COMMUNICATION ELECTRONICS

Principles and Applications

COMMUNICATION ELECTRONICS

Principles and Applications

Third Edition

Louis E. Frenzel

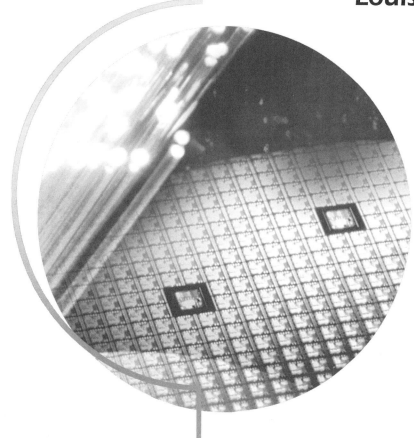

 **Glencoe McGraw-Hill**

New York, New York   Columbus, Ohio   Woodland Hills, California   Peoria, Illinois

Cover photos: Kuroda/Lee/SuperStock; background: David McLean/FPG

Library of Congress Cataloging-in-Publication Data

Frenzel, Louis E.
 Communication electronics / Louis E. Frenzel. —3rd ed.
  p.   cm.
 Includes index.
 ISBN 0-02-804837-7
 1. Electronics. 2. Radio—Receivers. 3. Telecommunication.
 I. Title.
 TK7816.F678   1999
 621.382—dc21   99-40937
                                    CIP

Glencoe/McGraw-Hill

A Division of The McGraw·Hill Companies

Communications Electronics, Third Edition

Send all inquiries to:
Glencoe/McGraw-Hill
936 Eastwind Drive
Westerville, Ohio 43081

ISBN 0-02-804837-7

1 2 3 4 5 6 7 8 9   079   06 05 04 03 02 01 00

Contents

The Glencoe *Basic Skills in Electricity and Electronics* series has been designed to provide entry-level competencies in a wide range of occupations in the electrical and electronic fields. The series consists of coordinated instructional materials designed especially for the career-oriented student. Each major subject area covered in the series is supported by a textbook, an experiments manual, and an instructor's productivity center. All the materials focus on the theory, practices, applications, and experiences necessary for those preparing to enter technical careers.

There are two fundamental considerations in the preparation of materials for such a series: the needs of the learner and needs of the employer. The materials in this series meet these needs in an expert fashion. The authors and editors have drawn upon their broad teaching and technical experiences to accurately interpret and meet the needs of the student. The needs of business and industry have been identified through personal interviews, industry publications, government occupational trend reports, and reports by industry associations.

The processes used to produce and refine the series have been ongoing. Technological change is rapid and the content has been revised to focus on current trends. Refinements in pedagogy have been defined and implemented based on classroom testing and feedback from students and instructors using the series. Every effort has been made to offer the best possible learning materials.

The widespread acceptance of the *Basic Skills in Electricity and Electronics* series and the positive responses from users confirm the basic soundness in content and design of these materials as well as their effectiveness as learning tools. Instructors will find the texts and manuals in each of the subject areas logically structured, well-paced, and developed around a framework of modern objectives. Students will find the materials to be readable, lucidly illustrated, and interesting. They will also find a generous amount of self-study and review materials and examples to help them determine their own progress.

Both the initial and ongoing success of this series are due in large part to the wisdom and vision of Gordon Rockmaker, who was a magical combination of editor, writer, teacher, electrical engineer and friend. Gordon has retired but he is still our friend. The publisher and editors welcome comments and suggestions from instructors and students using the materials in this series.

Charles A. Schuler,
Project Editor
and
Brian P. Mackin,
Editorial Director

Basic Skills in Electricity and Electronics

Charles A. Schuler, Project Editor

New Editions in the Series:

Electricity: Principles and Applications, Fifth Edition, Richard J. Fowler
Electronics: Principles and Applications, Fifth Edition, Charles A. Schuler
Digital Electronics: Principles and Applications, Fifth Edition, Roger L. Tokheim
Communication Electronics: Principles and Applications, Third Edition, Louis E. Frenzel

Other Series Titles Available:

Microprocessors: Principles and Applications, Second Edition, Charles M. Gilmore
Industrial Electronics, Frank D. Petruzella
Mathematics for Electronics, Harry Forster, Jr.

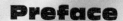

This book is written to accompany a first course in electronic communications. To get the most from this text, students should have the appropriate prerequisites. These include an understanding of ac circuits including resonance and filters, basic electronic circuits, and digital techniques.

In recent years, the study of microprocessors and personal computers has overshadowed study in other areas of electronics, including communications. Yet communications plays a vital role in our lives. Developments in communications technology have increased its applications in allied fields of electronics, including computers and industrial control. Regardless of a student's ultimate area of specialization, knowledge of communications concepts and applications is no longer optional; it is essential to understanding today's multidisciplinary applications.

We are currently experiencing a wireless revolution. Radio has been rediscovered and is being applied more widely than ever. In addition to its widespread use in cellular and cordless telephones, it is being employed in wireless local and networks, wireless Internet access, wireless remote control, and wireless identification (ID) systems. This textbook takes a look at all of the most recent wireless applications.

This third edition begins by introducing basic communications concepts and circuits, including modulation techniques as well as radio transmitters and receivers. Antennas and microwave techniques are also discussed. In addition, every effort has been made to include the latest in high-tech communication components, circuits, and equipment. Data communication techniques (including modems and local area networks, fiber optics, spread spectrum, and satellite communication) and advanced applications (such as cellular telephones, facsimile, television, and radar) are some of the topics covered.

Included in this edition is a new chapter (Chapter 15) on the telephone system. Besides being the largest and most widely used communications system, the telephone system is expanding and changing as it tries to accommodate an ever-larger percentage of our communication needs. The explosion of the Internet during the past few years has burdened the telephone system far more than expected. This new chapter will give you a working knowledge of telephones and their interconnecting systems.

All the chapters have been revised and updated to include the latest communication systems and applications. Some of the new material includes the Internet and World Wide Web, an update of cell phones with the new digital TDMA and CDMA systems, data communications with the latest modems (ADSL) and network components, and high-definition or digital TV systems.

My sincere hope is that this book will not only stimulate interest in the exciting field of electronic communications but will also help students learn the latest techniques they will surely encounter on the job.

Louis E. Frenzel

Safety

Electric and electronic circuits can be dangerous. Safe practices are necessary to prevent electrical shock, fires, explosions, and mechanical damage, and injuries resulting from the improper use of tools.

Perhaps the greatest hazard is electrical shock. A current through the human body in excess of 10 milliamperes can paralyze the victim and make it impossible to let go of a "live" conductor or component. Ten milliamperes is a rather small amount of electrical flow: It is only *ten one-thousandths* of an ampere. An ordinary flashlight uses more than 100 times that amount of current!

Flashlight cells and batteries are safe to handle because the resistance of human skin is normally high enough to keep the current flow very small. For example, touching an ordinary 1.5-V cell produces a current flow in the microampere range (a microampere is one-millionth of an ampere). This amount of current is too small to be noticed.

High voltage, on the other hand, can force enough current through the skin to produce a shock. If the current approaches 100 milliamperes or more, the shock can be fatal. Thus, the danger of shock increases with voltage. Those who work with high voltage must be properly trained and equipped.

When human skin is moist or cut, its resistance to the flow of electricity can drop drastically. When this happens, even moderate voltages may cause a serious shock. Experienced technicians know this, and they also know that so-called low-voltage equipment may have a high-voltage section or two. In other words, they do not practice two methods of working with circuits: one for high voltage and one for low voltage. They follow safe procedures at all times. They do not assume protective devices are working. They do not assume a circuit is off even though the switch is in the OFF position. They know the switch could be defective.

As your knowledge and experience grow, you will learn many specific safe procedures for dealing with electricity and electronics. In the meantime:

1. Always follow procedures.
2. Use service manuals as often as possible. They often contain specific safety information. Read,

and comply with, all appropriate material safety data sheets.
3. Investigate before you act.
4. When in doubt, *do not act.* Ask your instructor or supervisor.

General Safety Rules for Electricity and Electronics

Safe practices will protect you and your fellow workers. Study the following rules. Discuss them with others, and ask your instructor about any you do not understand.

1. Do not work when you are tired or taking medicine that makes you drowsy.
2. Do not work in poor light.
3. Do not work in damp areas or with wet shoes or clothing.
4. Use approved tools, equipment, and protective devices.
5. Avoid wearing rings, bracelets, and similar metal items when working around exposed electric circuits.
6. Never assume that a circuit is off. Double-check it with an instrument that you are sure is operational.
7. Some situations require a "buddy system" to guarantee that power will not be turned on while a technician is still working on a circuit.
8. Never tamper with or try to override safety devices such as an interlock (a type of switch that automatically removes power when a door is opened or a panel removed).
9. Keep tools and test equipment clean and in good working condition. Replace insulated probes and leads at the first sign of deterioration.
10. Some devices, such as capacitors, can store a *lethal* charge. They may store this charge for long periods of time. You must be certain these devices are discharged before working around them.
11. Do not remove grounds and do not use adaptors that defeat the equipment ground.

12. Use only an approved fire extinguisher for electrical and electronic equipment. Water can conduct electricity and may severely damage equipment. Carbon dioxide (CO_2) or halogenated-type extinguishers are usually preferred. Foam-type extinguishers may also be desired in *some* cases. Commercial fire extinguishers are rated for the type of fires for which they are effective. Use only those rated for the proper working conditions.

13. Follow directions when using solvents and other chemicals. They may be toxic, flammable, or may damage certain materials such as plastics. Always read and follow the appropriate material safety data sheets.

14. A few materials used in electronic equipment are toxic. Examples include tantalum capacitors and beryllium oxide transistor cases. These devices should not be crushed or abraded, and you should wash your hands thoroughly after handling them. Other materials (such as heat shrink tubing) may produce irritating fumes if overheated. Always read and follow the appropriate material safety data sheets.

15. Certain circuit components affect the safe performance of equipment and systems. Use only exact or approved replacement parts.

16. Use protective clothing and safety glasses when handling high-vacuum devices such as picture tubes and cathode-ray tubes.

17. Don't work on equipment before you know proper procedures and are aware of any potential safety hazards.

18. Many accidents have been caused by people rushing and cutting corners. Take the time required to protect yourself and others. Running, horseplay, and practical jokes are strictly forbidden in shops and laboratories.

Circuits and equipment must be treated with respect. Learn how they work and the proper way of working on them. Always practice safety: your health and life depend on it.

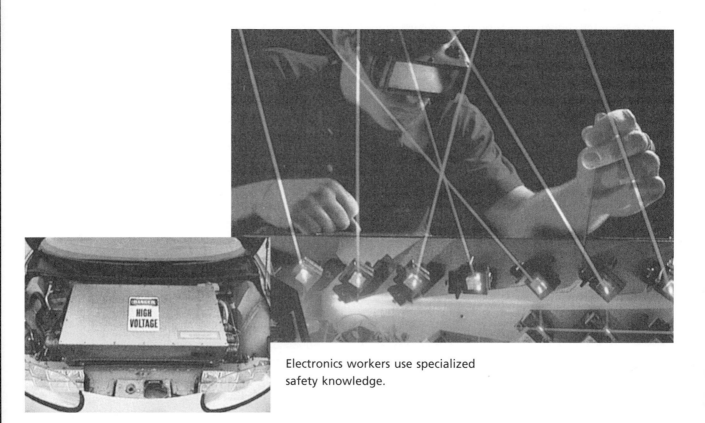

Electronics workers use specialized safety knowledge.

Chapter 1

Introduction to Electronic Communications

Chapter Objectives

This chapter will help you to:

1. *Name* the four main elements of any communications system and *state* what each does.
2. *Define* modulation and *explain* why it is necessary and important.
3. *List* six examples of common simplex and duplex communications systems in wide use today.
4. *List* the twelve major segments of the electromagnetic spectrum and *give* the frequency ranges of each.
5. *Convert* between measures of frequency and wavelength.
6. *Define* bandwidth and *compute* the bandwidth of a given piece of spectrum given the upper and lower frequency ranges.

One of the most pervasive human activities during the late twentieth and early twenty-first centuries is communications. Although human beings have always communicated face-to-face, communications has become even more widespread as new technologies have been developed. The printing press and the airplane greatly contributed to our ability to communicate from a distance. Human communication took a great leap forward in the nineteenth century with the development of the telephone and the telegraph.

In the twentieth century, radio, TV, radar, cell phones, fax machines, and dozens of other developments have further increased our ability to communicate. We continue to discover and use new communication technologies. The satellite and the pager are good examples. Yet nothing has increased our communications ability more than computers. Although computers still carry out their computational activities, they are now mostly interconnected to one another in local and long-distance networks so that they can exchange data. But this basic communications capability has led to amazingly popular new services such as e-mail and on-line shopping. And it has given us the Internet, with its enormous information resources. All of these recent systems are changing the way we live, work, and communicate. To top it off, more new radio or wireless applications have been developed in recent years than in the past 50 years.

We can thank the semiconductor industry for developing smaller and faster integrated circuits (ICs) that let us create sophisticated portable equipment for computing and wireless communicating.

This first chapter introduces you to the vast and varied communication services available today and the basic concepts upon which they are built.

1-1 The Importance of Communications

Communication is the basic process of exchanging information. It is what human beings do to convey their thoughts, ideas, and feelings to one another. Humans have been communicating with one another from the beginning of humankind. Most humans communicated through the spoken word, as they do today. Yet

Communication

a considerable amount of communication was nonverbal. Body movements and facial expressions were effective communication tools. Later, written communication was developed. Humans wrote letters to one another and eventually invented newspapers and books. Although the bulk of human communication today is still oral, a huge volume of information is exchanged by means of the written word. If anything, there is a glut of information in printed form.

Two of the main barriers to human communication are language and distance. When humans of different tribes, nations, or races come together, they often find that they do not speak the same language. This language barrier continues today, but it can be overcome as humans learn the languages of others and can serve as interpreters.

Communicating over long distances is another problem. Most human communication in the beginning was limited to face-to-face conversations. However, long-distance communication attempts were made as tribesmen signaled one another with drums or smoke signals. Other early forms of long-distance communication were blowing a horn, lighting a signal fire, or waving a flag (semaphore). But despite these long-distance communication attempts, transmission distances were limited. If the signal could be launched from a hill, mountain, or chain of high towers, distances of several miles could usually be achieved. Beyond that, other forms of communication were necessary.

Long-distance communication was extended by the written word. Messages and letters were transported from one place to another. For many years, long-distance communication was limited to the sending of verbal or written messages by human runner, horseback, ship, and later by trains.

Human communication took a dramatic turn in the late nineteenth century, when electricity was discovered and its many applications were explored. The telegraph was invented in 1844, and the telephone in 1876. In 1887 radio was

1440	Gutenberg invents the printing press.
1844	Morse patents the telegraph.
1866	First successful use of a transatlantic telegraph cable.
1876	Bell invents and patents the telephone.
1879	Eastman develops photographic film.
1887	Hertz discovers radio waves.
1895	Marconi demonstrates wireless telegraphy.
1901	Marconi makes first transatlantic radio transmission.
1903	The Fleming "valve" is invented.
1906	De Forest invents the triode vacuum tube and the first radiotelephone broadcast.
1923	Television is invented.
1931	Radio astronomy is discovered.
1940–45	Radar is perfected and helps win World War II.
1948	The transistor is invented.
1950s	Cable television first appears.
1954	Color television broadcasting begins.
1959	The integrated circuit is invented.
1962	First communications satellite.
1969	The Internet is invented.
1975–81	Personal computers come into use.
1981–85	Modems in PCs become widespread.
1983	First cellular telephone system becomes operational.
1989	The Global Positioning System (GPS) is used for commercial and personal (non-military) applications.
1989	The World Wide Web is invented.
1998	The first commercial use of digital/high-definition television takes place.

Fig. 1-1 Time line of milestones in human and electronic communications.

discovered, and it was demonstrated in 1895. From that point on, the exchange of information took a great leap forward. Figure 1-1 is a time line of important milestones in the evolution of human communications.

Well-known forms of electronic communications, such as the telephone, radio, and television, have increased our ability to share information. Today, they are a major part of our lives.

An unexpected development is the role that computers play in communications. E-mail now allows individuals with PCs to communicate with one another within and between messages transmitted over networks with office buildings, nationwide or worldwide. And interestingly, e-mail has not replaced the telephone or the fax. It has simply added a new way for people to interact with one another.

The Internet, with its overwhelming variety of data resources, is brought to us via data communications networks. The Internet not only puts more information than ever before at our fingertips but also is changing our buying habits and methods as well as the way we get information.

It is hard to imagine what our lives would be like without the knowledge and information that arrive from around the world by electronic

About ⟸⟹ Electronics

Electromagnetic radiation does not propagate underwater the way it does in air. Long-range communication underwater has mostly been based on sound waves.

communications. The way that we do things and the success of our work and personal lives are directly related to how well we communicate. It has been said that the emphasis in our society has now shifted from that of manufacturing and mass production of goods to the accumulation, packaging, and exchange of information. Ours is an information society, and a key part of it is communications. Without electronic communications, we could not access and apply the available information in a timely way. The so-called information superhighway of the future is the epitome of electronic communications technology.

■ TEST

Supply the missing word(s) in each statement:

1. Another name for radio is _____.
2. Communication is defined as the process of _____.
3. Most human communication is _____ even though there is a glut of _____ communication.
4. Two major barriers to human communication are _____.
5. Electronic communications came into being in the _____.

1-2   The Elements of a Communications System

All electronic communications systems have the basic form shown in Fig. 1-2. The basic components are a transmitter, a communications channel or medium, and a receiver. In most systems, a human generates a message that we call the information, or intelligence, signal. This signal is inputted to the transmitter which then transmits the message over the communications channel. The message is picked up by the receiver and is relayed to an-other human. Along the way, noise is added to the message in the communications channel. Noise is the general term applied to any interference that degrades the transmitted information. Let's take a closer look at each of these basic elements.

Transmitter

The *transmitter* is a collection of electronic components and circuits designed to convert the information into a signal suitable for transmission over a given communications medium. It may be as simple as a microphone or as complex as a microwave radio transmitter.

Transmitter

Communications Channel

The *communications channel* is the *medium* by which the electronic signal is sent from one place to another. In its simplest form, the medium may simply be a pair of wires that carry a voice signal from a microphone to a headset. The communications medium may also be a fiber-optic cable or "light pipe" that carries the message on a light wave.

Medium

On the other hand, the medium may be wireless or radio. *Radio* is the broad general term applied to any form of wireless communication from one point to another. Radio makes use of the electromagnetic spectrum where signals are communicated from one point to another by converting them into electric and magnetic fields that propagate readily over long distances.

Radio

Although the medium supports the transmission of information, it also attenuates it. Any type of media degrades the signal and causes it to appear much lower in amplitude at the receiver. Considerable amplification of the signal, both at the transmitter and the receiver, is required for successful communication.

*inter*NET
CONNECTION
Acronyms, words created using the first letter (or letters) of the words they represent, can quickly make the study of communications electronics an alphabet soup. When an acronym has you stumped, check out these listings on the Web for clarification:

⟨www.mot.com⟩
⟨www.sci.siemens .com⟩

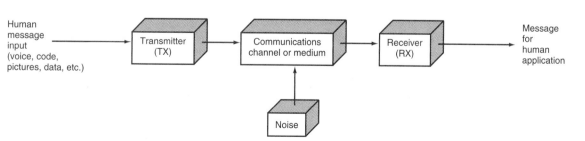

Fig. 1-2   The basic elements of any communications system.

Receiver

The *receiver* (RX) is another collection of electronic components and circuits that accept the transmitted message from the channel and convert it back into a form understandable by humans. Again it may be a simple earphone or a complex electronic receiver.

Noise

Noise is random, undesirable electric energy that enters the communication system via the communication medium and interferes with the transmitted message. However, some noise is also produced in the receiver. Noise comes from the atmosphere (e.g., from lightning which produces static), from outer space where the sun and other stars emit various kinds of radiation that can interfere with communications, and from electrical interference created by manufactured equipment. The electric ignition systems of cars, electric motors, fluorescent lights, and other types of equipment generate signals that can also interfere with the transmission of the message.

Finally, many electronic components generate noise internally due to thermal agitation of the atoms. Although such noise signals are low level, they can often seriously interfere with the extremely low-level signals that appear greatly attenuated at the receiver after being transmitted over a long distance. In some cases, noise completely obliterates the message. At other times, it simply causes interference, which, in turn, means some of the message is completely missed or misinterpreted.

Noise is one of the more serious problems of electronic communications. For the most part, it cannot be completely eliminated. However, there are ways to deal with it, as you will discover.

■ TEST

Supply the missing word(s) in each statement:

6. The three main elements of any communications system are _____.
7. The three major types of communications paths are _____.
8. The _____ converts the message into a form compatible with the selected medium.
9. The _____ converts the message from the medium into a form understandable by a human.
10. Undesirable interference in communications is _____ which is added to the signal in the _____.
11. The communications media greatly _____ and _____ the information signal.
12. Three common sources of interference are _____.

1-3 Types of Electronic Communications

There are three ways in which electronic communications is classified: one-way or two-way transmissions, analog versus digital signals, and baseband or modulated signals. Let's consider each of these categories in more detail.

There are two basic types of electronic communications. The simplest is one-way communications, normally referred to as *simplex*. Two-way communication is known as *duplex*.

Simplex

In simplex communications, the information travels in one direction only. A common example of simplex communications is radio and TV broadcasting. Another example is the information transmitted by the telemetry system of a satellite to earth. The telemetry system transmits information about the physical status of the satellite including its position and temperature.

Duplex

The bulk of electronic communications, however, is two-way. For example, when individuals communicate with one another over the telephone, each can transmit and hear simultaneously. Such two-way communications is referred to as *full duplex*.

Another form of two-way communications is where only one party transmits at a time. This is known as *half duplex*. The communication is two-way, but the direction alternates. Examples of half-duplex communication are most radio communications such as those used in the military, fire, police, and other services. Citizens band (CB) and amateur radio are half duplex in operation.

Analog Signals

Another way to categorize electronic communications is by the types of intelligence signals transmitted. There are two types of signals:

analog and digital. An *analog* signal is a continuously varying voltage or current. A typical analog signal is a sine wave tone. Voice and video voltages are analog signals.

Digital Signals

The other type of transmitted signal falls under the broad general category of *digital*. The earliest forms of both wire and radio communications used a type of on/off digital code. The telegraph used Morse code, whereas radio telegraphy used an international code of dots and dashes. Data used in computers is also digital where binary codes representing numbers, letters, and special symbols are transmitted by wire or radio. The most commonly used digital code in communications is the American Standard Code for Information Interchange (ASCII, pronounced "ass key").

Although digital transmission can be made up of signals that originated in digital form, such as telegraphy messages or computer data, analog signals may also be transmitted in digital form. It is very common today to take voice or video analog signals and "digitize" them with an analog-to-digital converter. There are many benefits to transmitting data in digital form as you will learn later.

Baseband Signals

Regardless of whether the original information signals are analog or digital, they are all referred to as *baseband signals.*

In a communications system, the information signals may be transmitted by themselves over the medium or may be used to modulate a carrier for transmission over the medium. Putting the original voice, video, or digital signals directly into the medium is referred to as *baseband transmission.* For example, in many telephone and intercom communications systems, it is the voice itself that is placed on the wires and transmitted over some distance to the receiver. In some computer networks, the digital signals are applied directly to coaxial cables for transmission to another computer.

Modulation

There are many instances when the baseband signals are incompatible with the media. For example, voice signals cannot be transmitted directly by radio. Although, theoretically this

is possible, realistically it is impractical. To transmit baseband signals by radio, modulation techniques must be used. Techniques using modulation are referred to as *broadband.*

Modulation is the process of having a baseband voice, video, or digital signal modify another, higher-frequency signal called the *carrier.* The information to be sent is said to be impressed upon the carrier.

The carrier is usually a sine wave that is higher in frequency than the highest intelligence signal frequency. Three basic characteristics of the carrier can be changed by the information signal: amplitude, frequency, and phase. The information or intelligence signal is also called the modulating signal or wave.

The two most common methods of modulation are amplitude modulation *(AM)* and frequency modulation *(FM)*. In AM, the baseband signal varies the amplitude of the higher-frequency carrier signal as shown in Fig. 1-3(a).

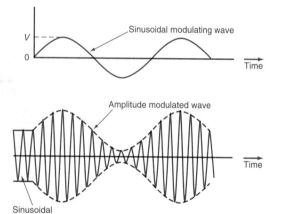

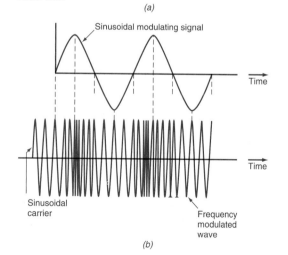

Fig. 1-3 Types of modulation. (a) Amplitude modulation and (b) frequency modulation.

Analog

**Broadband
Modulation**

Carrier

Digital

AM

FM

Baseband

*inter***NET**
CONNECTION

For a comprehensive source of telecommunications information aimed at basic understanding, see the following site:
⟨http:telecom .tbi.net⟩

PM

In FM, the baseband signal varies the frequency of the carrier as Fig. 1-3(b) shows.

Shifting the phase of the carrier in accordance with the intelligence signal produces phase modulation *(PM)*. Phase modulation, in turn, produces FM, so a PM signal looks like the FM signal in Fig. 1-3(b).

The higher-frequency carrier signal is more compatible with the medium (such as free space) and propagates through it with greater efficiency. At the receiver the carrier is demodulated to extract the original baseband information signal and transmitted message. Another name for the demodulation process is *detection*.

Detection

Multiplexing

Multiplexing

Multiplexing is the process of transmitting two or more signals simultaneously over the same channel. There are analog and digital as well as baseband and broadband methods of multiplexing, most of which are covered in this book.

■ TEST

Supply the missing word(s) in each statement:

13. One-way communications is called _____. An example is _____.
14. Simultaneous two-way communications is called _____. An example is _____.
15. Two-way communications where each party takes turns transmitting is referred to as _____.
16. Voice and video signals are continuous _____ voltages.
17. On/off or coded signals are referred to as _____ signals.
18. Voice and video signals may be transmitted digitally if they are first passed through a(n) _____.
19. An original voice, video, or data voltage is called the _____ signal.

20. To make the transmitted signal compatible with the medium, the process of _____ must be used. In this process the information signal is impressed upon a higher-frequency signal called the _____. This is called _____ transmission.
21. Recovering the originally transmitted signal is called _____.
22. The process of transmitting two or more baseband signals simultaneously over a common medium is called _____.

1-4 A Survey of Communications Applications

The applications of electronic techniques to communications are so common and pervasive that you are already familiar with most of them. You use the telephone, listen to the radio, and watch TV. You may also use other forms of communications such as cellular telephones, ham radios, CB radios, pagers, or remote-control garage door openers. But there are many more. The purpose of this section is to summarize all the various kinds of electronic communications, to show you both the breadth of the field and its importance.

Figure 1-4 lists all the various major applications of electronic communications. The list is divided into simplex and duplex categories. Read through the list to refresh your memory or possibly to discover some new applications you may not know about.

■ TEST

Answer the following questions.

23. Two methods of transmitting visual data over the telephone network are _____.
24. A common household remote-control unit is the _____.
25. The signaling of individuals at remote locations is called _____.
26. Performing, recording, and analyzing measurements at a distance is done with _____ equipment.
27. Radio astronomy is based on the fact that stars and other heavenly bodies emit _____.

28. List four ways radio is used in the telephone system.
29. Radar is based on the use of _____ radio signals.
30. Underwater radar is called _____.
31. The two types of sonar are _____.
32. The radio communications hobby is called _____.

33. Computers exchange digital data over the telephone network by using devices called _____.
34. Limited interconnections of PCs and other computers in offices or buildings are called _____.

SIMPLEX (One Way)

1. **AM and FM radio broadcasting.** Commercial stations broadcast music, news, and weather reports, as well as other programs and information.

2. **TV broadcasting.** Commercial stations broadcast a wide variety of entertainment, informational, and educational programs.

3. **Cable television.** The distribution of movies, sports events, and other programs by local cable companies to subscribers by coaxial cable. Cable stations originate some programming, but they primarily distribute "packaged" programming received by satellite from services such as HBO, CNN, and major networks.

4. **Wireless remote control.** A mechanism that controls missiles, satellites, robots, and other vehicles or remote plants or stations. A garage door opener is a special form of remote control that uses a tiny battery-operated transmitter in the car to operate a receiver in the garage that activates a motor to open or close the door. TV remote controls, remote keyless entry on cars, and radio-controlled models are others examples.

5. **Paging services.** A radio system for paging individuals, usually in connection with their work. Persons carry tiny battery-powered receivers or "beepers" that can pick up signals from a local paging station that receives telephone requests to locate and page individuals when they are needed.

6. **Navigation and direction-finding services.** Transmission by special stations of signals that can be picked up by receivers with highly directional antennas for the purpose of identifying exact location (latitude and longitude) or determining direction and/or distance from a station. Such systems employ both land-based and satellite stations. The services are used primarily by boats, ships, and airplanes, although systems for cars and trucks are being developed.

7. **Telemetry.** The transmission of measurements over a long distance. Telemetry systems use sensors to determine physical conditions (temperature, pressure, flow rate, voltages, frequency, etc.) at a remote location. The sensors modulate a carrier signal that is sent by wire or radio to a remote receiver that stores and/or displays the data for analysis. Telemetry systems permit remote equipment and systems to be monitored for the purpose of determining their status and condition. Examples are satellites, rockets, pipelines, plants, and factories.

8. **Radio astronomy.** Radio signals, including infrared, are emitted by virtually all heavenly bodies such as stars and planets. With the use of large directional antennas and sensitive high-gain receivers, these signals may be picked up and used to plot star locations and study the universe. Radio astronomy is an alternative and supplement to traditional optical astronomy.

9. **Surveillance.** *Surveillance* means discreet monitoring or "spying." Electronic techniques are widely used by police forces, governments, the military, business and industry, and others to gather information for the purpose of gaining some competitive advantage. Techniques include phone taps, tiny wireless "bugs," clandestine listening stations, as well as reconnaissance airplanes and satellites.

Fig. 1-4 Applications of electronic communications.

Continued

interNET CONNECTION

Cellular phones can interfere with pacemakers. For more information, see the following site: ⟨www.medtronic.com⟩

10. **Radio-frequency Identification.** RFID systems are tiny integrated circuit transmitters that send digital codes to a nearby receiver for the purpose of identifying, detecting, or locating packages, crates, or various types of vehicles (train cars, trucks, and so on).
11. **Music services.** The transmission of continuous background music for doctors' offices, stores, elevators, and so on, by local FM radio stations on special high-frequency subcarriers that cannot be picked up by conventional FM receivers.

DUPLEX (Two Way)

12. **Telephones.** One-on-one verbal communications over the vast worldwide telephone networks employing wire, radio relay stations, and satellites. Cordless telephones provide short-distance convenience communication without restrictive wires. Cellular radio systems provide telephone service in vehicles and for portable use.
13. **Facsimile.** The transmission of printed visual material over the telephone lines. A facsimile, or fax, machine scans a photo or other document and converts it into electronic signals that are sent over the telephone system for reproduction in original printed form by another fax machine at the receiving end.
14. **Two-way radio.** Commercial, industrial, and governmental communications between vehicles or between vehicles and a base station. Examples of users are police departments, fire departments, taxi services, forestry services, and trucking companies. Other forms of two-way radios are used in aircraft, marine, military, and space applications. Government applications are broad and diverse and include embassy communications, Treasury, Secret Service, and CIA.
15. **Radar.** A special form of communications that makes use of reflected microwave signals for the purpose of detecting ships, planes, and missiles and for determining their range, direction, and speed. Most radar is used in military applications, but civilian aircraft and marine services also use it. Police use radar in speed detection and enforcement.
16. **Sonar.** Underwater communications in which audible baseband signals use water as the transmission medium. Submarines and ships use sonar to detect the presence of enemy submarines. Passive sonar uses audio receivers to pick up water, propeller, and other noises. Active sonar is like an underwater radar in which reflections from a transmitted audio pulse are used to determine direction, range, and speed of an underwater target.
17. **Amateur radio.** A hobby for individuals interested in radio communications. Individuals may become licensed "hams" to build and operate two-way radio equipment for personal communication with other hams.
18. **Citizens and family radio.** Citizens band (CB) radio is a special service that any individual may use for personal communication with others. Most CB radios are used in trucks and cars for exchanging information about traffic conditions, speed traps, and emergencies. Family radio is a service designed to help family members keep in touch over short distances with small handheld radios on family outings, car trips, shopping, and so on.
19. **Data communications.** The transmission of binary data between computers. Computers frequently use the telephone system as the medium. Devices called modems make the computers and telephone networks compatible. Data communications also take place over terrestrial microwave relay links and satellites.
20. **Local area networks (LANs).** Wired (or wireless) interconnections of PCs or PCs and mini or mainframe computers within an office or building for the purpose of sharing mass storage, peripherals, and data.
21. **Internet.** The Internet is a worldwide network of computers that provides information and services to any computer user. The interconnections include the telephone system and many special wire, fiber-optic cable, and switched networks.

Fig. 1-4 *Continued*

1-5 The Electromagnetic Spectrum

Before it can be transmitted, information must be converted into electronic signals compatible with the medium. For example, a microphone changes voice into a voltage of varying frequency and amplitude. This baseband signal is then passed over wires to a receiver or headphone. This is the way the telephone system works. A tremendous amount of information is transmitted in this way.

Instead of using wires, free space can be used. The information is converted into electronic signals which radiate into space. Such signals consist of both electric and magnetic fields. These so-called electromagnetic signals travel through space for long distances. Electromagnetic signals are also referred to as *radio-frequency (RF) waves*.

Electromagnetic waves are signals that oscillate; that is, the amplitudes of the electric and magnetic fields vary at a specific rate. The field intensities fluctuate up and down a given number of times per second. The electromagnetic waves vary sinusoidally. Their frequency is measured in cycles per second (cps) or hertz (Hz). These oscillations may occur at a very low frequency or at an extremely high frequency. This entire range of frequencies is referred to as the *electromagnetic spectrum*. It includes signals such as the 60-Hz power line frequency and audio (voice) signals at the low end. In the midrange are the most commonly used radio frequencies for two-way communications, television, and other applications. At the upper end of the spectrum are infrared and visible light. Figure 1-5 shows the entire electromagnetic spectrum. Both frequency and wavelength are given.

Remember the relationship between frequency f and wavelength λ.

$$\lambda = \frac{300}{f}$$

where λ is in meters and f is in megahertz (MHz). For instance, if $f = 21$ MHz, then

$$\lambda = \frac{300}{21} = 14.29 \text{ m}$$

If you know the wavelength in meters, you can compute the corresponding frequency with the expression

$$f = \frac{300}{\lambda}$$

A wavelength of 2.4 m expressed as a frequency is:

$$f = \frac{300}{2.4} = 125 \text{ MHz}$$

Why Use a Carrier?

Most information signals to be transmitted occur at the lower frequencies in the spectrum. They are used to modulate a carrier of a higher frequency. The high-frequency signal gives the lower-frequency information

Radio-frequency (RF)

Electromagnetic spectrum

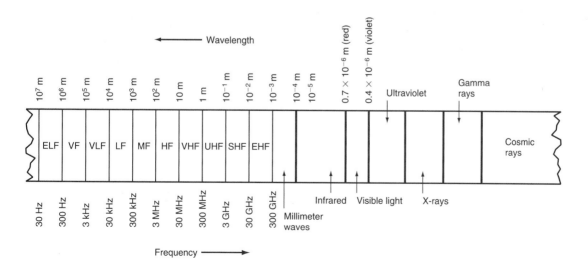

Fig. 1-5 The electromagnetic spectrum used in electronic communications.

FREQUENCY AND WAVELENGTH

Frequency is simply the number of times a particular phenomenon occurs in a given period of time. It may be the number of voltage polarity alternations or electromagnetic field oscillations that take place in one second. Each alternation or oscillation is called a cycle and the frequency is measured in cycles per second (cps). In electronics, the term *hertz* is used to express frequency in cps. For example, 60 cps=60 Hz.

Figure A shows a sine wave variation of voltage. One positive and one negative alternation form a cycle. If 1000 of the cycles take place in one second, the frequency is 1000 Hz.

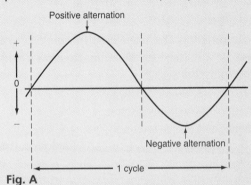

Fig. A

To express higher frequencies, the suffixes k, M, and G are used. The suffix k stands for kilo, M for mega, and G for giga. Also, k=1000, M=1,000,000, and G=1,000,000,000. A frequency of 1000 Hz is usually expressed as 1 kHz or 1 kilohertz. A frequency of 7,000,000 Hz is expressed as 7 MHz or 7 megahertz. A very high frequency such as 4,000,000,000 Hz is given as 4 GHz or 4 gigahertz.

Period is the distance between two points of similar cycles of a periodic wave. Figure B shows period at several points on a sine wave.

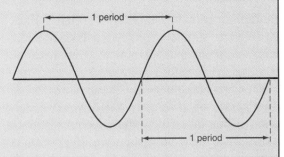

Fig. B

Wavelength is the distance traveled by an electromagnetic wave during one period. Electromagnetic waves travel at a speed of 300,000,000 meters per second or about 186,000 miles per second. The wavelength (λ) can be computed by dividing the speed by the frequency (f) in Hz.

$$\lambda = \frac{300,000,000}{f}$$

For example, the wavelength of a 4,000,000 Hz signal is

$$\lambda = \frac{300,000,000}{4,000,000} = 75 \text{ meters (m)}$$

If the frequency is expressed in megahertz, the formula can be simplified to

$$\lambda = \frac{300}{f}$$

A 150-MHz signal has a wavelength of

$$\lambda = \frac{300}{150} = 2 \text{ m}$$

A 430-MHz signal has a wavelength of

$$\lambda = \frac{300}{430} = 0.697 \text{ m}$$

or about 70 cm.

signals a ride through space. A receiver then picks up the modulated electromagnetic field, converts it back into a signal voltage, amplifies it, and demodulates it to recover the original information.

The inevitable question is, "Why can't the information signals be radiated directly?" Well, in theory they can be, but it is generally impractical. Let's take voice signals as an example. First, these occur in the 300- to 3000-Hz frequency range. After being amplified, they could be sent to a long wire antenna instead of a speaker. The resulting electromagnetic waves would be propagated through space to a receiver. In order for this to be done effectively, the antenna would have to be enormous. Antenna length is usually one-quarter or one-half the length of the waves it is set up to transmit. For audio signals, the antenna would be many miles long. That, of course, would be impractical because of the size and cost.

Second, simultaneously transmitted audio signals would interfere with one another since they occupy the same frequency range. The audio portion of the spectrum would be nothing but a jumble of hundreds or thousands of simultaneous communications.

For these reasons, the information signal, be it audio, video, or data, modulates a high-frequency carrier. At higher frequencies antenna sizes are much smaller and signals propagate farther. There is also more room at the higher frequencies, so many channels can be formed to carry thousands of simultaneous communications without interference.

The frequency spectrum is divided into segments for the purpose of classifying the various portions and how they are used. These segments are designated in Fig. 1-6. The frequency and wavelength range for each segment is given. Let's summarize briefly the signal characteristics and applications of each segment.

Extremely Low Frequencies

Extremely low frequencies (ELFs) are those in the 30- to 300-Hz range. These include ac power line frequencies (50 and 60 Hz are common) as well as those frequencies in the low end of the human hearing range.

Name	Frequency	Wavelength
Extremely low frequencies (ELF)	30–300 Hz	10^7–10^6 m
Voice frequencies (VF)	300–3000 Hz	10^6–10^5 m
Very low frequencies (VLF)	3–30 kHz	10^5–10^4 m
Low frequencies (LF)	30–300 kHz	10^4–10^3 m
Medium frequencies (MF)	300 kHz–3 MHz	10^3–10^2 m
High frequencies (HF)	3–30 MHz	10^2–10^1 m
Very high frequencies (VHF)	30–300 MHz	10^1–1 m
Ultra high frequencies (UHF)	300 MHz–3 GHz	1–10^{-1} m
Super high frequencies (SHF)	3–30 GHz	10^{-1}–10^{-2} m
Extremely high frequencies (EHF)	30–300 GHz	10^{-2}–10^{-3} m
Infrared	—	0.7–10 μm
The visible spectrum (light)	—	0.4×10^{-6}–0.8×10^{-6} m

Units of Measurement and Abbreviations:
kHz = 1000 Hz
MHz = 1000 kHz = 1×10^6 = 1,000,000 Hz
GHz = 1000 MHz = 1×10^6 = 1,000,000 kHz
= 1×10^9 = 1,000,000,000 Hz
m = meter
μm = micron = $\frac{1}{1,000,000}$ m = 1×10^{-6} m

Fig. 1-6 Segments of the electromagnetic spectrum.

Did You Know?

Communications Signals
- The ionosphere refracts radio waves, making long-distance communications possible for some frequencies at certain times.
- Secondary information may be encoded onto radio and television signals, and extra circuits are needed to take advantage of them.

Voice Frequencies

Voice frequencies (VFs) are those in the range of 300 to 3000 Hz. This is the normal range of human speech. Although human hearing extends from approximately 20 to 20,000 Hz, most intelligible sound occurs in the VF range.

Very Low Frequencies

Very low frequencies (VLFs) include the higher end of the human hearing range up to about 15 to 20 kilohertz (kHz). Many musical instruments also make sounds in this range as well as in the ELF and VF ranges. The VLF range is also used in some government and military communications. For example, VLF radio transmission is used by the navy to communicate with submarines.

Low Frequencies

Low frequencies (LFs) are those in the 30- to 300-kHz range. The primary communications services in this range are those used in aeronautical and marine navigation. Frequencies in this range are also used as subcarriers. *Subcarriers* are signals that carry the baseband modulating information but which, in turn, modulate another higher-frequency carrier.

Subcarriers

Medium Frequencies

Medium frequencies (MFs) are in the 300- to 3000-kHz [3-megahertz (MHz)] range. The major application of frequencies in this range is AM radio broadcasting (535 to 1605 kHz). Other services in this range include various marine and aeronautical communications applications.

High Frequencies

High frequencies (HFs) are those in the 3- to 30-MHz range. These are the frequencies generally known as short waves (SWs). All kinds of two-way radio communications take place in this range as well as some shortwave radio broadcasting. Voice of America and British Broadcasting Corporation (BBC) broadcasts occur in this range. Government and military services use these frequencies for two-way communications. Amateur radio and CB communications also occur in this part of the spectrum.

Very High Frequencies

Very high frequencies (VHFs) cover the 30- to 300-MHz range. This is an extremely popular frequency range and is used by many services including mobile radio, marine and aeronautical communications, FM radio broadcasting (88 to 108 MHz), and television channels 2 through 13. Radio amateurs also have numerous bands in this frequency range.

Ultrahigh Frequencies

Infrared

Ultrahigh frequencies (UHF) cover the 300- to 3000-MHz range. This too is an extremely widely used portion of the frequency spectrum. It includes the UHF television channels 14 through 67. It is also widely used for land mobile communications and services such as cellular telephones. The military services widely use these frequencies for communications. In addition, some radar and navigation services occupy this portion of the frequency spectrum. Radio amateurs also have bands in this part of the spectrum. Incidentally, frequencies above the 1000-MHz [1 gigahertz (GHz)] range are called *microwaves.*

Microwaves

Superhigh Frequencies

The superhigh frequencies (SHFs) are those in the 3- to 30-GHz range. These are microwave frequencies that are widely used for satellite communications and radar. Some specialized forms of two-way radio communications also occupy this region.

Extremely High Frequencies

Extremely high frequencies (EHFs) extend from 30 to 300 GHz. Equipment used to generate and receive signals in this range is extremely complex and expensive. Presently there is only a limited amount of activity in this range, but it does include satellite communications and some specialized radar. As technological developments permit equipment advances, this frequency range will be more widely used. Signals directly above this range are generally referred to as millimeter waves.

Infrared

Those electromagnetic signals whose frequencies are higher than 300 GHz are not referred to as radio waves. Special names are given to the various bands in the spectrum beyond that point. The infrared region is sandwiched between the highest radio frequencies and the visible portion of the electromagnetic spectrum. It occupies the range between approximately 0.01 millimeter (mm) and 700 nanometers (nm) or 0.7 to 10 microns (μm). Infrared frequencies are often given in microns, where a micron is one millionth of a meter (m), or in nanometers or one billionth of a meter (10^{-9} m). Infrared is divided into two areas, long infrared (0.01 mm to 1000 nm) and short infrared (1000 to 700 nm).

Infrared refers to radiation generally associated with heat. Anything that produces heat generates infrared signals. Infrared is produced by light bulbs, our bodies, and any physical equipment that generates heat. Infrared signals can also be generated by special types of light-emitting diodes.

Infrared signals are used for various special kinds of communications. For example, infrared is used in astronomy to detect stars and other physical bodies in the heavens. Infrared is also used for guidance in weapons systems where the heat radiated from airplanes or missiles can be picked up by infrared detectors and used to guide missiles toward these targets. Infrared is also used in most new TV remote-control units where special coded signals are transmitted by infrared to the TV receiver for the purpose of changing channels, setting the volume, and other functions.

Infrared signals also have many of the same properties as light. Optical devices such as lenses and mirrors are often used to process and manipulate infrared signals.

The Visible Spectrum

Just above the infrared region is the visible spectrum we ordinarily refer to as light. Light is a special type of electromagnetic radiation that has a wavelength in the 0.4- to 0.8-μm range. Light wavelengths are usually expressed in terms of angstroms (Å). An angstrom is one ten-thousandth of a micron. The visible range is approximately 8000 Å (red) to 4000 Å (violet).

Light is widely used for various kinds of communications. Light waves can be modulated and transmitted through glass fibers just as electrical signals can be transmitted over wires. Fiber optics is one of the fastest growing specialties of communications electronics. The great advantage of light wave signals is that their very high frequency gives them the ability to handle a tremendous amount of information. That is, the bandwidth of the baseband signals may be very wide.

Light signals can also be transmitted through free space. Various types of communications systems have been created using a laser that generates a light beam at a specific visible frequency. Lasers generate an extremely narrow beam of light which is easily modulated with voice, video, and data information.

Beyond the visible region are the x-rays, gamma rays, and cosmic rays. These are all forms of electromagnetic radiation, but they do not figure into communications systems and, as a result, we will not cover them here.

■ TEST

Answer the following questions.

35. Signals that travel free space for long distances are called _____.
36. Radio waves are made up of _____ fields.
37. A signal with a frequency of 18 MHz has a wavelength of _____ m.
38. Common power line frequencies of _____ and _____ Hz are in the _____ range.
39. Audio signals are not transmitted by electromagnetic waves because
    a. Antennas would be too long.
    b. Audio signals do not radiate.
    c. Simultaneous transmissions would interfere.
    d. The frequency is too low.
    (Choose all that apply.)

40. The human hearing range is approximately _____ to _____ Hz.
41. The frequency range of the human voice is _____ to _____ Hz.
42. True or false. Radio transmissions do not occur in the VLF and LF ranges.
43. AM broadcast stations are in the _____ range.
44. HF signals are also called _____.
45. TV (channels 2 to 13) and FM broadcasting is in the _____ part of the spectrum.
46. List five major uses of the UHF band.
47. A frequency of 1 GHz is the same as _____ MHz.
48. Frequencies above 1 GHz are called _____.
49. The SHF and EHF ranges are primarily used by _____ communications.
50. The frequencies just beyond the EHF range are called _____ waves.
51. One micron is the same as _____ m.
52. Infrared signals are usually derived from _____ sources.
53. The spectrum range of infrared signals is _____ to _____ μm.
54. One angstrom is equal to _____ μm.
55. The visible light range is from _____ to _____ Å.
56. Light signals use two mediums in electronic communications: _____.

1-6 Bandwidth

Bandwidth is that portion of the electromagnetic spectrum occupied by a signal. It is also the frequency range over which an information signal is transmitted or over which a receiver or other electronic circuit operates. More specifically, bandwidth (BW) is the difference between the upper and lower frequency limits of the signal or the equipment operation range. Figure 1-7 shows the bandwidth of the voice frequency range from 300 to 3000 Hz. The upper frequency is f_2 and the lower frequency is f_1. The bandwidth then is

$$BW = f_2 - f_1$$
$$= 3000 - 300$$
$$= 2700 \text{ Hz}$$

When information is modulated onto a carrier in the electromagnetic spectrum, the

Bandwidth

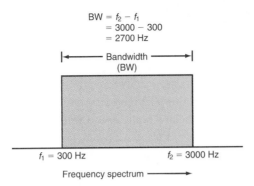

BW = $f_2 - f_1$
= 3000 − 300
= 2700 Hz

|← Bandwidth →|
(BW)

f_1 = 300 Hz f_2 = 3000 Hz

Frequency spectrum ⟶

Fig. 1-7 Bandwidth is the frequency range over which equipment operates or that portion of the spectrum occupied by the signal. This is the voice frequency bandwidth.

resulting signal occupies a small portion of the spectrum surrounding the carrier frequency. For example, in AM broadcasting, audio signals up to 5 kHz may be transmitted. The modulation process causes other signals, called *sidebands,* to be generated at frequencies above and below the carrier frequency by an amount equal to the modulating frequency. If the carrier frequency is 1000 kHz, or 1 MHz, and the modulating frequency is 5 kHz, then sidebands will be produced at 1000 − 5 = 995 kHz and at 1000 + 5 = 1005 kHz. In other words, the modulation process generates other signals which take up *spectrum space.* It is not just the carrier at 1000 kHz that is transmitted. The range of frequencies that contain the information is called the bandwidth. The term *channel bandwidth* is used to describe the range of frequencies required to transmit the desired information.

The bandwidth of the AM signal described above is the difference between the highest and lowest transmitting frequencies.

BW = 1005 kHz − 995 kHz

= 10 kHz

In this case, the channel bandwidth is 10 kHz. An AM broadcast signal, therefore, takes up a 10-kHz chunk of spectrum space. Signals transmitting on the same frequency or on overlapping frequencies will, of course, interfere with one another. For this reason, you can see that only a limited number of signals can be transmitted in the frequency spectrum.

Sidebands

Spectrum space

Channel bandwidth

More Space at the Higher Frequencies

As communications activities have grown over the years, there has been a continuous demand for more frequency channels over which communications can be conducted.

This has caused a push for the development of equipment that operates at the higher frequencies. Prior to World War II, frequencies above 1 GHz were virtually unused since there were no electronic components suitable for generating signals at those frequencies. But technological developments over the years have given us many microwave components such as klystrons, magnetrons, traveling-wave tubes, and today even transistors that work in the microwave range.

The benefit of using the higher frequencies is that a signal of a given bandwidth represents a smaller percentage of the spectrum at the higher frequencies than at the lower frequencies. For example, the 10-kHz-wide AM signal discussed earlier represents 1 percent of the spectrum at 1000 kHz.

$$\text{Percent} = \frac{10 \text{ kHz}}{1000 \text{ kHz}} \times 100 = 1\%$$

But at 1 GHz, or 1,000,000 kHz, it only represents a thousandth of 1 percent.

$$\text{Percent} = \frac{10 \text{ kHz}}{1,000,000 \text{ kHz}} \times 100 = 0.001\%$$

What this really means is that there are many more 10-kHz channels at the higher frequencies than at the lower frequencies. In other words, there is more spectrum space for information signals at higher frequencies.

The higher frequencies also permit wider bandwidth signals to be used. A TV signal, for example, occupies a bandwidth of 6 MHz. Such a signal cannot be used to modulate a carrier in the MF or HF ranges. Television signals are transmitted in the VHF and UHF portions of the spectrum where sufficient space is available.

Today, virtually all the frequency spectrum between approximately 30 kHz and 1 GHz has been spoken for. Granted, there are some open areas and portions of the spectrum that are not heavily used, but for the most part, the spectrum is filled with communications activities of all kinds. There is tremendous competition for these frequencies, not only between companies, individuals, and government services but also between the different nations of the world. The electromagnetic spectrum is one of our most precious natural resources. Because of this, communications engineering is devoted to making the best use of that finite spectrum. A considerable amount of effort goes into developing communications techniques that will minimize the bandwidth required to transmit given information. By minimizing the bandwidth, spectrum space is conserved and this gives other services or users an opportunity to take advantage of it. Many of the techniques we will discuss later in this book evolved in an effort to minimize transmission bandwidth.

Spectrum Assignment and Regulation

The United States and other countries recognized early on that the frequency spectrum was indeed finite and a valuable resource. For that reason, governments set up agencies to control the use of the spectrum. In the United States, Congress created the Communications Act of 1934. Along with its various amendments, the act determined how the spectrum space was used. It also established the *Federal Communications Commission (FCC),* which is a regulatory body whose sole purpose is allocating spectrum space, issuing licenses, setting standards, and policing the airwaves. The FCC controls all telephone and radio communications in this country and, in general, regulates all electromagnetic emissions. The *National Telecommunications and Information Administration (NTIA)* performs a similar function for government and military services. Other countries have similar organizations.

There is also an international organization to which all countries belong. This is the *International Telecommunications Union (ITU),* an agency of the United Nations. It has 154 member countries that meet at regular intervals to promote cooperation and negotiate national interests. Various committees of the ITU set standards for various areas within the communications field. The ITU brings together the various countries to discuss how the frequency spectrum is to be divided up and shared. Because many of the signals generated in the spectrum do not carry for long distances, countries may use these frequencies simultaneously without interference. On the other hand, there are some ranges of frequency spectrum that can literally carry signals around the world. High-frequency shortwave signals are an example. As a result, countries must negotiate with one another for various portions of the high-frequency and spectrum and coordinate usage.

■ TEST

Supply the missing information in each statement.

57. The spectrum space occupied by a signal is called the _____.
58. The new signals above and below the carrier frequency produced by the modulation process are called _____.
59. A signal occupies the frequency range from 1.050 to 1.175 MHz. Its bandwidth is _____ kHz.
60. Wide-bandwidth signals must be transmitted at _____ frequencies.
61. Percentagewise, there is less spectrum space at the _____ frequencies.
62. Many communications electronics techniques are designed in order to conserve _____.
63. Electronic communications in the United States is regulated by a set of laws called the _____.
64. The regulatory body for electronic communications in the United States is the _____.
65. Government and military communications are coordinated by the _____.
66. The electromagnetic spectrum is managed worldwide by the _____ organization.

Federal Communications Commission (FCC)

National Telecommunications and Information Administration (NTIA)

About ⬤ Electronics

Extreme Temperatures and Components/Digital Signal Processing

• Temperature changes can have a greater influence on circuit performance than component tolerance does. When troubleshooting equipment, be sure to ask the operator what temperature conditions the components have been exposed to.

• DSP can be used to compress video signals and to identify fingerprints.

Summary

1. The three major fields of electronics are computers, communications, and control. The computer segment is the largest; communications is the second largest.
2. Communication is the process of exchanging information.
3. Most human communication is oral, but a great deal of it is also in written or printed form.
4. The two main barriers to communication are language and distance.
5. Major electrical discoveries in the mid- and late nineteenth century made possible the development of electronic communications over long distances.
6. The telegraph (1844) and telephone (1876) were the first two long-distance communications systems.
7. Radio was discovered in 1887, and wireless telegraphy was demonstrated in 1895.
8. Electronic communications plays a vital role in all our lives and is essential to the success of our information society.
9. The major elements of a communications system are a transmitter to send a message, a communications medium, a receiver to pick up the message, and noise.
10. The three primary communications media are wires, free space, and fiber-optic cable.
11. Radio waves are signals made up of electric and magnetic fields that propagate over long distances.
12. Noise is any interference that disturbs the legible transmission of a signal. Noise is produced by the atmosphere, heavenly bodies, manufactured electrical equipment, and thermal agitation in electronic components.
13. The transmission medium greatly attenuates and degrades the transmitted signal.
14. Electronic communications may be either one-way or two-way. One-way transmission is called simplex or broadcasting.
15. Two-way communications is called duplex. In half-duplex communications, only one of the

two parties can transmit at a time. In full duplex, both parties may transmit and receive simultaneously.
16. Information signals may be either analog or digital. Analog signals are smooth, continuous voltage variations such as voice or video. Digital signals are binary pulses or codes.
17. The information signal, called the baseband signal, is often transmitted directly over the communications medium.
18. In most communications systems, the baseband signal is used to modulate a higher-frequency carrier signal than is transmitted by radio.
19. Modulation is the process of having an information signal modify a carrier signal in some way. Common examples are AM and FM.
20. The baseband signal cannot usually be transmitted through space by radio because the antennas required are too long and because multiple baseband signals transmitting simultaneously would interfere with one another.
21. Multiplexing is the process of transmitting two or more signals simultaneously over the same channel or medium.
22. Besides TV, there are several other methods of transmitting visual or graphical information; they are facsimile, videotex, and teletext.
23. Simplex transmission of special signals from land-based or satellite stations is used by ship and airplanes for navigation.
24. Telemetry is measurement at a distance. Sensors convert physical characteristics to electric signals which modulate a carrier transmitted to a remote location.
25. Radio astronomy supplements optical astronomy by permitting the location and mapping of stars by the radio waves they emit.
26. Radar uses the reflection of radio waves from remote objects for the detection of their presence, direction, and speed.
27. Underwater radar is called active sonar. Passive sonar is simply listening underwater for the detection of objects of interest.

28. Two forms of personal communications services are CB radio and amateur or "ham" radio, which is a technical hobby as well as a communications service.
29. Data communications is the transmission of computer and other digital data via the telephone system, microwave links, or satellite.
30. Devices called modems permit digital data to be transmitted over the analog telephone networks.
31. Interconnections of PCs for the exchange of information are called local area networks.
32. The electromagnetic spectrum is that range of frequencies from approximately 30 Hz to visible light over which electronic communications take place.
33. The greatest portion of the spectrum covers radio waves, which are oscillating electric and magnetic fields that radiate for long distances.
34. Wavelength (λ) is the distance (in meters) between corresponding points on successive cycles of a periodic wave: $\lambda = 300/f$ (f is in megahertz). It is also the distance that an electromagnetic wave travels in the time it takes for one cycle of oscillation.
35. The range of human hearing is approximately 20 to 20,000 Hz. The voice frequency range is 300 to 3000 Hz.
36. Amplitude-modulated broadcasting occurs in the MF range from 300 kHz to 3 MHz.
37. The high-frequency range (3 to 30 MHz), or shortwave, is used for worldwide two-way communications and broadcasting.
38. Television broadcasting occurs in the VHF and UHF ranges.
39. Frequencies above 1 GHz are called microwaves.
40. The SHF and EHF bands are used primarily for satellite communications and radar.
41. Those frequencies directly above 300 GHz are called millimeter waves.
42. Electromagnetic signals produced primarily by heat sources are called infrared. They cover the 0.7- to 100-μm range.
43. A micron is one millionth of a meter.
44. Visible light occupies the region above infrared. Its wavelength is 4000 to 8000 Å.
45. An angstrom is one ten-thousandth of a micron.
46. Bandwidth is the spectrum space occupied by a signal, the frequency range of a transmitted signal, or the range of frequencies accepted by a receiver. It is the difference between the upper and lower frequencies of the range in question.
47. There is more spectrum space available at the higher frequencies. For a given bandwidth signal, more channels can be accommodated at the higher frequencies.
48. Spectrum space is a precious natural resource.
49. In the United States, the FCC regulates the use of the spectrum and most forms of electronic communications according to the Communications Act of 1934.
50. Most countries belong to the ITU, an organization devoted to worldwide cooperation and negotiation on spectrum usage.
51. The NTIA coordinates government and military communications in the United States.

Chapter Review Questions

Choose the letter that best answers each question.

1-1. Communication is the process of
   a. Keeping in touch.
   b. Broadcasting.
   c. Exchanging information.
   d. Entertainment by electronics.
1-2. Two key barriers to human communication are
   a. Distance.
   b. Cost.
   c. Ignorance.
   d. Language.
1-3. Electronic communications was discovered in which century?
   a. Sixteenth
   b. Eighteenth
   c. Nineteenth
   d. Twentieth
1-4. Which of the following is *not* a major communications medium?
   a. Free space
   b. Water
   c. Wires
   d. Fiber-optic cable
1-5. Random interference to transmitted signals is called
   a. Adjacent channel overlap.

b. Cross talk.

c. Garbage-in, garbage-out.

d. Noise.

1-6. The communications medium causes the signal to be

a. Amplified.

b. Modulated.

c. Attenuated.

d. Interfered with.

1-7. Which of the following is *not* a source of noise?

a. Another communications signal

b. Atmospheric effects

c. Manufactured electrical systems

d. Thermal agitation in electronic components

1-8. One-way communications is called

a. Half duplex.

b. Full duplex.

c. Monocomm.

d. Simplex.

1-9. Simultaneous two-way communications is called

a. Half duplex.

b. Full duplex.

c. Bicomm.

d. Simplex.

1-10. The original electrical information signal to be transmitted is called the

a. Modulating signal.

b. Carrier.

c. Baseband signal.

d. Source signal.

1-11. The process of modifying a high-frequency carrier with the information to be transmitted is called

a. Multiplexing.

b. Telemetry.

c. Detection.

d. Modulation.

1-12. The process of transmitting two or more information signals simultaneously over the same channel is called

a. Multiplexing.

b. Telemetry.

c. Mixing.

d. Modulation.

1-13. Continuous voice or video signals are referred to as being

a. Baseband.

b. Analog.

c. Digital.

d. Continuous waves.

1-14. Recovering information from a carrier is known as

a. Demultiplexing.

b. Modulation.

c. Detection.

d. Carrier recovery.

1-15. Transmission of graphical information over the telephone network is accomplished by

a. Television.

b. CATV.

c. Videotext.

d. Facsimile.

1-16. Measuring physical conditions at some remote location and transmitting this data for analysis is the process of

a. Telemetry.

b. Instrumentation.

c. Modulation.

d. Multiplexing.

1-17. Receiving electromagnetic emissions from stars is called

a. Astrology.

b. Optical astronomy.

c. Radio astronomy.

d. Space surveillance.

1-18. A personal communication hobby for individuals is

a. Ham radio.

b. Electronic bulletin board.

c. CB radio.

d. Cellular radio.

1-19. Radar is based upon

a. Microwaves.

b. A water medium.

c. The directional nature of radio signals.

d. Reflected radio signals.

1-20. A frequency of 27 MHz has a wavelength of approximately

a. 11 m.

b. 27 m.

c. 30 m.

d. 81 m.

1-21. Radio signals are made up of

a. Voltages and currents.

b. Electric and magnetic fields.

c. Electrons and protons.

d. Noise and data.

1-22. The voice frequency range is

a. 30 to 300 Hz.

b. 300 to 3000 Hz.

c. 20 Hz to 20 kHz.

d. 0 Hz to 15 kHz.

1-23. Another name for signals in the HF range is

a. Microwaves.

b. RF waves.

c. Shortwaves.

d. Millimeter waves.

1-24. Television broadcasting occurs in which ranges?

a. HF

b. EHF

c. VHF

d. UHF

1-25. Electromagnetic waves produced primarily by heat are called

a. Infrared rays.

b. Microwaves.

c. Shortwaves.

d. X-rays.

1-26. A micron is

a. One-millionth of a foot.

b. One-millionth of a meter.

c. One-thousandth of a meter.

d. One ten-thousandth of an inch.

1-27. The frequency range of infrared rays is approximately

a. 30 to 300 GHz.

b. 4000 to 8000 Å.

c. 1000 to 10,000 Å.

d. 0.7 to 100 μm.

1-28. The approximate wavelength of red light is

a. 1000 μm.

b. 7000 Å.

c. 3500 Å.

d. 4000 Å.

1-29. Which of the following is *not* used for communications?

a. X-rays

b. Millimeter waves

c. Infrared

d. Microwaves

1-30. A signal occupies the spectrum space from 1.115 to 1.122 GHz. The bandwidth is

a. 0.007 MHz.

b. 7 MHz.

c. 237 MHz.

d. 700 MHz.

1-31. In the United States, the electromagnetic spectrum is regulated and managed by

a. Business and industry.

b. ITU.

c. FCC.

d. The United Nations.

1-32. For a given bandwidth signal, more channel space is available for signals in the range of

a. VHF.

b. UHF.

c. SHF.

d. EHF.

Critical Thinking Questions

1-1. Considering the wide range of communications applications outlined in Fig. 1-4, which one of these interests you the most as a career? Why?

1-2. Could light be used as a carrier for an intelligence signal? Explain.

1-3. What if all the available spectrum space from ELF to microwave was used up or otherwise assigned? How could communications capability be added?

1-4. List all the electronic communications techniques and services that you use on a regular basis. Which additional services would you like to use?

1-5. Discuss how each of the following could be used as the transmission medium for intelligence signals: (a) the earth, (b) a rubber hose, and (c) a piece of string.

1-6. Radio waves are made up of electric and magnetic fields that move through space. These fields can also be made to move along a trans-

mission line of two parallel conductors connecting a transmitter to its antenna load. With a measuring instrument, you can determine the points of maximum (or minimum) signal strength along the line. If the distance between two maximum points of the sinusoidal signal along the line is 36 cm, what is the signal frequency?

1-7. Based upon your general knowledge of AM and FM broadcasting by listening to the radio, name the frequency ranges occupied by AM and FM broadcast stations.

1-8. The U.S. government transmits precise time signals on frequencies of 5, 10, 15, and 20 MHz. What kind of receiver would you need to receive these signals?

1-9. What is the wavelength of the 60-Hz power line frequency?

1-10. Explain how you could use the ordinary ac power line wiring as a transmission medium for voice or digital data.

1. wireless
2. exchanging information
3. oral, written
4. distance, language
5. late nineteenth century
6. transmitter, receiver, channel or medium
7. wire, radio, fiber-optic cable
8. transmitter
9. receiver
10. noise, communications channel
11. degrades, attenuates
12. the atmosphere, manufactured equipment, thermal agitation in components
13. simplex, radio and TV broadcasting
14. full duplex, telephone communications
15. half duplex
16. analog
17. digital
18. analog-to-digital converter
19. baseband
20. modulation, baseband, carrier
21. demodulation or detection
22. multiplexing
23. facsimile, teletext
24. garage door opener
25. paging
26. telemetry
27. radio waves
28. microwave relay, satellites, cordless phones, cellular phones
29. reflected
30. sonar (active)
31. active, passive
32. amateur or "ham" radio
33. modems
34. local area networks
35. electromagnetic waves or radio-frequency (RF) waves
36. electric, magnetic
37. 16.67 ($300/_{18} = 16.67$)
38. 50, 60, ELF
39. *a, c*
40. 20, 20,000
41. 300, 3000
42. false
43. MF
44. shortwaves
45. VHF
46. land mobile, cellular telephones, military, radar and navigation, amateur radio
47. 1000
48. microwaves
49. radar, satellite
50. millimeter
51. $1/_{1,000,000}$
52. heat
53. 0.7, 10
54. $1/_{10,000}$
55. 4000, 8000
56. fiber-optic cables, free space
57. bandwidth
58. sidebands
59. 125 ($1.175 - 1.050 = 0.125$ MHz $= 125$ kHz)
60. higher
61. lower
62. spectrum space
63. Communications Act of 1934
64. Federal Communications Commission
65. National Telecommunications and Information Administration
66. International Telecommunications Union

Amplitude Modulation
and Single-Sideband Modulation

Chapter Objectives

This chapter will help you to:

1. *Recognize* an AM signal in the time domain (oscilloscope display), the frequency domain (spectral display), or in trigonometric equation form.

2. *Calculate* the percentage of modulation of an AM signal given waveform measurements.

3. *Calculate* the upper and lower sidebands of an AM signal given the carrier and modulating signal frequencies.

4. *Calculate* the sideband power in an AM wave given the carrier power and the percentage of modulation.

5. *Define* the terms DSB and SSB and *state* the benefits of SSB over an AM signal.

One of the principal techniques used in electronic communications is modulation. Modulation is the process of having the information to be transmitted alter a higher-frequency signal for the purpose of transmitting the information somewhere in the electromagnetic spectrum via radio, wire, or fiberoptic cable. Without modulation, electronic communications would not exist as we know it today. Communications electronics is largely the study of various modulation techniques and of the modulator and demodulator circuits that make modulation possible. The three principal types of electronic communications are amplitude modulation (AM), frequency modulation (FM), and phase modulation (PM). The oldest and simplest form of modulation is AM. In this chapter we will cover AM along with a derivation known as single-sideband modulation. Chapter 3 will cover amplitude modulator and demodulator circuits.

2-1 Amplitude Modulation Principles

Information signals such as voice, video, or binary data are sometimes transmitted directly from one point to another over some communications medium. For example, voice signals are transmitted by way of wires in the telephone system. Coaxial cables carry video signals between two points, and twisted-pair cable is often used to carry binary data from one point to another in a computer network. However, when transmission distances are far, cables are sometimes impractical. In such cases, ratio communications is used. To carry out reliable long-distance radio communication, a high-frequency signal must be used. It is simply impractical to convert the information signal directly to electromagnetic radiation. Excessively long antennas and interference between signals would result if information signals were transmitted directly. For this reason, it is desirable to translate the information signal to a point higher in the electromagnetic frequency spectrum. It is the process of modulation that creates a higher-frequency signal containing the original information.

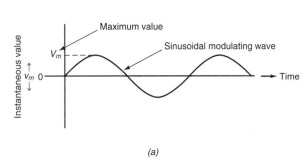

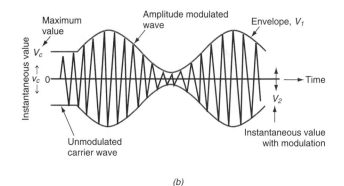

(a)

(b)

Fig. 2-1 Amplitude modulation. (a) The modulating or information signal. (b) The modulated carrier.

Modulation Defined

Modulation

Envelope

Modulation is the process of modifying the characteristic of one signal in accordance with some characteristic of another signal. In most cases, the information signal, be it voice, video, binary data, or some other information, is normally used to modify a higher-frequency signal known as the carrier. The information signal is usually called

Modulating signal

Carrier

the *modulating signal,* and the higher-frequency signal which is being modulated is called the *carrier* or *modulated wave.* The carrier is usually a sine wave, while the information signal can be of any shape, permitting both analog and digital signals to be transmitted. In most cases, the carrier frequency is considerably higher than the highest information frequency to be transmitted.

Time domain

Amplitude Modulation with Sine Waves

AM

In *AM,* the information signal varies the amplitude of the carrier sine wave. In other words, the

instantaneous value of the carrier amplitude changes in accordance with the amplitude and frequency variations of the modulating signal. Figure 2-1 shows a single-frequency sine wave modulating a higher-frequency carrier signal. Note that the carrier frequency remains constant during the modulation process but that its amplitude varies in accordance with the modulating signal. An increase in the modulating signal amplitude causes the amplitude of the carrier to increase. Both the positive and negative peaks of the carrier wave vary with the modulating signal. An increase or decrease in the amplitude of the modulating signal causes a corresponding increase or decrease in both the positive and negative peaks of the carrier amplitude.

If you interconnect the positive and negative peaks of the carrier waveform with an imaginary line (shown dashed in Fig. 2-1), then you re-create the exact shape of the modulating information signal. This imaginary line on the carrier waveform is known as the *envelope,* and it is the same as the modulating signal.

Because complex waveforms like that shown in Fig. 2-1 are difficult to draw, they are usually simplified by representing the high-frequency carrier wave as simply many equally spaced vertical lines whose amplitudes vary in accordance with a modulating signal. Figure 2-2 shows a sine wave tone modulating a higher-frequency carrier. We will use this method of representation throughout this book.

The signals illustrated in Figs. 2-1 and 2-2 show the variation of the carrier signal with respect to time. Such signals are said to be in the *time domain.* Time-domain signals are the actual variation of voltage over time. They are what you would see displayed on the screen of an oscilloscope. In this section we show the time-domain signals created by the various

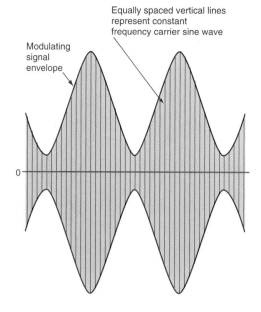

Equally spaced vertical lines represent constant frequency carrier sine wave

Modulating signal envelope

0

Fig. 2-2 A simplified method of representing an AM high-frequency sine wave.

types of modulation. Later you will see that modulated signals can also be expressed in the frequency domain.

Mathematical Representation of AM

Sinusoidal alternating current (ac) signals can be represented mathematically by trigonometric functions. For example, we can express the sine wave carrier with the simple expression

$$v_c = V_c \sin 2\pi f_c t$$

In this expression, v_c represents the instantaneous value of the sine wave voltage at some specific time in the cycle. The V_c represents the peak value of the sine wave as measured between zero and the maximum amplitude of either the positive- or negative-going alternations. See Fig. 2-1. The term f_c is the frequency of the carrier sine wave. Finally, t represents some particular point in time during the ac cycle.

In the same way, a sine wave modulating signal can also be expressed with a similar formula:

$$v_m = V_m \sin 2\pi f_m t$$

where f_m = the frequency of the modulating signal

Referring back to Fig. 2-1, you can see that the modulating signal uses the peak value of

---

DEGREE VERSUS RADIAN MEASURE OF ANGLES

You may recall that a sine wave reaches 70.7 percent of its maximum value at a phase angle of 45°. In general, the instantaneous value of a sine wave can be found by

$$v = V_m \times \sin \theta$$

where v = the instantaneous value
V_m = the maximum value
θ = the phase angle

Example: A 1-MHz sine wave has a peak or maximum value of 18 V. What is its instantaneous value at a phase angle of 45°?

$$\sin 45° = 0.707$$
$$v = 18 \text{ V} \times 0.707 = 12.7 \text{ V}$$

In communications, the phase angle may be stated in an equivalent way using the frequency of the signal and some time of interest. This is known as *radian* measure. For example, the instantaneous value of a signal can be found with

$$v = V_m \times \sin(2\pi f t)$$

where f = the frequency of the signal
t = the time of interest

A 1-MHz signal has a period of 1 μs. One period equals one cycle with 360°. A phase angle of 45° corresponds to 45/360 or 1/8 cycle, which is a time of 1μs divided by 8(0.125 μs).

Example: Use the second equation for finding the instantaneous value of a 1-MHz sine signal with a peak value of 18 volts (V) at a time of 0.125 μs. (Your calculator must be in the *radian* mode.)

$$v = 18 \text{ V} \times \sin (6.28 \times 1 \text{ MHz} \times 0.125 \text{ } \mu s)$$
$$= 12.7 \text{ V}$$

Conclusion: The two equations for finding v are equivalent. The first is based on angular measure, and the second is based on radian measure. Keep in mind that 1 rad = 57.3°. In our example, 45° = 45/57.3 = 0.785 rad.

$$\sin 0.785 = 0.707$$

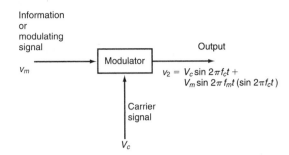

Fig. 2-3 Amplitude modulator showing input and output signals.

the carrier rather than zero as its reference point. The envelope of the modulating signal varies above and below the peak carrier amplitude. That is, the zero reference line of the modulating signal coincides with the peak value of the unmodulated carrier. Because of this, the relative amplitudes of the carrier and modulating signal are important. In general, the amplitude of the modulating signal should be less than the amplitude of the carrier. If the amplitude of the modulating signal is greater than the amplitude of the carrier, distortion will occur. Distortion causes incorrect information to be transmitted. It is important in AM that the peak value of the modulating signal be less than the peak value of the carrier.

Multiplication

Using the mathematical expressions for the carrier and the modulating signal, we can create a new mathematical expression for the complete modulated wave. First, keep in mind that the peak value of the carrier is the reference point for the modulating signal. The modulating signal value adds to or subtracts from the peak value of the carrier. This instantaneous value of either the top or bottom voltage envelope can be computed by the simple expression

Modulator

$$v_1 = V_c + v_m$$

Substituting the trigonometric expression for v_m, we get

$$v_1 = V_c + V_m \sin 2\pi f_m t$$

All this expression says is that the instantaneous value of the modulating signal algebraically adds to the peak value of the carrier. As you can see, the value of v_1 is really the envelope of the carrier wave. For that reason, we can write the instantaneous value of the complete modulated wave v_2 as

$$v_2 = v_1 \sin 2\pi f_c t$$

In this expression, the peak value of carrier wave V_c from the first equation given is replaced by v_1. Now, substituting the previously derived expression for v_1 and expanding, we get

$$v_2 = (V_c + V_m \sin 2\pi f_m t) \sin 2\pi f_c t$$
$$= \underbrace{V_c \sin 2\pi f_c t}_{} + \underbrace{(V_m \sin 2\pi f_m t)}_{} \underbrace{(\sin 2\pi f_c t)}_{}$$
$$\textit{Carrier} \quad + \quad \textit{modulation} \times \textit{carrier}$$

This expression consists of two parts: the first part is simply the carrier waveform, and the second part is the carrier waveform *multiplied* by the modulating signal waveform. It is this second part of the expression that is characteristic of AM. A circuit must be able to produce mathematical multiplication of analog signals in order for AM to occur.

The circuit used for producing AM is called a *modulator*. Its two inputs, the carrier and the modulating signal, and the resulting output are shown in Fig. 2-3. Amplitude modulators compute the product of the carrier and modulating signals.

Amplitude Modulation with Digital Signals

Digital, usually binary, signals may also be used to amplitude modulate a carrier. Figure 2-4 shows a binary signal modulating a sine wave carrier. In Fig. 2-4(*a*), the binary 1 level

Common AM Applications

1. AM radio broadcasting
2. TV picture (video)
3. Two-way radio
   a. Aircraft
   b. Amateur radio (SSB)
   c. Citizens' band radio
   d. Military
4. Digital data transmissions
5. Computer modems (used in combination with phase modulation QAM)
6. NIST time signals

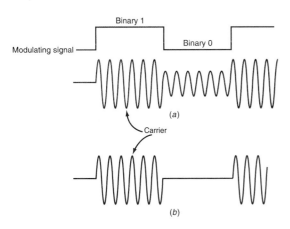

Fig. 2-4 Amplitude modulation of a carrier with a binary signal. (a) Amplitude shift keying (ASK). (b) On-off keying (OOK).

produces maximum carrier amplitude and the binary 0 level produces a lower-value carrier. Amplitude modulation in which the carrier is switched between two different carrier levels is known as *amplitude shift keying (ASK)*.

A special form of ASK is one in which the carrier is simply switched on or off. See Fig. 2-4(b). The binary 1 level turns the carrier on, and the binary 0 level turns the carrier off. This is called *on-off keying (OOK)*.

Some digital signals have more than two levels. As long as a signal varies in discrete steps, it is considered digital. Figure 2-5 shows a four-level digital signal and the resulting AM

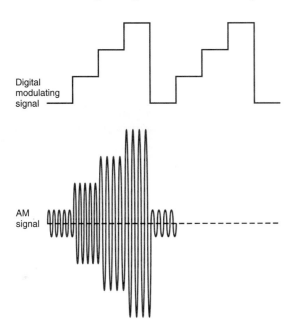

Fig. 2-5 Multilevel digital AM.

signal. To improve the speed of digital transmission in computer modems, 4-, 8-, 16- and 32-level digital signals are commonly used. Amplitude modulation is usually combined with simultaneous phase modulation of a carrier to produce *quadrature amplitude modulation (QAM)*.

Quadrature amplitude modulation (QAM)

■ TEST

Answer the following questions.

1. Modulation causes the information signal to be _____ to a higher frequency for more efficient transmission.
2. During modulation, the information signal _____ the amplitude of a high-frequency signal called the _____.
3. The circuit used to produce modulation is called a _____. Its two inputs are _____ and _____.
4. In AM, the instantaneous _____ of the carrier varies in accordance with the information signal.
5. True or false. The carrier frequency is usually lower than the modulating frequency.
6. The outline of the peaks of the carrier signal is called the _____, and it has the same shape as the _____.
7. Voltages varying over time are said to be _____ signals.
8. The trigonometric expression for the carrier is $v_c =$ _____.
9. True or false. The carrier frequency remains constant during AM.
10. An amplitude modulator performs the mathematical operation of
    a. Addition.   *c.* Multiplication.
    b. Subtraction.   *d.* Division.
11. AM with binary signals is called _____.
12. AM using the presence and absence of a carrier is called _____.

Amplitude shift keying (ASK)

On-off keying (OOK)

2-2 Modulation Index and Percentage of Modulation

In order for proper AM to occur, the modulating signal voltage V_m must be less than the carrier voltage V_c. Therefore, the relationship between the amplitudes of the modulating signal and carrier is important. This relationship is expressed in terms of a ratio known as the modulation

Did You Know?

Stray and distributed capacitances and inductances can greatly alter the operation and performance of a circuit.

index m (also called modulation factor, modulation coefficient, or the degree of modulation).

Modulation Index

Modulation index

Modulation index is simply the ratio of the modulating signal voltage to the carrier voltage:

$$m = \frac{V_m}{V_c}$$

The modulation index should be a number between 0 and 1. If the amplitude of the modulating voltage is higher than the carrier voltage, m will be greater than 1. This will cause severe distortion of the modulated waveform. This is illustrated in Fig. 2-6. Here a sine wave information signal modulates a sine wave carrier, but the modulating voltage is much greater than the carrier voltage. This condition is called *overmodulation.* As you can see, the waveform is flattened near the zero line. The received signal will produce an output waveform in the shape of the envelope, which in this case is a sine wave whose negative peaks have been clipped off. By keeping the amplitude of the modulating signal less than the carrier amplitude, no distortion will occur. The ideal condition for AM is where $V_m = V_c$ or $m = 1$, since this will produce the greatest output at the receiver with no distortion.

Overmodulation

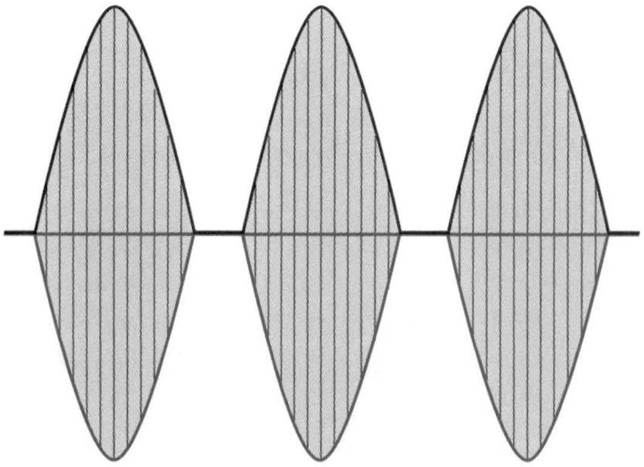

Fig. 2-6 Distortion of the envelope caused by overmodulation where the modulating signal amplitude V_m is greater than the carrier signal V_c.

interNET CONNECTION

For information on e-mail and Web services, visit the following Internet site:

⟨www.gl.umbc.edu⟩

The modulation index can be determined by measuring the actual values of the modulation voltage and the carrier voltage and computing the ratio. However, it is more common to compute the modulation index from measurements taken on the composite modulated wave itself. Whenever the AM signal is displayed on an oscilloscope, the modulation index can be computed from V_{max} and V_{min} as shown in Fig. 2.7.

The peak value of the modulating signal V_m is one-half the difference of the peak and trough values and is computed with the expression

$$V_m = \frac{V_{max} - V_{min}}{2}$$

By observing Fig. 2-7, you can see V_{max} is the peak value of the signal during modulation, while V_{min} is the lowest value, or trough, of the modulated wave. The V_{max} is one-half the peak-to-peak value of the AM signal or $V_{max(p\text{-}p)}/2$. Subtracting V_{min} from V_{max} produces the peak-to-peak value of the modulating signal. One-half of that, of course, is simply the peak value.

The peak value of the carrier signal V_c is the average of the V_{max} and V_{min} values and is computed with the expression

$$V_c = \frac{V_{max} + V_{min}}{2}$$

Substituting these values in our original formula for the modulation index produces the result

$$m = \frac{V_{max} - V_{min}}{V_{max} + V_{min}}$$

The values for V_{max} and V_{min} can be read directly from an oscilloscope screen and plugged into the formula to compute the modulation index.

For example, suppose that the V_{max} value read from the graticule on the oscilloscope screen is 4.6 divisions and V_{min} is 0.7 divisions. The modulation index is then

$$m = \frac{4.6 - 0.7}{4.6 + 0.7} = \frac{3.9}{5.3} = 0.736$$

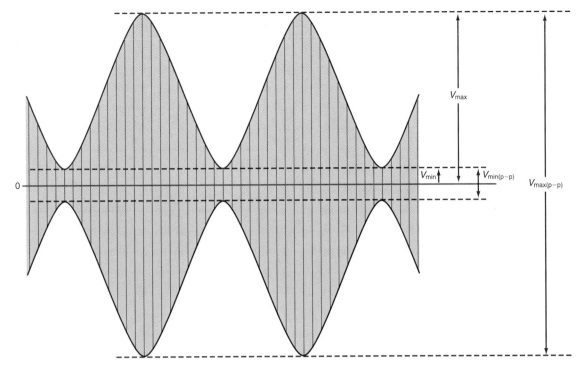

Fig. 2-7 An AM wave showing peaks (V_{max}) and troughs (V_{min}).

Percentage of Modulation

Whenever the modulation index is multiplied by 100, the degree of modulation is expressed as a percentage. The amount or depth of AM is more commonly expressed as percent modulation rather than as a fractional value. In the example above, the *percent of modulation* is 100% $\times$ *m* or 73.6 percent. The maximum amount of modulation without signal distortion, of course, is 100 percent where V_c and V_m are equal. At this time, $V_{min} = 0$ and $V_{max} = 2V_m$, where V_m is the peak value of the modulating signal.

In practice, it is desirable to operate with as close to 100 percent modulation as possible. In this way, the maximum information signal amplitude is transmitted. More information signal power is transmitted, thereby producing a stronger, more intelligible signal. When the modulating signal amplitude varies randomly over a wide range, it is impossible to maintain 100 percent modulation. A voice signal, for example, changes amplitude as a person speaks. Only the peaks of the signal produce 100 percent modulation.

■ TEST————————————————

Choose the letter that best answers each question.

13. Which of the following is the most correct?
    a. V_m should be greater than V_c.
    b. V_c should be greater than V_m.
    c. V_m should be equal to or less than V_c.
    d. V_c must always equal V_m.

14. Which of the following is *not* another name for modulation index?
    a. Modulation reciprocal
    b. Modulation factor
    c. Degree of modulation
    d. Modulation coefficient

15. The degree or depth of modulation expressed as a percentage is computed using the expression
    a. 2 V_m.
    b. 100/*m*.
    c. *m*/100.
    d. 100% $\times$ *m*.

Supply the missing information in each statement.

16. The modulation index is the ratio of the peak voltage of the _____ to the _____.

17. An AM wave displayed on an oscilloscope has values of $V_{max} = 3.8$ and $V_{min} = 1.5$ as read from the graticule. The percentage of modulation is _____ percent.

Percentage of modulation

18. The ideal percentage of modulation for maximum amplitude of information transmission is _____ percent.
19. To achieve 85 percent modulation of a carrier of $V_c = 40$ volts (V), a modulating signal of $V_m =$ _____ is needed.
20. The peak-to-peak value of an AM signal is 30 V. The peak-to-peak value of the modulating signal is 12 V. The percentage of modulation is _____ percent.
21. In Fig. 2-4(a), the carrier maximum value is 600 mV, and the carrier minimum is 300 mV. The percentage of modulation is _____.

2-3 Sidebands and the Frequency Domain

Sidebands

Whenever a carrier is modulated by an information signal, new signals at different frequencies are generated as part of the process. These new frequencies are called *side frequencies* or *sidebands* and occur in the frequency spectrum directly above and directly below the carrier frequency.

Sidebands

Side frequencies

If the modulating signal is a single-frequency sine wave, the resulting new signals produced by modulation are called *side frequencies.* If the modulating signal contains multiple frequencies such as voice, video, or digital signals, the result is a range of multiple side frequencies. These are referred to as sidebands.

The sidebands occur at frequencies that are the sum and difference of the carrier and modulating frequencies. Assuming a carrier frequency of f_c and a modulating frequency of f_m, the upper sideband f_{USB} and lower sideband f_{LSB} are computed as follows:

$$f_{USB} = f_c + f_m$$

$$f_{LSB} = f_c - f_m$$

The existence of these additional new signals that result from the process of modulation can also be proven mathematically. This can be done by starting with the equation for an AM signal v_2 described previously.

$$v_2 = V_c \sin 2\pi f_c t + (V_m \sin 2\pi f_m t)(\sin 2\pi f_c t)$$

There is a trigonometric identity that says that the product of two sine waves is

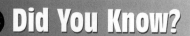

$$\sin A \sin B = \frac{\cos (A - B)}{2} - \frac{\cos (A + B)}{2}$$

By substituting this identity into the expression for our modulated wave, the instantaneous amplitude of the signal becomes

$$e_2 = \overbrace{V_c \sin 2\pi f_c t}^{\text{Carrier}} + \overbrace{\frac{V_m}{2} \cos 2\pi t(f_c - f_m)}^{\text{LSB}}$$
$$\underbrace{- \frac{V_m}{2} \cos 2\pi t(f_c + f_m)}_{\text{USB}}$$

As you can see, the second and third terms of this expression contain the sum $f_c + f_m$ and difference $f_c - f_m$ of the carrier and modulating signal frequencies. The first element in the expression is simply the carrier wave to which is added the difference frequency and the sum frequency.

By algebraically adding the carrier and the two sideband signals together, the standard AM waveform described earlier is obtained. This is illustrated in Fig. 2-8. This is solid proof that an AM wave contains not only the carrier but also the sideband frequencies. Observing an AM signal on an oscilloscope, you can see the amplitude variations of the carrier with respect to time. This is called a *time-domain display.* It gives no indication of the existence of the sidebands, although the modulation process does indeed produce them.

The Frequency Domain

Another method of showing the sideband signals is to plot the carrier and sideband amplitudes with respect to frequency. This is illustrated in Fig. 2-9. Here the horizontal axis represents frequency, and the vertical axis represents the amplitudes of the signals. A plot of

Adding these amplitudes

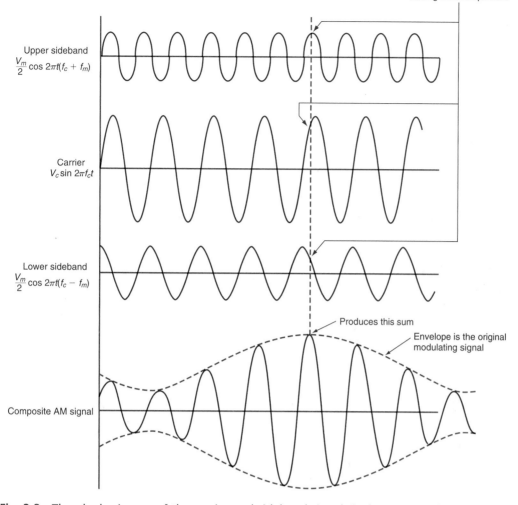

Upper sideband
$\frac{V_m}{2} \cos 2\pi t(f_c + f_m)$

Carrier
$V_c \sin 2\pi f_c t$

Lower sideband
$\frac{V_m}{2} \cos 2\pi t(f_c - f_m)$

Produces this sum

Envelope is the original modulating signal

Composite AM signal

Fig. 2-8 The algebraic sum of the carrier and sideband signals is the AM signal.

signal amplitude versus frequency is referred to as a *frequency-domain display*. A test instrument known as a spectrum analyzer will display the frequency domain of a signal.

Whenever the modulating signal is more complex than a single sine wave tone, multiple upper and lower side frequencies will be produced. For example, a voice signal consists of

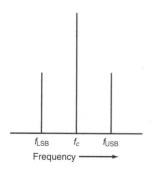

Fig. 2-9 A frequency-domain display of an AM signal.

many different-frequency sine wave components mixed together. Recall that voice frequencies occur in the 300- to 3000-Hz range. Therefore, voice signals will produce a range of frequencies above and below the carrier frequency as shown in Fig. 2-10. These sidebands take up spectrum space. You can compute the total bandwidth of the AM signal by computing the maximum and minimum sideband frequencies. This is done by finding the sum and difference of the carrier frequency and maximum modulating frequency, 3000 Hz, or 3 kHz, for voice transmission. If the carrier frequency is 2.8 MHz, or 2800 kHz, then the maximum and minimum sideband frequencies are

$$f_{\text{USB}} = 2800 + 3 = 2803 \text{ kHz}$$

$$f_{\text{LSB}} = 2800 - 3 = 2797 \text{ kHz}$$

The *total bandwidth* (*BW*), then, is simply the difference between the upper and lower sideband frequencies or

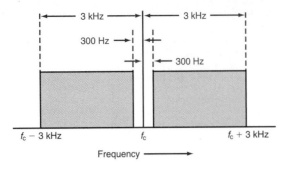

Fig. 2-10   The upper and lower sidebands of a voice modulated AM signal.

$$BW = f_{USB} = f_{LSB}$$

$$= 2803 - 2797$$

$$= 6\ kHz$$

As it turns out, the bandwidth of the AM signal is simply twice the highest frequency in the modulating signal. With a voice signal whose maximum frequency is 3 kHz, the total bandwidth would simply be twice this, or 6 kHz.

Sidebands Produced by Digital Signals

When other complex signals such as pulses or rectangular waves modulate a carrier, again a broad spectrum of sidebands is produced. According to the Fourier theory, complex signals such as square waves, triangular waves, sawtooth waves, or distorted sine waves are simply made up of a fundamental sine wave and numerous harmonic signals at different amplitudes. The classic example is that of a square wave which is made up of a fundamental sine wave and all odd harmonics. A modulating square wave will produce sidebands at frequencies of the fundamental square wave as well as at the third, fifth, seventh, etc., harmonics. The resulting frequency-domain plot would appear like that shown in Fig. 2-11(a). Pulses generate extremely wide bandwidth signals. In order for the square wave to be transmitted and received without distortion or degradation, all the sidebands must be passed by the antennas and the transmitting and receiving circuits.

Figure 2-11(b) shows the relationship between the time and frequency domain presentations of the modulating square wave. The time domain shows the individual sine wave harmonics that, when added together, produce the square wave. The frequency domain shows

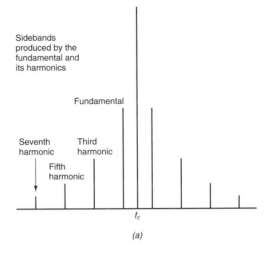

(a)

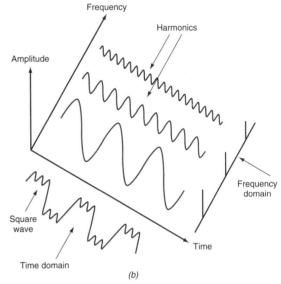

Fig. 2-11   (a) Frequency spectrum of an AM signal modulated by a square wave. (b) Relationship between time and frequency domain displays of the modulating signal.

the signal amplitudes of the harmonics that modulate the carrier and produce sidebands.

■ TEST

Answer the following questions.

22. New signals generated by the modulation process that appear directly above and below the carrier frequency are called

_____.

23. An AM radio station operating at 630 kHz is permitted to broadcast audio frequencies up to 5 kHz. The upper and lower side frequencies are

$$f_{USB} = _____\ kHz$$

$$f_{LSB} = _____\ kHz$$

24. The total bandwidth of the signal in the above example is _____ kHz.

25. A signal whose amplitude is displayed with respect to time is called a _____-domain display. The test instrument used to present such a display is the _____.

26. A signal whose amplitude is displayed with respect to frequency is called a _____-domain display. The test instrument used to present such a display is the _____.

27. Complex modulating signals containing many frequencies produce multiple _____ thus occupying more spectrum space.

28. The AM signal that occupies the greatest bandwidth is the one modulated by a
    a. 1-kHz sine wave.
    b. 5-kHz sine wave.
    c. 1-kHz square wave.
    d. 5-kHz square wave.

29. The composite AM signal can be re-created by algebraically adding which three signals?

30. True or False. Digital modulating signals typically produce an AM signal that has a greater bandwidth than an AM signal produced by an analog modulating signal.

2-4   Amplitude Modulation Power Distribution

To communicate by radio, the AM signal is amplified by a power amplifier and fed to the antenna with a characteristic impedance R. The total transmitted power divides itself between the carrier and the upper and lower sidebands. The total transmitted power P_T is simply the sum of the carrier power P_c and the power in the two sidebands P_{USB}, and P_{LSB}. This is expressed by this simple equation:

$$P_T = P_c + P_{LSB} + P_{USB}$$

Sideband Powers

The power in the sidebands depends upon the value of the modulation index. The greater the percentage of modulation, the higher the sideband power. Of course, maximum power appears in the sidebands when the carrier is 100 percent modulated. The power in each sideband P_s is given by the expression

$$P_s = P_{LSB} = P_{USB} = \frac{P_c (m^2)}{4}$$

Assuming 100 percent modulation where the modulation factor $m = 1$, the power in each sideband is one-fourth, or 25 percent, of the carrier power. Since there are two sidebands, their power together represents 50 percent of the carrier power. For example, if the carrier power is 100 watts (W), then at 100 percent modulation, 50 W will appear in the sidebands, 25 W in each. The total transmitted power then is the sum of the carrier and sideband powers or 150 W.

As you can see, the carrier power represents two-thirds of the total transmitted power assuming 100 percent modulation. With a carrier power of 100 W and a total power of 150 W, the carrier power percentage can be computed.

$$Carrier\ power\ percentage = \frac{100}{150}$$

$$= 0.667\ (or\ 66.7\%)$$

The percentage of power in the sidebands can be computed in a similar way:

$$Sideband\ power\ percentage = \frac{50}{150}$$

$$= 0.333\ (or\ 33.3\%)$$

The carrier itself conveys no information. The carrier can be transmitted and received, but unless modulation occurs, no information will be transmitted. When modulation occurs, sidebands are produced. It is easy to conclude, therefore, that all the transmitted information is contained within the sidebands. Only one-third of the total transmitted power is allotted to the sidebands, while the remaining two-thirds of it is literally wasted on the carrier. Obviously, although it is quite effective and still widely used, AM is a very inefficient method of modulation.

At lower percentages of modulation, the power in the sidebands is even less. You can compute the amount of power in a sideband with the previously given expression. Assume a carrier power of 500 W and a modulation of 70 percent. The power in each sideband then is

About ⬅️▬▬▶️ Electronics

Some cars rented in Miami come with cellular phones, GPS navigation, and a panic button that contacts police about the car's location.

$$P_s = \frac{P_c(m^2)}{4}$$

$$= \frac{500(0.7)^2}{4}$$

$$= \frac{500(0.49)}{4}$$

$$= 61.25 \text{ W}$$

At 70 percent modulation, only 61.25 W appears in each sideband for a total sideband power of 122.5 W. The carrier power, of course, remains unchanged at 500 W.

One way to calculate the total AM power is to use the formula

$$P_T = P_c(1 + m^2/2)$$

where P_c = unmodulated carrier power
m = modulation index

For example, if the carrier power is 1200 W and the percentage of modulation is 90 percent, the total power is

$$P_T = 1200(1 + 0.9^2/2) = 1200(1.405) = 1686 \text{ W}$$

If you subtract the carrier power, this will leave the power in both sidebands.

$$P_T = P_c + P_{\text{LSB}} + P_{\text{USB}}$$

$$P_{\text{LSB}} + P_{\text{USB}} = P_T - P_c$$

$$= 1686 - 1200 = 486 \text{ W}$$

Since the sideband powers are equal, the power in each sideband is 486/2 = 243 W. In practice, 100 percent modulation is difficult to maintain. The reason for this is that typical information signals, such as voice and video, do not have constant amplitudes. The complex voice and video signals vary over a wide amplitude and frequency range, so 100 percent modulation only occurs on the peaks of the modulating signal. For this reason, the average sideband power is considerably less than the ideal 50 percent produced by full 100 percent modulation. With less sideband power transmitted, the received signal is weaker and communication is less reliable.

Despite its inefficiency, AM is still widely used because it is simple and effective. It is used in AM radio broadcasting, CB radio, TV broadcasting, aircraft communications, and computer modems.

 that there are three basic ways to calculate the power dissipated in a load. These are:

$$P = V(I)$$

$$P = V^2/R$$

$$P = I^2R$$

Simply select the formula for which you have the values of current, voltage, or resistance. In an AM radio transmitting station, R is the load resistance, which is an antenna. To a transmitter, the antenna looks like a resistance. Although an antenna is not actually a physical resistor, it does appear to be one. This resistance is referred to as the *characteristic resistance of the antenna*. You will learn more about it in a later chapter.

Power Calculations

A common way to determine modulated power is to measure antenna current. Current in an antenna can be measured because accurate radio-frequency current meters are available. For example, if you know that the unmodulated carrier produces a current of 2.5 A in an antenna with a characteristic resistance of 73 Ω, the power is:

$$P = I^2R = (2.5)^2(73) = 6.25(73) = 456.25 \text{ W}$$

If the carrier is modulated, the antenna current will be higher because of the additional power in the sidebands. The total antenna current I_T is

$$I_T = I_c \sqrt{(1 + m^2/2)}$$

where I_c = unmodulated carrier current
m = modulation index

If the unmodulated carrier current is 4 A and the percentage of modulation is 70 percent, the total output current is

$$I_T = 4 \sqrt{(1 + 0.7^2/2)} = 4\sqrt{1.245}$$

$$= 4(1.116) = 4.46 \text{ A}$$

The total AM power then is

$$P_T = (I_T)^2 R$$

To determine the total power, monitor the total modulated antenna current and make the calculation above, given the antenna resistance.

If you measure both the modulated and the unmodulated carrier antenna currents, you can compute the percentage of modulation by using this formula:

$$m = \sqrt{2[(I_T/I_c)^2 - 1]}$$

Assume that you measured the unmodulated carrier current and found it to be 1.8 A. With modulation, the total current was 2 A. The percentage of modulation is:

$$m = \sqrt{2[(2/1.8)^2 - 1]} = \sqrt{2[1.234 - 1]}$$
$$= \sqrt{2(0.234)} = \sqrt{0.468} = 0.684 \text{ or } 68.4\%$$

■ TEST_____

Choose the letter which best answers each question.

31. The total sideband power is what percentage of the carrier power for 100 percent modulation?
    a. 25 percent
    b. 50 percent
    c. 100 percent
    d. 150 percent
32. Information in an AM signal is conveyed in the
    a. Carrier.
    b. Sidebands.
    c. Both together.

Supply the missing information in each statement.

33. The load into which the AM signal power is dissipated is a(n) _____.
34. The total transmitted power is the sum of the _____ and _____ powers.
35. A 5-kW carrier with 60 percent modulation produces _____ kW in each sideband.
36. In an AM signal with a carrier of 18 W and a modulation percentage of 75 percent, the total power in the sidebands is _____ W.
37. An AM signal with a carrier of 1 kW has 100 W in each sideband. The percentage of modulation is _____ percent.
38. An AM transmitter has a carrier power of 200 W. The percentage of modulation is

60 percent. The total signal power is _____ W.
39. The total AM signal power is 2800 W. The carrier power is 2000 W. The power in one sideband is _____ W. The percentage of modulation is _____.
40. The unmodulated carrier current in an antenna is 1.5 A. When the carrier is modulated by 95 percent, the total antenna current is _____ A.

2-5 Single-Sideband Communications

It is obvious from the previous discussion that AM is an inefficient and, therefore, wasteful method of communications. Two-thirds of the transmitted power appears in the carrier which itself conveys no information. The real information is contained within the sidebands. One way to overcome this problem is simply to suppress the carrier. Since the carrier does not provide any useful information, there is no reason why it has to be transmitted. By suppressing the carrier, the resulting signal is simply the upper and lower sidebands. Such a signal is referred to as a *double-sideband suppressed carrier* (*DSSC or DSB*) signal. The benefit, of course, is that no power is wasted on the carrier and that the power saved can be put into the sidebands. Double-sideband suppressed carrier modulation is simply a special case of AM with no carrier.

Double-sideband suppressed carrier

Double and Single Sidebands

Amplitude modulation generates two sets of sidebands, each containing the same information. The information is redundant in an AM or a DSB signal. Therefore, all the information can be conveyed in just one sideband. Eliminating one sideband produces a single-sideband (SSB) signal. Eliminating the carrier and one sideband produces a more efficient AM signal.

A typical DSB signal is shown in Fig. 2-12. This signal is simply the algebraic sum of the two sinusoidal sidebands. This is the signal produced when a carrier is modulated by a single-tone sine wave information signal. During the modulation process, the carrier is suppressed, but the two sideband signals remain. Even though the carrier is suppressed,

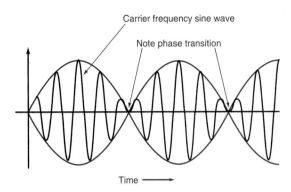

Fig. 2-12 A time-domain display of a DSB AM signal.

the time-domain DSB signal is a sine wave at the carrier frequency varying in amplitude as shown. Note that the envelope of this waveform is not the same as the modulating signal as it is in a pure AM signal with carrier.

A frequency-domain display of a DSB signal is given in Fig. 2-13. Note that the spectrum space occupied by a DSB signal is the same as that for a conventional AM signal.

Double-sideband suppressed carrier signals are generated by a circuit called a *balanced modulator*. The purpose of the balanced modulator is to produce the sum and difference frequencies but to cancel or balance out the carrier. You will learn more about balanced modulators in Chap. 3.

When DSB AM is used, considerable power is saved by eliminating the carrier. This power can be put into the sidebands for stronger signals over longer distances. Although a DSB AM signal is relatively easy to generate, DSB

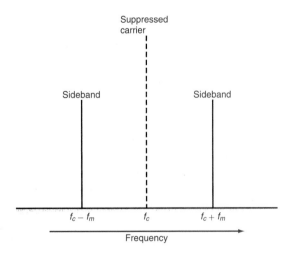

Fig. 2-13 A frequency-domain display of a DSB signal.

is rarely used because the signal is difficult to recover at the receiver.

Single Sideband Benefits

In a DSB signal, the basic information is transmitted twice, once in each sideband. Since the sidebands are the sum and difference of the carrier and modulating signals, the information must be contained in both of them. As it turns out, there is absolutely no reason to transmit both sidebands in order to convey the information. One sideband may be suppressed. The remaining sideband is called a *single-sideband suppressed carrier* (*SSSC* or *SSB*) signal.

The SSB signal offers four major benefits. First, the spectrum space occupied by the SSB signal is only half that of AM and DSB signals. This greatly conserves spectrum space and allows more signals to be transmitted in the same frequency range. It also means that there should be less interference between signals.

The second benefit is that all the power previously devoted to the carrier and other sideband can be channeled into the signal sideband, thereby producing a stronger signal that should carry farther and be more reliably received at greater distances.

The third benefit is that there is less noise on the signal. Noise gets added to all signals in the communications medium or in the receiver itself. Noise is a random voltage made up of an almost infinite number of frequencies. If the signal bandwidth is restricted and the receiver circuits are made with a narrower bandwidth, a great deal of the noise is filtered out. Since the SSB signal has less bandwidth than an AM or a DSB signal, logically there will be less noise on it. This is a major advantage in weak signal long-distance communications.

The fourth advantage of SSB signals is that they experience less fading than an AM signal. Fading means that a signal alternately increases and decreases in strength as it is picked up by the receiver. It occurs because the carrier and sidebands may reach the receiver shifted in time and phase with respect to one another. The carrier and sidebands, which are on separate frequencies, are affected by the ionosphere in different ways. The ionosphere bends the carrier and sideband signals back to earth at slightly different angles so that sometimes they reach the receiver in such a way that they cancel out one another rather than adding up to the desired AM wave. The result is fading. With SSB there is only one sideband, so this kind of fading does not occur.

An SSB signal has some unusual characteristics. First, when no information or modulating signal is present, no RF signal is transmitted. In a standard AM transmitter, the carrier is still transmitted even though it may not be modulated. This is the condition that might occur during a voice pause on an AM broadcast station. But since there is no carrier transmitted in an SSB system, no signals are present if the information signal is zero. Sidebands are generated only during the modulation process, such as when someone speaks into a microphone.

Figure 2-14 shows the frequency- and time-domain displays of an SSB signal produced when a steady 2-kHz sine wave tone modulates a 14.3-MHz carrier. Amplitude modulation would produce sidebands of 14.298 and 14.302 MHz. In SSB, only one sideband is used. Figure 2-14(a) shows that only the upper sideband is generated. The RF signal is simply a constant-power 14.302-MHz sine wave. A time-domain display of the SSB signal appears in Fig. 2-14(b).

Of course, most information signals transmitted by SSB are not pure sine waves. A more common modulation signal is voice with its varying frequency and amplitude content. The voice signal will create a complex RF SSB signal which varies in frequency and amplitude over the narrow spectrum defined by the voice signal bandwidth.

SSB Power

A voice signal with a frequency range of 200 to 4000 Hz modulates a 14.3-MHz carrier. A DSB AM modulator produces the following sidebands:

$$14.3 \text{ MHz} = 14,300 \text{ kHz} = 14,300,000 \text{ Hz}$$

Upper side frequencies:

$$14,300,000 + 200 = 14,300,200 \text{ Hz}$$
$$14,300,000 + 4000 = 14,304,000 \text{ Hz}$$

Lower side frequencies:

$$14,300,000 - 200 = 14,299,800 \text{ Hz}$$
$$14,300,000 - 4000 = 14,296,000 \text{ Hz}$$

The upper sideband extends from 14,300,200 to 14,304,000 Hz and occupies a bandwidth of

$$BW = 14,304,000 - 14,300,200 = 3800 \text{ Hz}$$

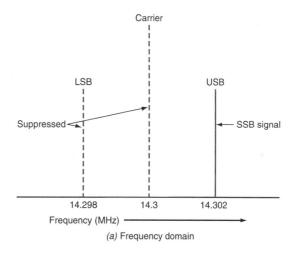

(a) Frequency domain

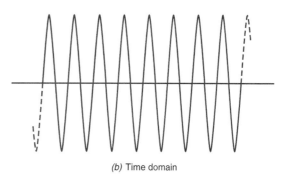

SSB signal
14.302 MHz sine wave

(b) Time domain

Fig. 2-14 An SSB signal produced by a 2-kHz sine wave modulating a 14.3-MHz sine wave carrier.

The lower sideband extends from 14,296,000 to 14,299,000 Hz and occupies a bandwidth of

$$BW = 14{,}299{,}800 - 14{,}296{,}000 = 3800 \text{ Hz}$$

The DSB signal consumes a total bandwidth of $4000 \times 2 = 8000$ Hz (8 kHz). If we transmit only one sideband, the bandwidth is 3800 Hz.

The SSB signal may be either the upper sideband (USB) or the lower sideband (LSB). In practice, an SSB transmitter generates both sidebands and a switch is used to select either the USB or the LSB for transmission.

When the voice or other modulating signal is zero, no SSB signal is produced. An SSB RF signal is produced only when modulation occurs. In AM, with no modulating signal, the carrier would still be transmitted. This explains why SSB is so much more efficient than AM.

In conventional AM, the transmitter power is distributed among the carrier and two sidebands. If we assume a carrier power of 100 W and 100 percent modulation, each sideband will contain 25 W of power. The total transmitted power will be 150 W. The communication effectiveness of this conventional AM transmitter is established by the combined power in the sidebands, or 50 W in this example.

An SSB transmitter sends no carrier, so the carrier power is zero. Such a transmitter will have the same communication effectiveness as a conventional AM unit running much more power. A 50-W SSB transmitter will equal the performance of an AM transmitter running a total of 150 W, since they both show 50 W of total sideband power. The power advantage of SSB over AM is 3:1.

Peak envelope power (PEP)

In SSB, the transmitter output is expressed in terms of *peak envelope power* (*PEP*). This is the maximum power produced on voice amplitude peaks. The PEP is computed by the familiar expression

$$P = \frac{V^2}{R}$$

where P = output power
V = root mean square (rms) output voltage
R = load resistance (usually antenna characteristic impedance)

As an example, assume that a voice signal produces a 120-volt (V) peak-to-peak signal across a 50-ohm (Ω) load. The rms voltage is 0.707 times the peak value. The peak value is one-half the peak-to-peak voltage. In this example, the rms voltage is

$$0.707\left(\frac{120}{2}\right) = 42.42 \text{ V}$$

The peak envelope power then is

$$PEP = \frac{V^2}{R}$$
$$= \frac{(42.42)^2}{50}$$
$$= 36 \text{ W}$$

the PEP input power is simply the direct-current (dc) input power of the transmitter final amplifier stage at the instant of the voice envelope peak. It is the final amplifier stage dc supply voltage multiplied by the maximum amplifier current that occurs at the peak or

$$PEP = V_s I_{max}$$

where V_s = the amplifier supply voltage
I_{max} = the current peak

A 300-V supply with a peak current of 0.6 ampere (A) produces a PEP of

$$PEP = 300(0.6) = 180 \text{ W}$$

It is important to point out that the PEP occurs only occasionally.

Voice amplitude peaks are produced only when very loud sounds are generated during certain speech patterns or when some word or sound is emphasized. During normal speech levels, the input and output power levels are much less than the PEP level. The average power is typically only one-fourth to one-third of the PEP value with typical human speech.

$$P_{avg} = \frac{PEP}{3} \quad or \quad P_{avg} = \frac{PEP}{4}$$

With a PEP of 180 W, the average power is only 45 to 60 W. Typical SSB transmitters are designed to handle only the average power level on a continuous basis, not the PEP.

The transmitted sideband will, of course, change in frequency and amplitude as a complex voice signal is applied. This sideband will occupy the same bandwidth as one sideband in a fully modulated AM signal with carrier.

Incidentally, it does not matter whether the upper or lower sideband is used since the information is contained in either. A filter is typically used to remove the unwanted sideband.

DSB and SSB Applications

Both DSB and SSB techniques are widely used in communications. Pure SSB signals are used in telephone systems as well as in two-way radio. Two-way SSB communications is used in the military, in CB radio, and by hobbyists known as radio amateurs.

The DSB signals are used in FM and TV broadcasting to transmit two-channel stereo signals. They are also used in some types of phase-shift keying which is used for transmitting binary data.

An unusual form of AM is that used in television broadcasting. A TV signal consists of the picture (video) signal and the audio signal which have different carrier frequencies. The audio carrier is frequency-modulated, but the video information amplitude-modulates the picture carrier. The picture carrier is transmitted, but one sideband is partially suppressed.

Video information typically contains frequencies as high as 4.2 MHz. A fully amplitude-modulated television signal would then occupy $2(4.2) = 8.4$ MHz. This is an excessive amount of bandwidth that is wasteful of spectrum space because not all of it is required to reliably transmit a TV signal. To reduce the bandwidth to the 6-MHz maximum allowed by the FCC or TV signals, a portion of the lower sideband of the TV signal is suppressed leaving only a small vestige of the lower sideband. Such an arrangement is know as a *vestigial sideband signal*. It is illustrated in Fig. 2-15. Video signals above 0.75 MHz (750 kHz) are suppressed in the lower sideband, and all video frequencies are transmitted in the upper sideband.

■ TEST

Answer the following questions.

41. An AM signal without a carrier is called a(n) _____ signal.
42. True or false. Two sidebands must be transmitted to retain all the information.
43. The acronym SSB means

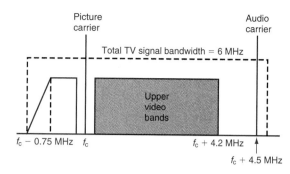

Fig. 2-15 Vestigial sideband transmission of a TV picture signal.

    a. Single sideband with carrier.
    b. Single sideband with suppressed carrier.
    c. Double sideband with no carrier.
    d. Double sideband with carrier.
44. A major benefit of DSB and SSB is
    a. Higher power can be put into the sideband(s).
    b. Greater power consumption.
    c. More carrier power.
    d. Double the sideband power.
45. List four benefits of SSB over AM and DSB.
46. List two common uses of SSB.
47. A common use of DSB is
    a. Two-way communication.
    b. Telephone systems.
    c. FM/TV stereo.
    d. Satellite communications.
48. True or false. In SSB, no signal is transmitted unless the information signal is present.
49. The type of AM signal that is used in TV broadcasting is called _____ transmission.
50. An SSB signal produces a peak-to-peak voltage of 720 V on voice peaks across a 75-Ω antenna. The PEP output is _____ W.
51. An SSB transmitter has a 150-V supply. Voice peaks produce a current of 2.3 A. The PEP input is _____ W.
52. The average output power of an SSB transmitter rated at 12 W PEP is in the _____ to _____ range.

Vestigial sideband

Summary

1. Modulation is the process of having the information to be communicated modify a higher-frequency signal called a carrier.
2. Amplitude modulation (AM) is the oldest and simplest form of modulation.
3. In AM, the amplitude of the carrier is changed in accordance with the amplitude and frequency or the characteristics of the modulating signal. The carrier frequency remains constant.
4. The amplitude variation of the carrier peaks has the shape of the modulating signal and is referred to as the envelope.
5. A time-domain display shows amplitude versus time variation of AM and other signals.
6. Amplitude modulation is produced by a circuit called a modulator which has two inputs and an output.
7. The modulator performs a mathematical multiplication of the carrier and information signals. The output is their analog product.
8. The carrier in an AM signal is a sine wave that may be modulated by either analog or digital information signals.
9. Amplitude modulation of a carrier by a binary signal produces amplitude shift keying (ASK).
10. The ratio of the peak voltage value of the modulating signal V_m to the peak value of the carrier V_c is called the modulation index m ($m = V_m/V_c$). It is also referred to as the modulation coefficient or factor and the degree of modulation.
11. The ideal value for m is 1. Typically m is less than 1. The condition where m is greater than 1 should be avoided as it introduces serious distortion of the modulating signal. This is called overmodulation.
12. When the modulation index is multiplied by 100, it is called the percentage of modulation.
13. The percentage of modulation can be computed from AM waveforms displayed on an oscilloscope by using the expression

$$\% \text{ mod.} = \frac{100(V_{max} - V_{min})}{V_{max} + V_{min}}$$

where V_{max} is the maximum peak carrier amplitude and V_{min} is the minimum peak carrier amplitude.

14. The new signals generated by the modulation process are called sidebands and occur at frequencies above and below the carrier frequency.
15. The upper f_{USB} and lower f_{LSB} sideband frequencies are the sum and difference of the carrier frequency f_c and the modulating frequency f_m and are computed with the expressions

$$f_{USB} = f_c + f_m$$
$$f_{LSB} = f_c - f_m$$

16. A display of signal amplitudes with respect to frequency is called a frequency-domain display.
17. An AM signal can be viewed as the carrier signal added to the sideband signals produced by AM.
18. The total transmitted power in an AM signal is the sum of the carrier and sideband powers $P_T = P_c + P_{USB} + P_{LSB}$) and is distributed among the carrier and sidebands. This power distribution varies with the percentage of modulation. The total power is

$$P_T = P_c(1 + m^2/2)$$

The power in each sideband is

$$P_s = \frac{P_c(m^2)}{4}$$

19. The higher the percentage of modulation, the greater the sideband power and the stronger and more intelligible the transmitted and received signal.
20. Despite its simplicity and effectiveness, AM is a highly inefficient method of modulation.
21. In an AM signal, the carrier contains no information. Any transmitted information lies solely in the sideband. For that reason, the carrier may be suppressed and not transmitted.
22. An AM signal with suppressed carrier is called a double-sideband (DSB) signal.

23. Since the same transmitted information is contained in both upper and lower sidebands, one is redundant. Full information can be transmitted using only one sideband.
24. An AM signal with no carrier and one sideband is called a single-sideband (SSB) signal. The upper and lower sidebands contain the same information, and one is not preferred over the other.
25. The main advantage of an SSB AM signal over an AM or DSB signal is that it occupies one-half the spectrum space.
26. Both DSB and SSB signals are more efficient in terms of power usage. The power wasted in the useless carrier is saved, thereby allowing more power to be put into the sidebands.
27. Power in an SSB transmitter is rated in terms of peak envelope power (PEP), the power that is produced on voice peaks. PEP output is computed using the expression

$$PEP = \frac{V^2}{R}$$

where PEP is in watts and V is the rms voltage across the antenna load impedance R. The PEP input is computed using the expression

$$PEP = V_s \times I_{max}$$

where V_s is the dc supply voltage of the final amplifier stage and I_{max} is the amplifier current on voice peaks.
28. The average output of an SSB transmitter is one-fourth to one-third of the PEP value.
29. DSB AM is not widely used. However, SSB is widely used in two-way radio communications.
30. A special form of amplitude modulation is used in TV transmission. Known as vestigial sideband, this method filters out a portion of the lower video sidebands to decrease the overall bandwidth of the AM picture signal to 6 MHz.

Chapter Review Questions

Choose the letter that best answers each question.

2-1. Having an information signal change some characteristic of a carrier signal is called
  a. Multiplexing.
  b. Modulation.
  c. Duplexing.
  d. Linear mixing.
2-2. Which of the following is *not* true about AM?
  a. The carrier amplitude varies.
  b. The carrier frequency remains constant.
  c. The carrier frequency changes.
  d. The information signal amplitude changes the carrier amplitude.
2-3. The opposite of modulation is
  a. Reverse modulation.
  b. Downward modulation.
  c. Unmodulation.
  d. Demodulation.
2-4. The circuit used to produce modulation is called a
  a. Modulator.
  b. Demodulator.
  c. Variable gain amplifier.
  d. Multiplexer.
2-5. A modulator circuit performs what mathematical operation on its two inputs?
  a. Addition
  b. Multiplication

  c. Division
  d. Square root
2-6. The ratio of the peak modulating signal voltage to the peak carrier voltage is referred to as
  a. The voltage ratio.
  b. Decibels.
  c. The modulation index.
  d. The mix factor.
2-7. If m is greater than 1, what happens?
  a. Normal operation.
  b. Carrier drops to zero.
  c. Carrier frequency shifts.
  d. Information signal is distorted.
2-8. For ideal AM, which of the following is true?
  a. $m = 0$
  b. $m = 1$
  c. $m < 1$
  d. $m > 1$
2-9. The outline of the peaks of a carrier has the shape of the modulating signal and is called the
  a. Trace.
  b. Waveshape.
  c. Envelope.
  d. Carrier variation.
2-10. Overmodulation occurs when
  a. $V_m > V_c$
  b. $V_m < V_c$

c. $V_m = V_c$

d. $V_m = V_c = 0$

2-11. The values of V_{max} and V_{min} as read from an AM wave on an oscilloscope are 2.8 and 0.3. The percentage of modulation is

a. 10.7 percent

b. 41.4 percent.

c. 80.6 percent.

d. 93.3 percent.

2-12. The new signals produced by modulation are called

a. Spurious emissions.

b. Harmonics.

c. Intermodulation products.

d. Sidebands.

2-13. A carrier of 880 kHz is modulated by a 3.5-kHz sine wave. The LSB and USB are, respectively,

a. 873 and 887 kHz.

b. 876.5 and 883.5 kHz.

c. 883.5 and 876.5 kHz.

d. 887 and 873 kHz.

2-14. A display of signal amplitude versus frequency is called the

a. Time domain.

b. Frequency spectrum.

c. Amplitude spectrum.

d. Frequency domain.

2-15. Most of the power in an AM signal is in the

a. Carrier.

b. Upper sideband.

c. Lower sideband.

d. Modulating signal.

2-16. An AM signal has a carrier power of 5 W. The percentage of modulation is 80 percent. The total sideband power is

a. 0.8 W.

b. 1.6 W.

c. 2.5 W.

d. 4.0 W.

2-17. For 100 percent modulation, what percentage of power is in each sideband?

a. 25 percent

b. 33.3 percent

c. 50 percent

d. 100 percent

2-18. An AM transmitter has a percentage of modulation of 88. The carrier power is 440 W. The power in one sideband is

a. 85 W.

b. 110 W.

c. 170 W.

d. 610 W.

2-19. An AM transmitter antenna current is measured with no modulation and found to be 2.6 amperes. With modulation, the current rises to 2.9 amperes. The percentage of modulation is

a. 35 percent.

b. 70 percent.

c. 42 percent.

d. 89 percent.

2-20. What is the carrier power in the problem above if the antenna resistance is 75 ohms?

a. 195 W.

b. 631 W.

c. 507 W.

d. 792 W.

2-21. In an AM signal, the transmitted information is contained within the

a. Carrier.

b. Modulating signal.

c. Sidebands.

d. Envelope.

2-22. An AM signal without the carrier is called a(n)

a. SSB.

b. Vestigial sideband.

c. FM signal.

d. DSB.

2-23. What is the minimum AM signal needed to transmit information?

a. Carrier plus sidebands

b. Carrier only

c. One sideband

d. Both sidebands

2-24. The main advantage of SSB over standard AM or DSB is

a. Less spectrum space is used.

b. Simpler equipment is used.

c. Less power is consumed.

d. A higher modulation percentage.

2-25. In SSB, which sideband is the best to use?

a. Upper

b. Lower

c. Neither

d. Depends upon the use

2-26. The typical audio modulating frequency range used in radio and telephone communications is

a. 50 Hz to 5 kHz.

b. 50 Hz to 15 kHz.

c. 100 Hz to 10 kHz.

d. 300 Hz to 3 kHz.

2-27. An AM signal with a maximum modulating signal frequency of 4.5 kHz has a total bandwidth of

a. 4.5 kHz.   c. 9 kHz.

b. 6.75 kHz.   d. 18 kHz.

2-28. Distortion of the modulating signal produces harmonics which cause an increase in the signal
   a. Carrier power.
   b. Bandwidth.
   c. Sideband power.
   d. Envelope voltage.
2-29. In Fig. 2-4(b), the peak carrier value is 7 V. What is the percentage of modulation?
   a. 0%
   b. 50%
   c. 75%
   d. 100%
2-30. The bandwidth of an SSB signal with a carrier frequency of 2.8 MHz and a modulating signal with a frequency range of 250 Hz to 3.3 kHz is
   a. 500 Hz
   b. 3050 Hz
   c. 6.6 kHz
   d. 7.1 kHz
2-31. The output of an SSB transmitter with a 3.85-MHz carrier and a 1.5-kHz sine wave modulating tone is

   a. a 3.8485-MHz sine wave.
   b. a 3.85-MHz sine wave.
   c. 3.85-, 3.8485-, and 3.8515-MHz sine waves.
   d. 3848.5- and 3851.5-MHz sine waves.
2-32. An SSB transmitter produces a 400-V peak-to-peak signal across a 52-Ω antenna load. The PEP output is
   a. 192.2 W.
   b. 384.5 W.
   c. 769.2 W.
   d. 3077 W.
2-33. The output power of an SSB transmitter is usually expressed in terms of
   a. Average power.
   b. RMS power.
   c. Peak-to-peak power.
   d. Peak envelope power.
2-34. An SSB transmitter has a PEP rating of 1 kW. The average output power is in the range of
   a. 150 to 450 W.
   b. 100 to 300 W.
   c. 250 to 333 W.
   d. 3 to 4 kW.

Critical Thinking Questions

2-1. Explain why an overmodulated AM signal occupies a lot of bandwidth.
2-2. Would it be possible to transmit one intelligence signal in the upper sideband and a different intelligence signal in the lower sideband of an AM or a DSB signal? Explain.
2-3. Explain how a potentiometer could be connected to demonstrate AM.
2-4. What are the side frequencies produced by a carrier modulated with a signal equal to the carrier frequency?
2-5. During a weak AM signal transmission, will talking louder produce a stronger and more intelligible signal? Explain.
2-6. An AM communication system consists of 30 channels spaced 5 kHz from one another.

Name two ways that can be used to prevent one station from interfering with adjacent channel stations.
2-7. An AM signal is restricted to a channel 4.5 kHz wide. What is the highest frequency that can be transmitted without going out of the channel?
2-8. Could a voice signal (300–3000 Hz) be transmitted without modulation by amplifying the signal audio power amplifier and connecting its output to an antenna? Explain. What are the problems with this system?
2-9. A constant-amplitude signal of 8.361 MHz is received. It is known that the source of the signal uses AM or SSB. Name three possible conditions the signal may represent.

Answers to Tests

1. translated
2. varies, carrier
3. modulator, carrier, modulating signal
4. amplitude

5. false
6. envelope, modulating signal
7. time-domain
8. $V_c \sin 2\pi f_c t$

9. true
10. *c*
11. amplitude shift keying
12. on-off keying
13. *c*
14. *a*
15. *d*
16. modulating signal, carrier
17. 43.4
18. 100
19. 34 V
20. 66.67
21. 33.33 percent
22. sidebands
23. 635, 625
24. 10
25. time, oscilloscope
26. frequency, spectrum analyzer
27. sidebands
28. *d*
29. Carrier, upper sideband, lower sideband
30. true
31. *b*

32. *b*
33. antenna
34. carrier, sideband
35. 0.45
36. 5.06
37. 63.25
38. 236 W
39. 400 W, 89.44 percent
40. 1.8
41. DSB
42. false
43. *b*
44. *a*
45. less spectrum space, more power in the sidebands with greater efficiency, less noise, little or no fading
46. telephone systems, two-way radio
47. *c*
48. true
49. vestigial sideband
50. 863.7
51. 345
52. 3-, 4-W

Chapter 3

Amplitude Modulation Circuits

Chapter Objectives

This chapter will help you to:

1. *Explain* the operation of low-level diode modulators and high-level collector modulators.

2. *Explain* the operation of a diode detector circuit.

3. *State* the function of a balanced modulator and *explain* the operation of diode and IC balanced modulators.

4. *Draw* a block diagram of a filter-type SSB generator and *name* three types of filters used to eliminate one sideband.

O ver the years, hundreds of circuits have been developed to produce AM. These modulator circuits cause the amplitude of the carrier to be varied in accordance with the modulating information signal. There are circuits to produce AM, DSB, and SSB. In this chapter, you will examine some of the more common and widely used discrete-component and integrated-circuit (IC) amplitude modulators. This chapter also covers demodulators. A demodulator is the circuit that recovers the original information signal from the modulated wave. Demodulators for AM, DSB, and SSB will also be discussed.

3-1 Amplitude Modulators

There are two basic ways to produce amplitude modulation. The first is to *multiply* the carrier by a gain or attenuation factor that varies with the modulating signal. The second is to linearly mix or algebraically add the carrier and modulating signals and then apply the composite signal to a nonlinear device or circuit. All amplitude modulators are based upon one of these two methods.

Analog Multiplication

You can see how the first method works by referring to the basic AM equation in Chap. 2. It is:

$$V_{am} = V_c \sin 2\pi f_c t + (V_m \sin 2\pi f_m t)(\sin 2\pi f_c)$$

We know that the modulation index is

$$m = V_m/V_c$$

Therefore

$$V_m = mV_c$$

Substituting this in the equation above and factoring gives:

$$V_{am} = \sin 2\pi f_c t(V_c + mV_c \sin 2\pi f_m t)$$
$$= V_c \sin 2\pi f_c t(1 + m \sin 2\pi f_m t)$$

You can see in this equation that AM is accomplished by multiplying the carrier by a factor equal to 1 plus the modulating sine wave. If we can create a circuit with a gain or attenuation that can be varied in accordance with the modulating signal, AM is produced by passing the carrier through it. Certain types of amplifiers and voltage dividers (for example, PIN diodes) can be created to do this. Better still, simply use one of the many available analog multiplier or modulator integrated circuits.

Multiplication

Nonlinear Mixing

With this method, we linearly mix the carrier and modulating signals. Then we use this composite voltage to vary the current in a nonlinear device. A nonlinear device is one whose current is proportional to but does not vary linearly with the applied voltage. One nonlinear device is the field-effect transistor (FET), whose response curve is approximately a parabola, indicating that the current in the device is proportional to the square of the input voltage (see Fig. 3-1). We say that such a device or circuit has a *square law response*. Squaring the sum of the carrier and modulating signals produces the classic AM equation described earlier. This exponential relationship is the basis for all amplitude modulation, mixing, and heterodyning. Other unwanted signals such as the second harmonics (also called second-order products) are also produced by this circuit. But a filter or tuned circuit used on the output will reject these unwanted products and leave only the carrier and sidebands.

Diodes and bipolar transistors are also nonlinear devices. Their response is not square law, but they also are capable of producing amplitude modulation. The advantage of using a true square-law device (such as an FET) is that only second-order products are produced. Other nonlinear devices, such as diodes and BJTs, produce third-order products as well. Since filters can remove the unwanted high-order products, this is often not a significant problem. So diodes and BJTs are useful as amplitude modulators.

One of the oldest and simplest amplitude modulators is shown in Fig. 3-2. It consists of a resistive mixing network, a diode rectifier, and an *LC* tuned circuit. The carrier is applied to one input resistor and the modulating signal to the other. The mixed signals appear across R_3. This network causes the two signals to be *linearly mixed,* that is, *algebraically added.* If both the carrier and the modulating signal are sine waves, the waveform resulting at the junction of the two resistors is shown in Fig. 3-3(c). The result is the carrier wave riding on the modulating signal. The important thing to remember about this signal is that it is not AM.

Square law response

Linear mixing (algebraically adding)

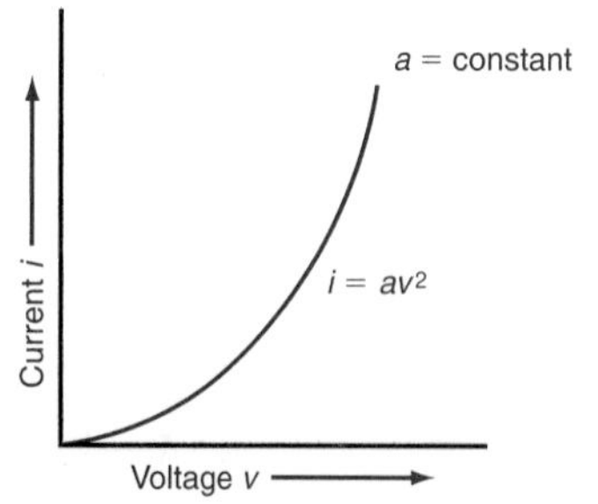

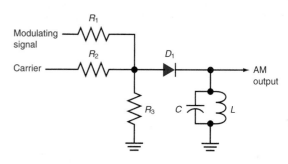

Fig. 3-2 Amplitude modulation with a diode.

The two signals have simply been added together or linearly mixed. Modulation is a multiplication process and not an addition process.

The composite waveform is applied to a diode rectifier that has an exponential response. The diode is connected so that it is forward-biased by the positive-going half cycles of the input wave. During the negative portions of the wave, the diode is cut off and no signal passes. The current through the diode is a series of positive-going pulses whose amplitude varies in proportion with the amplitude of the modulating signal. See Fig. 3-3(d).

These positive-going pulses are applied to the parallel tuned circuit made up of *L* and *C.* Both *L* and *C* resonate at the carrier frequency. Each time the diode conducts, a pulse of current flows through the tuned circuit. The coil and capacitor repeatedly exchange energy, causing an

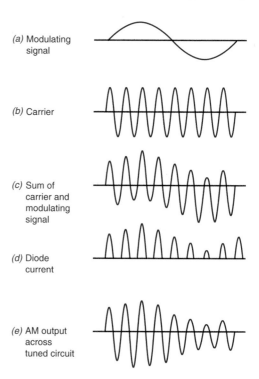

(a) Modulating signal

(b) Carrier

(c) Sum of carrier and modulating signal

(d) Diode current

(e) AM output across tuned circuit

Fig. 3-3 Waveforms in the diode modulator.

![Square law response curve graph]

$a = \text{constant}$

$i = av^2$

Current i →

Voltage v →

Fig. 3-1 A square law response curve produced by a diode or appropriately biased transistor.

oscillation or "ringing" at the resonant frequency. What happens is that the ringing or flywheel action of the tuned circuit creates a negative half cycle for every positive input pulse. High-amplitude positive pulses cause the tuned circuit to produce high-amplitude negative pulses. Low-amplitude positive pulses produce corresponding low-amplitude negative pulses. The resulting waveform across the tuned circuit is AM, as Fig. 3-3(e) illustrates. Of course, the Q of the tuned circuit should be high enough to produce a clean sine wave but also low enough so that its bandwidth will accommodate the sidebands generated. Another way to look at this is to view the

LC circuit as a bandpass filter that passes the carrier and sidebands but rejects the unwanted products of the exponential response.

Differential Amplifier Modulators

Most low-level amplitude modulators in use today are implemented with differential amplifiers in integrated-circuit (IC) form. A typical circuit is shown in Fig. 3-4(a). Transistors Q_1 and Q_2 form the

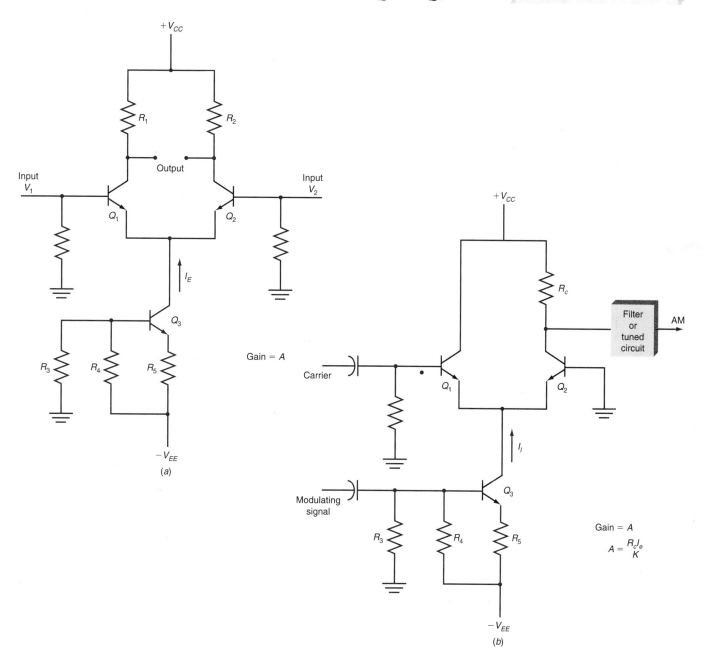

Fig. 3-4 (a) Basic differential amplifier. (b) Differential amplifier modulator.

differential pair, and Q_3 is a constant-current source. Transistor Q_3 supplies a fixed emitter current I_E to Q_1 and Q_2, half of which flows in each transistor. The output is developed across the collector resistors R_1 and R_2.

The output is a function of the difference between inputs V_1 and V_2; that is, $V_o = A(V_2 - V_1)$, where A is the circuit gain. The amplifier can also be operated with a single input. When this is done, the other input is grounded or set to zero. In Fig. 3-4(a), if V_1 is zero, the output at the collector of Q_1 is $V_o = A(V_2)$. If V_2 is zero, the output at the collector of Q_1 is $V_o = A(-V_1) = -AV_1$. This means that the circuit inverts V_1. The output at the collector of Q_2 is an inverted version of that at the collector of Q_1.

The output voltage can be taken between the two collectors, producing a *balanced*, or *differential, output*. The output can also be taken from the output of either collector to ground, producing a *single-ended output*. The two outputs are 180° out of phase with one another. If the balanced output is used, the output voltage across the load is twice the single-ended output voltage.

No special biasing circuits are needed, since the correct value of collector current is supplied directly by the constant-current source Q_3 in Fig. 3-4(a). Resistors R_3, R_4, and R_5, along with V_{EE}, bias the constant-current source Q_3. With no inputs applied, the current in Q_1 equals the current Q_2, which is $I_E/2$. The balanced output at this time is zero. The circuit formed by R_1 and Q_1 and Q_2 and R_2 is a *bridge circuit*. When no inputs are applied, V_{R_1} equals V_{R_2}, and Q_1 and Q_2 conduct equally. Therefore, the bridge is balanced and the output between the collectors is zero.

Now, if an input signal V_1 is applied to Q_1, the conduction of Q_1 and Q_2 is affected. Increasing the voltage at the base of Q_1 increases the collector current in Q_1 and decreases the collector current in Q_2 by an equal amount, so that the two currents sum to I_E. Decreasing the input voltage on the base of Q_1 decreases the collector current in Q_1 but increases it in Q_2. The sum of the emitter currents is always equal to the current supplied by Q_3.

The gain of a differential amplifier is a function of the emitter current and the value of the collector resistors. An approximation of the gain is given by the expression $A = (R_C I_E)/K$, where K is some value in the 10 to 100 range and is determined by the transistor. This is the single-ended gain, where the output is taken from one

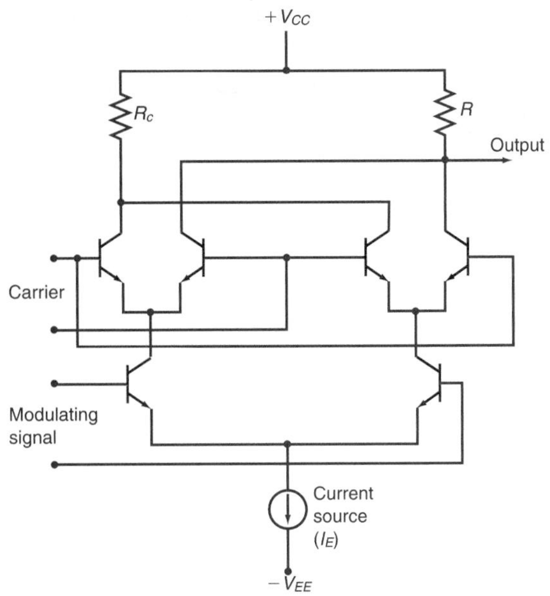

Fig. 3-5 Two parallel differential amplifiers form a high-quality amplitude modulator.

of the collectors with respect to ground. If the output is taken from between the collectors, the gain is two times the above value.

R_C is the collector resistor value in ohms, and I_E is the emitter current in milliamperes. If $R_C = R_1 = R_2 = 4.7$ kΩ, $I_E = 1.5$ mA, and $K = 75$, the gain will be about $A = 4700 (1.5)/75 = 7050/75 = 94$.

In most differential amplifiers, both R_C and I_E are fixed, providing a constant gain. But as the formula above shows, the gain is directly proportional to the emitter current. Thus, if the emitter current can be varied in accordance with the modulating signal, the circuit will produce AM. This is easily done by changing the circuit only slightly, as in Fig. 3-4(b). The carrier is applied to the base of Q_1, and the base of Q_2 is grounded. The output, taken from the collector of Q_2, is single-ended. Since the output from Q_1 is not used, its collector resistor can be omitted with no effect on the circuit. The modulating signal is applied to the base of the constant-current source Q_3. As the intelligence signal varies, it varies the emitter current. This changes the gain of the circuit, and the modulating signal amplitude determines the carrier amplitude. The result is AM in the output.

This circuit, like the basic diode modulator, has the modulating signal in the output in addition to the carrier and sidebands. The modulating signal can be removed by using a simple high-pass filter on the output, since the carrier and sideband frequencies are usually much higher than the modulating signal. A bandpass filter centered on the carrier with sufficient bandwidth to pass the sidebands can also be used. A parallel tuned circuit in the collector of Q_2 replacing R_C can also be used.

An improved version of the differential amplifier amplitude modulator is shown in Fig. 3-5. It uses two differential amplifiers operated in parallel. This is the circuit implemented in most IC AM circuits. It operates as an analog multiplier as long as the input signals are small enough to ensure linear operation. If either input is too large, the transistors will operate as switches and AM with a carrier will not be produced.

The differential amplifier makes an excellent amplitude modulator. It has high gain and good linearity and can be modulated 100 percent. And if high-frequency transistors or a high-frequency IC differential amplifier is used, this circuit can be used to produce low-level modulation at frequencies well into the tens of megahertz region.

PIN Diode Modulator

Some circuits for producing AM at very high frequencies are shown in Fig. 3-6. These circuits use PIN diodes to produce AM at VHF, UHF, and microwave frequencies. The *PIN diodes* are a special type of silicon junction diode designed to be used at frequencies above approximately 100 MHz. When forward-biased, these diodes act as variable resistors. The resistance of the diode varies linearly with the amount of current flowing through it. A high current produces a low resistance, whereas a low current produces a high resistance. By using the modulating signal to vary the forward-bias current through the PIN diode, AM is produced.

In Fig. 3-6(*a*), two PIN diodes are connected back to back and are forward-biased by a fixed negative dc voltage. The modulating signal is applied to the diodes through capacitor C_1. This ac modulating signal rides on the dc bias and, therefore, adds to and subtracts from it. In doing so, it varies the resistance of the PIN diodes. These diodes appear in series with the carrier oscillator and the load. A positive-going modulating signal will reduce the bias on the PIN diodes, causing their resistance to go up. This causes the amplitude of the carrier to be reduced across the load. A negative-going modulating signal will add to the forward bias,

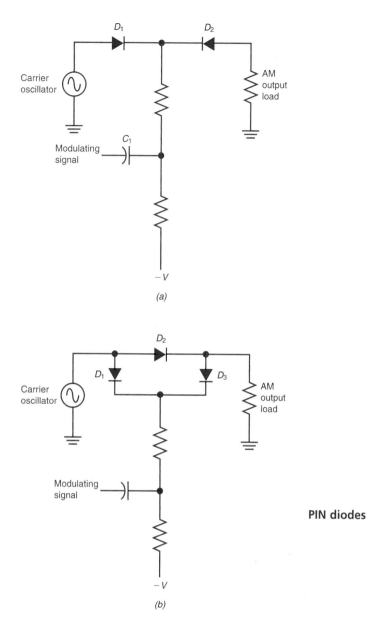

PIN diodes

Fig. 3-6 High-frequency amplitude modulators using PIN diodes.

causing the resistance of the diodes to go down, thereby increasing the carrier amplitude.

A variation of the PIN diode modulator circuit is shown in Fig. 3-6(*b*). The diodes are arranged in a pi network. This configuration is used when it is necessary to maintain a constant circuit impedance even under modulation.

In both circuits of Fig. 3-6, the PIN diodes form a variable attenuator circuit whose attenuation varies with the amplitude of the modulating signal. Such modulator circuits introduce a considerable amount of attenuation and, therefore, must be followed by amplifiers to increase the AM signal to a usable level. Despite this disadvantage, PIN modulators are widely used

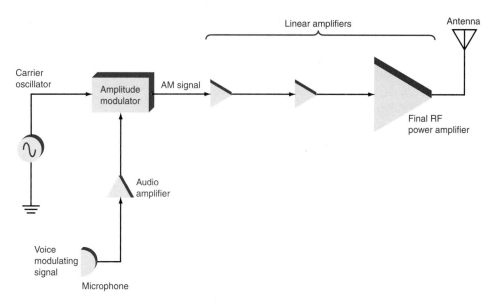

Fig. 3-7 Low-level modulation systems use linear power amplifiers to increase the AM signal level before transmission.

because they are one of the only methods available to produce AM at microwave frequencies.

High-Level Amplitude Modulation

The modulator circuits we have been discussing are known as low-level modulator circuits. "Low level" refers to the fact that the signals are generated at relatively low voltage and power amplitudes. Before an AM signal is transmitted, its power level must be increased. In systems using *low-level modulation,* the AM signal is applied to one or more linear amplifiers as shown in Fig. 3-7. These may be class A, class AB, or class B linear amplifier circuits. These raise the output level of the circuit to the desired power level before the AM signal is fed to the antenna. The key point here is that linear amplifiers must be used so as not to distort the AM signal.

High-level modulation is also possible. In high-level modulation, the modulator varies the voltage and power in the final RF amplifier stage of the transmitter. One example of a high-level modulator circuit is the collector modulator shown in Fig. 3-8. The output stage of the transmitter is a high-power class C amplifier. Class C amplifiers conduct for only a portion of the positive half cycle of their input signal. The collector current pulses cause the tuned circuit to oscillate or ring at the desired output frequency. The tuned circuit, therefore, reproduces the negative portion of the carrier signal.

The modulator is a linear power amplifier that takes the low-level modulating signal and

amplifies it to a high power level. The modulating output signal is coupled through modulation transformer T_1 to the class C amplifier. The secondary winding of the modulation transformer is connected in series with the collector supply voltage V_{cc} of the class C amplifier.

With a zero modulation input signal, there will be zero modulation voltage across the secondary of T_1. Therefore, the collector supply voltage will be applied directly to the class C amplifier, and the output carrier will be a steady sine wave.

When the modulation signal occurs, the ac voltage across the secondary of the modulation transformer will be added to and subtracted from the collector supply voltage. This varying supply voltage is then applied to the class C amplifier. Naturally, the amplitude of the current pulses

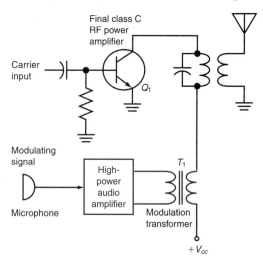

Fig. 3-8 A high-level collector modulator.

through transistor Q_1 will vary. As a result, the amplitude of the carrier sine wave varies in accordance with the modulating signal. For example, when the modulation signal goes positive, it adds to the collector supply voltage, thereby increasing its value and causing higher current pulses and a higher amplitude carrier. When the modulating signal goes negative, it subtracts from the collector supply voltage making it less. For that reason, the class C amplifier current pulses are smaller, thereby causing a lower amplitude carrier output.

For 100 percent modulation, the peak of the modulating signal across T_1 must be equal to the supply voltage. When the positive peak occurs, the voltage applied to the collector is twice the collector supply voltage. When the modulating signal goes negative, it subtracts from the collector supply voltage. When the negative peak is equal to the supply voltage, the effective voltage applied to the collector of Q_1 is zero, producing zero carrier output. This is illustrated in Fig. 3-9.

In practice, 100 percent modulation cannot be achieved with the high-level collector mod-

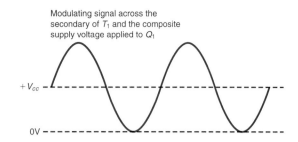

Modulating signal across the secondary of T_1 and the composite supply voltage applied to Q_1

Fig. 3-9 For 100 percent modulation the peak of the modulating signal must be equal to V_{cc}.

ulator circuit shown in Fig. 3-8. To overcome this problem, the driver amplifier stage driving the final class C amplifier is also collector-modulated simultaneously. This process is shown in Fig. 3-10. The output of the modulation transformer is connected in series with the collector supply voltage both to the driver transistor Q_1 and to the final class C amplifier Q_2. With this arrangement, solid 100 percent modulation is possible. This technique is widely used in low-power CB transmitters.

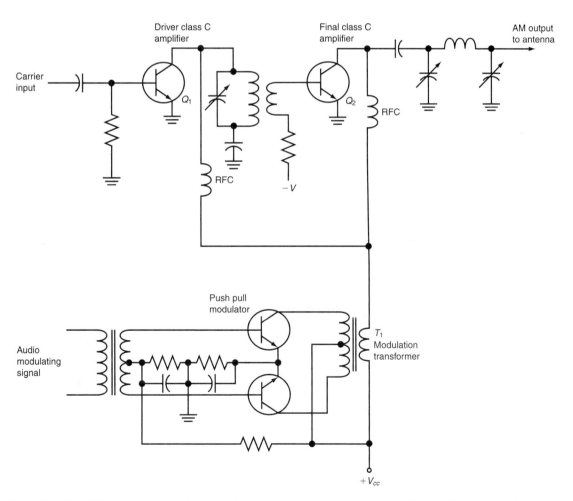

Fig. 3-10 For 100 percent modulation both the driver and the final class C amplifier are modulated.

High-level modulation produces the best type of AM, but it requires an extremely high-power modulator circuit. In fact, the power supplied by the modulator must be equal to one-half the total class C power amplifier rating for 100 percent modulation. If the class C amplifier has an input power of 1000 W, the modulator must be able to deliver one-half this amount, or 500 W.

■ TEST

Choose the letter which best answers each statement.

1. In the modulator circuit of Fig. 3-2, the carrier and modulating signals are
   a. Added.
   b. Subtracted.
   c. Multiplied.
   d. Divided.
2. In Fig. 3-2, D_1 acts as a(n)
   a. Capacitor.
   b. Rectifier.
   c. Variable resistor.
   d. Adder.
3. In Fig. 3-4, D_1 acts as a(n)
   a. Capacitor.
   b. Rectifier.
   c. Variable resistor.
   d. Adder.

Supply the missing information in each statement.

4. AM can be produced by passing the carrier through a circuit whose _____ or _____ can be varied in accordance with the modulating signal.
5. The name of the nonlinear response of a device that produces AM with only second-order products is _____.
6. A component that has an exponential response ideal for producing AM is the _____.

Demodulator

7. In Fig. 3-2, the negative peaks of the AM signal are supplied by the _____.

Detector

8. A differential amplifier used as an amplitude modulator performs the mathematical function of _____.
9. The gain of the differential amplifier is changed by varying the _____.
10. To produce AM, the differential amplifier must have input signals small enough to ensure _____ operation.
11. In Fig. 3-5, the FET acts like a(n) _____.

12. In Fig. 3-5, AM is produced by varying the _____ of the op-amp circuit.
13. The output of the circuit in Fig. 3-5 is usually connected to a _____.
14. When forward-biased, a PIN diode acts like a(n) _____.
15. PIN diode modulators are used only at frequencies above about _____.
16. A PIN diode modulator is a variable _____.
17. High current in a PIN diode means that its resistance is _____.
18. The AM signals generated by low-level modulating circuits must have their power level increased by a(n) _____ before being transmitted.
19. In a high-level AM transmitter, the output stage is usually a class _____ amplifier.
20. A high-level modulator like that in Fig. 3-8 is referred to as a _____ modulator.
21. The output of a high-level modulator causes the _____ applied to the final RF amplifier to vary with the amplitude of the modulating signal.
22. The final amplifier of a high-level modulation CB transmitter has an input power of 5 W. The modulator must be able to supply a power of _____ W for 100 percent modulation.
23. The final RF power amplifier has a supply voltage of 12 V. For 100 percent AM using a high-level modulator, the peak ac output of the modulation transformer must be _____ V.
24. To achieve 100 percent high-level modulation of an RF power amplifier, its _____ must also be modulated.

3-2   Amplitude Demodulators

A *demodulator* is a circuit that accepts a modulated signal and recovers the original modulating information. Also know as a *detector,* a demodulator circuit is the key circuit in any radio receiver. In fact, the demodulator circuit may be used alone as the simplest form of radio receiver.

Diode Detector

The simplest and most widely used amplitude demodulator is the diode detector shown in Fig. 3-11. The AM signal is usually transformer-coupled as indicated. It is applied to a basic

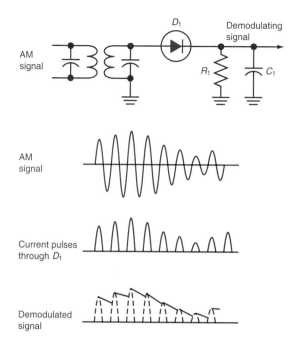

Fig. 3-11 The diode detector AM demodulator.

half-wave rectifier circuit consisting of D_1 and R_1. The diode conducts when the positive half cycles of the AM signals occur. During the negative half cycles, the diode is reverse-biased and no current flows through it. As a result, the voltage across R_1 is a series of positive pulses whose amplitude varies with the modulating signal.

To recover the original modulating signal, a capacitor is connected across resistor R_1. Its value is critical to good performance. The value of this capacitor is carefully chosen so that it has a very low impedance at the carrier frequency. At the frequency of the modulating signal, it has a much higher impedance. The result is that the capacitor effectively shorts or filters out the carrier, thereby leaving the original modulating signal.

Another way to look at the operation of the diode detector is to assume that the capacitor charges quickly to the peak value of the pulses passed by the diode. When the pulse drops to zero, the capacitor retains the charge but discharges into resistor R_1. The time constant of C and R_1 is chosen to be long compared to the period of the carrier. As a result, the capacitor discharges only slightly during the time that the diode is not conducting. When the next pulse comes along, the capacitor again charges to its peak value. When the diode cuts off, the capacitor will again discharge a small amount into the resistor. The resulting waveform across the capacitor is a close approximation to the original modulating signal. Because of the capacitor

charging and discharging, the recovered signal will have a small amount of ripple on it. This causes distortion of the demodulated signal. However, because the carrier frequency is usually many times higher than the modulating frequency, these ripple variations are barely noticeable. In Fig. 3-11, the ripple is quite pronounced because the carrier frequency is low.

The output of the detector is the original modulating signal. Because the diode detector recovers the envelope of the AM signal, which is the modulating signal, the circuit is sometimes referred to as an *envelope detector*.

The Crystal Radio

The basic diode detector circuit is really a complete radio receiver in its own right. In fact, this circuit is the same as that used in the *crystal radio receivers* of the past. The crystal refers to the diode. In Fig. 3-12, the diode detector circuit is redrawn, showing an antenna connection and headphones. A long wire antenna picks up the radio signal, which is inductively coupled to the tuned circuit. The variable capacitor C_1 is used to select a station. The diode detector D_1 recovers the original modulating information which causes current flow in the headphones. The headphones serve as the load resistance, whereas capacitor C_2 removes the carrier. The result is a simple radio receiver with very weak reception because no amplification is provided. Typically a germanium diode is used because its voltage threshold is lower than that of a silicon diode and permits reception of weaker

Envelope detector

Crystal radio receiver

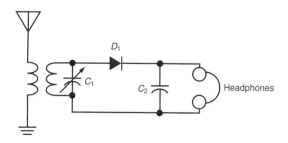

Fig. 3-12 A crystal radio receiver.

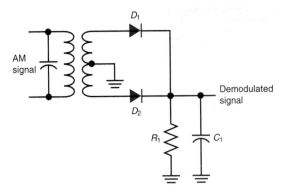

Fig. 3-13 A full-wave diode detector for AM signals.

Carrier suppression

Diode ring modulator

signals. Such a receiver can easily be built to receive standard AM broadcasting stations.

The performance of the basic diode detector can be improved by using a full-wave rectifier circuit as shown in Fig. 3-13. Here, two diodes and a center-tapped secondary on the RF transformer are used to form a standard full-wave rectifier circuit. With this arrangement, diode D_1 will conduct on the positive half cycle, while D_2 will conduct on the negative half cycle. This diode detector produces a higher average output voltage which is much easier to filter. The capacitor value necessary to remove the carrier can be half the size of the capacitor value used in a half-wave diode detector. The primary benefit of this circuit is that the higher modulating frequencies will not be distorted by ripple or attenuated as much as in the half-wave detector circuit.

■ **TEST**

Answer the following questions.

25. The purpose of a _____ is to recover the original modulating signal from an AM wave.
26. The most widely used amplitude demodulator is called a _____.
27. The most critical component in the circuit of Fig. 3-11 is _____.
28. The charging and discharging of C_1 in Fig. 3-11 produces _____ which causes _____ of the modulating signal.
29. Another name for the demodulator in Fig. 3-11 is _____ detector.
30. List the two main benefits of the full-wave amplitude demodulator over the half-wave circuit.
31. True or false. An amplitude demodulator is a complete radio receiver.

3-3 Balanced Modulators

A balanced modulator generates a DSB signal. The inputs to a balanced modulator are the carrier and a modulating signal. The output of a balanced modulator is the upper and lower sidebands. The balanced modulator *suppresses the carrier,* leaving only the sum and difference frequencies at the output. The output of a balanced modulator can be further processed by filters or phase-shifting circuitry to eliminate one of the sidebands, thereby resulting in an SSB signal.

Diode Lattice Modulator

One of the most popular and widely used balanced modulators is the *diode ring* or *lattice*

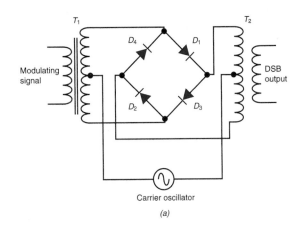

Carrier oscillator

(a)

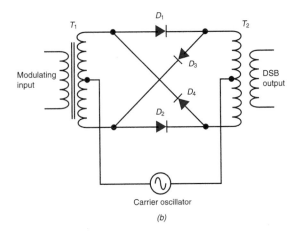

Carrier oscillator

(b)

Fig. 3-14 Lattice-type balanced modulator.

modulator illustrated in Fig. 3-14(*a*). It consists of an input transformer T_1, an output transformer T_2, and four diodes connected in a bridge circuit. The carrier signal is applied to the center taps of the input and output transformers. The modulating signal is applied to the input transformer T_1. The output appears across the secondary of the output transformer T_2.

Sometimes you will see the lattice modulator drawn as shown in Fig. 3-14(*b*). Physically the connections are the same, but the operation of the circuit can be more easily visualized with this circuit.

The operation of the lattice modulator is relatively simple. The carrier sine wave, which is usually considerably higher in frequency and amplitude than the modulating signal, is used as a source of forward and reverse bias for the diodes. The carrier turns the diodes off and on at a high rate of speed. The diodes act like switches which connect the modulating signal at the secondary of T_1 to the primary of T_2.

To see how the circuit works, refer to Fig. 3-15. Assume that the modulating input is zero. When the polarity of the carrier is as illustrated in Fig. 3-15(*a*), diodes D_1 and D_2 are forward-biased. At this time, D_3 and D_4 are reverse-biased and act like open circuits. As you can see, current divides equally in the upper and lower portions of the primary winding of T_2. The current in the upper part of the winding produces a magnetic field that is equal and opposite to the magnetic field produced by the current in the lower half of the secondary. Therefore, these magnetic fields cancel each other out and no output is induced in the secondary. Thus, the carrier is effectively suppressed.

When the polarity of the carrier reverses as shown in Fig. 3-15(*b*), diodes D_1 and D_2 are reverse-biased and diodes D_3 and D_4 conduct. Again, the current flows in the secondary winding of T_1 and the primary winding of T_2. The equal and opposite magnetic fields produced in T_2 cancel each other out and thus result in zero carrier output. The carrier is effectively balanced out. The degree of carrier suppression depends upon the degree of precision with which the transformers are made and the placement of the center tap to ensure perfectly equal upper and lower currents and magnetic field cancellation. The degree of the carrier attenuation also depends upon the diodes. The greatest carrier suppression will occur when the diode characteristics are perfectly matched. A carrier suppression of 40 decibels (dB) is achievable with well-balanced components.

Now assume that a low-frequency sine wave is applied to the primary of T_1 as the modulating signal. The modulating signal will appear across the secondary of T_1. The diode switches will connect the secondary of T_1 to the primary of T_2 at different times depending upon the carrier polarity. Refer to Fig. 3-15. When the carrier polarity is as shown in Fig. 3-15(*a*), diodes D_1 and D_2 conduct and act as closed switches. At this time, D_3 and D_4 are reverse-biased and are effectively not in the circuit. As a result, the modulating signal at the secondary of T_1 is applied to the primary of T_2 through D_1 and D_2.

When the carrier polarity reverses, D_1 and D_2 cut off and D_3 and D_4 conduct. Again, a portion of the modulating signal at the secondary of T_1 will be applied to the primary of T_2, but his time the leads have been effectively reversed because of the connections of D_3 and D_4. The result is a 180° phase reversal. If the modulating signal is positive, the output will be negative with this connection and vice versa.

The carrier is operating at a considerably higher frequency than the modulating signal

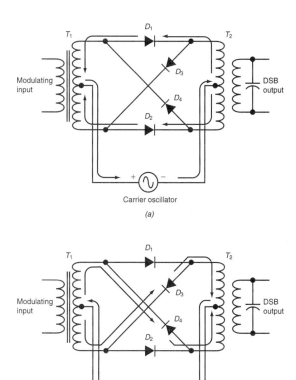

Fig. 3-15 Operation of the lattice modulator.

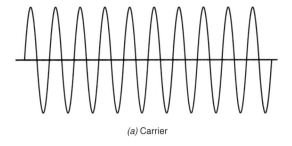

(a) Carrier

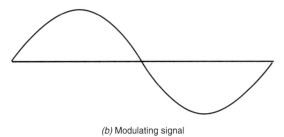

(b) Modulating signal

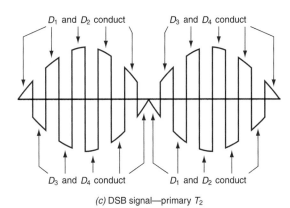

D_1 and D_2 conduct D_3 and D_4 conduct

D_3 and D_4 conduct D_1 and D_2 conduct

(c) DSB signal—primary T_2

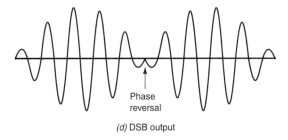

Phase
reversal

(d) DSB output

Fig. 3-16 Waveforms in the lattice-type balanced modulator.

(Fig. 3-16). Therefore, the diodes will switch off and on at a high rate of speed, causing portions of the modulating signal to be passed through the diodes at different times. The DSB signal appearing across the primary of T_2 is illustrated in Fig. 3-16(*c*). The steep rise and fall of the waveform is caused by the rapid switching of the diodes. The waveform contains harmonics of the carrier because of the switching action. Ordinarily, the secondary of T_2 is a resonant circuit as shown in Fig. 3-15, and thus the high-frequency harmonic con-

tent is filtered out, leaving a signal that appears like that in Fig. 3-16(*d*). This is a DSB signal.

There are several important things to notice about this signal. First, the output waveform is occurring at the carrier frequency. This is true even though the carrier has been removed. If you take two sine waves occurring at the sideband frequencies and algebraically add them together, the result is a sine wave signal at the carrier frequency but with the amplitude variation shown in Fig. 3-16(*c*) or (*d*). Observe that the envelope of the output signal is not the shape of the modulating signal. Note also the phase reversal of the signal in the very center of the waveform. This is one way you can tell whether the signal being observed is a true DSB signal.

Although diode lattice modulators can be constructed of discrete components, they are usually available in a single module containing the transformers and diodes in a sealed package. The unit can be used as an individual component. The transformers are carefully balanced, and matched hot-carrier diodes are used to provide a wide operating frequency range and superior carrier suppression.

The diode lattice modulator shown in Fig. 3-14 uses one low-frequency iron-core transformer for the modulating signal and an aircore transformer for the RF output. This is an inconvenient arrangement because the low-frequency transformer is large and expensive. More commonly, two RF transformers are used in the configuration shown in Fig. 3-17. Here the modulating signal is applied to the center taps of the RF transformers. The operation of the circuit is similar to that of the previously discussed circuit.

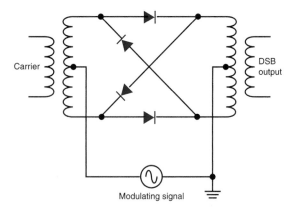

Carrier

DSB output

Modulating signal

Fig. 3-17 A modified version of the lattice modulator not requiring an iron-core transformer for the low-frequency modulating signal.

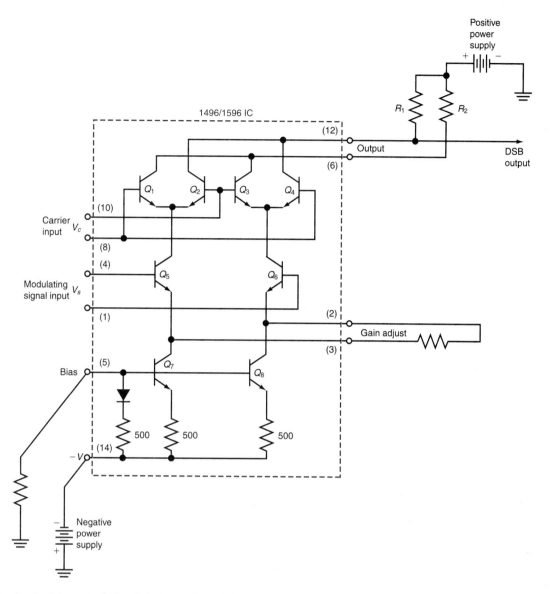

Fig. 3-18 Integrated-circuit balanced modulator.

IC Balanced Modulator

Another widely used balanced modulator circuit uses differential amplifiers. A typical circuit is shown in Fig. 3-18. This is the circuit of the popular 1496/1596 *IC balanced modulator.* This circuit can work at carrier frequencies up to approximately 100 MHz and can achieve a carrier suppression of 50 to 65 dB.

In Fig. 3-18, transistors Q_7 and Q_8 are constant current sources. The constant current sources are biased with a single external resistor and the negative supply. These current sources supply equal values of current to the two differential amplifiers. One differential amplifier is made up of Q_1, Q_2, and Q_5, and the other is made up of Q_3, Q_4, and Q_6. The modulating signal is applied to the bases of Q_5 and

Q_6. These transistors are connected in the current paths to the differential transistors and, therefore, will vary the amplitude of the current in accordance with the modulating signal. The currents in Q_5 and Q_6 will be 180° out of phase with each other. As the current in Q_5 increases, the current through Q_6 decreases, and vice versa.

The differential transistors Q_1 through Q_4 operate as switches. These transistors are controlled by the carrier. When the carrier input is such that the power input terminal is positive with respect to the upper input terminal, transistors Q_1 and Q_4 will conduct and act as closed switches and Q_2 and Q_3 will be cut off. When the polarity of the carrier signal reverses, Q_1 and Q_4 will be cut off and Q_2 and Q_3 will conduct and act as closed switches. These differential transistors, therefore,

IC balanced modulator

serve the same switching purpose as the diodes in the lattice modulator circuit in Fig. 3-15. That is, they switch the modulating signal off and on at the carrier rate.

To see how the circuit works, assume that a high-frequency carrier wave is applied to the switching transistors Q_1 through Q_4 and that a low-frequency sine wave is applied to the modulating signal input at Q_5 and Q_6. Assume that the modulating signal is positive-going so that the current through Q_5 is increasing while the current through Q_6 is decreasing.

When the carrier polarity is positive, Q_1 and Q_4 conduct. Since the current through Q_5 is increasing, the current through Q_1 and R_2 will increase proportionately; therefore, the output voltage at the collector of Q_1 will go in a negative direction. The current through Q_6 is decreasing; therefore, the current through Q_4 and R_1 is decreasing. The output voltage at the collector of Q_4 is hence increasing.

When the carrier polarity reverses, Q_2 and Q_3 conduct. Now the increasing current of Q_5 is passed through Q_2 and R_1. Therefore, the output voltage begins to decrease. The decreasing current through Q_6 is now passed through Q_3 and R_2. This decreasing current causes an increasing output voltage. The result of the carrier switching off and on and the modulating signal varying as indicated produces the classical DSB output signal shown in Fig. 3-16(c). The signal at R_1 is the same as the signal at R_2, but the two are 180° out of phase.

You may have noticed that the circuit in Fig. 3-18 is virtually identical to the circuit in Fig. 3-5. When the carrier signal is large, it forces the differential transistors to act as switches. In this mode of operation, the circuit produces suppressed-carrier AM or DSB AM. If the carrier and input signals are small, the differential amplifiers operate in the linear mode and produce true AM. The 1496/1596 can be used this way.

Filter method of SSB generation

■ TEST

Choose the letter which best answers each statement.

32. A balanced modulator eliminates which of the following from its output?
    a. Upper sideband
    b. Lower sideband
    c. Carrier
    d. Both sidebands

33. The output signal of a balanced modulator is
    a. AM.
    b. SSB.
    c. FM.
    d. DSB.

34. Which has better carrier suppression?
    a. Lattice modulator
    b. IC balanced modulator

Supply the missing information in each statement.

35. A balanced modulator using a diode bridge is called a(n) _____.

36. In the balanced modulator of Fig. 3-14, the diodes are used as _____.

37. A balanced modulator has a carrier frequency of 1.9 MHz and a modulating sine wave of 2.6 kHz. The output signals are _____ and _____ kHz.

38. In Fig. 3-18, transistors _____ supply a constant current.

39. In Fig. 3-18, transistors Q_1 to Q_4 operate as _____.

Determine whether the statement is true or false.

40. An IC balanced modulator may be used for AM signal generation.

3-4   SSB Circuits

There are two primary methods of generating SSB signals. These are the filter method and the phasing method. The filter method is by far the simplest and most widely used, but we will discuss both types here.

The Filter Method of SSB

Figure 3-19 shows a general block diagram of an SSB transmitter using the *filter method*. The modulating signal, usually voice from a microphone, is applied to the audio amplifier whose output is fed to one input of a balanced modulator. A crystal oscillator provides the carrier signal, which is also applied to the balanced modulator. The output of the balanced modulator is a DSB signal. An SSB signal is produced by passing the DSB signal through a highly selective band-pass filter. This filter selects either the supper or the lower sideband.

The filter, of course, is the critical component in the filter method SSB generator. Its primary requirement is that it have high selectivity so that it passes only the desired sideband and rejects the other. The filters are usually designed

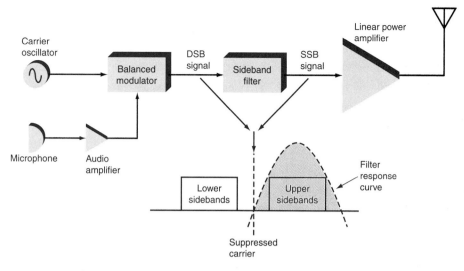

Fig. 3-19 An SSB transmitter using the filter method.

with a bandwidth of approximately 2.5 to 3 kHz, making them only wide enough to pass standard voice frequencies. The sides of the filter response curve are extremely steep, providing excellent rejection of the other sideband.

The filter is a fixed tuned device; that is, the frequencies that it can pass cannot be changed. Therefore, the carrier oscillator frequency must be chosen so that the sidebands fall within the filter bandpass. Usually, the filter is tuned to a frequency in the 455-kHz, 3.35-MHz, or 9-MHz range. Other frequencies are also used, but many commercially available filters are in these frequency ranges.

It is also necessary to select either the upper or the lower sideband. Since the same information is contained in both sidebands, it generally makes no difference which one is selected. However, various conventions in different communications services have chosen either the upper or the lower sideband as a standard. These vary from service to service, and it is necessary to know whether it is an upper or lower sideband to properly receive an SSB signal.

There are two methods of selecting the sideband. Many transmitters simply contain two filters, one that will pass the upper sideband and the other that will pass the lower sideband. A switch is used to select the desired sideband. See Fig. 3-20(a).

The other method of selecting the sideband is to provide two carrier oscillator frequencies. Two crystals change the carrier oscillator frequency to force either the upper sideband or the lower sideband to appear in the filter bandpass.

Assume a simple example in which the bandpass filter is fixed at 1000 kHz. The modulating signal f_m is 2 kHz. The balanced modulator generates the sum and difference frequencies. Therefore, the carrier frequency f_c must be chosen so that the USB or LSB is at 1000 kHz.

The balanced modulator outputs are USB $= f_c + f_m$ and LSB $= f_c - f_m$. To put the USB at 1000 kHz, the carrier must be

$$f_c + f_m = 1000$$
$$f_c + 2 = 1000$$
$$f_c = 1000 - 2 = 998 \text{ kHz}$$

To set the LSB at 1000 kHz, the carrier must be

$$f_c - f_m = 1000$$
$$f_c - 2 = 1000$$
$$f_c = 1000 + 2 = 1002 \text{ kHz}$$

Crystals and Crystal Filters

Crystal filters are by far the most commonly used filters in SSB transmitters. They are low in cost and relatively simple to design. Their very *high Q* provides extremely good selectivity.

Crystal filters are made from the same type of *quartz crystals* normally used in crystal oscillators. When a voltage is applied across a crystal, it will vibrate at a specific resonant frequency. This resonant frequency is a function of the size,

High Q

Quartz crystals

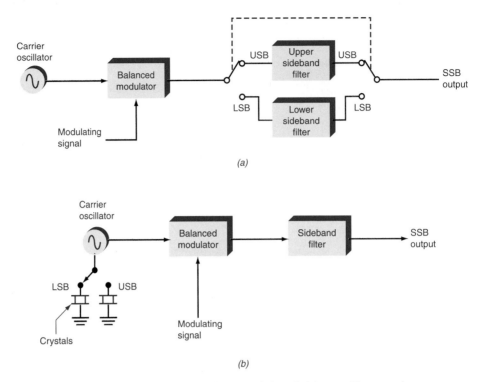

Fig. 3-20 Methods of selecting the upper or lower sideband. (*a*) Two filters and (*b*) two carrier frequencies.

thickness, and direction of cut of the crystal. Crystals can be cut and ground for almost any frequency in the 100-kHz to 100-MHz range. Its frequency of vibration is extremely stable, and, therefore, crystals are widely used to supply signals on exact frequencies with good stability.

The schematic symbol and the equivalent circuit of a quartz crystal are shown in Fig. 3-21. The crystal acts as a resonant *LC* circuit. The series *LCR* part of the equivalent circuit represents the crystal itself, whereas the parallel capacitance C_p is the capacitance of the metal mounting plates with the crystal as the dielectric.

Figure 3-22 shows the impedance variations of the crystal as a function of frequency. At frequencies below the crystal's resonant frequency, the circuit appears capacitive and has a high impedance. However, at some frequency, the reactances of the equivalent inductance *L* and the series capacitance C_s are equal, and the circuit will resonate. You should remember that a series circuit is at resonance when $X_L = X_C$. At this series resonant frequency f_s, the circuit is resistive. The resistance of the crystal is low compared to the equivalent inductance *L*, thereby giving the circuit an extremely high *Q*. Values of *Q* in the 10,000 to 100,000 range are common. This makes the crystal a highly selective series resonant circuit.

If the frequency of the signal applied to the crystal is above f_s, the crystal appears inductive.

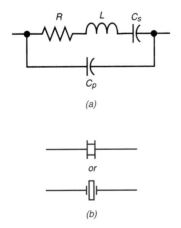

Fig. 3-21 (*a*) Quartz-crystal–equivalent electric circuit, and (*b*) schematic symbol.

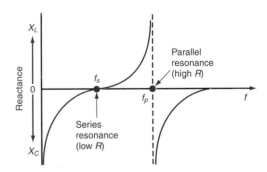

Fig. 3-22 Impedance variation of a quartz crystal as a function of frequency.

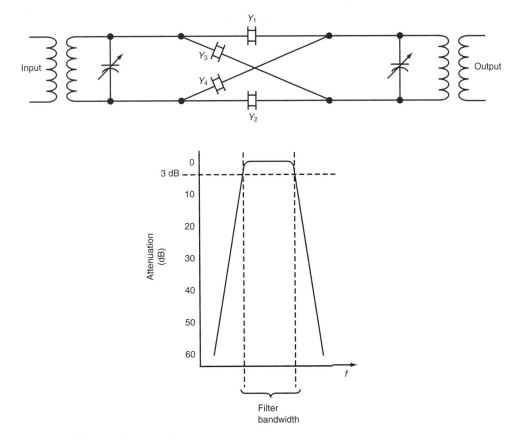

Fig. 3-23 Crystal lattice filter and its response curve.

At some frequency, the reactance of the parallel capacitance C_p equals the reactance of the net inductance. When this occurs, a parallel resonant circuit is formed. At this parallel resonant frequency f_p, the impedance of the circuit is resistive but extremely high.

Because the crystal has both series and parallel resonant frequencies that are close together, it makes an ideal component for use in filters. By combining crystals with selected series and parallel resonant points, highly selective filters with any desired bandpass can be constructed.

The most commonly used *crystal filter* is the full crystal lattice shown in Fig. 3-23. Crystals Y_1 and Y_2 resonate at one frequency, whereas crystals Y_3 and Y_4 resonate at another frequency. The difference between the two crystal frequencies determines the bandwidth of the filter. The 3-dB down bandwidth will be approximately 1.5 times the crystal frequency spacing. For example, if the Y_1-Y_2 frequency is 9 MHz and the Y_3-Y_4 frequency is 9.002 MHz, the difference is $9.002 - 9.000 = 0.002$ MHz $= 2$ kHz. The 3-dB bandwidth then is 1.5×2 kHz $= 3$ kHz.

The crystals are also chosen so that the parallel resonant frequency of Y_3-Y_4 equals the series resonant frequency of Y_1-Y_2. The series res-

onant frequency of Y_3-Y_4 is equal to the parallel resonant frequency of Y_1-Y_2. The result is a passband with extremely steep attenuation. Signals outside the passband are rejected as much as 50 to 60 dB below those inside the passband. See the response curve in Fig. 3-23. Such a filter can easily separate one sideband from another.

A popular variation of the crystal lattice filter is shown in Fig. 3-24. It uses only a center-tapped inductor rather than two transformers as in

Crystal filters

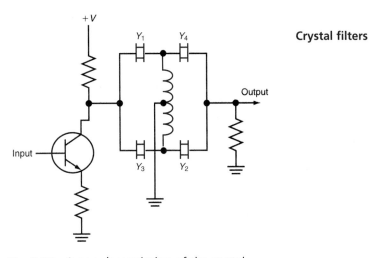

Fig. 3-24 A popular variation of the crystal lattice filter.

Fig. 3-23. Its primary advantage is that it is easily inserted between transistor amplifier stages.

Another type of crystal filter is the *ladder filter* shown in Fig. 3-25. All the crystals in this filter are cut for exactly the same frequency. The number of crystals used and the values of the shunt capacitors set the bandwidth. Crystals with frequencies near 9 MHz are commonly used. Because of the low cost and wide availability of crystals for CB radios with frequencies near 27 MHz, very simple low-cost filters with these frequencies are easily constructed. At least six crystals must usually be cascaded to achieve the kind of selectivity needed in SSB applications.

Other types of filters are also used to remove the unwanted sideband. These include both ceramic and mechanical filters. Ceramic is a manufactured crystal-like compound. It has the same piezoelectric qualities as quartz. Ceramic disks can be made so that they vibrate at a fixed frequency, thereby providing filtering actions. *Ceramic filters* are very small and inexpensive and are, therefore, widely used in transmitters and receivers. Typical center frequencies are 455 kHz and 10.7 MHz. These are available in different bandwidths up to 350 kHz.

Mechanical filters are also used in SSB-generating equipment. These mechanical filters consist of small metal disks coupled together with rods to form an assembly that vibrates or resonates over a narrow frequency range. The diameter and thickness of the disks determine the resonant frequency, whereas the number of disks and their spacing and method of coupling determine the bandwidth. The ac signal to be filtered is applied to a coil that creates a magnetic field. This magnetic field works against a permanent magnet to produce mechanical motion in the disks. If the input signal is within the bandpass resonant frequency range of the disks, they will vibrate freely. This vibration is mechanically coupled to a coil. The moving coil cuts the field of a permanent magnet inducing a voltage in the coil. This is

the output signal. If the input signal is outside of the resonant frequency range of the disks, they will not vibrate and little or no output will be produced. Such mechanical assemblies are extremely effective bandpass filters. Most are designed to operate over the 200- to 500-kHz range. A 455-kHz mechanical filter is commonly used.

The Phasing Method of SSB

The *phasing method of SSB generation* uses a phase-shift technique that causes one of the sidebands to be canceled out. A block diagram of a phasing-type SSB generator is shown in Fig. 3-26. It uses two balanced modulators instead of one. The balanced modulators effectively eliminate the carrier. The carrier oscillator is applied directly to the upper balanced modulator along with the audio modulating signal. Then both the carrier and the modulating signal are shifted in phase by 90° and applied to the second, lower, balanced modulator. The two balanced modulator outputs are then added together algebraically. The phase-shifting action causes one sideband to be canceled out when the two balanced modulator outputs are combined.

The carrier signal is $V_c \sin 2\pi f_c t$. The modulating signal is $V_m \sin 2\pi f_m t$. Balanced modulator 1 produces the product of these two signals:

$$(V_m \sin 2\pi f_m t)(V_c \sin 2\pi f_c t)$$

Applying a trigonometric identity,

$$(V_m \sin 2\pi f_m t)(V_c \sin 2\pi f_c t) = $$
$$0.5[\cos (2\pi f_c - 2\pi f_m)t - \cos (2\pi f_c + 2\pi f_m)t]$$

Note that these are the sum and difference frequencies or the upper and lower sidebands.

It is important to remember that a cosine wave is simply a sine wave shifted by 90°. A cosine wave has exactly the same shape as a sine wave, but it occurs 90° earlier in time. The cosine wave leads a sine wave by 90° or the sine wave lags a cosine wave by 90°.

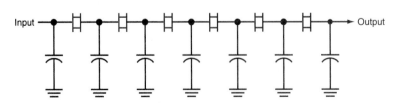

Fig. 3-25 A crystal ladder filter. All crystals are ground for the same frequency.

Ladder filter

Phasing method of SSB generation

Ceramic filters

Mechanical filters

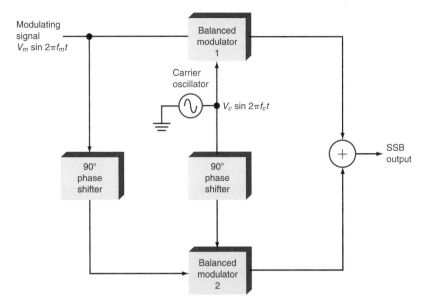

Fig. 3-26 The phasing method of SSB signal generation.

The 90° phase shifters in Fig. 3-26 create cosine waves of the carrier and modulating signals that are multiplied in balanced modulator 2 to produce

$$(V_m \cos 2\pi f_m t)(V_c \cos 2\pi f_c t)$$

Another common trigonometric identity translates this to

$$(V_m \cos 2\pi f_m t)(V_c \cos 2\pi f_c t) =$$
$$0.5 \left[\cos(2\pi f_c - 2\pi f_m)t + \cos(2\pi f_c + 2\pi f_m)t\right]$$

Now, if you add these two expressions together, the sum frequencies cancel whereas the difference frequencies add, producing only the lower sideband:

$$\cos (2\pi f_c - 2\pi f_m)t$$

A *phase shifter* is usually an *RC* network that causes the output to either lead or lag the input by 90°. Many different kinds of circuits have been devised for producing this phase shift. A simple RF phase shifter for the carrier is shown in Fig. 3-27. It consists of two *RC* sections set to produce a phase shift of 45°. The section made up of R_1 and C_1 produces an output that lags the input by 45°. The section made up of C_2 and R_2 produces a phase shift that leads the input by 45°. The total phase shift then between the two outputs is 90°. One output goes to one balanced modulator, and the other goes to the second balanced modulator.

The most difficult part of creating a phasing-type SSB generator is designing a circuit that maintains a constant 90° phase shift over a wide range of modulating frequencies. Keep in mind that the definition of phase shift is a time shift between sine waves of the same frequency. An *RC* network produces a specific amount of phase shift at only one frequency because the capacitive reactance varies with frequency. In the carrier phase shifter, this is not a problem since the carrier is maintained at a constant frequency. But the modulating signal is usually a band of frequencies, typically in the audio range from 300 to 3000 Hz.

One of the circuits commonly used to produce a 90° phase shift over a wide bandwidth is shown in Fig. 3-28. The phase-shift difference between output 1 and output 2 is 90°

Phase shifter

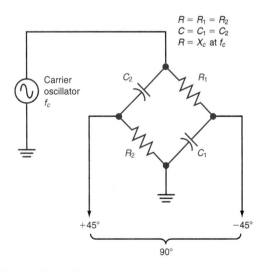

Fig. 3-27 A fixed-frequency, 90° *RC* phase shifter.

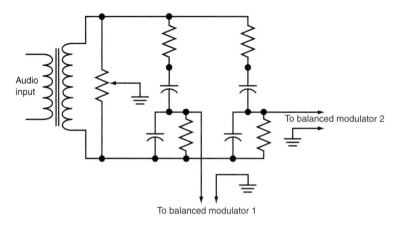

Fig. 3-28 A phase shifter that produces a 90° shift over the 300- to 3000-Hz range.

± 1.5° over the 300- to 3000-Hz range. The resistor and capacitor values are carefully selected to ensure this phase accuracy. Phase-shift inaccuracies will cause incomplete cancellation of the undesired sideband.

The phasing method can be used to select either the upper or the lower sideband. This is done by changing the phase shift of either the audio or carrier signals to the balanced modulator inputs. For example, applying the direct audio signal to balanced modulator 2 in Fig. 3-26 and the 90° phase-shifted signal to balanced modulator 1 will cause the upper sideband to be selected instead of the lower sideband. The phase relationship of the carrier could also be switched to make this change.

The output of the phasing generator is a low-level SSB signal. The degree of suppression of the carrier depends upon the quality of the balanced modulators. The precision of the phase shifting determines the degree of suppression of the unwanted sideband. The design and adjustment of phasing-type SSB generators is critical in order to ensure complete suppression of the undesired sideband. The SSB output is then applied to linear RF amplifiers, where its power level is increased before being applied to the transmitting antenna.

To demodulate an SSB signal, you must reinsert the carrier at the receiver. Assume that you generate an SSB signal by modulating a 9-MHz carrier with a 2-kHz sine wave intelligence signal. A balanced modulator suppresses the 9-MHz carrier but generates the upper and lower sideband frequencies of 9.002 and 8.998 MHz respectively. Assume that a sharp bandpass filter selects the upper sideband of 9.002 MHz and suppresses the lower sideband. At

Product detector

the receiver you will get only the 9.002-MHz signal. But what you want is the 0.002-MHz or 2-kHz intelligence signal.

Recovering the original modulating signal is a matter of mixing the received signal with a locally generated carrier. Inside the receiver is an oscillator that is set to 9 MHz. The oscillator signal is applied to a mixer circuit along with the incoming signal. The mixer forms the sum and difference frequencies of 9.002 + 9.000 = 18.002 MHz and 9.002 − 9.000 = 0.002 MHz or 2 kHz. The 18.002-MHz signal is filtered out, leaving the desired 2-kHz intelligence signal.

The demodulator for SSB signals is therefore a mixer. Typically, a balanced modulator is used for this purpose. Any of the previously described circuits will work. A balanced modulator used for this purpose is generally referred to as a *product detector.*

Because it is difficult to set the internal local oscillator to the exact frequency of the original carrier, the frequency of the recovered intelligence signal may be slightly higher or lower than the original 2 kHz. For voice transmissions, this means that the recovered voice may be higher or lower in pitch. To correct for this effect, the internal oscillator is usually made variable so that the operator of the receiver can adjust it and tune for the most intelligible and natural sounding output.

■ TEST

Supply the missing information in each statement.

41. The most common way of generating an SSB signal is the _____ method.

42. A filter capable of passing the desired sideband while rejecting the other sideband must have good _____.

43. A balanced modulator has a 3-MHz carrier input and a modulating signal input of 1.5 kHz. To pass the lower sideband, a filter must have a center frequency of _____ MHz.

44. The most popular filter used to select the desired sideband in an SSB generator uses _____ for selectivity.

45. Name the two ways of generating either the upper or lower sideband in a filter-type SSB generator.

46. A quartz crystal acts like a highly selective _____ circuit.

47. In a filter-type SSB generator, a crystal lattice filter is used. The two crystal frequencies are 3.0 and 3.0012 MHz. The filter bandwidth is approximately _____ kHz.

48. Mechanical filters provide selectivity because they _____ at a specific frequency.

49. A ceramic filter is similar in operation to a(n) _____ filter.

50. The operating frequency range of a mechanical filter is _____ to _____ kHz.

51. In the phasing method of SSB generation, _____ is used to cancel the undesired sideband.

52. A _____ circuit is commonly used to demodulate or recover an SSB signal.

53. The circuit used to demodulate an SSB signal is typically called a(n) _____.

Chapter 3 Review

Summary

1. Amplitude modulation can be accomplished by multiplying the carrier sine wave by a gain or attenuation factor that varies in accordance with the intelligence signal.
2. Amplitude modulation can be carried out by linearly combining the carrier and intelligence signals and then applying the result to a nonlinear component or circuit. A diode is an example.
3. The simplest AM circuit uses resistors to linearly mix the carrier and information signal, a diode to rectify the result, and a tuned circuit to complete the waveform.
4. The most widely used method of generating low-level AM is to use an integrated-circuit analog multiplier (modulator) or differential amplifier.
5. Low-level modulation is the process of generating the AM signal at low voltage and/or power levels and then using linear amplifiers to increase the power level.
6. High-level modulation is the process of amplitude modulating the final power amplifier of a transmitter.
7. High-level modulation is accomplished with a collector (plate in vacuum tubes) modulator that varies the collector supply voltage in accordance with the modulating signal.
8. For 100 percent high-level modulation, the modulation amplifier must produce an output wave whose peak-to-peak is 2 times the collector supply voltage.
9. For 100 percent high-level modulation, the modulation amplifier must generate an output power that is one-half of the final RF power amplifier input power ($P_i = V_{cc} \times I_c$).
10. The simplest and best amplitude demodulator is the diode detector. The AM signal is rectified by a diode and then filtered by a capacitor to recover the envelope, which is the original modulating information.
11. Balanced modulators are AM circuits that cancel or suppress the carrier but generate a DSB output signal that contains the upper (sum) and lower (difference) sideband frequencies.
12. A popular balanced modulator is the lattice modulator that uses a diode bridge circuit as a switch. The carrier turns the diodes off and on, letting segments of the modulating signal through to produce a DSB output signal. A carrier suppression of 40 dB is possible.
13. Another widely used balanced modulator is an integrated circuit (IC) using differential amplifiers as switches to switch the modulating signal at the carrier frequency. A popular device is the 1496 or 1596. Carrier suppression can be as high as 50 to 65 dB.
14. The most common way of generating an SSB signal is to use the filter method which incorporates a balanced modulator followed by a highly selective filter that passes either the upper or lower sideband.
15. To make both sidebands available, SSB generators use two filters, one for each sideband, or switch the carrier frequency to put the desired sideband into the fixed filter bandpass.
16. Most SSB filters are made with quartz crystals.
17. A quartz crystal is a frequency-determining component that acts like an LC circuit with a very high Q.
18. Crystals have series and parallel resonant modes. These can be combined into a lattice (bridge) circuit that provides extremely sharp selectivity over a desired bandwidth.
19. Ceramic filters use ceramic resonators that act like crystals but are smaller and lower in cost.
20. Mechanical bandpass filters are also used in SSB generators. These devices use multiple resonant disks that vibrate at some frequency in the 200- to 500-kHz range.
21. The phasing method of SSB generation uses two balanced modulators and 90° phase shifters for the carrier and modulating signal to produce two DSB signals that, when added, cause one sideband to be canceled out.

Chapter Review Questions

Choose the letter that best answers each question.

3-1. Amplitude modulation is the same as
   a. Linear mixing.
   b. Analog multiplication.
   c. Signal summation.
   d. Multiplexing.
3-2. In a diode modulator, the negative half of the AM wave is supplied by a(n)
   a. Tuned circuit.
   b. Transformer.
   c. Capacitor.
   d. Inductor.
3-3. Amplitude modulation can be produced by
   a. Having the carrier vary a resistance.
   b. Having the modulating signal vary a capacitance.
   c. Varying the carrier frequency.
   d. Varying the gain of an amplifier.
3-4. Amplitude modulators that vary the carrier amplitude with the modulating signal by passing it through an attenuator work on the principle of
   a. Rectification.
   b. Resonance.
   c. Variable resistance.
   d. Absorption.
3-5. A key requirement in using a differential amplifier as an amplitude modulator is that
   a. The input signals should be small enough to ensure linear operation.
   b. The transistors should operate as switches.
   c. Large input signals should be used.
   d. The gain should be constant.
3-6. In Fig. 3-6, D_1 is a
   a. Variable resistor.
   b. Mixer.
   c. Clipper.
   d. Rectifier.
3-7. The component used to produce AM at very high frequencies is a
   a. Varactor.
   b. Thermistor.
   c. Cavity resonator.
   d. PIN diode.
3-8. Amplitude modulation generated at a very low voltage or power amplitude is known as
   a. High-level modulation.
   b. Low-level modulation.
   c. Collector modulation.
   d. Minimum modulation.
3-9. A collector modulator has a supply voltage of 48 V. The peak-to-peak amplitude of the modulating signal for 100 percent modulation is
   a. 24 V.
   b. 48 V.
   c. 96 V.
   d. 120 V.
3-10. A collector-modulated transmitter has a supply voltage of 24 V and a collector current of 0.5 A. The modulator power for 100 percent modulation is
   a. 6 W.
   b. 12 W.
   c. 18 W.
   d. 24 W.
3-11. The circuit that recovers the original modulating information from an AM signal is known as a
   a. Modulator.
   b. Demodulator.
   c. Mixer.
   d. Crystal set.
3-12. The most commonly used amplitude demodulator is the
   a. Diode mixer.
   b. Balanced modulator.
   c. Envelope detector.
   d. Crystal filter.
3-13. A circuit that generates the upper and lower sidebands but no carrier is called a(n)
   a. Amplitude modulator.
   b. Diode detector.
   c. Class C amplifier.
   d. Balanced modulator.

3-14. The inputs to a balanced modulator are 1 MHz and a carrier of 1.5 MHz. The outputs are
a. 500 kHz.
b. 2.5 MHz.
c. 1.5 MHz.
d. All the above.
e. a and b

3-15. A widely used balanced modulator is called the
a. Diode bridge circuit.
b. Full-wave bridge rectifier.
c. Lattice modulator.
d. Balanced bridge modulator.

3-16. In a diode ring modulator, the diodes act like
a. Variable resistors.
b. Switches.
c. Rectifiers.
d. Variable capacitors.

3-17. The output of a balanced modulator is
a. AM.
b. FM.
c. SSB.
d. DSB.

3-18. The principal circuit in the popular 1496/1596 IC balanced modulator is a
a. Differential amplifier.
b. Rectifier.
c. Bridge.
d. Constant current source.

3-19. The most commonly used filter in SSB generators uses
a. LC networks.
b. Mechanical resonators.
c. Crystals.
d. RC networks and op amps.

3-20. The equivalent circuit of a quartz crystal is a
a. Series resonant circuit.
b. parallel resonant circuit.
c. Neither a nor b
d. Both a and b

3-21. A crystal lattice filter has crystal frequencies of 27.5 and 27.502 MHz. The bandwidth is approximately
a. 2 kHz.
b. 3 kHz.
c. 27.501 MHz.
d. 55.502 MHz.

3-22. An SSB generator has a sideband filter centered at 3.0 MHz. The modulating signal is 3 kHz. To produce both upper and lower sidebands, the following carrier frequencies must be produced:
a. 2.7 and 3.3 MHz.
b. 3.3 and 3.6 MHz.
c. 2997 and 3003 kHz.
d. 3000 and 3003 kHz.

3-23. In the phasing method of SSB generation, one sideband is canceled out because of
a. Phase shift.
b. Sharp selectivity.
c. Carrier suppression.
d. Phase inversion.

3-24. A balanced modulator used to demodulate a SSB signal is called a(n)
a. Transponder.
b. Product detector.
c. Converter.
d. Modulator.

Critical Thinking Questions

3-1. If AM can be achieved by varying the gain or attenuation of the carrier, name one or more devices or circuits that can be used for this purpose.

3-2. Is it possible for one AM signal to amplitude-modulate a carrier on another frequency? If so, describe what would happen. What would the output spectrum look like?

3-3. Will the circuit in Fig. 3-14 demodulate an AM signal? Draw the input and output voltage waveforms to make this determination.

3-4. Draw a simplified diagram of how an enhancement mode MOSFET could produce OOK or ASK modulation.

3-5. A voice signal with a 300- to 3000-Hz range modulates a carrier of 3.125 MHz to produce LSB SSB. At the receiver, the reinserted carrier has a frequency of 3.1256 MHz. Will the signal be received at all? If so, what will it sound like?

3-6. A 2-kHz sine wave tone modulates a 175-kHz carrier to produce a USB SSB signal that, in turn, modulates a 28-MHz carrier producing LSB SSB. Describe the final output signal and state its frequency.

3-7. A received OOK signal is on for 2 ms and then off for 2 ms. A diode detector is used to demodulate the signal. Describe the recovered demodulator output.

3-8. In an SSB modulator like that in Fig. 3-26, what determines the degree of carrier suppression?

1. *a*
2. *b*
3. *c*
4. gain, attenuation
5. square law
6. diode
7. tuned circuit
8. switches
9. emitter current
10. tuned circuit or filter
11. variable resistor
12. gain
13. increase, decrease, increase
14. resistor
15. 100 MHz
16. attenuator
17. low
18. linear amplifier
19. C
20. collector
21. supply voltage
22. 2.5
23. 12
24. driver
25. demodulator
26. diode detector
27. C_1
28. ripple, distortion

29. envelope
30. smaller filter capacitor, less ripple and distortion
31. true
32. *c*
33. *d*
34. *b*
35. lattice modulator or diode ring
36. switches
37. 1897.4, 1902.6
38. Q_7, Q_8
39. switches
40. true
41. filter
42. selectivity
43. 2.9985
44. crystals
45. Use one filter for each sideband; select the carrier frequency so that the desired sideband is in the filter passband.
46. tune, resonant, or *LC*
47. 1.8 (3.0012 − 3.0 = 0.0012 MHz = 1.2 kHz; 1.2 × 1.5 = 1.8 kHz)
48. vibrate or resonate
49. crystal
50. 200 to 500
51. phase shift
52. mixer or balanced modulator
53. product detector

Frequency Modulation

Chapter Objectives

This chapter will help you to:

1. *Define and explain* the processes of frequency modulation (FM) and phase modulation (PM) and state their differences.
2. *Calculate* the modulation index given the maximum deviation and maximum modulating frequency, *determine* the significant number of sidebands in an FM signal, and *calculate* the bandwidth of an FM signal.
3. *Define* pre-emphasis and de-emphasis, *state* their benefits, and *show* how they are accomplished.
4. *Name* the advantages and disadvantages of FM and PM compared to AM.

Angle modulation

Modulation is the process of modifying a carrier wave in accordance with an information signal to be transmitted. Changing the amplitude of the carrier produces AM. It is also possible to impress an information signal on a carrier by changing its frequency. Although not immediately obvious, another characteristic of a carrier that can be changed is its phase shift. If the amount of phase shift that a carrier experiences is varied, information can be impressed upon the carrier. This is known as phase modulation (PM). As it turns out, varying the phase shift of a carrier produces FM. Therefore, both FM and PM are closely related to one another. Together they are collectively referred to as types of *angle modulation*. Since FM is generally superior in performance to AM, it is widely used in many areas of communications electronics. In this chapter, we will introduce you to the fundamentals of FM and PM. Circuits for producing FM and PM and frequency demodulators are covered in Chap. 5.

4-1   Frequency Modulation Principles

Frequency deviation

In FM, the carrier amplitude remains constant, while the carrier frequency is changed by the modulating signal. As the amplitude of the information signal varies, the carrier frequency shifts in proportion. As the modulating signal amplitude increases, the carrier frequency increases. If the amplitude of the modulating signal decreases, the carrier frequency decreases. The reverse relationship can also be implemented. A decreasing modulating signal will increase the carrier frequency above its center value, whereas an increasing modulating signal will decrease the carrier frequency below its center value. As the modulating signal amplitude varies, the carrier frequency varies above and below its normal center frequency with no modulation. The amount of change in carrier frequency produced by the modulating signal is known as the *frequency deviation*. Maximum frequency deviation occurs at the maximum amplitude of the modulating signal.

The frequency of the modulating signal determines how many times per second the carrier frequency deviates above and below its nominal center frequency. If the modulating signal is a 100-Hz sine wave, then the carrier frequency will shift above and below the center frequency 100 times per second. This is called the *frequency deviation rate*.

An FM signal is illustrated in Fig. 4-1(c). Normally the carrier [Fig. 4-1(a)] is a sine wave, but it is shown as a triangular wave here to simplify the illustration. With no modulating signal applied, the carrier frequency is a constant-amplitude sine wave at its normal constant center frequency.

The modulating information signal [Fig. 4-1(b)] is a low-frequency sine wave. As the sine wave goes positive, the frequency of the carrier increases proportionately. The highest frequency occurs at the peak amplitude of the modulating signal. As the modulating signal amplitude decreases, the carrier frequency decreases. When the modulating signal is at zero amplitude, the carrier will be at its center frequency point.

When the modulating signal goes negative, the carrier frequency will decrease. The carrier frequency will continue to decrease until the peak of the negative half cycle of the modulating sine wave is reached. Then, as the modulating signal increases toward zero, the frequency will again increase. Note in Fig. 4-1(c) how the carrier sine waves seem to be first "compressed" and then "stretched" by the modulating signal.

Assume a carrier frequency of 50 MHz. If the peak amplitude of the modulating signal causes a maximum frequency shift of 200 kHz,

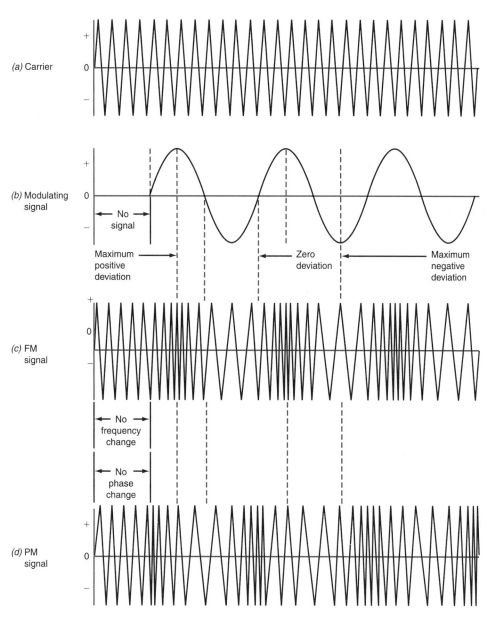

Fig. 4-1 Frequency modulation and phase modulation signals. The carrier is drawn as a triangular wave for simplicity, but in practice it is a sine wave.

About ⟨▭⟩ Electronics

Increasing the amplitude or frequency of the modulating signal in FM does not cause overmodulation or distortion.

the carrier frequency will deviate up to 50.2 MHz and down to 49.8 MHz. The total frequency deviation is $50.2 - 49.8 = 0.4$ MHz $= 400$ kHz. In practice, however, the frequency deviation is expressed as the amount of frequency shift of the carrier above or below the center frequency. Therefore, the frequency deviation in the example above is said to be ± 200 kHz. This means that the modulating signal varies the carrier above and below its center frequency by 200 kHz. The frequency of the modulating signal determines the rate of frequency deviation but has no effect on the amount of deviation, which is strictly a function of the amplitude of the modulating signal.

JOB TIP

Informally contacting companies for information—through e-mail, regular mail, or a phone call can help you narrow down the right career for you.

Phase modulation

Indirect FM

■ TEST

Answer the following questions.

1. The general name given to both FM and PM is _____ modulation.
2. True or false. In FM, the carrier amplitude remains constant with modulation.
3. The amount of frequency shift during modulation is called the _____.
4. The amount of frequency shift in FM is directly proportional to the _____ of the modulating signal.
5. As the modulating signal amplitude goes positive, the carrier frequency _____. As the modulating signal amplitude goes negative, the carrier frequency _____.

4-2   Phase Modulation

Another way to produce angle modulation is to vary the amount of phase shift of a constant-frequency carrier in accordance with a modulating signal. The resulting output is a PM signal. Imagine a modulator circuit whose basic function is to produce a phase shift.

Remember that a phase shift refers to a time separation between two sine waves of the same frequency. Assume that we can build a phase shifter that causes the amount of phase shift to vary with the amplitude of the modulating signal. The greater the amplitude of the modulating signal, the greater the phase shift. Assume further that positive alternations of the modulating signal produce a lagging phase shift and negative signals produce a leading phase shift.

If a constant-amplitude–constant-frequency carrier sine wave is applied to the phase shifter, the output of the phase shifter will be a PM wave. As the modulating signal goes positive, the amount of phase lag increases with the amplitude of the modulating signal. This means that the carrier output is delayed. That delay increases with the amplitude of the modulating signal. The result at the output is as if the constant-frequency carrier signal had been stretched out or its frequency lowered.

When the modulating signal goes negative, the phase shift becomes leading. This causes the carrier sine wave to be effectively speeded up or compressed. The result is as if the carrier frequency had been increased.

Phase modulation produces frequency modulation. Since the amount of phase shift is varying, the effect is as if the carrier frequency is changed. Since PM produces FM, PM is often referred to as *indirect FM*.

How FM and PM Differ

It is important to point out that it is the dynamic nature of the modulating signal that causes the frequency variation at the output of the phase shifter. In other words, FM is produced only as long as the phase shift is being varied. One way to understand this better is to assume a modulating signal like that shown in Fig. 4-2(*a*). It is a triangular wave whose positive and negative peaks have been clipped off at a fixed amplitude. During time t_0, the signal is zero so the carrier is at its center frequency.

Applying this modulating signal to a frequency modulator will produce the signal shown in Fig. 4-2(*b*). During the time that the waveform is rising (t_1), the frequency is increasing. During the time that the positive

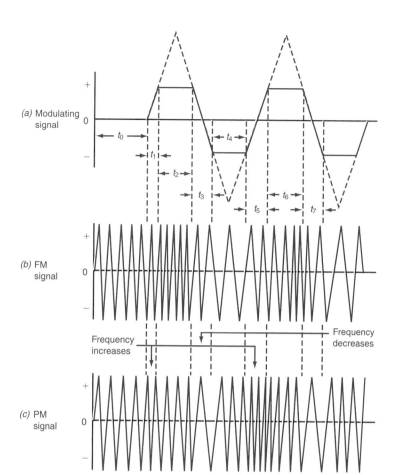

Fig. 4-2 A frequency shift occurs in PM only when the modulating signal amplitude varies.

amplitude is constant (t_2), the FM output frequency is constant. During the time the amplitude decreases and goes negative (t_3), the frequency will decrease. Then, during the constant-amplitude negative alternation (t_4), the frequency remains constant at a lower frequency. During t_5, the frequency increases.

When the modulating signal is applied to a phase shifter, the output frequency will change only during the time that the amplitude of the modulating signal is varying. Refer to the PM signal in Fig. 4-2(*c*). During increases or decreases in amplitude (t_1, t_3, and t_5), a varying frequency will be produced. However, during the constant-amplitude positive and negative peaks no frequency change takes place. The output of the phase shifter will simply be the carrier frequency which has been shifted in phase. This clearly illustrates that frequency variations take place only if the modulating signal amplitude is varying.

As it turns out, the maximum frequency deviation produced by a phase modulator occurs

during the time that the modulating signal is changing at its most rapid rate. For a sine wave modulating signal, the rate of change of the modulating signal is greatest when the modulating wave changes from plus to minus or from minus to plus. The maximum rate of change of modulating voltage occurs exactly at the zero crossing points. You can see this in Fig. 4-2(*c*). In contrast, note that in an FM wave the maximum deviation occurs at the peak positive and negative amplitude of the modulating voltage. So although a phase modulator does indeed produce FM, maximum deviation occurs at different points of a modulating signal. Of course, this is irrelevant since both the FM and the PM waves contain exactly the same information and, when demodulated, will reproduce the original modulating signal.

In FM, maximum deviation occurs at the peak positive and negative amplitudes of the modulating signal. In PM, the maximum amount of leading or lagging phase shift

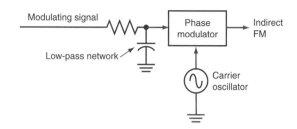

Fig. 4-4 A low-pass filter compensates for higher phase shift and frequency deviation at the higher modulating frequencies to produce indirect FM.

interNET CONNECTION

For definitions of the new key words used in this chapter, you might want to reference the following site:

⟨telecom.tbi.net⟩

occurs at the zero crossings of the modulating signal. Recall that we said that the frequency deviation at the output of the phase shifter depends upon the rate of change of the modulating signal. The faster the modulating signal voltage varies, the greater the frequency deviation produced. Because of this, the frequency deviation produced in PM increases with the frequency of the modulating signal. The higher the modulating signal frequency, the shorter its period and the faster the voltage changes. Higher modulating voltages produce greater phase shift which, in turn, produces greater frequency deviation. However, higher modulating frequencies produce a faster rate of change of the modulating voltage and, therefore, also produce greater frequency deviation. In PM then, the carrier frequency deviation is proportional to both the modulating frequency and the amplitude. In FM, frequency deviation is proportional only to the amplitude of the modulating signal regardless of its frequency. The relationship between carrier deviation and

modulating signal characteristics is summarized in Fig. 4-3.

Converting PM to FM

To make PM compatible with FM, we must compensate for the deviation produced by the frequency changes in the modulating signal. This is illustrated in Fig. 4-4. This low-pass filter causes the higher modulating frequencies to be attenuated in amplitude. Although the higher modulating frequencies will produce a greater rate of change and thus a greater frequency deviation, this is offset by the lower amplitude of the modulating signal, which will produce less phase shift and less frequency deviation. This network compensates for the excess frequency deviation caused by higher modulating frequencies. The result is an output that is the same as an FM signal. The FM produced by a phase modulator is called indirect FM.

Although both FM and PM are widely used in communications systems, most angle modulation is PM. The reason for this is that a

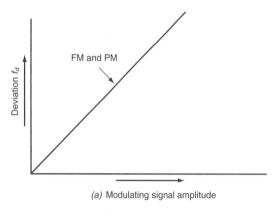

(a) Modulating signal amplitude

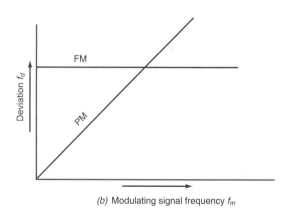

(b) Modulating signal frequency f_m

Fig. 4-3 Relationship between frequency deviation and modulating signal amplitude and frequency for FM and PM.

crystal oscillator with higher frequency accuracy and stability can be used to produce the carrier. In FM, crystal oscillators cannot, in general, be frequency-modulated over a very wide range. However, the crystal oscillator can drive a phase modulator that can produce the desired FM. Further, most phase modulators are simpler to implement than frequency modulators. You will see this when practical FM and PM circuits are discussed in Chap. 5.

■ TEST

Answer the following questions.

6. In PM, the carrier _____ is varied in proportion to the amplitude of the modulating signal.
7. A varying phase shift produces a(n) _____.
8. A phase modulator produces a frequency deviation only when the modulating signal is _____.
9. In PM, the frequency deviation is proportional to both the modulating signal _____ and _____.
10. When the modulating signal amplitude crosses zero, the phase shift and frequency deviation in a phase modulator are
    a. At a maximum.
    b. At a minimum.
    c. Zero.
11. A phase modulator may use a low-pass filter to offset the effect of increasing carrier frequency deviation for increasing modulating
    a. Amplitude.
    b. Frequency.
    c. Phase shift.
12. The FM produced by a phase modulator is known as _____.

Common FM Applications

1. FM radio broadcasting
2. TV sound broadcasting
3. Two-way mobile radio
   a. Police, fire, public service
   b. Marine
   c. Amateur radio
   d. Family radio
4. Cellular telephone (analog phones)
5. Digital data transmission

4-3 Sidebands and the Modulation Index

Any modulation process produces sidebands. As you saw in AM, when a constant-frequency sine wave modulates a carrier, two side frequencies are produced. The side frequencies are the sum and difference of the carrier and the modulating frequency. In FM and PM too, sum and difference sideband frequencies are produced. In addition, a theoretically infinite number of pairs of upper and lower sidebands are also generated. As a result, the spectrum of an FM/PM signal is usually wider than an equivalent AM signal. A special narrowband FM signal whose bandwidth is only slightly wider than that of an AM signal can also be generated.

Figure 4-5 shows an example of the spectrum of a typical FM signal produced by modulating a carrier with a single-frequency sine wave. Note that the sidebands are spaced from the carrier f_c and are spaced from one another by a frequency equal to the modulating frequency f_m. If the modulating frequency is 500 Hz, the first pair of sidebands are above and below the carrier by 500 Hz. The second pair of sidebands are above and below the carrier by 2×500 Hz = 1000 Hz, or 1 kHz, and so on. Note also that the amplitudes of the sidebands vary. If each sideband is assumed to be a sine wave with a frequency and amplitude as indicated in Fig. 4-5 and all these sine waves were added together, then the FM signal producing them would be created.

As the amplitude of the modulating signal varies, of course, the frequency deviation will change. The number of sidebands produced, their amplitude, and their spacing depend upon the frequency deviation and modulating frequency. Keep in mind that an FM signal has a constant amplitude. If that FM signal is a summation of the sideband frequencies, then you can see that the sideband amplitudes must vary with frequency deviation and modulating frequency if their sum is to produce a fixed-amplitude FM signal.

Although the FM process produces an infinite number of upper and lower sidebands, only those with the largest amplitudes are significant in carrying the information. Typically any sideband with an amplitude less than 1 percent of the unmodulated carrier is considered insignificant. As

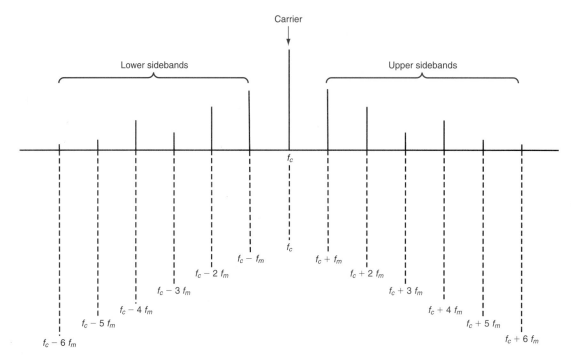

Carrier

Lower sidebands

Upper sidebands

f_c

f_c

$f_c - f_m$

$f_c - 2 f_m$

$f_c - 3 f_m$

$f_c - 4 f_m$

$f_c - 5 f_m$

$f_c - 6 f_m$

$f_c + f_m$

$f_c + 2 f_m$

$f_c + 3 f_m$

$f_c + 4 f_m$

$f_c + 5 f_m$

$f_c + 6 f_m$

Fig. 4-5 Frequency spectrum of an FM signal.

a result, this markedly narrows the bandwidth of an FM signal.

Modulation Index and Deviation Ratio

Deviation ratio

Modulation index

As indicated earlier, the number of significant sidebands and their amplitudes are dependent upon the amount of frequency deviation and the modulating frequency. The ratio of the frequency deviation to the modulating frequency is known as the *modulation index m.*

$$m = \frac{f_d}{f_m}$$

Bessel functions

where f_d is the frequency deviation and f_m is the modulating frequency.

For example, assume that the maximum frequency deviation of the carrier is ±25 kHz and the maximum modulating frequency is 10 kHz. The modulation index, therefore, is

$$m = \frac{25}{10} = 2.5$$

In most communications systems using FM, maximum limits are put on both the frequency deviation and the modulating frequency. For example, in standard FM broadcasting, the maximum permitted frequency deviation is 75 kHz, and the maximum permitted modulating

frequency is 15 kHz. This produces a modulation index of

$$m = \frac{75}{15} = 5$$

Whenever the maximum allowable frequency deviation and the maximum modulating frequency are used in computing the modulation index, m is known as the *deviation ratio.*

Determining the Number of Significant Sidebands

Knowing the modulation index, you can compute the number and amplitudes of the significant sidebands. This is done through a complex mathematical process known as the *Bessel functions.* This mathematical computation is beyond the scope of this text. In general, it is not necessary to know how to make the calculation as the Bessel functions have been computed and tabulated for a wide range of modulation indexes. An example is given in Fig. 4-6. The left-hand column gives the modulation index. The remaining columns indicate the relative amplitudes of the carrier and the various pairs of sidebands. Any sideband with a relative carrier amplitude of less than 1 percent (0.01) has been eliminated. Note that some of the carrier and sideband amplitudes have negative signs. This means that the signal

Modulation Index	Carrier	Sidebands (Pairs)															
		1st	2d	3d	4th	5th	6th	7th	8th	9th	10th	11th	12th	13th	14th	15th	16th
0.00	1.00	—	—	—	—	—	—	—	—	—	—	—	—	—	—	—	—
0.25	0.98	0.12	—	—	—	—	—	—	—	—	—	—	—	—	—	—	—
0.5	0.94	0.24	0.03	—	—	—	—	—	—	—	—	—	—	—	—	—	—
1.0	0.77	0.44	0.11	0.02	—	—	—	—	—	—	—	—	—	—	—	—	—
1.5	0.51	0.56	0.23	0.06	0.01	—	—	—	—	—	—	—	—	—	—	—	—
2.0	0.22	0.58	0.35	0.13	0.03	—	—	—	—	—	—	—	—	—	—	—	—
2.5	−0.05	0.50	0.45	0.22	0.07	0.02	—	—	—	—	—	—	—	—	—	—	—
3.0	−0.26	0.34	0.49	0.31	0.13	0.04	0.01	—	—	—	—	—	—	—	—	—	—
4.0	−0.40	−0.07	0.36	0.43	0.28	0.13	0.05	0.02	—	—	—	—	—	—	—	—	—
5.0	−0.18	−0.33	0.05	0.36	0.39	0.26	0.13	0.05	0.02	—	—	—	—	—	—	—	—
6.0	0.15	−0.28	−0.24	0.11	0.36	0.36	0.25	0.13	0.06	0.02	—	—	—	—	—	—	—
7.0	0.30	0.00	−0.30	−0.17	0.16	0.35	0.34	0.23	0.13	0.06	0.02	—	—	—	—	—	—
8.0	0.17	0.23	−0.11	−0.29	−0.10	0.19	0.34	0.32	0.22	0.13	0.06	0.03	—	—	—	—	—
9.0	−0.09	0.24	0.14	−0.18	−0.27	−0.06	0.20	0.33	0.30	0.21	0.12	0.06	0.03	0.01	—	—	—
10.0	−0.25	0.04	0.25	0.06	−0.22	−0.23	−0.01	0.22	0.31	0.29	0.20	0.12	0.06	0.03	0.01	—	—
12.0	−0.05	−0.22	−0.08	0.20	0.18	−0.07	−0.24	−0.17	0.05	0.23	0.30	0.27	0.20	0.12	0.07	0.03	0.01
15.0	−0.01	0.21	0.04	0.19	−0.12	0.13	0.21	0.03	−0.17	−0.22	−0.09	0.10	0.24	0.28	0.25	0.18	0.12

Fig. 4-6 A table showing carrier and sideband amplitudes for different modulation indexes of FM signals. Based on the Bessel functions.

represented by that amplitude is simply shifted in phase 180° (phase inversion).

As you can see, the spectrum of an FM signal varies considerably in *bandwidth* depending upon the modulation index. The higher the modulation index, the wider the bandwidth of the FM signal. When spectrum conservation is necessary, the bandwidth of an FM signal may be deliberately restricted by putting an upper limit on the modulation index.

Figure 4-7 shows several examples of an FM signal spectrum with different modulation indexes. Compare the examples to the entries in the table of Fig. 4-6. The unmodulated carrier has a relative amplitude of 1.0. With modulation, the carrier amplitude decreases while the amplitudes of the various sidebands increase. With some values of modulation index, the carrier can disappear completely.

The Bandwidth of an FM Signal

The total bandwidth of an FM signal can be determined by knowing the modulation index and using the table in Fig. 4-6. For example, assume that the modulation index is 2. Referring to the table, you can see that this produces four significant pairs of sidebands. The bandwidth can then be determined with the simple formula

$$BW = 2Nf_{m\ max}$$

where N is the number of significant sidebands.

Using the example above and assuming a highest modulating frequency of 2.5 kHz, the bandwidth of the FM signal is

$$BW = 2(4)(2.5)$$
$$= 20\ kHz$$

An FM signal with a modulation index of 2 and a highest modulating frequency of 2.5 kHz will then occupy a bandwidth of 20 kHz.

An alternative way to calculate the bandwidth of an FM signal is to use *Carson's rule*. This rule takes into consideration only the power in the most significant sidebands whose amplitudes are greater than 2 percent of the carrier. These are the sidebands whose values are 0.02 or more in Fig. 4-6. Carson's rule is given by the expression

$$BW = 2(f_{d\ max} + f_{m\ max})$$

In this expression, $f_{d\ max}$ is the maximum frequency deviation, and $f_{m\ max}$ is the maximum modulating frequency.

Assuming a maximum frequency deviation of 5 kHz and a maximum modulating frequency of 2.5 kHz, the bandwidth would be

$$BW = 2(5\ kHz + 2.5\ kHz)$$
$$= 2(7.5\ kHz) = 15\ kHz$$

Comparing the bandwidth with that computed in the preceding example, you can see that Carson's rule gives a smaller bandwidth. It has been determined that, if a circuit or system has that bandwidth (per Carson's rule),

Bandwidth

Carson's rule

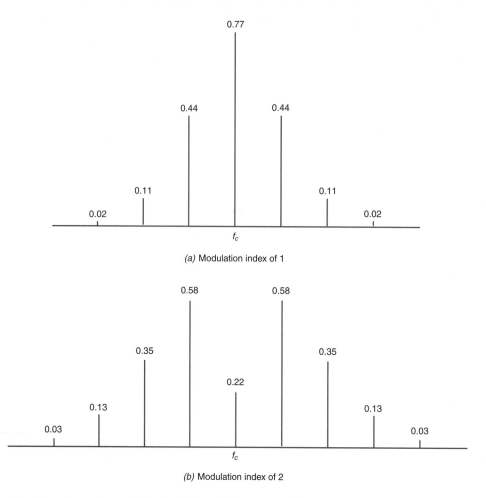

0.77

0.44   0.44

0.11   0.11

0.02   0.02

f_c

(a) Modulation index of 1

0.58   0.58

0.35   0.35

0.22

0.13   0.13

0.03   0.03

f_c

(b) Modulation index of 2

Fig. 4-7 Significant numbers of sidebands for different modulation indexes.

sufficient sideband power will be passed to ensure full intelligibility of the information signal.

Percentage of Modulation

In AM, the amount or degree of modulation is usually given as a *percentage of modulation.* The percentage of modulation is the ratio of the amplitude of the modulating signal to the amplitude of the carrier. When the two factors are equal, the ratio is 1 and we say that 100 percent modulation occurs. If the modulating signal amplitude becomes greater than the carrier amplitude, then overmodulation and distortion occur.

Such conditions do not exist with FM or PM. Since the carrier amplitude remains constant during modulating with FM and PM, the percentage-of-modulation indicator used in AM has no meaning. Further, increasing the amplitude or the frequency of the modulating signal will not cause overmodulation or distortion. Increasing the modulating signal ampli-

tude simply increases the frequency deviation. This, in turn, increases the modulation index, which simply produces more significant sidebands and a wider bandwidth. For practical reasons of spectrum conservation and receiver performance, there is usually some limit put on the upper frequency deviation and the upper modulating frequency. As indicated earlier, the ratio of the maximum frequency deviation permitted to the maximum modulating frequency is referred to as the deviation ratio.

The audio in broadcast TV is transmitted by FM. The maximum deviation permitted is 25 kHz, and the maximum modulating frequency is 15 kHz. This produces a deviation ratio of

$$d = \frac{25}{15}$$

$$= 1.6666$$

In standard two-way mobile radio communications using FM, the maximum permitted deviation is usually 5 kHz. The upper modulating frequency is usually limited to 2.5 kHz, which is high enough for intelligible voice transmission. This produces a deviation ratio of

$$d = \frac{5}{2.5}$$
$$= 2$$

The maximum deviation permitted can be used in a ratio with the actual carrier deviation to produce a percentage of modulation for FM. Remember, in commercial FM broadcasting the maximum allowed deviation is 75 kHz. If the modulating signal is producing only a maximum deviation of 60 kHz, then the FM percentage of modulation is

FM percent modulation

$$= \frac{\text{actual carrier deviation}}{\text{maximum carrier deviation}} \times 100$$

$$= \frac{60}{75} \times 100$$

$$= 80\%$$

When maximum deviations are specified, it is important that the percentage of modulation be held to less than 100 percent. The reason for this is that FM stations operate in assigned frequency channels. These are adjacent to other channels containing other stations. If the deviation is allowed to exceed the maximum, a greater number of pairs of sidebands will be produced and the signal bandwidth may be excessive. This can cause undesirable *adjacent channel interference.*

■ TEST_____

Answer the following questions.

13. True or false. An FM signal produces more sidebands than an AM signal.
14. The bandwidth of an FM signal is proportional to the _____.
15. The maximum frequency deviation of an FM signal is 10 kHz. The maximum modulating frequency is 3.33 kHz. The deviation ratio is _____.
16. An FM signal has a modulation index of 2.5. How many significant pairs of sidebands are produced? (See Fig. 4-6.)
17. In an FM signal, the modulating frequency is a 1.5-kHz sine wave. The carrier frequency is 1000 kHz. The frequencies of the third significant sidebands are _____ and _____ kHz.

18. Refer to Fig. 4-6. What is the relative amplitude of the fourth significant pair of sidebands in an FM signal with a deviation ratio of 4?
19. The amplitudes of the sidebands in an FM signal are dependent upon a mathematical process known as _____.
20. An FM signal has a deviation ratio of 3. The maximum modulating signal is 5 kHz. The bandwidth of the signal is _____ kHz.
21. If the maximum allowed deviation is 5 kHz but the actual deviation is 3.75 kHz, the percentage of modulation is _____ percent.
22. A negative sign on the carrier and sideband amplitudes in Fig. 4-6 means a(n) _____.
23. True or false. The carrier in an FM signal can never drop to zero amplitude.
24. Calculate the bandwidth of an FM signal with a maximum deviation of 10 kHz and a maximum modulating signal frequency of 4 kHz. Use the two methods given in the text (Fig. 4-6 and Carson's rule), and compare your answers. Explain the difference.

4-4 Frequency Modulation vs. Amplitude Modulation

In general, FM is considered to be superior to AM. Although both modulation types are suitable for transmitting information from one place to another and both are capable of equivalent fidelity and intelligibility, FM typically offers some significant benefits over AM, as follows: FM offers better noise immunity; it rejects interfering signals because of the capture effect; and it provides better transmitter efficiency. Its disadvantage lies in the fact that it uses an excessive amount of spectrum space. Let's consider each of these points in more detail.

Noise Immunity

The primary benefit of FM over AM is its superior *noise immunity.* Noise is interference to a signal generated by lightning, motors, automotive ignition systems, and any power line switching that produces transients. Such noise is typically narrow spikes of voltage with very broad frequency content. They add to a signal

Adjacent channel interference

Noise immunity

and interfere with it. If the noise signals are strong enough, they can completely obliterate the information signal.

Noise is essentially amplitude variations. An FM signal, on the other hand, has a constant carrier amplitude. Because of this, FM receivers contain limiter circuits that deliberately restrict the amplitude of the received signal. Any amplitude variations occurring on the FM signal are effectively clipped off. This does not hurt the information content of the FM signal, since it is contained solely within the frequency variations of the carrier. Because of the clipping action of the limiter circuits, noise is almost completely eliminated.

Another major benefit of FM is that interfering signals on the same frequency will be effectively rejected. Because of the limiters built into FM receivers, a peculiar effect takes place when two or more FM signals occur simultaneously on the same frequency. If the signal of one is more than twice the amplitude of the other, the stronger signal will "capture" the channel and will totally eliminate the weaker, interfering signal. This is known as the *capture effect* in FM. When two AM signals occupy the same frequency, both signals will generally be heard regardless of their relative signal strengths. When one AM signal is significantly stronger than the other, naturally the stronger signal will be intelligible; however, although the weaker signal will be unintelligible, it will still be heard in the background. When the signal strengths of the AM signals are nearly the same, they will interfere with one another making both of them nearly unintelligible. In FM, the capture effect allows the stronger signal to dominate while the weaker signal is eliminated.

However, when the strengths of the two FM signals begin to be nearly the same, the capture effect may cause the signals to alternate in their domination of the frequency. At some time one signal will be stronger than the other, and it will capture the channel. At other times, the

Capture effect

signal strengths will reverse, and the other signal will capture the channel. You may have experienced this effect yourself when listening to the FM radio in your car while driving on the highway. You may be listening to a strong station on a particular frequency, but as you drive, you move away from that station. At some point, you may begin to pick up the signal from another station on the same frequency. When the two signals are approximately the same amplitude, you will hear one station dominate and then the other as the signal amplitudes vary during your driving. However, at some point, the stronger signal will eventually dominate. In any case, once the strong signal dominates, the weaker is not heard at all on the channel.

Pre-emphasis and De-emphasis

Despite the fact that FM has superior noise rejection qualities, noise still interferes with an FM signal. This is particularly true for the high-frequency components in the modulating signal. Since noise is primarily sharp spikes of energy, it contains a considerable number of harmonics and other high-frequency components. These high frequencies can at times be larger in amplitude than the high-frequency content of the modulating signal. This causes a form of frequency distortion that can make the signal unintelligible.

Most of the content of a modulating signal, particularly voice, is at lower frequencies. In voice communications systems, the bandwidth of the modulating signal is deliberately limited to a maximum of approximately 3 kHz. The voice is still intelligible despite the bandwidth limitations. After all, telephones cut off at 3 kHz and give good voice quality. However, music would be severely distorted by such a narrow bandwidth because it contains high-frequency components necessary to high fidelity. Typically, however, these high-frequency components are of a lower amplitude. For example, musical instruments typically generate their signals at low frequencies but contain many lower level harmonics that give them their unique sound. If their unique sound is to be preserved, then the high-frequency components must be passed. This is the reason for such a wide bandwidth in high-fidelity sound systems. Since these high-frequency components are at a very low level, noise can obliterate them.

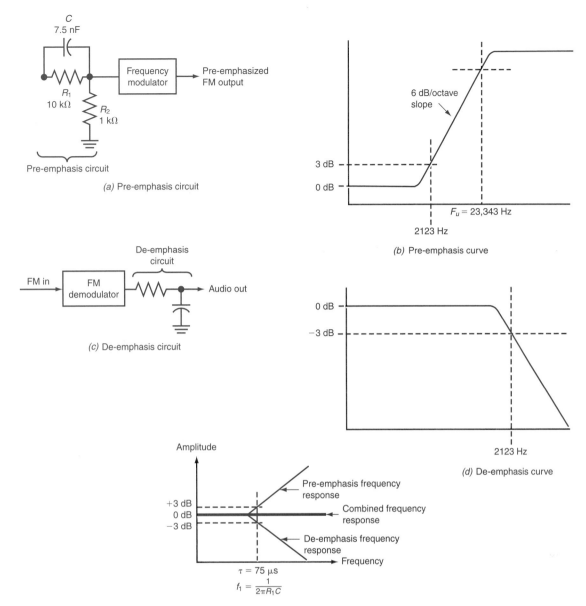

Fig. 4-8 Pre-emphasis and de-emphasis.

To overcome this problem, most FM systems use a technique known as *pre-emphasis,* which helps offset high-frequency noise interference. At the transmitter, the modulating signal is passed through a simple network which amplifies the high-frequency components more than the low-frequency components. The simplest form of such a circuit is the high-pass filter of the type shown in Fig 4.8(a). Specifications dictate a *time constant t* of 75 microseconds (μs) where $t = R_1 C$. Any combination of resistor and capacitor (or resistor and inductor) giving this time constant will be satisfactory. Such a circuit has a lower break frequency f_1 of 2123 Hz. This means that frequencies higher than 2123 Hz will be linearly

enhanced. The output amplitude increases with frequency at a rate of 6 dB per octave. The pre-emphasis curve is shown in Fig. 4-8(b). This pre-emphasis circuit increases the energy content of the higher-frequency signals so that they tend to become stronger than the high-frequency noise components. This improves the signal-to-noise ratio and increases intelligibility and fidelity.

The pre-emphasis circuit also has an upper break frequency f_u, where the signal enhancement flattens out. See Fig. 4-8(b). This upper break frequency is computed with the expression

$$f_u = \frac{R_1 + R_2}{2\pi R_1 R_2 C}$$

Pre-emphasis

75-μs time constant

De-emphasis

Narrowband FM (NBFM)

Transmission efficiency

It is usually set at some high value beyond the audio range. An f_u of greater than 20 kHz is typical.

To return the frequency response to its normal level, a *de-emphasis* circuit is used at the receiver. This is a simple low-pass filter with a time constant of 75 μs. See Fig. 4-8(*c*). It features a cutoff of 2123 Hz and causes signals above this frequency to be attenuated at the rate of 6 dB per octave. The response curve is shown in Fig. 4-8(*d*). As a result, the pre-emphasis at the transmitter is exactly offset by the de-emphasis circuit in the receiver, providing a normal frequency response. The combined effect of pre-emphasis and de-emphasis is to increase the high-frequency components during transmission so that they will be stronger and not masked by noise.

Transmission Efficiency

The third advantage of FM over AM is in *transmitting efficiency*. Recall that AM can be produced by both low-level and high-level modulation techniques. The most effective is high-level modulation, in which a class C amplifier is used as the final RF power stage and is modulated by a high-power modulation amplifier. The AM transmitter must produce both very high RF and modulating signal power. In addition, at very high power levels large modulation amplifiers are impractical. Under such conditions, low-level modulation must be used. The AM signal is generated at a lower level and then amplified with linear amplifiers to produce the final RF signal. Because linear amplifiers operate class A or class B, they are far less efficient than class C amplifiers. However, linear amplifiers must be used if the AM information is to be preserved.

An FM signal has a constant amplitude, and, therefore, it is not necessary to use linear amplifiers to increase its power level. In fact, FM signals are always generated at a lower level and then amplified by a series of class C amplifiers to increase their power. The result is greater use of available power because class C amplifiers are far more efficient.

Disadvantage of FM

Perhaps the greatest disadvantage of FM is that it simply uses too much spectrum space. The bandwidth of an FM signal is considerably wider than that of an AM signal transmitting similar information. Although the modulation index can be kept low to minimize the bandwidth used, still the bandwidth is typically larger than that of an AM signal. Further, reducing the modulation index also reduces the noise immunity of the FM signal. In commercial two-way FM radio systems, the maximum allowed deviation is 5 kHz, with a maximum modulating frequency of 3 kHz. This produces a deviation ratio of 5/3 = 1.67. This is usually referred to as *narrowband FM (NBFM)*.

Since FM occupies so much bandwidth, it typically has been used only at the very high frequencies. In fact, it is seldom used in communications below frequencies of 30 MHz. Most FM communications work is done at the VHF, UHF, and microwave frequencies. It is only in these portions of the spectrum where adequate bandwidth is available for FM signals and where line-of-sight transmission is prevalent. This means that the range of communication is more limited.

■ TEST

Answer the following questions.

25. The main advantage of FM over AM is its immunity from _____.
26. Noise is primarily a variation in
    a. Amplitude.
    b. Frequency.
    c. Phase.
27. FM receivers reject noise because of built-in _____ circuits.
28. The _____ in an FM receiver causes a stronger signal to dominate a weaker signal on the same frequency.
29. Typically FM transmitters are more efficient than AM transmitters because they use class _____ amplifiers.
30. The biggest disadvantage of FM is its excessive use of _____.
31. True or false. An AM circuit is usually more complex and expensive than an FM circuit.
32. Noise interferes primarily with _____ modulating frequencies.
33. The method used to offset the effect of noise in FM transmissions by boosting high frequencies is known as _____.
34. To boost high frequencies a(n) _____ circuit is used.

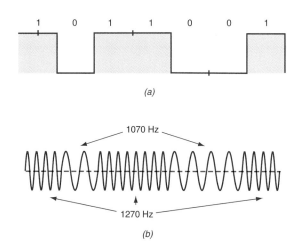

(a)

1070 Hz

1270 Hz

(b)

Fig. 4-9 Frequency-shift keying. (*a*) Binary signal. (*b*) FSK signal.

35. To correct for the high-frequency boost, a(n) _____ circuit is used at the receiver.
36. The time constant of an FM broadcast pre-emphasis circuit is _____ μs.
37. In an FM receiver, frequencies above _____ Hz are attenuated 6 dB per octave.

4-5 FM with Binary Signals

FM is also used to transmit digital data. When the modulating signal is binary, a binary 1 input produces one carrier frequency, and a binary 0 input produces another carrier frequency. This modulation technique is known as *frequency-shift keying (FSK)*.

Frequency-shift keying (FSK)

Figure 4-9 shows an FSK signal. A binary 0 input produces a 1070-Hz carrier, and a binary 1 produces a 1270-Hz carrier.

At one time, FSK was used in computer modems that transmit data through the telephone system. Today, FSK is used primarily to transmit binary data by radio.

■ TEST_____

Answer the following questions.

38. True or false. A binary 1 usually produces a higher carrier frequency than a binary 0.
39. True or false. FSK maintains a constant-amplitude carrier as in FM.

Chapter 4 Review

Summary

1. In FM, the information signal varies the frequency of the carrier.
2. The amount of frequency change from the carrier center frequency is called the frequency deviation.
3. In FM, the deviation is proportional to the amplitude of the modulating signal.
4. During FM, the carrier amplitude remains constant.
5. Both FM and PM are types of angle modulations.
6. In PM, the phase shift of the carrier is varied by the amplitude of the modulating signal.
7. Phase modulation produces frequency modulation.
8. The FM produced by PM is called indirect FM.
9. Maximum frequency deviation in a PM signal occurs where the rate of change of the modulating signal amplitude is greatest, which is at its zero-crossing points.
10. Frequency deviation does not occur at the output of a phase modulator unless the modulating signal amplitude varies.
11. The amount of frequency deviation produced by a phase modulator increases with the modulating frequency.
12. To produce true FM from a PM signal, the amplitude of the modulating signal must be decreased with frequency so that frequency deviation does not change with modulating frequency.
13. In PM, a low-pass filter on the modulating signal compensates for increased frequency deviation at the higher modulating frequencies.
14. Frequency modulation produces pairs of sidebands spaced from the carrier in multiples of the modulating frequency.
15. The modulation index m of an FM signal is the ratio of the frequency deviation f_d to the modulating frequency f_m ($m = f_d/f_m$).
16. The deviation ratio is the maximum frequency deviation divided by the maximum modulating frequency.
17. The modulation index determines the number of significant pairs of sidebands in an FM signal.
18. The amplitudes of the carrier and sidebands vary with the modulation index and can be calculated with a mathematical procedure known as the Bessel functions.
19. The carrier or sideband amplitudes are zero at some modulation indexes.
20. The bandwidth of an FM signal is proportional to the modulation index.
21. There are two ways to calculate the bandwidth of an FM signal.

$$BW = 2Nf_{m\ max}$$
$$BW = 2(f_{d\ max} + f_{m\ max})$$

22. For FM, the percentage of modulation is the ratio of the actual frequency deviation and the maximum allowed frequency deviation multiplied by 100.
23. The primary advantage of FM over AM is its immunity to noise.
24. Noise is short-duration amplitude variations caused by lightning, motors, auto ignitions, power transients, and other sources.
25. Limiter circuits in FM receivers clip off noise signals.
26. Another benefit of FM over AM is the capture effect that allows the strongest signal on a frequency to dominate without interference from the other signal.
27. A third benefit of FM over AM is greater transmitter efficiency since class C amplifiers may be used.
28. A major disadvantage of FM is that its bandwidth is wider than the bandwidth of AM.
29. The spectrum space taken up by an FM signal may be limited by carefully controlling the deviation ratio.
30. Another disadvantage of FM is that the circuits to produce and demodulate it are usually more complex and expensive than AM circuits.
31. Noise occurs primarily at high frequencies; therefore, noise interferes more with high modulating frequencies.
32. Interference from high-frequency noise can be minimized by boosting the amplitude of high-frequency modulating signals prior to modulation. This is called pre-emphasis.

33. Pre-emphasis is accomplished by passing the modulating signal through an *RC* network that linearly boosts the amplitude of frequencies above 2123 Hz in proportion to frequency. This increases the signal-to-noise ratio at the higher frequencies.

34. The effect of pre-emphasis is corrected for in an FM receiver by de-emphasizing the higher frequencies by passing them through an *RC* low-pass filter.

35. The pre-emphasis and de-emphasis networks have a time constant of 75 μs and a cutoff frequency of 2123 Hz.

36. A variation of FM called frequency-shift keying (FSK) is used to transmit digital data. A binary 1 input produces one carrier frequency, and a binary 0 produces another (usually lower) carrier frequency.

Chapter Review Questions

Choose the letter that best answers each question.

4-1. The amount of frequency deviation from the carrier center frequency in an FM transmitter is proportional to what characteristics of the modulating signal?
   a. Amplitude
   b. Frequency
   c. Phase
   d. Shape

4-2. Both FM and PM are types of what kind of modulation?
   a. Amplitude
   b. Phase
   c. Angle
   d. Duty cycle

4-3. If the amplitude of the modulating signal decreases, the carrier deviation
   a. Increases.
   b. Decreases.
   c. Remains constant.
   d. Goes to zero.

4-4. On an FM signal, maximum deviation occurs at what point on the modulating signal?
   a. Zero-crossing points
   b. Peak positive amplitude
   c. Peak negative amplitude
   d. Both *b* and *c*

4-5. In PM, a frequency shift occurs while what characteristic of the modulating signal is changing?
   a. Shape
   b. Phase
   c. Frequency
   d. Amplitude

4-6. Maximum frequency deviation of a PM signal occurs at
   a. Zero crossing points.
   b. Peak positive amplitude.
   c. Peak negative amplitude.
   d. Peak positive or negative amplitudes.

4-7. In PM, carrier frequency deviation is *not* proportional to:
   a. Modulating signal amplitude.
   b. Carrier amplitude and frequency.
   c. Modulating signal frequency.
   d. Modulator phase shift.

4-8. To compensate for increases in carrier frequency deviation with an increase in modulating signal frequency, what circuit is used between the modulating signal and the phase modulator?
   a. Low-pass filter
   b. High-pass filter
   c. Phase shifter
   d. Bandpass filter

4-9. The FM produced by PM is called
   a. FM.
   b. PM.
   c. Indirect FM.
   d. Indirect PM.

4-10. If the amplitude of the modulating signal applied to a phase modulator is constant, the output signal will be
   a. Zero.
   b. The carrier frequency.
   c. Above the carrier frequency.
   d. Below the carrier frequency.

4-11. A 100-MHz carrier is deviated 50 kHz by a 4-kHz signal. The modulation index is
   a. 5.
   b. 8.
   c. 12.5.
   d. 20.

4-12. The maximum deviation of an FM carrier is 2 kHz by a maximum modulating signal of 400 Hz. The deviation ratio is
   a. 0.2.
   b. 5.
   c. 8.
   d. 40.

4-13. A 70-kHz carrier has a frequency deviation of 4 kHz with a 1000-Hz signal. How many significant sideband pairs are produced?
a. 4
b. 5
c. 6
d. 7

4-14. What is the bandwidth of the FM signal described in question 4-13 above?
a. 4 kHz
b. 7 hKz
c. 14 kHz
d. 28 kHz

4-15. What is the relative amplitude of the third pair of sidebands of an FM signal with $m = 6$?
a. 0.11
b. 0.17
c. 0.24
d. 0.36

4-16. A 200-kHz carrier is modulated by a 2.5-kHz signal. The fourth pair of sidebands are spaced from the carrier by
a. 2.5 kHz.
b. 5 kHz.
c. 10 kHz.
d. 15 kHz.

4-17. An FM transmitter has a maximum deviation of 12 kHz and a maximum modulating frequency of 12 kHz. The bandwidth by Carson's rule is
a. 24 kHz.
b. 33.6 kHz.
c. 36.8 kHz.
d. 48 kHz.

4-18. The maximum allowed deviation of the FM sound signal in TV is 25 kHz. If the actual deviation is 18 kHz, the percent modulation is
a. 43 percent.
b. 72 percent.
c. 96 percent.
d. 139 percent.

4-19. Which of the following is *not* a major benefit of FM over AM?
a. Greater efficiency
b. Noise immunity
c. Capture effect
d. Lower complexity and cost

4-20. The primary disadvantage of FM is its
a. Higher cost and complexity.
b. Excessive use of spectrum space.
c. Noise susceptibility.
d. Lower efficiency.

4-21. Noise is primarily
a. High-frequency spikes.
b. Low-frequency variations.
c. Random level shifts.
d. Random frequency variations.

4-22. The receiver circuit that rids FM of noise is the
a. Modulator.
b. Demodulator.
c. Limiter.
d. Low-pass filter.

4-23. The phenomenon of a strong FM signal dominating a weaker signal on a common frequency is referred to as the
a. Capture effect.
b. Blot out.
c. Quieting factor.
d. Domination syndrome.

4-24. The AM signals generated at a low level may only be amplified by what type of amplifier?
a. Op amp
b. Linear
c. Class C
d. Push-pull

4-25. Frequency modulation transmitters are more efficient because their power is increased by what type of amplifier?
a. Class A
b. Class B
c. Class C
d. All the above

4-26. Noise interferes mainly with modulating signals that are
a. Sinusoidal.
b. Nonsinusoidal.
c. Low frequencies.
d. High frequencies.

4-27. Pre-emphasis circuits boost what modulating frequencies before modulation?
a. High frequencies
b. Mid-range frequencies
c. Low frequencies
d. All the above

4-28. A pre-emphasis circuit is a
a. Low-pass filter.
b. High-pass filter.
c. Phase shifter.
d. Bandpass shifter.

4-29. Pre-emphasis is compensated for at the receiver by a
a. Phase inverter.
b. Bandpass filter.
c. High-pass filter.
d. Low-pass filter.

4-30. The cutoff frequency of pre-emphasis and de-emphasis circuits is
  a. 1 kHz.
  b. 2.123 kHz.
  c. 5 kHz.
  d. 75 kHz.

4-31. In FSK, how many carrier frequencies does a binary modulating signal produce?
  a. 1
  b. 2
  c. 4
  d. An infinite number

Critical Thinking Questions

4-1. Discuss the effects and implications of using FM on the standard AM broadcast bands that use intelligence signal modulating frequencies up to 5 kHz and a 10 kHz standard station spacing. Assume a maximum deviation of 10 kHz.

4-2. The standard FM broadcast band has a frequency range of 88 to 108 MHz. Stations are spaced every 200 kHz beginning at 88.1 MHz and up to 107.9 MHz. The maximum permitted deviation is 75 kHz, with modulating frequencies up to 15 kHz. Compute the bandwidth of an FM station using two methods and then discuss how that bandwidth compares with the channel spacing.

4-3. TV sound is FM. The maximum deviation is 25 kHz for monaural and 50 kHz for stereo. The maximum modulating frequency is 15 kHz.

What is the maximum bandwidth of monaural and stereo broadcasts using Carson's rule?

4-4. Explain how the bandwidth of an FM signal is affected by transmitting a square-wave modulating signal.

4-5. What modulation index would produce the narrowest-bandwidth FM signal according to the Bessel functions?

4-6. Estimate the bandwidth of an 80-MHz carrier that is frequency-modulated by a 1-kHz square wave with a deviation of 2 kHz. Assume that the highest frequency content in the square wave is the seventh harmonic. Use Carson's rule.

4-7. Describe the signal resulting when a binary square-wave signal phase modulates a carrier.

Answers to Tests

1. angle
2. true
3. deviation
4. amplitude
5. increases, decreases (the reverse could also be true)
6. phase shift
7. frequency shift or deviation
8. changing or varying
9. amplitude, frequency
10. *a*
11. *b*
12. indirect FM
13. true
14. modulation index
15. 3 ($m = 10/3.33 = 3$)
16. 5
17. 995.5, 1004.5 [3(1.5) = 4.5 KHz; 1000 ± 4.5 = 995.5 and 1004.5 kHz]
18. 0.28
19. Bessel functions
20. 60 [2(5)(6) = 60 kHz]

21. 75 [(3.75/5)100 = 75%]
22. phase inversion or 180° shift
23. false
24. BW = 40 kHz; BW = 28 kHz using Carson's rule. Carson's rule gives narrower bandwidths because sidebands of less than 2 percent amplitude are not considered.
25. noise
26. *a*
27. limiter (or clipper)
28. capture effect
29. C
30. spectrum space
31. false
32. high
33. pre-emphasis
34. high-pass filter
35. de-emphasis
36. 75
37. 2123
38. true
39. true

Frequency Modulation Circuits

Chapter Objectives

This chapter will help you to:

1. *Describe* the operation of a voltage-variable capacitor and *show* how it is used to make frequency and phase modulator circuits in FM transmitters.
2. *Explain* the operation of a reactance modulator.
3. *Explain* the operation of varactor and transistor phase modulators.
4. *Describe* the operation of the Foster-Seeley discriminator, ratio detector, pulse-averaging discriminator, quadrature detector, and differential peak detector FM demodulator circuits.
5. *Explain* the operation and characteristics of a phase-locked loop and *show* how it is used for FM demodulation.

An incredible variety of circuits have been devised to produce FM and PM signals. There are direct FM circuits and circuits that produce FM indirectly by PM techniques. Direct FM circuits make use of techniques for actually varying the frequency of the carrier oscillator in accordance with the modulating signal. The PM techniques use a phase shifter after the carrier oscillator. In this chapter, some of the more commonly used FM and PM circuits will be discussed. Frequency demodulator circuits convert the FM signal back into the original modulating signal. Dozens of different circuits have been devised. The most popular and widely used demodulator circuits will be covered in detail here. Special emphasis will be given to the phase-locked loop demodulator, which provides the best overall frequency demodulation.

5-1 Frequency Modulators

The basic concept of FM is to vary the carrier frequency in accordance with the modulating signal. The carrier is generated by either an *LC* or a crystal oscillator circuit. The object then is to find a way to change the frequency of oscillation. In an *LC* oscillator, the carrier frequency is fixed by the values of the inductance and capacitance in a tuned circuit. The carrier frequency, therefore, may be changed by varying either this inductance or the capacitance. The idea is to find a circuit or component that converts a modulating voltage into a corresponding change in capacitance or inductance.

When the carrier is generated by a crystal oscillator, the frequency is fixed by the crystal.

However, keep in mind that the equivalent circuit of a crystal is an *LCR* circuit with both series and parallel resonant points. By connecting an external capacitor to the crystal, minor variations in operating frequency can be obtained. Again, the objective is to find a circuit or component whose capacitance will change in response to the modulating signal.

Voltage-Variable Capacitor

The component most frequently used in this application is a *varactor* or *voltage-variable capacitor (VVC)*. Also known as a variable capacitance diode, or varicap, this component is basically a semiconductor junction diode that is operated in a reverse-bias mode.

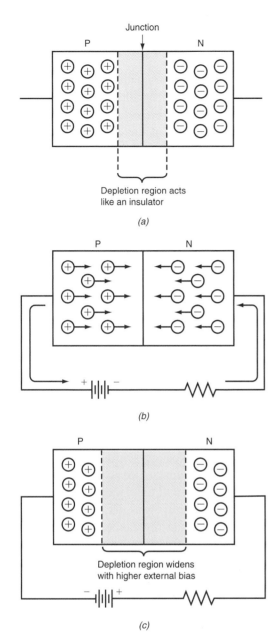

Fig. 5-1 Depletion region in a junction diode.

The depletion layer simply disappears. See Fig. 5-1b.

If an external reverse bias is applied to the diode, as in Fig. 5-1(c), no current will flow. The bias actually increases the width of the depletion layer. The width of this depletion mode depends upon the amount of the reverse bias. The higher the reverse bias, the wider the depletion layer and the less chance for current flow.

A reverse-biased junction diode appears to be a small capacitor. The P- and N-type materials act as the two plates of the capacitor, and the depletion region acts as the dielectric. The width of the depletion layer determines the width of the dielectric and, therefore, the amount of capacitance. If the reverse bias is high, the depletion region will be wide and the dielectric will cause the plates of the capacitor to be widely spaced, producing a low value of capacitance. Decreasing the amount of reverse bias narrows the depletion region, and, therefore, the plates of the capacitor will be effectively closer together and produce a higher capacitance.

All junction diodes exhibit this characteristic of variable capacitance as the reverse bias is changed. However, varactors or VVCs have been designed to optimize this particular characteristic. Such diodes are made so that the capacitance variations are as wide and linear as possible.

Voltage-variable capacitors are made with a wide range of capacitance values. Most units have a nominal capacitance in the range of 1 to 200 picofarads (pF). The capacitance variation range can be as high as 12 to 1. Figure 5-2 shows the curve for a typical diode. A maximum capacitance of 80 pF is obtained at 1 V. With 60 V applied, the capacitance drops to 20 pF, a

Refer to Fig. 5-1. When a junction diode is formed, P- and N-type semiconductors are joined to form a junction. Some electrons in the N-type material drift over into the P-type material and neutralize the holes there. Thus a thin region where there are no free carriers, holes, or electrons is formed. This is called the *depletion region*. It acts like a thin insulator that prevents current from flowing through the device. See Fig. 5-1(a).

If you apply a forward bias to the diode, it will conduct. The external potential forces the holes and electrons toward the junction, where they combine and cause a continuous current inside the diode as well as externally.

Depletion region

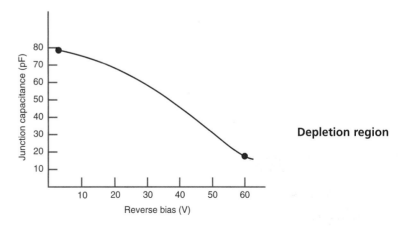

Fig. 5-2 Capacitance versus reverse junction voltage for a typical varactor.

4 to 1 range. The operating range is usually restricted to the linear center portion of the curve.

Varactor Modulator

Figure 5-3 shows the basic concept of a varactor frequency modulator. The L_1 and C_1 represent the tuned circuit of the carrier oscillator. Varactor diode D_1 is connected in series with capacitor C_2 across the tuned circuit. The value of C_2 is made very large at the operating frequency so that its reactance is very low. As a result, when C_2 is connected in series with the lower capacitance of D_1, the effect is as if D_1 were connected directly across the tuned circuit. The total effective circuit capacitance then is the capacitance of D_1 in parallel with C_1. This fixes the center carrier frequency.

The capacitance of D_1, of course, is controlled by two factors: a fixed dc bias and the modulating signal. In Fig. 5-3, the bias on D_1 is set by the voltage divider which is made up of R_1 and R_2. Usually either R_1 or R_2 is made variable so that the center carrier frequency can be adjusted over a narrow range. The modulating signal is applied through C_3 and the RFC. The C_3 is a blocking capacitor that keeps the dc bias out of the modulating signal circuits. The RFC is a radio frequency choke whose reactance is high at the carrier frequency to prevent the carrier signal from getting into the modulating signal circuits.

The modulating signal derived from the microphone is amplified and applied to the modulator. As the modulating signal varies, it adds to and subtracts from the fixed-bias voltage. Thus, the effective voltage applied to D_1 causes its capacitance to vary. This, in turn, produces a deviation of the carrier frequency as desired. A positive-going signal at point A adds to the reverse bias, decreasing the capacitance and increasing the carrier frequency. A negative-

going signal at A subtracts from the bias, increasing the capacitance and decreasing the carrier frequency.

The main problem with the circuit in Fig. 5-3 is that most *LC* oscillators are simply not stable enough to provide a carrier signal. Despite the quality of the components and the excellence of design, the *LC* oscillator frequency will vary because of temperature changes, circuit voltage variations, and other factors. Such instabilities cannot be tolerated in most modern electronic communications systems. As a result, crystal oscillators are normally used to set the carrier frequency. Not only do *crystal oscillators* provide a highly accurate carrier frequency, but their *frequency stability* is superior over a wide temperature range.

Frequency-Modulating a Crystal Oscillator

It is possible to vary the frequency of a crystal oscillator by changing the value of capacitance in series or in parallel with the crystal. Figure 5-4 shows a typical crystal oscillator. When a small value of capacitance is connected in series with the crystal, the crystal frequency can be "pulled" slightly from its natural resonant frequency. By making the series capacitor a varactor diode, frequency modulation of the crystal oscillator can be achieved. The modulating signal is applied to the varactor diode D_1 which changes the oscillator frequency.

The important thing to note about an FM crystal oscillator is that only a very small frequency deviation is possible. Greater deviations can be achieved with *LC* oscillators. Rarely can the frequency of a crystal oscillator be changed more than several hundred hertz from the nominal crystal value. The resulting deviation may be less than the total deviation

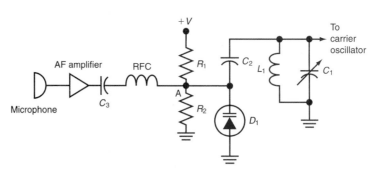

Fig. 5-3 Frequency modulation with a VVC.

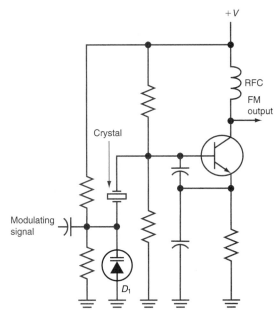

Fig. 5-4 Frequency modulation of a crystal oscillator with a VVC.

desired. To achieve a total frequency shift of 75 kHz as in commercial FM broadcasting, other techniques must be used. In two-way (narrowband) FM communications systems, the narrower deviations are acceptable.

Although a deviation of only several hundred cycles may be possible at the crystal oscillator frequency, the total deviation can be increased by using frequency multiplier circuits after the carrier oscillator. When the FM signal is applied to a *frequency multiplier,* both the frequency of operation and the amount of

deviation is increased. Typical frequency multipliers may increase the base (oscillator) frequency by 24 to 32 times.

For example, assume that the desired output frequency from an FM transmitter is 168 MHz and that the carrier is generated by a 7-MHz crystal oscillator. This is followed by frequency multiplier circuits that increase the frequency by a factor of 24 (7 MHz × 24 = 168 MHz).

Assume further that the desired maximum frequency deviation is 5 kHz. Frequency modulation of the crystal oscillator may produce a maximum deviation of only 200 Hz. When multiplied by the factor of 24 in the frequency multiplier circuits, this deviation, of course, is increased to 200 × 24 = 4800 Hz, or 4.8 kHz. Frequency multiplier circuits will be discussed in more detail in the chapter on transmitters.

Reactance Modulator

Another way to produce direct frequency modulation is to use a *reactance modulator.* This circuit uses a transistor amplifier to act like either a variable capacitor or an inductor. When the circuit is connected across the tuned circuit of an oscillator, the oscillator frequency can be varied by applying the modulating signal to the amplifier.

A reactance modulator is illustrated in Fig. 5-5. It is basically a standard common-emitter class A amplifier. Resistors R_1 and R_2 form a voltage divider to bias the transistor into the linear region. R_3 is an emitter bias resistor

Reactance modulator

Frequency multipliers

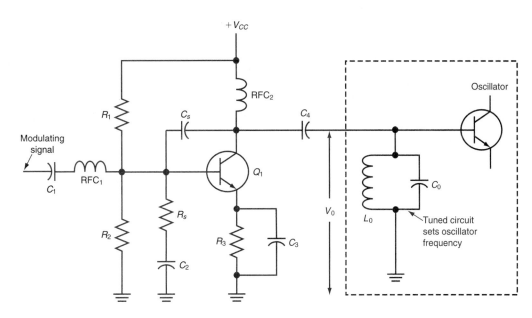

Fig. 5-5 A reactance modulator.

which is bypassed with capacitor C_3. Instead of a collector resistor, a radio frequency choke (RFC_2) is used to provide a high impedance load at the operating frequency.

Now, note that the collector of the transistor is connected to the top of the tuned circuit in the oscillator. Capacitor C_4 has a very low impedance at the oscillator frequency. Its main purpose is to keep the direct current from the collector of Q_1 from being shorted to ground through the oscillator coil L_o. As you can see, the reactance modulator circuit is connected directly across the parallel tuned circuit that sets the oscillator frequency.

The oscillator signal from the tuned circuit V_o is connected back to an RC phase-shift circuit made up of C_s and R_s. Capacitor C_2 in series with R_s has a very low impedance at the operating frequency, so it does not affect the phase shift. However, it does prevent R_s from disturbing the dc bias on Q_1. The value of C_s is chosen so that its reactance at the oscillator frequency is about 10 or more times the value of R_s. If the reactance is much greater than the resistance, the circuit will appear predominantly capacitive; therefore the current through the capacitor and R_s will lead the applied voltage by about 90°. This means that the voltage across R_s that is applied to the base of Q_1 leads the voltage from the oscillator.

Since the collector current is in phase with the base current, which in turn is in phase with the base voltage, the collector current in Q_1 leads the oscillator voltage V_o by 90°. Of course, any circuit whose current leads its applied voltage by 90° looks capacitive to the source voltage. This means that the reactance modulator looks like a capacitor to the oscillator-tuned circuit.

The modulating signal is applied to the modulator circuit through C_1 and RFC_1. The RFC helps keep the RF signal from the oscillator out of the audio circuits from which the modulating signal usually comes. The audio modulating signal will vary the base voltage and current of Q_1 according to the intelligence to be transmitted. The collector current will also vary in proportion. As the collector current amplitude varies, the phase-shift angle changes with respect to the oscillator voltage, which is interpreted by the oscillator as a change in the capacitance. So, as the modulating signal changes, the effective capacitance of the cir-

cuit varies and the oscillator frequency is varied accordingly. An increase in capacitance lowers the frequency, whereas a lower capacitance increases the frequency. The circuit produces direct frequency modulations.

If you reverse the positions of R_s and C_s in the circuit of Fig. 5-5, the current in the phase shifter will still lead the oscillator voltage by 90°. However, it is voltage from across the capacitor that is now applied to the base of the transistor. This voltage lags the oscillator voltage by 90°. With this arrangement, the reactance modulator acts like an inductor. The equivalent inductance changes as the modulating signal is applied. Again, the oscillator frequency is varied in proportion to the amplitude of the intelligence signal.

The reactance modulator is one of the best FM circuits because it can produce frequency deviation over a wide frequency range. It is also highly linear; that is, distortion is minimal. The circuit can also be implemented with a field-effect transistor (FET) in place of the NPN bipolar shown in Fig. 5-5.

Voltage-Controlled Oscillator

Oscillators whose frequencies are controlled by an external input voltage are generally referred to as *voltage-controlled oscillators (VCOs)*. A voltage-controlled crystal oscillator is generally referred to as a VXO. Although VCOs are used primarily for FM, there are other applications in which some form of voltage-to-frequency conversion is required.

In high-frequency communication circuits, VCOs are ordinarily implemented with discrete-component transistor and varactor diode circuits. However, there are many different types of lower-frequency VCOs in common use. These include IC VCOs using RC multi-vibrator-type oscillators whose frequency can be controlled over a wide range by an ac or dc input voltage. These VCOs typically have an operating range of less than 1 Hz to approximately 1 MHz.

The output is either a square or triangular wave rather than a sine wave. A typical IC VCO is shown in Fig. 5-6(*a*). This is a general block diagram of the popular NE566. External resistor R_1 at pin 6 sets the value of current produced by the internal current sources. The current sources linearly charge and discharge external

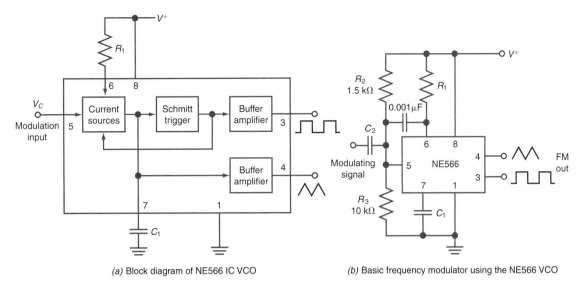

(a) Block diagram of NE566 IC VCO   (b) Basic frequency modulator using the NE566 VCO

Fig. 5-6 Frequency modulation with an IC VCO.

capacitor C_1 at pin 7. An external voltage V_C applied at pin 5 may also be used to vary the amount of current produced by the current sources.

The Schmitt trigger circuit is a level detector which switches when the capacitor charges or discharges to a specific voltage level. The Schmitt trigger controls the current source by switching the current sources between charging and discharging. A linear sawtooth of voltage is developed across the capacitor by the current source. This is buffered and made available at pin 4. The Schmitt trigger output is a square wave at the same frequency available at pin 3. If a sine wave output is desired, usually the triangle wave is filtered with a tuned circuit resonant to the desired carrier frequency.

A complete FM circuit is shown in Fig. 5-6(b). The current sources are biased with a voltage divider made up of R_2 and R_3. The modulating signal is applied through C_2 to the voltage divider at pin 5. The 0.001-μF capacitor between pins 5 and 6 is used to prevent unwanted spurious oscillations.

The center carrier frequency of the circuit is set by the values of R_1 and C_1. Carrier frequencies up to 1 MHz may be used with this IC. The outputs may be filtered or used to drive other circuits such as a frequency multiplier if higher frequencies and deviations are necessary. The modulating signal can vary the carrier frequency over a nearly 10:1 range, making very large deviations possible. The deviation is linear with respect to the input amplitude over the entire range.

■ TEST

Choose the letter which best answers each statement.

1. Increasing the reverse bias on a voltage-variable capacitor causes its capacitance to
   a. Increase.
   b. Decrease.
2. Connecting a VVC across a parallel *LC* circuit causes the resonant frequency to
   a. Increase.
   b. Decrease.
3. In the circuit of Fig. 5-3, a negative-going modulating signal causes the carrier frequency to
   a. Increase.
   b. Decrease.
4. A crystal is operating in its series resonant mode. A VVC is connected in series with it. The crystal frequency
   a. Increases.
   b. Decreases.
5. Which is capable of greater frequency deviation?
   a. *LC* oscillator
   b. Crystal oscillator

Supply the missing information in each statement.

6. Another name for voltage-variable capacitor is _____.
7. Most VVCs have a nominal capacitance in the _____ to _____ pF range.
8. A crystal oscillator has superior_____ over an *LC* oscillator.

9. The acronym VCO means _____.
10. A voltage-variable crystal oscillator is referred to as a(n) _____.
11. Carrier frequency and frequency deviation may be increased by using a(n) _____ after the carrier oscillator.
12. A reactance modulator is set up to act like an inductive reactance. If the modulating signal increases in amplitude, the effective inductance decreases. This causes the oscillator frequency to _____.
13. An IC VCO normally uses a combination of _____ and _____ to set the center operating frequency.

Determine whether each statement is true or false.

14. Voltage-variable capacitors should not be forward-biased.
15. For highly stable carrier generators, *LC* oscillators are preferred over crystal oscillators.
16. A reactance modulator is used with crystal oscillators.
17. IC VCOs operate primarily at frequencies below 1 MHz.

5-2 Phase Modulators

Most modern FM transmitters use some form of PM to produce indirect FM. The reason for using PM instead of direct FM is that the carrier oscillator can be optimized for frequency accuracy and stability. Crystal oscillators or crystal-controlled frequency synthesizers can be used to set the carrier frequency accurately and maintain solid stability.

The output of the carrier oscillator is fed to a phase modulator where the phase shift is made to vary in accordance with the modulating signal. Since phase variations produce frequency variations, indirect FM is the result.

Basic Phase-Shift Circuits

The simplest phase shifters are *RC* networks like those shown in Fig. 5-7(*a*) and (*b*). Depending upon the values of *R* and *C*, the output of the phase shifter can be set to any phase angle between 0° and 90°. In Fig. 5-7(*a*) the output leads the input by

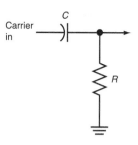

(*a*) Output leads input: 0°–90°

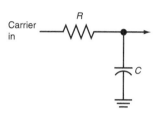

(*b*) Output lags input: 0°–90°

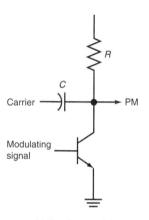

(*c*) Transistor modulator

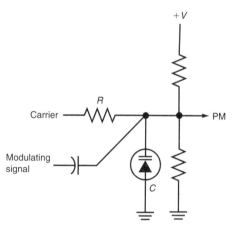

(*d*) Varactor modulator

Fig. 5-7 Simple PM circuits.

some angle between 0° and 90°. For example, when X_C equals *R*, the phase shift is 45°.

A low-pass version of the same *RC* filter can also be used, as shown in Fig. 5-7(*b*). Here the output is taken from across the capacitor, so it lags the input voltage by some angle between 0° and 90°.

One of these simple phase-shift circuits can be used as a phase modulator if the resistance or capacitance can be made to vary with the modulating signal. One way to do this is to substitute a transistor for the resistor in the circuit in Fig. 5-7(a). The resulting phase-shift circuit is shown in Fig. 5-7(c). The transistor simply acts as a variable resistor that varies in response to the modulating signal. If the modulating signal increases, the transistor base current and collector current increase. Therefore, the effective transistor resistance decreases. Lowering the resistance increases the amount of phase shift. This causes a corresponding frequency increase. If the amplitude of the modulated signal decreases, the base and collector currents decrease and the effective transistor resistance increases. This decreases the amount of phase shift, and, as a result, the amount of frequency shift decreases. An FET can be substituted for the bipolar transistor in Fig. 5-7(c) with comparable results.

Fig. 5-7(d) shows how a varactor can be used to implement a simple low-pass phase-shift modulator. Here the modulating signal causes the capacitance of the varactor to change. If the modulating signal amplitude increases, it adds to the varactor bias from R_1 and R_2, thereby causing the capacitance to decrease. This causes the reactance to increase; thus, the circuit produces less phase shift. A decreasing modulating signal subtracts from the reverse bias on the varactor diode, thereby increasing the capacitance or decreasing the capacitive reactance. This increases the amount of phase shift.

Practical Phase Modulators

A common phase modulator is shown in Fig. 5-8. It uses a phase shifter made up of a capacitor and the variable resistance of a field-effect transistor Q_1. The carrier signal from a crystal oscillator or a phase-locked loop frequency synthesizer is applied directly to the output through C_1 and C_2. The carrier signal is also applied to the gate of the FET through C_1. The series capacitance of C_1 and C_2 and the FET source to drain resistance produce a leading phase shift of current in the FET and a

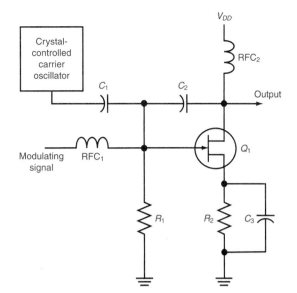

Fig. 5-8 An improved phase modulator.

For definitions of the new key words used in this chapter, you might want to reference the following site: ⟨telecom.tbi.net⟩.

leading voltage at the output. The carrier signal applied to the gate of the FET also varies the FET current. C_1 and R_1 produce a leading phase shift less than 90°. The leading voltage across R_1 also controls the current in Q_1. With two signals controlling the FET current, the result is a phasor sum of the two currents.

The modulating signal is applied to the gate of the FET. RFC_1 keeps the carrier RF out of the audio circuits. The audio signal now also controls the FET current. This changes the amplitude relationships of the other two controlling inputs, thereby producing a phase shift that is directly proportional to the amplitude of the modulating signal. The carrier output at the FET drain varies in phase and amplitude. The signal is then usually passed on to a class C amplifier or frequency multiplier which removes the amplitude variations but preserves the phase and frequency variations.

The problem with such simple phase modulators is that they are capable of producing only a small amount of phase shift because of the narrow range of linearity of the transistor or varactor. The total amount of phase shift is essentially limited to ±20°. Such a limited amount of phase shift produces, in turn, only a limited frequency shift.

Using a Tuned Circuit for Phase Modulation

An improved method of PM is to use a parallel tuned circuit to produce the phase shift. At resonance, the parallel resonant circuit will have a very high resistance. Off resonance, the circuit will act inductively or capacitively and, as a result, will produce a phase shift between its current and applied voltage.

Figure 5-9 shows the basic impedance Z response curve of a parallel resonant circuit. Also shown is the phase variation A. At the resonant frequency f_r, the inductive and capacitive reactances are equal, and, therefore, their effects cancel one another. The result is an extremely high impedance at f_r. The circuit acts resistive at this point, and, therefore, the phase angle between the current and the applied voltage is zero.

At frequencies below resonance, X_L decreases and X_C increases. This causes the circuit to act like an inductor. Therefore, the current will lag the applied voltage. Above resonance, X_L increases while X_C decreases. This causes the circuit to act capacitively, and the current leads the applied voltage. If the Q of the resonant circuit is relatively high, the phase shift will be quite pronounced, as shown in Fig. 5-9. The same effect is achieved if f is constant and either L or C is varied. For a relatively small change in L or C, a significant phase shift can be produced. The idea then is

to cause the inductance or capacitance to vary with the modulating voltage and thus produce a phase shift.

A variety of circuits have been developed based on this technique, but one illustrating its operation is shown in Fig. 5-10. The parallel tuned circuit made up of L, C_1, and C_2 is part of the output circuit of an RF amplifier driven by the carrier oscillator. Capacitor C_1 is large so that its reactance at the carrier frequency is low. Therefore, the resonant frequency is controlled by C_2. A varactor diode D_1 is connected in parallel with C_2 in the tuned circuit and, therefore, will provide a capacitance change with the modulating signal. The voltage divider made up of R_1 and R_2 sets the reverse bias on D_1. The value of C_3 is very large and simply acts as a dc blocking capacitor, preventing bias from being applied to the tuned circuit. Its value is very large, so it is essentially an ac short at the carrier frequency. Therefore, it is the capacitance of D_1 and C_2 that controls the resonant frequency.

The modulating signal is first passed through a low-pass network R_3-C_5 that provides the amplitude compensation necessary to produce FM. The modulating signal appears across potentiometer R_4. In this way, the desired amount of modulating signal can be tapped off and applied to the phase-shift circuit. The potentiometer acts as a deviation control. The higher the modulating voltage, the greater the frequency deviation. The modulating signal is applied to the varactor diode through capacitor C_4. The RFC has a high impedance at the carrier frequency to minimize the loading of the tuned circuit which reduces Q. With zero modulating voltage, the value of the capacitance of D_1 along with capacitor C_2 and the inductor L set the resonant frequency of the tuned circuit. The PM output across L is inductively coupled to the output.

When the modulating signal goes negative, it subtracts from the reverse bias of D_1. This increases the capacitance of the circuit and lowers the reactance, making the circuit appear capacitive. Thus a leading phase shift is produced. The parallel LC circuit looks like a capacitor to the output resistance of the RF amplifier, so the output lags the input. A positive-going modulating voltage will decrease the capacitance, and thus the tuned circuit will become inductive and produce a lagging phase shift. The LC circuit looks like an inductor to

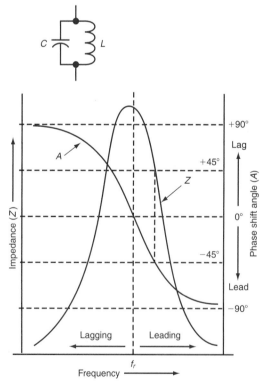

Fig. 5-9 Impedance and phase shift versus frequency of a parallel resonant circuit.

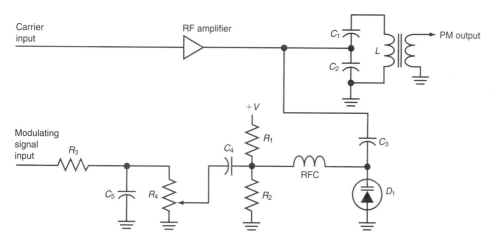

Fig. 5-10 One form of phase modulator.

output resistance of the RF amplifier, so the output leads the input. The result at the output is a relatively wide phase shift which, in turn, produces excellent linear frequency deviation.

Although phase modulators are relatively easy to implement, they have two main disadvantages. First, the amount of phase shift they produce and the resulting frequency deviation is relatively low. For that reason, the carrier is usually generated at a lower frequency and frequency multipliers are used to increase the carrier frequency and the amount of frequency deviation. Second, all the phase-shift circuits described above produce amplitude variations as well as phase changes. In the simple phase-shift circuits of Fig. 5-7, the phase shifters are all voltage dividers. When the value of one of the components is changed, the phase shifts but the output amplitude changes as well. This is also true of the tuned-circuit phase shifter. As a result, some means must be used to remove the amplitude variations.

Both these problems are solved by feeding the output of the phase modulator to class C amplifiers used as frequency multipliers. The class C amplifiers eliminate any amplitude variations, while at the same time they increase the carrier frequency and the deviation to the desired final values. Class C frequency multipliers will be discussed in the next chapter.

■ TEST_____

Answer the following questions.

18. Phase modulation is called _____ FM.
19. True or false. Phase modulation produces frequency variation as well as amplitude variation of the carrier.

20. In a simple *RC* network, the phase shift is between _____ and _____ degrees.
21. In the circuits of Fig. 5-7, phase is varied by changing the _____ or _____.
22. Larger linear phase shifts are obtained with a(n) _____ circuit.
23. The control element in a phase modulator is usually a(n) _____.
24. In Fig. 5-10, the component which adjusts the deviation is _____.
25. If a parallel *LC* circuit is at resonance, increasing *C* will cause the current to _____ (lead, lag) the applied voltage.
26. In Fig. 5-7(*b*), if *C* is decreased, will the phase shift increase or decrease?
27. In the phase modulator of Fig. 5-8, the FET current is the phasor sum of the currents produced by _____ and _____.
28. The small phase shifts produced by indirect FM are increased by sending the PM signal to a(n) _____ circuit.

5-3 Frequency Demodulators

There are literally dozens of circuits used to demodulate or detect FM and PM signals. The well-known *Foster-Seeley discriminator* and the ratio detector were among the most widely used frequency demodulators at one time, but today these circuits have been replaced with more sophisticated IC demodulators. Of course, they are still found in older equipment. The most widely used detectors today include the pulse-averaging discriminator, the quadrature detector, and the phase-locked loop. In this section, we will take a look at all these widely used demodulator circuits.

Foster-Seeley discriminator

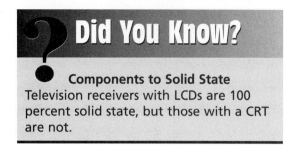

Foster-Seeley Discriminator

One of the best frequency demodulators is the Foster-Seeley discriminator shown in Fig. 5-11. The FM signal is applied to the primary of the RF transformer T_1. The primary and secondary windings are resonated at the carrier frequency with C_1 and C_2. The parallel tuned circuit in the primary of T_1 is connected in the collector of a limiter amplifier Q_1 that removes amplitude variations from the FM signal.

The signal across the primary of T_1 is also passed through capacitor C_3 and appears directly across an RFC. The voltage appearing across the RFC is exactly the same as that appearing across the primary winding simply because C_3 and C_5 are essentially short circuits at the carrier frequency. The voltage across this RFC is designated V_3.

The current flowing in the primary winding of T_1 induces a voltage in the secondary winding. Because the secondary winding is center-tapped, the voltage across the upper portion V_1 will be 180° out of phase with the voltage across the lower portion V_2. The voltage induced into the secondary winding is 90° out of phase with the voltage across the primary winding. When both the primary and the secondary windings of an air-core transformer are tuned resonant circuits, the phase relationship between the voltages across the primary and secondary will be 90°. This means that the voltages V_1 and V_2 will also be 90° out of phase with V_3, the voltage across the RFC. This phase relationship is shown in the vector diagrams in Fig. 5-11. In Fig. 5-11(a), the input is the unmodulated carrier frequency.

The remainder of the circuit consists of two diode detector circuits similar to those used for AM detection. The voltage $V_{1\text{-}3}$ applied to D_1, R_1, and C_4 is the sum of voltages V_1 and V_3. The voltage $V_{2\text{-}3}$ applied to D_2, R_2, and C_5 is the sum of voltages V_2 and V_3. Since voltages V_1 and V_2 are out of phase with voltage V_3, their respective sums $V_{1\text{-}3}$ and $V_{2\text{-}3}$, are vector sums, as illustrated in Fig. 5-11(a).

On one half cycle of the primary voltage, D_1 conducts and current flows through R_1 and

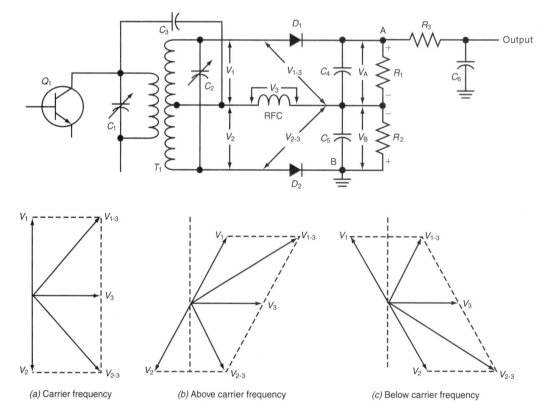

(a) Carrier frequency   (b) Above carrier frequency   (c) Below carrier frequency

Fig. 5-11 The Foster-Seeley discriminator.

charges C_4. On the next half cycle, D_2 conducts and current flows through R_2 and charges C_5. The voltages across R_1 and R_2, designated V_A and V_B, are identical because V_{1-3} and V_{2-3} are the same. Since these two voltages are equal but of the opposite polarity, the voltage between point A and ground is zero. At the carrier center frequency with no modulation, the modulator output is therefore zero.

The secondary of T_1 and capacitor C_2 form a series resonant circuit. The reason for this is that the voltage induced into the secondary appears in series with that winding. At resonance, the inductive reactance of the secondary winding equals the capacitive reactance of C_2. At that time, the current flowing in the circuit is exactly in phase with the voltage induced into the secondary. The output is derived from a portion of the secondary winding which is an inductor. The voltage across the secondary winding, therefore, is 90° out of phase with the current in the circuit.

Should the input frequency change as would be the case with FM, the secondary winding will no longer be at resonance. For example, if the input frequency increases, the inductive reactance will be higher than the capacitive reactance, making this circuit inductive. This causes the phase relationship between V_1 and V_2 to change with respect to V_3. If the input is above the resonant frequency, then V_1 will lead V_3 by a phase angle less than 90°. Since V_1 and V_2 remain 180° out of phase, V_2 will then lag V_3 by an angle of more than 90°. This change in phase relationship is shown in Fig. 5-8(b). When the new vector sums of V_1 and V_3 and V_2 and V_3 are computed, it is found that the voltage V_{1-3} applied to D_1 is greater than the voltage V_{2-3} applied to D_2. Therefore, the voltage across R_1 will be greater than the voltage across R_2 and the net output voltage will be positive with respect to ground.

If the frequency deviation is lower than the center frequency, voltage V_1 will lead by an angle of more than 90° while voltage V_2 will lag by an angle less than 90°. The resulting vector additions of V_1 and V_3 and V_2 and V_3 are shown in Fig. 5-11(c). This time, V_{2-3} is greater than V_{1-3}. As a result, the voltage across R_2 will be greater than the voltage across R_1 and the net output voltage will be negative.

As you can see, as the frequency deviates above and below the center frequency, the output at point A increases or decreases, and,

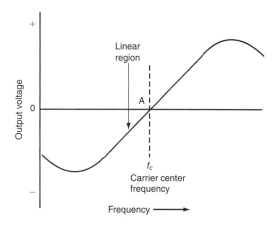

Fig. 5-12 Output voltage of the discriminator.

therefore, the original modulating signal is recovered. This signal is passed through the de-emphasis network consisting of R_3 and C_6. The output is then applied to an amplifier or to other circuits as required. Figure 5-12 shows the output voltage occurring across point A and ground with respect to frequency deviation. It is the linear range in the center that produces the accurate reproduction of the original modulating signal.

The discriminator circuit is sensitive to input amplitude variations. Higher inputs will produce a greater amount of signal at the output, whereas lower inputs create less output. These output variations will be interpreted as frequency changes, and the output will be an incorrect reproduction of the modulating signal. For that reason, all amplitude variations must be removed from the FM signal before being applied to the discriminator. This is usually handled by limiter circuits that will be discussed later.

Ratio Detector

Another widely used demodulator is the *ratio detector*. It is similar in appearance to the discriminator but has several important differences as is shown in Fig. 5-13. The FM signal is applied to the RF transformer T_1 with its center tapped secondary. The FM signal is also passed through capacitor C_3 and applied across the RFC as in the discriminator. The circuit uses two diodes, but note that the direction of D_2 is reversed from that in the discriminator. The voltages V_{1-3} and V_{2-3} applied to D_1 and D_2, respectively, are again a composite of V_1 and V_3 (V_{1-3}) and V_2 and V_3 (V_{2-3}) as before.

Ratio detector

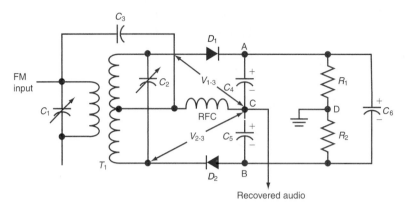

Fig. 5-13 A ratio detector.

Another major difference in the ratio detector is the use of a very large capacitor C_6 connected across the output. The load resistors R_1 and R_2 are equal in value, and their common connection is at ground. The output is taken from between points C and ground in the circuit. Capacitors C_4 and C_5 and resistors R_1 and R_2 form a bridge circuit. The voltage across capacitors C_4 and C_5 is the bridge input voltage, while the output is taken between points C and D.

With no modulation on the carrier, the voltage $V_{1\text{-}3}$ applied to D_1 is the same as the voltage $V_{2\text{-}3}$ applied to D_2. Therefore, capacitors C_4 and C_5 charge to the same voltage with the polarity shown. Since C_6 is connected across these two capacitors, it will charge to the sum of their voltages. Because C_6 is a very large capacitor, usually tantalum or electrolytic, it takes several cycles of the input signal for the capacitor to charge fully. However, once it charges, it will maintain a relatively constant voltage. Since R_1 and R_2 are equal, their voltage drops will be equal. Also, the voltage drops across C_4 and C_5 are equal. The bridge circuit, therefore, is balanced. Looking between points C and D, you will see 0 V because the potential is the same.

Assume that at the center carrier frequency, the voltage drops across C_4 and C_5 are each 2 V. This means the charge on C_6 is 4 V. Then the voltages across R_1 and R_2 are each 2 V.

If the frequency increases, the phase relationship in the circuit will change as described previously for the discriminator circuit [Fig. 5-11(a)]. This will cause the voltage across C_4 to be greater than the voltage across C_5. Assume that the voltage across C_4 is 3 V and the voltage across C_5 is 1 V. The voltages across R_1 and R_2 remain the same at 2 V each because

the charge on C_6 does not change. The bridge is now unbalanced, and an output voltage will appear between points C and D in the circuit. Using point B as a reference, the voltage at point C is 1 V positive, and the voltage across R_2 is 2 V positive. Therefore, the voltage difference at C is -1 V.

If the frequency decreases, then the phase relationship will be such that the charge on C_5 will be greater than the charge on C_4. If the voltage across C_5 is $+3$ V with respect to B and the voltage across R_2 remains 2 V, then point C is $+1$ V. The bridge is unbalanced, but in the opposite direction, and the output voltage is of the opposite polarity.

The primary advantage of the ratio detector over the discriminator is that it is essentially insensitive to noise and amplitude variations. The reason for this is the very large capacitor C_6. Since it takes a long time for this capacitor to charge or discharge, short noise pulses or minor amplitude variations are totally smoothed out. However, the average dc voltage across C_6 is the same as the average signal amplitude. This voltage can, therefore, be used in automatic gain control applications.

As indicated earlier, the ratio detector and Foster-Seeley discriminator are no longer widely used because they are difficult to implement in integrated-circuit form. Besides, the quadrature demodulator and PLL offer far superior performance for comparable cost.

Pulse-Averaging Discriminator

A simplified block diagram of a *pulse-averaging discriminator* is shown in Fig. 5-14. The FM signal is applied to a zero-crossing detector or a clipper/limiter which generates a binary voltage

Pulse-averaging discriminator

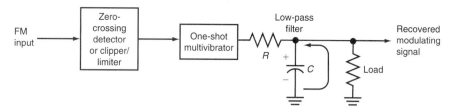

Fig. 5-14 A pulse-averaging discriminator for frequency demodulation.

level change each time the FM signal varies from minus to plus, or plus to minus. The result is a rectangular wave containing all the frequency variations of the original signal but without amplitude variations.

The FM square wave is then applied to a one-shot multivibrator. The one-shot or monostable multivibrator generates a fixed-amplitude, fixed-width dc pulse on the leading edge of each FM cycle. The duration of the one shot is set so that it is less than the period of the highest frequency expected during maximum deviation.

The one-shot output pulses are then fed to a simple *RC* low-pass filter that averages the dc pulses to recover the original modulating signal.

The waveforms for the pulse-averaging discriminator are shown in Fig. 5-15. Note how at the low frequencies, the one-shot pulses are widely spaced. At the higher frequencies, the one-shot pulses occur very close together. When these pulses are applied to the averaging filter, a dc output voltage is developed. The amplitude of this dc voltage is directly proportional to the frequency deviation.

When a one-shot pulse occurs, the capacitor in the filter charges to the amplitude of the

pulse. When the pulse turns off, the capacitor discharges into the load. If the *RC* time constant is high, the charge on the capacitor will not decrease much. If the time interval between pulses is long, however, the capacitor will lose some of its charge into the load, and so the average dc output will be low. When the pulses occur rapidly, the capacitor has little time to discharge between pulses, and therefore, the average voltage across it remains higher. As you can see, the filter output voltage varies in amplitude with the frequency deviation. Note in Fig. 5-15 how the voltage rises linearly as the frequency increases. The original modulating signal is developed across the filter output. The filter components are carefully selected to minimize the ripple caused by the charging and discharging of the capacitor while at the same time providing the necessary high-frequency response for the original modulating signal.

The pulse-averaging discriminator is a very high quality frequency demodulator. Prior to the availability of low-cost ICs, its use was limited to expensive telemetry and industrial control applications. Today, the pulse-averaging discriminator is easily implemented with low-cost

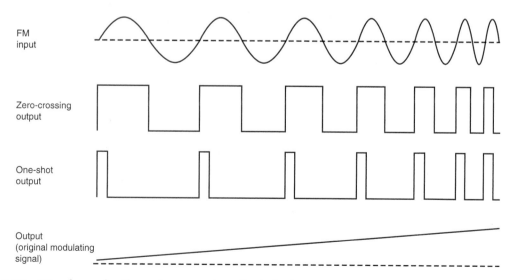

Fig. 5-15 Waveforms for the pulse-averaging discriminator.

ICs, and has, therefore, found its way into many electronic products.

Quadrature Detector

Quadrature detector

Another popular frequency demodulator is the *quadrature detector*. Its primary application is in TV audio demodulation, although it is also used in some FM radio systems. Most IC frequency demodulators are of the quadrature type.

The term *quadrature* refers to a 90° phase shift between two signals. The quadrature detector uses a phase-shift circuit to produce a phase shift of 90° at the unmodulated carrier frequency. The most commonly used phase-shift arrangement is shown in Fig. 5-16. The FM signal is applied through a very small capacitor C_1 to the parallel tuned circuit which is adjusted to resonance at the center carrier frequency. The tuned circuit appears as a high value of pure resistance at resonance. The small capacitor has a very high reactance compared to the tuned circuit impedance. The output across the tuned circuit then at the carrier frequency is very close to 90° leading the input. Now, when FM occurs, the carrier frequency will deviate above and below the resonant frequency of the tuned circuit. The result will be an increasing or decreasing amount of phase shift between the input and the output.

The two quadrature signals are then fed to a phase detector circuit. The phase detector is nothing more than a circuit whose output is a function of the amount of phase shift between two input signals. The most commonly used phase detector is a balanced modulator using differential amplifiers like those discussed in Chap. 3. The output of the phase detector is a series of pulses whose width varies with the amount of phase shift between the two signals. These signals are averaged in an *RC* low-pass filter to re-create the original modulating signal.

Differential peak detector

Normally the sinusoidal FM input signals to the phase detector are at a very high level. Therefore, they will drive the differential amplifiers in the phase detector into cutoff and saturation. The differential transistors will act as switches, so the output will be as a series of pulses. No limiter is needed if the input signal is large enough. The output pulse duration is determined by the amount of phase shift. You might think of the phase detector as simply an AND gate whose output is on only when the two input pulses are on and is off if either one or both of the inputs is off.

Figure 5-17 shows the typical waveforms involved in a quadrature detector. When there is no modulation, the two input signals are exactly 90° out of phase and, therefore, provide an output pulse width as indicated. When the frequency increases, the amount of phase shift decreases and, therefore, causes the output pulse width to be greater. The wider pulses averaged by the *RC* filter produce a higher average output voltage. This, of course, corresponds to the higher amplitude required to produce the higher carrier frequency.

When the FM signal frequency decreases, there is more phase shift and, as a result, the output pulses will be narrower. When averaged, the narrower pulses will produce a lower average output voltage, thus corresponding to the original lower modulating signal.

Differential Peak Detector

Another integrated-circuit FM demodulator is the *differential peak detector*. It is found in TV sets and other consumer electronic products. Usually the demodulator is only one of many other circuits on the chip. A general schematic diagram of the circuit is shown in Fig. 5-18(*a*).

The circuit is an enhanced differential amplifier. Transistors Q_3 and Q_4 form the

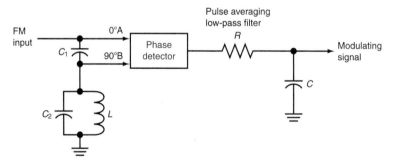

Fig. 5-16 A quadrature FM detector.

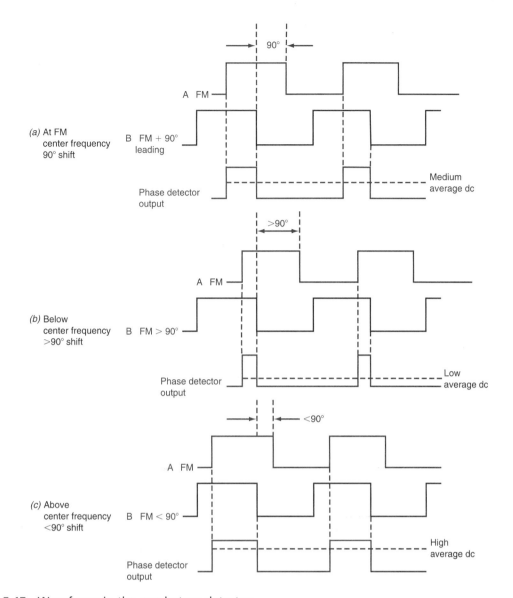

(a) At FM center frequency 90° shift

A FM

B FM + 90° leading

Phase detector output

Medium average dc

(b) Below center frequency >90° shift

A FM

B FM > 90°

Phase detector output

Low average dc

(c) Above center frequency <90° shift

A FM

B FM < 90°

Phase detector output

High average dc

Fig. 5-17 Waveforms in the quadrature detector.

differential amplifier, and the other transistors are emitter followers. Transistor Q_7 is a current source. Note that a tuned circuit is connected between the bases of Q_1 and Q_6. L_1, C_1, and C_2 are discrete components that are external to the chip. The FM input is applied to the base of Q_1. A typical input is a 4.5-MHz FM sound carrier in a TV set.

Q_1 and Q_6 are emitter followers that provide high input impedance and power amplification to drive two other emitter followers, Q_2 and Q_5. Q_2 and Q_5, along with on-chip capacitors C_a and C_b, are peak detectors. C_a and C_b charge and discharge as the voltages at the two inputs vary. For example, assume that the input is a carrier sine wave. When the input to Q_1 goes positive, the voltage at the emitter of Q_2 also goes positive, and C_a charges to the peak ac

value that appears across the tuned circuit. When the peak of the sine wave is past, the voltage declines to zero and then reverses polarity for the negative half-cycle of the sine wave. As the voltage drops, capacitor C_a retains the positive peak value as a charge. The input impedance to Q_3 is relatively high and so does not significantly discharge C_a. Capacitor C_a simply acts as a temporary storage cell for the peak voltage that cycle.

Capacitor C_b on Q_5 also forms a peak detector circuit. It too will charge up to the peak of the input from the opposite end of the tuned circuit. If the 4.5-MHz carrier is unmodulated, the two capacitors will charge up to the same value. These equal voltages appear at the inputs to the differential amplifier. Remember that the output of a differential amplifier is

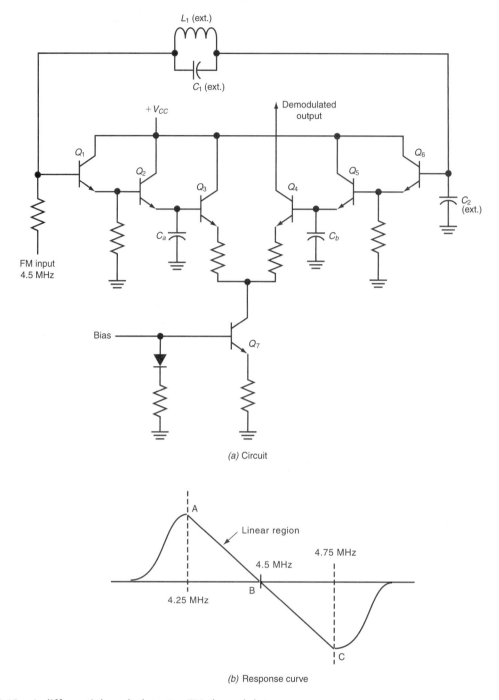

(a) Circuit

(b) Response curve

Fig. 5-18 A differential peak detector FM demodulator.

proportional to the difference of the two inputs. If the inputs are equal, the difference is zero, and so is the output. The output current from Q_4 when this condition exists is used as the zero reference. A lower or higher current will represent a varying frequency.

The external tuned circuit is designed so that L_1 forms a series-resonant circuit with C_2 at a frequency lower than the center frequency, usually 4.5 MHz. A typical series-resonant frequency is 4.25 MHz. L_1 forms a parallel reso-

nant circuit, with C_1 at a frequency above the center frequency. This frequency is usually 4.75 MHz.

To see how the circuit works, assume that an input is gradually increasing in frequency from below 4.25 MHz to above 4.75 MHz. At very low deviation frequencies, the effect on the tuned circuit is minimal and the inputs to the differential amplifiers are about equal; therefore the output is near zero. Now, as the input frequency increases, at some point it reaches the series-resonant point of L_1 and C_2. When this occurs, the total impedance of the series circuit is resistive and very low. This effectively shorts the input of Q_1 to ground, making it near zero. At the same time, the voltage across C_2 is at its peak value because of resonance, so the input to Q_6 is maximum. The result is a very high output current from Q_4. This condition is illustrated at point A in Fig. 5-18(*b*).

As the input frequency continues to rise, at some point the center frequency of 4.5 MHz is reached, at which time the output will be zero as described earlier. This is shown at point B in Fig. 5-18(*b*). The frequency continues to increase. At some point, L_1 will be parallel resonant with C_1. This produces a very high impedance at the base of Q_1, so the voltage is maximum at this time. The lower reactance of C_1 at this higher frequency makes the input to Q_6 very low. As a result, the difference between the two inputs is very high and the output current at Q_4 becomes very high in the opposite (negative) direction. This is point C in Fig. 5-18(*b*). As you can see, the variation in output current from point A to point C is very linear as the frequency increases. This change in output amplitude with a change in frequency results in excellent demodulation of FM signals.

Phase-Locked Loop Demodulator

The best frequency demodulator is the phase-locked loop (PLL). A *PLL* is a frequency- or phase-sensitive feedback control circuit. It is used not only in frequency demodulation but also in frequency synthesizers, as you will see in a later chapter, and in various filtering and signal detection applications.

All PLLs have the three basic elements illustrated in Fig. 5-19. A phase detector or mixer is used to compare the input or reference signal to the output of a VCO. The VCO frequency is varied by the dc output voltage from a low-pass filter. It is the output of the phase detector that the low-pass filter uses to produce the dc control voltage.

The primary job of the phase detector is to compare the two input signals and generate an output signal that when filtered will control the VCO. If there is a phase or frequency difference between the input and VCO signals, the phase detector output will vary in proportion. The filtered output will adjust the VCO frequency in an attempt to correct for the original frequency or phase change. This dc control voltage is called the *error signal* and is also the feedback in this circuit.

To examine the operation of the PLL, assume initially that no input signal is applied. At this time, the phase detector and low-pass filter outputs are zero. The VCO then operates at what is called its *free-running frequency*. This is the normal operating frequency of the VCO as determined by its internal frequency-determining components.

Now assume that an input signal close to the frequency of the VCO is applied. The phase detector will compare the VCO free-running frequency to the input frequency and produce an output voltage proportional to the frequency difference. The resulting dc voltage is applied to the VCO. This dc voltage is such that it forces the VCO frequency to move in a direction that reduces the dc error voltage. The error voltage forces the VCO frequency to change in the direction that reduces the amount of phase or frequency difference between the VCO and the input. At some point, the error voltage will eventually cause the VCO frequency to equal the input frequency. When this happens, the PLL is said to be in a *locked condition*. Although the input and VCO frequencies are equal, there will be a phase difference between them that produces the dc output voltage which causes the VCO to produce the frequency that will keep the circuit locked.

If the input frequency changes, the phase detector and low-pass filter will produce a new

Error signal

Free-running frequency

PLL demodulator

Locked condition

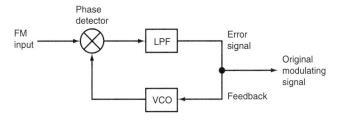

Fig. 5-19 The elements of a phase-locked loop (PLL).

Tracking

value of dc control voltage that will force the VCO output frequency to change until it is equal to the new input frequency. Because of this action, the PLL is said to *track* the input signal. Any input frequency variation will be matched by a VCO frequency change so that the circuit remains locked.

The VCO in a PLL, therefore, is capable of tracking the input frequency over a wide range. The range of frequencies over which the PLL will track the input signal and remain locked is

Lock range

known as the lock range. The *lock range* is usually a band of frequencies above and below the free-running frequency of the VCO. If the input signal frequency is out of the lock range, the PLL will not lock. When this occurs, the VCO output frequency jumps to its free-running frequency.

If an input frequency within the lock range is applied to the PLL, the circuit will immediately adjust itself and remain in a locked condition. The phase detector will determine the phase difference between the VCO free-running and input frequencies and generate the error signal that will force the VCO to equal the input frequency. Once the input signal is captured, the PLL remains locked and will track any changes in the input signal as long as it remains within the lock range.

The range of frequencies over which the PLL will capture an input signal is known as

Capture range

the *capture range*. It is much narrower than the lock range, but it is generally centered around the free-running frequency of the VCO as is the lock range. See Fig. 5-20.

The characteristic that causes the PLL to capture signals within a certain frequency range also causes it to act like a bandpass filter. Phase-locked loops are often used in signal conditioning applications where it is desirable to pass signals only in a certain range and reject signals outside of that range. The PLL is highly effective in eliminating the noise and interference on a signal.

Since the PLL responds to input frequency variations, naturally you can understand why it

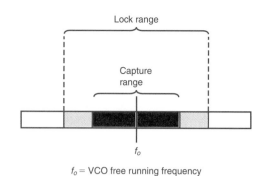

f_o = VCO free running frequency

Fig. 5-20 Capture and lock ranges of a PLL.

is useful in FM applications. If an FM signal is applied to the input, the VCO will track it. The VCO is, in effect, a frequency modulator that produces exactly the same FM signal as the input. In order for this to happen, however, the VCO input must be identical to the original modulating signal.

The VCO output follows the FM input signal because the error voltage produced by the phase detector and low-pass filter forces the VCO to track it. For that reason, the VCO output must be identical to the input signal if the PLL is to remain locked, and the error signal must be identical to the original modulating signal of the FM input. The low-pass filter cutoff frequency is designed such that it is capable of passing the original modulating signal.

The PLL is the best of all frequency demodulators in use because of the characteristics described earlier. Its ability to provide frequency selectivity and filtering give it a signal-to-noise ratio superior to any other type of FM detector. The linearity of the VCO ensures a highly accurate reproduction of the original modulating signal. Although PLLs are complex, they are easy to apply because they are readily available in low-cost IC form.

■ TEST

Supply the missing information in each statement.

29. In the Foster-Seeley discriminator and ratio detector circuits, an input frequency variation produces a(n) _____ that causes an output voltage variation.

30. In the pulse-averaging discriminator, a low-pass filter averages the fixed-width, fixed-amplitude pulses from a(n) _____.

About ⟺ Electronics

The ability of a phase-locked loop to provide frequency selectivity and filtering gives it a signal-to-noise ratio superior to that of any other type of FM detector.

31. Should the input frequency increase or decrease for the average voltage output of the low-pass filter in a pulse-averaging discriminator to increase?
32. Quadrature refers to a _____° phase difference.
33. In Fig. 5-16, does input B lag or lead input A?
34. The phase detector in a quadrature detector is usually a(n) _____.
35. The quadrature detector input circuit produces a varying _____ as the input frequency changes.
36. The varying-width pulses produced by the quadrature detector are converted into the original modulating signal by a(n) _____.
37. The circuit in a differential peak detector IC that temporarily stores the peak value of an input sine wave is called a(n) _____.
38. In the differential peak detector, the components _____ convert frequency variations into the amplitude variations that ultimately become the output.
39. The three main elements of a PLL are _____.
40. The _____ circuit in a PLL recognizes _____ or _____ changes between the input and VCO signals.

41. The _____ part of a PLL is a simple frequency modulator.
42. It is the _____ signal in the PLL that varies the VCO frequency.
43. The range of frequencies over which a PLL will cause the input and VCO signals to remain synchronized is known as the _____ range.
44. If the PLL input is zero, the VCO will operate at its _____ frequency.
45. The range of frequencies over which a PLL will latch onto or recognize an input signal is called the _____ range.
46. Since a PLL will only respond to signals over a narrow frequency range, it acts like a(n) _____.
47. In a PLL frequency demodulator, the error signal is the _____.
48. In a PLL demodulating an FM signal, the VCO output is an exact reproduction of the _____.

Determine whether each statement is true or false.

49. The Foster-Seeley discriminator is sensitive to input amplitude as well as frequency variations.
50. The ratio detector requires a limiter.
51. The lock range of a PLL is narrower than the capture range.

Summary

1. The component most widely used for FM or PM is the varactor diode or voltage-variable capacitor (VVC).

2. A VVC is a specially designed silicon junction diode optimized for large capacitive variations.

3. A reverse-biased junction diode will act as a small capacitor where the depletion region is the dielectric.

4. The capacitance of a varactor is inversely proportional to the reverse-biased voltage amplitude.

5. The most common frequency modulators use a varactor to vary the frequency of an LC circuit or crystal in accordance with the modulating signal.

6. A reactance modulator is an amplifier that is made to appear inductive or capacitive by phase shift. It is used to produce wide-deviation direct FM.

7. Crystal oscillators are preferred for their frequency stability over LC oscillators, but only very small frequency deviation is possible with crystal oscillators.

8. An IC VCO produces excellent deviation FM at frequencies below 1 MHz.

9. In a phase modulator, the carrier is shifted in phase in accordance with the modulating signal. This produces indirect FM.

10. One of the best phase modulators is a parallel tuned circuit controlled by a varactor.

11. Most phase modulators produce very small amounts of frequency deviation.

12. Frequency deviation and carrier frequency can be increased by passing them through a frequency multiplier.

13. One of the oldest and best frequency demodulators is the Foster-Seeley discriminator that is a phase detector whose output voltage increases or decreases with phase changes produced by input frequency deviation.

14. A Foster-Seeley discriminator is sensitive to input amplitude variations and, therefore, must be preceded by a limiter.

15. A variation of the Foster-Seeley discriminator is the ratio detector widely used in older TV re-

ceiver designs. A primary advantage of the ratio detector is that no limiter is needed.

16. A pulse-averaging discriminator converts an FM signal into a square wave of identical frequency variation using a zero-crossing detector, comparator, or limiter circuit. This circuit triggers a one shot that produces pulses that are averaged in a low-pass filter to reproduce the original modulating signal.

17. A quadrature detector uses a unique phase-shift circuit to provide quadrature (90°) FM input signals to a phase detector. The phase detector produces a different pulse width for different phase shifts. These pulse width variations are averaged in a low-pass filter to recover the modulating signal.

18. Quadrature detectors are available in IC form and are one of the most widely used TV audio demodulators.

19. A differential peak detector is an IC FM demodulator that uses a differential amplifier and capacitive storage peak detectors plus tuned circuits to translate frequency variations into voltage variations.

20. A phase-locked loop (PLL) is a feedback control circuit made up of a phase detector, voltage-controlled oscillator (VCO), and low-pass filter. The phase detector compares an input signal to the VCO signal and produces an output that is filtered by a low-pass filter into an error signal that controls the VCO frequency.

21. The PLL is synchronized or locked when the input and VCO frequencies are equal. Input frequency changes cause a phase or frequency shift which, in turn, produces an error signal that forces the VCO to track the input and reduce their difference to zero.

22. The range of frequencies over which a PLL will track an input is called the lock range. If the input strays outside the lock range, the PLL will go out of lock and the VCO will operate at its free-running frequency.

23. The capture range of a PLL is that narrow band of frequencies over which a PLL will

recognize and lock onto an input signal. The capture range is narrower than the lock range and it makes the PLL look like a bandpass filter.

24. The PLL is the best frequency demodulator because its filtering action removes noise and interference and its highly linear output faithfully reproduces the original modulating signal.

Chapter Review Questions

Choose the letter that best answers each question.

5-1. Another name for a VVC is
  a. PIN diode.
  b. Varactor diode.
  c. Snap diode.
  d. Hot-carrier diode.

5-2. The depletion region in a junction diode forms what part of a capacitor?
  a. Plates
  b. Leads
  c. Package
  d. Dielectric

5-3. Increasing the reverse bias on a varactor diode will cause its capacitance to
  a. Decrease.
  b. Increase.
  c. Remain the same.
  d. Drop to zero.

5-4. The capacitance of a varactor diode is in what general range?
  a. pF
  b. nF
  c. μF
  d. F

5-5. In Fig. 5-3, the varactor diode is biased by which components?
  a. R_1, R_2
  b. R_1, C_2
  c. L_1, C_1
  d. RFC, C_3

5-6. In Fig. 5-3, if the reverse bias on D_1 is reduced, the resonant frequency of C_1
  a. Increases.
  b. Decreases.
  c. Remains the same.
  d. Cannot be determined.

5-7. The frequency change of a crystal oscillator produced by a varactor diode is
  a. Zero.
  b. Small.
  c. Medium.
  d. Large.

5-8. A phase modulator varies the phase shift of the
  a. Carrier.
  b. Modulating signal.
  c. Both *a* and *b*.
  d. Neither *a* or *b*.

5-9. The widest phase variation is obtained with a(n)
  a. RC low-pass filter.
  b. RC high-pass filter.
  c. LR low-pass filter.
  d. LC resonant circuit.

5-10. In Fig. 5-10, R_4 is the
  a. Pre-emphasis circuit.
  b. De-emphasis circuit.
  c. Deviation control.
  d. Frequency determining component in the tuned circuit.

5-11. The small frequency change produced by a phase modulator can be increased by using a(n)
  a. Amplifier.
  b. Mixer.
  c. Frequency multiplier.
  d. Frequency divider.

5-12. A crystal oscillator whose frequency can be changed by an input voltage is called a
  a. VCO.
  b. VXO.
  c. VFO.
  d. VHF.

5-13. Which oscillators are preferred for carrier generators because of their good frequency stability?
  a. LC
  b. RC
  c. LR
  d. Crystal

5-14. Which of the following frequency demodulators requires an input limiter?
  a. Foster-Seeley discriminator
  b. Pulse-averaging discriminator
  c. Quadrature detector
  d. PLL

5-15. Which discriminator averages pulses in a low-pass filter?
  a. Ratio detector
  b. PLL
  c. Quadrature detector
  d. Foster-Seeley discriminator

5-16. Which frequency demodulator is considered the best overall?
  a. Ratio detector
  b. PLL
  c. Quadrature
  d. Pulse-averaging discriminator

5-17. In Fig. 5-11, the voltage at point A when the input frequency is below the FM center frequency is
  a. Negative.
  b. Positive.
  c. Zero.
  d. Indeterminant.

5-18. In Fig. 5-11, R_3 and C_6 form which kind of circuit?
  a. Carrier filter
  b. Pulse-averaging filter
  c. Pre-emphasis
  d. De-emphasis

5-19. In Fig. 5-3, the voltage across C_6 is
  a. Inversely proportional to signal amplitude.
  b. Directly proportional to signal amplitude.
  c. Directly proportional to frequency deviation.
  d. Constant.

5-20. In a pulse-averaging discriminator, the pulses are produced by a(n)
  a. Astable multivibrator.
  b. Zero-crossing detector.
  c. One shot.
  d. Low-pass filter.

5-21. A reactance modulator looks like a capacitance of 35 pF in parallel with the oscillator-tuned circuit whose inductance is 50 μH and capacitance is 40 pF. What is the center frequency of the oscillator prior to FM?
  a. 1.43 MHz
  b. 2.6 MHz
  c. 3.56 MHz
  d. 3.8 MHz

5-22. Which of the following is true about the NE566 IC?
  a. It is a VCO.
  b. Its output is sinusoidal.
  c. It is an FM demodulator.
  d. It uses LC-tuned circuits.

5-23. An FM demodulator that uses a differential amplifier and tuned circuits to convert frequency variations into voltage variations is the
  a. Quadrature detector.
  b. Foster-Seeley discriminator.
  c. Differential peak detector.
  d. Phase-locked loop.

5-24. The output amplitude of the phase detector in a quadrature detector is proportional to
  a. Pulse width.
  b. Pulse frequency.
  c. Input amplitude.
  d. The phase shift value at center frequency.

5-25. The input to a PLL is 2 MHz. In order for the PLL to be locked, the VCO output must be
  a. 0 MHz.
  b. 1 MHz.
  c. 2 MHz.
  d. 4 MHz.

5-26. Decreasing the input frequency to a locked PLL will cause the VCO output frequency to
  a. Decrease.
  b. Increase.
  c. Remain constant.
  d. Jump to the free-running frequency.

5-27. The range of frequencies over which a PLL will track input signal variations is known as the
  a. Circuit bandwidth.
  b. Capture range.
  c. Band of acceptance.
  d. Lock range.

5-28. The band of frequencies over which a PLL will acquire or recognize an input signal is called the
  a. Circuit bandwidth.
  b. Capture range.
  c. Band of acceptance.
  d. Lock range.

5-29. Over a narrow range of frequencies, the PLL acts like a
  a. Low-pass filter.
  b. Bandpass filter.
  c. Tunable oscillator.
  d. Frequency modulator.

5-30. The output of a PLL frequency demodulator is taken from the
  a. Low-pass filter.
  b. VCO.
  c. Phase detector.
  d. None of the above.

Critical Thinking Questions

5-1. A standard diode detector AM demodulator like that in Fig. 3-11 can be used to demodulate frequency modulation. Explain how this is possible. (Hint: Consider how the FM signal is affected by the tuned circuit before the diode.)

5-2. Describe what you think might be the result of demodulating an FM signal that had been passed by a tuned circuit whose bandwidth was not wide enough to pass all the FM sidebands.

5-3. Can a square-wave VCO be used as a carrier oscillator in an FM transmitter? If so, explain how.

5-4. An FM transmitter has an output frequency of 160 MHz. The maximum frequency deviation is 5 kHz. If the carrier oscillator uses a crystal whose maximum possible deviation is 200 Hz, what is the crystal frequency?

5-5. A frequency demodulator performs:
   a. voltage-to-frequency conversion.
   b. frequency-to-voltage conversion.

5-6. A standard analog cell phone transmitter uses FM in the 825- to 845-MHz range. Each channel is 30 kHz wide. If the maximum voice frequency is 3 kHz, what is the maximum deviation allowed?

Answers to Tests

1. *b*
2. *b*
3. *b*
4. *a*
5. *a*
6. varactor diode (or varicap)
7. 1,200
8. frequency stability
9. voltage-controlled oscillator
10. VXO
11. frequency multiplier
12. increase
13. resistance, capacitance
14. true
15. false
16. false
17. true
18. indirect
19. true
20. 0, 90
21. resistance, capacitance
22. parallel resonant (or tuned)
23. varactor
24. R_4
25. lead
26. decrease
27. C_1, C_2, Q_1, and C_1, R_1, Q_1
28. frequency multiplier
29. phase shift
30. one-shot or monostable multivibrator
31. increase
32. 90
33. lead
34. differential amplifier
35. phase shift
36. low-pass filter
37. peak detector
38. C_1, C_2, L_1
39. phase detector, VCO, low-pass filter
40. phase detector, frequency, phase
41. VCO
42. error
43. lock
44. free-running
45. capture
46. bandpass filter
47. modulating signal or information signal
48. FM input
49. true
50. false
51. false

Chapter

6

Radio Transmitters

Chapter Objectives

This chapter will help you to:

1. *Draw* a block diagram of a CW, an AM, an FM, or an SSB transmitter and *explain* the operation of each major circuit.
2. *State* the four major classes of RF power amplifiers and the advantages and disadvantages of each.
3. *Explain* the biasing and operation of a class C power amplifier.
4. *Explain* the operation of a frequency multiplier.
5. *Explain* the operation and benefits of Class D and Class E power amplifiers.
6. *Calculate* the component values in an *LC* impedance-matching network to match an output impedance to an input impedance.
7. *Show* how transformers and baluns are used in impedance matching.
8. *State* the benefits of speech processing and how it is achieved.

A radio transmitter takes the information to be communicated and converts it into an electronic signal compatible with the communications medium. Typically this process involves carrier generation, modulation, and amplification. The signal is then fed by wire, coaxial cable, or waveguide to an antenna that launches it into free space. In this chapter you will learn about the circuits commonly used in radio transmitters. These include amplifiers, frequency multipliers, impedance-matching networks, and speech-processing circuits.

6-1   Introduction to Transmitters

Transmitter

Oscillator

Continuous wave (CW) transmission

The *transmitter* is the electronic unit that accepts the information signal to be transmitted and converts it into an RF signal capable of being transmitted over long distances. Every transmitter has three basic functions. First, the transmitter must generate a signal of the correct frequency at a desired point in the spectrum. Second, it must provide some form of modulation that causes the information signal to modify the carrier signal. Third, it must provide sufficient power amplification to ensure that the signal level is high enough so that it will carry over the desired distance.

CW Transmitter

The simplest form of transmitter is the *oscillator* as shown in Fig. 6-1. The oscillator generates a carrier signal of the desired frequency. The frequency here is determined by a crystal. Information to be transmitted is expressed in a special form of code using dots and dashes to represent letters of the alphabet and numbers. Information transmitted in this way is referred to as *continuous wave (CW) transmission*. A key that is simply a convenient hand-operated switch is used in the emitter to turn the oscillator off and on to produce the dots and dashes. The oscillator produces a short burst of RF energy for a dot and a longer RF burst for a dash.

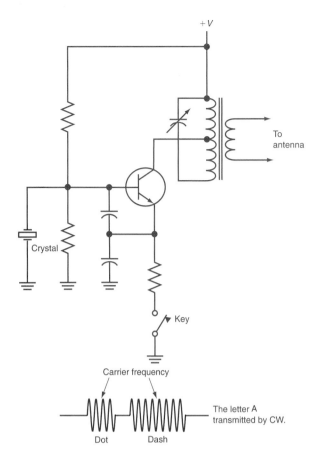

Fig. 6-1 The simplest transmitter, A CW oscillator.

Although such a simple transmitter may have a power of 1 W or less, at the right frequency and with a good antenna it is capable of sending signals halfway around the world.

The basic CW transmitter just described can be greatly improved by simply adding a power amplifier to it. The result is shown in Fig. 6-2. The oscillator is still keyed off and on to produce dots and dashes, and the amplifier increases the power level of the signal. The result is a stronger signal that will carry farther and produce more reliable communications.

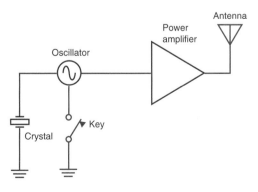

Fig. 6-2 A more powerful CW transmitter.

The basic oscillator-amplifier combination shown in Fig. 6-2 is the basis for virtually all radio transmitters. Many other circuits are added depending upon the type of modulation used, the power level, and other considerations. Now let's take a look at some of these transmitters which use various types of modulation.

AM Transmitter

Figure 6-3 shows an AM transmitter. An oscillator generates the final carrier frequency. In most applications, this is a crystal oscillator. Usually, transmitters operate on assigned frequencies or channels. Crystals provide the best way to obtain the desired frequency with good stability. In general, *LC* oscillators do not have the frequency stability required to stay on frequency. Temperature variations and other conditions will cause the frequency to drift outside of the limits imposed by the FCC.

The carrier signal is then fed to a buffer amplifier whose primary purpose is to isolate the oscillator from the remaining power amplifier stages. The buffer amplifier usually operates class A and provides a modest increase in power output. The main purpose of the buffer amplifier is to prevent load changes from causing frequency variations in the oscillator.

The signal from the buffer amplifier is applied to a driver amplifier. This is a class C amplifier designed to provide an intermediate level of power amplification. The purpose of this circuit is to generate sufficient output power to drive the final power amplifier stage.

The final power amplifier, normally just referred to as the *final,* also operates class C at very high power. The actual amount of power depends upon the application. For example, in a CB transmitter, the power input is only 5 W. However, AM radio stations operate at the much higher powers of 250, 500, 1000, 5000, or 50,000 W.

Final amplifier

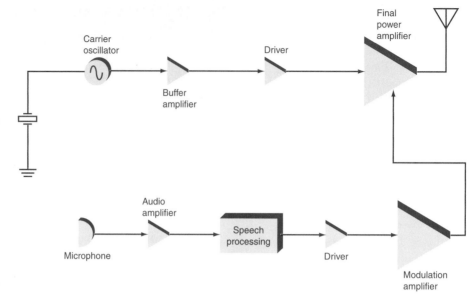

Fig. 6-3 An AM transmitter using high-level collector modulation.

All the RF circuits in the transmitter are usually solid-state; that is, they are implemented with either bipolar or field-effect transistors. Although bipolar transistors are by far the most common, MOSFETs are being more widely used as they are now capable of handling high power. Transistors are also typically used in the final as long as the power level does not exceed several hundred watts. Individual RF power transistors can handle up to about 300 W. Several of these may be connected in parallel or in push-pull configurations to increase the power-handling capability further. However, transistors are rarely used in final amplifiers where the power output exceeds 5000 W. For higher power levels, vacuum tubes are still used.

Refer again to Fig. 6-3. Assume that this is a voice transmitter. The input from the microphone is applied to a low-level class A audio amplifier. This amplifier boosts the small signal from the microphone to a higher voltage level. One or more stages of amplification may be used.

The voice signal is then fed to some form of *speech-processing circuit.* "Speech processing" refers to filtering and amplitude control. The filtering ensures that only voice frequencies in a certain range are passed. This helps to minimize the bandwidth occupied by the signal. Most communications transmitters limit the voice frequency to the 300- to 3000-Hz range. This is adequate for intelligible communications. However, AM broadcast stations

offer higher fidelity and allow frequencies up to 5 kHz to be used.

The speech processor also contains some kind of circuit used to hold the amplitude to a particular level. High-level amplitude signals are compressed and often lower-level signals are given more amplification. The result is that overmodulation is prevented. This reduces the possibility of signal distortion and harmonics which produce wider sidebands that can cause adjacent channel interference.

After the speech processor, a driver amplifier is used. The driver amplifier increases the power level of the signal so that it is capable of driving the high-power modulation amplifier. In the AM transmitter of Fig. 6-3, high-level or collector modulation is used. As you saw earlier, the power output of the modulation amplifier must be one-half the input power of the RF amplifier. The high-power modulation amplifier usually operates class AB or class B push-pull to achieve such power levels.

FM Transmitter

Figure 6-4 shows the typical configuration for an FM or PM transmitter. The indirect method of FM generation is used. A stable crystal oscillator is used to generate the carrier signal, and a buffer amplifier is used to isolate it from the remainder of the circuitry. The carrier signal is then applied to a phase modulator similar to those previously discussed. The voice input is

**Speech
processor**

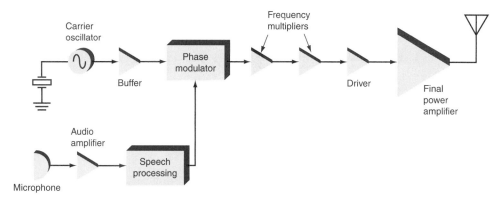

Fig. 6-4 A typical FM transmitter using indirect FM with a phase modulator.

amplified and processed to limit the frequency range and prevent overdeviation. The output of the modulator is the desired FM signal.

Most FM transmitters are used in the VHF and UHF range, and crystals are not available to generate those frequencies directly. As a result, the carrier is usually generated at a frequency considerably lower than the final output frequency. To achieve the desired output frequency, one or more frequency multiplier stages are used. A *frequency multiplier* is a class C amplifier whose output frequency is some integer multiple of the input frequency. Most frequency multipliers increase the frequency by a factor of 2, 3, 4, or 5, and because they are class C amplifiers, they also provide a modest amount of power amplification.

The frequency multiplier not only increases the carrier frequency to the desired output frequency but also multiplies the frequency deviation produced by the modulator. Many frequency and phase modulators generate only a small frequency shift, much lower than the desired final deviation. The design of the transmitter is such that the frequency multipliers will provide the correct amount of multiplication not only for the carrier frequency but also for the modulation deviation.

After the frequency multipliers, a class C driver amplifier is used to increase the power level sufficiently to operate the final power amplifier. The final power amplifier also operates class C.

Most FM communications transmitters operate at relatively low power levels, typically less than 100 W. All the circuits use transistors even in the VHF and UHF range. For power levels beyond several hundred watts, vacuum tubes must be used. The final amplifier stages in FM broadcast transmitters typically use large vacuum tube class C amplifiers. In FM transmitters

operating in the microwave range, klystrons, magnetrons, and traveling wave tubes are used to provide the final power amplification.

SSB Transmitter

A typical SSB transmitter is shown in Fig. 6-5. An oscillator signal generates the carrier which is then fed to the buffer amplifier. The buffer amplifier supplies the carrier input signal to the balanced modulator. The audio amplifier and speech-processing circuits described previously provide the other input to the balanced modulator. The balanced modulator output is then fed to a sideband filter which selects either the upper or lower sideband.

The SSB signal is then fed to a mixer circuit which is used to translate the signal to its final operating frequency. Typically, the SSB signal is generated at a much lower frequency. This makes the balanced modulator and filter circuits simpler and easier to design. The mixer translates the SSB signal to a higher desired frequency. The other input to the mixer is derived from an LO set at a frequency that, when mixed with the SSB signal, will produce the desired operating frequency. The mixer may be set up so that the tuned circuit at its output selects either the sum or difference frequency. The oscillator frequency must be set to provide the desired output frequency. This LO may use crystals for fixed-channel operation. However, in some equipment such as that used by amateur radio operators, a variable-frequency oscillator (VFO) is used to provide continuous tuning over a desired range. In most modern communications, a frequency synthesizer is used to set the final output frequency.

The output of the mixer is the desired final carrier frequency containing the SSB modulation.

Frequency multiplier

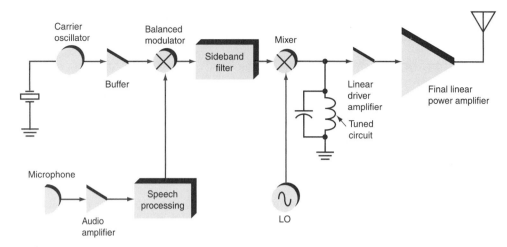

Fig. 6-5 An SSB transmitter.

It is then fed to linear driver and power amplifiers to increase the power level as required.

■ TEST

Supply the missing word(s) in each statement.

1. The simplest transmitter is a(n) _____ circuit.

2. Information sent in the form of coded dots and dashes is called _____ transmission and is abbreviated _____.

3. Fixed-frequency or channel operation of a transmitter is obtained by using a(n) _____.

4. An amplifier that isolates the carrier oscillator from other transmitter amplifiers is known as a _____ amplifier.

5. Intermediate power amplifier stages in a transmitter are commonly referred to as _____.

6. The output RF power amplifier in a transmitter is sometimes called the _____.

7. Audio circuits that prevent excessive signal bandwidth and overmodulation are called _____ circuits.

8. In an FM transmitter, special amplifiers called _____ are used to increase the carrier frequency and deviation to the desired output values.

9. In an SSB transmitter, the final output frequency is usually produced by a(n) _____ circuit before being amplified.

10. The power level of SSB signals must be increased by _____ amplifiers to prevent distortion.

11. Frequency-modulated signals use _____ amplifiers for power amplification.

6-2 Power Amplifiers

Three basic types of power amplifiers are used in transmitters: linear, class C, and switching. Linear amplifiers provide an output signal that is an identical, enlarged replica of the input. Their output is directly proportional to their input; therefore, they faithfully reproduce an input but at a higher power level. All audio amplifiers are linear. Linear RF amplifiers must be used to increase the power level of varying-amplitude RF signals such as low-level AM or SSB signals. Frequency-modulated signals do not vary in amplitude and, therefore, may be amplified with more efficient, nonlinear class C or switching amplifiers.

Linear Amplifiers

Linear amplifiers operate class A, AB, or B. The class of an amplifier indicates how it will be biased. A class A amplifier is biased so that it conducts continuously. The bias is set so that the input varies the collector (or drain) current over a linear region of the transistor's characteristic. In this way, its output is an amplified linear reproduction of the input. Usually we say that the class A amplifier conducts for 360° of an input sine wave.

A class B amplifier is biased at cutoff so that no collector current flows with zero input. The transistor conducts on only one-half of the

Linear amplifiers

sine wave input. In other words, it conducts for 180° of a sine wave input. This means that only one-half of the sine wave is amplified. Normally, two class B amplifiers are connected in a *push-pull* arrangement so that both the positive and negative alternations of the input are amplified simultaneously.

A class AB amplifier is biased near cutoff with some continuous collector current flow. It will conduct for more than 180° but for less than 360° of the input. It too is used primarily in push-pull amplifiers and provides better linearity than a class B amplifier but with less efficiency.

Class A amplifiers are linear but not very efficient. For that reason, they make poor power amplifiers. As a result, they are used primarily as small-signal voltage amplifiers or for low-power amplifiers. The buffer amplifiers described previously operate class A.

Class B and class C amplifiers are more efficient because current flows for only a portion of the input signal. They make good power amplifiers, the class C being the most efficient. Since both class B and class C amplifiers distort an input signal, special techniques are used to eliminate or compensate for the distortion. For example, class B amplifiers are operated in a push-pull configuration, while class C amplifiers use a resonant *LC* load to eliminate the distortion.

Class A Amplifiers

A simple class A buffer amplifier is shown in Fig. 6-6. The carrier oscillator signal is capacitively coupled to the input. The bias is derived from R_1, R_2, and R_3. The collector is tuned with

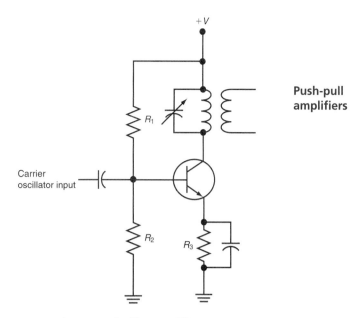

Push-pull amplifiers

Fig. 6-6 Class A RF buffer amplifier.

a resonant *LC* circuit at the operating frequency, and an inductively coupled secondary loop transfers power to the next stage. Buffers like this usually operate at a power level of well less than 1 W.

A high-power class A linear amplifier is shown in Fig. 6-7. Base bias is supplied by a constant-current circuit that is temperature-compensated. The RF input from a 50-Ω source is connected to the base via an impedance-matching circuit made up of C_1, C_2, and L_1. The output is matched to a 50-Ω load by the impedance-matching network made up of L_3, C_3, and C_4. When connected to a proper heat sink, the transistor can generate up to 100 W of power up to about 30 MHz. The

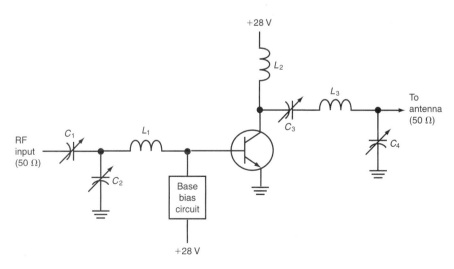

Fig. 6-7 High-power class A linear RF amplifier.

amplifier is designed for a specific frequency which is set by the input and output tuned circuits.

Keep in mind that a class A amplifier can achieve a maximum efficiency of 50 percent. That means only 50 percent of the dc power is converted to RF. The remaining power is dissipated in the transistor. For 100-W RF output, the transistor dissipates 100 W.

Neutralization

One problem that all RF amplifiers have is self-oscillation. When some of the output voltage finds its way back to the input of the amplifier with the correct amplitude and phase, the amplifier can oscillate. The amplifier may oscillate at its tuned frequency or, in some cases, at a much higher frequency. When the circuit oscillates at a higher frequency unrelated to the tuned frequency, the oscillation is referred to as *parasitic oscillation*. In either case, the oscillation is undesirable and prevents amplification from taking place or, in the case of parasitic oscillation, reduces the power amplification and introduces distortion of the signal.

Self-oscillation of an amplifier is usually the result of positive feedback that occurs because of the interelement capacitance of the amplifying device, whether it is a bipolar transistor, an FET, or a vacuum tube. In a bipolar transistor this is the collector-to-base capacitance C_{bc} as shown in Fig. 6-8(a). Transistor

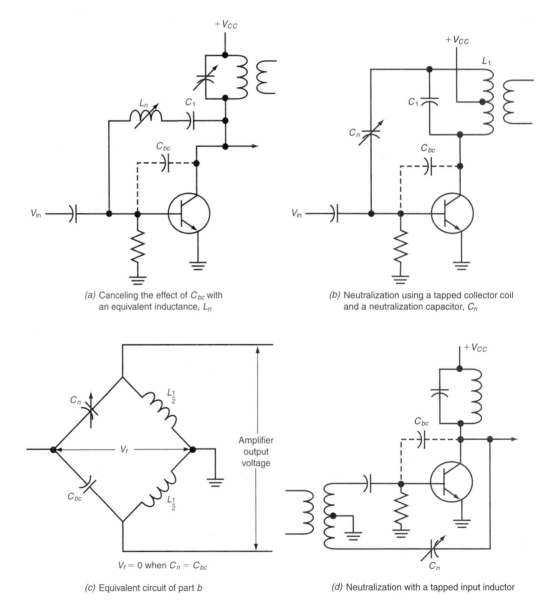

(a) Canceling the effect of C_{bc} with an equivalent inductance, L_n

(b) Neutralization using a tapped collector coil and a neutralization capacitor, C_n

$V_f = 0$ when $C_n = C_{bc}$

(c) Equivalent circuit of part b

(d) Neutralization with a tapped input inductor

Fig. 6-8 Neutralization circuits.

amplifiers are biased so that the emitter-base junction is forward-biased and the base-collector junction is reverse-biased. As you saw in a previous chapter, a reverse-biased diode or transistor junction acts like a capacitor. This small capacitance permits output from the collector to be fed back to the base. Depending upon the frequency of the signal, the value of the capacitance, and the values of stray inductances and capacitances in the circuit, the signal fed back may be in phase with the input signal and high enough in amplitude to cause oscillation.

This interelement capacitance cannot be eliminated; therefore its effect must be compensated for. This process is called *neutralization*. The concept of neutralization is to feed back another signal equal in amplitude to the signal fed back through C_{bc} and 180° out of phase with it. The result is that the two signals will cancel one another.

Several methods of neutralization are shown in Fig. 6-8. In Fig. 6-8(*a*), a signal of equal and opposite phase is provided by the inductor L_n. Capacitor C_1 is a high-value capacitor used strictly for dc blocking to prevent collector voltage from being applied to the base. L_n is made adjustable so that its value can be set to make its reactance equal to the reactance of C_{bc}

at the oscillation frequency. As a result, C_{bc} and L_n form a parallel resonant circuit that acts like a very high resistance at the resonant frequency. The result is effective cancellation of the positive feedback.

Figure 6-8(*b*) shows another type of neutralization using a tapped collector inductor and a neutralization capacitor C_n. The two equal halves of the collector inductance, the junction capacitance C_{bc}, and C_n form a bridge circuit as illustrated in Fig. 6-8(*c*). When C_n is adjusted to equal to C_{bc}, the bridge is balanced and no feedback signal V_f occurs. A variation of this is shown in Fig. 6-8(*d*), where a center-tapped base input inductor is used.

Neutralization

Class B Amplifiers

Most RF power transistors have an upper power limit of several hundred watts. To produce more power, two or more devices may be connected in parallel. A class B linear power amplifier using push-pull is shown in Fig. 6-9. The RF driving signal is applied to Q_1 and Q_2 through input

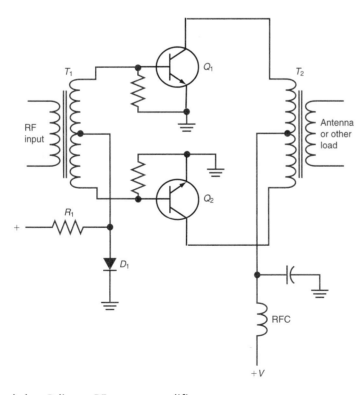

Fig. 6-9 Broadband class B linear RF power amplifier.

transformer T_1. It provides impedance matching and base-drive signals to Q_1 and Q_2 that are 180° out of phase. An output transformer T_2 couples the power to the antenna or load. Bias is provided by R_1 and D_1.

For class B operation, Q_1 and Q_2 must be biased right at the cutoff point. The emitter-base junction of a transistor will not conduct until about 0.6 to 0.8 V of forward bias is applied because of the built-in potential barrier. This effect causes the transistors to be biased beyond cutoff, not right at it. A forward-biased silicon diode D_1 has about 0.7 V across it. This is used to put Q_1 and Q_2 right on the conduction threshold.

Now, when the positive half cycle of input RF occurs, the base of Q_1 will be positive and the base of Q_2 will be negative. Q_2 will be cut off, but Q_1 will conduct and amplify the positive half cycle. Collector current flows in the upper half of T_2 which induces an output voltage in the secondary.

On the negative half cycle of the RF input, the base of Q_1 is negative, so it is cut off. The base of Q_2 is positive, so Q_2 amplifies the negative half cycle. Current flows in Q_2 and the lower half of T_2, thereby completing a full cycle. The power is split between the two transistors.

The circuit in Fig. 6-9 is a broadband circuit, that is, untuned. It will amplify signals over a broad frequency range. A typical range may be 2 to 30 MHz. A low-power AM or SSB signal would be generated at the desired frequency, and then applied to this power amplifier prior to being sent to the antenna. With push-pull circuits, power levels of up to several hundred watts are possible.

Figure 6-10 shows another push-pull RF power amplifier. It uses two power *MOSFETs* and can produce an output up to 1 kilowatt (kW) over the 10- to 90-MHz range. It has a 12-dB power gain. The RF input driving power must be 63 W to produce the full 1-kW output. Toroidal transformers are used at the input and output for impedance matching. They provide broadband operation over the 10- to 90-MHz range without tuning. The 20-nanohenry (nH) chokes and 20-Ω resistors form neutralization circuits that provide out-of-phase feedback from output to input to prevent self-oscillation.

Power Combiners. With a class B amplifier using higher-power MOSFETs, it is possible to achieve power levels up to about 1 kW. If more power is needed, there are two solutions: Use a high-power vacuum tube, or combine the power from two or more solid-state power amplifiers. Although vacuum tubes are still used for very high power levels, the power combiner method has become the method of choice for achieving very high power levels in modern communications equipment.

A power combiner is a broadband transformer made with several turns of heavy wire on a toroid ferrite core. The primary and secondary windings are made with two pieces of

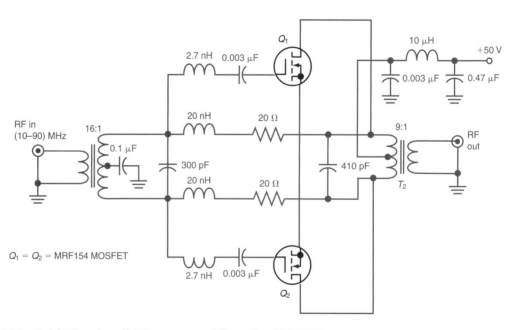

Fig. 6-10 A 1-kW push-pull RF power amplifier using MOSFETs.

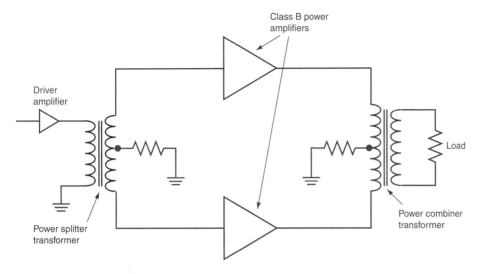

Fig. 6-11 Achieving high output power with solid-state amplifiers and power combiner transformers.

wire side by side or twisted together to form a transmission line. The wires are wound together on the core. This is called a *bifilar winding*. The resulting transformer is called a *transmission line transformer* since it uses the transmission line effects to achieve impedance matching.

Figure 6-11 shows how the power combiner is used. The triangles represent two class B power amplifiers. Their outputs are applied to the combiner transformer 180° out of phase. The output signals add in phase, and the combined power appears across the load.

Note also in Fig. 6-11 that a power splitter transformer is used at the inputs of the power amplifiers. The power from the driver stage is divided equally by the transformer.

The splitter transformer is also a broadband transmission line transformer like the output combiner. For even higher power, four or more amplifiers can be combined in this way.

Class C Amplifiers

The key circuit in most AM and FM transmitters is the *class C amplifier.* It is used for power amplification in the form of drivers, frequency multipliers, and final amplifiers.

A class C amplifier is biased so that it conducts for less than 180° of the input. It will typically have a conduction angle of 90° to 150°. This means that current flows through it in short pulses. The question then is, "How is a complete signal amplified?" As you will see, a resonant tuned circuit is used for that purpose.

Figure 6-12(*a*) shows one way of biasing a class C amplifier. The base of the transistor is simply connected to ground through a resistor. No external bias voltage is applied. An RF signal to be amplified is applied directly to the base. The transistor will conduct on the positive half cycles of the input wave and will be cut off on the negative half cycles. Ordinarily, you would think that this would be a class B amplifier. However, this is not the case. Recall that the emitter-base junction of a bipolar transistor has a forward voltage threshold of approximately 0.7 V. In other words, the emitter-base junction will not really conduct until the base is more positive than the emitter by +0.7 V. Because of this, the transistor has an inherent built-in bias. When the input signal is applied, the collector current will not flow until the base is positive by 0.7 V. See Fig. 6-12(*b*). The result is that collector current flows through the transistor in positive pulses for less than the full 180° of the positive ac alternation.

In many low-power driver and multiplier stages, no special biasing provisions other than the inherent emitter-base junction voltage are required. The resistor between base and ground simply provides a load for the driving circuit.

In some cases, a narrower conduction angle than that provided by the circuit in Fig. 6-12(*a*) may be required. In this case, some form of bias must be applied. A simple way of supplying bias is with the *RC* network shown in Fig. 6-13(*a*). Here the signal to be amplified is applied through capacitor C_1. When the emitter-base junction conducts on the positive half

Bifilar winding

Transmission line transformer

Class C amplifiers

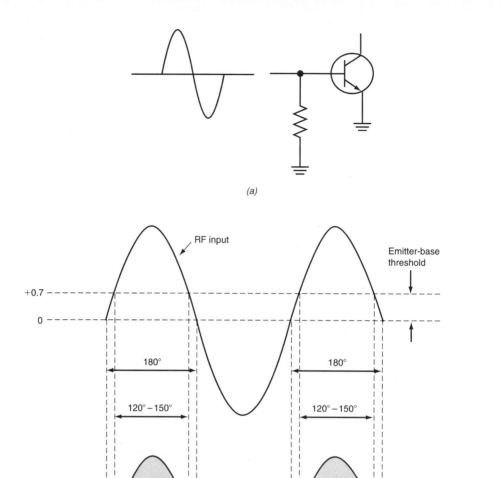

(a)

(b)

Fig. 6-12 Using the internal emitter-base threshold for class C biasing.

cycle, C_1 will charge to the peak of the applied voltage less the forward drop across the emitter-base junction. On the negative half cycle of the input, the emitter-base junction, of course, will be reverse-biased, so the transistor does not conduct. During this period of time, however, capacitor C_1 will discharge through R_1. This produces a negative voltage across R_1 which serves as a reverse bias on the transistor. By properly adjusting the time constant of R_1 and C_1, an average dc reverse-bias voltage will be established. The applied voltage will cause the transistor to conduct but only on the peaks. The higher the average dc bias voltage, the narrower the conduction angle and the shorter the duration of the collector current pulses. This method is referred to as *signal bias.*

Of course, negative bias can also be supplied to a class C amplifier from a fixed dc supply voltage as shown in Fig. 6-13(b). After

External bias

Self-bias

Signal bias

the desired conduction angle is determined, the value of the reverse voltage can be determined. It is applied to the base through the RFC. The incoming signal is then coupled to the base and causes the transistor to conduct only on the peaks of the positive input alternations. This is called *external bias* but requires a separate negative dc supply.

Another biasing method is shown in Fig. 6-13(c). The bias is also derived from the signal as in Fig. 6-13(a). This arrangement is known as the *self-bias method.* When current flows in the transistor, a voltage is developed across R_1. Capacitor C_1 is charged and holds the voltage constant. This makes the emitter more positive than the base, which is the same as a negative voltage on the base. A strong input signal is required for proper operation.

All class C amplifiers have some form of tuned circuit connected in the collector as

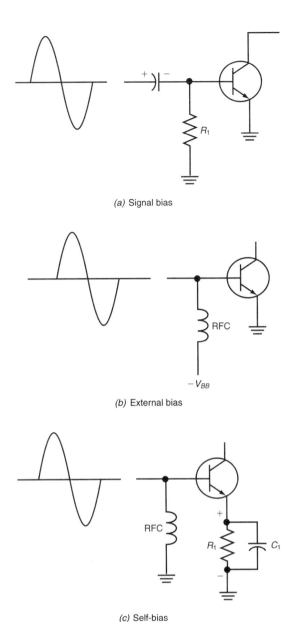

(a) Signal bias

(b) External bias

(c) Self-bias

Fig. 6-13 Methods of biasing a class C amplifier.

age across the tuned circuit will be a constant-amplitude sine wave at the resonant frequency. Even though the current flows through the transistor in short pulses, the class C amplifier output will be a continuous sine wave.

The tuned circuit in the collector also has another purpose, which is to filter out unwanted *harmonics*. Any nonsinusoidal signal, such as a square wave or the short pulses that flow through the class C amplifier, consists of a fundamental sine wave and multiple harmonics. The short pulses in a class C amplifier are made up of second, third, fourth, fifth, etc., harmonics. In a high-power transmitter, signals will be radiated at these harmonic frequencies as well as at the fundamental resonant frequency. Such harmonic radiation can cause out-of-band

Harmonics

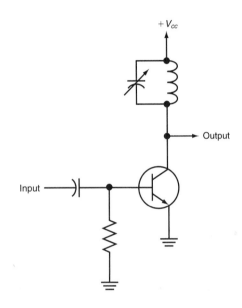

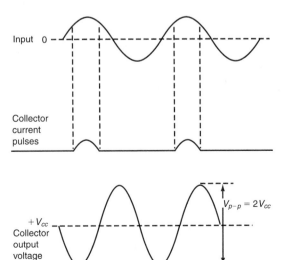

Fig. 6-14 Class C amplifier operation.

shown in Fig. 6-14. The primary purpose of this tuned circuit is to form the complete ac sine wave output. A parallel tuned circuit will ring or oscillate at its resonant frequency whenever it receives a dc pulse. The pulse will charge the capacitor which, in turn, will discharge into the inductor. The magnetic field in the inductor will increase and then collapse, causing a voltage to be induced. This induced voltage then recharges the capacitor in the opposite direction. This exchange of energy between the inductor and the capacitor is called the *flywheel effect* and produces a damped sine wave at the resonant frequency. If the resonant circuit receives a pulse of current every cycle, the volt-

Flywheel effect

interference. The purpose of the tuned circuit is to act as a selective filter that will eliminate these higher-order harmonics. If the Q of the tuned circuit is made high enough, the harmonics will be adequately suppressed.

It is important to point out that the Q of the tuned circuit is an important consideration in a class C amplifier. Remember that the bandwidth (BW) of a tuned circuit and its Q are related by the expression

$$\text{BW} = \frac{f_r}{Q} \quad Q = \frac{f_r}{\text{BW}}$$

The Q of the tuned circuit in the class C amplifier should be selected so that it provides adequate attenuation of the harmonics but also has adequate bandwidth to pass the sidebands produced by the modulation process. If the Q of the tuned circuit is too high, the bandwidth will be very narrow and some of the higher-frequency sidebands will be eliminated. This will cause a form of frequency distortion called *sideband clipping* and may make some signals unintelligible or will at least limit the fidelity of reproduction.

One of the main reasons why class C amplifiers are preferred for RF power amplification over class A and class B amplifiers is that class C amplifiers have high *efficiency*. Because the current flows for less than 180° of the ac input cycle, the average current in the transistor is fairly low, meaning that the power dissipated by the device is low. A class C amplifier functions almost as a transistor switch that is off for over 180° of the input cycle. The switch conducts for approximately 90° to 150° of the input cycle. During the time that it conducts, its emitter-to-collector impedance is very low. Even though the peak current may be high, the total power dissipation is much less than that in class A and class B circuits. For this reason, more of the dc power is converted to RF energy and passed on to the load, usually by an antenna. The efficiency of most class C amplifiers is in the 60 to 85 percent range.

In discussing power in a class C amplifier, there is input power and output power. The input power is the average power consumed by the circuit. This is simply the product of the supply voltage and the average collector current.

$$P_{in} = V_{cc}I_C$$

For example, if the supply voltage is 13.5 V and the average dc collector current is 0.7 A, the input power is

$$P_{in} = 13.5(0.7) = 9.45 \text{ W}$$

The output power, of course, is the power actually transmitted to the load. The actual amount of power depends upon the efficiency of the amplifier. The output power can be computed with the familiar power expression

$$P_{out} = \frac{V^2}{R_L}$$

Here V is the RF output voltage at the collector of the amplifier, and R_L is the load impedance. When a class C amplifier is set up and operating properly, the peak-to-peak RF output voltage is 2 times the supply voltage, or $2V_{cc}$.

Frequency Multipliers

A special form of class C amplifier is the *frequency multiplier.* Any class C amplifier is capable of performing frequency multiplication if the tuned circuit in the collector resonates at some integer multiple of the input frequency.

For example, a frequency doubler can be constructed by simply connecting a parallel tuned circuit in the collector of a class C amplifier that resonates at twice the input frequency. When the collector current pulse occurs, it excites the tuned circuit, which then rings at twice the input frequency.

A tripler circuit is constructed in the same way except that the tuned circuit resonates at three times the input frequency. In this way, the tuned circuit receives one input pulse for every three cycles of oscillation it produces. Multipliers can be constructed to increase the input frequency by any integer factor up to approximately 10. As the multiplication factor gets higher, the power output of the multiplier decreases. For most practical applications, the best result is obtained with multipliers of 2 and 3.

Another way to look at the operation of class C frequency multipliers is to remember that the nonsinusoidal current pulse is rich in harmonics. Each time the pulse occurs, the second, third, fourth, fifth, and higher harmonics are generated. The purpose of the tuned circuit in the collector is to act as a filter to select the desired harmonic.

Q

Sideband clipping
Frequency multipliers

Efficiency

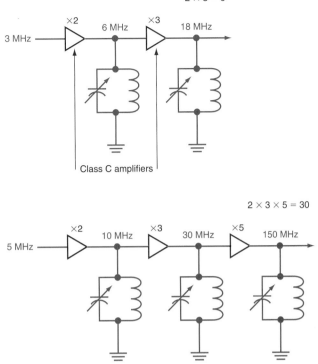

Fig. 6-15 Frequency multiplication with class C amplifiers.

In many applications, a multiplication factor greater than that achievable with a single multiplier stage is required. In such cases, two or more multipliers are cascaded. Figure 6-15 shows two multiplier examples. In the first case, multipliers of 2 and 3 are cascaded to produce an overall multiplication of 6. In the second example, three multipliers provide an overall multiplication of 30. The total multiplication factor is simply the product of the individual stage multiplication factors.

Switching Power Amplifiers

A *switching amplifier* is a transistor that is simply used as a switch and is either conducting or nonconducting. Both bipolar transistors and enhancement-mode MOSFETs are widely used in switching-amplifier applications. A bipolar transistor as a switch is either cut off or saturated. When it is cut off, no power is dissipated. When it is saturated, current flow is maximum, but the emitter-collector voltage is extremely low, usually less than 1 V. As a result, power dissipation is extremely low.

When enhancement-mode MOSFETs are used, the transistor is either cut off or turned on. In the cutoff state, no current flows, and so

no power is dissipated. When the transistor is conducting, its on resistance between source and drain is usually very low—again, no more than several ohms and typically far less than 1 Ω. As a result, power dissipation is extremely low even with high currents.

The use of switching power amplifiers permits efficiencies of over 90 percent. The current variations in a switching power amplifier are square waves, and thus harmonics are generated. However, these are relatively easy to filter out by the use of tuned circuits and filters between the power amplifier and the antenna.

The most commonly used switching amplifiers in RF application are class D and E amplifiers. A class D amplifier uses a pair of transistors to produce square-wave drive to a tuned circuit.

Figure 6-16 shows a class D amplifier implemented with enhancement-mode MOSFETs. The carrier is applied to the MOSFET gates 180° out of phase by the use of a transformer with a center-tapped secondary. When the input to the gate of Q_1 is positive, the input to the gate of Q_2 is negative. Thus Q_1 conducts and Q_2 is cut off. On the next half cycle of the input, the gate of Q_2 goes positive and the gate of Q_1 goes negative. Q_2 conducts, applying a negative pulse

Switching amplifiers

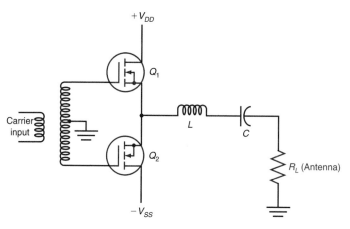

Fig. 6-16 A class D amplifier made with enhancement-mode MOSFETs.

to the tuned circuit. Recall that enhancement-mode MOSFETs are normally nonconducting until a gate voltage higher than a specific threshold value is applied, at which time the MOSFET conducts. The on resistance is very low. The transistors supply a bipolar square wave to the series resonant circuit, which filters out the odd harmonics and produces a sine wave across the load. In practice, efficiencies of up to 90 percent can be achieved using a circuit like that in Fig. 6-16.

In class E amplifiers, only a single transistor is used. Both bipolar and MOSFETs can be used, although the MOSFET is preferred because of its low drive requirements. Further, since MOSFETs switch faster (there is no carrier storage as in BJTs), they are more efficient. Figure 6-17 shows a typical class E RF amplifier. The carrier, which is initially a sine wave, is applied to a shaping circuit that effectively converts it into a square wave. The carrier is usually frequency-modulated. The square-wave carrier signal is then applied to the base of the class E bipolar power amplifier. Q_1 switches off and on at the carrier rate. The signal at the collector is applied to a low-pass filter and tuned impedance-matching circuit made up of C_1, C_2, and L_1. The harmonics are filtered out, leaving a fundamental sine wave that is applied to the antenna. A high level of efficiency is achieved with this arrangement.

■ TEST

Answer the following questions.

12. Linear power amplifiers are used to raise the power level of _____ and _____ signals.
13. A _____ power amplifier is used to increase the power level of an FM signal.
14. Linear power amplifiers operate class _____ and _____.
15. A class A transistor power amplifier has an efficiency of 50 percent. The output power is 27 W. The power dissipated in the transistor is _____ W.
16. Class A amplifiers conduct for _____ degrees of a sine wave input.
17. The cause of self-oscillation in a transistor RF amplifier is _____.

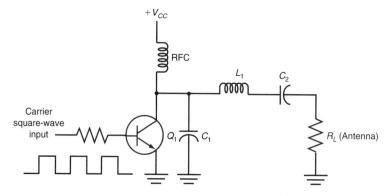

Fig. 6-17 A class E RF amplifier.

POWER GAIN IN DECIBELS

The gain of an amplifier is expressed as the ratio of the output to the input. Power gain G is:

$$G = \frac{P_o}{P_i}$$

where P_o and P_i are the output and input powers of an amplifier in watts, respectively.

If the output is 40 W and the input power is 5 W, the power gain is:

$$G = \frac{40\ W}{5\ W} = 8$$

Typically, power gain is more commonly expressed in decibels (dB). Recall that dB is calculated with the following formula:

$$dB = 10 \log \frac{P_o}{P_i}$$

The power gain of the amplifier described above is:

$$dB = 10 \log (8)$$

$$dB = 10\ (0.903) = 9.03\ dB$$

Power amplifier gains are usually stated in dB. If the gain and either the input or the output power are known, the remaining unknown quantity can be calculated by rearranging the above formula:

$$dB = 10 \log \frac{P_o}{P_i}$$

$$\frac{dB}{10} = \log \frac{P_o}{P_i}$$

$$\text{antilog } \frac{dB}{10} = \text{antilog of log } \frac{P_o}{P_i}$$

$$\text{antilog } \frac{dB}{10} = \frac{P_o}{P_i}$$

$$\text{antilog } \frac{dB}{10} = 10^{dB/10}$$

$$10^{dB/10} = \frac{P_o}{P_i}$$

Therefore:

$$P_o = 10^{db/10} P_i$$

$$P_i = \frac{P_o}{10^{db/10}}$$

If a power amplifier has a gain of 15 dB and the input power is 3 W, the output power is:

$$P_o = 10^{15/10}\ (3) = 10^{1.5}\ (3) = 31.62\ (3)$$

$$P_o = 94.86 = 95\ W$$

18. Self-oscillation in an RF amplifier that does not occur at the tuned frequency of the amplifier is called _____.

19. Self-oscillation of an RF amplifier can be eliminated by the process of _____.

20. True or false. Elimination of self-oscillation can be accomplished with either a feedback inductor or a capacitor.

21. Combining a feedback capacitor with the internal capacitance of the transistor forms a(n) _____ circuit which is used to cancel the self-oscillation.

22. True or false. With no input, a class B amplifier does not conduct.

23. Class B RF power amplifiers normally use a(n) _____ configuration.

24. A class C amplifier conducts for approximately _____ degrees to _____ degrees of the input signal.

25. In a class C amplifier, collector current flows in the form of _____.

26. In a class C amplifier, a complete sinusoidal output signal is produced by a(n) _____.

27. The efficiency of a class C amplifier is in the range of _____ to _____ percent.

28. The tuned circuit in the collector of a class C amplifier acts as a filter to eliminate _____.

29. A class C amplifier whose output tuned circuit resonates at some integer multiple of the input frequency is called a(n) _____.

30. Frequency multipliers with factors of 2, 3, 4, and 5 are cascaded. The input is 1.5 MHz. The output is _____ MHz.

31. A class C amplifier has a dc supply voltage of 28 V and an average collector current of 1.8 A. The power input is _____ W.

32. Without filtering, the output of a class D or E amplifier would be a(n) _____.

33. The most efficient amplifier is the class _____.

34. Higher output power can be obtained by adding the outputs of two or more

solid-state power amplifiers in a device
called a(n) _____.
35. The power gain of an amplifier with 6
mW in and 3 W out is _____ dB.
36. The output power of an amplifier with a
gain of 22 dB is 60 W. The input power
is _____ W.

6-3 Impedance-Matching Networks

One of the most important parts of any transmitter is the matching networks that connect one stage to another. In a typical transmitter, the oscillator generates the basic carrier signal which is then amplified usually by multiple stages before reaching the antenna. Since the idea is to increase the power of the signal, the interstage coupling circuits must permit an efficient transfer of power from one stage to the next. Finally, some means must be provided to connect the final amplifier stage to the antenna, again for the purpose of transferring the maximum possible amount of power.

Impedance-matching networks

The circuits used to connect one stage to another are known as *impedance-matching networks*. In most cases, they are either *LC* circuits, transformers, or some combination of the two. The basic function of a matching network is to provide for an optimum transfer of power through impedance-matching techniques.

Selectivity

Another important function of the matching network is to provide filtering and *selectivity*. Transmitters are designed to operate on a single frequency or on selectable narrow ranges of frequencies. The various amplifier stages in the transmitter must confine the RF generated to these frequencies. In class C amplifiers, a considerable number of high-amplitude harmonics are generated. These must be eliminated to prevent spurious radiation from the transmitter. The impedance-matching networks used for interstage coupling also provide this desired filtering and selectivity.

The basic problem of coupling is illustrated in Fig. 6-18(a). The driving stage appears as a signal source with an internal impedance of Z_i. The stage being driven represents a load to the generator with its internal resistance of Z_l. Ideally, Z_i and Z_l are resistive. As you recall, maximum power transfer takes place when Z_i equals Z_l. Although this basic relationship is essentially true in RF as in dc circuits, it is a much more

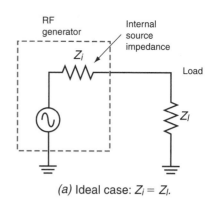

(a) Ideal case: $Z_i = Z_l$.

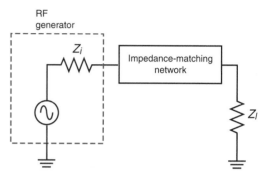

(b) The impedance-matching network makes Z_l "look like" Z_i.

Fig. 6-18 Impedance matching in RF circuits.

complex relationship. In RF circuits, Z_i and Z_l are seldom purely resistive and, in fact, usually include a reactive component of some type. Further, it is not always necessary to transfer maximum power from one stage to the next. The goal is to transfer a sufficient amount of power to the next stage so that it can provide the maximum amount of output of which it is capable.

The appropriate resistive output impedance (R_o) of a bipolar transistor amplifier is determined by the following expression. This output impedance is also the internal impedance (R_i) of the amplifier as seen by the load.

$$R_i = R_o$$

$$R_i = \frac{V^2}{KP}$$

where V is the dc power supply voltage, P is the desired output power, and K is a constant determined by the class of amplifier.

Amplifier class	K
A	1.3
AB	1.5
B	1.57
C	2

For example, the output impedance of a class B amplifier with an output power of 70 W and a supply voltage 36 V is:

$$R_i = R_0 = \frac{(36)^2}{1.57(70)} = \frac{1296}{109.9}$$

$$R_i = R_0 = 11.8 \ \Omega$$

This would be matched to the desired output load impedance (usually 50 Ω) with an *LC* matching network.

In most cases, the two impedances to be matched are considerably different from one another, and, therefore, a very inefficient transfer of power takes place. To overcome this problem, an impedance-matching network is introduced between the two as illustrated in Fig. 6-18(*b*). This impedance-matching network is usually an *LC* circuit or transformer as indicated before.

L-Networks

One of the simplest forms of impedance-matching networks is the *L network*. It consists of an inductor and a capacitor connected in various L-shaped configurations as shown in Fig. 6-19. The circuits shown in Fig. 6-19(*a*) and (*b*) are low-pass filters, while those shown in Fig. 6-19(*c*) and (*d*) are high-pass filters. Typically, the low-pass networks are used so that harmonic frequencies are filtered out.

By proper design of the L-matching network, the load impedance can be "matched" to the source impedance. For example, the network in Fig. 6-19(*a*) causes the load resistance to appear larger than it actually is. The load resistance Z_l appears in series with the inductor of the L network. The inductor and the capacitor are chosen to resonate at the transmitter frequency. When the circuit is at resonance, X_L equals X_C. To the generator impedance (Z_i), the complete circuit simply appears as a parallel resonant circuit. At resonance, the impedance represented by the circuit is very high. The actual value of the impedance depends upon the *L* and *C* values and the *Q* of the circuit. The higher the *Q*, the higher the impedance. The *Q* in this circuit is basically determined by the value of the load impedance. By proper selection of the circuit values, the load impedance can be made to appear as any desired value to the source impedance as long as Z_i is greater than Z_l.

By using the L network shown in Fig. 6-19(*b*), the impedance can be stepped down.

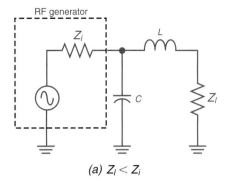

(a) $Z_l < Z_i$

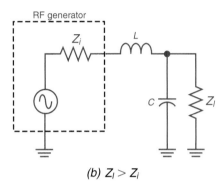

(b) $Z_l > Z_i$

L network

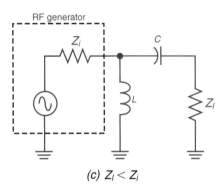

(c) $Z_l < Z_i$

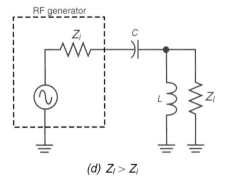

(d) $Z_l > Z_i$

Fig. 6-19 Four L-type impedance-matching networks.

In other words, the load impedance can be made to appear as if it were much smaller than it actually is. With this arrangement, the capacitor is connected in parallel to the load impedance. The parallel combination of C and Z_l

can be transformed to appear as an equivalent series RL combination. The result is that the overall network appears as a series resonant circuit. Recall that a series resonant circuit has a very low impedance at resonance. The impedance is, in fact, the equivalent load impedance.

The design equations for L networks are given in Fig. 6-20. We assume that the internal source and load impedances are resistive, so $Z_i = R_i$ and $Z_l = R_l$. The network of Fig. 6-20(a) assumes that $R_l < R_i$, while the network of Fig. 6-20(b) assumes that $R_i < R_l$.

Suppose we wish to match a 5-Ω transistor amplifier impedance to a 50-Ω antenna load at 120 MHz. In this case, $R_i < R_l$, so we use the formulas in Fig. 6-20(b).

$$X_L = \sqrt{R_i R_l - R_i^2}$$
$$= \sqrt{5(50) - 5^2}$$
$$= \sqrt{225}$$
$$= 15 \ \Omega$$
$$Q = \sqrt{\frac{R_l}{R_i} - 1}$$
$$= \sqrt{10 - 1} = \sqrt{9}$$
$$= 3$$
$$X_c = \frac{R_i R_l}{X_L} = \frac{5(50)}{15} = \frac{250}{15} = 16.67 \ \Omega$$

To find the values of L and C at 120 MHz, we rearrange the basic reactance formulas.

$$X_L = 2\pi f L$$
$$L = \frac{X_L}{2\pi f} = \frac{15}{6.28 \times 120 \times 10^6} = 20 \ \text{nH}$$
$$X_c = \frac{1}{2\pi f C}$$
$$C = \frac{1}{2\pi f X_c} = \frac{1}{6.28 \times 120 \times 10^6 \times 16.67}$$
$$= 80 \ \text{pF}$$

In most cases, the internal impedance and load impedances are not purely resistive. Internal and stray reactances make the impedances complex. Figure 6-21 shows an example. Here the internal resistance is 5 Ω, but it includes an internal inductance L_i of 8 nH. There is a stray capacitance C_l of 12 pF across the load. The way to deal with these reactances is simply to combine them with the L network values. In the example above, the calculation calls for an inductance of 20 nH. Since the stray inductance is in series with the L network inductance in Fig. 6-21, naturally the values will add. As a result, the L network inductance can be less than the computed value by an amount equal to the stray inductance of 8 nH.

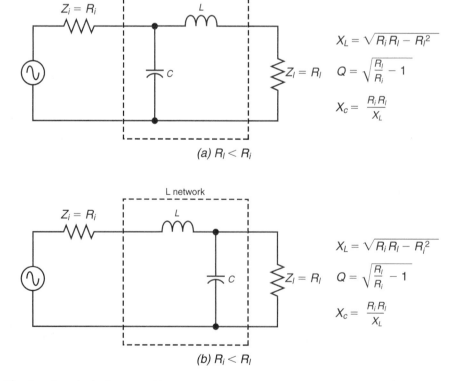

(a) $R_l < R_i$

(b) $R_i < R_l$

Fig. 6-20 The L network design equations.

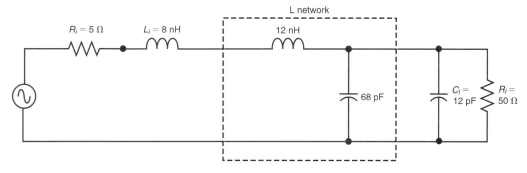

Fig. 6-21 Incorporating the internal and stray reactances into the matching network.

$$L = 20 - 8$$
$$= 12 \text{ nH}$$

By making the L network inductance 12 nH, the total circuit inductance will be correct when it adds to the stray inductance.

A similar thing occurs with the capacitance. The circuit calculations above call for a total of 80 pF. The L network capacitance and the stray capacitance will add as they are in parallel. Therefore, the L network capacitance can be less than the calculated value by the amount of the stray capacitance or

$$C = 80 - 12$$
$$= 68 \text{ pF}$$

Making the L network capacitance 68 pF gives the total correct capacitance when it is added to the stray capacitance.

Pi and T Networks

Although L networks are widely used in impedance matching, they are inflexible with regard to their selectivity. When L networks are designed, there is very little control over the Q of the circuit. The Q of the circuit is defined by the values of the internal and load impedances. The resulting Q is what you get but may not always be what you need to achieve the desired selectivity.

To overcome this problem, matching networks using three reactive elements can be used. The three most widely used impedance-matching networks containing three reactive components are illustrated in Fig. 6-22. The network in Fig. 6-22(a) is known as a *pi network* because its configuration resembles the Greek character pi. The circuit in Fig. 6-22(b) is known as a *T network*, again because the cir-

cuit elements essentially resemble the letter T. The circuit in Fig. 6-22(c) is a T network using two capacitors. Note that all are low-pass filters that provide maximum harmonic attenuation. The pi and T networks can be designed to either step up or step down the impedances as required by the circuit. The capacitors are usually made variable so that the circuit can be tuned to resonance and adjusted for maximum power output.

The most popular circuit is the T network of Fig. 6-22(c). Often called an *LCC network*, it is widely used to match the low output impedance of a transistor power amplifier to the higher impedance of another amplifier or an antenna. The design formulas are given in Fig. 6-23. Using our previous example of matching a 5-Ω source R_i to a 50-Ω load R_l at 120 MHz, let's calculate the *LCC* network. Assume a Q of 10. First calculate the inductance.

$$X_L = QR_i = 10(5) = 50 \ \Omega$$

$$L = \frac{X_L}{2\pi f} = \frac{50}{6.28(120 \times 10^6)} = 66.3 \text{ nH}$$

Next, calculate C_1.

$$X_{C_1} = R_l \sqrt{\frac{R_i(Q^2 + 1)}{R_l} - 1}$$

$$= 50 \sqrt{\frac{5(10^2 + 1)}{50} - 1}$$

$$= 50 \sqrt{\frac{505}{50} - 1} = 50\sqrt{9.1} = 150 \ \Omega$$

$$C_1 = \frac{1}{2\pi f X_c} \ \frac{1}{6.28(120 \times 10^6)(150)} \qquad \textbf{Pi network}$$

$$= 8.8 \times 10^{-12} = 8.8 \text{ pF}$$

Finally, calculate C_2. **T network**

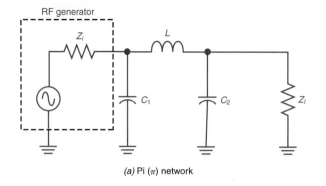

(a) Pi (π) network

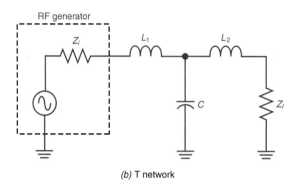

(b) T network

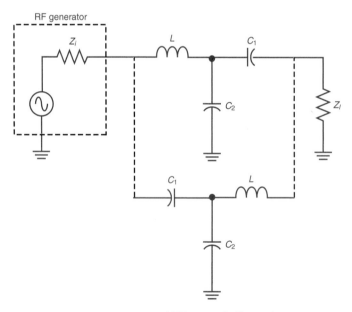

(c) Two-capacitor T network

Fig. 6-22 Three-element matching networks.

$$X_{C_2} = \frac{R_i(Q^2 + 1)}{Q} \times \frac{1}{1 - X_{C_1}/QR_l}$$

$$= \frac{5(10^2 + 1)}{10} \times \frac{1}{1 - 150/(10 \times 50)}$$

$$= 50.5 \frac{1}{1 - 0.3} = 50.5\ (1.43) = 72\ \Omega$$

$$C_2 = \frac{1}{2\pi f X_c} = \frac{1}{6.28(120 \times 10^6)(72)}$$

$$= 18.4\ \text{pF}$$

Transformer Impedance Matching

One of the best impedance-matching components is the transformer. Recall that iron-core transformers are widely used at lower frequencies to match one impedance to another. Any load impedance can be made to look like a desired load impedance if the correct value of the transformer turns ratio is selected.

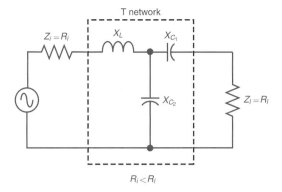

T network

$Z_i = R_i$ X_L X_{C_1}

X_{C_2}

$Z_l = R_l$

$R_i < R_l$

Design Procedure:
1. Select a desired circuit Q
2. Calculate $X_L = QR_i$
3. Calculate X_{C_1}:

$$X_{C_1} = R_i \sqrt{\frac{R_l(Q^2 + 1)}{R_l} - 1}$$

4. Calculate X_{C_2}:

$$X_{C_2} = \frac{R_l(Q^2 + 1)}{Q} \times \frac{1}{\left(1 - \frac{X_{C_1}}{QR_i}\right)}$$

5. Compute final L and C values:

$$L = \frac{X_L}{2\pi f}$$

$$C = \frac{1}{2\pi f X_C}$$

Fig. 6-23   Design equations for an *LCC* T network.

Refer to Fig. 6-24. The relationship between the turns ratio and the input and output impedances is

$$\frac{Z_i}{Z_l} = \left(\frac{N_p}{N_s}\right)^2 \quad or \quad \frac{N_p}{N_s} = \sqrt{\frac{Z_i}{Z_l}}$$

What this formula says is that the ratio of the input impedance Z_i to the load impedance Z_l is equal to the square of the ratio of the number of turns on the primary N_p to the number of turns on the secondary N_s. As an example, to match a 5-Ω generator impedance to a 50-Ω load impedance, the turns ratio would be

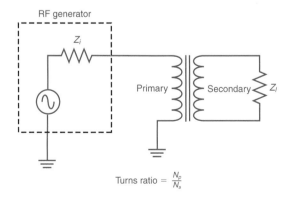

RF generator

Z_i

Primary   Secondary Z_l

Turns ratio = $\frac{N_p}{N_s}$

Fig. 6-24   Impedance matching with an iron-core transformer.

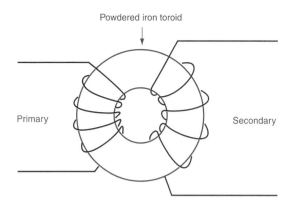

Powdered iron toroid

Primary

Secondary

Fig. 6-25   A toroid transformer.

$$\frac{N_p}{N_s} = \sqrt{\frac{Z_i}{Z_l}} = \sqrt{\frac{5}{50}} = \sqrt{0.1}$$

$$\frac{N_p}{N_s} = 0.316 \quad or \quad \frac{N_s}{N_p} = \frac{1}{3.16} = 0.316$$

This means that there are 3.16 times as many turns on the secondary as on the primary.

This relationship holds true only on iron-core transformers. When air-core transformers are used, the coupling between primary and secondary windings is not complete and, therefore, the impedance ratio is not as indicated. Although air-core transformers are widely used at RF frequencies and can be used for impedance matching, they are less efficient than iron-core transformers.

Special types of core materials have been created for use at very high frequencies so that iron-core transformers can be made. The core material is a ferrite or powdered iron. Both the primary and secondary windings are wound on a core of this material.

The most widely used core for RF transformers is the popular toroid. A *toroid* is a circular, doughnut-shaped core usually made of a special type of powdered iron. Copper wire is wound on the toroid to create the primary and secondary windings. A typical arrangement may appear like that shown in Fig. 6-25.

Single-winding tapped coils called *autotransformers* are also used for impedance matching between RF stages. Figure 6-26 shows both impedance step-down and step-up arrangements. Toroids are commonly used.

Unlike air-core transformers, toroid transformers cause the magnetic field produced by the primary to be completely contained within the core itself. This has several important advantages. First, a toroid will not radiate RF energy. Air-core coils radiate because the magnetic field produced

Toroid

Autotransformers

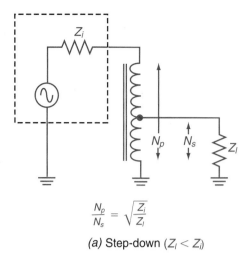

$$\frac{N_p}{N_s} = \sqrt{\frac{Z_i}{Z_l}}$$

(a) Step-down $(Z_l < Z_i)$

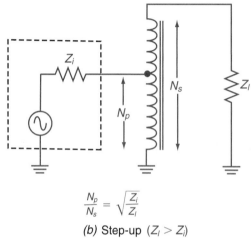

$$\frac{N_p}{N_s} = \sqrt{\frac{Z_i}{Z_l}}$$

(b) Step-up $(Z_l > Z_i)$

Fig. 6-26 Impedance matching in an autotransformer.

around the primary is not contained. Transmitter and receiver circuits using air-core coils must be contained by magnetic shields to prevent them from interfering with other circuits. The toroid, on the other hand, confines the entire magnetic field, and, therefore, no shields are necessary.

Another benefit is that most of the magnetic field produced by the primary cuts the turns of the secondary winding. For this reason, the basic turns ratio, input-output voltage, and impedance formulas for standard low-frequency transformers also apply to high-frequency toroid transformers.

In most new RF designs, toroid transformers are used for RF impedance matching between stages. Further, the primary and secondary windings are sometimes used as inductors in tuned circuits. Alternately, toroid inductors can also be built. Toroid inductors have an advantage over air-core inductors for RF applications; that is, the high permeability of the core causes the inductance to be high. Recall that whenever an iron core is inserted into a coil of wire, the inductance will increase dramatically. For RF applications, this means that the desired values of inductance can be created by using fewer turns of wire. This results in smaller inductors. Also, fewer turns have less resistance, giving the coil a higher Q than that obtainable with air-core coils.

Powdered-iron toroids are so effective that they have virtually replaced air-core coils in most modern transmitter designs. They are available in sizes from a fraction of an inch to several inches in diameter. In most applications, a minimum number of turns are required to create the desired inductance.

Figure 6-27 shows a toroid transformer used for interstage coupling between two class C amplifiers. The primary of the driver transformer is tuned to resonance by capacitor C_1. The capacitor is adjustable so that the exact frequency of operation can be set. The relatively high output impedance of the transistor is coupled with the low input impedance of the next class C stage by a step-down transformer which provides the desired impedance-matching effects. Usually, the secondary winding is only a few turns of wire and is not resonated. The circuit in Fig. 6-27 also

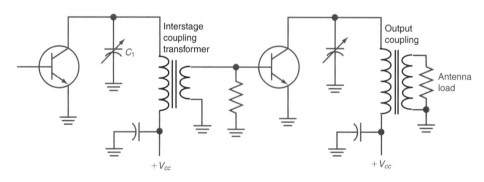

Fig. 6-27 Using toroid transformers for coupling and impedance matching in class C amplifier stages.

shows a similar transformer used for output coupling to the antenna.

Baluns for Impedance Matching

A *balun* is a transmission line transformer connected to perform impedance matching over a wide range of frequencies. One of the most widely used configurations is shown in Fig. 6-28. This transformer is usually wound on a toroid in which the number of primary and secondary turns are equal, giving the transformer a 1:1 turns ratio and a 1:1 impedance-matching ratio. The dots indicate the phasing of the windings. Note the unusual way in which the windings are connected. A transformer connected in this way is generally known as a balun. The name balun is derived from the first letters of the words balanced and unbalanced since such transformers are normally used to connect a balanced source to an unbalanced load or vice versa. In the circuit of Fig. 6-28(a), a balanced generator is connected to an unbalanced (grounded) load. In Fig. 6-28(b), an unbalanced (grounded) generator can be connected to a balanced load.

Figure 6-29 shows how a 1:1 turns ratio balun can be used for impedance matching.

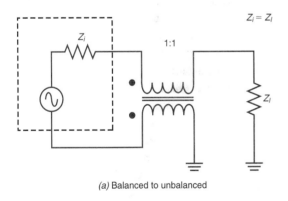

(a) Balanced to unbalanced

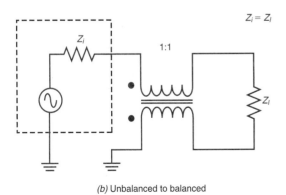

(b) Unbalanced to balanced

Fig. 6-28 Balun transformers used for connecting balanced and unbalanced loads or generators.

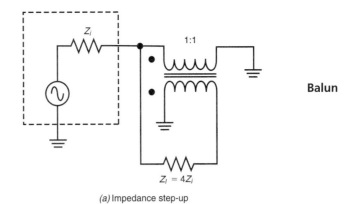

(a) Impedance step-up

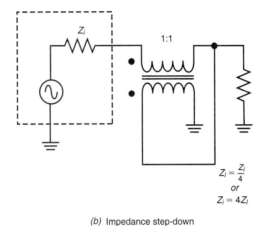

(b) Impedance step-down

Fig. 6-29 Using a balun for impedance matching.

With the arrangement shown in Fig. 6-29(a), an impedance step-up is obtained. A load impedance of four times the source impedance Z_i provides a correct match. The balun makes the load "look like" $Z_l/4$ to match Z_i. In Fig. 6-29(b), an impedance step-down is obtained. The balun makes the load Z_l "look like" $4Z_i$.

Many other balun configurations are possible with different impedance ratios. Several common 1:1 baluns can be interconnected for both 9:1 and 16:1 impedance transformation ratios. In addition, baluns may also be cascaded so that the output of one appears as the input to the other, and so on. By cascading baluns, impedances may be stepped up or stepped down by wider ratios.

An important point to note here is that the windings in a balun are not resonated with capacitors to a particular frequency. As a result, they operate over a wide frequency range. The winding inductances are made such that the coil reactances are four or more times that of the highest impedance being matched. In this way, the transformer will provide the designated impedance matching over a tremendous range of frequencies.

This broadband characteristic of balun transformers allows designers to construct broadband RF power amplifiers. Such amplifiers provide a specific amount of power amplification over a *wide bandwidth*. These amplifiers are preferred particularly in communications equipment that must operate in more than one frequency range. Rather than having one transmitter for each desired band, a single transmitter can be used.

When using conventional tuned amplifiers, some method of switching the correct tuned circuit into the circuit must be provided. Such switching networks are complex and expensive. Further, they introduce problems, particularly at the high frequencies. In order for them to perform effectively, the switches must be located very close to the tuned circuits so that stray inductances and capacitances are not introduced by the switch and the interconnecting leads.

One way to overcome the switching problem is simply to use a *broadband amplifier*. No switching or tuning is needed. The broadband amplifier provides the necessary amplification as well as impedance matching.

The primary problem with the broadband amplifier is that it does not provide the filtering necessary to get rid of harmonics. One way to overcome this problem is to generate the desired frequency at a lower power level, allowing tuned circuits to filter out the harmonies and then to provide final power amplification with the broadband circuit. The broadband power amplifier operates either a class A or a class B push-pull so that the inherent harmonic content of the output is very low.

Figure 6-30 shows a typical broadband linear amplifier. Note that two 4:1 balun transformers are cascaded at the input so that the low base input impedance is made to look like an impedance 16 times higher than the input. The output uses a 1:4 balun that steps up the very low output impedance of the final amplifier to an impedance 4 times higher to equal the antenna load impedance. In some transmitters, broadband amplifiers may be followed by low-pass filters which are used to eliminate undesirable harmonics in the output.

■ TEST

Answer the following questions.

37. Maximum power transfer occurs when the generator impedance _____ the load impedance.
38. A commonly used two-element (*LC*) circuit called a(n) _____ is widely used for impedance matching.
39. List the two popular types of three-element (*LC*) matching networks.
40. Most impedance-matching networks are _____ filters, so they eliminate harmonics.
41. The main advantage of pi and T networks over the L network is that the circuit _____ can be chosen.
42. The output impedance of a class A amplifier with a supply voltage of 15 V and an output power of 400 mW is _____ Ω.
43. Calculate the L network components to match a 3-Ω internal resistance in series with an internal inductance of 19 nH to a

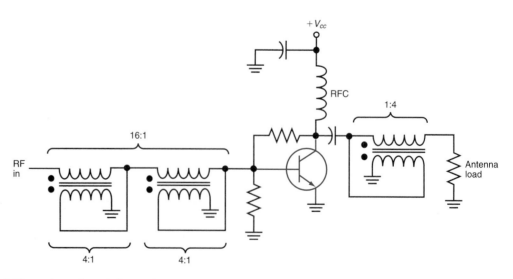

Fig. 6-30 A broadband class A linear power amplifier.

72-Ω load impedance in parallel with a stray capacitance of 32 pF at a frequency of 18 MHz.

44. Calculate the components of an *LCC* T network to match a 4-Ω internal resistance to a 52-Ω load at 72 MHz. Assume a *Q* of 12.

45. A transformer has 6 turns on the primary and 18 turns on the secondary. If the generator (source) impedance is 50 Ω, the load impedance should be _____ Ω.

46. A transformer must match a 200-Ω generator to a 50-Ω load. The turns ratio must be _____.

47. A single-winding transformer is known as a(n) _____.

48. A doughnut-shaped powdered-iron core widely used to make RF transformers and inductors is called a(n) _____.

49. An RF transformer with a 1:1 turns ratio called a(n) _____ is connected in a special way to provide 1:4 or 4:1 impedance matching.

50. Untuned RF transformers permit _____ operation over a wide frequency range.

6-4 Speech Processing

Speech processing refers to the ways that the voice signal used in communications is dealt with prior to being applied to the modulator. The voice signal from the microphone is usually amplified and applied to circuits that deliberately limit its amplitude and frequency response.

Benefits of Speech Processing

The primary purpose of speech-processing circuits is to ensure that overmodulation does not occur and to restrict the bandwidth of the signal. In AM, an excessively high voice signal will cause *overmodulation* and severe distortion of the signal. If the amplitude is deliberately limited, overmodulation will not occur.

In an AM transmitter, the distortion caused by overmodulation causes harmonics of the primary frequencies in the voice signal to be generated. As a result, these harmonics will also modulate the carrier and produce sidebands that extend far beyond the assigned band-

width of the AM signal. These harmonics may interfere with signals on adjacent channels. This interference is generally known as *splatter*. Excessive modulation causes splatter which, in turn, produces interference to other signals. The distortion also causes the signal to be generally less intelligible.

Splatter

Most transmitters are designed to operate in a specific frequency range. This means that the sidebands may not extend beyond certain assigned limits. In AM transmitters, it is the highest modulating frequency that determines the total bandwidth of the AM signal. For example, in AM broadcasting, the channel width is limited to 10 kHz. For this reason, the upper audio frequency must be limited to one-half of this, or 5 kHz. Speech-processing circuits are designed to limit both the upper amplitude and the frequency.

Another purpose of speech processing in AM and SSB transmitters is that it helps keep the average transmitted power higher. Recall that when an AM transmitter is modulated, the power from the modulator that appears in the sidebands increases with the percentage of modulation. The more power transmitted in the sidebands, the greater the transmission distance is and the more reliable communications are. In other words, the strength of the RF signal is determined by the strength of the modulating signal. The same is true of SSB signals. The strength of the modulating signal determines the strength of the transmitted sideband.

Speech processing

Voice signals vary over an extreme amplitude range. It has been determined in the study of speech that the highest peaks of intensity come from vowel sounds. On the other hand, vowel sounds tend to do nothing for intelligibility. As it turns out, it is the consonants, such as B, K, L, S, T, and V, that provide the best detail for good intelligibility. Yet consonant sounds are typically spoken at a much lower amplitude level than vowel sounds. If the vowel sounds could be de-emphasized while consonant sounds were emphasized through increased amplification, the intelligibility of the signal would improve tremendously. At the same time, the total average voice signal would be higher and would, therefore, provide a higher transmitted power. Special-speech processing circuits have been developed to provide for increased amplification of low-level audio content while at the same time limiting

Overmodulation

the peaks. This process is generally known as *dynamic compression.*

In FM, it is also necessary to deliberately limit the bandwidth and amplitude of the modulating signal. Recall that the frequency deviation of an FM carrier is directly proportional to the amplitude of the modulating signal. Most FM transmitters are allowed to operate up to a certain maximum deviation limit. In mobile communications transmitters, this frequency deviation is ± 5 kHz. In order to ensure that the deviation limit is not exceeded, some means must be provided to prevent the modulating signal amplitude from causing overdeviation. This is comparable to overmodulation in AM transmitters.

In addition, the frequency content of the modulating signal must also be limited. Recall that an FM signal produces multiple sidebands above and below the carrier. These are spaced from the carrier by the frequency of the modulating signal. The higher the modulating signal frequency, the wider the spacing of the sidebands and the greater the spectrum space occupied by the FM signal. As in most applications, the allowed bandwidth of an FM signal is specified. For this reason, the upper frequencies of the modulating signal are usually limited to prevent the bandwidth from becoming too wide.

Clipping causes harmonics

Analog Speech Processing

Clipper circuit

The basic speech-processing circuit is made up of an amplifier operating in conjunction with a *clipper circuit* that will limit the amplitude swings of the modulating signal and a filter that cuts off modulating frequencies above a certain point. Such a basic circuit is illustrated in Fig. 6-31. The low-level signal from the microphone is amplified by two stages of audio amplification to raise the signal to a high enough amplitude to make it compatible with the modulator. The output of the amplifier is applied to a diode clipper. Diodes D_1 and D_2 are low-level silicon diodes that conduct when about $+0.7$ V is applied across them. Notice the diodes are connected in an inverse parallel arrangement; that is, one diode is connected to conduct on positive-going ac signals, while the other will conduct on negative-going ac signals.

As long as the peak-to-peak amplitude of the voice signal is below approximately 1.4 V peak-to-peak, no clipping will occur. However, if voice peaks should cause the signal to be greater, diode D_1 or D_2 will conduct, thereby clipping off the peak and holding the amplitude to ± 0.7 V. Regardless of the amplitude of the amplifier output, the signal across the diodes will never be greater than 1.4 V peak-to-peak. Thus with the amplitude deliberately limited, an AM transmitter will not be overmodulated and an FM transmitter will not be overdeviated.

Clipping, however, does cause signal distortion. The deliberate flattening of the signal peaks introduces harmonics. One way to get rid of these harmonics is to pass the clipped signal through a low-pass filter. In Fig. 6-31 this low-pass filter is made up of resistors R_1 and R_2 and capacitors C_1 and C_2. This filter gets rid of those harmonics and thus compensates for the clipping distortion.

The other purpose of the low-pass filter is to deliberately restrict the modulating signal bandwidth. Typically, the cutoff frequency of the filter is chosen so that it is somewhere in the 2.5- to 3-kHz range. This is the upper limit for most two-way communications systems. As you can see then, the filter has a dual purpose: restricting the bandwidth of the signal and reducing the distortion caused by clipping.

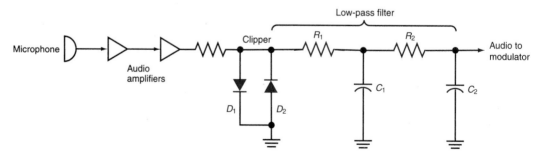

Fig. 6-31 A basic speech-processing circuit incorporating clipping and filtering.

Such circuits are almost always incorporated in FM transmitters and in many AM transmitters as well. However, in AM and SSB transmitters, *speech compression* is generally preferred over clipping. Speech compression permits the average modulating signal to be higher, thus increasing the average power output. However, this technique is not applicable to FM.

Figure 6-32 shows a typical speech-compression circuit. It consists of two stages of amplification whose gain is controlled automatically by the amplitude of the audio signal. Some of the audio signal is tapped off and applied to a diode rectifier D_1. The audio signal is rectified and filtered by C_2 to produce a dc output. It is amplified by a dc amplifier. The average dc output is proportional to the amplitude of the modulating signal. This dc voltage is fed back to an earlier amplifier stage to control its gain. The gain of a transistor amplifier stage can be controlled by adjusting collector current. This is done by modifying the base bias current. The higher the collector current, the higher the gain. If the modulating signal amplitude is high, the dc output from the rectifier will be high and it, in turn, will reduce the gain of the amplifier thereby limiting the peaks. In this regard, the compression circuit effectively limits the peaks, preventing overmodulation and splatter.

If the amplitude of the modulating signal is very low, then the rectifier will produce a very low average dc output. This causes the amplifier gain to be very high, thereby providing extra amplification for low-level signals. This compression circuit is said to have *automatic gain control (AGC)* simply because the gain of the circuit automatically adjusts itself to a level appropriate to the amplitude of the incoming signal. High-level amplitude signals are deliberately limited, while low-level amplitude signals are provided with extra amplification.

The clipping and compression can also be performed in the RF stages of the transmitter rather than in the modulating signal stages. A common technique is to use RF compression for this purpose. This is usually done in SSB transmitters and is commonly referred to as *automatic level control (ALC)*. The technique for RF compression is similar to that used in audio compression. A rectifier circuit is used to tap off a sample of the RF signal at the final power amplifier stage. The RF signal is rectified and converted into a dc voltage that is fed back to control the gain of an earlier stage used to amplify the SSB signal. Again, if the output level exceeds a certain limit, the fed-back dc voltage will reduce the gain of the stage and thus automatically control the amplitude. Low-level signals will receive full amplification.

Digital Speech Processing

In many modern transmitters, speech processing is accomplished with digital circuits. Virtually all of the various speech-processing techniques such as limiting, filtering, and compression can be performed using a *digital signal processor (DSP)*. A DSP chip is a powerful microprocessor that executes special programs that implement the speech processing methods on digital data. The analog voice signal is first digitized by an *analog-to-digital converter (ADC)*. The ADC converts the voice into a stream of binary numbers that is processed by the DSP chip. The DSP output is another stream of digital data representing the processed signal. This data is sent to a *digital-to-analog converter (DAC)* that reproduces the processed analog signal.

■ TEST

Answer the following questions.

51. The two techniques used in speech processing are _____ and _____.
52. Name the two primary purposes of speech processing.
53. A diode _____ circuit is frequently used to restrict voice signal amplitude.
54. Name the two main purposes of a low-pass filter in a speech-processing circuit.
55. Overmodulation in an AM transmitter causes distortion and out-of-band operation which is referred to as _____.

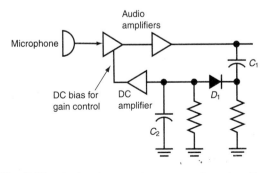

Fig. 6-32 A simple speech-compression circuit.

To continue studying the topics in this chapter, see the following university sites:

⟨www.speech.cs .cmu.edu⟩
⟨www.isip .msstate.edu⟩

56. Audio compression is achieved by controlling the _____ of an audio amplifier.
57. Audio compression is used to increase the average _____ in an AM or SSB signal.
58. _____ and _____ circuits are used in a speech compressor to create a dc feedback control signal.
59. The gain of a transistor can be controlled by varying the _____.

60. The process of adjusting the amplification of a signal based on its amplitude is known as _____.
61. In an audio compressor, do low-level signals receive more or less amplification?
62. True or false. Speech processing may be done on the RF signal or the audio signal.
63. The three circuits used in digital speech processing are the _____, _____, and _____.

Summary

1. A radio transmitter generates the carrier signal, provides power amplification, and applies modulation.
2. The simplest transmitter is an oscillator that is keyed off and on to produce cw Morse code.
3. Most transmitters consist of a crystal oscillator used to generate an accurate and stable carrier frequency, a buffer amplifier that isolates the carrier oscillator from its load, one or more driver amplifiers to increase the RF power level, and a final power amplifier that applies the signal to the antenna.
4. In AM transmitters, class C amplifiers are used to increase the RF power level. The final stage is modulated.
5. Frequency modulation transmitters use class C amplifiers to increase the RF power level.
6. Single-sideband transmitters generate the carrier and modulation at a low frequency and then translate it up with a mixer. Linear amplifiers are used to increase the power level.
7. In a class A amplifier, collector current flows continuously. The output is directly proportional to the input; therefore, it is a linear amplifier.
8. Class B amplifiers are biased at cutoff so that collector current flows for 180° of the input cycle.
9. Class B amplifiers are usually connected in a push-pull circuit where one transistor amplifies each half of the input signal.
10. Class C amplifiers are biased beyond cut-off. Current flows for 90° to 150° of the input cycle.
11. Collector current pulses in a class C amplifier are converted into a continuous sine wave by a resonant circuit.
12. The collector current pulses in a class C amplifier contain many harmonics which are filtered out by the tuned output circuit.
13. A class C amplifier can be used as a frequency multiplier by connecting a resonant circuit tuned to some integer multiple of the input frequency in the output.
14. RF amplifiers may oscillate because of feedback from internal transistor capacitance. This can be eliminated, or prevented, by neutralization, a process that cancels the feedback with out-of-phase feedback.
15. Frequency multipliers can be cascaded to produce higher output frequencies.
16. In an FM transmitter, the frequency multipliers increase the deviation as well as the carrier frequency.
17. Linear amplifiers operating class A or B are used to increase the power level of low-level AM and SSB signals.
18. Class D and class E power amplifiers use transistors as switches to produce a square-wave output at the carrier frequency. Tuned circuits are used to eliminate the harmonics and match impedances. Switching power amplifiers are the most efficient because they can achieve efficiencies of 90 to 98 percent.
19. Higher power can be achieved with solid-state amplifiers by adding the outputs of two or more amplifiers in a power combiner transformer.
20. The gain of a power amplifier is usually expressed in decibels (dB). It is calculated with the formula:

$$dB = 10 \log \frac{P_o}{P_i}$$

where P_o is the output power and P_i is the input power.
21. Impedance-matching networks are used to interconnect RF amplifiers and to couple power to the antenna to ensure the optimum transfer of power.
22. Common impedance-matching circuits and components include LC L networks, LC pi networks, LC T networks, transformers, and baluns. The pi and T networks are preferred since Q can be controlled.
23. Maximum power transfer occurs when the load impedance equals the generator source impedance.
24. Radio-frequency transformers are typically constructed with doughnut-shaped powdered-iron cores called toroids.

25. The impedance-matching ability of a transformer is determined by its turns ratio: $(N_p/N_s)^2 = Z_p/Z_s$.
26. Inductors made with toroids for a given inductance are smaller, use fewer turns of wire, have a higher Q, and do not require shielding.
27. A balun is a transformer connected in a special way to transform circuits from balanced to unbalanced or vice versa and to provide impedance matching.
28. Toroid transformers and baluns are broadband devices that operate over a wide bandwidth.
29. Broadband, linear, untuned RF power amplifiers provide amplification over a broad frequency range.
30. Speech-processing circuits in a transmitter prevent overmodulation, prevent excessive signal bandwidth, and increase the average transmitted power in AM and SSB systems.
31. A voice clipper uses diodes to limit the amplitude of the audio modulating signal. A low-pass filter smooths out any clipping distortion and prevents excessive sidebands.
32. Voice compressors use automatic-gain-control (AGC) circuits to limit the audio amplitude. The gain of the circuit is inversely proportional to the audio signal amplitude.
33. In AGC circuits, a rectifier and filter convert the audio or RF into a dc voltage that controls the gain of an audio or RF amplifier to prevent overmodulation.
34. The gain of a transistor amplifier can be varied by changing the collector current.
35. Speech processing is often performed digitally by converting the audio signal to digital form, manipulating it in a digital signal processor, and reconverting it to analog.

Chapter Review Questions

Choose the letter that best answers each question.

6-1. Which of the following circuits is *not* typically part of every radio transmitter?
a. Carrier oscillator
b. Driver amplifier
c. Mixer
d. Final power amplifier

6-2. Class C amplifiers are *not* used in which type of transmitter?
a. AM   c. CW
b. SSB   d. FM

6-3. A circuit that isolates the carrier oscillator from load changes is called a
a. Final amplifier.
b. Driver amplifier.
c. Linear amplifier.
d. Buffer amplifier.

6-4. A class B amplifier conducts for how many degrees of an input sine wave?
a. 90° to 150°
b. 180°
c. 180° to 360°
d. 360°

6-5. Bias for a class C amplifier produced by an input RC network is known as
a. Signal bias.
b. Self-bias.
c. Fixed external bias.
d. Threshold bias.

6-6. An FM transmitter has a 9-MHz crystal carrier oscillator and frequency multipliers of 2, 3, and 4. The output frequency is
a. 54 MHz.
b. 108 MHz.
c. 216 MHz.
d. 288 MHz.

6-7. The most efficient RF power amplifier is which class amplifier?
a. A
b. E
c. B
d. C

6-8. Collector current in a class C amplifier is a
a. Sine wave.
b. Half sine wave.
c. Pulse.
d. Square wave.

6-9. The maximum power of typical transistor RF power amplifiers is in what range?
a. Milliwatts
b. Watts
c. Hundreds of watts
d. Kilowatts

6-10. Self-oscillation in a transistor amplifier is usually caused by
a. Excessive gain.
b. Stray inductance.
c. Internal capacitance.
d. Unmatched impedances.

6-11. Neutralization is the process of
a. Canceling the effect of internal device capacitance.
b. Bypassing undesired alternating current.
c. Reducing gain.
d. Eliminating harmonics.

6-12. Low-level AM signals must be amplified by which type of amplifier?
a. Push-pull
b. Class C
c. Switching
d. Linear

6-13. In which class of amplifier is the transistor used as a switch?
a. Class D
b. Class C
c. Class B
d. Class A

6-14. A power combiner is most like which component?
a. Transistor
b. Transformer
c. Capacitor
d. Inductor

6-15. The output impedance of a class C amplifier with a dc supply voltage of 24 V and an output power of 30 W is
a. 6.8 Ω
b. 7.3 Ω
c. 9.6 Ω
d. 18.2 Ω

6-16. Maximum power transfer occurs when what relationship exists between the generator impedance Z_i and the load impedance Z_l?
a. $Z_i = Z_l$
b. $Z_i > Z_l$
c. $Z_i < Z_l$
d. $Z_i = 0\ \Omega$

6-17. Which of the following is *not* a benefit of a toroid RF inductor?
a. No shielding required
b. Fewer turns of wire
c. Higher Q
d. Self-supporting

6-18. A toroid is a
a. Type of inductor.
b. Transformer.

c. Magnetic core.
d. Coil holder.

6-19. Which of the following is *not* commonly used for impedance matching in a transmitter?
a. Resistive attenuator
b. Transformer
c. L network
d. T network

6-20. To match a 6-Ω amplifier impedance to a 72-Ω antenna load, a transformer must have a turns ratio N_p/N_s of
a. 0.083.
b. 0.289.
c. 3.46.
d. 12.

6-21. Impedance matching in a broadband linear RF amplifier is handled with a(n)
a. L network.
b. Parallel tuned circuit.
c. Pi network.
d. Balun.

6-22. A class C amplifier has a supply voltage of 24 V and a collector current of 2.5 A. Its efficiency is 80 percent. The RF output power is
a. 24 W.
b. 48 W.
c. 60 W.
d. 75 W.

6-23. Which of the following is *not* a benefit of speech-processing circuits?
a. Improved frequency stability
b. Increased average output power
c. Limited bandwidth
d. Prevention of overmodulation

6-24. In an AM transmitter, a clipper circuit eliminates
a. Harmonics.
b. Splatter.
c. Overdeviation.
d. Excessive gain.

6-25. In a speech-processing circuit, a low-pass filter prevents
a. Overdeviation.
b. Overmodulation.
c. High gain.
d. Excessive signal bandwidth.

6-26. What values of L and C in an L network are required to match a 10-Ω transistor amplifier impedance to a 50-Ω load at 27 MHz?
a. $L = 47$ nH, $C = 185$ pF
b. $L = 118$ nH, $C = 236$ pF
c. $L = 0.13\ \mu$H, $C = 220$ pF
d. $L = 0.3\ \mu$H, $C = 330$ pF

Critical Thinking Questions

6-1. Refer to Sec. 7-6, in Chap. 7, which covers phase-locked loops (PLLs) and frequency synthesizers. Explain how you could use a PLL in a transmitter. What other circuits would it replace?

6-2. If transistor RF amplifiers are limited to several hundred watts of power, how is the high power in AM, FM, and TV broadcasting stations produced?

6-3. Explain how a small low-power transmitter could be powered (by direct current) by just the voice signal to be transmitted.

6-4. A CW transmitter can be made similar to an AM transmitter by using what type of modulation?

6-5. A transmitter with 1.5-kW output has an antenna load of 50 Ω. How much voltage is applied to the antenna? Is it dangerous to touch the antenna?

6-6. If a class A amplifier is only 50 percent efficient, only half of the total power generated gets to the load (usually the antenna). If a transmitter puts 175 W into the load, what is the total power consumed? Where does the remaining power go?

6-7. High-power amplifiers use class C, D, and E modes of operation that generate harmonics. Name two ways in which these harmonics are minimized.

6-8. The impedance-matching circuits given in Figs. 6-19 and 6-20 are used to match the final amplifier impedance of the transistor to an output load, usually an antenna. Show how you can use one of these L networks to match a transmitter with a 50-Ω output impedance to an antenna with an equivalent resistance of 4 Ω in series with 20 Ω of capacitive reactance. Which circuit do you use? Calculate L and C values at 13.56 MHz.

6-9. Show how you can create a variable inductance by using a fixed inductor in series with a variable capacitor. Demonstrate this with a variable capacitor that tunes over the 20- to 200-pF range. Assume a frequency of 12 MHz. What value of inductance will give a minimum inductance value of 8 μH for the combination?

6-10. Could a digital logic gate be used to make a simple transmitter? Explain.

6-11. A power amplifier receives an input driving power of 3 W. The amplifier power gain is 26 dB. What is the output power?

Answers to Tests

1. oscillator
2. continuous-wave, CW
3. crystal oscillator
4. buffer
5. drivers
6. final
7. speech-processing
8. frequency multipliers
9. mixer
10. linear
11. class C
12. AM, SSB
13. class C
14. A, B, AB
15. 27
16. 360
17. base-collector capacitance
18. parasitic oscillation
19. neutralization
20. true
21. bridge
22. true
23. push-pull
24. 90, 150
25. pulses
26. tuned or resonant circuit
27. 60, 85
28. harmonics
29. frequency multiplier
30. 180 [2(3)(4)(5) = 120; 120(1.5) = 180 MHz]
31. 50.4 [28(1.8) = 50.4 W]
32. square wave
33. D or E
34. power combiner
35. 27
36. 0.378
37. equals
38. L network
39. pi (π), T
40. low-pass
41. Q
42. 433
43. $L = 108$ nH, $C = 558$ pF
44. $L = 106$ nH, $C_1 = 13.5$ pF, $C_2 = 34$ pF

45. 450
46. 2:1
47. autotransformer
48. toroid
49. balun
50. broadband
51. clipping, compression
52. prevent overmodulation and distortion, eliminate out-of-band operation
53. clipper
54. smooth out distortion caused by clipping, restrict signal bandwidth
55. splatter
56. gain
57. power
58. rectifying, filtering
59. collector current
60. automatic gain control
61. more
62. true
63. analog-to-digital converter, DSP chip, digital-to-analog converter

Chapter 7

Communications Receivers

Chapter Objectives

This chapter will help you to:

1. *Define* the terms selectivity and sensitivity as they apply to receivers.
2. *Calculate* the *Q*, bandwidth, and shape factor of a tuned circuit.
3. *Explain* the basic concept and benefits of a superheterodyne receiver.
4. *Explain* the process of frequency conversion and *recognize* common mixer circuits.
5. *Calculate* the intermediate frequency (IF) and image frequencies of a superheterodyne receiver.
6. *Draw* block diagrams of single- and dual-conversion receivers and *explain* the function and operation of each major section.
7. *Define* noise and *state* the three basic types of noise.
8. *Calculate* the thermal noise, noise figure, and noise temperature in a receiver.
9. *Explain* the need for and operation of AGC circuits.
10. *Define* squelch and *explain* the operation of a basic squelch circuit.
11. *Draw* a block diagram of a phase-locked loop frequency synthesizer, *name* the major circuits and *explain* their operation, and *calculate* the output frequencies given the input frequencies and frequency divider ratio.
12. *Explain* the concept of digital and software radios.

In radio communications systems, the transmitted signal is very weak when it reaches the receiver, particularly if it has traveled any distance. Further, the signal has had to share the free-space transmission media with thousands of other radio signals. Various kinds of noise also get added to the signal. The job of the radio receiver, then, is to have the sensitivity and selectivity to fully reproduce the modulating signal at its output. The kind of radio receiver best suited to this task is known as the superheterodyne. The superheterodyne receiver was conceived in the early 1900s, and today virtually every communications receiver is of that design. The various circuits in the superheterodyne receiver will be discussed here in detail. Many special circuits including frequency synthesizers will also be covered. Finally, we will take a look at transceivers that combine both a transmitter and a receiver in a single package.

7-1   The Superheterodyne Receiver

The primary requirement for any communications receiver is that it have the ability to select the desired signal from among thousands of others present and to provide sufficient amplification to recover the modulating signal.

Selectivitity

These two requirements are generally referred to as selectivity and sensitivity.

Selectivity

Selectivity refers to the ability of a receiver to select a signal of a desired frequency while rejecting those on closely adjacent frequencies.

A receiver with good selectivity will isolate the desired signal in the RF spectrum and eliminate all other signals.

Selectivity in a receiver is obtained by using tuned circuits. These are *LC* circuits tuned to resonate at a desired signal frequency. The *Q* of these tuned circuits determines the selectivity. Recall that *Q* is the ratio of inductive reactance to resistance ($Q = X_L/R$). The *bandwidth* of a tuned circuit is a measure of its selectivity. The bandwidth is the difference between the upper f_2 and lower f_1 cutoff frequencies that are located at the 3-dB-down or 0.707 points on the selectivity curve as shown in Fig. 7-1. This bandwidth is determined by the resonant f_r and the *Q* according to the relationship BW = f_r/Q. The higher the *Q*, the narrower the bandwidth and the better the selectivity. High-*Q* tuned circuits are used to keep the bandwidth narrow to ensure that only the desired signal is passed.

Some examples will illustrate these concepts. Assume that a 10-μH coil with a resistance of 20 Ω is connected in parallel with a 101.4-pF variable capacitor. The circuit resonates at

$$f_r = \frac{1}{2\pi\sqrt{LC}}$$

$$= \frac{1}{6.28\sqrt{(10 \times 10^{-6})(101.4 \times 10^{-12})}}$$

$$= 5 \text{ MHz}$$

The response curve is as shown in Fig. 7-1. The peak occurs at 5 MHz. The *Q* of the coil and circuit is

$$Q = \frac{X_L}{R}$$

The inductive reactance is

$$X_L = 2\pi f L$$

$$= 6.28(5 \times 10^6)(10 \times 10^{-6})$$

$$= 314 \text{ }\Omega$$

The coil resistance was given as 20 Ω. Therefore, the *Q* is

$$Q = \frac{X_L}{R}$$

$$= \frac{314}{20}$$

$$= 15.7$$

Now we can compute the bandwidth.

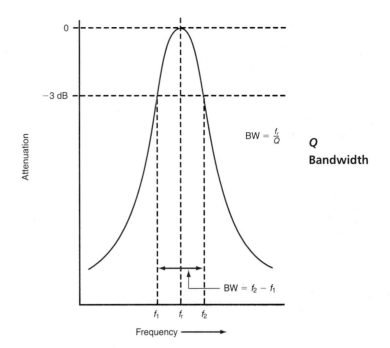

Fig. 7-1 Selectivity curve of a tuned circuit.

$$\text{BW} = \frac{f_r}{Q}$$

$$= \frac{5 \times 10^6}{15.7}$$

$$= 318{,}471.3 \text{ Hz } or \text{ 318 kHz}$$

This bandwidth is centered around 5 MHz, one-half of it appearing on each side of 5 MHz. The upper and lower cutoff frequencies are 5 MHz plus or minus one-half the bandwidth.

$$\frac{318 \text{ kHz}}{2} = 159 \text{ kHz } or \text{ 0.159 MHz}$$

The upper cutoff frequency is

$$f_2 = 5 + 0.159$$

$$= 5.159 \text{ MHz}$$

The lower cutoff frequency is

$$f_1 = 5 - 0.159$$

$$= 4.841 \text{ MHz}$$

The bandwidth can be verified by computing it with the cutoff frequencies.

$$\text{BW} = f_2 - f_1$$

$$= 5.159 - 4.841$$

$$= 0.318 \text{ MHz } or \text{ 318 kHz}$$

To improve the selectivity of the circuit, we need to narrow the bandwidth. Let's say we want a bandwidth of 40 kHz. With this information, we can calculate the needed circuit *Q*.

$$Q = \frac{f_r}{\text{BW}}$$

$$= \frac{5 \times 10^6}{40 \times 10^3}$$

$$= 125$$

Skirts

To increase the Q, we must lower the coil resistance. One way to do this is to use much larger gage wire. Alternately, a higher inductance could be used, but the capacitor value would also have to be changed. Assume we simply wind a new 10-μH coil with larger wire. If $Q = X_L/R$, then $R = X_L/Q$, and the resis-

Shape factor tance then must be

$$R = \frac{X_L}{Q}$$

$$= \frac{314}{125}$$

$$= 2.512 \ \Omega$$

Although very high Q circuits can be obtained, keep in mind that the bandwidth of the receiver must be such that it will pass not only the carrier but also the sideband frequencies that contain the transmitted information. It is possible to make a tuned circuit too selective where the higher-frequency sidebands are eliminated or greatly attenuated. This, of course, distorts the transmitted information. As you can see, the bandwidth is a compromise and must be adjusted to pass all signal components.

The ideal receiver selectivity curve would have perfectly vertical sides as shown in Fig. 7-2(a).

Such a curve cannot be obtained with tuned circuits or any other electronic circuit. The response curve of a tuned circuit, as Fig. 7-1 shows, has gradual attenuation on either side of the center frequency. The sides of the curve are referred to as *skirts*. The objective is to obtain steep skirts and thus better selectivity. Improved selectivity is achieved by cascading tuned circuits or by using crystal or mechanical filters. Both methods are widely used in communications receivers.

A measure of the steepness of the skirts, or the skirt selectivity of a receiver, is the shape factor. The *shape factor* is the ratio of the 60-dB-down bandwidth to the 6-dB-down bandwidth of a tuned circuit or filter. This is illustrated in Fig. 7-2(b). The bandwidth at the 60-dB-down points is $f_4 - f_3$. The bandwidth of the 6-dB-down points is $f_2 - f_1$. The shape factor is their ratio:

$$\text{Shape factor} = \frac{f_4 - f_3}{f_2 - f_1}$$

Assume the 60-dB bandwidth is 12 kHz and the 6-dB bandwidth is 3 kHz. Then,

$$\text{Shape factor} = \frac{12}{3}$$

$$= 4 \ or \ 4{:}1$$

The lower the shape factor, the better the skirt selectivity. The ideal, of course, is 1, which cannot be obtained in practice. Only the ideal curve in Fig. 7-2(a) has a shape factor of 1.

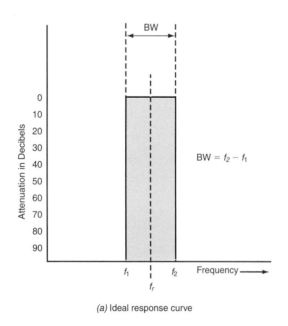

(a) Ideal response curve

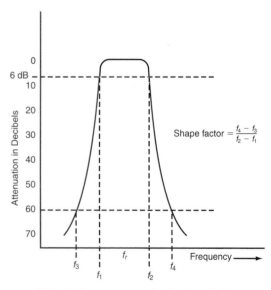

(b) Practical response curve showing shape factor

Fig. 7-2 Receiver selectivity.

Sensitivity

The *sensitivity* of a communications receiver refers to the receiver's ability to pick up weak signals. Sensitivity is primarily a function of the overall receiver gain. *Gain,* of course, is the factor by which an input signal is multiplied to produce the output signal. In general, the higher the gain of a receiver, the better its sensitivity. The more gain that a receiver has, the smaller the input signal necessary to produce a desired level of output. High gain in communications receivers is obtained by using multiple stages of amplification.

The sensitivity of a communications receiver is usually expressed as the minimum amount of signal voltage input that will produce an output signal that is 10 dB higher than the receiver background noise. Some specifications state a 20-dB signal-to-noise ratio. A typical sensitivity figure might be 1-μV input. The lower this figure, the better the sensitivity. Good communications receivers typically have a sensitivity of 0.2 to 1 μV. Consumer AM and FM receivers designed for receiving strong local stations have much lower sensitivity. Sensitivities of 5 to 10 μV are typical for FM receivers, whereas the sensitivity of an AM receiver could be 100 μV or more.

As we discuss the various receiver configurations and circuits, keep in mind that selectivity and sensitivity are the primary requirements of any receiver and all designs attempt to enhance these two capabilities.

Crystal Radio

The simplest radio receiver is the crystal set consisting of a tuned circuit, a diode (crystal) detector, and earphones. Refer to Fig. 7-3(*a*).

The antenna picks up the signal and causes current to flow in the primary winding of coupling transformer T_1. This induces a voltage into the secondary winding that is the inductance in a series resonant circuit. When signal current flows in the primary, a varying magnetic field cuts the turns of the secondary winding, inducing a voltage in them. Each turn of the secondary, therefore, acts like a tiny voltage generator. The total effect is the same as having a signal generator in series with the secondary. See Fig. 7-3(*b*). The generator, secondary winding, and C_1 together form a series resonant circuit. Current flows in the secondary winding causing a voltage to be developed across capacitor C_1. When the circuit is tuned to resonance, a high voltage is developed across the capacitor. This is known as the *resonant rise* or *resonant step-up voltage*. This voltage is significantly higher than the actual voltage induced into the secondary winding.

It is the tuned circuit that provides both selectivity and sensitivity. The tuned circuit, if it has a high Q, can select a desired frequency when it is tuned to resonance. The resonant step-up voltage across the capacitor provides some gain. It is the high Q of the tuned circuit that will permit good selectivity and, in this case, also good sensitivity.

The diode or crystal rectifies the signal, and capacitor C_2 filters out the carrier, leaving only the original information, which is heard in the earphones.

The simple crystal receiver in Fig. 7-3 does not provide the kind of selectivity and sensitivity necessary for modern communications. Only the strongest of signals will produce an output, and selectivity is often insufficient to separate incoming signals. The use of

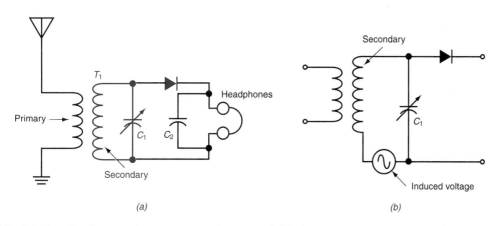

Fig. 7-3 (*a*) The simplest receiver—a "crystal" set, and (*b*) the series resonant circuit formed by the secondary.

earphones is also inconvenient. However, this simple receiver does make one very important point. The minimum communications receiver is a demodulator. The demodulator strips the RF carrier from the incoming signal, leaving only the original modulating information. Most other circuits used in a receiver are designed to improve the sensitivity and selectivity so that the demodulator may perform better.

The TRF Radio

TRF receiver

The earliest communications receivers were, of course, improvements upon the basic crystal receiver. The obvious way to improve sensitivity is to add stages of amplification, both before and after the demodulator. The result is a *tuned radio frequency* (*TRF*) *receiver* shown in Fig. 7-4. Here three stages of RF amplification are used between the antenna and the detector. This is followed by two stages of audio amplification. The RF amplifier stages increase the amplitude of the received signal tremendously before it is applied to the detector. The recovered signal is amplified further by the audio amplifiers which provide enough gain to operate a loudspeaker.

Another important factor is that the RF amplifiers use tuned circuits. Whenever resonant *LC* circuits tuned to the same frequency are cascaded, the overall selectivity is improved. The greater the number of tuned stages cascaded, the narrower the bandwidth and the steeper the skirts. This is illustrated in Fig. 7-5. A signal above or below the resonant frequency will be attenuated by one tuned circuit. It will be further attenuated by a second tuned circuit and even more by a third. The effect is to steepen the skirts.

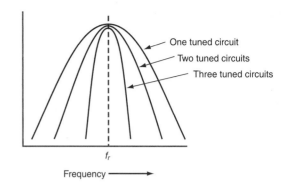

Fig. 7-5 How the cascading of tuned circuits improves selectivity.

The TRF receivers were widely used for many years, but they suffered from a number of annoying problems. The main problem was tracking of the tuned circuits. In a receiver, the tuned circuits must be made variable so that they can be set to the frequency of the desired signal. In most receivers, the capacitors in the tuned circuits are made variable. By rotating a knob, the capacitance is varied, thereby changing the resonant frequency. In early receivers, each tuned circuit had a separate capacitor and, therefore, multiple dials had to be adjusted to tune in a signal. Needless to say, this was a difficult and trying process.

The solution to the problem was to "gang" all the tuning capacitors so that they all would be changed simultaneously when the tuning knob was rotated. See Fig. 7-4. Although this essentially solved the problem, tracking errors still occurred. Differences in capacitors caused the resonant frequency of each tuned circuit to be slightly different, thereby increasing the bandwidth. This problem was offset by connecting a small "trimmer" capacitor in parallel

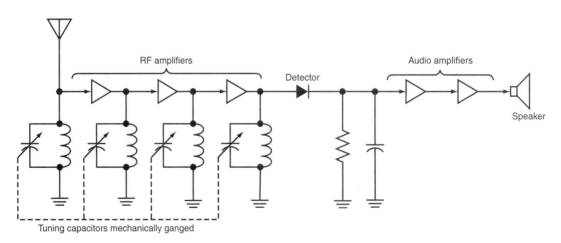

Fig. 7-4 Tuned radio-frequency (TRF) receiver.

with each of the tuning capacitors. The trimmers could then be adjusted to ensure that these minor differences were compensated for.

Another problem with TRF receivers was that the selectivity varied with frequency. In general, the bandwidth of a tuned circuit will increase with its resonant frequency. Selectivity was good at the low frequencies but less at the higher frequencies. These basic problems ultimately led to the development of a receiver that eliminated them. This receiver configuration is known as the *superheterodyne* or *superhet* (both terms are used interchangeably throughout this book).

Superheterodyne Receivers

The basic process of a superheterodyne receiver is to convert all incoming signals to a lower frequency known as the *intermediate frequency* (*IF*) where fixed tuned amplifiers can be used to provide a fixed level of sensitivity and selectivity. Most of the gain and selectivity in a superheterodyne receiver is obtained in the *IF amplifiers.*

Figure 7-6 shows a general block diagram of a superheterodyne receiver. The key circuit, of course, is the *mixer,* which produces the frequency translation of the incoming signal down to the IF. The incoming signal is mixed with an LO signal to produce this conversion. Now let's take a look at each of the circuits in the receiver to understand their basic function.

The antenna picks up the weak radio signal and feeds it to the *RF amplifier.* The purpose of the RF amplifier is to provide some initial gain and selectivity. Because the RF amplifier provides some initial selectivity, it is sometimes referred to as a *preselector.* Tuned circuits help select the desired signal or at least the frequency range in which the signal resides. In fixed-tuned receivers, the tuned circuits can be given a very high Q so that excellent selectivity is obtained. However, in receivers that must tune over a broad frequency range, selectivity is somewhat more difficult to obtain. The tuned circuits must resonate over a wide frequency range. Therefore, the Q and the selectivity of the amplifier change with frequency.

Some communications receivers do not use an RF amplifier. Instead, the antenna is connected directly to a tuned circuit at the input to the mixer. The input tuned circuit will provide the desired initial selectivity.

The reason for omitting the RF amplifier is that in many designs, extra gain is simply not needed, particularly at low frequencies. Most of the receiver gain is in the IF amplifier section, and if relatively strong signals are to be received, additional RF gain is not necessary.

In general, it is better to have an RF amplifier. It improves sensitivity because of the extra gain and improves selectivity because of the added tuned circuits. The signal-to-noise ratio is also improved. Further, image (covered in Sec. 7-3) and spurious signals are better rejected, thereby minimizing unwanted heterodyning effects in the mixer.

One major reason for including the RF amplifier is to minimize radiation of the LO signal through the receiving antenna. The LO signal is relatively strong, and some of it can leak through and appear at the input to the mixer. If the mixer input is connected directly to the antenna, some of the LO signal will be radiated, and it could cause interference to other nearby

Superheterodyne receiver

Intermediate frequency (IF)

IF amplifier

Mixer

Local oscillator

RF amplifier

Preselector

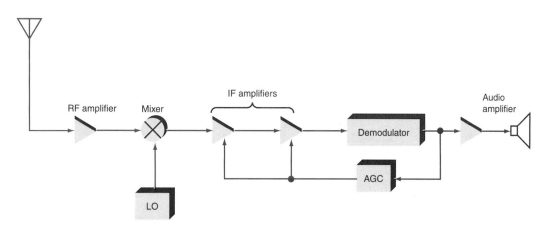

Fig. 7-6 Block diagram of a superheterodyne receiver.

receivers. The RF amplifier between the mixer and the antenna isolates the two and thereby effectively eliminates any LO radiation.

Although both bipolar and field-effect transistors can be used as RF amplifiers, in most modern receivers MOSFETs are used. The reason for this is that they contribute less noise than bipolar transistors.

The output of the RF amplifier is applied to the input of the mixer. The mixer also receives an input from an LO. The output of the mixer will be the input signal, the LO signal, as well as their sum and difference frequencies. Usually a tuned circuit at the output of the mixer selects the difference frequency. This frequency is known as the IF. The mixer may be a diode, a balanced modulator, or a transistor. A MOSFET is the preferred mixer because of its low noise characteristics.

The LO is made tunable so that its frequency can be adjusted over a relatively wide range. By tuning the LO frequency, the mixer will translate a wide range of input frequencies to the IF.

In most receivers, the mixer and LO are separate circuits. But in some simple low-frequency receivers, the mixer and LO functions are combined within a single stage called a *converter*. Here, a bipolar transistor is normally used as the mixer.

The output of the mixer is a signal at the IF containing the same modulation that appeared on the input RF signal. This signal is amplified by one or more IF amplifier stages, and most of the receiver gain is obtained in these stages. Tuned circuits provide fixed selectivity. Since the IF is usually much lower than the input signal frequency, the IF amplifiers are easier to design and good selectivity is easier to obtain with tuned circuits. Crystal, mechanical, and ceramic filters are used in some IF sections to obtain better selectivity.

The highly amplified IF signal is finally applied to the demodulator or *detector*. This circuit recovers the original modulating information. The output of the detector is then usually fed to an audio amplifier with sufficient voltage and power gain to operate a speaker.

Another important circuit in the superheterodyne receiver is the *AGC circuit*. The output of the detector is usually the original modulating signal whose amplitude is directly proportional to the amplitude of the received signal. The recovered signal, which is usually ac, is rectified and filtered into a dc voltage by the AGC circuit. This dc voltage is fed back to the IF amplifiers, and sometimes to the RF amplifier, to control the gain of the receiver. The whole purpose of the AGC circuit is to help maintain a constant output voltage level over a wide range of RF input signal levels.

The amplitude of the RF signal at the antenna of a receiver can vary from a fraction of a microvolt to thousands of microvolts. This represents a very wide signal range, known as the *dynamic range*. Typically, a receiver is designed with very high gain so that weak signals may be reliably received. However, when a very high amplitude signal is applied to a receiver, the circuits will be overloaded, causing distortion and reducing intelligibility.

By using AGC, the overall gain of the receiver can be automatically adjusted depending upon the input signal level. The signal amplitude at the output of the detector will be proportional to the amplitude of the input signal. If it is very high, the AGC circuit will produce a high dc output voltage. This will cause the gain of the IF amplifiers to be reduced. With reduced gain, the distortion normally produced by a large input signal will be eliminated.

When the incoming signal is weak, the detector output will be low. The output of the AGC will then be only a very small dc voltage. This will cause the gain of the IF amplifiers to remain high thereby providing maximum amplification.

Virtually all superheterodyne receivers use some form of AGC. In low-frequency broadcast receivers, it is sometimes referred to as *automatic volume control* (*AVC*). In all cases, the gain of the IF amplifiers is controlled.

■ TEST

Answer the following questions.

1. The ability of a receiver to choose a desired signal frequency while rejecting closely adjacent signal frequencies is known as _____.
2. Decreasing the Q of a resonant circuit causes its bandwidth to _____.
3. Good selectivity usually means _____ (wide, narrow) bandwidth.
4. True or false. A tuned circuit can provide voltage gain.

Dynamic range

Converter

AVC

Detector

AGC circuit

5. Cascading tuned circuits cause the selectivity to _____ (increase, decrease).

6. If the selectivity of a tuned circuit is too sharp, the _____ of the received signal may be attenuated.

7. A tuned circuit has a Q of 100 at its resonant frequency of 500 kHz. Its bandwidth is _____ kHz.

8. A parallel LC tuned circuit has a coil of 3 μH and a capacitance of 75 pF. The coil resistance is 10 Ω. The circuit bandwidth is _____ MHz.

9. A tuned circuit has a resonant frequency of 10 MHz and a bandwidth of 100 kHz. The upper and lower cutoff frequencies are f_1 (upper) = _____ MHz, f_2 (lower) = _____ MHz.

10. To achieve a bandwidth of 3 kHz at 4 MHz, a Q of _____ is required.

11. To narrow the bandwidth of a tuned circuit, the coil resistance must be _____ (increased, decreased).

12. A filter has a 6-dB bandwidth of 500 Hz and a 60-dB bandwidth of 1200 Hz. The shape factor is _____.

13. The greater the gain of the receiver, the better its _____.

14. A receiver that uses only tuned amplifiers and a detector is known as a(n) _____ receiver.

15. A receiver that uses a mixer to convert the received signal to a lower frequency is called a(n) _____.

16. The acronym IF means _____.

17. Tuning a superheterodyne is done by varying the frequency of its _____.

18. Most of the gain and selectivity in a superhet is obtained in the _____.

19. The _____ circuit in a receiver compensates for a wide range of input signal levels.

20. The mixer output is usually the difference between the _____ frequency and the _____ frequency.

21. The AGC voltage controls the gain of the _____.

7-2 Frequency Conversion

The key concept in superheterodyne receivers is frequency conversion. *Frequency conversion* is the process of translating a modulated signal to a higher or lower frequency while retaining all the originally transmitted information. Frequency conversion is often performed before or after transmission or reception to provide some benefit. There are many occasions when it is desirable to convert a signal to a higher or lower frequency for more convenient processing. In radio receivers, high-frequency radio signals are regularly translated to a lower frequency where gain and selectivity can be readily obtained. In satellite communications, the original signal is more conveniently generated at a lower frequency but is then translated (converted) to a higher frequency for transmission. Any modulation on the original signal is retained.

Frequency conversion is a form of AM. It is carried out by a mixer circuit. In some applications, the mixer is referred to as a converter. The function performed by the mixer is called *heterodyning*.

Figure 7-7 shows the simplified symbol used to represent a mixer. The mixer, like a modulator, accepts two inputs. The signal f_s, which is to be translated to another frequency, is applied to one input, whereas the sine wave from an oscillator f_o is applied to the other input. The signal to be translated may be a simple sine wave or any complex modulated signal containing sidebands.

The mixer, like an amplitude modulator, essentially performs a mathematical multiplication of its two input signals. The oscillator is the carrier, and the signal to be translated is the modulating signal. The output will contain not only the carrier signal but also the sidebands formed when the local oscillator and input signal are mixed. The output of the mixer, therefore, consists of these signals:

$$f_s$$
$$f_o$$
$$f_o + f_s$$
$$f_o - f_s \quad \text{or} \quad f_s - f_o$$

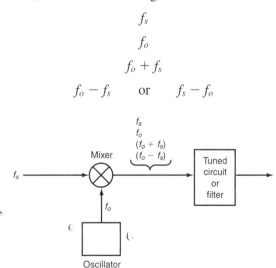

Fig. 7-7 A mixer used for frequency translation.

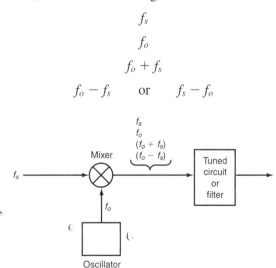

For information on wireless local loop (WWL) and other receivers, see the following Internet sites:

⟨www.galtronics.com⟩
⟨www.optaphone.com⟩
⟨www.xertex.com⟩

Heterodyning

Frequency conversion

Only one of these signals is the desired one. For example, to translate the input signal to a lower frequency, the lower sideband or difference signal $f_o - f_s$ will be chosen. The local oscillator frequency will be chosen such that when the information signal is subtracted from it, a signal with the desired lower frequency is obtained. When translating upward in frequency, the upper sideband or sum signal $f_o + f_s$ is chosen. Again the local oscillator frequency will determine what the new, higher frequency will be. A tuned circuit or filter is normally used at the output of the mixer to select the desired signal.

For example, an FM signal at 107.1 MHz is translated by an FM radio receiver to an intermediate frequency of 10.7 MHz for amplification and detection. A local oscillator frequency of 96.4 MHz is used. The mixed output signals are

$$f_s = 107.1 \text{ MHz}$$

$$f_o = 96.4 \text{ MHz}$$

$$f_o + f_s = 96.4 + 107.1$$

$$= 203.5 \text{ MHz}$$

$$f_s - f_o = 107.1 - 96.4$$

$$= 10.7 \text{ MHz}$$

A tuned circuit selects the 10.7-MHz signal and rejects the others.

Keep in mind that although the original signal is translated in frequency, any modulation regardless of type will continue to appear on this signal. In the above example, the 10.7-MHz output signal contains the original frequency modulation. The result is as if the carrier frequency of the input signal were changed as well as all the sideband frequencies. Such frequency conversion makes it possible to shift a signal from one part of the spectrum to another as required by the application.

■ **TEST**───────────────────

Answer the following questions.

22. The process of translating a signal to a higher or lower frequency for more convenient processing is called _____.
23. The circuit used for translating the frequency of a signal is referred to as a(n) _____ or _____.
24. The input signals to a frequency translation circuit are $f_s = 3.7$ MHz and $f_o = 4.155$ MHz. The output signals are _____.

25. In the example given in the text, what other local oscillator frequency could be used to produce the 10.7-MHz output with a 107.1-MHz input?
26. Unwanted mixer output signals are eliminated by a _____.
27. True or false. The local oscillator is modulated.
28. True or false. Any modulation on the input signal to be translated is retained.
29. The operation carried out by a mixer is known as _____.

7-3 Intermediate Frequency Selection and Images

The choice of the IF is usually a design compromise. The main objective is to obtain good selectivity. Narrowband selectivity is best obtained at lower frequencies, particularly when conventional *LC* tuned circuits are used. A typical IF is 455 kHz. This has become a standard in AM broadcast receivers and in many other radio receivers. Intermediate frequencies as low as 50 kHz have been used successfully to obtain extremely narrow bandwidths.

The benefits of using a low IF are that the design is not only simpler but also less troublesome. At low frequencies, the circuits will be far more stable with high gain. At higher frequencies, the design is complicated by stray inductances and capacitances as well as the need for shielding. This makes the layouts of the circuits critical if undesired feedback paths are to be avoided. With very high circuit gain, some of the signal can be fed back in phase and cause oscillation. Oscillation is not as much of a problem at lower frequencies.

Images

When low IFs are selected, another problem occurs, particularly if the signal to be received is very high in frequency. This is the problem of images. An *image* is an RF signal that is spaced from the desired incoming signal by a frequency that is 2 times the IF above or below the incoming frequency. Expressed mathematically, this is

$$f_i = \begin{cases} f_s + 2f_{IF} \\ f_s - 2f_{IF} \end{cases}$$

where f_i = image frequency
$\quad f_s$ = desired signal frequency
$\quad f_{IF}$ = intermediate frequency

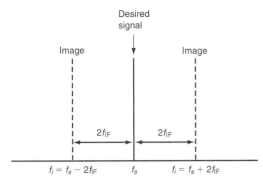

Fig. 7-8 Relationship of the signal and image frequencies.

This is illustrated graphically in Fig. 7-8. Only one of these images will occur, depending upon whether the LO frequency f_o is above or below the signal frequency.

Recall that the mixer in a superhet produces the sum and difference frequencies of the incoming and LO signals. Normally, it is the difference frequency that is selected as the IF. The frequency of the LO is usually chosen to be higher than the incoming signal by an amount equal to the IF. However, the LO frequency could also be made lower than the incoming signal frequency by the same amount. Either choice will work and will produce the desired difference frequency. Let's assume that our LO frequency is higher than the incoming signal frequency.

Now, what happens if an image signal ($f_i = f_s + 2f_{IF}$) appears at the input of the mixer? The mixer will, of course, produce the sum and difference frequencies regardless of the inputs. Therefore, the mixer output will again be the difference frequency at the IF value. Let's look at an example.

Assume a desired signal frequency f_s of 90 MHz and an LO frequency f_o of 100 MHz. The IF is the difference $100 - 90 = 10$ MHz. The image frequency is

$$f_i = f_s + 2f_{IF}$$
$$= 90 + 2(10)$$
$$= 90 + 20$$
$$= 110 \text{ MHz}$$

If the image appears at the mixer input, the output will be the difference $110 - 100 = 10$ MHz. The IF amplifier would pass it. Now look at Fig. 7-9. This shows that the mixer will generate the difference between the LO and

the received signals, if it is the desired one or an image, since each will produce the same IF.

What this means is that a signal spaced from the desired signal by two times the IF will also be picked up by the receiver and converted to the IF. When this occurs, the image signal will interfere with the desired signal.

What are the chances that there will be a signal on the image frequency? In the modern crowded RF spectrum, the chances are quite large. The interference from the image could be so great as to make the desired signal unintelligible.

Images are a real problem in superheterodyne receivers. The design of a superhet, therefore, must take this problem into consideration. And, as you will see, the choice of the IF is critical to minimizing the problem.

The only reason why image interference occurs is that the image signal is allowed to appear at the mixer input. This is the reason for using high-Q tuned circuits ahead of the mixer or a selective RF amplifier. If the selectivity of the RF amplifier and tuned circuits is good enough, then the image will be rejected. In a fixed-tuned receiver designed for a specific frequency, it is possible to optimize the receiver front end for the good selectivity necessary to eliminate images. But many receivers have broadband RF amplifiers that pass many frequencies within a

Image interference

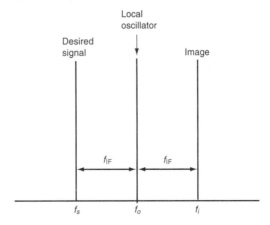

Fig. 7-9 Signal, LO, and image frequencies in a superheterodyne.

specific band. Other receivers must be made tunable over a wide frequency range. In such cases, the selectivity is a problem. An example will illustrate this.

Assume that the receiver is designed to pick up a signal at 20 MHz. The IF is 500 kHz, or 0.5 MHz. The LO is adjusted to a frequency above the incoming signal by an amount equal to the IF, or $20 + 0.5 = 20.5$ MHz. When the LO and signal frequencies are mixed, the difference, of course, is 0.5 MHz as desired.

The image frequency can be computed using the expression given earlier:

$$f_i = f_s + 2f_{IF}$$
$$= 20 + 2(0.5)$$
$$= 21 \text{ MHz}$$

The image frequency of 21 MHz will cause interference to the desired signal at 20 MHz unless it is rejected. The signal, LO, and image frequencies are shown in Fig. 7-10.

Now, assume that a tuned circuit ahead of the mixer has a Q of 10. Knowing this and the resonant frequency, we can calculate the bandwidth of the resonant circuit from the familiar expression:

$$BW = \frac{f_R}{Q}$$

where f_R is the resonant frequency of the tuned circuit.

$$BW = \frac{20}{10}$$
$$= 2 \text{ MHz}$$

The response curve for this tuned circuit is shown in Fig. 7-10.

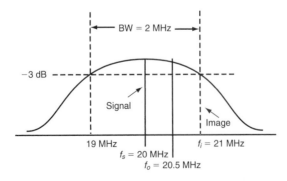

Fig. 7-10 The tuned-circuit bandwidth is such that it allows the image to get to the mixer.

As you can see, the bandwidth of the resonant circuit is relatively wide. The bandwidth is centered on the signal frequency of 20 MHz and will, therefore, extend from 19 to 21 MHz. Remember that the bandwidth is measured at the 3-dB-down points on the tuned circuit response curve. The upper cutoff frequency is 21 MHz, and the lower cutoff frequency is 19 MHz.

Note that the upper cutoff frequency is equal to the image frequency. What this means is that the image frequency would essentially be passed by the tuned circuit and would, therefore, cause interference.

Cascading tuned circuits and making them with higher Q would indeed help solve the problem. However, this usually not only complicates the design, but in some cases, compromises it. To overcome this problem, the usual solution is to choose a higher IF.

Let's assume that we choose an IF of 5 MHz. Now, the image frequency will be

$$f_i = 20 + 2(5)$$
$$= 30 \text{ MHz}$$

The image frequency is now 30 MHz. A signal at this frequency would interfere with our desired signal at 20 MHz if it was allowed to pass into the mixer, but as you can see, 30 MHz is well out of the bandpass of our tuned circuit. The selectivity, even though it is poor, is sufficient to adequately reject the image. Of course, by choosing the higher IF, we make the design more difficult as indicated earlier.

The selection of an IF is usually a compromise. The IF is made as high as possible to effectively eliminate the image problem, while at the same time it is made as low as possible to simplify the design. The IF in most receivers generally varies in proportion to the frequencies that must be covered. At low frequencies, low values of IF are used. A value of 455 kHz is common for AM broadcast band receivers and for others covering that general frequency range. At higher frequencies, up to about 30 MHz, 3385 kHz, and 9 MHz are common IF frequencies. In FM radios that receive frequencies in the range of 88 to 108 MHz, 10.7 MHz is a standard IF. In TV receivers, an IF in the 40- to 50-MHz range is common. Radar receivers typically use an IF in the 6-MHz range, and satellite communications gear uses 70-MHz and 140-MHz IFs.

Dual Conversion

Another way to get the desired selectivity while at the same time eliminating the image problem is to use what is known as a *dual-conversion superheterodyne receiver.* A block diagram of a dual-conversion receiver is shown in Fig. 7-11. Notice that it uses two mixers and two LOs. The result is that the receiver has two IFs. The first mixer converts the incoming signal to a relatively high IF for the purpose of eliminating the images. Then a second mixer converts that IF down to a much lower frequency where good selectivity is easier to obtain.

Following the frequencies given in Fig. 7-11, you can see how each is obtained. Each mixer produces the difference frequency. Note that the first LO is variable and provides the tuning for the receiver. The second LO is fixed in frequency. Since it must convert only one fixed IF to a lower IF, the LO need not be tunable. In most cases, its frequency is set by a quartz crystal. In some receivers, the first mixer is driven by the fixed-frequency LO, while tuning is done with the second LO.

In some special situations, triple-conversion receivers can be used to further minimize the image problem. A triple-conversion receiver uses three mixers and three different IF values. Such receivers are not common, but they have been used in unusual or critical applications. Dual-conversion receivers are relatively common. Most shortwave receivers and those at VHF, UHF, and microwave frequencies use dual conversion. For example, a CB receiver typically uses a first IF of 10.7 MHz and a second IF of 455 kHz.

In most dual-conversion receivers, the first IF is typically lower in frequency than the received signals. The second IF is even lower.

However, it is not uncommon for the first IF to be higher than the received signal. For example, some modern shortwave receivers (3 to 30 MHz) use 45 or 55.845 MHz for the first IF and 455 kHz for the second IF. This arrangement is used to eliminate the image problem usually associated with these receivers.

Dual-conversion receivers

■ TEST

Answer the following questions.

30. For best selectivity and stability, the IF should be
    a. High.
    b. Medium.
    c. Low.
31. An interfering signal that is spaced from the desired signal by twice the IF is called a(n) _____.
32. A superhet has an input signal of 15 MHz. The LO is tuned to 18.5 MHz. The IF is _____ MHz.
33. Images are caused by the lack of _____ before the mixer input.
34. A desired signal at 27 MHz is mixed with an LO frequency of 27.5 MHz. The image frequency is _____ MHz.
35. True or false. The LO may be above or below the signal frequency.
36. The main feature of a dual-conversion superhet is that it has two _____ circuits.
37. The image problem can be solved by proper choice of the _____.
38. A dual-conversion superheterodyne has an input frequency of 50 MHz and LO frequencies of 59 MHz and 9.6 MHz. The two IFs are _____ MHz and _____ kHz.

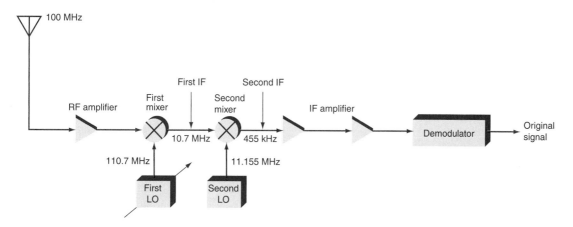

Fig. 7-11 A dual-conversion superheterodyne.

7-4 Noise

One of the most important considerations in any communications system is noise. Noise is random energy that interferes with the information signal. You can hear what noise sounds like by turning on any AM or FM receiver and tuning it to some position between stations. The hiss or static that you hear in the speaker is noise. Noise also shows up in a television picture as snow or colored snow known as confetti. If the noise level is too high or the signal is too weak, the noise may actually dominate and make reception unreliable. The noise may be so high as to completely obliterate the signal. Noise that occurs on transmissions of digital data makes itself known as bit errors. Thus the information being transmitted is lost or incorrect.

Noise is a problem in communications systems simply because the received signals are so low in amplitude. When the transmission is over short distances or when very high power transmitters are used, signal strength is good and noise is not a problem. But in most communications systems, weak signals are normal, so noise becomes a significant factor in the design of communications equipment. Noise is most problematic in the receiver because the receiver has the responsibility for amplifying the weak signal and recovering the information reliably.

Signal-to-noise (S/N) ratio

In any discussion of communications systems and receiver performance, the term signal-to-noise ratio is used. *Signal-to-noise (S/N) ratio* is simply a number that indicates the relative strengths of the signal and the noise. The stronger the signal and the weaker the noise, the higher the S/N ratio. If the signal is weak and the noise is strong, the S/N ratio will be low and reception will be less reliable. The design of communications equipment has as its objective to produce the highest S/N ratio possible.

Noise Sources

Noise comes from two basic sources. First, there is noise that is generated external to the receiver. Then there is internal noise that is produced within the receiver itself. Both types affect the S/N ratio.

External noise usually comes from industrial, atmospheric, or extraterrestrial or space sources. Industrial noise is produced by manufactured equipment, such as automotive ignition systems and electric motors and generators. Any electrical equipment that causes high voltages or currents to be abruptly switched produces "transients" that create noise. Fluorescent and other forms of gas-filled lights are another common noise source.

Regardless of its source, noise shows up as a random ac voltage. You can actually view it on an oscilloscope. The amplitude varies over a wide range as does the frequency. Industrial noise is mostly produced by voltage transients which in general are signals containing a tremendous amount of harmonic energy. You might say that noise in general contains all frequencies, varying randomly.

Another source of noise is electrical disturbances that occur naturally in the earth's atmosphere. You will often hear atmospheric noise referred to as static. Static usually comes from lightning. Lightning is the general term used to refer to electric discharges that occur between clouds or between the earth and clouds. Huge static charges build up, and when the potential difference is great enough, an arc occurs where electricity literally flows through the air. Lightning is very much like the static charges that you experience during a dry spell in the winter, only the voltages involved are enormous. This results in a transient electrical signal that generates harmonic energy that can travel extremely long distances.

Like industrial noise, atmospheric noise shows up primarily as amplitude variations that add to a signal and interfere with it. Atmospheric noise has its greatest impact on signals at frequencies less than 30 MHz.

Extraterrestrial noise comes from sources in space. For example, one of the primary sources of noise is the sun. The sun radiates a wide range of signals in a broad noise spectrum. The noise intensity produced by the sun varies with time. In fact, the sun has a repeating 11-year noise cycle. During the peak of the cycle, the sun produces an awesome amount of noise that causes tremendous radio signal interference, making many frequencies unusable for communications. During other years, the noise is at a minimum level.

Noise is also generated by stars, which are also suns. Noise from the stars is generally known as cosmic noise, and its level is not as great because of the great distances between the stars and earth. Nevertheless, it is an important source of noise that must be considered. It shows up primarily in the 10-MHz to 1.5-GHz range but causes the greatest disruptions in the 15- to 150-MHz range.

External noise is a fact of life and must be dealt with. Atmospheric and space noise simply cannot be eliminated. Industrial noise, because it is manufactured, can sometimes be eliminated at the source, if you have control over the source. In general, because there are so many sources of industrial noise, there is no way to control it. The key to reliable communications then is simply to generate the signal at a high enough power so that it overcomes this noise.

Thermal Noise

As if external noise wasn't enough, designers of communications systems must also contend with internal noise, that noise generated within the communications receiver. Electronic components such as resistors, diodes, and transistors are major sources of noise. Although this is a low-level noise, it is often great enough to interfere with weak signals. However, since the sources of internal noise are well known, the designer has some control over it.

Most internal noise is caused by a phenomenon know as *thermal agitation.* Thermal agitation refers to the random motion of the atoms and electrons in an electronic component caused by heat. Increasing the temperature causes this motion to increase. Since the components are conductors, the movement of electrons constitutes a current flow that causes a small voltage to be produced across that component.

You can actually observe this noise by simply connecting a resistor to a very high gain oscilloscope. The motion of the electrons (due to room temperature) in the resistor will cause a voltage to appear across it. The voltage variation is completely random and at a very low level. You will often hear thermal agitation referred to as *white noise* or *Johnson noise.*

The noise power developed across a resistor is directly proportional to the temperature. In a relatively large resistor at room temperature or higher, the noise voltage across it can be as high as several microvolts. This is the same order of magnitude as many weak RF signals. Some signals will have even less amplitude and thus will be totally masked by this noise.

Since noise is a very broad band signal containing a tremendous range of random frequencies, its level can be reduced by limiting the bandwidth. By feeding a noise signal into a selective tuned circuit, many of the noise frequencies will be rejected and the overall noise level will go down. The noise level is directly proportional to the bandwidth of any circuit to which it is applied. Although filtering does reduce the noise level, it does not eliminate it entirely.

The amount of noise voltage appearing across a resistor or the input impedance to a receiver can be calculated with the expression

$$v_n = \sqrt{4kTBR}$$

where v_n = rms noise voltage
k = Boltzmann's constant or
$\quad$ 1.38×10^{-23} joule/kelvin (J/K)
T = temperature, K
B = bandwidth, Hz
R = resistance, Ω

Let's look at an example. Assume that a receiver has a 50-Ω input resistance, a bandwidth of 6 MHz, and a temperature of 29°C or 29° + 273° = 302 K. (The relationship between degrees Celsius and degrees kelvin is discussed on page 158.) The input thermal noise voltage then is

$$v_n = \sqrt{4kTBR}$$
$$= \sqrt{4(1.38 \times 10^{-23})(302)(6 \times 10^6)(50)}$$
$$= 2.24 \ \mu V$$

This noise voltage would, of course, mask signals less than this value. The best way to reduce the noise voltage is to reduce the bandwidth to the minimum value acceptable to the application.

Electronic components such as transistors and tubes are also major contributors of noise. The most common type of noise is referred to as *shot noise,* which is produced by the random arrival of electrons or holes at the output element, at the plate in a tube, or at the collector or drain in a transistor. Shot noise is also produced by the random movement of electrons or holes across a PN junction. Even though current flow is established by external bias voltages, there will still be some random movement of electrons or holes due to discontinuities in the device. An example of such a discontinuity is the contact between the copper lead and the

Thermal agitation

Shot noise

semiconductor material. The interface between the two creates a discontinuity that causes random movement of the current carriers.

Another kind of noise that occurs in transistors is called transit time noise. *Transit time* is the duration of time that it takes for a current carrier such as a hole or electron to move from the input to the output. The devices themselves are very tiny, so the distances involved are minimal. Yet the time it takes for the current carriers to move even a short distance is finite. At low frequencies this time is negligible. But when the frequency of operation is high and the period of the signal being processed is the same order of magnitude as the transit time, then problems can occur. The transit time shows up as a kind of random noise within the device, and this noise is directly proportional to the frequency of operation.

Signal-to-Noise Ratio

Noise is usually expressed as a power because the received signal is also expressed in terms of power. If the signal and noise powers are known, the S/N ratio can be computed. Rather than express the S/N ratio as simply a number, you will usually see it expressed in terms of decibels.

A receiver has an input signal power of 1.2 μW. The noise power is 0.8 μW. The S/N ratio is

$$S/N = 10 \log \frac{1.2}{0.8}$$

$$= 10 \log 1.5$$

$$= 10(0.176)$$

$$= 1.76 \text{ dB}$$

A variety of methods are used to express the noise quality of the receiver. One of these methods is known as the noise figure, which is the ratio of the S/N power at the input to the S/N power at the output. The device under consideration can be the entire receiver or a single amplifier stage. The *noise figure F,* also called the *noise factor,* can be computed with the expression

$$F = \frac{\text{S/N input}}{\text{S/N output}}$$

You can express the noise figure as a number, but more often you will see it expressed in decibels.

An amplifier or receiver will always have more noise at the output than at the input. The reason for this is that the amplifier or receiver generates internal noise which will be added to the signal. And even though the signal may be amplified along the way, that noise will be amplified along with it. The S/N ratio at the output will be less than the S/N ratio of the input; therefore, the noise figure will always be greater than 1. A receiver that contributes zero noise to the signal would have a noise figure of 1, or 0 dB. Such a noise figure is not attainable in practice. A transistor amplifier in a communications receiver will usually have a noise figure of several decibels. The lower the noise figure, the better the amplifier.

Another method of expressing the noise in an amplifier or receiver is to use *noise temperature.* Most of the noise produced in a device is thermal noise that is directly proportional to temperature. Therefore, temperature is an appropriate unit of measurement.

The noise temperature is expressed in degrees kelvin. Normally we express temperature in terms of degrees Fahrenheit or Celsius. The kelvin temperature scale is related to the Celsius scale by the relationship

$$T_K = T_C + 273°$$

where T_K is the temperature in degrees kelvin and T_C is the temperature in degrees Celsius. The *kelvin temperature scale* is referred to as the absolute temperature scale. The letter K is used to designate that the temperature is expressed in terms of kelvin. If the temperature is in Celsius, all you have to do is add 273° to it to get the temperature in kelvin.

Remember too that the temperatures in Fahrenheit and Celsius are related by the expression

$$T_C = \frac{5}{9}(T_F - 32°)$$

To compute the noise temperature, the following expression is used.

$$T_N = 290(F - 1)$$

Here *F* is the noise factor discussed earlier. This expression shows that the noise temperature T_N is related to the noise figure *F.* In the expression above, the ratio rather than the decibel value is used for *F.* For example, if the noise factor is 1.5, the equivalent noise temperature is

Transit time

Noise temperature

Kelvin scale

Noise figure *F*

$$T_N = 290(1.5 - 1)$$
$$= 290(0.5)$$
$$= 145 \text{ K}$$

By examining the expression above, you can see that if the amplifier or receiver contributes no noise, then the noise factor F will be 1 as indicated before. Plugging this value into the expression gives an equivalent noise temperature of 0 K. If the noise figure is greater than 1, then an equivalent noise temperature will be produced.

Noise temperature is used only in circuits or equipment that operate at VHF, UHF, or microwave frequencies. The noise figure or factor is used at lower frequencies. A good low-noise transistor or amplifier stage will typically have a noise temperature of less than 100 K, the lower the better. Often you will see the noise temperature of a transistor given in the data sheet. Noise figures of less than about 3 dB are excellent.

Another method of comparing the quality of communications receivers is called *SINAD*, which is an acronym for *s*ignal plus *n*oise *a*nd *d*istortion. All receivers will contribute both noise and distortion to an incoming signal. By determining the ratio of the composite signal plus noise plus distortion to the noise plus distortion, a relative figure of merit is obtained.

$$\text{SINAD} = \frac{\text{signal} + \text{noise} + \text{distortion}}{\text{noise} + \text{distortion}}$$

Amplifiers and other circuits in a receiver are not perfectly linear and will, therefore, introduce some distortion. The distortion takes the form of very low levels of signal harmonics. The harmonics add to the signal along with the noise. The SINAD ratio makes no attempt to discriminate between or separate the noise and distortion signals.

To obtain the SINAD figure, an RF signal modulated by an audio signal usually of 400 Hz or 1 kHz is applied to the input of an amplifier or receiver. The composite output is then measured. This gives the numerator value.

Next, a highly selective notch (band reject) filter is used to eliminate the modulating audio signal from the output. The result is only the noise and the distortion. This gives the denominator value. The SINAD ratio can then be calculated. The SINAD ratio is also used to express the sensitivity of a receiver.

Although noise is an important consideration at all communications frequencies, it is particularly critical in the microwave region. The reason for this is that the noise increases with bandwidth and will impact high-frequency signals more than lower-frequency signals. The limiting factor in most microwave communications systems, such as satellites and radar, is internal noise.

The noise level in a system is proportional to the temperature and the bandwidth. It is also proportional to the amount of current flowing in a component, the gain of the circuit, and the resistance of the circuit. Increasing any of these factors will increase the noise. Therefore, low noise is best obtained by using low-gain circuits, low dc, low resistance values, and narrow bandwidths. Keeping the temperature low can also help. In fact, in some special microwave receivers, the noise level is reduced by cooling the input stages to the receiver.

Noise, of course, has its greatest effect at the input to a receiver simply because that is the point at which the signal level is lowest. The noise performance of a receiver is invariably determined in the very first stage of the receiver, usually an RF amplifier or mixer. Care should be taken in the design of these to ensure the use of very low noise components and to take into consideration the current, resistance, and gain figures in this circuit. Beyond the first stage, noise is basically no longer a problem.

SINAD

■ TEST

Answer the following questions.

39. List three sources of external noise.
40. List three main types of internal noise.
41. Noise from the sun and stars is called _____ noise.
42. Atmospheric noise comes primarily from _____.
43. List four sources of industrial noise.
44. The main source of internal noise is _____.
45. The S/N ratio is usually expressed in _____.
46. For best reception, the S/N ratio should be _____ (low, high).
47. Increasing the temperature of a component causes its noise power to _____.
48. Thermal noise is sometimes called _____ or _____ noise.

49. Narrowing the bandwidth of a circuit causes the noise level to _____.

50. The noise voltage produced across a 75-Ω input resistance at a temperature of 25°C with a bandwidth of 1.5 MHz is _____ μV.

51. Two types of noise caused by tubes or transistors are _____ noise and _____ noise.

52. True or false. The noise at the output of a receiver will be less than the noise at the input.

53. True or false. The receiver amplifies noise as well as the signal.

54. The ratio of the S/N power at the input to the S/N power at the output is called the _____.

55. Noise temperature in degrees kelvin is used to express the noise in a system at _____ (low, microwave) frequencies.

56. The noise figure of an amplifier is 2.6. The noise temperature is _____ K.

57. The SINAD method considers the signal, noise, and _____ levels in a receiver.

58. Noise is more of a problem at _____ frequencies.

59. The stages of a receiver that contribute the most noise are the _____ and the _____.

60. True or false. An amplifier with a noise temperature of 170 K is better than one with a rating of 235 K.

FET RF amplifiers

7-5 Typical Receiver Circuits

Now let's take a more detailed look at some of the circuits used in communications receivers. We have already discussed mixers and demodulators in previous chapters, so we will not discuss them here. Instead, we will focus on RF and IF amplifiers, mixers, AGC, and AFC circuits.

RF Amplifiers

RF amplifiers

The most critical part of a communications receiver is the front end. The front end usually consists of the *RF amplifier,* mixer, and related tuned circuits. This is the part of the receiver that processes the very weak input signal. It is essential that low-noise components be used to ensure a sufficiently high S/N ratio. Further, the selectivity should be such that it effectively eliminates images.

In many communications receivers, an RF amplifier is not used. This is particularly true in receivers designed for frequencies lower than about 30 MHz. The extra gain is not necessary, and its only contribution would be more noise. Therefore, the RF amplifier is usually eliminated, and the antenna is connected directly to the mixer input through one or more tuned circuits. The tuned circuits must provide the input selectivity necessary for image rejection. In a receiver of this kind, the mixer must also be of the low-noise variety. Today, most mixers are MOSFETs, which provide the lowest noise contribution.

Receivers used at frequencies above approximately 100 MHz, however, do typically use RF amplifiers. And RF amplifiers are found in some lower-frequency communications systems as well. The primary purpose of this amplifier is to increase the weak signal amplitude prior to mixing. The RF amplifier also provides some selectivity for image rejection.

In most of these receivers, a single RF stage is used, usually providing a voltage gain in the 10- to 30-dB range. This is easily obtainable with a single transistor. Bipolar transistors are used at the lower frequencies, and FETs are preferred at VHF, UHF, and microwave frequencies. Typically, FETs have a lower noise figure than bipolar transistors and, therefore, give better performance.

The RF amplifier is usually a simple class A circuit. A typical bipolar circuit is shown in Fig. 7-12(*a*), and a typical *FET circuit* is shown in Fig. 7-12(*b*). Note that the bipolar circuit does not have a tuned input. The antenna is connected directly to the base of the transistor. This makes the circuit very broad band as the transistor will amplify virtually any signal picked up by the antenna. However, note that the collector is tuned with a parallel resonant circuit which provides the initial selectivity for the mixer input. An untuned broadband transformer could also be used.

The FET circuits are particularly effective because their high input impedance minimizes loading on tuned circuits, thereby permitting the Q of the circuit to be higher and selectivity to be sharper. Most FETs also have a lower noise figure.

At microwave frequencies (those above 1 GHz), a newer type of FET is preferred. Known as a metal-semiconductor FET or MESFET,

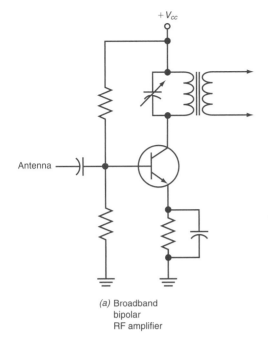

(a) Broadband bipolar RF amplifier

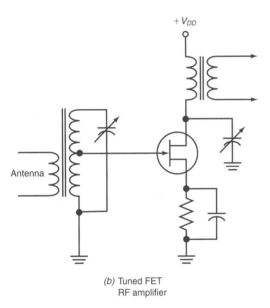

(b) Tuned FET RF amplifier

Fig. 7-12 The RF amplifiers used in receiver front ends.

the device is a junction FET (JFET) made with gallium arsenide (GaAs). It is also referred to as a GASFET. A cross section of a typical MESFET is shown in Fig. 7-13. The gate junction is a metal-to-semiconductor interface as it is in a Schottky or hot-carrier diode. Like in other JFET circuits, the gate-to-source is reverse-biased. Also, as in other depletion mode JFETs, the reverse-bias voltage between the source and the gate controls the conduction of current between the source and the drain. The actual transit time of electrons through

gallium arsenide is far shorter than in similar silicon transistors, and, therefore, the MESFET provides greater gain at microwave frequencies. In addition, MESFETs have an extremely low noise figure, typically 2 dB or less. Most MESFETs have a noise temperature of less than 200 K.

Mixers and Converters

The circuits used to perform frequency translation or conversion are called *mixers* or *converters*. The function they perform is referred to as *heterodyning*.

Mixing

Heterodyning

There are two basic types of frequency conversion, up conversion and down conversion. *Up conversion* means that the input signal along with any modulation is translated to a higher frequency. In *down conversion,* the input signal is translated to some lower frequency.

Up conversion

Down conversion

Figure 7-14 shows a simplified diagram of a mixer circuit. The two inputs to the mixer are the input signal along with any modulation and the input from a local oscillator (LO). The output will consist of the sum and difference frequencies of the two inputs. In some mixers, the LO and input signals may also appear in the output. Of course, the sum and difference frequencies are what we are interested in. A tuned circuit or filter at the output of the mixer

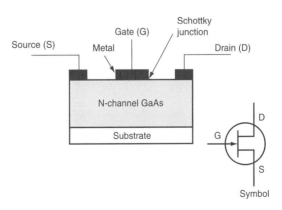

Fig. 7-13 The MESFET configuration and symbol.

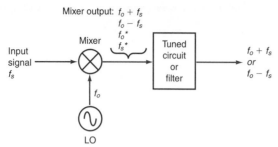

Mixer output: $f_o + f_s$
$f_o - f_s$
f_o*
f_s*

* May or may not be in the output depending upon the type of mixer.

Fig. 7-14 Basic mixer circuit.

selects the desired output signal while rejecting the others.

The mixer can be one of many different components or circuits. Its primary characteristic is *nonlinearity*. Any device or circuit whose output does not vary linearly with the input can be used as a mixer. A diode is a good example.

The mixer, because it is an amplitude modulator, basically performs the function of analog multiplication. It is this multiplication process that generates the sum and difference frequencies. For up conversion, the sum of the two inputs is the selected output, while for down conversion, the difference between the two inputs is selected.

Diode Mixers   One of the most widely used type of mixer is the diode type. Diode mixers are the most common type found in microwave applications.

A simple single-diode mixer circuit is shown in Fig. 7-15. The input signal is applied to the primary winding of transformer T_1. The signal is coupled to the secondary winding and applied to the diode mixer. The LO signal is cou-

pled to the diode by way of capacitor C_1. The output signals, including both inputs, are developed across the tuned circuit, which selects either the sum or difference frequency and eliminates the others.

In small-signal applications, germanium diodes, because of their relatively low turn-on voltage, are used in mixers. Silicon diodes also make excellent RF mixers. The best diode mixers at VHF, UHF, and microwave frequencies are hot-carrier or Schottky barrier diodes.

Doubly Balanced Mixers   Balanced modulators are also widely used as mixers. These circuits offer the advantage of having the carrier eliminated from the output, thereby making the job of filtering much easier. Any balanced modulator described previously can be used in mixing applications. Both the diode lattice balanced modulator and the integrated differential amplifier-type balanced modulator are quite effective in mixing applications. A version of the balanced modulator, known as a *doubly balanced mixer* and illustrated in Fig. 7-16, is one of the most popular mixers at VHF and UHF. It may be the best mixer available.

Transistor Mixers   A transistor can be used as a *mixer* if it is biased into the nonlinear range so that it produces analog multiplication. The primary benefit of transistor mixers over the diode mixers is that gain is obtained with a transistor stage. An example of such a mixer circuit is illustrated in Fig. 7-17. This bipolar transistor is typically biased as if it were a linear amplifier,

Nonlinearity

Doubly balanced mixer

Transistor mixers

Diode mixer

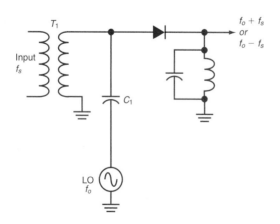

Fig. 7-15   A simple diode mixer.

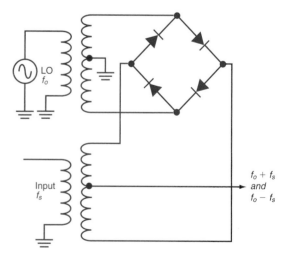

Fig. 7-16   A doubly balanced mixer very popular at high frequencies.

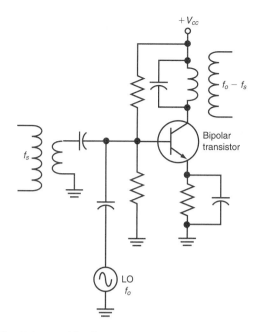

Fig. 7-17 A bipolar transistor mixer.

but the bias is adjusted so that the collector current does not vary linearly with variations in the base current. The class of operation is AB rather than A. This results in analog multiplication, which produces the sum and difference frequencies. In this circuit, both the incoming signal and the LO are applied to the base of the transistor. The tuned circuit in the collector usually selects the difference frequency.

Improved performance can be obtained by using an FET as shown in Fig. 7-18(*a*). Like the bipolar transistor, the FET is biased so that it operates in a nonlinear portion of its characteristic. The input signal is applied to the gate, while the LO signal is coupled to the source. Again the tuned circuit in the drain selects the difference frequency.

Another popular FET mixer is shown in Fig. 7-18(*b*). Here a dual-gate metal-oxide-semiconductor FET (MOSFET) is used. The input signal is applied to one gate, while the LO is coupled to the other gate. *Dual-gate MOSFETs* provide superior performance in mixing applications. Today, bipolar transistors are rarely used in mixing applications except at very low radio frequencies. In receivers built for VHF, UHF, and microwave applications, junction FETs and dual-gate MOSFETs are widely used as mixers because of their high gain and low noise. Gallium arsenide (GaAs) rather than silicon FETs are preferred at the higher frequencies because of their lower noise contribution and higher gain. Diode mixers are also used in many high-frequency receivers.

IC Mixer A typical integrated-circuit mixer is shown in Fig. 7-19. Known as the NE602, it consists of a double-balanced mixer circuit made up of two cross-connected differential

Dual-gate mixer

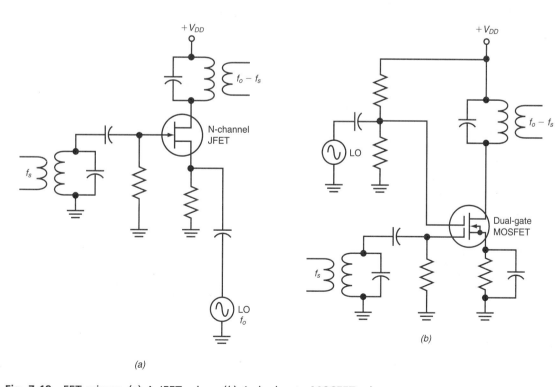

(a)

(b)

Fig. 7-18 FET mixers. (*a*) A JFET mixer. (*b*) A dual-gate MOSFET mixer.

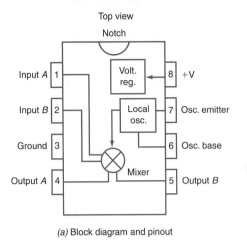

(a) Block diagram and pinout

Local oscillator

(b) Simplified schematic

Fig. 7-19   NE602 IC mixer.

amplifiers. Also on the chip is an NPN transistor that can be connected as a stable oscillator circuit and a dc voltage regulator. The device is housed in an 8-pin dual in-line IC package. It operates from a single dc power supply voltage of 4.5 to 8 V. The circuit can be used at frequencies up to 500 MHz, which makes it useful in HF, VHF, and low-frequency UHF applications. The oscillator will operate up to about 200 MHz. An external LC-tuned circuit

or a crystal is required to set the operating frequency. The oscillator is internally connected to one input of the mixer.

Figure 7-19 shows the circuit details of the mixer itself. Q_1 and Q_2 form a differential amplifier with current source Q_3, and Q_4 and Q_5 form another differential amplifier with current source Q_6. Note that the inputs are connected in parallel. The collectors are cross-connected; that is, the collector of Q_1 is connected to the collector of Q_4

instead of Q_3, as would be the case for a parallel connection. In the same way, the collector of Q_2 is connected to the collector of Q_5. This connection results in a balanced modulator-like circuit in which the internal oscillator signal is suppressed, leaving only the sum and difference signals in the output. The output may be balanced or single-ended as the application requires. A filter or tuned circuit must be connected to the output to select the desired sum or difference signal.

A typical circuit using the NE602 IC mixer is shown in Fig. 7-20. It is powered by 6 V dc. R_1 and C_1 are used for decoupling. A resonant transformer T_1 couples the 49-MHz input signal to the mixer. C_2 resonates with the transformer secondary at the input frequency. C_3 is an ac bypass connecting pin 2 to ground. External components C_4 and L_1 form a tuned circuit that sets the oscillator to 59 MHz. C_5 and C_6 form a capacitive voltage divider that connects the on-chip NPN transistor as a Colpitts oscillator circuit. C_7 is an ac coupling and blocking capacitor. The output is taken from pin 5 and connected to a ceramic bandpass filter. Note the special symbol for the ceramic filter. The output, in this case the difference signal, or $59 - 49 = 10$ MHz, appears across R_2. The balanced mixer circuit suppresses the 59-MHz oscillator signal. The sum signal of 108 MHz is eliminated by the ceramic filter.

IF Amplifiers

Another critical part of a superheterodyne receiver is the *IF amplifier*. This is where most

of the gain and selectivity is obtained. The choice of an IF is critical to the design of a receiver. It is a compromise between good selectivity and stability, which is best obtained at lower frequencies, and good image rejection, which is best obtained at the higher frequencies.

One of the most common IF values is 455 kHz. It is low enough to provide good selectivity and to make high gain with minimum instability. With input frequencies of up to about 10 MHz, the image rejection is satisfactory. But beyond that frequency, the input tuned circuits do not provide sufficient selectivity to reduce the images to an acceptable level. When operating above about 10 MHz, a higher IF frequency is selected. A value in the 1500- to 2000-kHz range is satisfactory for frequencies up to about 30 MHz.

When the input frequency is in the VHF range and beyond, a very high value of IF is chosen. For example, most FM receivers operating in the 88- to 108-MHz band use an IF of 10.7 MHz. In many communications receivers operating at higher frequencies, double conversion is used to solve the image and selectivity problems. In some high-frequency receivers operating in the 3- to 30-MHz range, single conversion is used, but a very high IF of 9 MHz is chosen because it effectively solves the image problem.

Like RF amplifiers, IF amplifiers are tuned class A amplifiers capable of providing a gain in the 10- to 30-dB range. Usually two or more

IF amplifiers

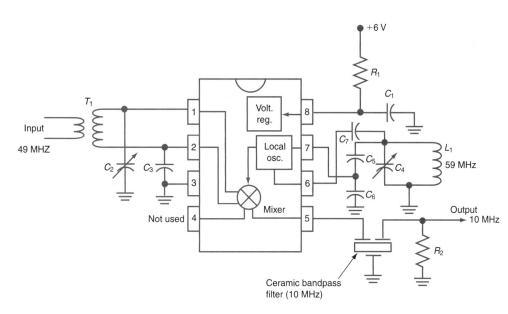

Fig. 7-20 NE602 mixer used for frequency translation.

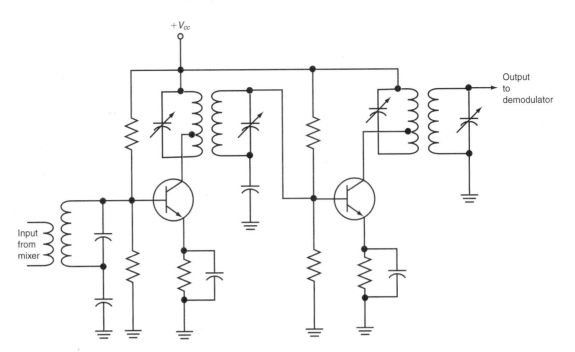

Fig. 7-21 A two-stage IF amplifier using double-tuned transformer coupling for selectivity.

IF amplifiers are used to provide adequate overall receiver gain. See Fig. 7-21. Ferrite-core transformers are used for coupling between stages. Most IF amplifiers use bipolar transistors, although FETs are used in some designs. In most new designs, IC differential amplifiers are used to implement IF amplifiers.

The selectivity in the IF amplifier is provided by the tuned circuits. As indicated, cascading tuned circuits cause the overall circuit bandwidth to be considerably narrowed. High-Q tuned circuits are used, but with multiple tuned circuits, the bandwidth is even narrower.

In designing an IF amplifier, care should be taken so that the selectivity is not too sharp. If the IF bandwidth is too narrow, it will cause sideband cutting. This means that the higher modulating frequencies will be greatly reduced in amplitude, thereby distorting the received signal. The exact nature of the kinds of signals to be received must be known so that the bandwidth of the IF amplifier can be properly set.

It is sometimes necessary when receiving very broad band signals to widen the bandwidth of the IF amplifier. There are several ways of doing this. First, a high value of resistance can be connected across the parallel tuned circuits thereby lowering their Q to a value that will produce the appropriate bandwidth.

Double-Tuned Transformers Another technique is to use overcoupled tuned circuits. The coupling between IF amplifier stages in some receivers is done with double-tuned ferrite-core transformers as shown in Fig. 7-21. Both the primary and the secondary windings are resonated with capacitors. The output voltage versus frequency for such a *double-tuned coupled circuit* is strictly dependent upon the amount of coupling or mutual inductance between the primary and secondary windings. That is, the spacing between the windings determines how much of the magnetic field produced by the primary will cut the turns of the secondary. This affects not only the amplitude of the output voltage, but also the bandwidth.

Figure 7-22 shows the effect of different amounts of *coupling* between the primary and secondary windings. When the windings are spaced far apart, the coils are said to be *undercoupled*. The amplitude will be low, and the bandwidth will be relatively narrow.

At some particular degree of coupling, the output will reach a peak value. This is known as *critical coupling*. In most IF designs, critical coupling provides the best gain if the bandwidth provided is adequate.

Moving the coils closer together and increasing the coupling causes the actual bandwidth to start to increase. The output signal amplitude is at a maximum and will not increase

Double-tuned transformer

Coupling

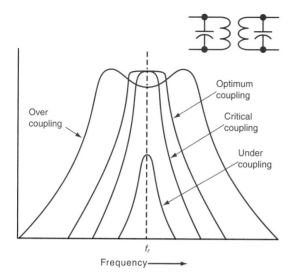

Fig. 7-22 Response curves of a double-tuned air-core transformer for various degrees of coupling.

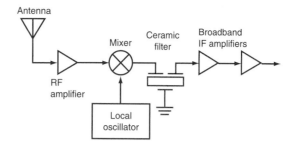

Fig. 7-23 Using a ceramic bandpass filter for IF selectivity.

beyond that obtained at critical coupling, but the bandwidth is considerably widened. This point is usually known as *optimum coupling*.

Increasing the amount of coupling further produces an effect known as *overcoupling*. It produces a doubled peak output response curve with a considerably wider bandwidth. By setting the amount of coupling between the windings in the IF coupling transformers, the desired amount of bandwidth can be obtained.

Crystal and Ceramic Filters   In most communications receivers where superior selectivity is required, very sharp crystal filters are used to obtain the desired selectivity. These crystal filters are usually of the lattice variety discussed previously in the chapter on SSB. *Ceramic filters* are also used.

Ceramic filters are similar to crystal filters in that they are based upon the piezoelectric principle. Ceramic has a resonant frequency like quartz that depends upon the size and thickness. Ceramic can be cut and shaped so that it is sharply resonant over a narrow band of frequencies centered on the resonant frequency. Because ceramic filters are designed for a specific bandwidth, they make ideal filters for obtaining IF selectivity.

Ceramic bandpass filters are small and considerably less expensive than crystal filters and therefore are widely used. In fact, virtually all modern receivers use some form of crystal or ceramic filter. Tuned circuits and transformers are no longer used as often as they once were.

Such filters are usually packaged as a unit and are connected directly at the output of the mixer but prior to the first IF stage. See Fig. 7-23. The desired selectivity is obtained with this filter. The remaining stages of the IF amplifier are, by comparison, broadband circuits. It is the filter alone that sets the bandwidth.

Limiters   In FM receivers, one or more of the IF amplifier stages are used as limiters. The limiters remove any amplitude variations on the FM signal prior to it being applied to the demodulator.

The *limiter* is not a special circuit. Typically it is nothing more than a conventional class A IF amplifier. However, any amplifier will act as a limiter if the input signal is high enough. With a very large input signal applied to a single transistor stage, the transistor will alternately be driven between saturation and cutoff. For example, in a bipolar class A amplifier, applying a very large positive input signal to the amplifier will cause the base bias to increase, thereby increasing the collector current. When a sufficient amount of input voltage is supplied, the transistor will reach maximum conduction where both the emitter-base and the base-collector junctions become forward-biased. At this point, the transistor is saturated and the voltage between the emitter and collector drops to some very small value, typically less than 0.1 V. At this time, the amplifier output is approximately equal to the dc voltage drop across any emitter resistor that may be used in the circuit.

When a very large negative-going signal is applied to the base, the transistor can be driven into cutoff. The collector current drops to zero, and the voltage seen at the collector is simply the supply voltage. Figure 7-24 shows the collector current and voltage for both extremes.

By driving the transistor between saturation and cutoff, the positive and negative peaks of

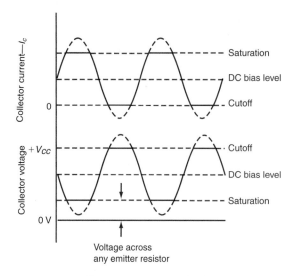

Fig. 7-24 Collector current and voltage in a bipolar limiter IF amplifier circuit.

the input signal are effectively flattened or clipped off. Any amplitude variations are essentially removed. The output signal at the collector, therefore, is a square wave. The most critical part of the limiter design is to set the initial base-bias level to an appropriate point where symmetrical clipping, that is, equal amounts of clipping on the positive and negative peaks, will occur. The square wave at the collector, which is made up of many undesirable harmonics, is effectively filtered back into a sine wave by the tuned circuit in the collector.

Automatic Gain Control The selection of the overall *gain* of a communications receiver is usually based on the weakest signal to be received. In most modern communications receivers the individual gains are as follows: the voltage gain between the antenna and the demodulator is usually in excess of 100 dB; the RF amplifier usually has a gain in the 5- to 15-dB range; the mixer gain is in the 6- to 10-dB range; the IF amplifiers have individual stage gains of 20 to 30 dB; the detector may introduce a typical loss of -2 to -5 dB if it is of the diode type; and the gain of the audio amplifier stage is in the 20- to 40-dB range. As an example, assume the following circuit gains:

RF amplifier	9 dB
Mixer	4 dB
IF amplifiers (3 stages)	27 dB (81 dB total)
Detector	-3 dB
Audio amplifier	30 dB

The total gain then is simply the algebraic sum of the individual stage gains, or

$$\text{Total gain} = 9 + 4 + 27 + 27 + 27 - 3 + 30$$
$$= 121 \text{ dB}$$

In many cases, the total gain is far greater than that required for adequate reception. Excessive gain will usually cause the received signal to be distorted and the transmitted information to be less intelligible. One solution to this problem is to provide a gain control in the receiver. Sometimes a potentiometer is connected at some point in an RF or IF amplifier stage to control the RF gain manually. Further, all receivers include a volume control in the audio circuit. The gain control allows the overall receiver gain to be reduced in order to deal with large signals.

A more effective way of dealing with large signals is to include *AGC circuits.* Automatic gain control is a feedback system that automatically adjusts the gain of the receiver based on the amplitude of the received signal. Very low level signals cause the gain of the receiver to be high. Large input signals cause the gain of the receiver to be reduced.

The use of AGC results in the receiver having a very wide dynamic range. *Dynamic range* refers to the measure of a receiver's ability to receive both very strong and very weak signals without introducing distortion and is the ratio of the largest signal that can be handled to the lowest, expressed in decibels. The dynamic range of a typical communications receiver with AGC is usually in the 60- to 100-dB range.

An AGC circuit takes the received signal either at the output of an IF amplifier or at the output of the demodulator and rectifies it into a direct current. The amplitude of the dc voltage is proportional to the level of the received signal. This dc voltage is applied to one or more IF amplifier stages for the purpose of controlling their gain.

The gain of a bipolar transistor amplifier is proportional to the amount of collector current flowing. Increasing the collector current from some very low level will cause the gain to increase proportionately. At some point, the gain flattens over a narrow collector current range and then begins to decrease as the current increases further. Figure 7-25 shows the typical gain variation versus collector current of a typical bipolar transistor. The gain peaks at 30 dB over the 6- to 12-mA range.

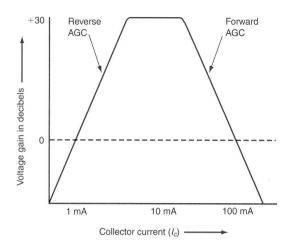

Fig. 7-25 Voltage gain of a bipolar transistor amplifier versus collector current.

The amount of collector current in the transistor, of course, is a function of the base bias applied. A small amount of base current will produce a small amount of collector current and vice versa. Usually the bias level is fixed by a voltage divider, but in IF amplifiers the bias level is usually controlled by the AGC circuit. In some circuits, a combination of fixed voltage divider bias plus a dc input from the AGC circuit controls the overall gain.

As you can see from Fig. 7-25, the gain can be adjusted in two ways. The gain can be decreased by decreasing the collector current. An AGC circuit that decreases the current flowing in the amplifier in order to decrease the gain is called *reverse AGC*. The gain of the IF amplifier can also be reduced by increasing the collector current. As the signal gets stronger, the AGC voltage increases, which increases the base current. The increase in base current, in turn, increases the collector current, which reduces the gain. This method of gain control is known as *forward AGC*. In general, reverse AGC is more common in communications receivers. However, forward AGC is widely used in TV sets and typically requires special transistors for optimum operation.

Figure 7-26 shows two typical ways of applying AGC to an *IF amplifier.* In Fig. 7-26(a), the common emitter IF amplifier bias is derived from the voltage divider made up of R_1 and R_2 and the emitter resistor R_3. Resistor R_4 applied to the base accepts a negative dc voltage from the AGC circuit. As the signal amplitude level increases, the negative dc voltage increases, which thereby decreases the base current. This in turn

decreases the collector current and lowers the circuit gain. This is a reverse AGC circuit.

The circuit in Fig. 7-26(b) is similar, but the bias for the stage is derived from the emitter

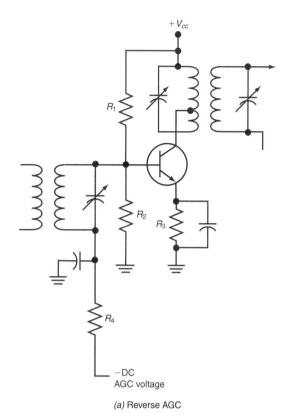

(a) Reverse AGC

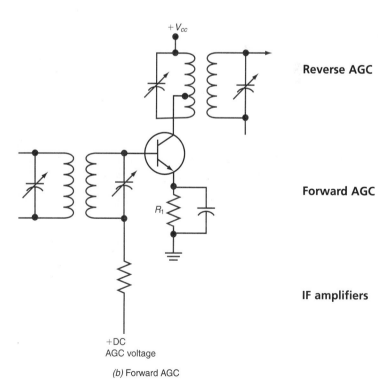

Reverse AGC

Forward AGC

IF amplifiers

(b) Forward AGC

Fig. 7-26 Methods of applying AGC to an IF amplifier.

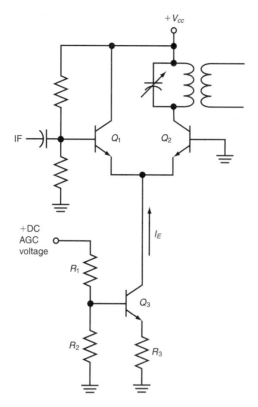

Fig. 7-27 An IF differential amplifier with AGC.

resistor R_1 and the AGC circuit itself. In this case, the AGC dc voltage is positive, which sets the bias level. A strong signal will increase the positive voltage and, therefore, will increase the base current and collector current. This will reduce the circuit gain. This is an example of forward AGC.

Integrated-circuit differential amplifiers are widely used as IF amplifiers. The gain of a differential amplifier is directly proportional to the amount of emitter current I_E flowing. Because of this, the AGC voltage can be conveniently applied to the constant-current source transistor in a differential amplifier. A typical circuit is shown in Fig. 7-27. The bias on constant-current source Q_3 is adjusted by R_1, R_2, and R_3 to provide a fixed level of emitter current I_E to differential transistors Q_1 and Q_2. Normally in a constant-gain stage, the emitter current value is fixed and the current divides between Q_1 and Q_2. The gain is easy to control by varying the bias on Q_3. In the circuit shown, increasing the positive AGC voltage will increase the emitter current and increase the gain. Decreasing the AGC voltage will decrease the gain so that reverse AGC action is obtained.

The dc voltage used to control the gain is usually derived by rectifying either the IF signal or the recovered information signal after the demodulator. One of the simplest and most widely used methods of AGC voltage generation in an AM receiver is simply to use the output from the diode detector as shown in Fig. 7-28. The diode detector recovers the original AM information. The voltage developed across R_1 is a negative dc voltage. Capacitor C_1 filters out the IF signal leaving the original modulating signal. The time constant of R_1 and C_1 is adjusted to eliminate the IF ripple but to retain the highest modulating signal. The recovered signal is passed through C_2 to remove the dc. The resulting ac signal is further amplified and applied to a loudspeaker. The dc voltage across R_1 and C_1 must be further filtered to provide a

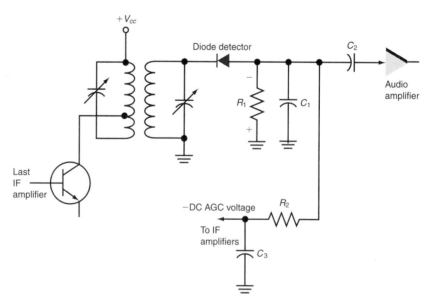

Fig. 7-28 Deriving the AGC voltage from the diode detector in an AM receiver.

pure dc voltage. This is done with R_2 and C_3. The time constant of these components is chosen to be very large so that the voltage at the output is pure dc. The dc level, of course, will vary with the amplitude of the received signal. The resulting negative voltage is then applied to one or more IF amplifier stages.

In an FM receiver, the dc voltage can usually be derived directly from the demodulator. Both Foster-Seeley discriminator and ratio detector circuits provide convenient takeoff points for a dc voltage proportional to the signal amplitude. With additional RC filtering, a dc level proportional to the signal amplitude is derived for use in controlling the IF amplifier gain.

In many receivers, a special rectifier circuit devoted strictly to deriving the AGC voltage is used. Figure 7-29 shows a typical circuit. The input can be either the recovered modulating signal or the IF signal. It is applied to an AGC amplifier. A voltage doubler rectifier circuit made up of D_1, D_2, and C_1 is used to increase the voltage level high enough for control purposes. The RC filter R_1-C_2 removes any signal variations and produces a pure dc voltage. In some circuits, further amplification of the dc control voltage may be necessary. A simple IC op amp like that shown in Fig. 7-29 can be used. The connection of the rectifier and any phase inversion in the op amp will determine the polarity of the AGC voltage. It may be either positive or negative depending upon the types of transistors used in the IF and their bias connections.

Squelch   Another circuit found in most communications receivers is a *squelch circuit.* Also known as a *mute circuit,* it is designed to keep the receiver audio turned off until an RF signal appears at the receiver input. Most two-way communications are short conversations that do not take place continuously. In most cases, the receiver is left on so that if a call is to be received, it will be heard. With no RF signal at the input to the receiver, the audio output will simply be background noise. In AM systems such as CB radios, the noise level is relatively high and very annoying. Even the noise level in FM systems is high. Most people do not want to listen to the background noise and would naturally turn the audio volume down. However, if this is done, when a signal is received, it may not be heard.

The squelch circuit provides a means of keeping the audio amplifier turned off during the time that noise is received in the background. When an RF signal appears at the input, the audio amplifier is enabled.

Most squelch circuits have a built-in level control that allows the circuit threshold to be adjusted to quiet very weak signals and allow only strong signals to turn the audio on.

The most popular type of squelch circuit amplifies the high-frequency background noise that exists when no signal is present and uses it to keep the audio turned off. When a signal is received, the noise circuit is overridden and the audio amplifier is turned on.

Figure 7-30 shows a noise-derived squelch circuit commonly used in most communications receivers. The background noise with no signal is taken from one of the IF amplifiers and passed through C_1 and potentiometer R_1, which form a high-pass filter. Only frequencies above 5 kHz are passed. Most noise is of the higher-frequency variety. Resistor R_1 also serves as a squelch level control.

The noise is then further amplified by two transistor stages and is then rectified into a dc control voltage by a rectifier-voltage doubler circuit made up of C_2, C_3, D_1, and D_2. The rectifier output causes squelch gate Q_1 to saturate. The base current of audio amplifier Q_2 is shunted away through Q_1, and therefore, no audio amplification takes place and the receiver is quiet.

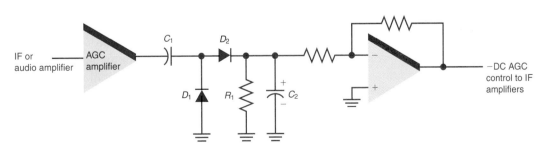

Fig. 7-29 An AGC rectifier and amplifier.

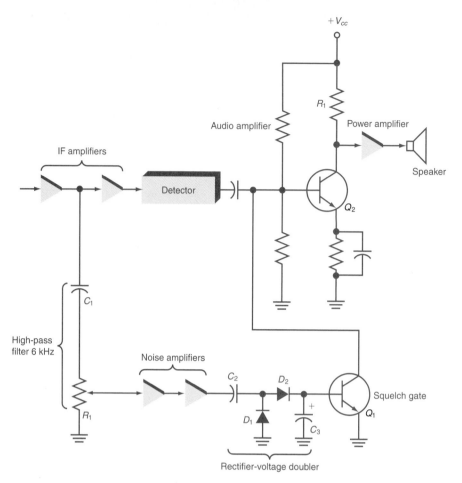

Fig. 7-30 A noise-derived squelch circuit.

If a voice signal occurs, it will blank or mask any high-frequency noise. Thus no noise signal will be applied to the squelch circuit. High-pass filter C_1-R_1 will not pass the audio signal since its frequency content is below 3 kHz. As a result, no squelch voltage is developed at the rectifier output. Q_1 cuts off so that Q_2 is biased normally, and the audio signal is passed through the audio power amplifier to the speaker.

A more sophisticated form of squelch used in some systems is known as *continuous tone control squelch* (*CTCS*). This is a squelch system that is activated by a low-frequency tone transmitted along with the audio. The purpose of CTCS is to provide some communications privacy on a particular channel. Although the squelch circuit keeps the speaker quiet when no input signal is received, in some communications systems a particular frequency channel may be active with many calls. Listening to a variety of transmissions may be tiring or undesirable. The objective is to activate the squelch only when the desired signal is received. This is done by having the desired signal transmit a sine wave code tone along with the audio. The code tone is usually a very low frequency sine wave (usually less than 50 Hz) that is added to the audio prior to being applied to the modulator. The low-frequency tone will appear at the output of the demodulator in the receiver but usually will not be heard in the speaker since the audio response of most communications systems cuts off below 300 Hz. However, the low-frequency tone can be used to activate the squelch circuit.

Figure 7-31 shows a general block diagram of the transmitter and receiver used in a CTCS system. In the transmitter, the low-frequency audio tone has a very specific frequency, and it is linearly mixed with the audio signal prior to being applied to the modulator.

At the receiver, a very selective bandpass filter tuned to the desired tone selects the tone at the output of the demodulator and applies it to a rectifier and filter to generate a dc voltage that will operate the squelch circuit.

Continuous tone control squelch (CTCS)

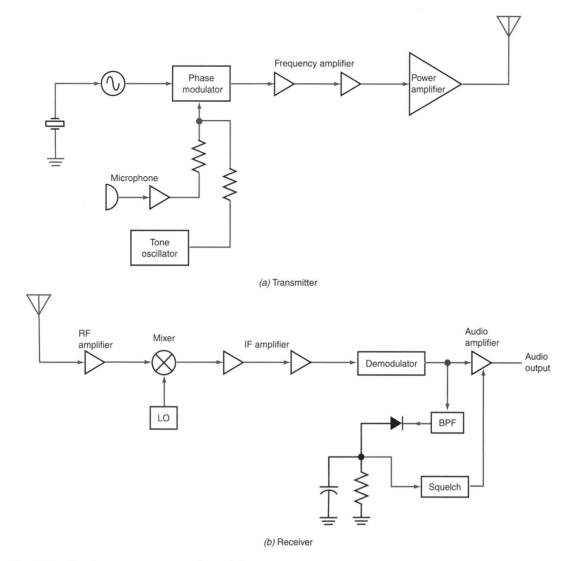

(a) Transmitter

(b) Receiver

Fig. 7-31 Continuous tone control squelch.

Signals that do not transmit the desired tone will, of course, not trigger the squelch. When the desired signal comes along, the low-frequency tone is received, is converted into a dc voltage that operates the squelch, and then turns on the receiver audio. In most systems, multiple tones are available so that the transmitter or receiver may be set up to send to a particular station or to receive a particular station. The overall result is a nearly private communications channel.

CW and SSB Detection

Communications receivers designed for receiving SSB or CW (Morse code) signals have a built-in oscillator that permits recovery of the transmitted information. This circuit is called the *beat frequency oscillator* (*BFO*). Usually it

is designed to operate near the IF and is applied to the demodulator along with the IF signal containing the modulation.

Recall from the discussion of SSB that the basic demodulator is a balanced modulator. See Fig. 7-31(*a*). The balanced modulator has two inputs: the incoming SSB signal at the IF, and a carrier which will mix with the incoming signal to produce the sum and difference frequencies, the difference being the original audio. The BFO is the oscillator that supplies the carrier signal to the balanced modulator. The term "beat" refers to the difference frequency output.

The BFO will be set to a value above or below the SSB signal frequency by an amount equal to the frequency of the modulating signal. Usually the BFO is made variable so that its frequency can be adjusted for optimum

Beat frequency oscillator (BFO)

reception. By varying the BFO over a narrow frequency range, the pitch of the received audio will change from low to high. It is typically adjusted for the sound of a natural voice.

The BFO is also used in receiving CW code. When dots and dashes are transmitted, the carrier is simply turned off and on for short and long periods of time. Neither the amplitude of the carrier nor its frequency varies. However, the on and off nature of the carrier does constitute a form of AM.

Consider for a moment what would happen if a CW signal were applied to a diode detector. The output of the diode detector would simply be pulses of dc voltage representing the dots and dashes. When applied to the audio amplifier, the dots and dashes would quiet the receiver (blank the noise), but they would not be discernible. To make the dots and dashes audible, the IF signal is mixed with the signal from a BFO. The BFO is usually injected directly into the diode detector as shown in Fig. 7-32(*b*). In this way, the CW signal at the IF signal output mixes with the BFO in the diode detector. This results in sum and difference frequencies at the output. The sum frequency, of course, is very high, nearly double the IF value. It is filtered out by capacitor C_1 in the detector. The difference frequency will be a low audio frequency. In fact, if the BFO is variable, the difference frequency can be adjusted to any desired audio tone, usually in the 400- to 900-Hz range. Now, when the dots and dashes appear at the input to the diode detector, the output will be an audio tone that is amplified and heard in a speaker or earphones. Of course, the BFO is turned off for standard AM signal reception.

■ TEST

Choose the letter which best answers each statement.

61. The mixing process is
    a. Linear.
    b. Nonlinear.
62. In up conversion, which signal is selected?
    a. $f_1 - f_2$
    b. $f_1 + f_2$
63. Which transistor mixer is preferred?
    a. Bipolar
    b. FET

Supply the missing information in each statement.

64. Another name for the mixing process is _____.
65. The purpose of a mixer is to perform _____.
66. The output signals produced by a mixer with inputs f_1 and f_2 are _____, _____, _____, and _____.
67. The type of mixer most used at microwave frequencies is a(n) _____.
68. The two mixers preferred at VHF and UHF are _____.
69. Transistor mixers provide _____, whereas diode mixers do not.
70. A mixer producing down conversion has an input of 1390 kHz and an LO frequency of 1845 kHz. The output is _____ kHz.
71. The mixer circuit in the popular NE602 IC is a(n) _____.
72. Mixers made with _____ or _____ diodes are best for small signals.
73. In addition to a mixer and a voltage regulator, the NE602 mixer IC contains a(n) _____.
74. The output of a NE602 mixer IC is 5.5 MHz. The local oscillator is set to 26.8 MHz. The input frequency is _____.

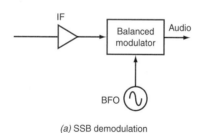

(a) SSB demodulation

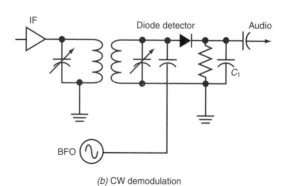

(b) CW demodulation

Fig. 7-32 The use of a BFO.

Determine whether each statement is true or false.

75. Mixing is the same as AM.
76. Any balanced modulator may be used as a mixer.

Supply the missing information in each statement.

77. RF amplifiers provide initial _____ and _____ in a receiver but also add _____.
78. A low-noise transistor preferred at microwave frequencies is the _____ made of _____.
79. Most of the gain and selectivity in a superhet is obtained in the _____ amplifier.
80. The selectivity in an IF amplifier is usually produced by using _____ between stages.
81. The bandwidth of a double-tuned transformer depends upon the degree of _____ between primary and secondary windings.
82. In a double-tuned circuit, minimum bandwidth is obtained with _____ coupling, maximum bandwidth with _____ coupling, and peak output with _____ or _____ coupling.
83. An IF amplifier that clips the positive and negative peaks of a signal is called a(n) _____.
84. Clipping occurs in an amplifier because the transistor is driven by a high-level signal into _____.
85. The gain of a bipolar class A amplifier can be varied by changing the _____ _____.
86. The overall RF-IF gain of a receiver is approximately _____ dB.
87. Using the amplitude of the incoming signal to control the gain of the receiver is known as _____.
88. AGC circuits vary the gain of the _____ amplifier.
89. The dc AGC control voltage is derived from a(n) _____ circuit connected to the _____ or _____ output.
90. Reverse AGC is where a signal amplitude increase causes a(n) _____ in the IF amplifier collector current.
91. Forward AGC uses a signal amplitude increase to _____ the collector current, which decreases the IF amplifier gain.

92. The AGC of a differential amplifier is produced by controlling the current produced by the _____ transistor.
93. In a dual-gate MOSFET IF amplifier, the dc AGC voltage is applied to the _____.
94. Another name for AGC in an AM receiver is _____.
95. In an AM receiver, the AGC voltage is derived from the _____.
96. Large input signals cause the gain of a receiver to be _____ by the AGC.
97. A circuit that blocks the audio until a signal is received is called a(n) _____ circuit.
98. Two types of signals used to operate the squelch circuit are _____.
99. In a CTCS system, a low-frequency _____ is used to trigger the _____ circuit.
100. A BFO is required to receive _____ and _____ signals.

7-6 A Typical Communications Receiver

Now let's combine all you have learned so far and apply it to real receivers. This will give you practice in analyzing and servicing communications receivers.

VHF AM Aircraft Receivers

A typical receiver circuit is shown in Fig. 7-33. This is a receiver designed to receive two-way communication between airplanes and airport controllers. Such communication from air to ground and ground to air takes place in the VHF range of 118 to 135 MHz. Amplitude modulation is used. Like most modern receivers, the circuit is a combination of discrete components and integrated circuits.

The signal is picked up by an antenna and fed through a transmission line to input jack J_1. The signal is coupled through C_1 to a tuned filter consisting of the series- and parallel-tuned circuits made up of L_1 through L_5 and C_2 through C_6. This broad bandpass filter passes the entire 118- to 135-MHz range.

The output of the filter is connected to an RF amplifier through C_7. The RF amplifier is

For information on the latest transceiver developments in Europe, go to the following French-English Web site:

⟨www.ers.fr⟩

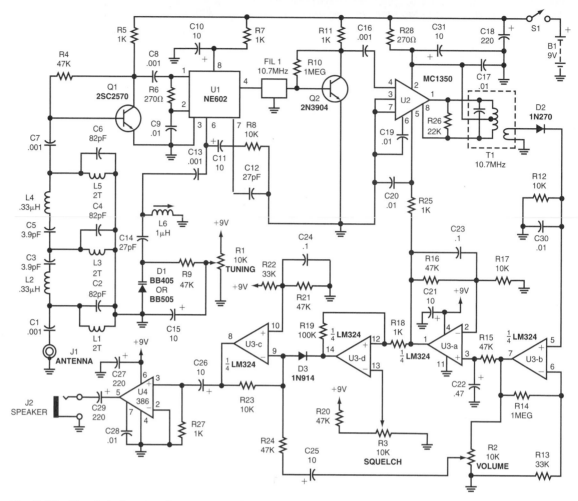

Fig. 7-33 The Aviation Receiver—a superheterodyne unit, built around four ICs—is designed to receive AM signals in the 118- to 135-MHz frequency range. (Popular Electronics, January 1991, Gernsback Publications, Inc.)

made up of transistor Q_1 and its bias resistor R_4 and collector load R_5. The signal is then applied to the NE602 IC, designated U_1, through C_8. As you will recall, the NE602 contains a balanced mixer and a local oscillator. The local oscillator frequency is set by the circuit made up of inductor L_6 and the related components. C_{14} and D_1 in series form the capacitor that resonates with L_6 to set the frequency of the local oscillator. Tuning of the oscillator is accomplished by varying the dc bias on a varactor D_1. Potentiometer R_1 sets the reverse bias, which in turn varies the capacitance to tune the oscillator.

Remember, to tune a superheterodyne receiver, you should vary the local oscillator frequency which is set to a frequency above the incoming signal by the amount of the intermediate frequency (IF). In this receiver, the IF is 10.7 MHz, a standard value for many VHF receivers. This means that to tune the receiver to

the 118 to 135 range, the local oscillator must be varied from 128.7 to 145.7 MHz.

The output of the mixer, which is the difference between the incoming signal frequency and the local oscillator frequency, appears at pin 4 of the NE602 and is fed to a ceramic bandpass filter set to the IF of 10.7 MHz. This filter provides most of the receiver's selectivity. The insertion loss of the filter is offset by an amplifier consisting of Q_2, its bias resistor R_{10}, and collector load R_{11}. The output of this amplifier drives an MC1350 IC through C_{16}. U_2 is an integrated IF amplifier that provides extra gain and selectivity. The selectivity comes from the tuned circuit made up of IF transformer T_1. The MC1350 also contains all the AGC circuitry.

The signal at the secondary of T_1 is then fed to a simple AM diode detector consisting of D_2, R_{12}, and C_{30}. The demodulated audio signal appears across R_{12} and is then fed to an op amp U_{3-b}. This amplifier is a noninverting circuit biased by R_{13} and R_{14}. It provides extra amplification for the demodulated audio and the average direct current at the detector output. This amplifier feeds the volume control, which is potentiometer R_2. The audio signal goes from there through C_{25} and R_{24} to another op amp U_{3-c}. Here the signal is further amplified. It is then fed to the 386 IC power amplifier U_4. This circuit drives the speaker, which is connected by means of jack J_2.

The audio signal from the diode detector contains the dc level resulting from detection (rectification). Both the audio and the direct current are amplified by U_{3-b} and are further filtered into almost pure direct current by the low-pass filter made of R_{15} and C_{22}. This dc signal is applied to op amp U_{3-a}, where it is amplified into a dc control voltage. The direct current at the output pin 1 of U_3 is fed back to pin 5 on the MC1350 IC to provide AGC control. This ensures a constant comfortable listening level despite wide variations in the signal strength.

The AGC voltage from U_{3-a} is also fed to an op-amp comparator circuit made from amplifier U_{3-d}. The other input to this comparator is a dc voltage from potentiometer R_3, which is used as a squelch control. Since the AGC voltage from U_{3-a} is directly proportional to the signal strength, it is used as the basis for setting the squelch to a level that will keep the receiver quiet until a signal of a predetermined strength comes along.

If the signal strength is very low or no signal is tuned in, the AGC voltage will be very low or nonexistent. This causes D_3 to conduct and effectively disable amplifier U_{3-c}, which in turn prevents the audio signal from the volume control from passing through to the power amplifier. If a strong signal exists, D_3 will be reverse-biased, and so will not interfere with amplifier U_{3-c}. As a result, the signal from the volume control passes to the power amplifier and is heard in the speaker.

IC FM Receiver

IC FM receiver

Figure 7-34 shows a diagram of a single-chip FM receiver like that used in a cordless telephone. All the circuitry except for the selective circuits and audio power amplifier are contained in a 28-pin integrated circuit. This device is the Motorola MC 3363 dual-conversion superheterodyne.

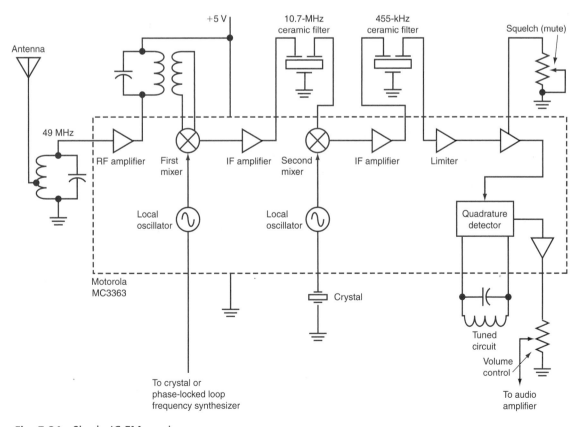

Fig. 7-34 Single IC FM receiver.

The incoming FM signal is picked up by a small antenna rod. Initial selectivity is provided by the tuned circuit set to the 49-MHz range. The tap on the coil provides impedance matching. An internal RF amplifier boosts the signal level. The amplified signal is coupled through the external RF transformer to the first mixer. The on-chip local oscillator can be used with an external crystal to set the frequency, or it can be used as the VCO in a phase-locked loop (PLL) frequency synthesizer for multichannel operation. The local oscillator is set so that the mixer output is 10.7 MHz, the first IF. Selectivity is obtained with an external ceramic filter.

The 10.7-MHz IF signal is amplified and sent to the second mixer, where it is combined with the signal from the second local oscillator, which is controlled by an external crystal. The output of the second mixer is the difference frequency 455 kHz, the second IF. Additional selectivity is obtained with another ceramic filter.

The 455-kHz IF is passed through a limiter to remove amplitude variations. An IF squelch or mute circuit follows. The original voice signal is then recovered by the quadrature detector. Its tuned circuit is external to the chip. The audio signal is amplified and sent to the off-chip volume control. A separate IC audio power amplifier is used to operate the speaker.

■ TEST

Answer the following questions.

101. What components or circuits in Fig. 7-33 determine the bandwidth of the receiver?

Transceiver

102. As R_1 is varied so that the voltage on the arm of the potentiometer increases toward +9 V, how does the frequency of the local oscillatory vary?
103. If there was no audio output from the speaker but you knew that a signal was present, what two controls would you check first?
104. What component provides most of the gain in this receiver?

Shared circuit

105. Is the squelch signal or noise derived?
106. Does this receiver contain a BFO? Could it receive CW or SSB signals?
107. If the dc AGC voltage on pin 5 of U_2 was decreased, what would happen to the gain of U_2?

108. Where would you inject an audio signal to test the complete audio section of this receiver?
109. What frequency signal would you use to test the IF section of this receiver, and where would you connect it?
110. What component would be inoperable if C_{31} became shorted?
111. Explain how you would connect a digital counter to this receiver so that it would read the frequency of the signal to which it was tuned.
112. Which circuit in Fig. 7-34 controls the frequency of operation or channel selection?
113. What is the frequency of the crystal on the second local oscillator?
114. What is the purpose of the limiter?
115. What type of FM demodulator is used in the Motorola MC3363?
116. IF selectivity is obtained with

_____ _____.

7-7 Transceivers and Frequency Synthesizers

At one time, all pieces of communications equipment were individually packaged in units based on their function. In general, transmitters and receivers were always separate units. Today, however, most pieces of two-way radio communications equipment are packaged so that all the functions are contained within a single housing. Typically, this means that both the transmitter and receiver are in the same package. For this reason, this combination of units is referred to as a *transceiver*. Transceivers vary in size and complexity from very large, high-power, desktop units to the very small, pocket-size, hand-held walkie-talkies.

Transceiver Organization

There are advantages to packaging communications equipment in a single housing. In addition to sharing housing and power supply, in many designs the transmitters and receivers can also *share circuits*. This results in less circuitry, lower cost, and, in some cases, smaller size. Some of the circuits that can often be shared are antennas, oscillators, power supplies, tuned circuits, filters, and various kinds of amplifiers. These circuits perform a dual

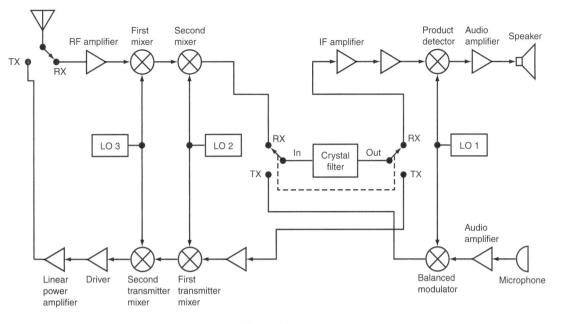

RX = receiver
TX = transmitter

Fig. 7-35 An SSB transceiver showing circuit sharing.

function, serving both the transmitter and receiver but not at the same time.

In FM transceivers, both the transmitter and the receiver operate from the same power supply, but usually this is the only shared circuit. The only exception is that in some designs the oscillators or frequency synthesizers used for generating the carrier and LO signals may sometimes be shared.

On the other hand, transceivers designed for AM, CW, and SSB can share many circuits. Figure 7-35 is a general block diagram of a high-frequency transceiver capable of CW and SSB operation. Because both the receiver and the transmitter make use of heterodyning techniques for generating the IF and the final transmission frequencies, respectively, proper design enables both transmitter and receiver to share common LOs. Local oscillator 1 is the BFO for the receiver product detector and the carrier for the balanced modulator for producing DSB. Later, crystal LO 2 drives the second mixer in the receiver and the first transmitter mixer used for up conversion. Local oscillator 3 supplies the first receiver mixer and the second transmitter mixer.

Another shared circuit is the crystal filter. In the transmission mode, the crystal filter provides sideband selection after the balanced modulator. In the receiver, this same filter provides selectivity for the IF section of the re-

ceiver. In some designs, tuned circuits can also be shared. A particular tuned circuit may serve as the tuned input for the receiver or as the tuned output for the transmitter. Circuit switching may be manual but often is done with relays or diode electronic switches.

Frequency Synthesizer

In many of the newer transceiver designs, both the transmitter and receiver typically share a circuit known as a frequency synthesizer. A *frequency synthesizer* is a variable-frequency signal generator that replaces one or more LOs or carrier oscillators in a transceiver, providing the frequency stability of a crystal oscillator but the convenience of continuous tuning over a broad frequency range in small, equal increments. Frequency synthesizers usually provide an output signal that varies in fixed frequency increments over a wide range. They are used in receivers as LOs and perform the receiver tuning function. In a transmitter, a frequency synthesizer provides basic carrier generation for channelized operation.

Over the years, many techniques have been developed for implementing frequency synthesizers with frequency multipliers and mixers. Today most frequency synthesizers are some variation of the phase-locked loop (PLL).

Frequency synthesizer

*inter***NET**
C O N N E C T I O N

The following Internet sites include information related to the material in this section:

⟨www.glb.com⟩
⟨www
.rf-connection
.com⟩

An elementary frequency synthesizer based on a PLL is shown in Fig. 7-36. As with all PLLs, it consists of a phase detector, a low-pass filter, and a VCO. The input to the phase detector is a reference oscillator. The reference oscillator is normally crystal-controlled to provide high-frequency stability. The frequency of the reference oscillator sets the increments in which the frequency may be changed, in this example, 100 kHz.

Frequency divider

Note that the VCO output is not connected directly back to the phase detector. The VCO is applied to a frequency divider first. A *frequency divider* is a circuit whose output frequency is some integer submultiple of the input frequency. A divide-by-10 frequency divider will produce an output frequency that is one-tenth of the input frequency. Frequency dividers can be easily implemented with digital circuits to provide any integer value of frequency division.

Refer to Fig. 7-36. The reference oscillator is set to 100 kHz, or 0.1 MHz. Initially assume that the frequency divider is set for a division by 10. Keep in mind that in order for a PLL to become locked or synchronized, the second input to the phase detector must be equal in frequency to the reference frequency. In other words, for this PLL to be locked, the frequency divider output must be 100 kHz. This means that the VCO output has to be 10 times higher than this, or 1 MHz. One way to look at this circuit is as a frequency multiplier. The 100-kHz input was multiplied by 10 to produce the 1 MHz. In the design of the synthesizer, the VCO frequency is set to 1 MHz so that when it is divided, it will provide the 100-kHz input

signal required for the locked condition. The synthesizer output is the output of the VCO. What we have created then is a 1-MHz signal source. Because the PLL is locked to the crystal reference source, the VCO output frequency has the same stability as the crystal oscillator.

To make the frequency synthesizer more useful, some means must be provided to vary its output frequency. This is done by varying the frequency division ratio. Through various switching techniques, the flip-flops in a frequency divider can be arranged to provide any desired frequency division ratio. The frequency division ratio is normally designed to be manually changed in some way. For example, a rotary switch may control logic circuits to provide the correct configuration. In another circuit, a thumbwheel switch may be used. Some designs incorporate a keyboard by means of which the desired frequency division ratio can be keyed in. In most sophisticated circuits, a microprocessor may actually provide the correct frequency division ratio while at the same time controlling a direct frequency readout display.

By varying the frequency division ratio, the output frequency is changed. For example, in our circuit of Fig. 7-36, if the frequency division ratio is changed from 10 to 11, the VCO output frequency must change to 1.1 MHz. In this way, the output of the divider will remain at 100 kHz as necessary to maintain a locked condition. Each increment of frequency division ratio change produces an output frequency change of 0.1 MHz. You can see now how the frequency increment is set by the reference oscillator.

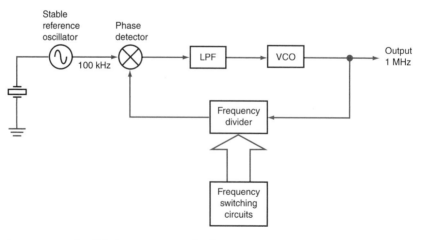

Fig. 7-36 Basic concept of a PLL frequency synthesizer.

All frequency synthesizers use the basic principles of the circuit in Fig. 7-36. However, more sophisticated versions are typical. Most frequency synthesizers also incorporate mixers for additional frequency conversion capabilities. Also, two or more PLLs may be used as part of a single synthesizer to provide multiple frequencies, say, for both LO and carrier frequencies. In most transceivers, different LOs and carrier frequency generators are needed. These can usually be supplied by a synthesizer where channel frequency selection is simultaneous. That is, one manual frequency control will change both LO and carrier frequencies to put them simultaneously on the correct transmit and receive frequencies.

Figure 7-37 shows a PLL synthesizer for a 40-channel *CB transceiver.* Using two crystal oscillators for reference and a single-loop PLL, it synthesizes the transmitter frequency and the two LO frequencies for a dual-conversion receiver for all 40 CB channels.

The reference crystal oscillator operates at 10.24 MHz. It is divided by a divide-by-2 flip-flop to 5.12 MHz. Then a binary frequency divider that divides by 1024 produces 5.12/1024 MHz = 5 kHz. This is applied to the phase detector. The channel spacing, therefore, is 5 kHz.

The phase detector drives the low-pass filter and a VCO that generates a signal in the 16.27- to 16.71-MHz range. This is the LO frequency for the first receiver mixer. Assume that we want to receive on CB channel 1, or 26.965 MHz. The programmable divider is set to the correct ratio to produce a 5-kHz output when the VCO is 16.27. The first receiver mixer produces the difference frequency of $26.965 - 16.27 = 10.695$ MHz. This is the first IF.

The 16.27-MHz VCO signal is also applied to mixer A, whose other input is 15.36 MHz, which is derived from the divided 10.24-MHz reference oscillator and a frequency tripler ($\times 3$) (5.12 MHz $\times 3 = 15.36$ MHz). The output of mixer A drives the programmable divider, which feeds the phase detector.

The 10.24-MHz reference output is also used as the LO signal for the second receiver mixer. With a 10.695-MHz first IF, the second IF then is the difference or $10.695 - 10.24$ MHz $= 0.455$ MHz, or 455 kHz.

CB transceiver

The VCO output is also applied to mixer B along with a 10.695-MHz signal from a second crystal oscillator. The sum output is selected to produce the transmit frequency $10.695 + 16.27 = 26.965$ MHz. This signal drives the class C drivers and power amplifiers.

Channel selection is done by changing the frequency division ratio on the programmable

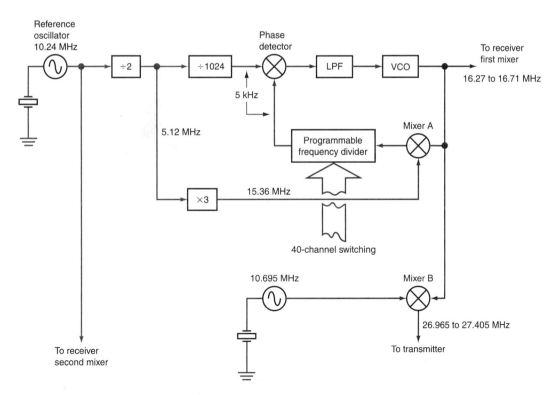

Fig. 7-37 A frequency synthesizer for a CB transceiver.

divider. It is usually done with a rotary switch. Most of the circuitry in the synthesizer is usually contained on one or two IC chips.

■ TEST

Answer the following questions.

117. In an FM transceiver, the only shared circuitry is usually the _____.

118. In an SSB transceiver, the shared circuits are the _____ and the _____.

119. The circuit used to generate one or more stable frequencies for the transmitter and receiver is known as a(n) _____.

120. A basic circuit of a frequency synthesizer is a(n) _____.

121. The output frequency of a synthesizer is changed by varying the _____ of the divider between the VCO and phase detector.

122. A frequency synthesizer has a phase detector input reference of 10 kHz. The divide ratio is 286. The output frequency is _____ MHz. The frequency change increment is _____ kHz.

123. The PLL is often combined with _____ circuits to produce multiple frequencies.

124. What three frequencies does the synthesizer in a transceiver usually generate?

Summary

1. The simplest form of communications receiver is the tuned radio-frequency (TRF) receiver which provides RF and AF gain, selectivity, and a demodulator.

2. The two primary characteristics of a receiver are selectivity and sensitivity.

3. Selectivity is the ability to separate signals on different frequencies.

4. The bandwidth (BW) of a tuned circuit is the difference between the upper f_2 and lower f_1 3-dB-down frequencies (BW $= f_2 - f_1$).

5. The bandwidth of a tuned circuit is determined by the ratio of the resonant frequency to the Q, where Q is X_L/R of the inductor (BW $= f_r/Q$). The Q can be computed from the bandwidth $Q = f_r/\text{BW} = f_r/(f_2 - f_1)$.

6. Sharper selectivity with steeper response curve skirts can be obtained by cascading tuned circuits.

7. The shape factor of a response curve is the ratio of the 60-dB-down bandwidth to the 6-dB-down bandwidth. The lower the shape factor, the better the selectivity.

8. Sensitivity is the ability of the receiver to pick up weak signals and is a function of gain.

9. The disadvantages of a TRF receiver are tuning difficulties and selectivity varying with frequency.

10. These problems of a TRF receiver are eliminated by using a superheterodyne (superhet) receiver.

11. A superheterodyne uses a mixer to translate the incoming signal to a lower frequency, known as the intermediate frequency (IF), where fixed gain and selectivity can be obtained.

12. Frequency conversion is a form of AM used to translate signals to higher or lower frequencies for improved processing.

13. Frequency conversion is carried out by a circuit called a mixer or converter.

14. The mixer performs analog multiplication of the input signal and a local oscillator signal.

15. The frequency conversion process faithfully retains any modulation (AM, FM, etc.) on the input. The translated signal contains the same modulation.

16. The output of a mixer consists of the local oscillator signal f_o and the sum and difference frequencies of the local oscillator and input frequencies:

$$f_s$$
$$f_o$$
$$f_o + f_s$$
$$f_o - f_s \quad \text{or} \quad f_s - f_o$$

17. Either the sum or the difference frequency is selected with a filter, while the others are rejected.

18. The process of mixing is also known as heterodyning.

19. Most of the gain and selectivity in a superheterodyne is obtained in the IF amplifier.

20. Low IFs are preferred because higher selectivity and better stability can be obtained with simple circuits.

21. Low IFs usually cause image interference problems.

22. An image is a signal on a frequency separated from the desired signal frequency by two times the IF value that interferes with reception.

23. The mixer will convert both the desired signal and the image to the IF.

24. Image interference is caused by poor receiver input selectivity that does not adequately reject the image.

25. Image interference can be reduced by using a double-conversion superhet that uses two mixers and IFs. The first IF is high to eliminate images, and the second is low to ensure good selectivity.

26. The most critical part of any receiver is the front end, which consists of the RF amplifier and mixer, because these circuits add the most noise to a weak signal.

27. Noise is any random interference to a weak signal.

28. A measure of a receiver's noise performance is its signal-to-noise (S/N) ratio.

29. External noise comes from industrial, atmospheric, and space sources.

30. Industrial noise sources are ignitions, motors and generators, switching devices, and fluorescent lights.

31. Atmospheric noise comes from lightning and other sources.

32. Space noise comes from the sun as well as other stars and extraterrestrial sources.

33. Internal noise from electronic components also interferes with reception and can totally mask weak signals.

34. Most internal noise comes from thermal agitation—the random movement of electrons and atoms in a material due to heat.

35. The thermal noise voltage across a resistor is proportional to the temperature, the bandwidth, and the resistance value and can be computed with the expression $v_n = \sqrt{4kTBR}$, where k is Boltzmann's constant 1.38×10^{-23}, T is the temperature in degrees kelvin, B is the bandwidth in hertz, and R is the resistance in ohms.

36. Other types of internal noise are shot and transit-time noise in semiconductors.

37. Noise cannot be eliminated, but its effect can be minimized.

38. Noise performance of a receiver or circuit is expressed in terms of the noise figure F, also called the noise factor, which is the ratio of the S/N input to the S/N output. It is usually given in decibels.

39. Noise performance of microwave components is usually expressed in terms of noise temperature in degrees kelvin $K = 290(F - 1)$.

40. Thermal noise is random and is often referred to as white noise or Johnson noise.

41. Since noise is a mixture of all random frequencies, its level can be reduced by narrowing the bandwidth.

42. Frequency translation or conversion is an AM process that converts a signal with any modulation to a higher (up conversion) or lower (down conversion) frequency.

43. Frequency conversion is produced by a circuit called a mixer.

44. Another name for frequency conversion is heterodyning.

45. Almost any low-level AM circuit can be used for mixing.

46. One of the most commonly used mixers is a single diode. Germanium or hot-carrier diodes are used for mixing at very high radio frequencies.

47. The inputs to a mixer are the signal to be translated f_s and a sine wave f_o from a local oscillator (LO). The outputs are f_o, f_s, $f_o + f_s$, and $f_o - f_s$. A tuned circuit at the output selects either the sum or difference frequency while suppressing the others. Any modulation on the input appears on the output signal.

48. Bipolar and field-effect transistors can be used as mixers by operating them in the nonlinear region of their characteristics.

49. Transistor mixers offer the benefit of gain over diode mixers.

50. Balanced modulators are widely used as mixers.

51. Doubly balanced modulators and GaAs FETs are the best mixers at VHF, UHF, and microwave frequencies.

52. A popular IC mixer is the NE602, which contains a cross-connected differential amplifier mixer circuit and an on-chip local oscillator.

53. Most receivers get their selectivity from ceramic or crystal filters.

54. Most receivers have AGC circuits so that a wide dynamic range of input signal amplitudes can be accommodated without distortion.

55. An AGC circuit rectifies the IF or demodulator output into dc to control the IF amplifier gain.

56. The gain of a bipolar transistor can be varied by changing its collector current.

57. In reverse AGC, an increase in the AGC voltage decreases the collector current.

58. In forward AGC, an increase in AGC voltage increases collector current.

59. The gain of a dual-gate MOSFET in an IF amplifier is controlled by varying the dc voltage on the second gate.

60. A squelch circuit is used to cut off the audio output to prevent annoying noise until a signal is received. Either the audio signal or background noise can be used to operate the squelch circuit.

61. Continuous tone control squelch (CTCS) circuits permit selective signaling by allowing only low-frequency tones to trigger the squelch.

62. A beat frequency oscillator (BFO) is used in SSB and CW receivers to provide a carrier that will mix with the input signal in the demodulator to generate the audio output.

63. A transceiver is a piece of communications equipment that combines a receiver and a transmitter in a common package where they share a common housing and power supply. Single-

64. Many new transceivers contain a frequency synthesizer that eliminates multiple crystal oscillators and *LC* tuned oscillators in the transmitter and receiver.
65. A frequency synthesizer is a signal generator usually implemented with a PLL that produces LO and transmitter carrier signals in a transceiver.
66. A frequency synthesizer has the stability of a crystal oscillator, but the frequency can be varied in small, equal increments over a wide range.
67. The frequency increments in a synthesizer are set by the frequency of the reference input to the phase detector.
68. The frequency of a synthesizer is changed by varying the divide ratio of the frequency divider between the VCO output and phase detector input.
69. Phase-locked loop synthesizers often incorporate mixers and multipliers to permit more than one frequency to be generated.

Chapter Review Questions

Choose the letter which best answers each question.

7-1. The simplest receiver is a(n)
   a. RF amplifier.
   b. Demodulator.
   c. AF amplifier.
   d. Tuned circuit.

7-2. The key conceptual circuit in a superhet receiver is the
   a. Mixer.
   b. RF amplifier.
   c. Demodulator.
   d. AF amplifier.

7-3. Most of the gain and selectivity in a superhet is obtained in the
   a. RF amplifier.
   b. Mixer.
   c. IF amplifier.
   d. AF amplifier.

7-4. The sensitivity of a receiver depends upon the receiver's overall
   a. Bandwidth.
   b. Selectivity.
   c. Noise response.
   d. Gain.

7-5. The ability of a receiver to separate one signal from others on closely adjacent frequencies is called the
   a. Sensitivity.
   b. S/N ratio.
   c. Selectivity.
   d. Gain.

7-6. A mixer has a signal input of 50 MHz and an LO frequency of 59 MHz. The IF is
   a. 9 MHz.
   b. 50 MHz.
   c. 59 MHz.
   d. 109 MHz.

7-7. A signal 2 times the IF from the desired signal that causes interference is a(n)
   a. Ghost.
   b. Image.
   c. Phantom.
   d. Inverted signal.

7-8. A receiver has a desired input signal of 18 MHz and an LO frequency of 19.6 MHz. The image frequency is
   a. 1.6 MHz.
   b. 18 MHz.
   c. 19.6 MHz.
   d. 21.2 MHz.

7-9. The main cause of image interference is
   a. Poor front-end selectivity.
   b. Low gain.
   c. A high IF.
   d. A low S/N ratio.

7-10. For best image rejection, the IF for a 30-MHz signal would be
   a. 455 kHz.
   b. 3.3 MHz.
   c. 9 MHz.
   d. 55 MHz.

7-11. A tuned circuit is resonant at 4 MHz. Its *Q* is 100. The bandwidth is
   a. 400 Hz.
   b. 4 kHz.
   c. 40 kHz.
   d. 400 kHz.

7-12. A crystal filter has a 6-dB bandwidth of 2.6 kHz and a 60-dB bandwidth of 14 kHz. The shape factor is
   a. 0.186.
   b. 5.38.
   c. 8.3.
   d. 36.4.

Communications Receivers   Chapter 7   185

7-13. Most internal noise comes from which of the following?
 a. Shot noise
 b. Transit-time noise
 c. Thermal agitation
 d. Skin effect

7-14. Which of the following is *not* a source of external noise?
 a. Thermal agitation
 b. Auto ignitions
 c. The sun
 d. Fluorescent lights

7-15. Noise can be reduced by
 a. Widening the bandwidth.
 b. Narrowing the bandwidth.
 c. Increasing temperature.
 d. Increasing transistor current levels.

7-16. Noise at the input to a receiver can be as high as several
 a. Microvolts.
 b. Millivolts.
 c. Volts.
 d. Kilovolts.

7-17. Which circuit contributes most to the noise in a receiver?
 a. IF amplifier
 b. Demodulator
 c. AF amplifier
 d. Mixer

7-18. Which noise figure represents the lowest noise?
 a. 1.6 dB
 b. 2.1 dB
 c. 2.7 dB
 d. 3.4 dB

7-19. Which filter shape factor represents the best skirt selectivity?
 a. 1.6
 b. 2.1
 c. 5.3
 d. 8

7-20. Which input signal below represents the best receiver sensitivity?
 a. 0.5 μV
 b. 1 μV
 c. 1.8 μV
 d. 2 μV

7-21. The transistor with the lowest noise figure in the microwave region is a(n)
 a. MOSFET.
 b. Dual-gate MOSFET.
 c. JFET.
 d. MESFET.

7-22. Frequency translation is done with a circuit called a
 a. Summer.
 b. Multiplier.
 c. Filter.
 d. Mixer.

7-23. The inputs to a mixer are f_o and f_m. In down conversion, which of the following mixer output signals is selected?
 a. f_o
 b. f_m
 c. $f_o - f_m$
 d. $f_o + f_m$

7-24. Mixing for frequency conversion is the same as
 a. Rectification.
 b. AM.
 c. Linear summing.
 d. Filtering.

7-25. Which of the following can be used as a mixer?
 a. Balanced modulator
 b. FET
 c. Diode modulator
 d. All the above

7-26. The desired output from a mixer is usually selected with a
 a. Phase-shift circuit.
 b. Crystal filter.
 c. Resonant circuit.
 d. Transformer.

7-27. The two inputs to a mixer are the signal to be translated and a signal from a(n)
 a. Modulator.
 b. Filter.
 c. Antenna.
 d. LO.

7-28. An NE602 mixer IC has a difference output of 10.7 MHz. The input is 146.8 MHz. The local oscillator frequency is
 a. 101.9 MHz.
 b. 125.4 MHz.
 c. 131.6 MHz.
 d. 157.5 MHz.

7-29. The AGC circuits usually control the gain of the
 a. Mixer.
 b. Detector.
 c. IF amplifiers.
 d. Audio amplifiers.

7-30. Selectivity is obtained in most receivers from
 a. Crystal filters.
 b. Mechanical filters.
 c. Double-tuned circuits.
 d. Audio filters.

7-31. Widest bandwidth in a double-tuned circuit is obtained with
a. Undercoupling.
b. Critical coupling.
c. Optimum coupling.
d. Overcoupling.

7-32. Automatic gain control permits a wide range of signal amplitudes to be accommodated by controlling the gain of the
a. RF amplifier.
b. IF amplifier.
c. Mixer.
d. AF amplifier.

7-33. In an IF amplifier with reverse AGC, a strong signal will cause the collector current to
a. Increase.
b. Decrease.
c. Remain the same.
d. Drop to zero.

7-34. Usually AGC voltage is derived by the
a. RF amplifier.
b. IF amplifier.
c. Demodulator.
d. AF amplifier.

7-35. A circuit that keeps the audio cut off until a signal is received is known as a(n)
a. Squelch.
b. AFC.
c. AGC.
d. Noise blanker.

7-36. A BFO is used in the demodulation of which types of signals?
a. AM
b. FM
c. SSB or CW
d. QPSK

7-37. Which of the following circuits are *not* typically shared in an SSB transceiver?
a. Crystal filter
b. Mixers
c. Power supply
d. LO

7-38. The basic frequency synthesizer circuit is a(n)
a. Mixer.
b. Frequency multiplier.
c. Frequency divider.
d. PLL.

7-39. The output frequency increment of a frequency synthesizer is determined by the
a. Frequency division ratio.
b. Reference input to the phase detector.
c. Percentage of output frequency.
d. Frequency multiplication factor.

7-40. The output frequency of a synthesizer is changed by varying the
a. Reference frequency input to phase detector.
b. Frequency division ratio.
c. Frequency multiplication factor.
d. Mixer LO frequency.

7-41. In Fig. 7-36, if the input reference is 25 kHz and the divide ratio is 144, the VCO output frequency is
a. 173.61 Hz.
b. 144 kHz.
c. 3.6 MHz.
d. 5.76 MHz.

7-42. The bandwidth of a parallel *LC* circuit can be increased by
a. Increasing X_C.
b. Decreasing X_L.
c. Decreasing coil resistance.
d. A resistor connected in parallel.

7-43. The upper and lower cutoff frequencies of a tuned circuit are 1.7 and 1.5 MHz respectively. The circuit Q is
a. 8.
b. 10.
c. 16.
d. 24.

7-44. The noise voltage across a 300-Ω input resistance to a TV set with a 6-MHz bandwidth and a temperature of 30°C is
a. 2.3 μV.
b. 3.8 μV.
c. 5.5 μV.
d. 6.4 μV.

7-45. The stage gains in a superheterodyne are as follows: RF amplifier, 10 dB; mixer, 6 dB; two IF amplifiers, each 33 dB; detector, −4 dB; AF amplifier, 28 dB. The total gain is
a. 73 dB.
b. 82 dB.
c. 106 dB.
d. 139 dB.

7-46. A tuned circuit resonates at 12 MHz with an inductance of 5 μH whose resistance is 6 Ω. The circuit bandwidth is
a. 98 kHz.
b. 191 kHz.
c. 754 kHz.
d. 1.91 MHz.

7-47. In a receiver with noise-derived squelch, the presence of an audio signal causes the audio amplifier to be
a. Enabled.
b. Disabled.

Critical Thinking Questions

7-1. Explain the operation of a superheterodyne receiver designed with a local oscillator that is set to the exact frequency of the received signal.

7-2. How could you design a receiver that can receive transmissions of AM, FM, and SSB signals?

7-3. Explain how you would make a receiver that could cover the frequency range from 500 kHz to 180 MHz. What special circuits and controls would you need?

7-4. How do scanners work? Could you design a receiver that would scan four frequencies? How would you proceed?

7-5. Using the principles of PLL frequency synthesizers described in this chapter, create a synthesizer that will generate all the AM radio broadcast frequencies from 540 to 1650 kHz in 10-kHz increments. This unit would be used as the carrier source in an AM broadcast transmitter.

7-6. Draw a simplified block diagram of what you think a pager receiver is like. What are the outputs? What part of the receiver is the greatest mystery to you?

7-7. Is it possible for a receiver to use an IF value higher than the received signal frequency? If so, why do you think this would be done?

7-8. Would it be possible to have a 100 percent digital software radio? What are the limitations in doing this?

7-9. A dual-conversion shortwave receiver is set to receive 18-MHz signals. The second IF is 455 kHz. The local first oscillator frequency is set to 27 MHz. What is the first IF? Compute all possibilities.

Answers to Tests

1. selectivity
2. increase
3. narrow
4. true (resonant step-up)
5. increase (narrower bandwidth)
6. sidebands
7. 5
8. 0.53 MHz or 530 kHz ($f_r = 10.6$ MHz, $X_L = 200 \ \Omega$, $Q = 20$)
9. 9.95, 10.05
10. 1333.3
11. decreased
12. 2.4
13. sensitivity
14. tuned radio frequency
15. superheterodyne
16. intermediate frequency
17. local oscillator
18. IF amplifier
19. automatic gain control
20. signal or input, local oscillator
21. IF amplifier
22. frequency conversion
23. mixer, converter
24. 4.155 MHz, 7.855 MHz, 455 kHz, 3.7 MHz
25. 117.8 MHz
26. filter or tuned circuit
27. false
28. true
29. heterodyning
30. c
31. image
32. 3.5
33. selectivity
34. 28
35. true
36. mixer
37. intermediate frequency
38. 9, 600
39. industrial, atmospheric, and space
40. thermal agitation, shot noise, transit-time noise
41. extraterrestrial or space
42. lightning
43. automobiles, motors, generators, fluorescent lights
44. thermal agitation
45. decibels
46. high
47. increase
48. white, Johnson
49. decrease
50. 1.36
51. shot, transit-time
52. false
53. true

54. noise figure
55. microwave
56. 464
57. distortion
58. high
59. RF amplifier, mixer
60. true
61. b
62. b
63. b
64. heterodyning
65. frequency conversion or translation
66. $f_1, f_2, f_1 + f_2, f_1 - f_2$
67. diode
68. doubly balanced mixer, GASFET
69. gain
70. 455 (1845 − 1390 = 455)
71. cross-connected differential amplifier
72. germanium, hot-carrier
73. local oscillator
74. 21.3 or 32.3 MHz
75. true
76. true
77. gain, selectivity, noise
78. MESFET or GASFET, gallium arsenide
79. IF
80. tuned circuits
81. mutual inductance
82. under, over, optimum, critical
83. limiter
84. cutoff, saturation
85. collector current
86. 100
87. automatic gain control
88. IF
89. rectifier, IF amplifier or detector
90. decrease
91. increase
92. constant-current source
93. control gate
94. automatic volume control
95. diode detector
96. reduced
97. squelch
98. audio, noise
99. tone, squelch
100. SSB, CW

101. the ceramic filter FIL_1
102. As the potentiometer arm voltage goes more positive, the reverse bias across D_1 increases, thus decreasing its capacitance and the overall capacitance of D_1 in series with C_{14}; this in turn increases the local oscillator frequency.
103. volume R_2 and squelch R_3
104. U_2, the MC1350 IF amplifier IC
105. The squelch is signal-derived.
106. With no BFO, the receiver cannot receive CW or SSB.
107. The gain increases.
108. pin 5 of the U_{3-b} op amp
109. 10.7 MHz injected to the ceramic filter FIL_1
110. U_2, the MC1350 IC, would lose supply voltage at pin 2.
111. A digital counter could be connected to monitor the local oscillator output signal at pin 6 or 7 of the NE602 mixer IC. This would read the local oscillator frequency, which is 10.7 MHz higher than the actual input signal frequency. Therefore, the counter would have to be preset to a value that is 10.7 MHz less than zero count. In this way, the IF value would effectively be subtracted from the input count to read the correct frequency. Assume that the counter is set up to count four digits, 000.0. The counter is preset to $100.0 - 10.7 = 89.3$, and the input signal frequency is 118 MHz. The local oscillator is tuned to $118 + 10.7 = 128.7$ MHz. The counter counts 10.7 to reach 000.0 and then counts 118 more to display a final value of 118.0.
112. first local oscillator (along with XTAL or synthesizer)
113. $10.7 - .455$ MHz $= 10.245$ MHz
114. Eliminate amplitude variations
115. Quadrature detector
116. ceramic filters
117. power supply
118. crystal filter, local oscillators
119. frequency synthesizer
120. phase-locked loop
121. frequency division ratio
122. 2.86, 10
123. mixer
124. LO for first mixer, LO for second mixer, transmitter

Chapter 8

Multiplexing

Chapter Objectives

This chapter will help you to:

1. *Define* the terms multiplexing, demultiplexing, codec, and telemetry.
2. *Explain* the processes of frequency division multiplexing and demultiplexing and *state* three modern applications.
3. *Explain* the processes of time division multiplexing and demultiplexing and *name* two common applications.
4. *Explain* the operation of a pulse-code modulation system and *name* a common application.
5. *Calculate* the lowest possible sampling rate for a pulse-amplitude modulation system.
6. *Define* the terms compression, expansion, and companding.
7. *Explain* the operation and application of a sample and hold amplifier.

A communications link is established whenever a cable is strung between two points or when radio transmitters and receivers at two or more locations make contact. With such arrangements, voice conversations can take place, data can be exchanged, and control functions can be performed. Generally speaking, only one of these functions can be performed at a time. With one communications link, only one transaction can take place. However, there are many occasions when it would be desirable for multiple functions to occur simultaneously using the established link. In fact, many applications require that multiple conversations take place, dozens of data signals be transmitted, or several control operations take place at the same time. Logically, multiple cables or radio links would have to be installed to carry out such applications. Unfortunately, that would usually be impractical because of the cost and complexity. As it turns out, a single cable or radio link can handle multiple signals. Thanks to a process called multiplexing, hundreds or even thousands of signals of any type can be combined and transmitted over a single medium. As a result, multiplexing has made communications more practical and economically feasible. It helps conserve spectrum space and has allowed new sophisticated applications to be implemented. In this chapter, the techniques of multiplexing are introduced and some real world uses explained.

8-1   Introduction

Multiplexing

Multiplexing is the process of simultaneously transmitting two or more individual signals over a single communications channel. Multiplexing has the effect of increasing the number of communications channels so that more information can be transmitted.

There are many instances in communications where it is necessary or desirable to transmit more than one voice or data signal. The application itself may require multiple signals, and money can be saved by using a single communications channel to send multiple information signals. Telemetry and telephone applications are good examples. In satellite communications, multiplexing is essential to making the system practical and for justifying the expense.

Telemetry is a good illustrative example. *Telemetry* is the process of measurement at a distance. Telemetry systems are used to monitor physical characteristics of some applications for the purpose of determining their status and operational conditions. This information may also be used as feedback in a closed-loop control system. Most spacecraft and many chemical plants, for example, use telemetry systems for monitoring their operation. Physical characteristics such as temperature, pressure, speed, light level, flow rate, and liquid level are monitored. Sensitive transducers convert these physical characteristics into electrical signals. These electrical signals are then processed in various ways and sent to a central monitoring location.

The most obvious way to send multiple signals from one place to another is to provide a single communications channel for each. For example, each signal could be sent over a single pair of wires. If long distances are involved, the signals will be degraded, and, therefore, special techniques must be used to prevent this. Using multiple wires is also an expensive process. When a very large number of signals must be monitored, many pairs of cables will be needed. This leads to extra cost and complexity. Ideally, it would be more economical if all the telemetry signals could somehow be combined and sent over a single cable.

In a spacecraft with multiple transducers, multiple transmitters would be required to send the signals back to earth. Again, this leads to incredible cost and complexity. In the case of a spacecraft, multiple transmitters would weigh much and consume an enormous amount of extra power, making them impractical. Again, the ideal situation would be to use a single transmitter and in some way combine all the various information signals and transmit them simultaneously over the single radio channel.

The telephone system is another example of the need for some means to increase the information-carrying capability of a single channel. There are hundreds of millions of telephones in this country, and each must be capable of being connected to any other telephone. This is what the telephone system is all about. If each telephone requires a two-wire path, imagine the enormous number of wires required to make all the various interconnections. The problem is further compounded if you want to add in the ability to

connect each telephone to any other telephone in the world. On such a large scale, cost is a major factor. Everything must be done to minimize the number of interconnecting wires. This leads to the use of some methods of combining multiple telephone conversations in such a way that they can be transmitted over a single pair of wires or a single radio communications channel.

The concept of a simple multiplexer is illustrated in Fig. 8-1. Multiple input signals are combined by the multiplexer into a single composite signal that is transmitted over the communications medium. Alternatively, the multiplexed signals may modulate a carrier before transmission. At the other end of the communications link, a demultiplexer is used to sort out the signals into their original form.

There are two basic types of multiplexing: *frequency division multiplexing (FDM)* and *time division multiplexing (TDM)*. Generally speaking, FDM systems are used to deal with analog information, and TDM systems are used for digital information. Of course, TDM techniques are found in many analog applications as well because the process of analog-to-digital (A/D) and digital-to-analog (D/A) is so common. The primary difference between these techniques is that in FDM, individual signals to be transmitted are assigned a different frequency within a common bandwidth. In TDM, the multiple signals are transmitted in different time slots. In the following sections, we will discuss FDM and TDM in more detail.

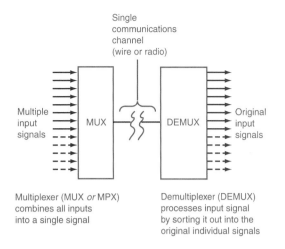

Fig. 8-1 Concept of multiplexing.

Telemetry

For technical reference information on multiplexing, see the following Web site:

⟨telecom.tbi.net⟩

JOB TIP

Do you enjoy working in a team environment? Many employers want to hire team members who work together to accomplish the goals of the company.

For information on about a large telecommunication corporation, see the following informative site:

⟨www.gte.com⟩

Frequency division multiplexing (FDM)

Time division multiplexing (TDM)

■ TEST_____

Answer the following questions.

1. Which statement is most correct?
   a. Multiplexing uses multiple channels to transmit multiple signals.
   b. Multiplexing uses multiple channels to transmit individual signals.
   c. Multiplexing uses a single channel to transmit multiple signals.
   d. Multiplexing uses a single channel to transmit a single signal.

2. Two common applications of multiplexing are _____.

3. The primary benefit of multiplexing is _____.

4. When multiplexing is used to transmit information, then a _____ must be used at the receiving end of the communications link.

5. The two types of multiplexing are _____.

8-2 Frequency Division Multiplexing

Frequency division multiplexing (FDM) is based on the idea that a number of signals can share the bandwidth of a common communications channel. The multiple signals to be transmitted over this channel are each used to modulate a separate carrier. Each carrier is on a different frequency. The modulated carriers are then added together to form a single complex signal that is transmitted over the single channel.

FDM Concept

Figure 8-2 shows a general block diagram of an FDM system. Each signal to be transmitted feeds a modulator circuit. The carrier for each modulation f_c is on a different frequency. The carrier frequencies are usually equally spaced from one another over a specific frequency

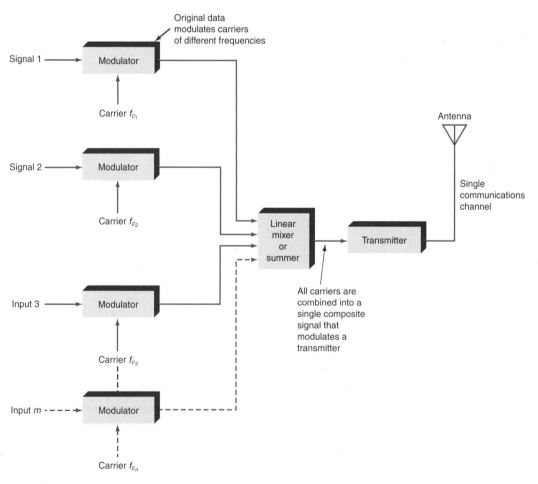

Fig. 8-2 The transmitting end of an FDM system.

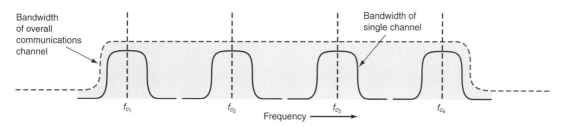

Fig. 8-3 Spectrum of an FDM signal.

range. Each input signal is given a portion of the bandwidth. The result is illustrated in Fig. 8-3. As for the type of modulation, any of the standard kinds can be used including AM, SSB, FM, or PM.

The modulator outputs containing the side-band information are added together in a *linear mixer*. In a linear mixer, modulation and the generation of sidebands do not take place. Instead, all the signals are simply added together algebraically. The resulting output signal is a composite of all carriers containing their modulation. This signal is then used to modulate a radio transmitter. Alternatively, the composite signal itself may be transmitted over the single communications channel. Another option is that the composite signal may become one input to another multiplexer system.

Demultiplexing FDM Signals

The receiving portion of the system is shown in Fig. 8-4. A receiver picks up the signal and demodulates it into the composite signal. This is sent to a group of *bandpass filters (BPF)*, each centered on one of the carrier frequencies. Each filter passes only its channel and rejects all others. A channel demodulator then recovers each original input signal.

To be specific about FDM systems, let's consider three practical examples, telemetry, telephone, and FM stereo.

FDM in Telemetry

Telemetry is one of the most common uses for multiplexing techniques. In telemetry systems, many different physical characteristics are monitored by sensors. These sensors generate

Bandpass filters (BPF)

Linear mixer

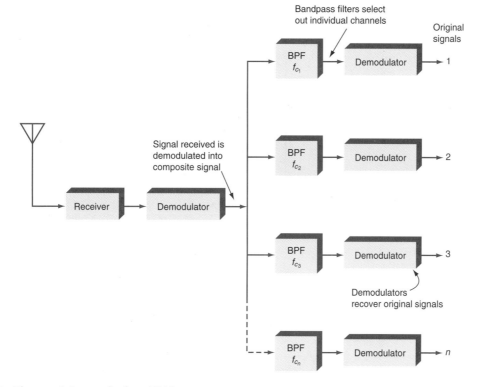

Fig. 8-4 The receiving end of an FDM system.

electrical signals that change in some way to indicate the amplitude or measurement of the physical characteristics. An example of a sensor is a thermistor used to measure temperature. A thermistor's resistance varies inversely with temperature: As the temperature increases, the resistance decreases. The thermistor is usually connected into some kind of a resistive network, such as a voltage divider or bridge, and it is connected to a dc voltage source. The result is a dc output from this network whose voltage varies in accordance with the temperature. That varying dc level must then be transmitted to a remote receiver for measurement, readout, and recording. In this case, the thermistor signal becomes one channel of an FDM system.

Other sensors have different kinds of outputs. Many simply have varying dc outputs, and others may be ac in nature. Each of these signals is typically amplified, filtered, and otherwise conditioned before being used to modulate a carrier. All the carriers are then added together to form a single multiplexed channel. In such systems, FM is normally used.

The conditioned transducer outputs are used to modulate a subcarrier. The varying direct or alternating current changes the frequency of an oscillator operating at the carrier frequency. Such a circuit is generally referred to as a *voltage-controlled oscillator (VCO)* or a *subcarrier oscillator (SCO)*. The outputs of the SCOs are added together. A diagram of such a system is shown in Fig. 8-5.

The output of the signal conditioning circuits is fed to the VCOs. Of course, to produce FDM, each VCO operates at a different frequency.

VCO Figure 8-6(*a*) shows a block diagram of a typical VCO circuit. The VCOs are available as single IC chips. One version is called the 566. It consists of a dual-polarity current source that linearly charges and discharges an external capacitor C. The current value is set by an external resistor R_1. Together R_1 and C set the operating or center carrier frequency, which can be any value up to about 1 MHz.

The current source may be varied by an external signal, either direct or alternating current. This is the modulating signal from a transducer or other source. The input signal varies the charging and current and, therefore, varies the carrier frequency, producing FM.

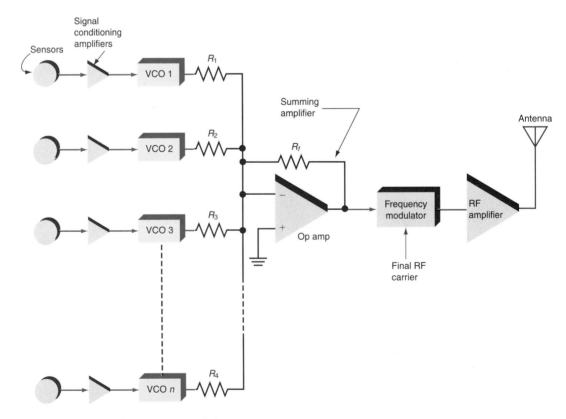

Fig. 8-5   An FDM telemetry transmitting system.

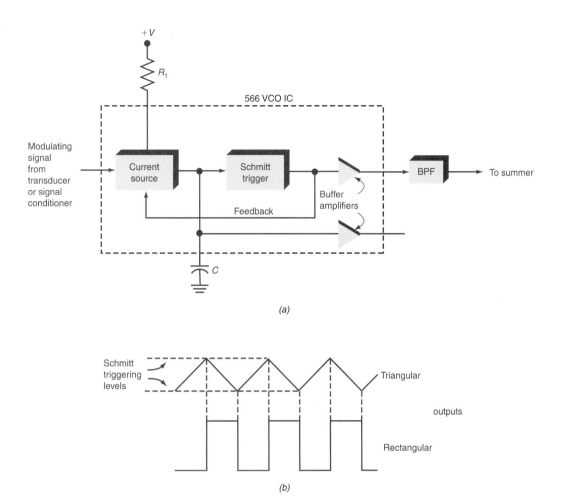

(a)

(b)

Fig. 8-6 (a) Typical IC VCO circuit and (b) waveforms.

The current source output is a linear triangular wave that is buffered by an amplifier for external use. This triangular waveform is also fed to an internal Schmitt trigger, which generates a rectangular pulse at the operating frequency. This is fed to a buffer amplifier for external use.

The Schmitt trigger output is also fed back to the current source, where it controls whether the capacitor is charged or discharged. For example, the VCO may begin by charging the capacitor. When the Schmitt trigger senses a specific level on the triangular wave, it switches the current source. Discharging then occurs. The waveforms in Fig. 8-6(b) show this action. It is this feedback that creates a free-running, astable oscillator.

Most VCOs are astable multivibrators whose frequency is controlled by the input from the signal conditioning circuits. The frequency of the VCO changes linearly in proportion to the input voltage. Increasing the input voltage causes the VCO frequency to increase.

The rectangular or triangular output of the VCO is usually filtered into a sine wave by a bandpass filter centered on the unmodulated VCO center frequency. This may be either a conventional *LC* filter or an active filter made with an op amp and *RC* input and feedback networks. The resulting sinusoidal output is applied to the linear mixer.

Linear Mixing   The linear mixing process in the FDM system can be accomplished with a simple resistor network as shown in Fig. 8-7. However, such networks greatly attenuate the signal. Typically, some voltage amplification is required in practical systems. A way to achieve the mixing and amplification at the same time is to use an op-amp summer like that shown in Fig. 8-5. Recall that the gain of each input is a function of the ratio of the feedback resistor R_f to the input resistor value (R_1, R_2, etc.). The output is given by the expression

$$V_o = V_1 \frac{R_f}{R_1} + V_2 \frac{R_f}{R_2}$$

$$+ V_3 \frac{R_f}{R_3} + \ldots + V_n \frac{R_f}{R_n}$$

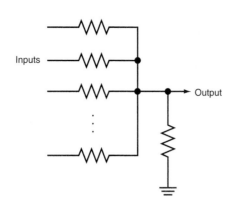

Fig. 8-7 Resistive summing network.

Inter-Range Instrumentation Group (IRIG)

In most cases, the VCO FM output levels are the same, and, therefore, all input resistors on the summer amplifier are equal. However, if variations do exist, amplitude corrections can be made by making the summer input resistors adjustable. The output of the summer amplifier does invert the signal, but this has no effect upon the content.

Modulating Schemes The composite output signal is then typically used to modulate a radio transmitter. Again, most telemetry systems use FM. A system that uses FM of the VCO subcarriers as well as FM of the final carrier is usually called an *FM/FM system*. However, keep in mind that other kinds of modulation schemes may be used.

FM/FM

Most FM/FM telemetry systems conform to standards established many years ago by an organization known as the *Inter-Range Instrumentation Group (IRIG)*. These standards define specific channels as indicated in Fig. 8-8. The center frequency of each channel is given along with the related channel number. A frequency deviation of ±7.5 percent is used on most of the channels, and this is increased to ±15 percent on the upper frequency channels. Also given in Fig. 8-8 is the upper frequency range that the modulating signal can have on each channel. On channel 1, for example, with a 400-Hz center frequency, the maximum signal frequency that can be used is 6 Hz. Most of the lower frequency channels are used for direct current or very low frequency ac signals.

This set of standards is known as the proportional bandwidth FM/FM system. Since a fixed percentage of frequency deviation is specified, this means that the bandwidth is proportional to the carrier frequency. The higher the carrier frequency, the wider the bandwidth over which the modulating signal can occur.

Constant-bandwidth FM telemetry channels are also used. Carrier frequencies in the same approximate range as those given in Fig. 8-8 are used. However, a fixed deviation of ±2 kHz is typically specified, thereby creating

Band Number	Center Frequency (Hz)	Lower Limit (Hz)	Upper Limit (Hz)	Maximum Deviation (%)	Frequency Response (cps)
1	400	370	430	±7.5	6.0
2	560	518	602	"	8.4
3	730	675	785	"	11
4	960	888	1,032	"	14
5	1,300	1,202	1,399	"	20
6	1,700	1,572	1,828	"	25
7	2,300	2,127	2,473	"	35
8	3,000	2,775	3,225	"	45
9	3,900	3,607	4,193	"	59
10	5,400	4,995	5,805	"	81
11	7,350	6,799	7,901		110
12	10,500	9,712	11,288	"	160
13	14,500	13,412	15,588	"	220
14	22,000	20,350	23,650	"	330
15	30,000	27,750	32,250	"	450
16	40,000	37,000	43,000	"	600
17	52,500	48,562	56,438	"	790
18	70,000	64,750	75,250	"	1,050
A.	22,000	18,700	25,300	±15	660
B.	30,000	25,500	34,500	"	900
C.	40,000	34,000	46,000	"	1,200
D.	52,500	44,625	60,375	"	1,600
E.	70,000	59,500	80,500	"	2,100

Fig. 8-8 The IRIG FM subcarrier bands and specifications.

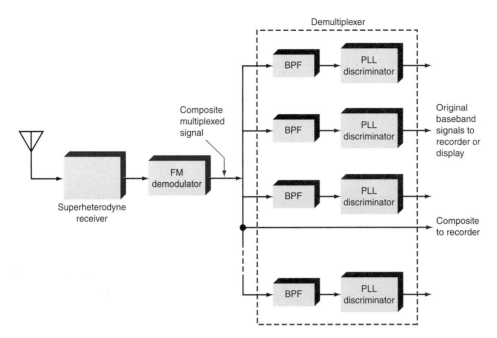

Fig. 8-9 An FM/FM telemetry receiver.

multiple channels with a 4-kHz bandwidth. These are spaced throughout the frequency spectrum with some guard space between channels to minimize interference.

The receiving end of a telemetry system appears as shown in Fig. 8-9. A standard superheterodyne receiver tuned to the RF carrier frequency is used to pick up the signal. An FM demodulator then reproduces the original composite multiplexed signal. This multiplexed signal is then fed to a demultiplexer that divides the signals and reproduces the original inputs.

The output of the first frequency demodulator is fed simultaneously to multiple BPFs, each of which is tuned to the center frequency of one of the specified channels. Each filter passes only its subcarrier and related sidebands and rejects all the others. As you can see, the demultiplexing process is essentially that of using filters to sort the composite multiplex signal back into its original components. The output of each filter is the VCO frequency with its modulation.

These signals then, in turn, are applied to frequency demodulators. Also known as discriminators, these circuits take the FM signal and recreate the original dc or ac signal produced by the transducer. These original signals are then measured and otherwise interpreted to provide the desired information from the remote transmitting source. In most systems, the multiplexed signal is sent to a data recorder

where it is stored for possible future use. The original telemetry output signals may be graphically displayed on a strip chart recorder or otherwise converted into usable outputs.

The demodulator circuits used in typical FM demultiplexers are either of the PLL or pulse-averaging type. Of these two types, the PLL circuit is generally preferred because of its superior noise performance. However, it is typically more complex and expensive than the simpler pulse-averaging type. A PLL discriminator is also used to demodulate the receiver output.

FDM in Telephone Systems

Another example of a commonly used FDM system is the telephone system. For years telephone companies have been using FDM to send multiple telephone conversations over a minimum number of cables. The concepts are the same as those previously discussed. Here the original signal is voice in the 300- to 3000-Hz range. The voice is used to modulate a subcarrier. Each subcarrier is on a different frequency. These subcarriers are then added together to form a single channel. This multiplexing process is repeated at several levels so that an enormous number of telephone conversations can be carried over a single communications channel, assuming its bandwidth is sufficient.

The frequency plan for a typical telephone multiplex system is shown in Fig. 8-10. The

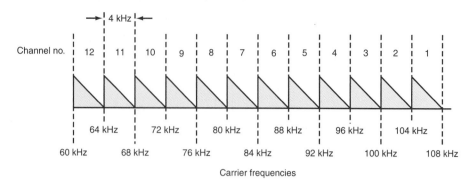

Fig. 8-10 Basic group frequency plan for FDM telephone system.

hertz, slightly higher than the highest frequency used in a typical voice communication.

Basic group

symbols are explained in Fig. 8-11. Here the voice signal amplitude modulates 1 of 12 channels in the 60- to 108-kHz range. The carrier frequencies begin at 60 kHz with a spacing of 4 kHz, slightly higher than the highest frequency used in a typical voice communication.

Single-sideband, suppressed-carrier modulation is used in telephone multiplex systems. Refer to Fig. 8-12. The voice signal is applied to a balanced modulator along with a carrier. The output of the balanced modulator consists of the upper and lower sideband frequencies. The carrier is suppressed by the balanced modulator, and a highly selective filter is used to pass either the upper or lower sideband. The upper sidebands are selected here. The output of the filter is the sideband containing the original voice signal. All 12 SSB signals are then summed in a linear mixer to produce a single frequency multiplexed signal. This set of 12

Super group

Master group

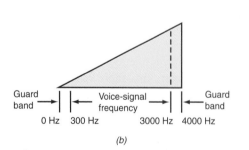

(a)

(b)

Fig. 8-11 Meaning of symbols in Fig. 8-10. (*a*) Upper and lower sidebands, and (*b*) sideband detail showing guard bands.

modulated carriers is generally referred to as a *basic group*.

If more than 12 voice channels are needed, multiple basic groups are used. The outputs of these basic groups can then be further multiplexed onto higher-frequency subcarriers. In the telephone system, as many as five 12-channel basic groups can be combined. Carrier frequencies in the 360- to 552-kHz range are used. These carriers are spaced 48 kHz apart. The frequency plan is shown in Fig. 8-13. The multiplexing process is similar, but here the output of each basic group modulates the higher-frequency carriers which are again summed to create an even more complex single-channel signal. Also SSB is used, and the lower sidebands are selected. Each of the 5 channels in this group carries 12 channels for a total of 60 voice signals. This composite signal is referred to as a *super group*.

The process may again be taken another step further. Up to 10 super groups can be used to modulate subcarriers in the 60- to 2540-kHz range. This allows a total of 5(12)(10) = 600 voice channels to be carried. The output of these 10 multiplexers is referred to as a *master group*. The process can continue with six master groups being further combined into one jumbo group for a total of 3600 channels. These three jumbo groups can then be multiplexed again into one final output to achieve a total of 10,800 voice channels.

Just keep in mind that as more and more levels of multiplexing are used, the bandwidth required to carry all the signals increases. A bandwidth of many megahertz is required to deal with the 10,800-channel composite signal mentioned above.

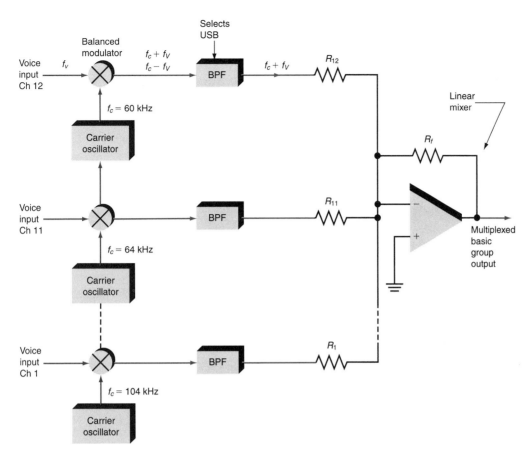

Fig. 8-12 An FDM telephone multiplexer using SSB.

The receiving end of the system is shown in Fig. 8-14. Bandpass filters select out the various channels, and balanced modulators are used to reinject the carrier frequency and produce the original voice input.

The telephone FDM system described here is no longer used in modern telephone systems. Instead, a digital multiplexing technique is used. It will be described later in this chapter.

FDM in Stereo FM

Another well-known example of FDM is broadcast *stereo FM*. All FM broadcast sta-

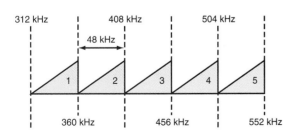

Fig. 8-13 Five basic groups are used to form a super group.

tions use frequency multiplexing to transmit two channels of audio to the FM receiver in your car or to your hi-fi system at home. Let's take a look at how it works.

In stereo, two microphones are used to generate two separate audio signals. The two microphones pick up sound from a common source, such as a voice or band, but from different directions. The separation of the two microphones provides sufficient difference in the two audio signals to provide more realistic reproduction of the original sound. These two independent signals must somehow be transmitted by a single transmitter. This is done by frequency multiplexing techniques.

Figure 8-15 is a general block diagram of a stereo FM multiplex modulator. The two audio signals generally called the left L and right R signals designate the positions of the microphones picking up the original sound. These two signals are fed to a circuit where they are combined to form sum $L + R$ and difference $L - R$ signals. The $L + R$ signal is a linear algebraic combination of the left and right channels. The composite signal it produces is the

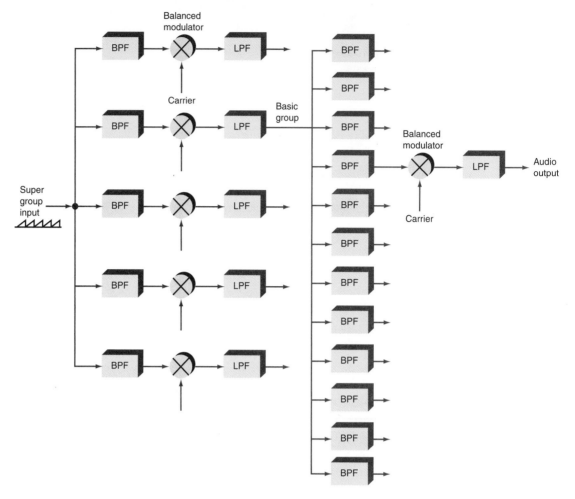

Fig. 8-14 Demultiplexing the telephone signals.

same as if a single microphone were used to pick up the sound. It is this signal that a monaural receiver will hear. See Fig. 8-16.

The combining circuit inverts the right channel signal thereby subtracting it from the left channel signal to produce the $L - R$ signal. These two signals, $L + R$ and $L - R$, will be transmitted independently and recombined later in the receiver to produce the individual right- and left-hand channels.

The $L - R$ signal is used to amplitude-modulate a 38-kHz carrier. This carrier is fed to a balanced modulator along with the $L - R$ signal. The balanced modulator suppresses the carrier but generates upper and lower sidebands as shown in Fig. 8-16. Since the audio response of an FM signal is in the 0.05- to 15-kHz range, the sidebands are in the frequency range of 38 kHz $\pm$ 15 kHz or in the range of 23 to 53 kHz. This DSB signal will be transmitted along with the standard $L + R$ audio signal.

Also transmitted with the $L + R$ and $L - R$ signals is a 19-kHz *pilot carrier.* This is generated by an oscillator whose output will also modulate the main transmitter. Note that the 19-kHz oscillator drives a frequency doubler to generate the 38-kHz carrier for the balanced modulator.

Some FM stations also broadcast another signal referred to as the *Subsidiary Communications Authorization (SCA)* signal. This is a separate subcarrier of 67 kHz that is frequency-modulated by audio signals, usually music. Special SCA FM receivers can pick up the music. The SCA portion of the system is generally used for broadcasting background music for stores, offices, and restaurants. If the SCA system is used, the 67-kHz subcarrier with its music modulation will also modulate the FM transmitter.

As in other FDM systems, all the subcarriers are added with a linear mixer to form a single signal. The spectrum of that composite

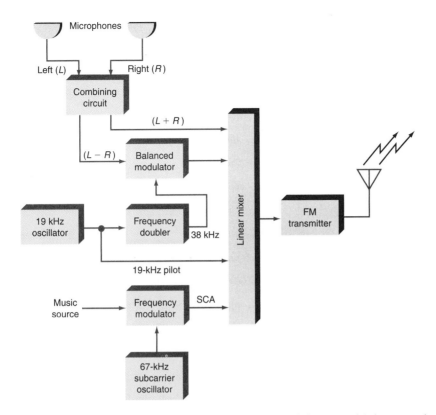

Fig. 8-15 General block diagram of an FM stereo multiplex modulator, multiplexer, and transmitter.

signal is shown in Fig. 8-16. This signal is used to frequency-modulate the carrier of the broadcast transmitter. Again note that FDM simply provides a portion of the frequency spectrum for such independent signals to be transmitted. In this case there is sufficient spacing between adjacent FM stations so that the additional information can be accommodated. Some FM stations now transmit computer data over other subcarriers.

At the receiving end, the demodulation is accomplished with a circuit similar to that illustrated in Fig. 8-17. The FM superheterodyne receiver picks up the signal, amplifies it, and translates it to an IF, usually 10.7 MHz. It is then demodulated. The output of the demodulator is the original multiplexed signal. The various additional circuits now sort out the various signals and reproduce them in their original form.

The original audio $L + R$ signal is extracted by passing the multiplex signal through a low-pass filter. Only the 50- to 15,000-Hz original audio is passed. This signal is fully compatible

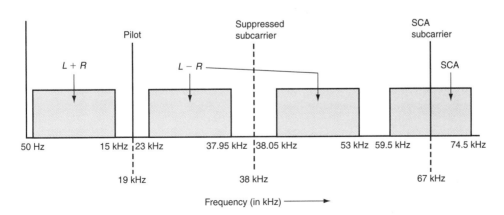

Fig. 8-16 Spectrum of FM stereo multiplex broadcast signal. This signal frequency-modulates the RF carrier.

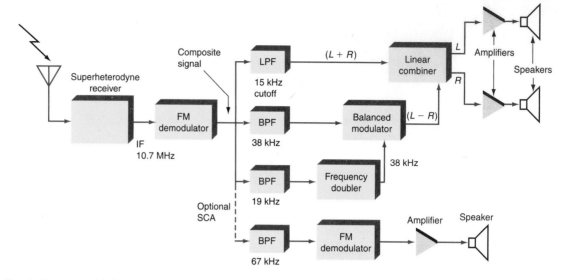

Fig. 8-17 Demultiplexing and recovering the FM stereo and SCA signals.

with monaural FM receivers without stereo capability. In a stereo system, the $L + R$ audio signal is fed to a linear matrix where it is mixed with the $L - R$ signal to create the two separate L and R channels.

The multiplexed signal is also applied to a bandpass filter that passes the 38-kHz suppressed subcarrier with its sidebands. This is the $L - R$ signal that modulates the 38-kHz carrier. This signal is fed to a balanced modulator for demodulation.

The 19-kHz pilot carrier on the multiplexed signal is extracted by passing the multiplexed signal through a narrow-bandpass filter. This 19-kHz subcarrier is then fed to an amplifier and frequency doubler circuit which produces a 38-kHz carrier signal. This is fed to the balanced modulator. The output of the balanced modulator, of course, is the $L - R$ audio signal. This is fed to the linear resistive matrix along with the $L + R$ signal.

If the SCA signal is used, a separate bandpass filter centered on the 67-kHz subcarrier would extract the signal and feed it to a frequency demodulator. The demodulator output would then be sent to an audio amplifier and speaker.

At this point both the $L + R$ and the $L - R$ audio signals have been recovered and fed to the linear matrix. The matrix simply performs an algebraic operation on the two signals. The matrix both adds and subtracts these two signals. Adding the signals produces the left-hand channel.

$$(L + R) + (L - R) = 2L$$

Subtracting the two signals produces the right-hand channel.

$$(L + R) - (L - R) = 2R$$

The left- and right-hand audio signals are then sent to separate audio amplifiers and ultimately to the speakers. Similar multiplex systems are used in AM stereo radio and TV stereo.

■ TEST_____

Supply the missing word(s) in each statement.

6. In FDM, multiple signals share a common channel _____.
7. The multiple signals in an FDM system are used to modulate separate _____.
8. To combine the multiple signals in FDM, a _____ is used.
9. The composite multiplex signal is normally used to _____ an RF carrier before being transmitted.
10. Most telemetry multiplex systems used _____ modulation.
11. Telemetry signals are first conditioned by _____ and/or _____ before modulating a subcarrier.
12. Telemetry signals typically frequency-modulate a _____.
13. The key circuit in a demultiplexer is a(n) _____.
14. The best demodulator for telemetry work is the _____ discriminator.

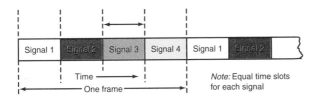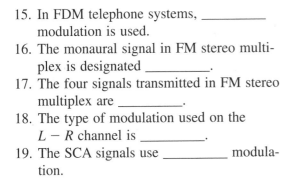

Fig. 8-18 The basic TDM concept.

15. In FDM telephone systems, _____ modulation is used.
16. The monaural signal in FM stereo multiplex is designated _____.
17. The four signals transmitted in FM stereo multiplex are _____.
18. The type of modulation used on the $L - R$ channel is _____.
19. The SCA signals use _____ modulation.

8-3 Time Division Multiplexing

In FDM, multiple signals are transmitted over a single channel by sharing the channel bandwidth. This is done by allocating each signal a portion of the spectrum within that bandwidth. In TDM, each signal can occupy the entire bandwidth of the channel. However, each signal is transmitted for only a brief period of time. In other words, the multiple signals take turns transmitting over the single channel. This concept is illustrated graphically in Fig. 8-18. Here, four signals are transmitted over a single channel. Each signal is allowed to use the channel for a fixed period of time, one after another. Once all the signals have been transmitted, the cycle repeats again and again.

Time division multiplexing may be used with both digital and analog signals. To transmit multiple digital signals, the data to be transmitted is formatted into serial data words. For example, the data may consist of sequential bytes. One byte of data may be transmitted during the time interval assigned to a particular channel. For example, in Fig. 8-18, each time slot might contain 1 byte from each channel. One channel transmits 8 bits and then halts while the next channel transmits 8 bits. The third channel then transmits its data word and so on. One transmission of each channel completes one cycle of operation called a *frame*. The cycle repeats itself at a high rate of speed. In this way, the data bytes of the individual channels are simply interleaved. The resulting

single-channel signal is a digital bit stream that must somehow be deciphered and reassembled at the receiving end.

Pulse-Amplitude Modulation

The transmission of digital data by TDM is straightforward in that digital data is incremental and can be broken up into words that can be easily assigned to different time slots. What is not obvious is how TDM can be used to transmit continuous analog signals. Yet, virtually any analog signal, be it voice, video, or telemetry measurements, can readily be transmitted by TDM techniques. This is accomplished by sampling the analog signal repeatedly at a high rate. *Sampling* is the process of "looking at" an analog signal for a brief instant of time. During this very short sampling interval, the amplitude of the analog signal is allowed to be passed or stored. By taking multiple samples of the analog signal at a periodic rate, most of the information contained in the analog signal will be passed. The resulting signal will be a series of samples or pulses that vary in amplitude according to the variation of the analog signal.

Modulation Figure 8-19 shows an analog signal. The resulting output is a series of pulses whose amplitudes are the same as those of the

Sampling

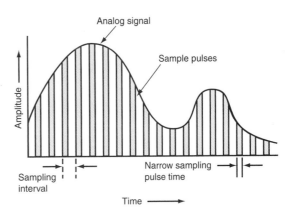

Fig. 8-19 Sampling an analog signal to produce pulse-amplitude modulation.

Sampling theorem

Pulse-amplitude modulation (PAM)

analog signal during the sample period. This process is known as *pulse-amplitude modulation (PAM)*.

The basic circuit for generating PAM is illustrated in Fig. 8-20. An astable clock oscillator drives a one-shot multivibrator which generates a narrow fixed-width pulse. This pulse is applied to a gate circuit that is essentially a switch that will open and close in accordance with the one-shot signal. When the one shot is off, the gate is closed and the analog signal applied to it will not pass. When the clock triggers the one shot once per cycle, the gate opens for a short period of time allowing the analog signal to pass through. The gate circuit may be constructed with diodes or can be an arrangement of bipolar or field-effect transistors.

Demodulation To recover the original information, the transmitted pulses are simply passed through a low-pass filter. The upper cutoff frequency of the low-pass filter is selected to pass the highest-frequency components contained within the analog signal. All higher frequencies are eliminated. Since the pulses themselves represent a composite of many high-frequency harmonics, these are effectively filtered out. The pulses, therefore, are smoothed into a continuous analog signal that is virtually identical in information content to the original transmitted signal. The process is similar to that used in a simple AM diode detector.

Sampling Rate In order for the recovered signal to be an accurate representation of the orig-

inal, the sampling rate must be high enough to ensure that rapid fluctuations are sampled a sufficient number of times. It has been determined that the sampling rate must be at least two times the highest-frequency component of the original signal in order for the signal to be adequately represented. This relationship between the original analog signal and the sampling frequency is known as the *sampling theorem*. If the upper bandwidth value of the analog signal is known, the minimum sampling rate can be found by simply multiplying it by 2.

If the sampled signal is a simple sine wave, then the minimum sampling frequency can be twice the sine wave frequency. A 2-kHz sine wave would have to be sampled a minimum of 2×2 kHz, or 4 kHz. A more complex signal containing harmonics up to 650 kHz would have to be sampled at a 2×650 kHz = 1300 kHz, or 1.3-MHz, rate or higher. The higher the sampling rate, the better the representation. Most systems sample at a rate higher than the minimum 2 times to ensure good fidelity.

In telephone communications, the upper frequency value of the voice content is assumed to be 3 kHz. This dictates a 2×3 kHz, or 6-kHz, sampling rate. In practice, the sampling rate for audio in telephone systems is 8 kHz. The higher sampling rate provides more faithful reproduction of the audio signal.

Although in many applications a sampling rate of twice the highest-frequency content is satisfactory, usually the sampling rate is made much higher. The actual value depends upon the application, but typically the sampling rate is 4 to 5 times the highest-frequency component in the analog signal. A sampling rate of 10 times the maximum analog bandwidth is ideal. This provides excellent representation of the signal.

Time Multiplexer

Now, by combining the concepts of TDM and PAM, you can see how multiple analog signals can be transmitted over a single channel. This is accomplished by a circuit called a multiplexer (usually abbreviated MUX or MPX). The *multiplexer* is simply a single-pole, multiple-position mechanical or electronic switch that sequentially samples the multiple analog inputs at a high rate of speed. The simple rotary switch shown in Fig. 8-21 is an

Multiplexer

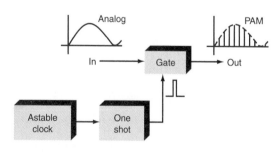

Fig. 8-20 A pulse-amplitude modulator.

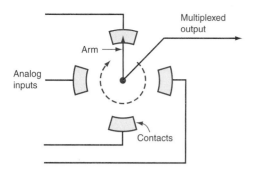

Fig. 8-21 Simple rotary-switch multiplexer.

example. The switch arm dwells momentarily on each contact allowing the input signal to be passed through to the output. It then switches quickly to the next channel and allows that channel to pass for a fixed duration. The remaining channels are sampled in the same way. After each signal has been sampled, the cycle repeats. The result is that four analog signals will be sampled, creating PAM signals that are interleaved with one another. Figure 8-22 illustrates how four different analog signals are sampled by this technique. Be sure that you study the figure so that you can recognize each of the four signals in the composite waveform.

Multiplexers used in early TDM/PAM systems used a form of rotary switch known as a commutator. Multiple switch segments were attached to the various incoming signals, and a high-speed brush rotated by a dc motor rapidly sampled the signal as it passed over the contacts. Such commutators were used in early telemetry systems but have now been totally replaced by electronic circuits.

In practice, the duration of the sample pulses is shorter than the time which is allocated to each channel. For example, assume it takes the commutator or multiplexer switch 1 ms to move from one contact to another. The contacts could be set up so that each sample is 1 ms long. Typically, the duration of that sample is usually made about half that period, or in this example, 0.5 ms.

One complete revolution of the commutator switch is referred to as a *frame*. In other words, during one frame, each input channel is sampled one time. The number of contacts on the multiplexer switch or commutator sets the number of samples per frame. The number of frames completed in 1 s is called the *frame rate*. If you multiply the number of samples per frame by the frame rate, you will get the commutation rate or multiplex rate. This is the frequency of the pulses in the final multiplexed signal.

In our example in Fig. 8-22, the number of samples per frame is four. Assume that the frame rate is 100 frames per second. The period for one frame, therefore, is $1/100 = 0.01 = 10$ ms. During that 10-ms frame period, each of the four channels will be sampled once. Assuming equal sample durations, each channel would be allowed $10/4 = 2.5$ ms. As indicated earlier, the full 2.5-ms period would not be used. Instead, the sample duration during that interval might only be 1 ms long.

Since there are four samples taken per frame, the commutation rate would be 4(100) or 400 pulses per second. This would be the basic frequency of the composite signal to be transmitted over the communications channel.

In practical TDM/PAM systems, electronic circuits are used instead of mechanical switches or commutators. The multiplexer itself is usually implemented with FETs, which are nearly ideal off/on switches that can turn off and on at very high speeds. A complete TDM/PAM

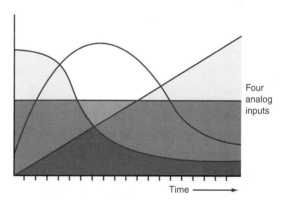

Four analog inputs

Time ⟶

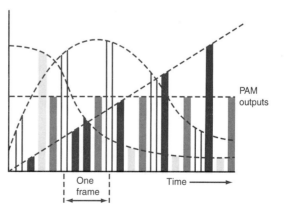

PAM outputs

One frame

Time ⟶

Fig. 8-22 Four-channel PAM time division multiplexer.

Frame

Frame rate

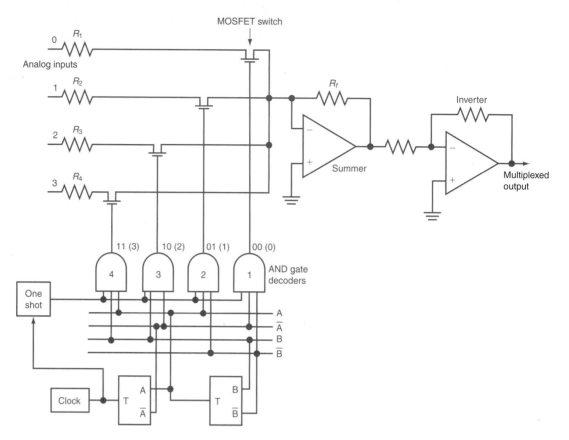

Fig. 8-23 A PAM multiplexer.

circuit is illustrated in Fig. 8-23. Only four channels are used so as to simplify the discussion.

The multiplexer is an op-amp summer circuit with FETs on each input resistor. When the FET is conducting, it has a very low resistance and, therefore, acts as a closed switch. When the transistor is cut off, no current flows through it and, therefore, it acts as an open switch. A digital pulse applied to the gate of the FET turns the transistor on. The absence of a pulse means that the transistor is cut off. The control pulses to the FET switches are such that only one FET is turned on at a time. These FETs are turned on in sequence by the digital circuitry illustrated.

All the FET switches are connected in series with resistors R_1 to R_4 that in combination with the feedback resistor R_f on the op-amp circuit determine the gain. For our discussion here, we will assume that the input and feedback resistor are all equal in value, meaning that the op-amp circuit has a gain of 1. Since this op-amp summing circuit inverts the polarity of the analog signals, it is followed by another op-amp inverter that again inverts and restores the proper polarity.

The digital control pulses are developed by the counter and decoder circuit shown in Fig. 8-23. Since there are four channels, four counter states are needed. Such a counter can be implemented with two flip-flops which can represent four discrete states. These are 00, 01, 10, and 11. These are the binary equivalents of the decimal numbers 0, 1, 2, and 3. We can, therefore, label our four channels as channels 0, 1, 2, and 3.

A clock oscillator circuit triggers the two flip-flop counters. The clock and flip-flow wave-forms are illustrated in Fig. 8-24. The flip-flop outputs are applied to the decoder gates. These are AND gates that are connected to recognize the four binary combinations 00, 01, 10, and 11. The output of each decoder gate is applied to one of the multiplexer FET gates.

The one-short multivibrator shown in Fig. 8-24 is used to trigger all the decoder AND gates at the clock frequency. This one-shot multivibrator produces an output pulse whose duration has been set to the desired sampling interval. Recall in our earlier discussion the sampling interval was 2.5 ms and the actual

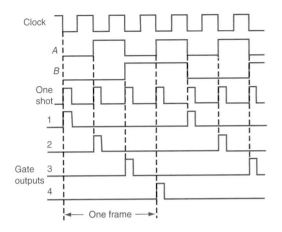

Fig. 8-24 Waveforms for a PAM multiplexer.

sample length was set at 1 ms. Here the one shot would have a 1-ms pulse duration.

Each time the clock pulse occurs, the one shot generates its pulse which is applied simultaneously to all four AND decoder gates. At any given time, only one of the gates is enabled. The output of the enabled gate will be a pulse whose duration is the same as that of the one shot.

When the pulse occurs, it turns on the associated MOSFET and allows the analog signal to be sampled and passed through the op amps to the output. The output of the final op amp is the multiplexed PAM signal like that in Fig. 8-22.

Using PAM to Modulate a Carrier

The varying-amplitude PAM signal is not transmitted as it is over the single channel. Instead, these varying-amplitude pulses are used to modulate a carrier that is then transmitted over the communications medium. In most systems, the PAM signal is used to frequency-modulate either an RF carrier for radio transmission or a subcarrier which, in turn, modulates a final RF carrier. Two such arrangements are shown in Fig. 8-25. In the first arrangement, the PAM signals phase-modulate a carrier. This system is, therefore, referred to as a PAM/PM system.

In the second arrangement, the PAM signals phase-modulate a subcarrier. In fact, two sets of PAM signals are shown, each phase-modulating a subcarrier. These subcarriers are then linearly mixed and used to phase-modulate the RF carrier, which is the final transmitted signal. This system is referred to as PAM/PM/PM. The second system uses a combination of TDM and FDM schemes to create the final composite signal.

Once the composite signal is received, it must be demodulated and demultiplexed. In a PAM/PM/PM system, the signal is picked up by the receiver which ultimately sends the signal to a phase demodulator which recovers the

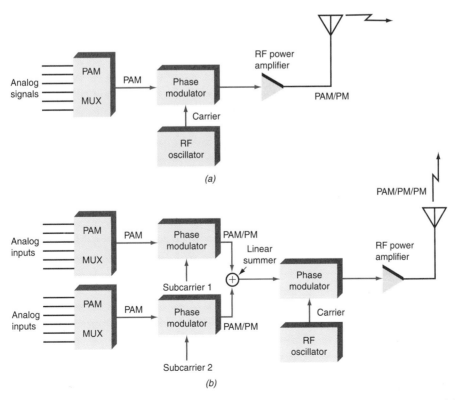

Fig. 8-25 Combining analog (frequency) and digital (PAM) multiplexing. (a) PAM/PM, and (b) PAM/PM/PM.

original PAM data. In a PAM/PM/PM system, two levels of phase demodulation are required before the PAM signal is available. Once the composite PAM signal is obtained, it is applied to a demultiplexer (usually abbreviated *DEMUX*). The DEMUX is, of course, the reverse of a multiplexer. It has a single input and multiple outputs, one for each original input signal. Following with our four-channel example, a DEMUX for this system would have a single input and four outputs. Again, most DEMUXes use FETs driven by a pulse counter arrangement like that shown earlier in Fig. 8-23.

Demultiplexing

The main problem encountered in demultiplexing is *synchronization*. That is, in order for the PAM signal to be accurately demultiplexed into the original sampled signals, some method must be used to ensure that the clock frequency used on the DEMUX is identical to that used at the transmitting multiplexer. Further, even though the clock frequencies may be identical, the sequence of the DEMUX must be identical to that of the multiplexer so that when channel 1 is being sampled at the transmitter, channel 1 will be turned on in the receiver DEMUX at the same time. Such synchronization is usually carried out by a special synchronizing pulse in-

cluded as a part of each frame. Let's take a look at some of the circuits used for clock frequency and frame synchronization.

Instead of using a free-running clock oscillator set to the identical frequency of the transmitter system clock, the clock for the DEMUX is derived from the received PAM signal itself. The circuits shown in Fig. 8-26 are typical of those used to generate the DEMUX clock pulses. They are called *clock recovery circuits*.

In Fig. 8-26(a), the PAM signal is first applied to an amplifier-limiter circuit. This amplifies all the received pulses to a high level and then clips them off at a fixed level. The result is that the output of the limiter is a constant-amplitude rectangular wave whose output frequency is equal to the commutation rate. This is the frequency at which the PAM pulses occur. This, of course, is determined by the transmitting multiplexer clock.

The rectangular pulses at the output of the limiter are applied to a bandpass filter. This bandpass filter eliminates all the upper harmonics, creating a sine wave signal at the transmitting clock frequency. This signal is applied to the phase detector circuit in a PLL along with the input from a VCO. The VCO is set to operate at the frequency of the PAM pulses. However, the VCO frequency is controlled by a dc error voltage applied to its input. This

DEMUX

Clock recovery

Synchronization

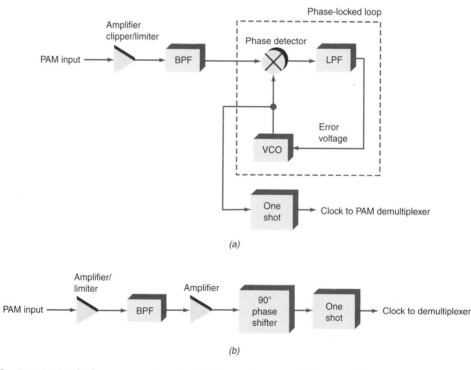

(a)

(b)

Fig. 8-26 Two PAM clock recovery circuits. (a) Closed loop and (b) open loop.

input is derived from the phase detector output which is filtered by a low-pass filter into a dc voltage.

The phase detector compares the phase of the incoming PAM sine wave to the VCO sine wave. If a phase error exists, the phase detector will produce an output voltage that is translated into direct current to vary the VCO frequency. The system is stabilized or locked when the VCO output frequency is identical to that of the sine wave frequency derived from the PAM input. When the PLL is locked, the two sine waves are shifted in phase by 90°.

If the PAM signal frequency changes for some reason, the phase detector picks up the variation and generates an error signal that is used to change the frequency of the VCO to match. Because of the closed-loop feature of the system, the VCO will automatically track frequency changes in the PAM signal. This means that the clock frequency used in the DE-MUX will always perfectly match that of the original PAM signal regardless of any frequency changes that occur.

The output signal of the VCO is applied to a one-shot pulse generator that creates rectangular pulses at the proper frequency. These are used to step the counter in the DEMUX from which are derived the gating pulses for the FET DEMUX switches.

A simpler open-loop clock pulse circuit is shown in Fig. 8-26(*b*). Again, the PAM signal is applied to an amplifier-limiter and then a bandpass filter, just as it was in the previous circuit. The sine wave output of the bandpass filter is then amplified and applied to a phase-shift circuit which produces a 90° phase shift at the frequency of operation. This phase-shifted sine wave is then applied to a pulse generator which, in turn, creates the clock pulses for the DEMUX. Although this circuit

works satisfactorily, the phase-shift circuit is fixed to create a 90° shift at only one frequency, and, therefore, minor shifts in input frequency will produce clock pulses whose timing is not perfectly accurate. In most systems where frequency variations are not great, the circuit operates reliably.

With clock pulses of the proper frequency, some means is now needed to synchronize the multiplexer channels. This is usually done with a special synchronizing pulse that is applied to one of the input channels at the transmitter. In our example of a four-channel system, only three actual signals would be transmitted. The fourth channel would be used to transmit a special pulse whose characteristics would be unique in some way so that it could be easily recognized. The amplitude of the pulse may be higher than the highest-amplitude data pulse, or the width of the pulse may be wider than those pulses derived by sampling the input signals. Special circuits can then be used to detect the *synchronizing (sync) pulse.*

Figure 8-27 shows an example of a sync pulse that is higher in amplitude than the maximum pulse value of any data signal. The sync pulse is also the last to occur in the frame. At the receiver, a comparator circuit is used to detect the sync pulse. One input to the comparator is set to a dc reference voltage equal to slightly higher than the maximum amplitude possible for the data pulses. When a pulse occurs that is greater than this amplitude, the comparator will generate an output pulse. The only time this occurs is when the sync pulse occurs. The output of the comparator will be a pulse that occurs at the same time as the sync pulse. This pulse can then be used for synchronization purposes.

Another method of providing sync is not to transmit a pulse during one channel

Synchronizing pulse

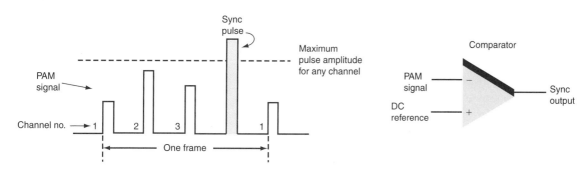

Fig. 8-27 Frame sync pulse and comparator detector.

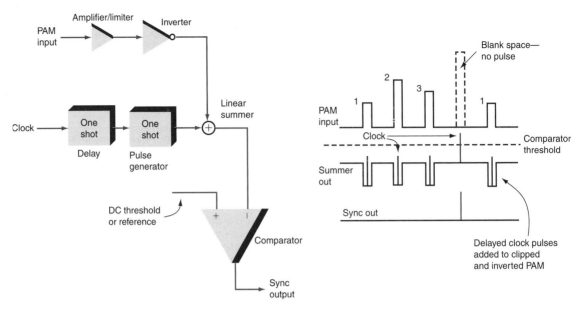

Fig. 8-28 A PAM sync detector circuit.

interval. This leaves a blank space in each frame. This blank space can then be detected and used for synchronizing purposes. Such a circuit for doing this is illustrated in Fig. 8-28. Here the PAM multiplex signal is applied to an amplifier-limiter as it was in the clock circuits. The output is a series of pulses occurring at the pulse repetition rate of the PAM signal. However, since no pulse is transmitted during the sync interval, a blank space occurs. The resulting signal is inverted, so the blank space actually appears as a wider pulse, as shown in Fig. 8-28. This signal is then added to a series of narrow clock pulses. These pulses are delayed for one-half the clock period before they are linearly added to the clipped PAM signal. This delay is accomplished with a one-shot multivibrator whose pulse duration is set to one-half the clock period. The first one shot generates a delay, while another one shot generates clock pulses at the same frequency but at a predetermined width. These are then added in a linear resistive circuit to form the composite shown. The output of the linear mixer is applied to a comparator. The comparator threshold is set so that only the clock pulse added to the blank portion of the original signal is passed. Again, this synchronizing signal is used by the DEMUX.

As indicated, the sync pulse is usually the last one transmitted within a given frame.

This sync pulse, when detected at the receiver, is used as a reset pulse for the counter in the DEMUX circuit. At the end of each frame, the counter is reset to zero, meaning that channel 0 is selected. Now when the next PAM pulse occurs, the DEMUX will be set to the proper channel. Clock pulses then step the counter in the proper sequence for demultiplexing.

Finally, at the output of the DEMUX, separate low-pass filters are applied to each channel to recover the original analog signals. Figure 8-29 shows the complete PAM DEMUX.

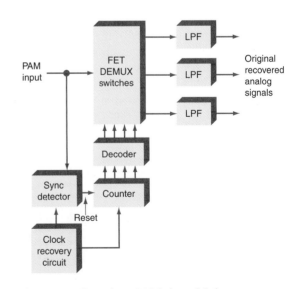

Fig. 8-29 Complete PAM demultiplexer.

Supply the missing information in each statement.

20. In TDM, multiple signals share a channel by transmitting in different _____.

21. The process of sampling an analog signal at a high rate of speed is called _____.

22. The minimum sampling rate for a signal with a 7-kHz bandwidth is _____ kHz.

23. The basic switching element in a PAM multiplexer is a _____.

24. The name of the circuit used to re-create the clock pulses at the receiver in a PAM system is called a _____.

25. The multiplexer and demultiplexer are kept in step with one another by a(n) _____.

26. The time period during which all channels are sampled once is called a(n) _____.

27. A PAM signal is demodulated with a(n) _____.

28. A multiplexer is an electronic multiposition _____.

29. The circuit that samples an analog signal in a PAM system is called a(n) _____.

30. In a PAM/FM system, the RF carrier modulation is _____.

31. A clock recovery circuit that tracks PAM frequency variations uses a(n) _____.

8-4 Pulse-Code Modulation

The most popular form of pulse modulation used in TDM systems is *pulse-code modulation (PCM)*. Pulse-code modulation is a form of digital modulation in which the code refers to a binary word that represents digital data. Multiple channels of serial digital data are transmitted with TDM by allowing each channel a time slot in which to transmit one binary word of data. The various channel data are interleaved and transmitted sequentially. Instead of transmitting a single pulse whose amplitude is the same as that of the analog signal being sampled, in PCM a binary number representing the amplitude of the analog waveform at the sampling point is transmitted. As this state-

About ⬛ Electronics

As a rule of thumb, the per-channel sample rate should be about five times the highest-frequency component of the signal.

ment implies, analog signals may be transmitted by PCM. The analog signal is sampled as in PAM and is then converted into digital format by an *analog-to-digital converter (ADC)*. The ADC converts the analog signal into a series of binary numbers where each number is proportional to the amplitude of the analog signal at the various sampling points. These binary words are converted from parallel to serial format and are then transmitted.

Analog-to-digital converter (ADC)

At the receiving end, the various channels are demultiplexed and the original sequential binary numbers are recovered. These are usually stored in a digital memory and then transferred to a *digital-to-analog converter (DAC)* which reconstructs the analog signal. Of course, the original data may be strictly digital in format, in which case no D/A conversion is required.

Digital-to-analog converter (DAC)

Pulse-code modulation systems allow the transmission of any form of digital data regardless of what it represents. Pulse-code modulation is used in telephone systems to transmit analog voice conversations, binary data for use in digital computers, and even video data. Most long-distance space probes such as the Mariner and Voyager have on-board video cameras whose output signals are digitized and transmitted back to earth in binary format. Such PCM video systems make possible the transmission of pictures over incredible distances.

Multiplexing

Pulse-code modulation (PCM)

Figure 8-30 shows a general block diagram of the major components in a PCM system. We will assume that analog voice signals are the initial inputs. These are applied to ADCs as shown. The output of each ADC is an 8-bit parallel binary word. Since the digital data must be transmitted serially, the ADC output is fed to a shift register that produces a serial data output from the parallel input. The clock oscillator circuit driving the shift register operates at the desired frequency.

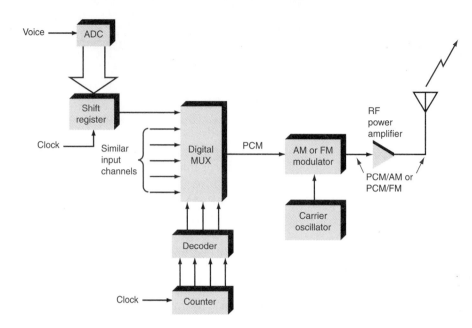

Fig. 8-30 A PCM system.

The multiplexing is done with a simple digital multiplexer. Since all the signals to be transmitted are binary in nature, a multiplexer constructed of standard AND or NAND gates can be used. A binary counter drives a decoder that selects the desired input channel.

The multiplexer output is a serial data waveform of the interleaved binary words. This binary signal is used to modulate a carrier. Either AM or FM may be used in typical systems. This creates either PCM/AM or PCM/FM. The output of the modulator is then fed to a transmitter for radio communications or can otherwise be transmitted by wire or fiber-optic cable. Additional levels of modulation may also be used.

Demultiplexing

At the receiving end of the communications link, the RF signal is picked up by a receiver and then demodulated. Refer to Fig. 8-31. The original serial PCM binary waveform is recovered. This is fed to a shaping circuit, such as a Schmitt trigger, to clean up and rejuvenate the binary pulses. The original signal is then demultiplexed. This is done with a digital DEMUX using AND or NAND gates. The binary counter and decoder driving the DEMUX are kept in step with the receiver through a combination of clock recovery and sync-pulse detector circuits similar to those used in PAM systems. The demultiplexed serial out-

put signals are fed to a shift register for conversion to parallel data and are then sent to a DAC followed by a low-pass filter. The result is a very accurate reproduction of the original voice signal.

Sample/Hold Circuit

Quantizing is the name given to the process of translating amplitude samples of an analog waveform into a binary code word. Quantizing is really the same thing as A/D conversion. You will also hear the term digitizing used to designate the same process.

The first step in the quantizing process is virtually identical to that in PAM. The analog signal is sampled at periodic intervals. This is done as described previously, with a gate. Another method of sampling is to use a *sample and hold (S/H) amplifier.* An S/H amplifier, also called a track/store circuit, accepts the analog input signal and passes it through, unchanged, during its sampling mode. In the hold mode, the amplifier remembers a particular voltage level at the instant of sampling. The output of the S/H amplifier is a fixed dc level whose amplitude is the value at the sampling time.

Figure 8-32 shows a simplified drawing of an S/H amplifier. A high-gain dc differential op amp is the basic element. The amplifier is connected as a follower with 100 percent feedback. Any signal applied to the noninverting

Quantizing

Sample and hold (S/H)

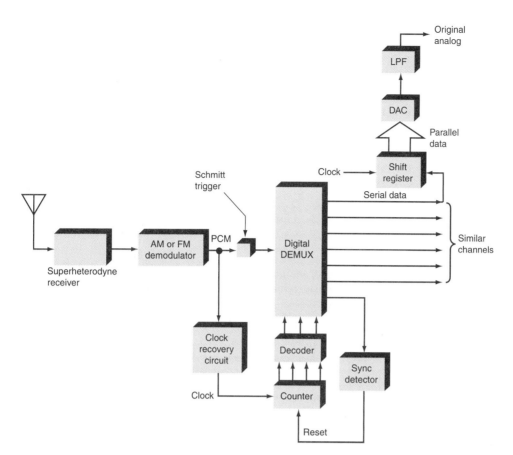

Fig. 8-31 A PCM receiver-demultiplexer.

*inter*NET
CONNECTION

For a fun tutorial on pulse-amplitude modulation, complete with sound, take a look at and listen to the following, which is a university site:

⟨smartpixel
.colorado.edu⟩

(+) input will be passed through unaffected. The amplifier has unity gain and no inversion.

A storage capacitor is connected across the very high input impedance of the amplifier. The input signal is applied to the storage capacitor and the amplifier input through a MOSFET gate. A depletion mode MOSFET is normally used. This MOSFET acts as an on/off switch. When the gate is at 0 V, the transistor turns on, acting as a very low resistance and connecting the input signal to the amplifier. The charge on the capacitor follows the input signal. This is the sample or track mode for the amplifier. The output is simply equal to the input. When the gate voltage is made positive with a pulse, the MOSFET cuts off. In this mode, it acts as an open switch.

During the sample mode, the charge on the capacitor and the op-amp output simply follows the input signal. When the S/H control

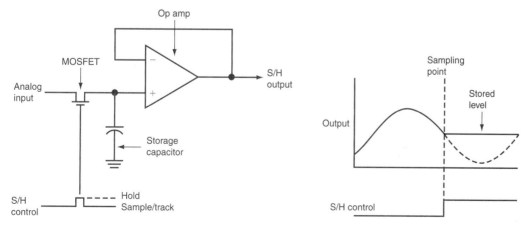

Fig. 8-32 An S/H amplifier.

Multiplexing   **Chapter 8**   213

signal goes high, the transistor is cut off. The charge on the capacitor remains. The very high input impedance of the amplifier allows the capacitor to retain the charge for a relatively long period of time. The output of the S/H amplifier then is the dc voltage value of the input signal at the instant the S/H control pulse switches from low (sample) to high (hold). It is this voltage that is applied to the ADC for conversion into a proportional binary number.

The primary benefit of an S/H amplifier is that it stores the analog voltage during the sampling interval. In some high-frequency signals, the analog voltage may actually change during the sampling interval, either increasing or decreasing. This is an undesirable condition since it will confuse the A/D converter and introduce some error. The S/H amplifier, however, stores the voltage on the capacitor, which remains constant during the sampling interval and thus ensures more accurate quantizing.

A/D Conversion

In the quantizing process, we are effectively dividing the total analog signal amplitude range into a number of equal amplitude increments. Each one of these increments will be represented by a specific binary code. For example, assume that the total analog amplitude voltage range is 0 to 15 V. We could represent each voltage increment by a 4-bit binary number where 0 V was represented by 0000 and 15 V was represented by 1111. During the sampling of the analog wave, the amplitudes of the samples can assume any one of a number of infinite values between 0 and 15 V. In the quantizing process, each of those values will be converted into an even or integer value. For example, one of the analog samples may be 9.2 V. This is closest to the integer 9 and, therefore, the 9.2-V value will be represented by the value 9 in binary form, or 1001. An analog value of 12.7 V might be represented as the integer 13, or 1101. As you can see, the quantizing process introduces some error. This is called *quantizing error*. The result is that the analog signal is somewhat distorted by the process. The quantized analog signal is only an approximation of the real thing.

Quantizing error

Although 15 levels provide only crude quantization, improved representations of the analog signal can be used by providing more quantizing increments. The greater the number of individual voltage increments or levels provided in the quantizer, the more closely the analog signal can be approximated.

It has been determined that the range of voice amplitude levels in the telephone system is approximately 1000 to 1. In other words, the largest-amplitude voice peak is approximately 1000 times the smallest voice signal. This voltage ratio of 1000:1 represents a 60-dB range. If a quantizer with 1000 increments were used, very high quality analog signal representation would be achieved. For example, an ADC with a 10-bit word can represent 1024 individual levels. A 10-bit ADC would provide excellent signal representation. If the maximum peak audio voltage were 1 V, then the smallest voltage increment would be one-thousandth of this, or 1 mV.

In practice, it has been found that it is not necessary to use this many quantizing levels for voice. In most practical PCM systems, a 7- or 8-bit ADC is used for quantizing. One popular format is to use an 8-bit code where 7 bits represent 128 amplitude levels and the eighth bit designates polarity ($0 = +$, $1 = -$). Overall, this provides 256 levels, one-half positive, the other half negative.

Companding

As indicated earlier, the analog voltage range of a typical voice signal is approximately 1000 to 1. It turns out, however, that lower-level signals predominate. Most of the conversations take place at a normal low level. Therefore, the upper end of the quantizing scale is not often used. It may be reached during momentary peaks of loud talking, shouting, or emotional outbursts, but for most general conversations, the lower-level signals will be more typical.

Since most of the signals are low level, the quantizing error will be larger. In other words, the smallest increment of quantization becomes a larger percentage of the lower-level signal. It is a smaller percentage of the peak amplitude value, of course, but that is irrelevant when the signals are much lower in amplitude. The increased quantizing error can produce garbled or distorted sound.

In addition to increased quantizing error, low-level signals are also more susceptible to noise. Noise represents random spikes or voltage impulses added to the signal. The result is static that interferes with the low-level signals and makes intelligibility difficult.

The most common means of overcoming the problems of quantizing error and noise is to use a process of signal compression and expansion known as *companding*. At the transmitting end, the voice signal to be transmitted is compressed. That is, its dynamic range is decreased. The lower-level signals are emphasized, and the higher-level signals are de-emphasized. This compression can take place prior to quantizing. But in some systems, companding is accomplished digitally in the analog-to-digital converter (ADC) by having unequal quantizing steps, small ones at low levels and larger ones at higher levels.

At the receiving end, the recovered signal is fed to an expander circuit that does the opposite, de-emphasizing the lower-level signals and emphasizing the higher-level signals, thereby returning the transmitted signal to its original condition. Companding greatly improves the quality of the signal being transmitted.

One type of compression circuit is a nonlinear amplifier that amplifies lower-level signals more than it does upper-level signals. Figure 8-33 shows a graph illustrating the companding process. The curve shows the relationship between the input and output of the compander. Note that at the lower input voltages, the gain of the amplifier is high and produces high output voltages. As the input voltage increases, the curve begins to flatten, producing proportionately lower gain. The nonlinear curve compresses the upper-level signals while bringing the lower-level signals up to a higher amplitude. Such compression greatly reduces the dynamic range of the audio signal. Instead of an

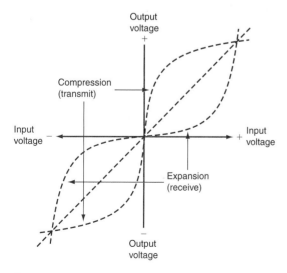

Fig. 8-33 Compression and expansion curves.

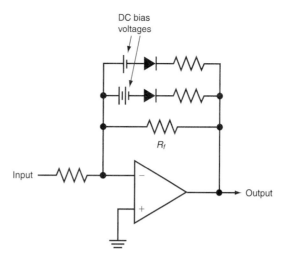

Fig. 8-34 Compression amplifier.

amplitude ratio of approximately 1000:1, compression reduces this to approximately 60:1. The actual degree of compression, of course, can be controlled by carefully designing the gain characteristics of the compression amplifier. Thus the 60-dB voice range is reduced to around a 36-dB range.

In addition to minimizing quantizing error and the effects of noise, compression also lowers the dynamic range so that fewer binary bits are required to digitize the audio signal. A 64:1 voltage ratio could be easily implemented with a 6-bit ADC. In practice, a 7-bit ADC is used.

Figure 8-34 shows a simplified diagram of a *compression amplifier.* Two biased diodes and their resistors are connected as the feedback elements in an op amp. Although only two diodes are shown, typically multiple diodes each biased to a different voltage level are used. At the lower-level signals, the amplifier output is basically linear. But as the input signals get larger, the amplifier output will grow larger and at some point the diodes will begin to turn on. When the diodes turn on, they begin to shunt the feedback resistor R_F and thereby reduce the amplifier gain. Once the audio signal has been compressed, it is then applied to the ADC for quantizing.

At the receiving end, the digital signals are translated into analog signals. An analog signal is then passed through an *expander amplifier* that performs the opposite function of the compressor. A typical circuit is shown in Fig. 8-35. Again, biased diodes are used to shape the amplifier gain curve. Here the biased diodes are used on the input resistance to an op amp.

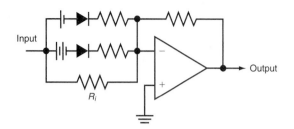

Fig. 8-35 Expansion amplifier.

Figure 8-33 shows the variation of the output voltage with respect to the input voltage during expansion. The lower-level signals are amplified less than the higher-level signals. If the compression and expansion curves are equal and opposite, the result will be a highly accurate reproduction of the original transmitted signal. In Fig. 8-35, as the input increases, the output increases up to a certain level. Beyond that point, the diodes cut in, shunting the input resistor R_i, thereby increasing the gain of the circuit in steps at the higher-input levels.

There are two basic types of companding used in telephone systems. One is called the μ-*law compander* and the other is called the *A-law compander.* The μ-law compander has a slightly different compression and expansion curve than the A-law compander. The μ-law compander is used in telephone systems in the United States and Japan, and the A-law compander is used in European telephone networks. The two are incompatible, but conversion circuits have been developed to make them compatible. These circuits convert μ-law to A-law and vice versa. According to international telecommunications regulations, those using μ-law are responsible for the conversions.

Although some companding is analog in nature as described, digital companding is far more widely used. The most common method is to use a nonlinear ADC. These converters provide a greater number of quantizing steps at the lower levels than they do at the higher levels, providing compression. On the receiving end, a matching nonlinear DAC is used to provide the opposite, compensating, expansion effect.

The T-1 System

Now let's take a look at a typical PCM system. The most commonly used PCM system is the *T-1 system,* developed by Bell Telephone for transmitting telephone conversations by high-speed digital links. The T-1 system multiplexes 24 voice channels onto a single line using TDM techniques. Each analog voice channel is sampled at an 8-kHz rate. In other words, the analog waveform is sampled every 125 μs. Those samples are then converted to serial digital words by the ADCs. These serial digital words from each of the 24 channels are then transmitted one after another. Each sample is an 8-bit word, 7 bits of magnitude and 1 bit representing the sign of polarity.

T-1 Frame A T-1 signal is show in Fig. 8-36. During the 125-μs interval between analog sampling on each channel, twenty-four 8-bit words occur, each representing a sample from each of the channels. The channel sampling rate is (125 μs)/24 = 5.2 μs per cycle or 192 kHz. This represents a total of 24(8) = 192 bits. One additional bit is then added to this stream for synchronization purposes. That bit is a sync pulse used to keep the transmitting and receiving signals in step with one another. This represents a total of 193 bits. The twenty-four 8-bit words and the synchronizing bit form one frame. The sequence is then repeated again and again. The 193 bits multiplied by the 8-kHz sampling rate produce a total bit rate for the multiplexed signal of 193 × 8 kHz = 1544 kHz = 1.544 MHz.

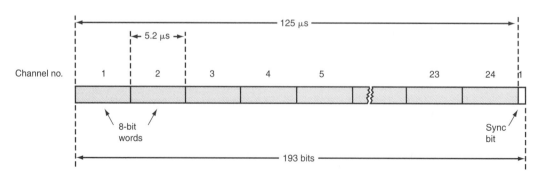

Fig. 8-36 The T-1 frame format, serial data.

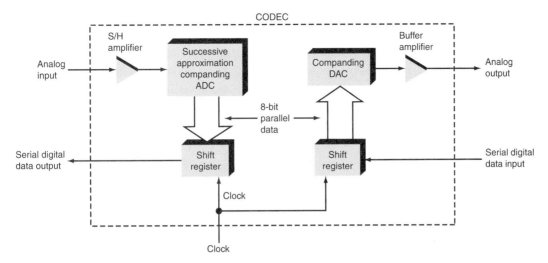

Fig. 8-37 Simplified block diagram of an IC codec.

Since telephone conversations are two-way in nature, both ends of the communication link have transmitting (multiplexing) and receiving (demultiplexing) capability. All the A/D and D/A conversion and related functions, such as serial-to-parallel and parallel-to-serial conversion as well as companding, are usually taken care of by a single large-scale IC chip known as a *codec*. One codec is used per channel at each end of the communication channel. The codecs are combined with the digital multiplexers and DEMUXes and the clock and synchronizing circuits to complete the system.

Figure 8-37 shows a simplified block diagram of a codec. The analog input is sampled by the S/H amplifier at an 8-kHz rate. The samples are quantized by the successive-approximation type of ADC. Compression is done digitally in the ADC. The parallel ADC output is sent to a shift register to create the serial data output which goes to one input of a digital multiplexer.

The serial digital input is derived from a digital DEMUX. The clock shifts it into a shift register for serial-to-parallel conversion. The 8-bit parallel word is sent to the DAC. This DAC has digital expansion built in. The analog output is then buffered. It may be filtered externally. Most codecs are made with CMOS and come in a 16-pin dual-in-line IC package.

Today PAM is rarely used in telephone, telemetry, or other systems requiring multiplexing. Neither are pulse-width modulation (PWM) or pulse-position modulation (PPM), both of which were once popular in TDM systems. Today, virtually all TDM systems use PCM. Because it is all digital, PCM is highly immune to noise, very reliable, and inexpensive.

In PCM, the transmitted binary pulses all have the same amplitude; therefore, like an FM signal, they may be clipped to reduce noise. Further, even if the signals are highly degraded due to noise, attenuation, or distortion, all the receiver has to do is determine whether a pulse was transmitted or not. It does not have to worry about amplitude, width, frequency, phase shape, or otherwise. It just determines whether the pulse is there or not. Such signals are easily recoverable and rejuvenated, even under the most severe attenuation, noise, or distortion.

Codec

T-1 Variations   The T-1 system is widely used to transmit telephone calls between local central offices over distances of 5 to 50 mi. For longer-distance transmission, T-1 signals are multiplexed into other, faster TDM systems. The hierarchy of TDM systems used in telecommunications is given below.

System	Number of Channels	Data Rate, MHz	Distance, mi
T1	24	1.544	5–50
T2	96	6.312	500
T3	672	46.304	500
T4	4032	274.176	500
T5	8064	560.16	500

T-1 systems are also frequently used to transmit fewer than 24 inputs at a faster rate. For example, a T-1 line can transmit a single source of computer data at a 1.544-MHz rate. It can also transmit two data sources at a 722-kHz rate or four sources at a 386-kHz rate, and so on. These are known as *fractional T-1 lines.*

■ TEST

Supply the missing information in each statement.

32. In PCM systems, analog signals are transmitted as _____.
33. Quantizer is another name for a(n) _____.
34. The standard audio sampling rate in a PCM telephone system is _____ kHz.
35. The standard word size in a PCM audio system is _____ bits.
36. The name of a widely used PCM telephone system is _____.
37. The process of emphasizing lower-level signals and de-emphasizing higher-level signals is called _____.
38. An LSI circuit containing an ADC, DAC, compander, expander, and serial/parallel converters is called a(n) _____.
39. The main benefit of PCM over PAM, FM, and other modulation techniques is its immunity to _____.
40. A T-1 multiplexer has _____ channels and a total bit rate of _____ MHz.

Summary

1. Multiplexing is the process of transmitting multiple signals over a single communications channel.

2. The primary benefit of multiplexing is economic since multiple communications can take place for the cost of a single link plus the multiplexing equipment.

3. The two major types of multiplexing are frequency division and time division multiplexing.

4. Some examples of multiplexing include telemetry, telephone systems, satellite communications, and radio/TV stereo broadcasting.

5. In frequency division multiplexing (FDM), multiple signals share the common bandwidth of a single communications channel, each occupying a separate portion of the bandwidth.

6. In FDM, each signal modulates a subcarrier on a different frequency. The subcarriers are then linearly mixed to form a composite signal that is usually used to modulate a final carrier for transmission.

7. In FDM telemetry systems, transducers frequency-modulate subcarrier oscillators whose outputs are combined and used to frequency-modulate a final carrier (FM/FM).

8. A subcarrier oscillator is a VCO whose frequency varies linearly in proportion to the amplitude of a modulating dc or ac signal.

9. Recovering the individual signals at the receiver is done with a demultiplexer whose main components are bandpass filters tuned to the individual subcarrier frequencies.

10. In FM/FM telemetry systems, PLL and pulse-averaging discriminators are used to demodulate the subcarriers; PLL discriminators are preferred because of their superior noise performance.

11. The telephone system routinely uses FDM systems to allow many conversations to be carried on a single cable.

12. Single-sideband suppressed-carrier modulation is used in telephone FDM systems to minimize bandwidth requirements.

13. Stereo broadcasts from FM stations use FDM techniques. Two channels of audio L and R are combined to form $L + R$ and $L - R$ signals. The $L - R$ signal modulates a 38-kHz subcarrier by DSB. The $L + R$ signal, DSB signal, and a 19-kHz pilot carrier are combined and used to frequency-modulate the final transmitter carrier.

14. In some FM broadcast systems, music is broadcast on a separate 67-kHz subcarrier which is frequency-modulated. This is referred to as a Subsidiary Communication Authorization (SCA) signal or channel.

15. In time division multiplexing (TDM), each channel is assigned a time slot and may transmit for a brief period using the entire bandwidth of the medium. Signal sources take turns transmitting.

16. Both digital and analog signals may be transmitted by TDM.

17. Analog signals are transmitted by converting them into a series of pulses whose amplitudes approximate the shape of the analog signal. This process is called pulse-amplitude modulation (PAM).

18. Pulse-amplitude modulation is produced by sampling the analog signal. This is done by periodically opening a gate for a brief period, allowing a narrow portion of the analog signal to pass through.

19. Samples must be taken fast enough in order for high-frequency components to be recognized and adequately represented.

20. A minimum sampling rate is two times the highest frequency component or upper bandwidth limit of the analog signal. This is called the sampling theorem.

21. Pulse-amplitude modulation signals may be multiplexed by allowing samples of several signals to be interleaved into adjacent time slots.

22. Field-effect transistor switches are commonly used in sampling gates and PAM multiplexers. They are controlled by digital circuits that set the sampling intervals and pulse rates.

23. The period of time during which each channel in a PAM system is sampled once is called a frame.
24. Pulse-amplitude modulation signals are normally used to frequency-modulate another carrier creating PAM/FM.
25. Demultiplexing PAM signals requires some means of synchronization to ensure matching clock frequencies and channel timing at the receiver.
26. Special clock recovery circuits use the PAM signal itself to derive the clock signal at the receiver rather than generating it independently. This ensures perfect frequency and phase relationships.
27. A special sync pulse with a unique shape to distinguish it from the PAM pulses is used to keep the demultiplexer in synchronization with the multiplexer. The sync pulse usually occurs as the last pulse in a frame.
28. Most TDM systems in use today use pulse-code modulation (PCM) to transmit analog signals.
29. Pulse-code modulation uses A/D conversion techniques to translate analog signals into binary form for serial transmission.
30. In a multichannel PCM system, each signal is furnished with an analog-to-digital converter (ADC) and parallel-to-serial converter. The resulting binary outputs are digitally multiplexed using TDM techniques.
31. Voice, telemetry data, video, and other analog signals may be transmitted via PCM.
32. Pulse-code modulation signals are recovered at the receiver by demultiplexing and D/A conversion plus appropriate filtering.
33. Pulse-code modulation is generated by periodically sampling the analog signal as in PAM systems. Then the varying amplitude pulses are converted into proportional binary words by an ADC. This process is called quantizing.
34. Quantizing means dividing a given signal voltage range into a number of discrete increments, each represented by a binary code. Each analog sample is matched to the nearest binary level.
35. Sampling in a PCM system is done by a sample/hold (S/H) amplifier that stores the analog value on a capacitor at the instant of measurement.
36. Most ADCs have an 8-bit word providing a quantizing resolution of 1 in 256.
37. Most PCM systems, especially for voice transmission, use companding to compress the voice signal dynamic range by emphasizing lower-level signals and de-emphasizing higher-level signals. Companding minimizes quantization error and improves noise immunity.
38. Companding may be done by analog or digital techniques. Analog companders use diode circuits in amplifiers for compression and expansion. Digital companders use special nonlinear ADCs and DACs.
39. A common PCM system using TDM techniques is the Bell T-1 used in telephone work.
40. A T-1 system multiplexes 24 voice channels, each represented by an 8-bit word. The sampling interval is 125 μs. A frame consists of twenty-four 8-bit words and a single sync pulse. The frame data rate is 1.544 MHz.
41. Most PCM systems like T-1 use a special LSI circuit called a codec to handle conversion and companding. A codec contains the ADC, DAC, serial-to-parallel conversion circuits, companders, and related timing circuits.
42. The sampling rate of a codec is usually 8 kHz.
43. Pulse-width modulation (PWM) and pulse-position modulation (PPM) systems are no longer used.
44. Most TDM systems use PCM, though some use PAM.
45. Pulse-code modulation is preferred over PAM because of its superior noise immunity.

Chapter Review Questions

Choose the letter which best answers each question.

8-1. Multiplexing is the process of
   a. Several signal sources transmitting simultaneously to a receiver on a common frequency.
   b. Sending the same signal over multiple channels to multiple destinations.
   c. Transmitting multiple signals over multiple channels.
   d. Sending multiple signals simultaneously over a single channel.

8-2. In FDM, multiple signals
   a. Transmit at different times.
   b. Share a common bandwidth.

c. Use multiple channels.

d. Modulate one another.

8-3. Each signal in an FDM system

a. Modulates a subcarrier.

b. Modulates the final carrier.

c. Is mixed with all the others before modulation.

d. Serves as a subcarrier.

8-4. Frequency modulation in FDM systems is usually accomplished with a

a. Reactance modulator.

b. Varactor.

c. VCO.

d. PLL.

8-5. Which of the following is *not* a typical FDM application?

a. Telemetry

b. Stereo broadcasting

c. Telephone

d. Secure communications

8-6. The circuit that performs demultiplexing in an FDM system is a(n)

a. Op amp.

b. Bandpass filter.

c. Discriminator.

d. Subcarrier oscillator.

8-7. Most FDM telemetry systems use

a. AM.

b. FM.

c. SSB.

d. PSK.

8-8. The best frequency demodulator is the

a. PLL discriminator.

b. Pulse-averaging discriminator.

c. Foster-Seeley discriminator.

d. Ratio detector.

8-9. The modulation used in FDM telephone systems is

a. AM.

b. FM.

c. SSB.

d. PSK.

8-10. The FDM telephone systems accommodate many channels by

a. Increasing the multiplexer size.

b. Using many final carriers.

c. Narrowing the bandwidth of each.

d. Using multiple levels of multiplexing.

8-11. In FM stereo broadcasting, the $L + R$ signal

a. Double-sideband-modulates a subcarrier.

b. Modulates the FM carrier.

c. Frequency-modulates a subcarrier.

d. Is not transmitted.

8-12. In FM stereo broadcasting, the $L - R$ signal

a. Double-sideband-modulates a subcarrier.

b. Modulates the FM carrier.

c. Frequency-modulates a subcarrier.

d. Is not transmitted.

8-13. The SCA signal if used in FM broadcasting is transmitted via

a. A 19-kHz subcarrier.

b. A 38-kHz subcarrier.

c. A 67-kHz subcarrier.

d. The main FM carrier.

8-14. In TDM, multiple signals

a. Share a common bandwidth.

b. Modulate subcarriers.

c. Are sampled at high speeds.

d. Take turns transmitting.

8-15. In TDM, each signal may use the full bandwidth of the channel.

a. True

b. False

8-16. Sampling an analog signal produces

a. PAM.

b. AM.

c. FM.

d. PCM.

8-17. The maximum bandwidth that an analog signal can use with a sampling frequency of 108 kHz is

a. 27 kHz.

b. 54 kHz.

c. 108 kHz.

d. 216 kHz.

8-18. Pulse-amplitude modulation signals are multiplexed by using

a. Subcarriers.

b. Bandpass filters.

c. A/D converters.

d. FET switches.

8-19. In PAM demultiplexing, the receiver clock is derived from

a. Standard radio station WWV.

b. A highly accurate internal oscillator.

c. The PAM signal itself.

d. The 60-Hz power line.

8-20. In a PAM/TDM system, keeping the multiplexer and DEMUX channels in step with one another is done by a

a. Clock recovery circuit.

b. Sync pulse.

c. Sampling.

d. Sequencer.

8-21. Transmitting data as serial binary words is called

a. Digital communications.

b. Quantizing.

c. PAM.

d. PCM.

8-22. Converting analog signals to digital is done by sampling and

    a. Quantizing.

    b. Companding.

    c. Pre-emphasis.

    d. Mixing.

8-23. A quantizer is a(n)

    a. Multiplexer.

    b. Demultiplexer.

    c. A/D converter.

    d. D/A converter.

8-24. Emphasizing low-level signals and compressing higher-level signals is called

    a. Quantizing.

    b. Companding.

    c. Pre-emphasis.

    d. Sampling.

8-25. Which of the following is *not* a benefit of companding?

    a. Minimizes noise

    b. Minimizes number of bits

    c. Minimizes quantizing error

    d. Minimizes signal bandwidth

8-26. A telephone system using TDM and PCM is called

    a. PBX.

    b. RS-232.

    c. T-1.

    d. Bell 212.

8-27. An IC that contains A/D and D/A converters, companders, and parallel-to-serial converters is called a

    a. Codec.

    b. Data converter.

    c. Multiplexer.

    d. Modem.

8-28. Pulse-code modulation is preferred to PAM because of its

    a. Resistance to quantizing error.

    b. Simplicity.

    c. Lower cost.

    d. Superior noise immunity.

Critical Thinking Questions

8-1 How does increasing the number of signals to be multiplexed affect the bandwidth of the transmitted composite signal? Explain your answer for both FDM and TDM.

8-2 Assume that AM and FM broadcast stations could transmit signals other than their main programming on higher-frequency subcarriers. Name some applications you can think of for this capability.

8-3 What type of multiplexing do you think is being used in cable TV systems to transmit many TV signals on a single cable?

8-4 Think of a way in which multiple transmitters and receivers can have a common frequency band without interference and without formal multiplexing techniques.

8-5 In color TV, two analog signals representing the color in a scene are transmitted by multiplexing them in a common bandwidth. Each color signal AM-DSB modulates a 3.58-MHz carrier, with one carrier being shifted 90 degrees from the other. The two DSB signals are added before transmission. What type of multiplexing is this? How do you recover the signals at the receiver? Draw a block diagram of the demultiplexer.

Answers to Tests

1. *c*
2. telephone, telemetry
3. economy
4. demultiplexer
5. time division, frequency division
6. bandwidth
7. subcarriers
8. linear mixer
9. modulate
10. frequency
11. amplifying, filtering
12. subcarrier oscillator or VCO
13. bandpass filter
14. PLL
15. SSB
16. $L + R$
17. $L + R$, $L - R$, 19-kHz pilot, SCA
18. DSB

19. frequency
20. time slots or intervals
21. pulse-amplitude modulation
22. 14
23. FET or MOSFET
24. clock recovery circuit
25. sync pulse
26. frame
27. low-pass filter
28. switch
29. gate

30. FM
31. phase-locked loop
32. binary codes
33. A/D converter
34. 8
35. 8
36. T-1
37. companding
38. codec
39. noise
40. 24, 1.544

Chapter 9

Antennas, Transmission Lines, and Radio Wave Propagation

Chapter Objectives

This chapter will help you to:

1. *Define* and *state* the importance of standing waves and compute standing-wave ratio.
2. *Explain* how quarter- and half-wave transmission lines can act as tuned circuits at high frequencies.
3. *State* the characteristics of a radio wave and *explain* how antenna polarization is determined.
4. *Compute* the length of quarter-, half-, and full-wavelength antennas and transmission lines given the operating frequency.
5. *Name* three widely used types of antenna arrays that exhibit gain.
6. *Name* the three main types of radio wave propagation and *state* how each behaves.

All electronic communications systems consist of a transmitter, a receiver, and a communications medium. In some systems, the transmission medium is a direct link such as a wire, cable, or fiber-optic cable. In radio communications systems, there are no direct connections. The system is "wireless." The RF signal generated by the transmitter is sent into free space and is eventually picked up by the receiver. The processes of launching the signal into space and receiving it are the functions of the antenna. The antenna is a device that acts as the interface between the transmitter and free space and between free space and the receiver. It converts the trans-mitter RF power into electromagnetic signals that can propagate over long distances, and it is also the device that picks up the electromagnetic signals and converts them into signals for the receiver. Since the antennas are typically located remotely from the transmitters and receivers, some means must be used to get the power to and from the antenna. This is the job of the transmission line, a special kind of cable. In this chapter we introduce you to antenna and transmission line principles. We will also briefly examine the propagation characteristics of electromagnetic signals in free space to see how they propagate over long distances.

9-1 Transmission Lines

A *transmission line* is a two-wire cable that connects the transmitter to the antenna or the antenna to the receiver. The purpose of the transmission line is to carry the RF energy for the desired distance. Ideally, the antenna should be connected directly to the transmitter or the receiver, but this is not always practical. Antennas

must be located out of doors as high as possible for best radiation or reception. The equipment, of course, is usually located inside a building or vehicle. Thus the transmission line is a very important link in the communications system.

Balanced Line

There are two basic types of wire transmission lines. These are balanced line and coaxial cable. *Balanced line* is made up of two parallel conductors spaced from one another by a distance of $^1/_2$ inch (in.) up to several inches. Insulating spacers may be used to keep the wires separated, or the spacing may be maintained by a continuous plastic insulator that is part of the conductor insulation (called *twin lead*). Figure 9-1(*a*) shows a two-wire balanced line.

Unbalanced Line

The term *balanced line* means that the same current flows in each wire with respect to ground, although the direction of current in one wire is 180° out of phase with the current in the other wire. Neither wire is connected to ground. In an unbalanced line, one conductor is connected to ground.

The other type of transmission line is *coaxial cable,* usually just called *coax.* Refer to Fig. 9-1(*b*). It consists of a solid-center conductor surrounded by a plastic insulator such as Teflon. Over the insulator is a second conductor, a tubular braid or shield made of fine wires. An outer plastic sheath protects and insulates the braid. Coax is an unbalanced line since the current in the center conductor is referenced to

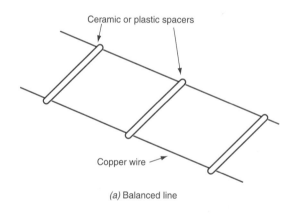

Ceramic or plastic spacers

Copper wire

(a) Balanced line

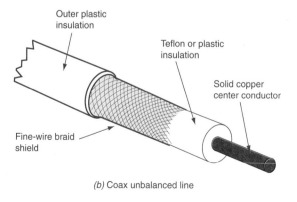

Outer plastic insulation

Teflon or plastic insulation

Solid copper center conductor

Fine-wire braid shield

(b) Coax unbalanced line

Fig. 9-1 The two most common types of transmission lines.

the braid which is connected to ground. Coax comes in a variety of sizes from approximately $^1/4$ in. to several inches in diameter. Of the two types of transmission lines, coax is the most widely used.

Wavelength Matters

Two-wire cables are very common in electricity and electronics. The cables that carry the 60-Hz power line signal into your house are a form of transmission line. The wires connecting the audio output of your stereo receiver to your speaker are also a form of transmission line. At these low frequencies, the transmission line acts simply as a carrier of the ac signal. In these applications, the only characteristic of the cable we are interested in is its resistive loss. Such cables are basically noncritical in their size and characteristics except for conductor size, which determines the current-carrying capability and the voltage drop over long distances.

However, when such cables are used to carry RF energy, strange things happen. The cables are no longer simply resistive conductors but instead appear to be complex interconnections of inductors and capacitors as well as resistors. Further, whenever the length of the cable becomes about one-tenth of the wavelength of the transmitted signal, or more, the transmission line takes on special characteristics that require a more complex analysis.

Wavelength (λ) is a distance. It is the length or distance of one cycle of an ac wave. See Fig. 9-2. Wavelength is also the distance that an ac wave travels in the time required for one cycle of that signal.

Mathematically, wavelength (λ) is expressed as the ratio of the speed of light to the frequency of the signal.

$$\lambda = \frac{300,000,000}{f}$$

In this formula, the large number is the speed of light in meters per second. This is also the

Coaxial cable

Wavelength

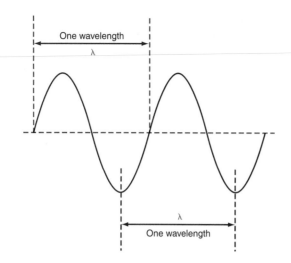

Fig. 9-2 One wavelength.

speed of a radio signal. If the distance is converted to miles, you get the familiar number 186,000 miles per second (mi/s). The frequency f is given in hertz. As you can see then, the wavelength of a 60-Hz power line signal is

$$\lambda = \frac{300,000,000}{60}$$

$$= 5,000,000 \text{ m}$$

That's an incredibly long distance and, in fact, represents a distance of several thousand miles. Power lines are typically much shorter than a wavelength at 60 Hz.

But at RF, say those frequencies of 3 MHz or more, the wavelength gets considerably shorter. The wavelength at 3 MHz is

$$\lambda = \frac{300,000,000}{3,000,000}$$

$$= 100 \text{ m}$$

That's a distance of a little more than 300 ft. This is a very practical distance. As the frequency gets higher, the wavelength gets shorter. The wavelength formula can be simplified at these higher frequencies.

$$\lambda = \frac{300}{f}$$

About ◁▦▷ Electronics

Most transmission lines come with standard fixed values of characteristic impedance.

Here, the frequency is expressed in megahertz. A 50-MHz signal will have a wavelength of 6 m. Converting meters to feet, the wavelength formula becomes

$$\lambda = \frac{984}{f} \quad (1 \text{ m} = 3.281 \text{ ft})$$

where f is again in megahertz.

If you know the wavelength and wish to compute frequency, the above formulas can be rearranged:

$$f = \frac{300}{\lambda} \quad or \quad f = \frac{984}{\lambda}$$

Velocity Factor

An important factor in transmission line applications is that the speed of the signal in the transmission line is slower than the speed of a radio signal in free space. Light and radio signals propagate through free space at a speed of 300,000,000 m/s or 186,000 mi/s as indicated earlier. This speed is considerably less in a transmission line. This leads to the development of a velocity factor F, which is the ratio of the transmission speed in the transmission line V_L and the transmission speed in free space V_S.

$$F = \frac{V_L}{V_S}$$

Velocity factors in transmission lines vary from approximately 0.8 to 0.6. The velocity factor of coax is typically in the 0.6 to 0.7 range. Open-wire line and twin lead have a velocity factor near 0.8. This velocity factor must be taken into consideration when computing the length of a transmission line in wavelengths.

One wavelength is given by the expression

$$\lambda = \frac{984}{f}$$

However, this is not the wavelength of a transmission line. The above formula must be modified by the velocity factor F in order to get the correct length. The formula for a transmission line is:

$$\lambda = 984 \frac{F}{f}$$

For example, what is the actual length in feet of one-quarter wavelength of a coax with a velocity factor of 0.66 at 30 MHz?

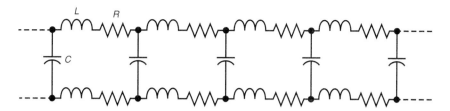

Fig. 9-3 Equivalent electric circuit of a two-wire transmission line.

$$\lambda = 984\,\frac{F}{f}$$

$$= 984\,\frac{0.66}{30}$$

$$= 21.65\text{ ft}$$

One wavelength is 21.65 ft. One-quarter wavelength then is one-fourth of this, or 21.65/4 = 5.41 ft.

It is often necessary to know the length of the transmission line in wavelengths; the above formula can give it to you. Just be sure to use the correct velocity factor for the desired transmission line. This can be obtained from the manufacturer's literature and various handbooks. Common values of F for coax are 0.66, 0.7, 0.75, and 0.8.

Characteristic Impedance

When the length of a transmission line is longer than several wavelengths at the signal frequency, the two parallel conductors of the transmission line appear as a complex impedance. The wires exhibit considerable series inductance at high frequencies. Further, there is a capacitance between the parallel conductors. The result is that to a high-frequency signal, the transmission line appears as a distributed low-pass filter, as shown in Fig. 9-3, consisting of series inductors, resistors, and shunt capacitors. An RF generator connected to the transmission line sees an impedance that is a function of the inductance and capacitance in the circuit. This impedance is known as the *characteristic impedance Z_o*. It is also referred to as the surge impedance. This impedance can be calculated with the simple expression

$$Z_o = \sqrt{\frac{L}{C}}$$

where Z_o = characteristic impedance of transmission line, Ω
    L = inductance of transmission line for a given length
    C = capacitance for that same length

Of course, knowing or measuring the inductance and capacitance for a given length of line is impractical. However, we do know that the inductance and capacitance depend upon the physical characteristics of the line just as the inductance of a coil and capacitance of a capacitor depend upon the physical characteristics of the coil and the capacitor, respectively. Therefore, we can easily express the characteristic impedance in terms of the physical sizes of the cable. The basic expression for the characteristic impedance of a parallel two-wire transmission line is

$$Z_o = 276\log\frac{2S}{D}$$

In this expression, S is the spacing between the centers of the parallel conductors and D is the diameter of the conductors. See Fig. 9-4.

For example, assume two wires are spaced 0.5 in. apart. Each wire has a diameter of 0.082 in. The impedance is found by substituting 0.5 in. and 0.082 in. in the formula above:

$$Z_o = 276\log\frac{2(0.5)}{0.082}$$

$$= 276\log 12.22$$

$$= 276(1.087)$$

$$= 300\ \Omega$$

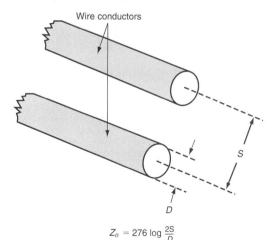

Wire conductors

$$Z_o = 276\log\frac{2S}{D}$$

Characteristic impedance

Fig. 9-4 The characteristic impedance Z_o of a two-wire transmission line depends upon its physical dimensions.

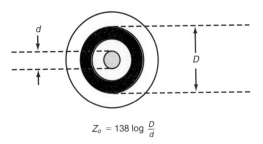

$$Z_o = 138 \log \frac{D}{d}$$

Fig. 9-5 The characteristic impedance of a coax.

The characteristic impedance of a coax is given by the expression

$$Z_o = 138 \log \frac{D}{d}$$

Here, D is the inside diameter of the outer braid shield, and d is the diameter of the inner conductor. See Fig. 9-5.

Assume that the shield inner diameter is 0.2 in. and that the center conductor has a diameter of 0.087 in. The impedance is

$$Z_o = 138 \log \frac{0.2}{0.087}$$
$$= 138 \log 2.3$$
$$= 138(0.3617)$$
$$= 50 \ \Omega$$

Most transmission lines come with standard fixed values of characteristic impedance. For example, a widely used balanced transmission line is the popular TV twin lead in which two stranded wires are separated by a flat continuous plastic insulator as shown in Fig. 9-6. The typical characteristic impedance is 300 Ω. Coaxial cables are commonly available with characteristic impedances of 50 and 75 Ω.

Figure 9-7 gives a summary of the characteristics of several popular types of transmis-

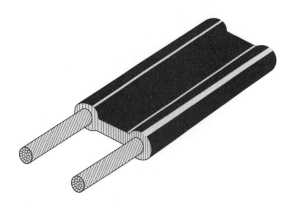

Fig. 9-6 A 300-Ω twin-lead balanced line widely used on TV antennas.

sion lines. Note that coaxes are designated by an alphanumeric code beginning with the letters RG. The primary specifications are characteristic impedance and attenuation. The attenuation is the amount of power lost per 100 ft of cable expressed in decibels. This attenuation is directly proportional to the cable length and increases with frequency. The loss is significant at very high frequencies. However, the larger the cable, the lower the loss. The characteristics of 300-Ω twin lead are also listed in Fig. 9-7 for contrast. Note the low loss compared to coax.

Calculating Cable Attenuation

Assume that a cellular telephone cell site antenna is mounted on a tower 65 ft high. The operating frequency is in the 800- to 900-MHz range. Type RG-214/U coax cable is used. What is the attenuation of the cable?

Figure 9-7 provides the attenuation of 7.8 dB/100 ft. The attenuation per foot is:

$$7.8/100 = 0.078 \ \text{dB/ft}$$

A cable of 65 ft has an attenuation of:

$$0.078 \times 65 = 5.07 \ \text{dB}$$

What does that dB loss mean in terms of power? You can determine this by using the familiar dB power formula:

$$dB = 10 \log \left(\frac{P_o}{P_i} \right)$$

where P_o is the output power
P_i is the input power.

You can rearrange this formula to calculate P_o or P_i:

$$P_o = 10^{dB/10} \ P_i$$

$$P_i = \frac{P_o}{10^{10/dB}}$$

Assume that the power line (P_i) into the 65-ft cable above is 20 W. The output power is:

$$P_o = 10^{dB/10} \ P_i$$
$$= 10^{-5.07/10} \ (20)$$
$$= 10^{-0.507} \ (20)$$
$$= 0.3111 \ (20) = 6.22 \ \text{W}$$

Note: A dB loss is indicated as a negative dB value (-5.07dB). This is a significant loss! Only 6.22 W of the 20-W input gets to the

Type	Characteristic Impedance (Ω)	Outside Diameter (in.)	Frequency (MHz)	Attenuation (dB/100 ft)	Applications
RG-59/U	73	0.242	100	3.4	TV antennas cable TV, HF antennas
			400	7.1	
RG-11/U	75	0.405	100	2.5	VHF antennas
			400	5.5	
RG-214/U	50	0.405	100	2.0	Satellite TV antennas, UHF and microwave
			400	4.7	
			900	7.8	
RG-58/U	53.5	0.195	100	5.3	CB antennas
Twin lead	300	—	100	0.55	TV antennas

Fig. 9-7 Common types and characteristics of transmission lines.

antenna. To have 20 W at the antenna, you would need to have an input power of:

$$P_i = \frac{P_o}{10^{dB/10}}$$

$$= \frac{20}{10^{-5.07/10}}$$

$$= \frac{20}{0.311} = 64.3 \text{ W}$$

One solution to this power loss problem is to use a better coaxial cable with less loss per foot or to shorten the length of cable as much as possible.

The Ideal Transmission Line Application

If a transmission line is used properly, it must be terminated in a resistive impedance equal in value to its characteristic impedance. For example, a 50-Ω coax must be terminated with a 50-Ω impedance as shown in Fig. 9-8. If the load is an antenna, then that antenna must look like a resistance of 50 Ω. When the load impedance and characteristic impedance match,

the transmission operates properly and maximum power transfer takes place—minus any resistive losses in the line. The line can be any length. If you could measure the voltage (or current) at any point on the line, you would see a constant value, disregarding losses. A correctly terminated transmission line, therefore, is said to be matched or flat.

Standing Waves

If the load impedance is different from the line characteristic impedance, then not all the power transmitted will be absorbed by the load. The power not dissipated in the load will actually be reflected back toward the source. The power sent down the line toward the load is called the *forward* or *incident power,* and any power that is not absorbed by the load is known as *reflected power.* The signal on the line is simply the algebraic sum of the forward and reflected signals.

The reflected power represents a loss. Not all the power generated at the transmitter is taken advantage of. Some of the reflected power is dissipated in the transmission line,

Forward power
Incident power

Reflected power

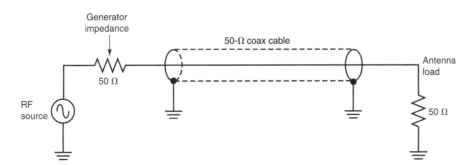

Fig. 9-8 A transmission line must be terminated in its characteristic impedance for proper operation.

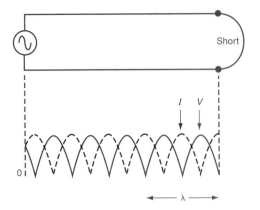

Fig. 9.9 Standing waves on a shorted transmission line.

and, in some cases when the mismatch is great, the reflected power can be high enough to actually cause damage to the transmitter or the line itself. For proper application then, the load impedance should equal the characteristic impedance as closely as possible. One of the key objectives in designing antenna and transmission line systems is to ensure this impedance match.

Standing wave

The forward and reflected signals on an incorrectly terminated transmission line produce what is referred to as a *standing wave*. A standing wave is the unique distribution of voltage and current along a transmission line that is not terminated in its characteristic impedance.

Shorted Load

The concept of standing waves is best illustrated by considering the two worst cases of impedance mismatch at the load or antenna end of the transmission line: an open circuit

and a short circuit. The condition for a short circuit is illustrated in Fig. 9-9. The graph below the transmission line shows a plot of the voltage and current at each point on the line. If you had a voltmeter and ammeter that you could move along the line and measure the voltage and current values and if you then plotted these values, you would obtain the curves shown.

As you would expect with a short at the end of the line, the voltage is zero while the current is maximum. All the power is reflected back toward the generator. Working back toward the generator, you see the voltage and current variations distribute themselves according to the wavelength of the signal. The pattern repeats every one-half wavelength. The voltage and current levels at the generator will be dependent upon the signal wavelength and the actual line length. Just remember that this standing wave pattern is fixed and is the result of a composite of the forward and reflected signals.

Open Load

Figure 9-10 shows the standing waves on an open-circuit line. With an infinite impedance, the voltage at the end of the line is maximum and the current is zero. All the energy is reflected, thereby setting up this stationary pattern of voltage and current standing waves.

Now, in practice you won't actually have a short or open. Instead, the load impedance will not equal the transmission line impedance. Further, the antenna will probably have a reactive component, either inductive or capacitive, as well as its resistance. The mismatch will

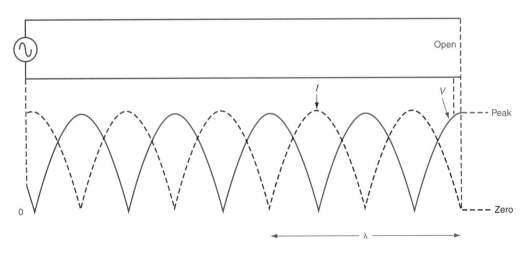

Fig. 9-10 Standing waves on an open-circuit transmission line.

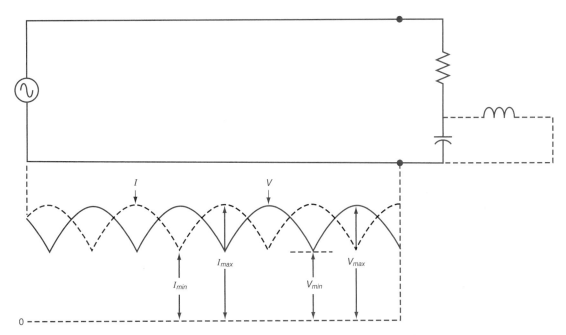

Fig. 9-11 Transmission line with mismatched load and the resulting standing waves.

produce standing waves, but their amplitude will be lower and their distribution will look something like that shown in Fig. 9-11. Note that the voltage never goes to zero.

Calculating SWR

The magnitude of the standing waves on a transmission line is determined by the ratio of the maximum current to the minimum current along the line, or the ratio of the maximum voltage to the minimum voltage. These ratios are referred as the *standing-wave ratio (SWR)*. See Fig. 9-11.

$$\text{SWR} = \frac{I_{\max}}{I_{\min}} = \frac{V_{\max}}{V_{\min}}$$

With the shorted and open conditions described earlier, the current or voltage minimums are zero. This produces an SWR of infinity. It means that no power is dissipated in the load since it is all reflected.

On a properly terminated line, there are no standing waves. The voltage and current are constant along the line, so there are no maximums or minimums, or else you could say the maximums and minimums are the same. Therefore, the SWR is 1. Of course, this is the ideal condition.

The SWR can also be computed by knowing the impedance of the transmission line and the actual impedance of the load. The SWR is the ratio of the load impedance Z_l to the characteristic impedance Z_o or vice versa.

$$\text{SWR} = \frac{Z_l}{Z_o} \text{ if } Z_l > Z_o$$

$$= \frac{Z_o}{Z_l} \text{ if } Z_l < Z_o$$

Since the standing wave is really the composite of the original incident wave added to the reflected wave, we can also define SWR in terms of these waves. If V_i is the incident voltage wave and V_r is the reflected voltage wave, then their ratio will tell us what is happening along the line. This ratio R is called the *reflection coefficient*.

$$R = \frac{V_r}{V_i}$$

If the line is terminated in its characteristic impedance, then there is no reflected voltage, so $R = 0$. If the line is open or shorted, then total reflection occurs. This means that V_r and V_i are the same. In this case, $R = 1$. The reflection coefficient really expresses the percentage of reflected voltage to incident voltage. If $R = 0.5$, it means that 50 percent of the incident power is reflected.

If the load is not matched but is also not an open or short, the line will have voltage minimums and maximums as described previously. These can also be used to obtain the reflection coefficient.

Standing-wave ratio (SWR)

Reflection coefficient

$$R = \frac{V_{\max} - V_{\min}}{V_{\max} + V_{\min}}$$

To obtain the SWR from the reflection co-efficient, we use the formula

$$SWR = \frac{1 + R}{1 - R}$$

If the load matches the line impedance, then $R = 0$. The formula above gives an SWR of 1 as you would expect. With an open or shorted load, $R = 1$. This produces an SWR of infinity.

The reflection coefficient can also be derived from the line and load impedances.

$$R = \frac{Z_l - Z_o}{Z_l + Z_o}$$

As before, Z_l is the load impedance and Z_o is the characteristic impedance of the line.

The SWR is always expressed as a number greater than 1. If the characteristic and load impedances are equal, the SWR is 1, as it should be.

The importance of the SWR is that it gives a relative indication of just how much power is lost in the transmission line. This assumes that none of the reflected power is re-reflected by the generator. In a typical transmitter, some power is reflected and sent to the load again. The curve in Fig. 9-12 shows the relationship between the percentage of reflected power and the SWR. Naturally when the SWR is 1, the percent reflected power is zero. But as a line and load mismatch grows, reflected power increases. When the SWR is 1.5, for example, the percent reflected power is 4 percent. For SWR values of less than about 2, the reflected power is less than 10 percent. For values greater than this, the percent increases dramatically.

Impedance Matching

The best way to prevent a mismatch between the antenna and transmission lines is to have a correct design. But, in practice, mismatches will occur. The antenna resistance may be other than the characteristic impedance of the line, or the antenna may be inductive or capacitive. One solution is to tune the antenna, usually by adjusting its length.

You can also insert an impedance-matching circuit or antenna tuner between the transmitter and the transmission line (matching or tuning at this point does *not* improve the SWR on the transmission line). This can be a balun or

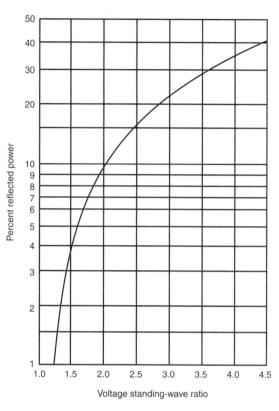

Fig. 9-12 Percent reflected power on a transmission line for different values of the SWR.

LC, L, T, or pi network like those described in a previous chapter. These circuits will make the transmitter behave properly but will not reduce the SWR on the transmission line.

Resonant Lines

The standing-wave conditions described earlier with open- and short-circuited loads must usually be avoided when working with transmission lines. On the other hand, the unique conditions that occur can be used for other purposes. For example, short lengths of transmission line, up to one-half wavelength long, with an open or short circuit can be used as resonant or reactive circuits. Consider the shorted quarter-wavelength ($\lambda/4$) line shown in Fig. 9-13. The voltage is zero and the current is maximum at the end, as you saw earlier. But at the generator, one-quarter wavelength back, the voltage is maximum while the current is zero. To the generator, the line appears as an open circuit or at least a very high impedance. The key point here is that this condition exists at only one frequency, the frequency at which the line length is exactly one-quarter of the

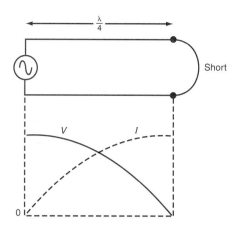

Fig. 9-13 A shorted quarter-wavelength line acts like a parallel resonant circuit.

wavelength. Because of this frequency sensitivity, the line acts like an *LC* tuned or resonant circuit, in this case, a parallel resonant circuit because of its very high impedance.

If you make the line one-half wavelength long, you get the standing-wave pattern shown in Fig. 9-14. The generator sees the same conditions as at the end of the line: zero voltage and maximum current. This represents a short or very low impedance. That condition occurs only if the line is exactly one-half wavelength long at the generator frequency. In this case, the line looks like a series resonant circuit to the generator.

If the line length is less than one-quarter wavelength at the operating frequency, the shorted line looks like an inductor to the generator. If the shorted line length is between one-quarter and one-half wavelength, the line looks like a capacitor to the generator. All these conditions repeat with multiple quarter or half wavelengths of shorted line.

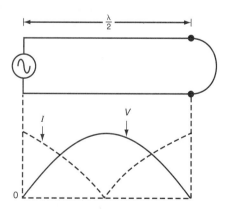

Fig. 9-14 A shorted half-wavelength line acts like a series resonant circuit.

A similar set of conditions is obtained with an open line as shown in Fig. 9-15. To the generator, a quarter-wavelength line looks like a series resonant circuit, whereas a half-wavelength line looks like a parallel resonant circuit, just the opposite of a shorted line. If the line is less than one-quarter wavelength, the generator sees a capacitance. If the line is between one-quarter and one-half wavelength, the generator sees an inductance. Again, all these characteristics repeat for multiple quarter or half wavelengths. Figure 9-16 is a complete summary of the conditions represented by open and shorted lines.

At low frequencies, these characteristics of open and shorted lines have minimum significance. However, at UHF (more than 300 MHz) and microwaves (more than 1 GHz), the length of one-half wavelengths becomes less than 1 ft. Because of this, we can actually use transmission lines as tuned circuits and reactances. The values of inductance and capacitance become so small at these frequencies that it is difficult to realize them physically with standard coils and capacitors. At UHF and microwave frequencies, transmission lines become useful components and circuits. They are used as tuned circuits, filters, phase shifters, reactors, and impedance-matching circuits.

Measuring Power and SWR

A variety of test instruments are available to measure RF power and SWR at a transmitter output. The most common type is inserted

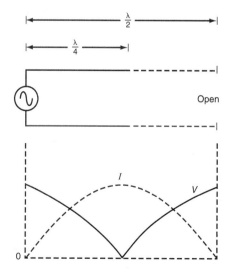

Fig. 9-15 One-quarter and one-half wavelength open lines also look like resonant circuits to a generator.

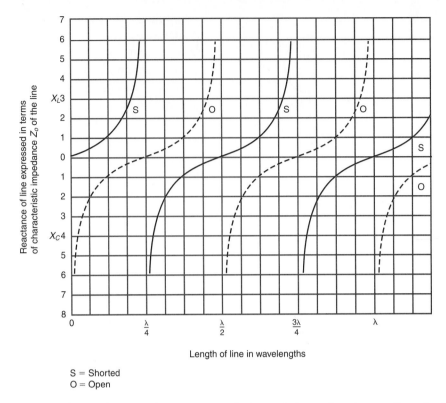

Length of line in wavelengths

S = Shorted
O = Open

Fig. 9-16 Summary of impedance and reactance variations of shorted and open lines for lengths up to one wavelength.

between the transmitter coax output connector and the input to the transmission line. A coupling circuit picks up some of the transmitted signal. The RF voltage is then rectified into direct current and used to operate a meter that is calibrated for power. A variation of this design uses a section of transmission line that produces two voltages, one the forward or incident voltage and the other the reflected voltage. The relative values of forward and reflected power can be determined, as can the SWR.

Figure 9-17 shows a representative circuit. The large copper strip in the center (2) forms a 50-Ω transmission line segment on a printed circuit board. The smaller strips, (1) and (3), are coupling elements that pick up signal from the main conductor. Signal pickup is effected by inductive and capacitive coupling. The lower coupling strip picks up a signal proportional to the forward power, and the upper strip picks up a signal proportional to the reflected power. These signals are rectified by germanium diodes D_1 and D_2 and filtered into direct current by C_1 and C_2. The direct current is read on the meter. Resistors R_3, R_4, and R_5 are adjusted to read power on the meter scale, which is usually calibrated in watts.

When the switch is in the forward position, the forward or incident power is measured. Putting the switch in the reflected position causes reflected power to be shown on the meter.

To use the meter for SWR measurement, the forward P_F and reflected P_R power values are read and then used in the formula:

$$\text{SWR} = \frac{1 + \sqrt{P_R/P_F}}{1 - \sqrt{P_R/P_F}}$$

To provide the best match to an antenna, the meter is set to read reflected power. Then antenna adjustments are made or impedance-matching circuits are tuned to produce the minimum reflected power that will give the lowest SWR.

For example, assume that you read a forward power of 450 W and a reverse power of 25 W. This SWR is:

$$\text{SWR} = \frac{1 + \sqrt{\dfrac{25}{450}}}{1 - \sqrt{\dfrac{25}{450}}} = \frac{1 + 0.2357}{1 - 0.2357}$$

$$= \frac{1 + 0.2357}{0.7643} = 1.62$$

Or, with 10 W forward and 2 W reflected:

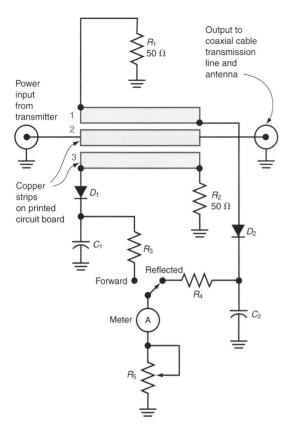

Fig. 9-17 Power/SWR meter.

$$\text{SWR} = \frac{1 + \sqrt{\dfrac{2}{10}}}{1 - \sqrt{\dfrac{2}{10}}} = 2.62 : 1$$

■ TEST_____

Supply the missing information in each statement.

1. The two basic types of transmission line are _____.
2. If one wire of a transmission line is connected to ground, the line is said to be _____.
3. The distance that a signal travels during one cycle is called the _____.
4. One wavelength at a frequency of 450 MHz is _____ m.
5. A line 4 in. long represents one-half wavelength at a frequency of _____ GHz.
6. The physical dimensions of a transmission line determine its _____.
7. To a generator, a transmission line looks like a(n) _____ made up of distributed _____ and _____.

8. A coax line has a shield braid with an inside diameter of 0.2 in. and a center conductor with a diameter of 0.057 in. The characteristic impedance is _____ Ω.
9. The attenuation of 250 ft of RG-11U coax at 100 MHz is _____ dB. (See Fig. 9-7.)
10. For optimum transfer of power from a generator to a load of 52 Ω, the transmission line impedance should be _____ Ω.
11. The current and voltage along a properly matched line are _____.
12. If a transmission line is not terminated in its characteristic impedance, _____ _____ will develop along the line.
13. If a load and line have mismatched impedances, power not absorbed by the load will be _____.
14. Patterns of voltage and current variations along a transmission line with a mismatched load are known as _____ _____.
15. A 52-Ω coax has a 36-Ω antenna load. The SWR is _____.
16. If the load and line impedances are matched, the SWR will be _____.
17. All incident power on a line will be reflected if the line is _____ or _____ at its end.
18. The ratio of the reflected voltage to the incident voltage on a transmission line is called the _____.
19. The maximum voltage along a transmission line is 150 V, and the minimum voltage is 90 V. The SWR is _____. The reflection coefficient is _____.
20. The reflection coefficient of a transmission line is 0.75. The SWR is _____.
21. An open or shorted transmission line will have a reflection coefficient of _____ and an SWR of _____.
22. Transmission lines, one-quarter or one-half wavelength, can be used as _____.
23. A transmission line has an SWR of 1.75. The power applied to the line is 90 W. The amount of reflected power is _____ W.
24. A shorted quarter-wave line looks like a(n) _____ impedance to the generator.
25. The following lines look like a series resonant circuit: a _____ λ/4 line; a _____ λ/2 line.
26. Transmission lines less than λ/4 or between λ/4 and λ/2 act like _____ or _____.

27. An open transmission line 6 in. long acts as a _____ resonant circuit at a frequency of 492 MHz.
28. A coax has a velocity factor of 0.68. One-half wavelength of this coax at 120 MHz is _____ ft long.
29. What is the power at the antenna driven by a transmitter of 300 W through a 175-ft coaxial line with an attenuation of 2.5 dB/100 ft?
30. If the forward power measured by the circuit in Fig. 9-17 is 25 W and the reflected power is 4 W, the SWR is

_____.

9-2 Antenna Fundamentals

An antenna, or aerial as it is sometimes called, is one or more electrical conductors of a specific length that radiate radio waves generated by a transmitter or that collect radio waves at the receiver. There are hundreds of different types of antennas in use today. In this section we will introduce you to the most common and popular types.

Radio Waves

A radio wave is generally called an *electromagnetic wave* because it is made up of a combination of both electric and magnetic fields. Whenever voltage is applied to the antenna, an electric field will be set up. At the same time, this voltage will cause current to flow in the antenna. This current flow will produce a magnetic field. The electric and magnetic fields are at right angles to one another. These electric and magnetic fields are emitted from the antenna and propagate through space over very long distances. Figure 9-18 shows the general concept of electric and magnetic fields making up a radio signal.

A closer approximation to what an electromagnetic field may look like is illustrated in Fig. 9-19. Note that the amplitude and direction of the magnetic and electric fields vary in a sinusiodal manner depending upon the frequency of the signal being radiated. The electric and magnetic fields are perpendicular to one another, and they are also both perpendicular to the direction of propagation of the wave.

Polarization

Electromagnetic wave

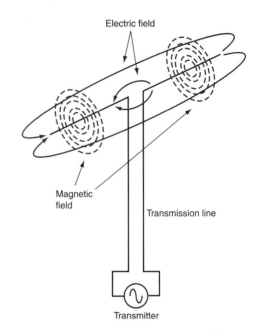

Fig. 9-18 The electric and magnetic fields produced by an antenna.

Polarization

An important consideration in the transmission and reception of radio waves is the orientation of the magnetic and electric fields with respect to the earth. The direction of the electric field specifies the *polarization* of the antenna. If the electric field is parallel to the earth as in Fig. 9-19(*a*), the electromagnetic wave is said to be horizontally polarized. However, if the electric field is perpendicular to the earth as in

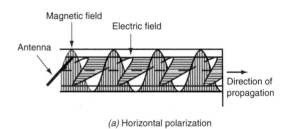

(a) Horizontal polarization

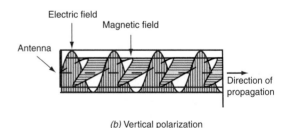

(b) Vertical polarization

Fig. 9-19 The orientation of the electric and magnetic fields determines the signal polarization.

Fig. 9-19(*b*), then the wave is vertically polarized. As you would expect, antennas that are horizontal to the earth produce horizontal polarization, and antennas that are vertical to the earth produce vertical polarization.

The most important factor is that both the transmitting and receiving antennas must be of the same polarization for optimum transmission and reception. Theoretically, a vertically polarized wave will produce 0 V in a horizontal antenna and vice versa. But during transmission over long distances, the waves have their polarization changed slightly due to various propagation effects. For this reason, even though the polarization of the transmitting and receiving antennas may be mismatched, a signal will typically be received, although optimum results will take place when the polarization of the two antennas is matched.

Dipole Antenna

As indicated earlier, an antenna is some form of electrical conductor. It may be a length of wire, a metal rod, or a piece of tubing. Many different sizes and shapes are used. The length of the conductor is dependent upon the frequency of transmission. Antennas radiate most effectively when their length is directly related to the wavelength of the transmitted signal. Most antennas have a length that is some fraction of a wavelength. The most common lengths are one-half and one-quarter wavelengths.

One of the most widely used antenna types is the *half-wave dipole* shown in Fig. 9-20. Also called a doublet, it is simply a piece of wire, rod, or tubing that is one-half wavelength long at the operating frequency. The antenna is

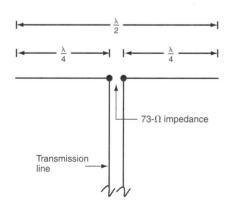

Fig. 9-20 A half-wave dipole antenna.

actually cut into two quarter-wavelength sections. The transmission line is connected at the center. The dipole has an impedance of 73 Ω at its center.

As you saw earlier, it is important for maximum power transfer that the impedance of the transmission line match the load. For the dipole, a 73-Ω coax like RG-59/U is a perfect transmission line; RG-11/U coax with an impedance of 75 Ω is also an excellent match.

The length of a half-wavelength dipole antenna is computed with the formula

$$L = \frac{468}{f}$$

As an example, an antenna for a frequency of 18 MHz would have a length of

$$L = \frac{468}{18}$$

$$= 26 \text{ ft}$$

To create a half-wave dipole for 18 MHz, you would cut two 13-ft lengths of wire. Copper wire size no. 12 or 14 gage is commonly used. Physically the antenna would be suspended between two points as high as possible off the ground. See Fig. 9-21. The wire conductors themselves would be connected to glass or ceramic insulators in each end and at the middle of the antenna to provide good insulation between the antenna itself and its supports. The transmission line would be attached to the two conductors at the center insulator. The transmission line should leave the antenna at a right angle so that it will not interfere with the antenna's radiation.

Half-wave dipole

Most half-wave dipole antennas are mounted horizontally to the earth and, therefore, will be horizontally polarized. Although the antenna in general will radiate energy in many different directions, typically that radiation is concentrated into a very specific geometric radiation pattern. The *radiation pattern* of a half-wave dipole has the shape of a doughnut. Figure 9-22 shows this pattern with half the doughnut cut away. The

Radiation pattern

Directional antenna

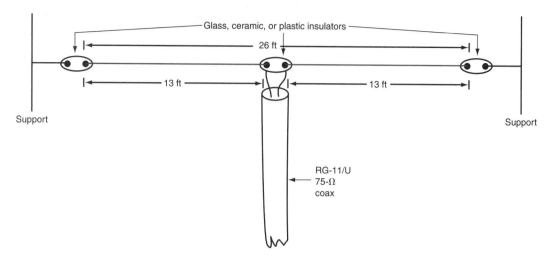

Fig. 9-21 A half-wave dipole for 18 MHz.

dipole is at the center hole of the doughnut, and the doughnut itself represents the radiated energy. If you look down on the top of the dipole, the radiation pattern appears to be a figure eight, as shown in Fig. 9-23. This is the horizontal radiation pattern. It is plotted on a polar coordinate graph. The center of the antenna is assumed to be at the center of the graph. The dipole is assumed to be aligned with the 90°–270° axis. As you can see, the maximum amount of energy is radiated at right angles to the dipole, at 0° and 180°. For that reason, a dipole is what is known as a *directional antenna*. For optimum transmission and reception, the antenna should be aligned broadside to the signal destination. The transmitting and receiving antennas must be parallel to one another for maximum signal transmission.

Whenever you point a dipole receiving antenna toward a transmitter or vice versa, it must be broadside to the direction of the transmitter. If the antenna is at some angle, the maximum signal will not be received. As you can see from the radiation pattern in Fig. 9-23, if the end of the receiving antenna is pointed directly at the transmitting antenna, then zero signal will be received. As indicated earlier, this is a theoretical condition as the radiated wave does undergo some shifts during propagation and, therefore, some signal will be received off the ends of the antenna, but it is a very minimum amount. In practice, if you observed the signal strength as you rotate the receiving antenna, the maximum signal will be obtained when the antenna is broadside, while the minimum signal will be obtained when the end of the antenna is pointed toward the transmitter.

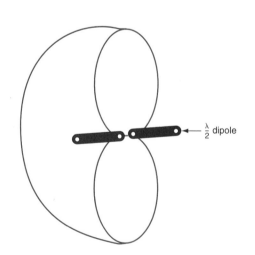

Fig. 9-22 Radiation pattern of a half-wave dipole.

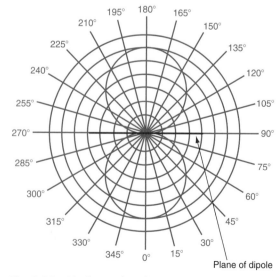

Fig. 9-23 Horizontal radiation pattern of a half-wave dipole.

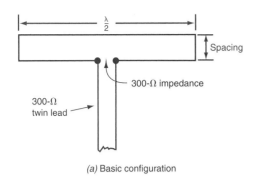

(a) Basic configuration

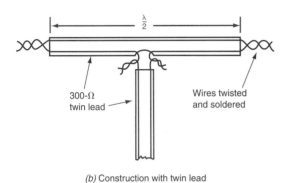

(b) Construction with twin lead

Fig. 9-24 Folded dipole.

Although most half-wave dipoles are mounted horizontally to the earth, they can also be mounted vertically. Now the antenna is vertically polarized. Its radiation pattern is still a doughnut, but now looking down upon the antenna you see a radiation pattern that is a perfect circle. This means that the antenna will transmit an equal amount of energy in all directions. Such an antenna is said to be *omnidirectional.*

Folded Dipole

A popular variation of the half-wave dipole is the *folded dipole* shown in Fig. 9-24(a). Like the standard dipole, it is one-half wavelength long. However, note that it consists of two parallel conductors connected at the ends with one side open at the center for connection to the transmission line. The reason for the popularity of this particular antenna is the fact that its impedance is 300 Ω, and, therefore, it makes an excellent match to the widely available 300-Ω twin lead. The actual spacing between the two parallel conductors is not particularly critical, although it is typically inversely proportional to the frequency. For very high frequency antennas, spacing is less than 1 in. For low-frequency antennas, the spacing may be 2 or 3 in.

An easy way to make a folded dipole antenna is to construct it entirely of 300-Ω twin lead. A piece of twin lead is cut to a length of one-half wavelength, and the two ends are soldered together as shown in Fig. 9-24(b). When figuring the half wavelength, be sure to include the velocity factor of twin lead (0.8) in the calculation. The center of one conductor is then cut open and a 300-Ω twin-lead transmission line is soldered to the two wires. The result is an effective, low-cost antenna that can be used for both transmitting and receiving purposes. Such antennas are commonly used for both TV and FM radio reception.

Ground Plane Antenna

If vertical polarization and omnidirectional characteristics are required, it is not necessary to use a half-wave length dipole. The same effect can be achieved by using a *quarter-wavelength vertical radiator* as illustrated in Fig. 9-25(a). This antenna is known as a quarter-

Quarter-wave vertical

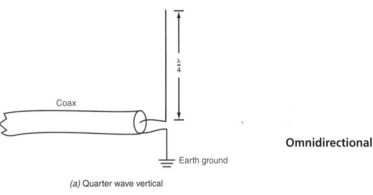

(a) Quarter wave vertical

Omnidirectional

Folded dipole

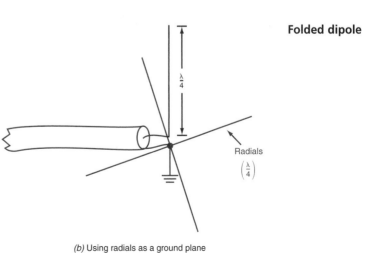

(b) Using radials as a ground plane

Fig. 9-25 Ground plane antenna.

wave vertical or ground plane antenna. It is usually fed with coax, the center conductor connected to the vertical radiator and the shield connected to earth ground. With this arrangement, the earth acts as a type of electrical "mirror," effectively providing the other quarter wavelength of the antenna and making it the equivalent of a vertical dipole. The result is a vertically polarized omnidirectional antenna.

The effectiveness of this antenna depends upon a good electrical ground. Making good electrical contact with the earth is a tricky process. Usually a reasonable ground can be obtained by driving a copper rod 5 to 10 ft long into the earth. In some cases, even this is insufficient. When a good electrical connection is made to the earth, the earth becomes what is known as a ground plane. If a good electrical connection cannot be made to the earth, then an artificial ground plane can be constructed on several quarter wavelength wires laid horizontally on the ground or buried in the earth. Usually four wires are sufficient, but in some antenna systems, more are used. See Fig. 9-25(*b*). These horizontal wires at the base of the antenna are referred to as *radials*. The entire collection of radials is often referred to as a *counterpoise* or *ground plane*.

Radials

Counterpoise

At very high frequencies where antennas are short, any large, flat metallic surface can serve as an effective ground plane. For example, vertical antennas are widely used on cars, trucks, boats, and other vehicles. The metallic roof of a car makes a superior ground plane for VHF and UHF antennas. In any case, the ground plane must be large enough so that it has a radius of greater than one quarter wavelength.

The impedance of a vertical ground plane antenna is exactly one-half the impedance of the dipole or approximately 36.5 Ω. Since there is no such thing as 36.5-Ω coax, 50-Ω coax is commonly used to feed power to the ground plane antenna. This represents a mismatch with an SWR of 1.39. This is a relatively low SWR and will not be the cause of any significant loss in power.

One way to adjust the antenna's impedance is to use "drooping" radials as shown in Fig. 9-26. At some angle, depending upon the height of the antenna above the ground, the antenna's impedance will be near 50 Ω, making it a near perfect match for most 50-Ω coax.

In addition to its vertical polarization and omnidirectional characteristics, another major

Directivity

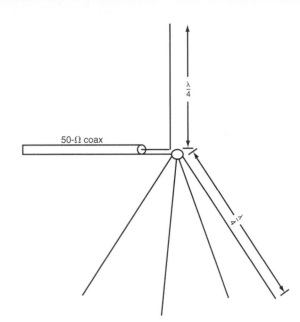

Fig. 9-26 Drooping radials adjust the antenna's impedance to near 50 Ω.

benefit of the quarter-wave vertical antenna is its length. It is one-half the length of a standard dipole; this represents a significant savings at lower RFs. For example, a half-wave antenna for a frequency of 5 MHz is 93.6 ft long. Creating a vertical antenna of that length represents a major structural problem as it requires a support at least that long. By using a quarter-wave vertical antenna, however, the necessary length is cut in half; in this case the length is 46.8 ft. Creating a structure 50 ft high instead of 100 ft high is a lot easier. Most low-frequency transmitting antennas use verticals for this reason. Amplitude modulation broadcast stations use vertical antennas because they are shorter, less expensive, and not as visually obtrusive. Additionally, they provide an equal amount of radiation in all directions.

Directional Antennas

One of the primary advantages of a vertical antenna or any antenna with omnidirectional characteristics is that it can send messages in any direction or receive them from any direction. In many communications systems this is a highly desirable advantage. On the other hand, there are communications systems in which it is more advantageous to restrict the direction in which signals are sent or received. This requires an antenna with directivity. *Directivity* refers to the ability of an antenna to send or receive signals over a narrow horizontal directional range. In other words, the antenna has a response or directivity curve that makes it highly directional based on its physical orientation. The advantage of a directional antenna is that it eliminates interference from other signals being received from all directions except the direction of the desired signal. A highly directional antenna acts as a type of filter to provide selectivity based on the direction of the signal. The receiving antenna is pointed directly at the station to be received, thereby effectively rejecting signals from transmitters in all other directions.

When a directional antenna is used, there is greater efficiency of power transmission. When a transmitter uses an omnidirectional antenna, the transmitted power goes off into all directions. Only a small portion of it is received by the desired station. In effect, all the other power is wasted. If the antenna is made directional, the transmitter power can be focused into a narrow beam that can be directed toward the station of interest.

The conventional half-wave dipole has some directivity in that it sends or receives signals in those directions perpendicular to the line of the antenna. You saw this earlier in the figure-eight response curve of the dipole in Fig. 9-23. The antenna is directional in that no signal is radiated from or picked up off the ends of the antenna. Such an antenna is referred to as *bidirectional* since it receives signals best in two directions. An antenna can be made *unidirectional* with proper design. Unidirectional means that the antenna sends or receives signals in only one direction.

The measure of an antenna's directivity is *beam width*. Beam width refers to the angle of the radiation pattern over which a transmitter's energy is directed or received. The beam width is measured on the antenna's radiation pattern. Figure 9-27 shows the horizontal radiation pattern of a typical directional antenna plotted on a polar coordinate graph. The antenna is assumed to be at the center of the graph. The concentric circles extending outward from the pattern indicate the relative strength of the signal as it moves away from the antenna. The beam width is measured between the points on the radiation curve that are down 3 dB from the maximum amplitude of the curve. The maximum amplitude of the pattern occurs at 0°.

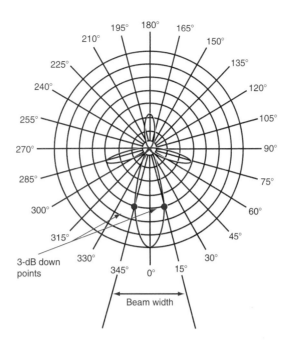

Fig. 9-27 Measuring the beam width of a directional antenna.

The 3-dB down points are 70.7 percent of the maximum voltage. Then an angle is formed with two lines extending from the center of the curve to the 3-dB down points. The angle formed by these two lines is the beam width. In this example, the beam width is 30°.

Note in Fig. 9-27 the three smaller patterns or loops in different directions from the main larger pattern we have been discussing. They are called *minor lobes*. Few antennas are perfectly directive. Because of various imperfections, some power is radiated in other directions. The goal is to eliminate or at least minimize these minor lobes through various antenna adjustments to put more power into the *main lobe*.

The beam width on a standard half-wave dipole is approximately 70°. This is not a highly directional antenna. The narrower the beam width, of course, the better the directivity and the more highly focused the signal becomes. At microwave frequencies, antennas with beam widths of less than a degree have been built for pinpoint communications accuracy.

When a highly directive antenna is used, all the transmitted power is focused in one direction. Because the power is concentrated into a small beam, the effect is as if the antenna amplified the transmitted signal. Directivity, because it focuses the power, causes the antenna to exhibit gain. Gain, as you know, is a form of amplification. And although the antenna does

Minor lobes

Bidirectional antenna

Unidirectional antenna

Main lobe

Beam width

Antenna gain

not have the capability of actually amplifying the signal, since it can focus the energy in a single direction, the effect is as if the amount of radiated power were substantially higher than the power output of the transmitter. The antenna effectively has power gain.

The gain of an antenna can be expressed as the ratio of the focused power transmitted P_t to the input power of the antenna P_i. This is the power gain. But usually, the gain is expressed in decibels of power gain.

$$\text{dB} = 10 \log \frac{P_t}{P_i}$$

Effective radiated power (ERP)

The total amount of power radiated by the antenna is called the *effective radiated power (ERP)* and is simply the power applied to the antenna multiplied by the antenna gain. Power gains of 10 or more are easily achieved, especially at the higher RFs. This means that a 100-W transmitter can be made to perform like a 1000-W transmitter when applied to an antenna with gain.

Array
Director

To create an antenna with directivity and gain, two or more antenna elements are combined to form what is known as an *array*. There are two basic types of antenna arrays used to achieve gain and directivity. These are parasitic arrays and driven arrays.

Parasitic Arrays

Parasitic array

A *parasitic array* consists of a basic antenna connected to a transmission line plus one or more additional conductors that are not connected to the transmission line. These extra conductors are referred to as parasitic elements. The basic antenna itself is referred to as the driven element. Typically the driven element is a half-wave dipole or some variation. The parasitic elements are slightly longer than and slightly less than one-half wavelength long. These parasitic elements are placed in parallel with and near the driven element. A common arrangement is illustrated in Fig. 9-28.

Yagi

Reflector

One type of parasitic element is known as a *reflector*. It is typically about 5 percent longer than the half-wave dipole driven element. The reflector is spaced from the driven element by a distance of 0.15 to 0.25 wavelength. The physical arrangement is shown in Fig. 9-28. When the signal radiated from the dipole reaches the reflector, it will induce a voltage into the reflector. The reflector will produce some radiation of its own. Because of the spacing, the re-

Beam antenna

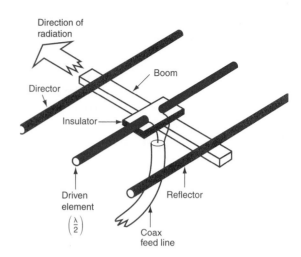

Fig. 9-28 A parasitic array known as a Yagi antenna.

flector's radiation will be mostly in phase with the driven-element radiation. As a result, the reflected signal will add to the dipole signal and create a stronger, more highly focused beam. The reflector minimizes the radiation to the rear of the driven element and reinforces the radiation to the front of the driven element.

Another kind of parasitic element is called a *director*. It is approximately 4 percent shorter than the half-wave driven element and is mounted in front of the driven element. See Fig. 9-28.

The directors are placed in front of the driven element and are spaced by some distance between approximately one-tenth to two-tenths of a wavelength from the driven element. The signal from the driven element will be induced into the director. The signal radiated by the director will then add in phase to that from the driven element. The result is increased focusing of the signal, a narrow beam width, and a higher antenna gain. Note the direction of radiation in Fig. 9-28.

An antenna made up of a driven element, a reflector, and one or more directors is generally referred to as a *Yagi antenna,* named after one of its Japanese inventors, Yagi and Uda. The antenna elements are usually made of aluminum tubing and are mounted on an aluminum cross member or boom. Since the center of the parasitic elements is neutral electrically, they may be connected directly to the boom. The boom itself may then be connected to a metal mast and electrical ground for the best lightning protection. The resulting antenna is often referred to as a *beam antenna* because it is very highly directional and has very high gain. Gains of 6 to 15 dB are possible with beam angles of 40° to 20°.

The three-element Yagi shown in Fig. 9-28 has a gain of about 8 dB when compared to a half-wave dipole. Most Yagis have a driven element, a reflector, and from 1 to 20 directors. The greater the number of parasitic elements, the higher the gain and the narrower the beam angle.

Yagis are widely used communications antennas because of their directivity and gain. At one time they were widely used for TV reception, but since they are tuned to only one frequency, they are not good for reception or transmission over a wide frequency range. Amateur radio operators are major users of beam antennas. Many other communications services use them because of their excellent performance and low cost.

Because of its physical size, a beam antenna is used mainly at VHF and UHF. For example, at a frequency of 450 MHz, the elements of a Yagi are only about 1 ft long. This makes the antenna relatively small and easy to handle. The lower the frequency, the larger the elements and the longer the boom. In general, such antennas are only practical above frequencies of about 15 MHz. At this frequency, the elements are large and difficult to work with, although they are still widely used in some communications services.

Driven Arrays

The other major type of directional antenna is the *driven array*. This is an antenna that has two or more driven elements. Each element receives RF energy from the transmission line. The elements may be arranged in various ways to produce different degrees of directivity and gain.

Collinear There are three basic types of driven arrays, the collinear, the broadside, and the end-fire. The *collinear antenna* usually uses two or more half-wave dipoles mounted end to end as shown in Fig. 9-29(*a*). The lengths of the transmission lines connecting the various driven elements are carefully selected so that

Driven array

Collinear antenna

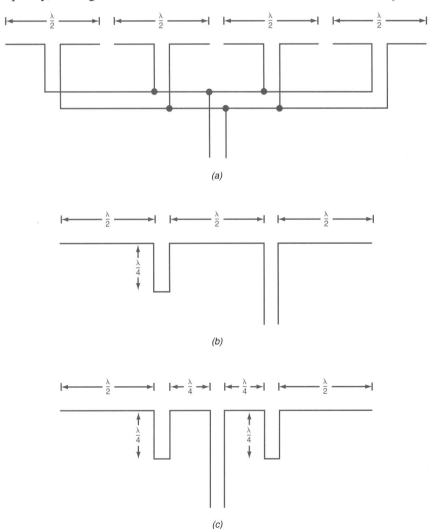

Fig. 9-29 Types of collinear antennas.

the energy reaching each antenna is in phase with all the other antennas. In this way, the individual antenna signals will combine and produce a more focused beam. Collinears still have a bidirectional radiation pattern like that of a dipole, but the two beam widths are much narrower, providing greater directivity and gain.

The collinears in Fig. 9-29(b) and (c) use half-wave sections separated by shorted quarter-wave matching stubs which ensure that the signals radiated by each half-wave section are in phase. The greater the number of half-wave sections used, the higher the gain and the narrower the beam width.

Collinears are usually used only on VHF and UHF bands because their length becomes prohibitive at the lower frequencies. Collinears can be mounted vertically at the high frequencies to provide an omnidirectional antenna with gain.

Broadside array

Broadside   A *broadside array* is essentially a stacked collinear antenna. It consists of half-wave dipoles spaced from one another by one-half wavelength as shown in Fig. 9-30. Two or more elements can be combined. Each is connected to the other and to the transmission line. The crossover transmission line ensures the correct signal phasing. The resulting antenna produces a highly directional radiation pattern, not in the line of the elements as in a Yagi, but broadside or perpendicular to the plane of the array. Like the collinear, the broadside is bidirectional in radiation, but the radiation pattern has a very narrow beam width and high gain.

End-fire array

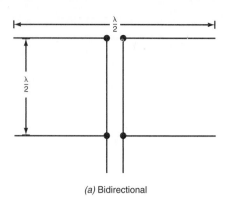

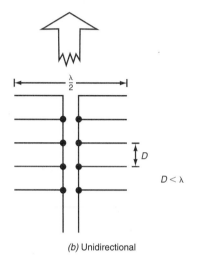

(a) Bidirectional

(b) Unidirectional

Fig. 9-31   End-fire antennas.

End-Fire   The *end-fire array* shown in Fig. 9-31(a) uses two half-wave dipoles spaced one-half wavelength apart. Both elements are driven by the transmission line. The antenna has a bidirectional radiation pattern but with narrower beam widths and gain. The radiation is in the plane of the driven elements as in a Yagi. The end-fire array in Fig. 9-31(b) uses five driven elements spaced some fraction of a wavelength D. By selecting the desired number of elements and a related spacing, a highly directional unidirectional antenna is created. The spacing causes the lobe in one direction to be canceled so that it adds to the other lobe, creating high gain and directivity.

Log-Periodic   A special version of the driven array is the wide bandwidth log-periodic antenna shown in Fig. 9-32. The lengths of the driven elements vary from long to short and are related logarithmically. The longest element has a length of one-half wavelength at the lowest frequency to be covered, and the shortest element is one-half wavelength at the highest frequency. Each element is fed with a special short transmission line

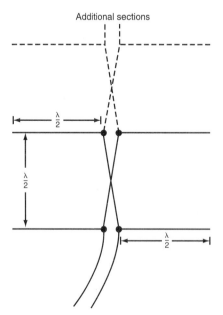

Fig. 9-30   A broadside array.

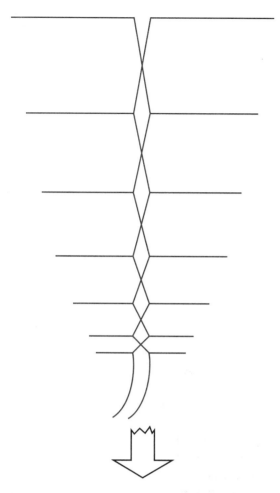

Fig. 9-32 Log-periodic antenna.

segment to properly phase the signal sent to each element. The result is a highly directional antenna with excellent gain. Its great advantage over a Yagi or other array is its very wide bandwidth. Most Yagis and other driven arrays are designed for a very specific frequency or a narrow band of frequencies. The lengths of the elements set the operating frequency. When more than one frequency is to be used, of course, multiple antennas can be used. Most TV antennas in use today are of the *log-periodic* variety so that they can provide high gain and directivity on both VHF and UHF TV channels. Log-periodic antennas are also used in other two-way communications systems where multiple frequencies must be covered.

All the antennas described in this section will operate effectively at their design frequency. They can be used in the 500-kHz to 1-GHz range. The higher the frequency, the shorter the antenna. At frequencies above 1 GHz (microwaves), different types of antennas are used. These include the horn, parabolic,

and helix. These will be discussed in the next chapter on microwaves.

Antenna Impedance Matching

One of the most critical aspects of any antenna system is ensuring that the maximum amount of power is transferred from the transmitter to the antenna. An important part of this, of course, is the transmission line itself. The characteristic impedance of the transmission line must match the output impedance of the transmitter and the impedance of the antenna itself. If this is the case, the SWR will be 1 and maximum power transfer will take place.

Inevitably, there will be situations in which a perfect match between an antenna, transmission line, and transmitter is not possible. In these applications, special techniques are used. Collectively, they are generally referred to as antenna tuning or antenna matching and involve numerous techniques for ensuring maximum power output.

Most of the techniques are aimed at *impedance matching,* that is, making one impedance look like another through the use of tuned circuits or other devices. For example, one technique is to use what is known as a matching stub or *Q section*. This is a quarter wavelength of coax or balanced transmission line of a specific impedance that is connected between a load and a source for the purpose of matching impedances. See Fig. 9-33. It has been found that a quarter-wave transmission line can be used to make one impedance look like another. This is expressed in the following relationship:

$$Z_Q = \sqrt{Z_L Z_O}$$

where Z_Q = characteristic impedance of quarter-wave matching stub or Q section

Z_O = characteristic impedance of transmission line or transmitter at input of Q section

Z_L = impedance of load, usually antenna feed point impedance

Impedance matching

Q section

Log-periodic antennas

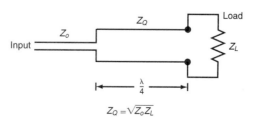

Fig. 9-33 A quarter-wave matching stub or Q section.

An example of such an application is the use of the Q section to match a standard 73-Ω coax transmission line to the 36.5-Ω impedance of a quarter-wave vertical antenna. To compute the impedance of the transmission line used as the matching stub, the expression above is used.

Antenna tuners

$$Z_Q = \sqrt{36.5(73)}$$
$$= \sqrt{2664.5}$$
$$= 51.6$$

This tells us that a quarter-wavelength section of 50-Ω coax will make the 73-Ω transmission line look like the 36.5-Ω antenna or vice versa. In this way maximum power is transferred.

Balun

Another commonly used impedance-matching technique is a *balun*. Recall from previous discussions that a balun is simply a type of transformer used to match impedances. Most baluns are made of a ferrite core, either a toroid or a rod, and windings of copper wire. Baluns have a very wide bandwidth and, therefore, are essentially independent of frequency. Baluns can be created for producing impedance-matching ratios of 4:1, 9:1, and 16:1.

A balun can also be constructed from coax as shown in Fig. 9-34. A half wavelength of coax is connected as shown between the antenna and the feed line. When figuring the length of this half-wave section, be sure to take into account the velocity factor of the coax.

Such a balun provides a 4:1 impedance transformation. For example, it can easily convert a 300-Ω antenna to a 75-Ω load and vice versa.

When baluns and matching sections cannot do the job, *antenna tuners* are used. An antenna tuner is a variable inductor, one or more variable capacitors, or a combination of these components connected in various configurations. Here L, T, and pi networks are all widely used. The inductor and capacitor values are adjusted until the impedances match as indicated by minimum SWR.

■ TEST

Answer the following questions.

31. Radio waves are made up of _____ and _____ fields.
32. The polarization of a radio wave depends upon the position of its _____ with respect to the earth's surface.
33. The antenna is connected to the transmitter or receiver by a(n) _____.
34. A radio wave has its magnetic field horizontal to the earth. It is, therefore, _____ polarized.
35. One of the most widely used and simplest antennas is the half-wave _____.
36. The length of a doublet antenna at 150 MHz is _____ ft.
37. The feed impedance of a dipole antenna is approximately _____ Ω.
38. The horizontal radiation pattern of a dipole looks like a(n) _____.
39. The measure of an antenna's directivity is _____.
40. An antenna that radiates equally well in all horizontal directions is said to be _____.
41. A quarter-wave vertical antenna is commonly know as a(n) _____ antenna.
42. The length of a quarter-wave vertical antenna at 890 MHz is _____ in.
43. For proper operation of a vertical antenna, the shield of the feed coax must make a good connection to _____ or a set of quarter-wave wires called _____.
44. The horizontal radiation pattern of a quarter-wave vertical is a(n) _____.
45. The feed impedance of a quarter-wave vertical is approximately _____ Ω.
46. An antenna that transmits or receives equally well in two opposite directions is said to be _____.

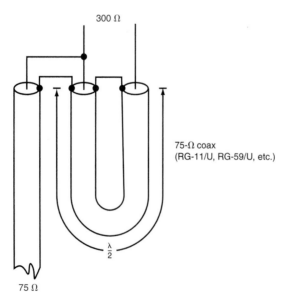

Fig. 9-34 A coax balun with a 4:1 impedance ratio.

47. A unidirectional antenna transmits best in _____ directions(s).

48. A directional antenna that focuses the energy into a narrow beam has _____ since it effectively amplifies the signal.

49. To have gain and directivity, an antenna must have two or more _____.

50. List two basic types of antenna arrays.

51. The three basic elements in a Yagi antenna are the _____.

52. The two parasitic elements in a beam antenna are the _____.

53. A Yagi may have one or more _____ elements.

54. The beam width of a Yagi is usually in the range of _____ to _____ degrees.

55. The length of the driven element in a Yagi at 222 MHz is _____ ft.

56. List three kinds of driven arrays.

57. True or false. Yagis and driven arrays may be operated either horizontally or vertically.

58. A popular wideband driven array is the _____ array.

59. An impedance-matching circuit used to make the antenna, transmission line, and transmitter impedances match is the _____.

60. A transformer used for impedance matching is the _____.

61. A quarter-wavelength section of transmission line used for impedance matching is called a(n) _____.

62. A coax balun has an impedance-matching ratio of _____.

63. A quarter-wavelength of coax with a velocity factor of 0.7 at 220 MHz is _____ in.

9-3 Radio-Frequency Wave Propagation

Once a radio signal has been radiated by the antenna, it will travel or propagate through space and will ultimately reach the receiving antenna. As you would expect, the energy level of the signal decreases rapidly as the distance from the transmitting antenna is increased. Further, the electromagnetic signal can take one or more of several different paths to the receiving antenna. The path that a radio signal takes depends upon many factors including the frequency of the signal, atmospheric conditions, and the time of day.

The three basic paths that a radio signal can take through space are the ground wave, the sky wave, and the space wave.

Ground Waves

The *ground* or *surface wave* leaves the antenna and remains close to the earth. See Fig. 9-35. The ground wave will actually follow the curvature of the earth and can, therefore, travel at distances beyond the horizon.

Ground wave

Ground-wave propagation is strongest at the low- and medium-frequency ranges. That is, ground waves are the main signal path for radio signals in the 30-kHz to 3-MHz range. The signals can propagate for hundreds and sometimes thousands of miles at these low frequencies. Amplitude modulation broadcast signals are propagated primarily by ground waves.

At the higher frequencies beyond 3 MHz, the earth begins to attenuate the radio signals. Objects on the earth and terrain features become the same order of magnitude in size as the wavelength of the signal and will, therefore, absorb and otherwise affect the signal. For this reason, the ground-wave propagation of signals above 3 MHz is insignificant except within several miles of the antenna.

Sky Waves

A *sky-wave signal* is one that is radiated by the antenna into the upper atmosphere where it is bent or reflected back to earth. This bending of the signal is caused by a region in the upper atmosphere known as the *ionosphere*. Ultraviolet radiation from the sun causes the upper atmosphere to ionize, that is, to become electrically charged. The atoms take on extra electrons or lose electrons to become positive and negative ions, respectively. Free electrons are also present. This results in a relatively thick but invisible layer that exists above the earth. The ionosphere is illustrated in Fig. 9-36. At

Sky-wave signal

Ionosphere

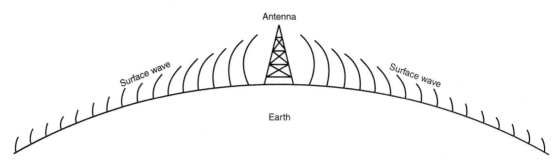

Fig. 9-35 Ground- or surface-wave radiation from an antenna.

its lowest point, the ionosphere is approximately 30 mi above the earth and extends to as much as 250 mi above the earth. The ionosphere is generally considered to be divided into three basic layers, designated the D layer, the E layer, and the F layer.

The D and E layers are only weakly ionized areas primarily because they are the farthest from the sun. Primarily they exist only during daylight hours. During daylight hours they tend to absorb radio signals in the medium-frequency range from 300 kHz to 3 MHz.

The F layer is generally considered to be divided into two sublayers designated F1 and F2. These are the most highly ionized because they are closest to the sun and, as a result, will have the most effect on radio signals. The F layer exists both during daylight and nighttime hours.

The primary effect of the F layer is to cause refraction of the radio signal. *Refraction* is the deflection or bending of electromagnetic waves such as radio waves, light, or even sound when the waves cross a boundary line between two mediums with different characteristics. In the case of the ionosphere, the boundary is the interface between different levels of ionization. When a radio signal goes into the ionosphere, the different levels of ionization will cause the radio waves to be gradually bent. The direction of bending depends upon the angle at which the radio wave enters the ionosphere and the different degrees of ionization of the layers. Figure 9-36 shows the effects of refraction with different angles of radio signals entering the ionosphere. When the angle is large with respect to the earth, the radio signals are bent slightly and pass on through the ionosphere and are lost in space. If the angle of entry is smaller, the radio wave will actually be bent and sent back to earth. Because of this effect, it actually appears as though the radio wave

Multiple hop

Refraction

Direct waves

has been reflected by the ionosphere. In reality, the ionosphere does not act like a mirror and cause reflection, but it does cause refraction, which produces an equivalent effect.

The radio waves are sent back to earth with minimum signal loss. The result is that the signal is propagated over an extremely long distance. This effect is most pronounced in the 3- to 30-MHz range, which permits extremely long distance communications.

In some cases, the signal refracted back from the ionosphere strikes the earth and is reflected back up to the ionosphere again to be bent and sent back to earth. When this occurs, this is known as *multiple-skip* or *multiple-hop transmission.* For strong signals and ideal ionospheric conditions, two, three, or more hops may be possible. This can extend the communications range by many thousands of miles. The maximum distance of a single hop is usually no more than about 2000 mi, but with multiple hops, transmissions halfway around the world are possible.

Radio signals can be reflected by large objects. The earth reflects signals upward as indicated above, but smaller metallic reflectors at least one-half wavelength long at the signal frequency can also reflect signals. Tall buildings, mountains, water towers, and even airplanes can reflect radio signals and thus change their direction. Such reflectors act just like mirrors.

Space Waves

The third method of radio signal propagation is by *direct* or *space waves.* A direct wave travels in a straight line directly from the transmitting antenna to the receiving antenna. You will often hear direct-wave radio signaling referred to as line-of-sight communications. Direct or space waves are not refracted, nor do they follow the curvature of the earth.

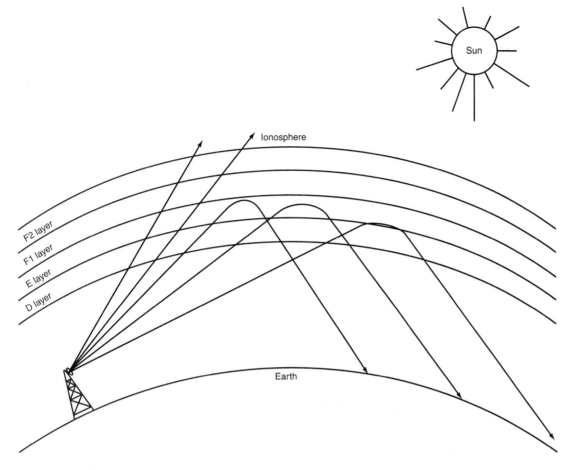

Fig. 9-36 Sky-wave propagation.

Because of their straight-line nature, direct waves will at some point be blocked because of the curvature of the earth. The signals will travel horizontally from the antenna until they reach the horizon at which point they are blocked. This is illustrated in Fig. 9-37. If the signal is to be received beyond the horizon, then the antenna must be high enough to intercept the straight-line radio waves.

Obviously, the transmitting distance with direct waves is limited to relatively short distances and is strictly a function of the height of the transmitting and receiving antennas. For example, the formula for computing the distance between a transmitting antenna and the horizon is

$$d = \sqrt{2h_t}$$

where h_t = the height of the transmitting antenna in feet

d = the distance from the transmitter to the horizon in miles

For a complete communications system, the height of both the transmitting and receiving

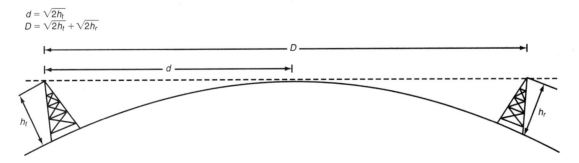

Fig. 9-37 Line-of-sight communications by direct or space waves.

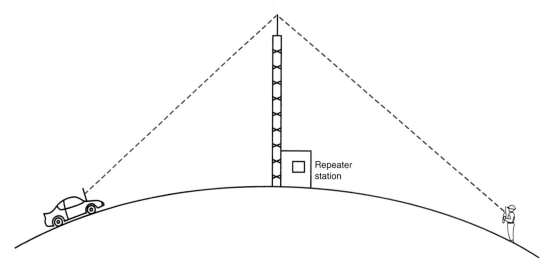

Fig. 9-38 How a repeater extends the communications distance of mobile radio units at VHF and UHF.

Repeater

antennas must be taken into consideration. The distance between transmitting and receiving antennas then is given by the formula

$$D = \sqrt{2h_t} + \sqrt{2h_r}$$

In this formula, h_r is the height of the receiving antenna in feet.

For example, if a transmitting antenna is 150 ft high and the receiving antenna is 40 ft high, the greatest transmission distance is

$$D = \sqrt{2(150)} + \sqrt{2(40)}$$
$$= \sqrt{300} + \sqrt{80}$$
$$= 17.32 + 8.9$$
$$= 26.22 \text{ mi}$$

Line-of-sight communications by direct wave is characteristic of most radio signals with a frequency above approximately 30 MHz. This is particularly true of VHF, UHF, and microwave signals. Such signals pass through the ionosphere and are not bent. Transmission distances at these frequencies are extremely limited. Antenna height is of utmost importance. You can see why FM and TV broadcasts cover only limited distances. This is the reason why very high transmitting antennas must be used. Transmitters and receivers operating at the very high frequencies typically locate their antennas on top of tall buildings or on mountains to greatly increase the range of transmission and reception.

Repeaters

To extend the communications distance at VHF, UHF, and microwave frequencies, special tech-

niques have been adopted. The most important of these is the use of *repeater stations*. The concept of a repeater is illustrated in Fig. 9-38. A repeater is a combination of a receiver and a transmitter. The function of the repeater is to pick up the signal from a transmitter, amplify it, and retransmit it on another frequency to the receiver. Usually the repeater will be located between the transmitting and receiving stations and will, therefore, extend the communications distance. Repeaters have extremely sensitive receivers, high-power transmitters, and antennas located at a high point.

Repeaters are extremely effective in increasing the communications range for mobile and hand-held radio units. The antennas on cars and trucks are naturally not very high off the ground, and, therefore, their transmitting and receiving range is extremely limited. However, by operating through a repeater located at some high point, the communications range can be increased considerably.

This concept can be taken further by using a whole series of repeater stations as shown in Fig. 9-39. Each repeater contains a receiver and a transmitter. The original signal is picked up, amplified, and retransmitted on a different frequency to a second repeater which does the same thing. A whole string of repeater stations can relay signals for great distances. Typically such relay stations are located 20 to 60 mi apart, mostly at high elevations to ensure reliable communications over long distances. The microwave relay stations used by many telephone companies for long-distance communications are good examples of such series of repeaters.

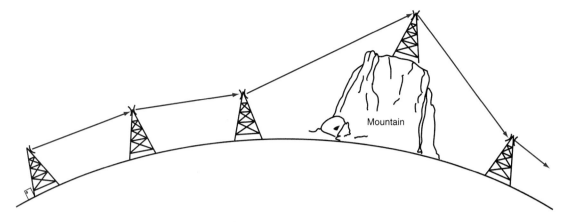

Satellite

Fig. 9-39 Using repeater stations to increase communications distances at microwave frequencies.

The ultimate repeater or relay station, of course, is the communications satellite. Most communications satellites are located in a geostationary orbit 22,300 mi out in space. At this distance, when rotating about the equator, they rotate in synchronism with the earth and, therefore, appear stationary above the earth. They act as fixed repeater stations. Signals sent to the satellite are amplified and retransmitted back to earth from long distances away. The receiver-transmitter combination within the satellite is known as a transponder. Most satellites have many transponders so that multiple signals can be relayed. With such an arrangement, worldwide communication is possible at microwave frequencies. You will learn more about satellite communications in a later chapter.

■ TEST

Answer the following questions.

64. List the three paths that a radio signal may take through space.
65. A radio wave that propagates near the surface of the earth is called a(n) _____ or _____ wave.
66. The radio wave that is refracted by the ionosphere is known as a(n) _____ wave.

67. A radio wave that propagates only over-line-of-sight distances is called a(n) _____ or _____ wave.
68. The surface wave is effective only at frequencies below about _____ MHz.
69. The upper part of the earth's atmosphere ionized by the sun that affects radio waves is called the _____.
70. The _____ layer has the greatest effect on a radio signal.
71. The ionized atmosphere causes radio waves at some frequencies to be _____.
72. True or false. Radio waves are easily reflected by large objects.
73. True or false. The ionosphere reflects radio waves.
74. Only signals in the _____ to _____ MHz range are significantly affected by the ionosphere.
75. Worldwide radio communications is possible thanks to _____ transmission.
76. The VHF, UHF, and microwave signals travel in a(n) _____.
77. To increase transmission distances at VHF and above, special stations called _____ are used.
78. A microwave relay station contains a(n) _____ and a(n) _____ operating on different frequencies.

Summary

1. A transmission line is a two-wire cable used to carry RF energy between two different pieces of communications equipment or between an antenna and a receiver or transmitter.

2. The two most common types of transmission lines are balanced and coaxial.

3. The primary feature of a transmission line is its characteristic or surge impedance Z_o which is a function of the distributed inductance L and the capacitance C per unit length ($Z_o = L/C$).

4. The characteristic impedance of a balanced line is determined by its physical dimensions. [$Z_o = 276 \log (2S/D)$ where S is the center-to-center spacing of the conductors and D is the diameter of the conductors.]

5. The characteristic impedance of coax also depends on its physical dimensions. [$Z_o = 138 \log (D/d)$ where D is the inside diameter of the shield and d is the diameter of the inner conductor.]

6. The proper use of a transmission line is to terminate it in a load impedance equal to its surge impedance. All the power applied to the line will be absorbed by the load.

7. Wavelength is the distance between adjacent peaks of an RF wave. It is also the distance traveled by a signal in one cycle. Wavelength (λ) is computed with the expression $\lambda = 300/f$ where f is the frequency in megahertz and λ is in meters, Or $\lambda = 984/f$ where λ is in feet.

8. If a transmission line is not terminated in its characteristic impedance, the load will not absorb all the power. Some of it will be reflected back toward the generator.

9. If the load on a transmission line is an open or short, all the power applied to the line will be reflected back to the generator.

10. The forward or incident power applied to the line combines with the reflected power to produce a pattern of voltage and current variations along the line known as standing waves.

11. If the load impedance matches the line impedance, there are no standing waves.

12. A measure of the mismatch between line and load impedances or the maximum and mini-mum voltage and current variations along the line is the standing-wave ratio (SWR) which is a number always greater than 1.

13. The SWR indicates how much power is delivered to the load and lost in the line. With SWR = 1, all power is delivered to the load.

14. The ratio of the reflected voltage V_r to the incident voltage V_i on a transmission line is called the reflection coefficient R ($R = V_r/V_i$). A properly terminated line will have $R = 0$. A shorted or open line will have $R = 1$.

15. The SWR in terms of the reflection coefficient is

$$\text{SWR} = \frac{1 + R}{1 - R}$$

The SWR in terms of power is

$$\frac{1 + \sqrt{P_R/P_F}}{1 - \sqrt{P_R/P_F}}$$

16. The SWR = Z_o/Z_l or Z_l/Z_o where Z_o is the characteristic impedance and Z_l is the load impedance.

17. Every effort is made to reduce the SWR by using impedance-matching circuits to ensure that maximum power is delivered to the load.

18. Transmission lines, one-quarter or one-half wavelength long and either shorted or open, act like resonant or reactive circuits.

19. At UHF and microwave frequencies where one-half wavelength is less than 1 ft, transmission lines are commonly used to replace conventional LC tuned circuits.

20. A shorted quarter wave and an open half wave act like a parallel resonant circuit.

21. Both an open quarter-wave circuit and a shorted half-wave circuit act like a series resonant circuit.

22. The velocity of propagation of a radio signal is slower in a transmission line than in free space. This difference is expressed as the velocity factor F for different types of lines. Coax has a velocity factor of 0.6 to 0.7. The velocity factor of open wire line or twin lead is in the 0.7 to 0.8 range. In computing the length of a transmission

line at a specific frequency, the velocity factor must be considered ($\lambda = 984\ F/f$).

23. An antenna or aerial is one or more conductors used to transmit or receive radio signals.

24. A radio signal is electromagnetic energy made up of electric and magnetic fields at right angles to one another and to the direction of signal propagation.

25. The polarization of a radio signal is defined as the orientation of the electric field with respect to the earth and is either vertical or horizontal.

26. The most common antenna is the half-wave dipole or doublet that has a characteristic impedance of approximately 73 Ω at the center. Its length in feet is $468/f$ where f is the frequency in megahertz.

27. The dipole has a bidirectional figure-eight radiation pattern and is usually mounted horizontally but may also be used vertically.

28. A popular variation of the dipole is the folded dipole which is one-half wavelength long and has an impedance of 300 Ω.

29. Another popular antenna is the quarter-wave vertical. The earth acts as the other quarter wave to simulate a half-wave vertical dipole.

30. The quarter-wave vertical is referred to as a ground plane antenna. It is fed with coax with the center conductor connected to the antenna and the shield connected to earth ground, to an array of quarter-wave wires called radials, or to a large, flat, metal surface. Its length in feet is $234/f$ where f is in megahertz.

31. The characteristic impedance of a ground plane is about 36.5 Ω. It has an omnidirectional radiation pattern that sends or receives equally well in all directions.

32. A directional antenna is one that transmits or receives over a narrow range in only one direction.

33. Directional antennas made up of two or more elements focus the radiation into a narrow beam, thus giving the antenna gain.

34. The gain of the antenna is the power amplification resulting from the concentration of power in one direction. The gain may be expressed as a power ratio or in decibels.

35. The effective radiated power (ERP) of an antenna is the power input multiplied by the antenna power gain.

36. Directional antennas with two or more elements are called arrays. There are two types of arrays: parasitic and driven.

37. Parasitic elements called reflectors and directors when spaced parallel to a half-wave dipole driven element help focus the signal into a narrow beam.

38. The measure of the directivity of an antenna is the beam width or beam angle measured in degrees.

39. A parasitic array made up of a driven element, reflector, and one or more directors is known as a Yagi or beam antenna and has a gain of 10 to 20 dB with a beam width of 40° to 20°.

40. Driven arrays consists of two or more half-wavelength elements, each receiving power from the transmission line.

41. The three most popular driven arrays are the collinear, end-fire, and broadside.

42. A widely used driven array is the log-periodic antenna which exhibits gain, directivity, and a wide operating frequency range.

43. A radio wave propagates through space in one of three ways: ground wave, sky waves, or direct waves.

44. The ground or surface wave leaves the antenna and follows the curvature of the earth. The ground wave is only effective on frequencies below 3 MHz.

45. The sky wave propagates from the antenna upward where it is bent back to earth by the ionosphere.

46. The ionosphere is a portion of the earth's atmosphere 30 to 250 mi above the earth that has been ionized by the sun.

47. The ionosphere is made up of three layers of different ionization density: the D, E, and F layers. The F layer is the most highly ionized and causes refraction or bending of radio waves back to earth.

48. The refraction of the ionosphere causes a radio signal to be bent back to earth with little or no attenuation long distances from the transmitter. This is known as a skip or hop.

49. Multiple skips or hops between the ionosphere and earth permit very long distances, even worldwide, communications. This effect is useful over the 3- to 30-MHz range.

50. At frequencies above 30 MHz, propagation is primarily by the direct or space wave which travels in a straight line between transmitting and receiving antennas. This is known as line-of-sight communications.

51. Radio waves are easily blocked or reflected by large objects. This is particularly true of VHF, UHF, and microwave signals.
52. The communications distance at VHF, UHF, and microwave frequencies is limited to the line-of-sight distance between transmitting and receiving antennas.
53. The line-of-sight distance D is limited by the curvature of the earth and is dependent upon the heights h_t and h_r of the transmitting and receiving antennas, respectively.

$$D = \sqrt{2h_t} + \sqrt{2h_r}$$

54. To extend transmission distances at VHF, UHF, and microwave frequencies, relay stations known as repeater stations receive and retransmit signals.

Chapter Review Questions

Choose the letter which best answers each question.

9-1. The most commonly used transmission line is a
   a. Two-wire balance line.
   b. Single wire.
   c. Three-wire line.
   d. Coax.
9-2. The characteristic impedance of a transmission line does *not* depend upon its
   a. Length.
   b. Conductor diameter.
   c. Conductor spacing.
   d. None of the above.
9-3. Which of the following is *not* a common transmission line impedance?
   a. 50 Ω
   b. 75 Ω
   c. 120 Ω
   d. 300 Ω
9-4. For maximum absorption of power at the antenna, the relationship between the characteristic impedance of the line Z_o and the load impedance Z_l should be
   a. $Z_o = Z_l$
   b. $Z_o > Z_l$
   c. $Z_o < Z_l$
   d. $Z_o = 0$
9-5. The mismatch between antenna and transmission line impedances *cannot* be corrected for by
   a. Using an *LC* matching network.
   b. Adjusting antenna length.
   c. Using a balun.
   d. Adjusting the length of transmission line.
9-6. A pattern of voltage and current variations along a transmission line not terminated in its characteristic impedance is called
   a. An electric field.
   b. Radio waves.
   c. Standing waves.
   d. A magnetic field.
9-7. The desirable SWR on a transmission line is
   a. 0.
   b. 1.
   c. 2.
   d. Infinity.
9-8. A 50-Ω coax is connected to a 73-Ω antenna. The SWR is
   a. 0.685.
   b. 1.
   c. 1.46.
   d. 2.92.
9-9. The most desirable reflection coefficient is
   a. 0.
   b. 0.5.
   c. 1.
   d. Infinity.
9-10. A ratio expressing the percentage of incident voltage reflected on a transmission line is known as the
   a. Velocity factor.
   b. Standing-wave ratio.
   c. Reflection coefficient.
   d. Line efficiency.
9-11. The minimum voltage along a transmission line is 260 V, while the maximum is 390 V. The SWR is
   a. 0.67.
   b. 1.0.
   c. 1.2.
   d. 1.5.
9-12. Three feet is one wavelength at a frequency of
   a. 100 MHz.
   b. 164 MHz.
   c. 300 MHz.
   d. 328 MHz.

9-13. At very high frequencies, transmission lines are used as
  a. Tuned circuits.
  b. Antennas.
  c. Insulators.
  d. Resistors.

9-14. A shorted quarter-wave line at the operating frequency acts like a(n)
  a. Series resonant circuit.
  b. Parallel resonant circuit.
  c. Capacitor.
  d. Inductor.

9-15. A shorted half-wave line at the operating frequency acts like a(n)
  a. Capacitor.
  b. Inductor.
  c. Series resonant circuit.
  d. Parallel resonant circuit.

9-16. A popular half-wavelength antenna is the
  a. Ground plane.
  b. End-fire.
  c. Collinear.
  d. Dipole.

9-17. The length of a doublet at 27 MHz is
  a. 8.67 ft.
  b. 17.3 ft.
  c. 18.2 ft.
  d. 34.67 ft.

9-18. A popular vertical antenna is the
  a. Collinear.
  b. Dipole.
  c. Ground plane.
  d. Broadside.

9-19. The magnetic field of an antenna is perpendicular to the earth. The antenna's polarization
  a. Is vertical.
  b. Is horizontal.
  c. Is circular.
  d. Cannot be determined from the information given.

9-20. An antenna that transmits or receives equally well in all directions is said to be
  a. Omnidirectional.
  b. Bidirectional.
  c. Unidirectional.
  d. Quasidirectional.

9-21. The horizontal radiation pattern of a dipole is a
  a. Circle.
  b. Figure eight.
  c. Clover leaf.
  d. Narrow beam.

9-22. The length of a ground plane vertical at 146 MHz is
  a. 1.6 ft.
  b. 1.68 ft.
  c. 2.05 ft.
  d. 3.37 ft.

9-23. The impedance of a dipole is about
  a. 50 Ω.
  b. 73 Ω.
  c. 93 Ω.
  d. 300 Ω.

9-24. A directional antenna with two or more elements is known as a(n)
  a. Folded dipole.
  b. Ground plane.
  c. Loop.
  d. Array.

9-25. The horizontal radiation pattern of a vertical dipole is a
  a. Figure eight.
  b. Circle.
  c. Narrow beam.
  d. Clover leaf.

9-26. In a Yagi antenna, maximum direction of radiation is toward the
  a. Director.
  b. Driven element.
  c. Reflector.
  d. Sky.

9-27. Conductors in multielement antennas that do not receive energy directly from the transmission line are known as
  a. Parasitic elements.
  b. Driven elements.
  c. The boom.
  d. Receptors.

9-28. A coax has an attenuation of 2.4 dB per 100 ft. The attenuation for 275 ft is
  a. 2.4 dB.
  b. 3.3 dB.
  c. 4.8 dB.
  d. 6.6 dB.

9-29. An antenna has a power gain of 15. The power applied to the antenna is 32 W. The effective radiated power is
  a. 15 W.   c. 120 W.
  b. 32 W.   d. 480 W.

9-30. Which beam width represents the best antenna directivity?
  a. 7°
  b. 12°
  c. 19°
  d. 28°

9-31. The radiation pattern of collinear and broad-side antennas is
a. Omnidirectional.
b. Bidirectional.
c. Unidirectional.
d. Clover-leaf shaped.

9-32. Which antenna has a unidirectional radiation pattern and gain?
a. Dipole
b. Ground plane
c. Yagi
d. Collinear

9-33. A wide-bandwidth multielement driven array is the
a. End-fire.
b. Log-periodic.
c. Yagi.
d. Collinear.

9-34. Ground-wave communications is most effective in what frequency range?
a. 300 kHz to 3 MHz
b. 3 to 30 MHz
c. 30 to 300 MHz
d. Above 300 MHz

9-35. The ionosphere causes radio signals to be
a. Diffused.
b. Absorbed.
c. Refracted.
d. Reflected.

9-36. The ionosphere has its greatest effect on signals in what frequency range?
a. 300 kHz to 3 MHz
b. 3 to 30 MHz
c. 30 to 300 MHz
d. Above 300 MHz

9-37. The type of radio wave responsible for long-distance communications by multiple skips is the
a. Ground wave.
b. Direct wave.
c. Surface wave.
d. Sky wave.

9-38. Microwave signals propagate by way of the
a. Direct wave.
b. Sky wave.
c. Surface wave.
d. Standing wave.

9-39. Line-of-sight communications is *not* a factor in which frequency range?
a. VHF
b. UHF
c. HF
d. Microwave

9-40. A microwave-transmitting antenna is 550 ft high. The receiving antenna is 200 ft high. The maximum transmission distance is
a. 20 mi.
b. 33.2 mi.
c. 38.7 mi.
d. 53.2 mi.

9-41. To increase the transmission distance of a UHF signal, which of the following should be done?
a. Increase antenna gain.
b. Increase antenna height.
c. Increase transmitter power.
d. Increase receiver sensitivity.

9-42. A coax has a velocity factor of 0.68. What is the length of a half wave at 30 MHz?
a. 11.2 ft
b. 12.9 ft
c. 15.6 ft
d. 16.4 ft

9-43. Which transmission line listed here has the lowest attenuation?
a. Twin lead
b. RG-11/U
c. RG-59/U
d. RG-214/U

9-44. Refer to Fig. 9-40. The beam width of this antenna pattern is approximately
a. 30°.
b. 38°.
c. 45°.
d. 60°.

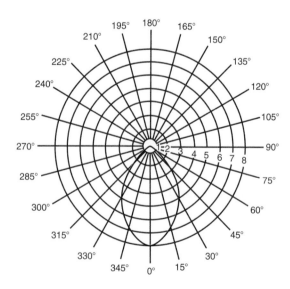

Fig. 9-40 Pattern for Chapter Review Question 9-44.

9-45. A receiver-transmitter station used to increase the communications range of VHF, UHF, and microwave signals is called a(n)
  a. Transceiver.
  b. Remitter.
  c. Repeater.
  d. Amplifier.

9-46. The forward or incident power from a transmitter is measured as 880 W. The reflected power is 60 W. The SWR is
  a. 0.26
  b. 1.07
  c. 1.36
  d. 1.7

Critical Thinking Questions

9-1. Assume that you wanted to transmit audio directly by antenna without modulation. If the signal to be transmitted has a frequency of 2 kHz, calculate the length of a quarter-wave vertical antenna. Explain why this is an impractical antenna.

9-2. What do you think happens to the gain of a directional antenna as its beam width is decreased? Explain.

9-3. Does the length of a properly terminated and matched transmission line affect the SWR?

9-4. How can a satellite work if radio signals are affected by the ionosphere, which extends only 250 miles out?

9-5. Name two sure things that you can do to improve transmission distance and communications reliability of VHF and UHF signals.

9-6. If a coaxial cable slows the signal down because of its velocity factor, this means that a cable introduces a time delay or a phase shift. Calculate the approximate phase shift between the input and output of a coaxial cable 25 ft long with a velocity factor of 0.66 at a frequency of 10 MHz.

9-7. The length and height of an antenna affect its resonant frequency and transmission line driving impedance. If the antenna impedance is not purely resistive but contains some reactance, how can the antenna be properly matched to the transmitter or receiver?

9-8. A low-power "bugging" transmitter is used for surveillance but cannot transmit over a distance of 400 ft. The receiver is a quarter of a mile away. Name some techniques that can be used to make this system work.

Answers to Tests

1. balanced, coax
2. unbalanced
3. wavelength
4. 0.667
5. 1.475
6. characteristic or surge impedance
7. low-pass filter, capacitors, inductors
8. 75.23
9. 6.25
10. 52
11. constant or flat
12. standing waves
13. reflected
14. standing waves
15. 1.44
16. 1
17. open, shorted
18. reflection coefficient
19. 1.67, 0.25
20. 7
21. 1, infinity
22. tuned or resonant circuits
23. 6.3 [0.07(90) = 6.3] (See Fig. 9-12.)
24. high or infinite
25. open, shorted
26. inductors, capacitors
27. series
28. 2.788
29. 109.55 W
30. 2.33:1
31. electric, magnetic
32. electric field
33. transmission or feed line
34. vertically

35. dipole or doublet
36. 3.12
37. 73
38. figure eight
39. beam width
40. omnidirectional
41. ground plane
42. 3.16
43. earth ground, radials
44. circle
45. 36.5
46. bidirectional
47. one
48. gain
49. elements
50. parasitic, driven
51. director, reflector, driven element
52. director, reflector
53. director
54. 20, 40
55. 2.1
56. collinear, end-fire, broadside
57. true

58. log-periodic
59. antenna tuner
60. balun
61. Q section, matching stub
62. 4:1
63. 9.39
64. ground or surface wave, sky wave, direct or space wave
65. ground, surface
66. sky
67. direct or space
68. 3
69. ionosphere
70. F
71. refracted or bent
72. true
73. false
74. 3, 30
75. multiple hop
76. straight line
77. repeaters
78. transmitter, receiver

Chapter 10

Microwave Techniques

Chapter Objectives

This chapter will help you to:

1. *State* the importance of microwaves to the future of communications.
2. *Define* the term waveguide, *explain* how a waveguide works, and *calculate* the cutoff frequency of a waveguide.
3. *Define* microstrip and stripline and *tell* how they are used.
4. *Explain* the physical structure of a cavity resonator and *tell* what cavity resonators are used for.
5. *Name* seven microwave semiconductor devices and *explain* how they operate and where they are used.

6. *Name* the three commonly used microwave vacuum tubes, *explain* their operation, and *state* where they are used.
7. *Name* four common types of microwave antennas and *calculate* the gain and beam width of horn and parabolic dish antennas.
8. *Explain* the physical construction of the helical, bicone, slot, lens, and patch antennas.
9. *Describe* the operation of a radar system.

The electromagnetic frequency spectrum is a finite natural resource. As the use of electronic communications has increased over the years, the frequency spectrum normally used for radio signals has become extremely crowded. Many new communications services have been added, and the increased number of users has crowded the airwaves. Thanks to technological developments, more efficient usage of the spectrum is being made. But the future holds only more growth and further overcrowding. One of the primary solutions to

this problem has been to extend radio communications into higher frequencies in the spectrum. Initially this expansion was in the VHF and UHF ranges, but now, the primary expansion of radio communications services is in those frequencies beyond 1 GHz. Known as microwaves, these frequencies offer tremendous bandwidth for communications and at least temporarily resolve the problem of spectrum crowding. In this chapter, we introduce you to microwaves and show you some of the special techniques and components unique to this field.

10-1 Microwaves in Perspective

Microwaves are signals with a frequency greater than 1 GHz. The microwave region is generally considered to extend to 30 GHz, although some definitions include frequencies up to 300 GHz. In the 1- to 30-GHz range, these signals have wavelengths of 30 cen-

timeters (cm) (about 1 ft) to 1 cm (or about 0.4 in.).

Operating in the microwave region solves many of the problems of the overcrowding in the radio spectrum. At the same time, it introduces additional benefits but also causes some unique problems. Working with microwave

Microwaves

equipment requires special knowledge and skills considerably different from those needed for conventional electronic equipment.

Bandwidth Benefits

Every electronic signal used in communications has a finite bandwidth. When a carrier is modulated by the information signal, sidebands are produced. The resulting signal occupies a certain amount of space in the RF spectrum. We call that small segment of spectrum a *channel.* As the number of communications signals and channels increase, more and more of the spectrum space is used up. Over the years as the need for electronic communication has increased, the number of radio communications stations has increased dramatically, and the radio spectrum has become extremely crowded.

Channel

Use of the RF spectrum is typically regulated by the government. In the United States, this job is assigned to the *Federal Communications Commission (FCC).* The FCC establishes various classes of radio communications and regulates the assignment of spectrum space. For example, certain areas of the spectrum are set aside for radio and TV broadcasting, and frequency assignments are given to the stations. For two-way radio communications, other portions of the spectrum are used. All the various classes of radio communications are given specific areas in the spectrum in which they can operate. Over the years, the available spectrum space has essentially been used up. In many cases, communications services share frequency assignments. In some areas, new licenses are no longer being granted since the current spectrum space for that service is full. Yet the demand for new electronic communications channels continues.

Federal Communications Commission (FCC)

One of the solutions to this problem has been to let technology help solve the problem. For example, great improvements in communications receivers have been made to improve selectivity so that adjacent channel interference will not be as great. This permits stations to operate on more closely spaced frequencies.

On the transmitting side, new techniques have helped squeeze more signals into the same frequency spectrum. A classic example is the use of SSB signals where only one sideband is used rather than two, thereby cutting the spectrum usage in half. Limiting the deviation of FM signals also helps to control the bandwidth. In data communications new modulation techniques such as phase-shift keying have been developed to help narrow the bandwidth of transmitted information. Multiplexing techniques help put more signals into a single channel.

Although excellent progress has been made, the available spectrum is still overcrowded and future expectations are that the problem will worsen. The spectrum space between 0 and approximately 300 MHz is absolutely full.

To solve this problem, higher and higher frequencies are being used. Initially, the VHF and UHF bands were tapped. Today, most new communications services are assigned to the microwave region simply because more channels and wider bandwidths are available.

To give you some idea why more bandwidth is available at the higher frequencies, let's take an example. A standard AM broadcast station operating on 1000 kHz has a maximum bandwidth of 10 kHz. The bandwidth represents only $10/1000 = 0.01 = 1$ percent of the spectrum space at that frequency. Now consider a microwave carrier frequency of 3 GHz. One percent of 3 GHz is $0.01(3,000,000,000) = 30,000,000$ Hz, or 30 MHz. A bandwidth of 30 MHz is incredibly wide. In fact, it represents the low-frequency, medium-frequency, and high-frequency portions of the spectrum. This is the space that might be occupied by a 3-GHz carrier modulated by a 15-MHz information signal. Obviously, most information signals do not require that kind of bandwidth. A voice signal would take up only a tiny fraction of that. Our 10-kHz bandwidth AM signal represents only $10,000/3,000,000,000 = 0.000333$

About ⬛⬛⬛ Electronics

Materials used for Microwave Devices, Dopants, and Semiconductors

- GaAs (gallium arsenide) and silicon-germanium (Si-Ge) work better than silicon in microwave devices because they allow faster movement of electrons.
- Materials other than boron and arsenic are used as dopants.
- It is theoretically possible to make semiconductor devices from crystalline carbon.
- Crystal radio receivers were an early application of semiconductors.

percent of 3 GHz. Up to 3000 AM broadcast stations with 10-kHz bandwidth could be accommodated.

Obviously then, the higher the frequency, the greater the bandwidth available for the transmission of information. This gives not only more space for individual stations, but also allows wide-bandwidth information signals such as video to be accommodated. The average TV signal has a bandwidth of approximately 6 MHz. It is impractical to transmit video signals on low frequencies because they use up entirely too much spectrum space. That's why most TV transmission is in the VHF and UHF ranges. There is even more space for video in the microwave region.

The wide bandwidth also makes it possible to use various multiplexing techniques to transmit more information. When signals are multiplexed, the resulting bandwidth is generally wide, but these multiplexed signals can be easily handled in the microwave region. Finally, data communications of binary information often requires relatively wide bandwidths. These are easily transmitted on microwave frequencies.

The biggest benefit of microwave radio transmission, then, is the enormous amount of spectrum space available for new communications services. Today most new communications equipment is being developed for the microwave region. It is the only solution to the growing need for more spectrum space.

The Challenges of Microwaves

Although the use of microwaves offers some advantages, it also presents numerous problems. For example, the higher the frequency, the more difficult it becomes to analyze electronic circuits. The analysis of electronic circuits at lower frequencies, say those below 30 MHz, is based upon current-voltage relationships. Such relationships are simply not usable at microwave frequencies. Instead, most components and circuits are analyzed in terms of electric and magnetic fields. As a result, those techniques commonly used for analyzing antennas and transmission lines also become effective for understanding microwave electronic circuits.

Measuring techniques are also different. In low-frequency electronics, we regularly measure currents and voltages. In microwave circuits, the measurements are those of electric and magnetic fields. Power measurements are more common than voltage and current measurements.

Another problem is that at microwave frequencies, conventional components become more difficult, if not impossible, to implement. For example, a common resistor may look like a pure resistance at lower frequencies, but at microwave frequencies, it exhibits other characteristics. The short leads of a resistor, although they may be less than an inch, represent a significant amount of inductive reactance at very high frequencies. A small capacitance also exists between the leads. As a result, at microwave frequencies a simple resistor looks like a complex *LCR* circuit. The same goes for any inductor or capacitor as well.

In addition, to create resonant circuits at microwave frequencies, the values of inductance and capacitance must be smaller and smaller. Physical limits become a problem. Even a half-inch piece of wire represents a significant amount of inductance at some microwave frequencies.

To overcome these problems, *lumped components* are not used at microwave frequencies. Instead, *distributed circuit elements* are used. Transmission lines are examples of distributed circuit elements where they constitute inductance and capacitance. When cut to the appropriate length, transmission lines can act as inductors, capacitors, and resonant circuits. Special versions of transmission lines known as strip lines, microstrips, waveguides, and cavity resonators are widely used to implement tuned circuits and reactances.

Conventional semiconductor devices also do not work properly at microwave frequencies. Because of inherent capacitances and inductances, ordinary diodes and transistors simply will not function. Sufficient gain is lacking.

Another serious problem is *transit time* in a transistor. Transit time refers to the amount of time it takes for the current carrier (holes or electrons) to move through a device. At low frequencies, the transit times are totally negligible, but at microwave frequencies, transit times become a high percentage of the actual signal period.

This problem has been solved by designing special smaller microwave diodes and transistors and using special materials such as gallium arsenide, in which transit time is

*inter*NET
CONNECTION

For information about microwave transmission, see the following company sites:
⟨www.cm-mrc.com⟩
⟨www.hdcom.com⟩
⟨www.nucomm.com⟩

Lumped components

Distributed circuit elements

Transit time

significantly less than in silicon. In addition, new and different components are used. This is particularly true for power amplification, where special vacuum tubes known as klystrons, magnetrons, and traveling-wave tubes are the primary components used for power amplification.

Another problem is that microwave signals like light waves travel in perfectly straight lines. This means that the communications distance is limited to line-of-sight range. Antennas must be very high for long-distance transmission. Further, because of the very short wavelength, microwave signals are easily reflected and otherwise diverted by both large and small objects. Even rain, snow, fog, and other atmospheric conditions can greatly absorb and attenuate microwave signals, especially those above about 20 GHz.

Typical Applications

Currently, the microwave frequency spectrum is used primarily for telephone communications and radar. Many long-distance telephone systems use microwave relay links for carrying telephone calls. By using multiplexing techniques, thousands of two-way communications are modulated on a single carrier and relayed from one station to another over long distances. Satellite repeater stations are also widely used for long-distance telephone communications. Some cell phones operate in the microwave region.

Radar

Another communications technique that is used primarily in the microwave region is *radar*. Radar is an acronym for *ra*dio *d*etection *a*nd *r*anging. It is a method of detecting the presence of a distant object and determining its distance and direction. Radar systems transmit a high-frequency signal that is then reflected from the distant object: an airplane, a missile, a ship, an automobile, or any other object. The reflected signal is picked up by the radar unit and compared to the transmitted signal. The time difference between the two gives the distance to the object. Since highly directional antennas are used, naturally the direction or azimuth of the object can also be determined.

The greatest use of radar, of course, is in military applications, but it is also used for safety purposes in aviation and marine fields. Another major use of radar is in weather fore-

casting. It has also found application in other areas such as in highway law enforcement.

Television stations and networks use microwave relay links to transmit TV signals over long distances instead of relying on long, lossy coaxial cables. Cable TV networks use satellite communications to transmit programs from one location to another. There are even consumer microwave satellite TV receivers capable of picking up TV transmissions directly from the satellite.

A growing application for microwave communications is space communications. Communications with satellites, deep-space probes, and other spacecraft is usually done by microwave transmission. The reason for this is that microwave signals are not reflected or absorbed by the ionosphere as are many lower-frequency signals. A microwave signal penetrates all the ionospheric layers directly into outer space with minimum attenuation.

Electromagnetic radiations from the stars are also primarily in the microwave region. Through the use of radio telescopes, astronomers can study the stars and other bodies in this galaxy and beyond. Sensitive radio receivers and large antennas operating in the microwave region are capable of plotting outer space even more effectively than can be done with visual (optical) telescopes.

Finally, microwaves are also used for heating. Microwave ovens are a common appliance in most kitchens today, and special microwave equipment known as diathermy machines are used in medicine for heating body muscles and tissues without hurting the skin. Microwave heating is also widely used in industry for a variety of heating and melting operations.

In the remainder of this chapter, we will introduce you to the major types of microwave components and circuits, including waveguides and cavity resonators, microstrips and stripline, microwave semiconductors, transmitting tubes, and antennas.

■ TEST

Answer the following questions.

1. Microwaves are frequencies above _____ GHz.
2. The main advantage of microwaves is that more of the _____ is available for signals.

3. List seven reasons why microwaves are more difficult to work with than lower-frequency signals.
4. The applications most commonly found in the microwave region are _____ and _____.
5. List four popular uses for microwaves.
6. The government agency that regulates radio communications in the United States is the _____.
7. Name four techniques that have helped squeeze more signals into the given spectrum.
8. Two reasons why conventional transistors won't work at microwave frequencies are _____ and _____.
9. At microwave frequencies, conventional components like resistors, capacitors, and inductors act like _____ circuits.
10. The TV channels 2 to 13 occupy a total bandwidth of about 72 MHz. At a frequency of 200 MHz, this represents a bandwidth of _____ percent of the spectrum space. At 2 GHz, it represents _____ percent.

10-2 Transmission Lines, Waveguides, and Cavity Resonators

Transmission lines are the means for carrying electromagnetic energy from one place to another. At frequencies below microwaves, coaxial cable is the primary means of carrying radio signals. But at microwave frequencies, this kind of transmission line is less effective.

Coax works well at microwave frequencies, but it has significant loss. The attenuation at microwave frequencies is so large as to make coax virtually unusable at all but the lower microwave frequencies (less than 10 GHz). Special types of coax using larger inner conductors and shields have been developed to minimize loss, although they are more difficult to use because in most cases they are rigid rather than flexible like most other cable. Special coax has been developed for the lower microwave frequencies and is used in some applications, such as consumer satellite TV. These special cables can be used up to approximately 6 GHz if the length is kept short. Above this frequency, coax loss is too great. Short lengths of coax, several feet or less, can be used to in-terconnect pieces of equipment that are close together, but for longer runs other methods of transmission must be used.

Stripline and Microstrip

As indicated in Chap. 9, transmission lines can be used as tuned circuits, filters, and even reactive components. And this is exactly how these circuits and components are implemented at microwave frequencies. Lengths of one-quarter and one-half wave are only inches or less, making them a practical size. But instead of coax or balanced line, special transmission lines are constructed by using printed circuit boards (PCBs) or IC manufacturing techniques.

Recall that a *PCB* is a flat insulating base made of phenolic or fiberglass to which is bonded copper on one or both sides and sometimes in several layers. Teflon is used as the base for some PCBs in microwave applications. The copper is etched away in patterns to form the interconnections for transistors, ICs, resistors, and other components. Thus point-to-point connections with wire are eliminated.

PCB

The PCB techniques can be used to create special microwave transmission lines called microstrip and stripline. *Microstrip* is a flat conductor separated from a large conducting ground plane by an insulating dielectric as shown in Fig. 10-1. The length of the microstrip is usually one-quarter or one-half wavelength. The ground plane is the circuit common.

Microstrip

Stripline is a flat conductor sandwiched between two ground planes as shown in Fig. 10-2. It is more difficult to make, but it will not radiate as microstrip does. The length is one-quarter or one-half wave.

Stripline

Both stripline and microstrip are widely used to form the tuned circuits used in microwave receiver front ends and in the amplifier

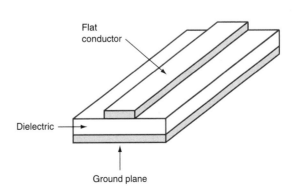

Fig. 10-1 Microstrip.

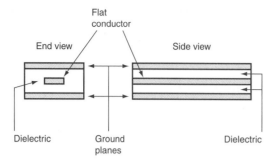

Fig. 10-2 Stripline.

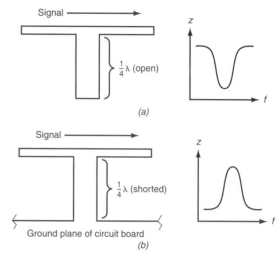

Note: Quarter-wave transmission lines act as impedance inverters

Fig. 10-3 Examples of λ/4 lines used as (*a*) band-reject or (*b*) bandpass filters.

sections of transmitters. Diodes, transistors, and other components are mounted right on the PCB and are connected directly to the formed microstrip or stripline. By shaping the stripline in specific ways, inductors and capacitors are also formed.

For example, at 1.9 GHz, a popular microwave frequency, a quarter-wavelength line would have a length of:

$$\lambda = 984/f_{\text{MHz}} = \frac{984}{1900} = 0.518 \text{ ft}$$

$$\lambda/4 = 0.518/4 = 0.129 \text{ ft or } 1.56 \text{ in.}$$

The velocity factor of microstrip or stripline is a function of the dielectric constant of the PC board material (*E*).

$$\text{VF} = \frac{1}{\sqrt{E}}$$

The value of *E* is usually in the 2 to 3 range. For an *E* of 2.7,

$$\text{VF} = \frac{1}{\sqrt{2.7}} = 0.61$$

The line length then is:

$$\lambda/4 = 1.56 \, (0.61) = 0.95 \text{ in.}$$

This λ/4 line could be a series or parallel resonant circuit depending upon whether it is shorted or open. Refer to Fig. 10-3, which shows quarter-wave lines used to pass or reject signals. In Fig. 10-3(*a*), an open λ/4 line acts

Waveguide

as a notch or band rejection filter. The line appears as a series resonant *RLC* circuit connected from the signal line to ground. In Fig. 10-3(*b*), the λ/4 wave line is connected to the ground plane of the printed-circuit board. The line is shorted so that it acts like a parallel resonant *RLC* circuit and a bandpass filter.

Even tinier microstrips and striplines can be made using monolithic, thin-film, and hybrid IC techniques. When combined with diodes, transistors, and other components, microwave integrated circuits (MICs) are formed.

Waveguides

Most microwave energy transmission is handled by waveguides. A *waveguide* is a hollow metal tube designed to carry microwave energy from one place to another. Waveguides may be used to carry energy between pieces of equipment or over longer distances to carry transmitter power to an antenna or microwave signals from an antenna to a receiver.

Waveguides are made from copper, aluminum, or brass. These metals are extruded into long rectangular or circular pipes. Often the insides of these waveguides are plated with silver to reduce their resistance to a very low level.

A microwave signal to be carried by a waveguide is injected into one end of the waveguide. This is done with an antennalike device which creates an electromagnetic wave that propagates through the waveguide. The electric

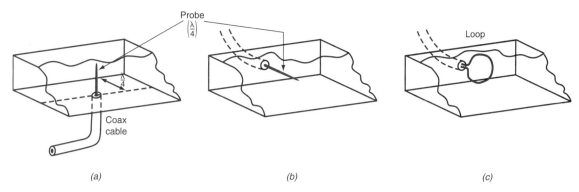

Fig. 10-4 Coupling RF energy to a waveguide.

and magnetic fields associated with the signal bounce off the inside walls back and forth as it progresses down the waveguide. The waveguide completely contains the signal so that none escapes by radiation.

Energy Coupling A common method of applying a microwave signal to a waveguide is done by using a *probe* as shown in Fig. 10-4(*a*). The probe is a quarter-wavelength vertical antenna at the signal frequency which is inserted in the waveguide one-quarter wavelength from the end that is closed. The signal is usually coupled to the probe through a coaxial connector. The probe sets up a vertically polarized electromagnetic wave in the waveguide, and the wave is then propagated down the line. The location of the probe one-quarter wavelength from the closed end of the waveguide causes the signal from the probe to be reflected from the closed end of the line back toward the open end. Over a quarter-wave distance, the reflected signal will appear back at the probe in phase to aid the signal going in the opposite direction. Remember that a radio signal consists of both electric and magnetic fields at right angles to one another. The electric and magnetic fields established by the probe propagate down the waveguide at a right angle to those two fields. An alternative probe position is shown in Fig. 10-4(*b*). The position of the probe determines whether the signal is horizontally or vertically polarized.

Figure 10-4(*c*) shows how a loop can be used to introduce electromagnetic radiation into the waveguide. Here the loop is mounted in the closed end of the waveguide. Microwave energy applied through a short piece of coax causes a magnetic field to be set up in the loop.

The magnetic field also establishes an electric field which is then propagated down the waveguide.

Extracting a signal from a waveguide is also done with probes and loops. When the signal strikes a probe or loop, a signal will appear across the coaxial connector which can then be fed to other circuitry.

Operating Frequency Range Most waveguides are of the rectangular variety. It is the size of the waveguide that determines its operating frequency range. Figure 10-5 shows the most important dimensions of a waveguide. The width of the waveguide is designated *a*, and the height is designated *b*. Note that these are the inside dimensions of the waveguide. The frequency of operation is determined by the *a* dimension. This dimension is usually made equal to one-half the wavelength at the lowest frequency of operation. This frequency is known as the waveguide *cutoff frequency*. At the cutoff frequency and below, the waveguide will not transmit energy. At frequencies above the cutoff frequency, the waveguide will propagate electromagnetic energy. As you can see

Probe

Cutoff frequency

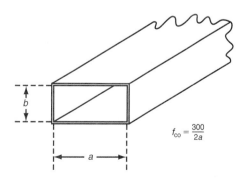

$$f_{co} = \frac{300}{2a}$$

Fig. 10-5 Dimensions of a waveguide determine its operating frequency range.

then, the waveguide is essentially a high-pass filter with a cutoff frequency equal to

$$f_{co} = \frac{300}{2a}$$

Here f_{co} is in megahertz and a is in meters.

As an example, a waveguide has an a width of 0.8 in. What is the cutoff? To convert 0.8 in. to meters, multiply by 2.54 to get centimeters and divide by 100 to get meters. So, 0.8 in. = 0.02032 m.

$$f_{co} = \frac{300}{2(0.02032)}$$

$$= 7382 \text{ MHz}$$

$$= 7.382 \text{ GHz}$$

Normally, the height of the waveguide is made equal to approximately one-half the a dimension, in this case about 0.4 in.

When a probe or loop launches energy into the waveguide, the electromagnetic fields will bounce off the side walls of the waveguide as illustrated in Fig. 10-6. The angles of incidence and reflection depend upon the operating frequency. At high frequencies, the angles are large, and, therefore, the path between the op-posite walls is relatively long as shown in Fig. 10-6(a). As the operating frequency gets lower, the angles decrease and the path between the sides shortens. When the operating frequency reaches the cutoff frequency of the waveguide, the signal simply bounces back and forth directly between the side walls of the waveguide and has no forward motion. At the cutoff frequency and below, no energy is propagated.

Rectangular waveguides usually come in a variety of standard sizes. The exact size is selected based on the desired operating frequency. The size of the waveguide is chosen so that its rectangular width is greater than one-half the wavelength but less than one wavelength at the operating frequency. This gives a cutoff frequency that is below the operating frequency, thereby ensuring that the signal will be propagated down the line. For example, a typical waveguide used in transmitting a 7-GHz signal has dimensions of $a = 1.37$ in. and $b = 0.62$ in. A waveguide of this size has a cutoff frequency of 4.3 GHz.

Modes of Operation   Whenever a microwave signal is launched into a waveguide by a probe

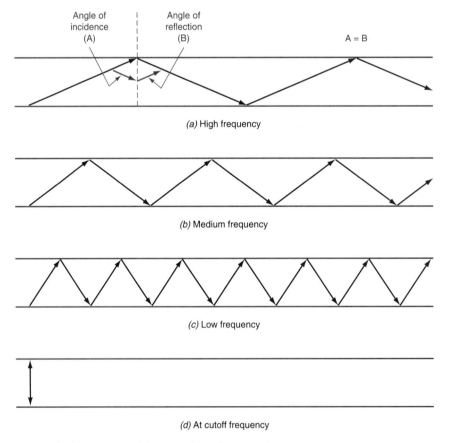

(a) High frequency

(b) Medium frequency

(c) Low frequency

(d) At cutoff frequency

Fig. 10-6 Wave paths in a waveguide at various frequencies.

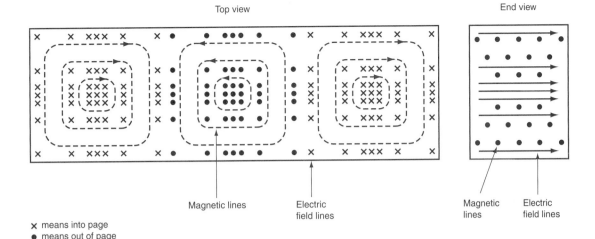

Top view   End view

Magnetic lines   Electric field lines   Magnetic lines   Electric field lines

✗ means into page
● means out of page

Fig. 10-7 Electric (E) and magnetic (H) fields in a rectangular waveguide.

or loop, the electric and magnetic fields are set up in various patterns depending upon the method of energy coupling, the frequency of operation, and the size of the waveguide.

Figure 10-7 shows typical fields in a waveguide. From the end view, you see the lines which represent the electric (E) field. The X's and dots represent the magnetic (H) field. The X's represent H field lines going into the page; the dots represent lines coming out of the page.

In the top view, the dashed lines represent the H field, and the X's and dots represent the E field lines.

The pattern of electromagnetic fields within a waveguide takes many forms. Each form is called an operating mode. Either the H or E field must be perpendicular to the direction of propagation of the wave. Modes are classified as either *transverse electric (TE)* or *transverse magnetic (TM)*. In the TE mode, the E field exists across the guide and no E lines extend lengthwise along the guide. In the TM mode, the H lines form loops in planes perpendicular to the walls of the guide, and no part of an H line is lengthwise along the guide.

Subscript numbers are used along with the TE and TM designations to further describe the E and H field patterns. A typical designation is $TE_{0,1}$. The first number indicates the number of half-wave patterns of transverse lines that exist along the short dimension of the guide through the center of the cross section. Transverse lines are those lines perpendicular to the walls of the guide. The second number indicates the number of transverse half-wave patterns that exist along the long dimen-

sion of the guide through the center of the cross section. If there is no change in the field intensity of one dimension, a zero is used.

The waveguide in Fig. 10-7 is TE because the E lines are perpendicular (transverse) to the sides of the guide. Looking at the end view in Fig. 10-7 along the short length, there is no field intensity change, so the first subscript is 0. Along the long dimension of the end view, the E lines spread out at the top and bottom but are close together in the center. The actual field intensity is sinusoidal, zero at the ends and maximum in the center. This is one-half of a sine wave variation. The second subscript, therefore, is 1. So the mode of the line in Fig. 10-7 is $TE_{0,1}$. This, by the way, is the main or dominant mode of most rectangular waveguides. Of course, there are many other patterns. Two others are shown in Fig. 10-8.

Coupling Hardware Waveguides are available in a variety of different lengths. As many sections as necessary can be interconnected to form a straight-line path between the microwave generator and its ultimate destination. When a waveguide is used as a transmission line between a transmitter or receiver and the antenna, relatively long straight sections are common. However, in most installations it is necessary for the waveguide to make turns. Special curved waveguide sections are available for making 90° bends. Curved sections introduce reflections and power loss, but

Transverse electric (TE)

Transverse magnetic (TM)

Wireless telecommunications standards may change often over the life of your career. Be sure to keep up to date by reading industry journals and publications from the FCC.

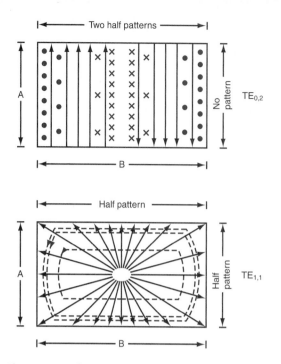

Fig. 10-8 Other waveguide operating modes.

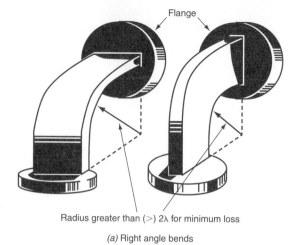

Radius greater than (>) 2λ for minimum loss

(a) Right angle bends

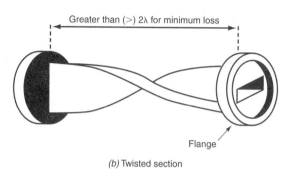

Greater than (>) 2λ for minimum loss

Flange

(b) Twisted section

Fig. 10-9 Curved waveguide sections.

these are kept small by proper design. When the radius of the curved section is greater than two wavelengths at the signal frequency, losses are minimized. Figure 10-9 shows several different configurations.

Most of these special sections are accompanied by flanges that are used to interconnect them with straight runs. Most waveguides and their fittings are precision-made so that the dimensions match perfectly. Any mismatch in dimensions or misalignment of pieces that fit together will introduce significant losses and reflections.

Cavity Resonators

A waveguide is a basic microwave component. Not only is it used as a transmission line, but special shorter segments are used for a variety of other purposes. These include simulating reactive components and forming resonant circuits. Other applications include the coupling of multiple units and impedance matching. Some of the special waveguide-derived components include cavity resonators, directional couplers, circulators, hybrids, and duplexers. Let's take a brief look at each of these.

One of the most important microwave devices is the *cavity resonator*. A cavity resonator is a short segment of waveguide that acts as a high-Q parallel resonant circuit. A simple cav-

ity resonator can be formed with a short piece of waveguide one-half wavelength long as illustrated in Fig. 10-10. A small probe at the center injects microwave energy. The short piece of waveguide is closed at each end with a metallic short. The result is that when microwave energy is injected into the cavity, the signal will bounce off the shorted ends of the waveguide and reflect back toward the probe. Because the probe is located a quarter wavelength from each shorted end, the reflected signal will reinforce the signal at the probe. The result is that the signal will bounce back and forth off the shorted ends. If the signal is removed, the wave will continue to bounce back and forth for a continuous length of time until losses cause it to die out.

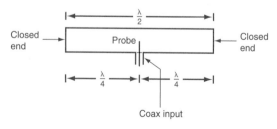

Fig. 10-10 Side view of a half-wavelength waveguide used as a cavity resonator.

This effect is pronounced at a frequency where the length of the waveguide is exactly one-half wavelength. The cavity is said to resonate and acts like a parallel resonant circuit. A brief burst of energy applied to the probe will cause the cavity to oscillate until losses cause it to die out. Cavities such as this have an extremely high Q, as high as 30,000. For this reason, they are commonly used to create resonant circuits and filters at microwave frequencies.

A cavity can also be formed by using a short section of circular waveguide as shown in Fig. 10-11. The diameter should be one-half wavelength at the operating frequency. Other cavity shapes are also possible. Further, a coupling loop rather than a probe is more commonly used in cavity resonators. In some cavity resonators two probes or two coupling loops are used, one for the input and one for the output. In this way, the cavity can serve as a filter.

Cavities can be constructed in a variety of ways. Although short segments of waveguide can be used, typically cavities are specially designed components. They are often hollowed out sections in a block of metal. By machining the cavity, very precise dimensions can be obtained for a specific frequency. The internal walls of the cavity are often plated with silver or some other low-loss material to ensure minimum losses and a very high Q.

Some cavities are also tunable. One wall of the cavity is made movable as shown in Fig. 10-12(a). An adjustment screw moves the end wall in and out to adjust the resonant frequency. The smaller the cavity, the higher the operating frequency. Cavities can also be tuned with adjustable plugs in the side of the cavity as Fig. 10-12(b) illustrates. As the plug is screwed in, more of it protrudes into the cavity and the operating frequency goes up.

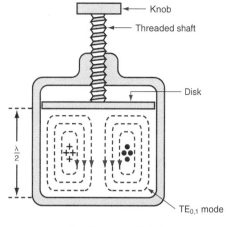

(a) Cylindrical cavity with adjustable disk

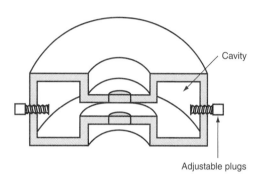

(b) Cavity with adjustable plugs

Fig. 10-12 Adjustable or tunable cavities.

■ **TEST**

Answer the following questions.

11. What is the length of $\lambda/2$ of microstrip line at 5.8 GHz? The dielectric constant of the PCB is 2.2. If the line is shorted, what type of resonant circuit is formed?

12. The main disadvantage of using coax for microwave signals is its high _____.

13. Coax is not used beyond frequencies of about _____ GHz.

14. A hollow metal pipe used to carry microwaves is called a(n) _____.

15. Two types of transmission line made of PCB material used to produce tuned circuits and filters are _____.

16. The two ways to couple or extract energy from a waveguide are by using a _____ or a _____.

17. A waveguide acts as a(n) _____ filter.

18. A rectangular waveguide has a width of 1.2 in. and a height of 0.7 in. The waveguide will pass all signals above _____ GHz.

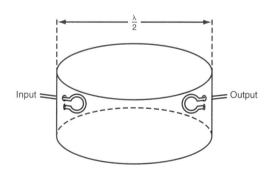

Fig. 10-11 Circular resonant cavity with input-output loops.

19. A waveguide has a cutoff frequency of 5 GHz. (a) True or false. It will pass a signal of 8 GHz. (b) True or false. It will pass a signal of 3 GHz.

20. The magnetic and electric fields in a waveguide are designated by the letters _____ and _____, respectively.

21. The basic operating mode of most waveguides is designated _____.

22. If the H field in the waveguide is perpendicular to the direction of signal travel, the mode is said to be _____.

23. A one-half wavelength section of waveguide shorted at both ends creates a(n) _____.

24. A cavity resonator acts as a(n) _____ circuit.

25. Resonant cavities are used as _____.

26. Mechanically varying the cavity's dimensions allows its _____ to be adjusted.

Point-contact diode

10-3 Microwave Semiconductor Devices

Most conventional diodes and transistors do not operate at microwave frequencies. The lengths of the diode and transistor leads present a significant amount of inductance at microwave frequencies and, therefore, introduce losses. The internal capacitances of diodes and transistors are also very high and will actually prevent operation at microwave frequencies. In addition, the transit time of the carriers in a diode or transistor is sufficiently high to prevent normal operation. Transit time refers to the amount of time it takes for an electron or hole in a diode or transistor to pass from cathode to anode, emitter to base, emitter to collector, or source to drain.

These problems have been overcome by designing special semiconductor devices for use at microwave frequencies. In addition to new semiconductor materials other than silicon, the geometry is different. In this section we will review briefly the most common semiconductor devices used in microwave equipment, including diodes, transistors, and integrated circuits.

Microwave Diodes

A wide variety of semiconductor diodes have been developed for microwave use. Those diodes used for signal detection and mixing are the most common. Two types are widely used: the point-contact diode and Schottky barrier (or hot-carrier) diode.

The typical semiconductor diode is a junction formed of P- and N-type semiconductor materials. Because of the relatively large surface area of the junction, these diodes exhibit a high capacitance that prevents normal operation at microwave frequencies. For this reason, standard PN junction diodes are not used in the microwave region. Instead, special microwave diodes are used.

Point Contact Diode Perhaps the oldest microwave semiconductor device is the *point-contact diode*. This is nothing more than a piece of semiconductor material and a fine wire which makes contact with the semiconductor material. Because the wire makes contact with the semiconductor over a very small surface area, the capacitance is extremely low. Current flows easily from the fine wire (the cathode) to the semiconductor material (the anode). However, current will not flow easily in the opposite direction.

Most early point-contact diodes used germanium as the semiconductor material, but today point-contact diodes are made of P-type silicon with a fine tungsten wire as the cathode. See Fig. 10-13. The forward threshold voltage is extremely low.

Point-contact diodes are ideal for low-signal applications and are widely used in microwave mixers and detectors. They are extremely delicate and cannot withstand high power. Also, they

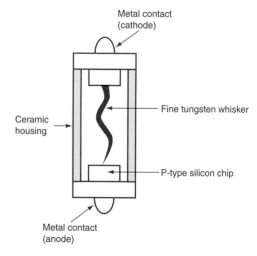

Fig. 10-13 A point-contact diode.

are easily damaged and, therefore, must be used in such a way as to avoid shock and vibration.

Hot Carrier Diodes   Perhaps the most widely used microwave diode is the *Schottky* or *hot-carrier diode*. You will see the terms Schottky and hot carrier both used, but they are the same thing. However, the term *Schottky diode* is the more widely used. A Schottky diode has a metal semiconductor junction.

Most Schottky diodes are made with N-type silicon on which has been deposited a thin metal layer. The semiconductor forms the cathode, and the metal forms the anode. Typical anode materials include nickel chromium and aluminum, although other metals such as gold are also used.

The Schottky diode, like the point-contact diode, is extremely small and, therefore, has a tiny junction capacitance. It also has a low bias threshold voltage. It too is ideal for mixing, signal detection, and other low-level signal operations. They are widely used in balanced modulators and mixers. Because of their very high frequency response, Schottky diodes are also used as fast switches at microwave frequencies.

The most important use of microwave diodes is as *mixers*. Some microwave diodes are installed as part of a waveguide or cavity resonator tuned to the incoming signal frequency. An LO signal is injected with a probe or loop. The diode mixes the two signals and produces the sum and difference output frequencies. The difference frequency is usually selected with another cavity resonator, stripline, or microstrip tuned circuit, or with a low frequency *LC* tuned circuit. Another common microwave mixer is the doubly balanced type shown in Fig. 7-16. It uses hot carrier diodes in a bridge configuration. The mixer is

sometimes the input circuit in a microwave receiver since it is desirable to convert the microwave signal to a lower frequency level as early as possible so that amplification and demodulation can take place with simpler, more conventional electronic circuits. Some microwave receivers also use an RF amplifier stage before the mixer to amplify very low level signals. But although the RF stage adds necessary gain, it also adds noise, which can degrade small signals.

Varactor Diodes   Another widely used microwave semiconductor device is the *varactor diode*. As you saw earlier, a varactor diode is basically a VVC. When a reverse bias is applied to the diode, it acts like a capacitor. Its capacitance depends upon the value of the reverse bias. Varactor diodes are VVCs made with gallium arsenide and optimized for use at microwave frequencies. Their main application is as frequency multipliers.

Figure 10-14 shows a *varactor frequency multiplier circuit*. When an input signal is applied across the diode, it alternately conducts and cuts off. The result is a nonlinear or distorted output containing many harmonics. By using a tuned circuit in the output, the desired harmonic is selected and others are rejected. Since the lower harmonics produce the greatest amount of energy, varactor multipliers are usually used only for doubling and tripling operations. In Fig. 10-14, the input tuned circuit L_1-C_2 resonates at the input frequency f_i, and the output tuned circuit L_2-C_3 resonates at two or three times the input frequency as desired. In practice, the tuned circuits are not actually made up of individual inductors or capacitors. Instead, they are microstrip, stripline, or cavity resonators. Both C_1 and C_4 are used for impedance matching.

Schottky diode
Hot-carrier diode

Varactor diode

Varactor frequency multiplier circuit

Mixers

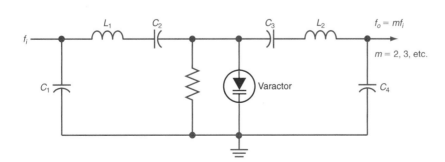

Fig. 10-14   A varactor frequency multiplier.

A varactor frequency multiplier does not have gain; in fact, it produces a signal power loss. But it is a relatively efficient circuit, and the output can be as high as 80 percent of the input. Typical efficiencies are in the 50 to 80 percent range. Note also in Fig. 10-14 that no external source of dc power is required for this circuit. Only the RF input power is required for proper operation. Outputs up to 20 W are obtainable with special high-power varactors.

Varactors are used in those applications where it is difficult to generate microwave signals. Usually it is a lot easier to generate a VHF or UHF signal and then use a series of frequency multipliers to put it into the desired microwave region. Varactor diodes are available for producing relatively high power outputs at frequencies up to 100 GHz.

Step-recovery diode

Step Recovery Diodes The *step-recovery diode* or *snap-off varactor* is a PN junction diode made with gallium arsenide or silicon. When it is forward-biased, it conducts the way any diode does, but a charge is stored in the depletion layer. When reverse bias is applied, the charge keeps the diode on momentarily. Then suddenly the diode turns off abruptly. This snap-off produces an extremely high intensity reverse-current pulse with a duration of about 100 picoseconds (ps), where 1 ps = 10^{-12} s. It is extremely rich in harmonics. Even the higher harmonics are of relatively high amplitude.

IMPATT diodes
TRAPATT diodes

Step-recovery diodes are used in the circuit of Fig. 10-14 to produce multipliers with factors up to 5 and 10. Power ratings of 50 W can be obtained. Operating frequencies up to 10 GHz are possible with an efficiency of 80 percent or more.

Tunnel diode

Gunn Diode A very popular microwave diode is the *Gunn diode.* It is also called a transferred-electron device (TED). The Gunn diode is not a diode in the usual sense as it does not have a junction. Instead, it is a thin piece of N-type gallium arsenide (GaAs) or indium phosphide (InP) semiconductor that forms a special resistor when voltage is applied to it. This device exhibits a negative resistance characteristic. That is, over some voltage range, an increase in voltage results in a decrease in current and vice versa, just the opposite of what Ohm's law predicts for a linear resistance. When it is so biased, the time it takes for electrons to flow

Gunn diode

Microwave transistors

across the material is such that the current will be 180° out of phase with the applied voltage. If the Gunn diode is so biased and connected to a cavity that is resonant near the frequency determined by the electron transit time, the resulting combination will oscillate. The Gunn diode, therefore, is used primarily as a microwave oscillator.

Gunn diodes are available to oscillate at frequencies up to about 50 GHz. In the lower microwave range, power outputs up to several watts are possible. It is the thickness of the semiconductor that determines the frequency of oscillation. However, if the cavity is made variable, the Gunn oscillator frequency can be adjusted over a narrow range.

Gunn diodes are widely used as LOs in receivers and as primary frequency sources in transmitters. Police radar guns use a Gunn diode oscillator.

IMPATT and TRAPATT Diodes

Two other microwave diodes are also widely used as oscillators. These are the *IMPATT* and *TRAPATT diodes.* Both are PN junction diodes made of silicon, GaAs, or InP. They are designed to operate with a high reverse bias that causes them to avalanche or break down. A high current flows. Over a narrow range, a negative resistance characteristic is produced. This will produce oscillation when the diode is mounted in a cavity. The IMPATT diodes are available with power ratings up to about 25 W in the low gigahertz range.

Tunnel Diode The *tunnel diode* is another negative resistance diode. It produces a narrow range of negative resistance when forward-biased. Tunnel diodes are used to produce low-power microwave oscillators.

Microwave Transistors

Over the years, transistor manufacturers have learned to make transistors that work at microwave frequencies. Both bipolar transistors and FETs are available for frequencies up to 40 GHz. There are two basic types of *microwave transistors:* small signal and RF power.

Small Signal These transistors, both bipolar and FET, are designed for low-level amplification.

Most are used as class A linear RF amplifiers in receivers or in low-power-level buffers in transmitters. In receivers, the key specifications are the gain-bandwidth product and the noise figure. The gain-bandwidth product (f_T) of a bipolar transistor is the upper cutoff frequency for unity gain. At higher gains, the cutoff frequency is proportionally less. For example, with $f_T = 5$ GHz, a transistor with a gain of 5 would have an upper cutoff frequency of 1 GHz. Transistors with f_T up to 40 GHz are available.

Noise figure is stated in dB. For bipolar transistors, a noise figure in the 2–3-dB range is typical.

FETs are even better at microwave frequencies. One of the best for small signal amplification is the GaAs MESFET described in Chap. 7 (Fig. 7-13). This is available for frequencies up to 40 GHz with a noise figure less than 2 dB. Several newer types of bipolar transistors have also been developed for microwave amplification. These include heterojunction bipolar transistors (HBTs), pseudomorphic high electron mobility transistors (PHEMTs), and bipolars made with a silicon-germanium (SiGe) compound. These are usually incorporated into microwave integrated circuits.

Power Microwave power transistors are also available. Bipolar power transistors can operate at class A power levels up to 150 W at frequencies as high as 2 GHz. MOSFET power transistors can operate at Class A or AB power levels up to 15 W at frequencies up to 3 GHz. If Class C operation is used, power levels to 300 W can be achieved. For power levels beyond about 200 W, special microwave tubes must be used.

Microwave transistors are usually soldered directly to a printed-circuit board where they are connected to microstrip or strip-line circuits used for frequency selectivity or impedance matching.

Microwave Integrated Circuits

Monolithic microwave integrated circuits (MMICs) are available for small signal amplification and frequency conversion.

Typical MMICs include linear amplifiers and mixers. Typical MMIC amplifiers can work up to about 20 GHz with gains of up to 10 dB. The mixers use hot-carrier diodes in a doubly balanced circuit like that described in Chap. 7. They can also be used as balanced modulators or phase modulators and detectors. They are available to work up to frequencies of 8–10 GHz.

■ TEST

Answer the following questions.

27. List three reasons why conventional diodes and transistors do not work in the microwave region.
28. A point-contact diode has a(n) _____ anode and a(n) _____ cathode.
29. A microwave diode with an N-type silicon cathode and a metal anode forming a junction is called a(n) _____ diode.
30. The most common application of microwave signal diodes is in _____ circuits.
31. A diode that acts like a VVC is a(n) _____ diode.
32. Two types of diodes widely used as frequency multipliers are the _____ and _____ diodes.
33. A diode with no junction that is widely used with a cavity resonator to form an oscillator is the _____ diode.
34. The _____, _____, and _____ diodes are also used as oscillators.
35. Microwave diodes that cause a current decrease for a voltage increase have what is known as a(n) _____ characteristic.
36. An IMPATT diode operates with _____ (forward, reverse) bias.
37. A tunnel diode operates with _____ (forward, reverse) bias.
38. The key specifications of a bipolar microwave transistor are _____ and _____.
39. One of the best small-signal microwave transistors is the _____.
40. The upper power level of a FET microwave power transistor for linear amplification is about _____ W.
41. The two commonly available MMICs are _____ and _____.

MMICs

10-4 Microwave Tubes

Before transistors were invented, all electronic circuits were implemented with vacuum tubes. Like transistors, tubes are devices used for controlling a large signal with a smaller signal

Klystron

to produce amplification, oscillation, switching, and other operations. Today, vacuum tubes are no longer used in most circuits. Yet they have survived in special applications. The cathode-ray tube (CRT) used in TV sets, oscilloscopes, and other display devices is a special form of vacuum tube that has no semiconductor equivalent. Vacuum tubes are also still found in microwave equipment. This is particularly true in microwave transmitters used for producing high output power. Although both bipolar and field-effect transistors are available to produce power in the microwave region up to approximately 50 W, many applications require more power. In satellite earth stations, TV stations, and in

some military equipment such as radar, very high output powers are needed. Special microwave tubes developed during World War II are still widely used for microwave power amplification. These tubes are the klystron, the magnetron, and the traveling-wave tube. In this section we take a brief look at each.

Klystron

A *klystron* is a microwave vacuum tube using cavity resonators to produce velocity modulation of the electron beam and to produce amplification. Figure 10-15 shows the concept of a two-cavity klystron. The vacuum tube contains a cathode that is heated by a filament. At a very high temperature, the cathode emits electrons. These negative electrons are attracted by a positive plate or collector. Thus, current flow is established between the cathode and the collector inside the evacuated tube.

The electrons emitted by the cathode are focused into a very narrow stream by electrostatic and electromagnetic focusing techniques. In

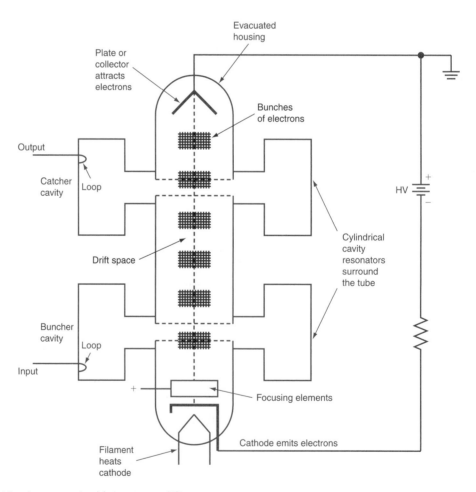

Fig. 10-15 A two-cavity klystron amplifier.

electrostatic focusing, special elements which are called focusing plates and which have high applied voltages force the electrons into a narrow beam. Electromagnetic focusing makes use of coils around the tube through which current is passed to produce a magnetic field. This magnetic field helps focus the electrons into a narrow beam.

The sharply focused beam of electrons is then forced to pass through the centers of two cavity resonators. The cavity is open in the center area. The microwave signal to be amplified is applied to the lower cavity through a coupling loop. This sets up electric and magnetic fields in the cavity which influence the electron beam. Specifically, they cause the electrons to speed up and slow down as they pass through the cavity. On one half cycle, the electrons are speeded up, and on the next half cycle of the input, they are slowed down. The effect of this is to create bunches of electrons that are one-half wavelength apart in the drift space between the cavities. This speeding up and slowing down of the electron beam is known as *velocity modulation*. Since the input cavity produces bunches of electrons, it is commonly referred to as the *buncher cavity*.

The bunched electrons move on through the tube since they are attracted by the positive collector. They then pass through the center of another cavity known as the catcher cavity. These bunched electrons move toward the collector in clouds of alternately dense and sparse areas. We can say that the electron beam is *density-modulated*.

As the bunches of electrons pass through the catcher cavity, the cavity is excited into oscillation at the resonant frequency. Therefore, the dc energy in the electron beam is converted into RF energy at the cavity frequency. Thus, amplification takes place. The output is extracted from the catcher cavity with a loop.

Klystrons are also constructed with additional cavities between the buncher and catcher cavities. These intermediate cavities produce further bunching which causes increased amplification of the signal. If the buncher cavities are tuned off the center frequency from the input and output cavities, they have the effect of broadening the bandwidth of the tube. The frequency of operation of a klystron is set by the sizes of the input and output cavities. Since cavities typically have high Q's, their bandwidth is limited. By lowering the Q's of the cavities and by introducing intermediate cavities, wider bandwidth operation can be achieved.

Klystrons are available in a wide range of sizes. Small hand-held units produce only milliwatts of power amplification, while larger, human-size klystrons produce many thousands of watts of power. Klystrons are used at frequencies as low as UHF and as high as 100 GHz in the microwave region.

A special variation of the basic klystron tube known as a *reflex klystron* is commonly used as a microwave oscillator. A reflex klystron uses only a single cavity as shown in Fig. 10-16. The cathode emits electrons which are accelerated forward by an accelerating grid with a positive voltage on it and focused into a narrow beam as previously described. The electrons pass through the cavity and undergo velocity modulation, which produces electron bunching. The electron bunches move toward the repeller plate, which is made negative with respect to the cathode. This causes the electrons to be repelled rather than attracted as they are by a positive collector in a standard klystron. The result is that the bunched electrons are forced back through the cavity. The distance between the repeller and the cavity is chosen such that the repelled electron bunches will reach the cavity at the proper time to be in synchronization. Because of this, they deliver energy to the cavity. The result is oscillation at the cavity frequency. The cavity itself is made positive so that the electrons are ultimately attracted by the cavity and cause direct current flow in the external circuit. A coupling loop in the cavity removes the RF energy.

Reflex klystrons are small, low-power devices. They typically generate output power from 100 mW to several watts of power. They are commonly used for microwave signal generators and for LOs in microwave receivers. Reflex klystrons can also be easily frequency-modulated by simply adding the ac modulating

Reflex klystron

Velocity modulation

Buncher cavity

Density modulation

About ⟺ Electronics

One university research group in Singapore is studying the effects of leaky microwave cables in subway tunnels. To find out more, check out a summary on the World Wide Web: ⟨www.ee.nus.edu.sg⟩.

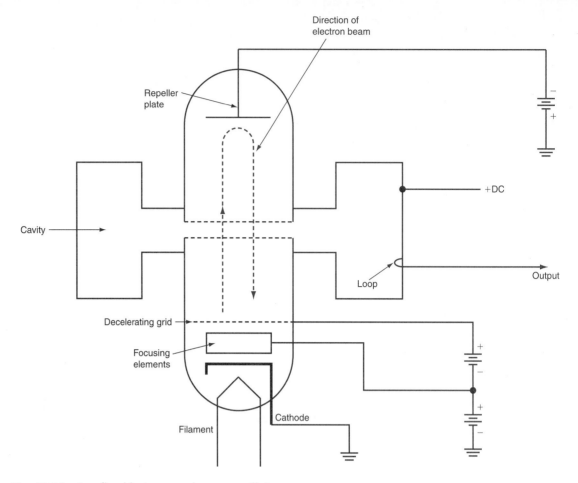

Fig. 10-16 A reflex klystron used as an oscillator.

signal in series with the repeller supply voltage. This causes the repelled electron bunches to reach the cavity at different times, thereby producing a variable output frequency (FM).

Klystrons are gradually being replaced in most microwave equipment. Gunn diodes are replacing the smaller reflex klystrons in signal-generating applications because they are smaller and lower in cost and do not require high dc supply voltages. The larger multicavity klystrons are being replaced by traveling-wave tubes in higher-power applications.

Magnetron

Magnetron

Another widely used microwave tube is the magnetron. The *magnetron* is the combination of a simple diode vacuum tube with built-in cavity resonators and an extremely powerful permanent magnet. The typical tube assembly is shown in Fig. 10-17. It consists of a circular anode into which has been machined an even number of resonant cavities. The diameter of

each cavity is equal to one-half wavelength of the desired operating frequency. The anode is usually made of copper and is connected to a high-voltage positive dc source.

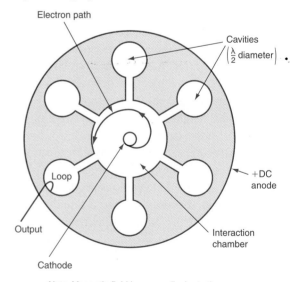

Note: Magnetic field is perpendicular to the page.

Fig. 10-17 A magnetron tube used as an oscillator.

In the center of the anode, called the interaction chamber, is a circular cathode that emits electrons when heated. As in a normal diode vacuum tube, the electrons would flow directly from the cathode straight to the anode, causing a high current to flow. However, the direction of the electrons is modified by surrounding the tube with a strong magnetic field. This is usually supplied by a C-shaped permanent magnet. In Fig. 10-17, the lines of force could be coming out of the page or going into the page depending upon the construction of the tube.

When an electron moves, it has a magnetic field associated with it. The magnetic fields of the electrons will, therefore, interact with the strong field supplied by the magnet. The result is that the path for electron flow from the cathode will not be directly to the anode. Instead, the path will be curved. By properly adjusting the anode voltage and the strength of the magnetic field, the electrons can be made to bend such that they rarely reach the anode and cause current flow. The path becomes circular loops as illustrated in Fig. 10-17. At some point, the electrons will eventually reach the anode and cause current flow. By adjusting the dc anode voltage and the strength of the magnetic field, the electron path is made circular. In making their circular passes in the interaction chamber, the electrons excite the resonant cavities into oscillation. The magnetron, therefore, is an oscillator, not an amplifier. A takeoff loop in one cavity provides the output.

Magnetrons are capable of developing extremely high levels of microwave power. Thousands and even millions of watts of power can be produced by a magnetron. When operated in a pulsed mode, magnetrons can generate several megawatts of power in the microwave region. Pulsed magnetrons are commonly used in radar systems. Continuous-wave magnetrons are also used and can generate hundreds and even thousands of watts of power. A typical application for a CW magnetron is for heating purposes in microwave ovens.

Traveling-Wave Tubes

One of the most versatile microwave RF power amplifiers is the *traveling-wave tube (TWT)*. The main virtue of the TWT is its extremely wide bandwidth of operation. In addition to its ability to generate hundreds and even thousands of watts of microwave power, it is not resonant at a single frequency.

Figure 10-18 shows the basic structure of a TWT. It consists of a cathode and filament heater plus an anode that is biased positively to accelerate the electron beam forward and to focus it into a narrow beam. The electrons are attracted by a positive plate called the collector to which is applied a very high dc voltage. The length of the tube can be anywhere from approximately 1 ft to several feet. In any case, the length of the tube is usually many wavelengths at the operating frequency. Surrounding the tube are either permanent magnets or electromagnets that keep the electrons tightly focused into a narrow beam.

Traveling-wave tube (TWT)

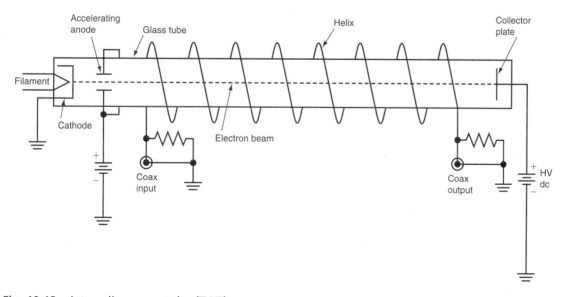

Fig. 10-18 A traveling-wave tube (TWT).

JOB TIP

If you have good people skills, pay attention to detail, and enjoy solving technical problems, you may want to look into the career of cable TV technician. Whether you work on cable transmission lines or you monitor transmission signal quality from cable headquarters, this career offers personal responsibility and a team atmosphere.

The unique feature of the TWT is a helix or coil that surrounds the length of the tube. The electron beam passes through the center or axis of the helix. The microwave signal to be amplified is applied to the end of the helix near the cathode, and the output is taken from the end of the helix near the collector. The purpose of the helix is to provide a path for the RF signal that will slow down its propagation. The propagation of the RF signal along the helix is made approximately equal to the velocity of the electron beam from cathode to collector. The structure of the helix is such that the wave traveling along it is slightly slower than that of the electron beam.

The passage of the microwave signal down the helix produces electric and magnetic fields that will interact with the electron beam. The effect on the electron beam is similar to that in a klystron. The electromagnetic field produced by the helix causes the electrons to be speeded up and slowed down. This produces velocity modulation of the beam which produces density modulation. Density modulation, of course, causes bunches of electrons to group together one wavelength apart. These bunches of electrons travel down the length of the tube toward the collector. Since the density-modulated electron beam is essentially in step with the electromagnetic wave traveling down the helix, the electron bunches induce voltages into the helix which reinforce the voltage already present there. The result is that the strength of the electromagnetic field on the helix increases as the wave travels down the tube toward the collector. At the end of the helix, the signal is considerably amplified. Coaxial cable or waveguide structures are used to extract energy from the helix.

As indicated earlier, the primary benefit of the TWT is its extremely wide bandwidth. Tubes can be made to amplify signals from UHF to hundreds of gigahertz. Most TWTs have a frequency range of approximately 2:1 in the desired segment of the microwave region to be amplified. The TWTs can be used in both continuous and pulsed modes of operation with power levels up to several thousand watts. One of the most common applications of TWTs is as power amplifiers in satellite transponders.

■ **TEST**

Determine whether each statement is true or false.

42. Klystrons can be used as amplifiers or oscillators.
43. A magnetron operates as an amplifier.

Supply the missing words in each statement.

44. A TWT is an _____ (amplifier, oscillator).
45. The operating frequency of klystrons and magnetrons is determined by _____.
46. The cavities in a klystron produce _____ modulation of the electron beam.
47. In a klystron amplifier, the input is applied to the _____ cavity and the output taken from the _____ cavity.
48. A one-cavity klystron that oscillates is known as a(n) _____ klystron.
49. Low-power klystrons are being replaced by _____, and high-power klystrons are being replaced by _____.
50. A(n) _____ causes the electrons in a magnetron to travel in circular paths.
51. Two major applications of magnetrons are in _____.
52. The TWT is a microwave _____.
53. Density modulation of the electron beam in a TWT is produced by a(n) _____.
54. The major benefit of a TWT is its wide _____.

10-5 Microwave Antennas

All the antennas we discussed in Chap. 9 can also be used at microwave frequencies. However, these antennas will be extremely small. A half-wave dipole at 5 GHz is slightly less than 1 in. long. A quarter-wave vertical at this frequency would be slightly less than 1/2 in. long. Naturally, these antennas will radiate microwave signals but inefficiently. Because of the line-of-sight transmission of microwave signals, highly directive antennas are preferred because they do not waste the radiated energy and provide an increase in gain which helps offset the path loss at these frequencies. For these important reasons, special high-gain,

highly directive antennas are normally used in microwave applications.

Horn Antenna

The most widely used microwave antenna is the horn. A *horn antenna* is nothing more than a flared waveguide. The horn exhibits gain and directivity. However, its performance is significantly improved when it is used in combination with a parabolic reflector. This is the familiar "dish" normally associated with microwave and satellite transmissions. In this section we introduce you to the horn antenna and parabolic reflectors. The popular helical antenna is also discussed.

Waveguides are the most predominant type of transmission line used with microwave signals. Below approximately 6 GHz, special coax can be used effectively if the distances are kept short. In most microwave systems, waveguides are preferred because of their low loss. Microwave antennas, therefore, must be some extension of or compatible with a waveguide.

Actually, a waveguide will act as an inefficient radiator if it is simply left open at the end as shown in Fig. 10-19. The problem with using a waveguide as a radiator is that it makes a poor impedance match with free space. This represents a mismatch that causes standing waves and reflected power. The result is tremendous power loss of the radiated signal.

It has been discovered that this mismatch can be overcome by simply flaring the end of the waveguide. Flaring of the waveguide ends creates a horn antenna as shown in Fig. 10-20. The longer and more gradual the flare, the better the impedance match and the lower the loss. Horn antennas have excellent gain and directivity. The longer the horn, the greater its gain and directivity.

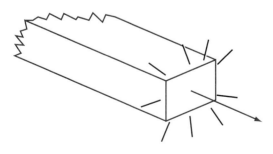

Fig. 10-19 A waveguide will act as an inefficient radiator.

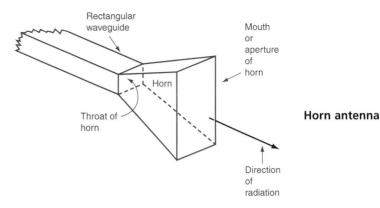

Fig. 10-20 Basic horn antenna.

Different kinds of horn antennas can be created by flaring the ends of the waveguide in different ways. For example, flaring the waveguide in only one dimension creates a *sectoral horn* as shown in Fig. 10-21(*a*) and (*b*). Two sides of the horn remain parallel with the sides of the waveguide, while the other dimension is flared. **Sectoral horn**

Flaring both dimensions of the horn produces a *pyramidal horn* as shown in Fig. 10-21(*c*). If a circular waveguide is used, the flare produces a *conical horn* as shown in Fig. 10-21(*d*). **Pyramidal horn** **Conical horn**

The gain and directivity of a horn are direct functions of its various dimensions. The important dimensions are illustrated in Fig. 10-22. They include horn length, aperture area, and flare angle.

The length of a typical horn is usually 2 to 15 wavelengths at the operating frequency. Assume an operating frequency of 10 GHz. The length of one wavelength at this frequency is

$$\lambda = \frac{300}{f}$$

$$= \frac{300}{10,000}$$

$$= 0.03 \text{ m}$$

The frequency f is in megahertz. A length of 0.03 m is the same as 3 cm. One inch equals 2.54 cm. Therefore, a wavelength at 10 GHz is

$$\lambda = \frac{3}{2.54}$$

$$= 1.18 \text{ in.}$$

or just a shade short of $1\frac{1}{4}$ in. This means that the length of a typical horn at this frequency could be anywhere from about $2\frac{1}{2}$ to 18 in.

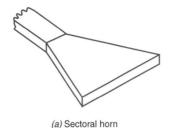

(a) Sectoral horn

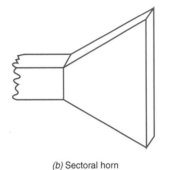

(b) Sectoral horn

Beam width

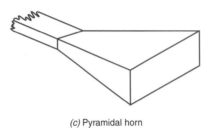

(c) Pyramidal horn

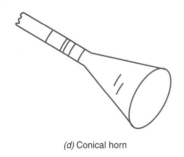

(d) Conical horn

Fig. 10-21 Types of horn antennas.

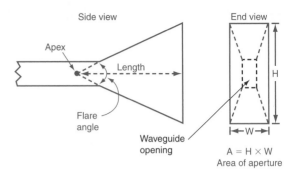

Fig. 10-22 Dimensions of a horn.

size of aperture area, decreasing the length increases the flare angle. Each of these dimensions is adjusted to achieve the desired design objective.

Recall that we indicated that the directivity of an antenna is measured in terms of its beam width. The *beam width* is the angle formed by extending lines from the center of the antenna response curve to the 3-dB-down points as shown in Fig. 10-23. In this example, the beam width is approximately 30°. Horn antennas typically have a beam angle somewhere in the 10° to 60° range.

Figure 10-24 shows another way of plotting the beam angle of an antenna. Instead of using polar coordinates, standard *x* and *y* coordinates are used. The horizontal, or *x* axis, represents the beam width in degrees, and the vertical, or *y* axis, represents the gain in decibels.

Remember that the signal radiated from an antenna is three-dimensional in nature. The directivity patterns indicate the horizontal radiation pattern of the antenna. The antenna also has a vertical radiation pattern. A typical plot

The longer horns are, of course, more difficult to mount and work with but otherwise provide higher gain and better directivity.

The aperture area is the area of the rectangle formed by the opening of the horn and is simply the product of the height and width of the horn as shown in Fig. 10-22. The greater this area, the higher the gain and directivity.

The flare angle also affects gain and directivity. Typical flare angles vary from about 20° to 60°. You can see how all these dimensions are interrelated. Increasing the flare angle, of course, increases the aperture area. For a given

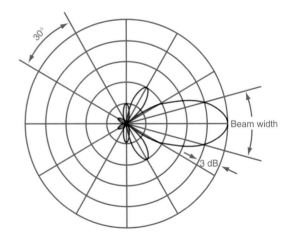

Fig. 10-23 Directivity of an antenna as measured by beam width.

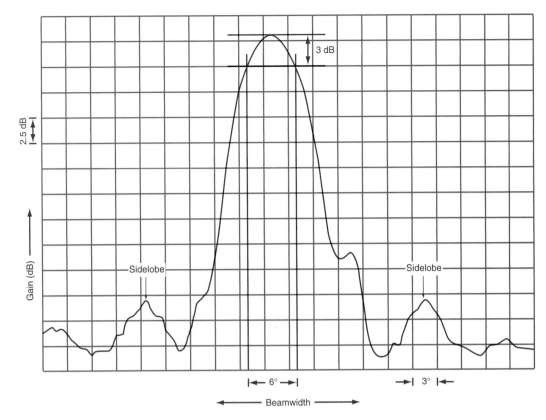

Fig. 10-24 Plot of beam width versus gain of a horn antenna.

of the vertical radiation pattern of a horn antenna is illustrated in Fig. 10-25. Usually the vertical beam width is approximately the same angle as the horizontal beam width on a horn.

The horizontal beam width of a pyramidal horn may be computed with the simple expression

$$B = \frac{80}{w/\lambda}$$

where B = beam width, degrees
w = width of horn, meters
λ = wavelength of operating frequency meters

Assume that the operating frequency is 10 GHz = 10,000 MHz. This gives us a wavelength as computed earlier of 0.03 m. Assume that the dimensions of the pyramidal horn are 10 cm high and 12 cm wide. The beam width then is

$$B = \frac{80}{0.12/0.03}$$

$$= \frac{80}{4}$$

$$= 20°$$

The gain of a pyramidal horn can also be computed from its dimensions. The expression below gives the approximate *power gain* of the antenna.

Power gain

$$G = \frac{4\pi KA}{\lambda^2}$$

where K = constant derived from how uniformly the phase and amplitude of the electromagnetic fields are distributed across the aperture (typical values are in the 0.5 to 0.6 range; we will assume 0.5 here)
A = aperture of horn, m²
λ = wavelength, meters

Using the previously described horn with a height dimension of 10 cm and a width of 12 cm, the aperture area is

$$A = \text{height} \times \text{width}$$

$$= 10(12)$$

$$= 120 \text{ cm}^2$$

About ⟨⟨⟨ Electronics

Most microwave amplifiers are designed to have input and output impedances of 50 Ω.

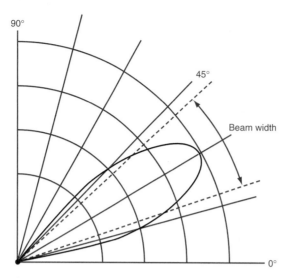

Fig. 10-25 Vertical directivity of an antenna.

Since there is 10,000 cm²/m², then 120 cm² is

$$\frac{120}{10,000} = 0.012 \text{ m}^2$$

Assuming an operating frequency of 10 GHz, one wavelength is equal to 0.03 m or 3 cm. The gain, therefore, is

$$G = \frac{4(3.14)(0.5)(0.012)}{(0.03)^2}$$

$$= \frac{0.07536}{0.0009}$$

$$= 83.7$$

To find the gain in decibels, use the standard power formula

$$\text{dB} = 10 \log G$$

where G is the power ratio or gain.

$$\text{dB} = 10 \log 83.7$$

$$= 10(1.923)$$

$$= 19.23$$

This is the power gain of the horn over a standard half-wave dipole or quarter-wave vertical.

An important aspect of a microwave antenna is its *bandwidth*. Most antennas, as you know, have a narrow bandwidth because they are resonant at only a single frequency. Their dimensions determine the frequency of operation. Bandwidth is an important consideration at microwave frequencies because the spectrum transmitted on the microwave carrier is usually very wide so that a considerable amount of information can be carried.

Parabolic reflector

Bandwidth

Luckily, horn antennas do have a relatively large bandwidth. Horns are essentially nonresonant or aperiodic, which means that they will operate over a wide frequency range. The bandwidth of a typical horn antenna is approximately 10 percent of the operating frequency. Using our 10-GHz example, the bandwidth of our horn would be approximately 1 GHz. This is an enormous bandwidth and certainly wide enough to accommodate almost any kind of complex modulating signal.

Parabolic Antenna

Horn antennas are used by themselves in many microwave applications, but many times, higher gain and directivity are desirable. This can easily be obtained by using a horn in conjunction with a *parabolic reflector*. A parabolic reflector is a large dish-shaped structure made of metal or screen mesh. The energy radiated by the horn is pointed at the reflector, which focuses the radiated energy into a narrow beam and reflects it toward its destination. Because of the unique parabolic shape, the electromagnetic waves are narrowed into an extremely small beam. Beam widths of only a few degrees are typical with parabolic reflectors. Of course, such narrow beam widths also represent extremely high gains.

A parabola, illustrated in Fig. 10-26, is a common geometric figure. A key dimension of the parabola is a line drawn from its center at point Z to a point on the axis labeled F, which is the focal point. The ends of the parabola could extend outwardly for an infinite distance, but usually they are limited. The limits are shown by the dashed vertical line with the end points labeled X and Y.

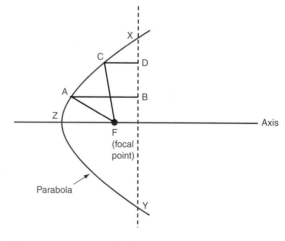

Fig. 10-26 A parabola.

The parabola has a unique characteristic. The distance between the focal point and the parabola and then to the vertical dashed line is a constant value. For example, in Fig. 10-26, the distance represented by the sum of the lines FA to AB and FC to CD are equal. This effect causes a parabolic-shaped surface to collimate electromagnetic waves into a narrow beam of energy. Placing an antenna at the focal point F will cause it to radiate waves from the parabola in parallel lines. If used as a receiver, the parabola will pick up the electromagnetic waves and reflect them to the antenna located at the focal point.

The key thing to remember about a parabola is that it is not a two-dimensional figure as shown in Fig. 10-26. If you rotate the parabola about its axis, a three-dimensional dish-shaped structure results.

Figure 10-27 shows how a parabolic reflector is used in conjunction with a horn antenna for both transmission and reception. The horn antenna is placed at the focal point. In transmitting, the horn radiates the signal toward the reflector which bounces the waves off and collimates them into a narrow parallel beam. When used for receiving, the reflector picks up the electromagnetic signal and bounces the waves toward the antenna at the focal point. The result is an extremely high-gain, narrow beam width antenna.

The gain of a parabolic antenna is directly proportional to the aperture of the parabola. The aperture, of course, is the area of the outer circle of the parabola. This area is

$$A = 3.14R^2$$

The gain of a parabolic antenna is given by the simple expression

$$G = 6\left(\frac{D}{\lambda}\right)^2$$

where G = gain expressed as a power ratio
D = diameter of dish, m
λ = wavelength, m

Most parabolic reflectors are designed so that the diameter is no less than 10 wavelengths at the lowest operating frequency.

Assume that we are using a 5-m-diameter dish at 10 GHz. The 10 GHz represent a wave-

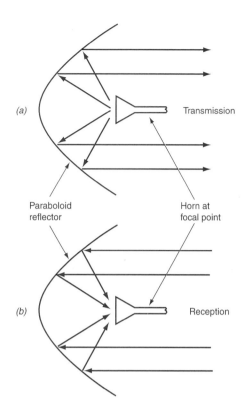

(a) Transmission

Paraboloid reflector

Horn at focal point

(b) Reception

Fig. 10-27 Sending and receiving with a parabolic reflector antenna.

length of 0.03 as computed earlier. The gain, therefore, is

$$G = 6\left(\frac{5}{0.03}\right)^2$$
$$= 6(166.67)^2$$
$$= 6(27778.9)$$
$$= 166{,}673$$

This is the *power gain,* which can be expressed in decibels.

Power gain

$$dB = 10 \log 166{,}673$$
$$= 10(5.22)$$
$$= 52.2$$

The beam width of a parabolic reflector can also be computed by knowing the parabolic reflector dimensions. The beam width is given by the expression

$$G = \frac{58}{D/\lambda}$$

Here the beam width is inversely proportional to the diameter.

Let's compute the beam width of our 5-m, 10-GHz antenna. Its beam width is

$$B = 58\left(\frac{0.03}{5}\right)$$

$$= \frac{1.74}{5}$$

$$= 0.348°$$

With a beam width of less than $1/2°$ the signal radiated from an antenna such as this is a pencil-thin beam. Naturally it must be pointed with great accuracy in order for the signal to be picked up. Usually both the transmitting and the receiving antennas are of a parabolic design and have extremely narrow beam widths. For that reason, they must be accurately pointed if contact is to be made. Naturally, the beam spreads out and grows in size with distance. Nevertheless, the directivity is good and the gain is high. The directivity helps to prevent interference from signals coming in at angles outside of the beam width.

Keep in mind that the parabolic dish is not the antenna, only a part of it. The antenna is the horn at the focal point. There are many physical arrangements used in positioning the horn. One of the most common is that shown in Fig. 10-28. The waveguide feeds through the center of the parabolic dish and is curved around so that the horn is positioned exactly at the focal point. Such a structure is inconvenient, but it is a widely used arrangement.

One of the more popular methods of feeding a parabolic antenna is to use a second reflector positioned at the focal point as shown in Fig. 10-29. Here the horn antenna is positioned at the center of the parabolic reflector. At the focal point is another small reflector with either a parabolic or a hyperbolic shape.

Cassegrain feed

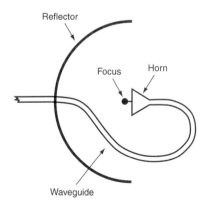

Fig. 10-28 Standard waveguide and horn feed.

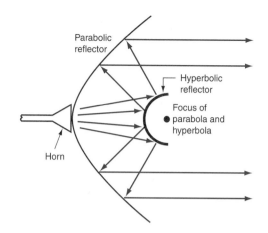

Fig. 10-29 Cassegrain feed.

The electromagnetic radiation from the horn strikes the small reflector, which then reflects the energy toward the large dish. The dish then radiates the signal in parallel beams. This arrangement is known as a *Cassegrain feed*.

The Cassegrain feed has several advantages over the more obvious feed arrangement shown earlier. The first is that the waveguide transmission line is shorter. In addition to being shorter, the radical bends in the waveguide are eliminated. Both add up to less signal attenuation. The noise figure is also improved somewhat. Most large earth station antennas use a feed arrangement of this type.

Many other feed arrangements have been developed for parabolic reflectors. In those antennas used for satellite TV reception, a waveguide is not used. Instead, a horn antenna mounted at the focal point is usually fed with large microwave coax. In other large antenna systems, various mechanical arrangements are used to permit the antenna to be rotated or its position otherwise physically changed. Many earth station antennas must be set up so that their azimuth and elevation can be changed to ensure proper orientation for the receiving antenna. This is particularly true of antennas used in satellite communications systems.

Some sophisticated systems use multiple horns on a single reflector. Such multiple feeds permit several signals on different frequencies to be either radiated or received with a single large reflecting structure.

Miscellaneous Microwave Antennas

Although the horn and the parabolic reflector are the most commonly used microwave antennas, two other antennas find wide use in the

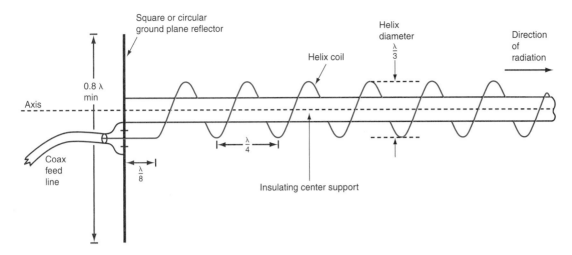

Fig. 10-30 The helical antenna.

microwave region. These are the helical antenna and the bicone antenna.

Helical Antenna   A *helical antenna,* as its name implies, is a wire helix as shown in Fig. 10-30. A center insulating support is used to hold a heavy wire or tubing formed into a circular coil or helix as shown. The diameter of the helix is typically one-third wavelength, and the spacing between turns is approximately one-quarter wavelength. Most helical antennas use from six to eight turns. A circular or square ground plane reflector is also used behind the helix. Figure 10-30 shows a coaxial feed line. This antenna is also widely used in the VHF and UHF range.

The gain of a helical antenna is typically in the 12- to 20-dB range. Beam widths vary from approximately 12° to 45°. Although the gain and beam width do not compare with horns and parabolic reflectors, the helical antenna is favored in many applications because of its simplicity and low cost.

A peculiar characteristic of a helical antenna is that the signal it radiates is *circularly polarized.* Most antennas transmit either a vertical or a horizontally polarized electromagnetic field. With the helix, the electromagnetic field is caused to rotate. This is known as circular polarization. Either right-hand (clockwise) or left-hand (counterclockwise) circular polarization may be produced depending upon the direction of winding of the helix. Because of the rotating nature of the magnetic field, a circularly polarized signal will be received by either a horizontally or vertically polarized receiving antenna, but with a 3-dB loss. A helical re-

ceiving antenna will also receive horizontally or vertically polarized signals, again with a 3-dB loss. Although this is true, a right-hand circularly polarized (RHCP) signal cannot be picked up by a left-hand circularly polarized (LHCP) antenna and vice versa. Therefore, when helical antennas are used at both transmitting and receiving ends of the communication link, both antennas must have the same polarization.

Bicone   Most microwave antennas are highly directional. But in some applications an omnidirectional antenna may be required. One of the most widely used omnidirectional microwave antennas is the *bicone,* shown in Fig. 10-31.

Helical antenna

Bicone antenna

Circular polarization

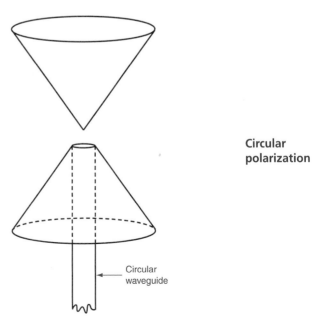

Fig. 10-31 The omnidirectional bicone antenna.

The signal is fed to the antenna through a circular waveguide which ends in a flaired cone. The upper cone acts as a reflector and causes the signal to be radiated equally in all directions with a very narrow vertical beam width.

Slot antennas

Slot Antennas A *slot antenna* is a radiator made by cutting a half-wavelength slot in a conducting sheet of metal (or into the side of a waveguide). It has the same characteristics as a standard dipole antenna, as long as the metal sheet is very large compared to 1λ at the operating frequency.

Several slots can be cut into the same waveguide to create a slot antenna array. Slot arrays, which are equivalent to driven arrays with many elements, have better gain and better directivity than single-slot antennas.

Slot antennas are widely used on high-speed aircraft. External antennas would be torn off at such high speeds or would slow the aircraft. The slot antenna can be integrated into the metallic skin of the aircraft. The slot itself is filled in with an insulating material to create a smooth skin surface.

Patch antennas

Dielectric (lens) antennas

Dielectric (Lens) Antennas As discussed previously, radio waves, like light waves, can be reflected, refracted, diffracted, and otherwise manipulated. This is especially true of microwaves, which are closer in frequency to light. Thus a microwave antenna can be created by constructing a device that serves as a lens for microwaves just as glass or plastic can serve as a lens for light waves. These *dielectric* or *lens antennas* use a special dielectric material to collimate or focus the microwaves from a source into a narrow beam.

An example of a lens antenna is one used in the millimeter-wave range. The microwave energy is coupled to a horn antenna through a waveguide. A dielectric lens is placed over the end of the horn, which focuses the waves into a narrower beam with greater gain and directivity. In technical terms, the lens takes the microwaves from a source with a spherical wavefront (e.g., a horn antenna) and concentrates them into a plane wavefront. A lens like that shown in Fig. 10-32(*a*) can be used. The shape of the lens ensures that all of the entering waves with a spherical wavefront are put into phase at the output to create the concentrated plane wavefront. However, the lens in Fig. 10-32(*a*) will work only when it is very thick at the center. This creates great signal loss, especially at the lower microwave frequencies. To get around this problem, a stepped or zoned lens, such as the one shown in Fig. 10-32(*b*), can be used. The spherical wavefront is still converted into a focused plan wavefront, but the thinner lens causes less attenuation.

Lens antennas are usually made of polystyrene or some other plastic, although other types of dielectric can be used. They are rarely used at the lower microwave frequencies. Their main use is in the millimeter range above 40 GHz.

Patch Antennas *Patch antennas* are made with microstrip on PCBs. The antenna is a circular or rectangular area of copper separated from the ground plane on the bottom of the board by the thickness of the PCB's insulating material (see Fig. 10-33). The width of the rectangular antenna is approximately one-half wavelength, and the diameter of the circular antenna is about 0.55 to 0.59 wavelength. In both cases the exact dimensions depend upon the dielectric constant and the thickness of the

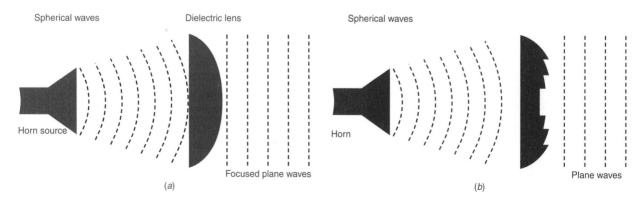

Fig. 10-32 Lens antenna operations. (*a*) Dielectric lens. (*b*) Zoned lens.

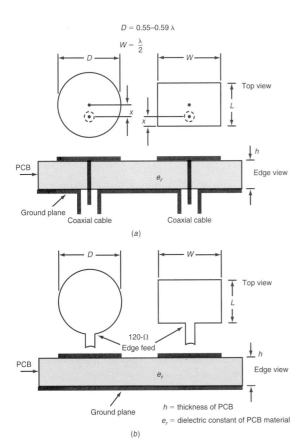

$D = 0.55–0.59\ \lambda$

$W = \dfrac{\lambda}{2}$

(a)

h = thickness of PCB
e_r = dielectric constant of PCB material

(b)

Fig. 10-33 Patch or microstrip antennas.
(a) Coaxial feed. (b) Edge feed.

PCB material. The most commonly used PC board material for patch antennas is a Teflon-fiberglass combination.

The feed method for patch antennas can be either coaxial or edge. With the coaxial method, the center conductor of a coaxial cable is attached somewhere between the center and the edge of the patch and the coaxial shield is attached to the ground plane [Fig. 10-33(a)]. If the antenna is fed at the edge, a length of microstrip is connected from the source to the edge, as shown in Fig. 10-33(b). The impedance of the edge feed is about 120 Ω at the edge. Correctly positioning the coaxial cable center on the patch [dimension x in Fig. 10-33(a)] allows extremely accurate impedance matching.

Patch antennas are small, inexpensive, and easy to construct. In many applications they can simply be integrated on the PCB with the transmitter or receiver. A disadvantage of patch antennas is their narrow bandwidth, which is usually no more than about 5 percent of the resonant frequency with circular patches and up to 10 percent with rectangular patches. The

bandwidth is directly related to the thickness of the PCB material, that is, the distance between the antenna and the ground plan [h in Fig. 10-33(b)]. The greater the thickness of the PCB dielectric, the greater the bandwidth.

The radiation pattern of a patch antenna is approximately circular in the direction opposite to that of the ground plane.

Phased Arrays A *phased array* is an antenna system made up of a large group of similar antennas on a common plane. Patch antennas on a common PCB can be used, or separate antennas like dipoles can be physically mounted together in a plane. The antennas are driven by transmission lines that incorporate impedance matching, power splitting, and phase-shift circuits. The basic purpose of an array is to improve gain and directivity. Arrays also offer better control of directivity, since individual antennas in an array can be turned off or on or driven through different phase shifters. The result is that the array can be "steered"—that is, its radiation pattern can be pointed over a wide range of different directions—without physically moving the antenna, as is necessary with Yagis or parabolic dish antennas.

There are two common arrangements for phased arrays. In one configuration, the multiple antennas are driven by a common transmitter or feed a common receiver. Another approach is to have a low-power transmitter amplifier or low-noise receiver amplifier associated with each dipole or patch in the array. In both cases, the switching and phase shifting are under the control of a microprocessor or computer. Different programs in the processor select the gain, directivity, and other factors as required by the application.

Most phased arrays are used in radar systems, but they are finding applications in some specialized wireless communications systems and in satellites.

Phased array

■ TEST―――――――――――――――――

Supply the missing information in each statement.

55. The most commonly used microwave antenna is the _____ antenna.
56. Increasing the length of a horn antenna causes its gain to _____ and its beam width to _____.

57. A horn antenna is six wavelengths long at 8 GHz. That length is _____ in.

58. The beam width of a horn antenna is usually in the _____ to _____ degree range.

59. The width of a horn antenna is 8 cm. The height is 6 cm. The operating frequency is 8 GHz. The beam angle is _____ degrees. The gain is _____ dB. Assume $k = 0.5$.

60. Horn antennas have _____ (narrow, wide) bandwidth.

61. The geometric shape of a microwave reflector is a(n) _____.

62. In order for a dish reflector to work, the antenna must be located at the _____ of the dish.

63. The antenna used with a dish reflector is usually a(n) _____.

64. A parabolic reflector antenna has a diameter of 9 m. The frequency of operation is 8 GHz. The gain is _____ dB. The beam width is _____.

65. The effect of a parabolic reflector on an antenna is to _____ gain and _____ beam width.

66. A dish with a horn at its center and a small reflector at the focal point is said to use _____ feed.

67. The benefits of a helical antenna are its _____ and _____.

68. The gain of a helical antenna is typically in the _____ to _____ -dB range.

69. The beam width of a helical antenna is in the _____ to _____ degree range.

70. The helical antenna radiates a(n) _____ polarized wave.

71. The acronym RHCP means _____.

72. A popular omnidirectional microwave antenna is the _____.

Determine whether each statement is true or false.

73. A helical antenna will receive either vertically or horizontally polarized signals.

74. An RHCP antenna can receive a signal from an LHCP antenna.

Answer the following questions:

75. What is the primary dimension of a slot antenna?

76. At what frequencies are lens antenna used? Why?

77. What are the base and construction of a patch antenna?

78. Describe the radiation pattern of a patch antenna.

79. State two benefits of phased arrays.

10-6 Radar

The communications applications in which microwaves are most widely used today are telephone communications and radar. However, there are many other significant uses of microwave frequencies in communications. For example, TV stations use microwave relay links instead of coaxial cables to transmit TV signals over long distances, and cable TV networks use satellite communications to transmit programs from one location to another. Communications with satellites, deep-space probes, and other spacecraft is usually done by microwave transmission because microwave signals are not reflected or absorbed by the ionosphere as are many lower-frequency signals. Electromagnetic radiation from the stars is also primarily in the microwave region. Radio telescopes made up of sensitive radio receivers and large antennas operating in the microwave region are used to map outer space with far greater precision than could be achieved with optical telescopes. Finally, microwaves are also used for heating—in the kitchen (microwave ovens), in medical practice (diathermy machines used to heat muscles and tissues without causing skin damage), and in industry.

This section is devoted to a discussion of radar, one of the most widely used microwave systems. Satellites are covered in detail in Chap. 11.

Radar is an electronic communications system used to detect objects at distances that cannot be observed visually. The range, direction, and elevation of a remote object can be precisely determined with radar.

Radar Concepts

Radar is based on the principle that high-frequency RF signals are reflected by conductive targets. The most usual targets are airplanes, missiles, ships, and automobiles. In a radar system, a signal is transmitted toward the target. The reflected signal is picked up by a receiver in the radar unit. This reflected or returned radio signal is called an *echo*. The radar unit can then determine the distance to the target, its direction or azimuth, and in some cases, its elevation or distance above the ground. See Fig. 10-34.

The ability of radar to determine the distance between a remote object and the radar unit is dependent upon knowing the exact speed of radio signal transmission. As you know, the speed of radio signals and light is 300,000,000 m/s. This is the familiar speed of 186,000 mi/s. In most applications where radar is used, the nautical mile is used instead of statute miles for measurement. Recall that a mile is equal to 5280 ft. A nautical mile (nmi) is equal to 6076 ft. The speed of a radio signal is 162,000 nmi/s. Sometimes, a special unit known as a radar mile is used. A radar mile is equal to 6000 ft.

Knowing the speed of a signal in miles per second, we can actually determine the length of time that it takes for the signal to travel 1 mi. For example, it takes a radio signal 5.376 μs to travel 1 mi. The time it takes for a signal to travel 1 nmi is 6.17 μs.

A radar signal must travel twice the distance between the radar unit and the remote object. The signal is transmitted, and a finite time passes before it reaches the remote object. The signal is then reflected and travels an equal distance back to the radar station. If an object is exactly 1 nmi away, the signal takes 6.18 μs to reach the target, and then 6.18 μs for the return trip. The total elapsed time from the instant of initial transmission to the reception of the echo is 12.34 μs.

The distance to the remote target, therefore, can be calculated using the following simple expression

$$D = \frac{T}{12.36}$$

Here, D is the distance between the radar unit and the remote object in nautical miles, and T is the total time that occurs between the transmission and reception of the signal in microseconds. For example, if the time is 33.7 μs, the distance is

Echo

$$D = \frac{33.7}{12.34}$$

$$= 2.73 \text{ nmi}$$

It is the job of the radar unit to measure this time and compute the distance for display.

In short-distance applications, the yard (yd) is the common unit of distance measurement. A radio signal travels 328 yd/μs. Knowing this fact, the distance to an object can be computed using the expression

$$D = \frac{328T}{2}$$

$$= 164T$$

A measured time of 4.8 μs corresponds to a distance of

$$D = 164(4.8)$$

$$= 787.2 \text{ yd}$$

In order to obtain a strong reflection or echo from a distant object, the wavelength of the radar signal should be small compared to the size of the object being observed. If the wavelength of the radar signal is long with respect to the distant object, only a small amount of energy will be reflected. At higher frequencies, the wavelength is shorter and, therefore, the reflected energy is greater. Ideally, the size of the target should be one-quarter wavelength or more at the transmitted frequency for the optimum reflected signal.

The shorter the wavelength of the signal compared to the observed object, the greater the resolution or definition of the remote object. In most cases, it is necessary to detect just the presence of a remote object, but if very short wavelengths are used, the actual shape of the object can often be clearly determined.

The earliest radars operated in the VHF and UHF ranges primarily because technology did

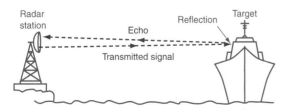

Fig. 10-34 Concept of radar.

About ⬌ Electronics

Transistor Applications—Then and Now
The earliest commercial transistors worked only at frequencies below 1 MHz. Transistors intended for harsh environments use metal, glass, and ceramic packages. Pseudomorphic high-electron-mobility transistors (pHEMTs) provide new application areas in radar and microwave communications.

not permit reliable higher-frequency operation. However, as microwave components and techniques were developed, they were quickly adopted for radar applications. Virtually all radar systems use microwave signals. Many systems use frequencies as high as millimeter waves for precision detection.

Since radar uses microwave frequencies, line-of-sight communications results. In other words, radar cannot detect objects which are beyond the horizon. Objects need not be physically visible, but they must be within line-of-sight radio distance in order for detection to occur.

Ranging

In locating a distant object by radar, there are three factors that must be determined. These are distance or range, azimuth or bearing, and elevation or altitude. These can be expressed by a right triangle as shown in Fig. 10-35. Assume that the radar is land-based and used to detect aircraft. The distance or range between the radar unit and the remote airplane is the hypotenuse of the right triangle. The angle of elevation is the angle between the hypotenuse and the base line which is a line tangent to the surface of the earth at the radar location. The altitude is defined by the angle of elevation. The greater the angle of elevation, the greater the altitude. By knowing the range and the angle of elevation, the altitude can be computed using standard trigonometric techniques.

Another important factor in locating a distant object is knowing its direction with respect to the radar set. If the radar station is fixed and land-based, the direction or azimuth is usually given as a compass direction. Recall that true north is 0° or 360°, east is 90°, south is 180°, and west is 270°. The direction or bearing of the remote object is usually given in degrees.

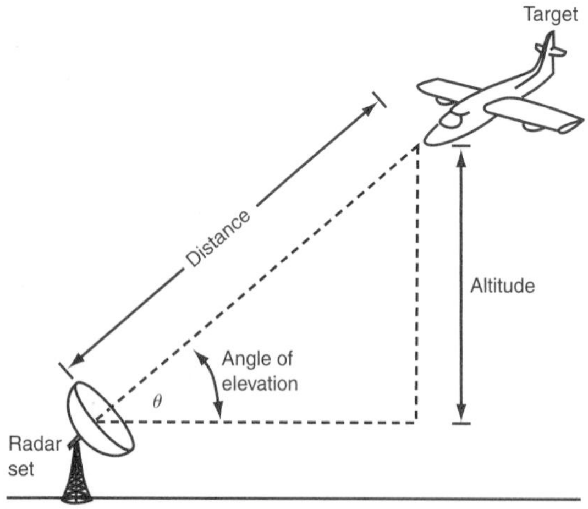

If the radar unit is located in a moving vehicle such as an airplane or ship, the azimuth or bearing is given as a relative bearing with respect to the forward direction of the vehicle. Straight ahead is 0° or 360°, directly to the right is 90°, directly behind is 180°, and directly to the left is 270°.

The ability of a radar unit to determine the direction of a remote object is strictly dependent upon a highly directional antenna. By using an antenna with an extremely narrow beam width, the radar unit will receive signals from only over a narrow angle. The narrower the beam width of the antenna, the more precisely the actual azimuth or bearing can be determined.

Since most radars operate in the microwave region, highly directional antennas can be easily obtained. Horn antennas with parabolic reflectors are the most common, and beam widths of less than 1° are readily attainable. These highly directional antennas are rotated continuously 360°. The same antenna is used for transmitting the original signal and receiving the reflected signal.

Circuits within the radar unit are calibrated so that the direction in which the antenna is pointing is accurately known. Then when the echo is received, it can be compared to the calibrated values and the precise direction can be determined.

The ability of a radar unit to determine the altitude of a remote target depends upon the vertical beam width of the radar antenna. The radar antenna may scan vertically while measuring the distance of the object during the scan. When the object is detected, the vertical elevation of the antenna is noted and the actual altitude can then be compared.

Pulsed Radar

There are two basic types of radar systems: pulsed and continuous wave (CW). There are also numerous variations of each. By far the most commonly used radar system is the *pulsed type*. Signals are transmitted in short bursts or pulses as shown in Fig. 10-36. The duration or width w of the pulse is very short and, depending upon the application, can be anywhere

Pulsed mode

Fig. 10-35 The trigonometry of radar.

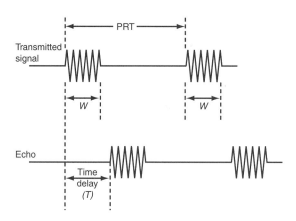

Fig. 10-36 Pulsed radar signals.

from less than a microsecond to several microseconds. The time between transmitted pulses is known as the pulse repetition time (PRT). From this can be determined the pulse repetition frequency (PRF).

$$PRF = \frac{1}{PRT}$$

If the PRT is 120 μs, the PRF is

$$PRF = \frac{1}{120 \times 10^{-6}}$$

$$= 8.33 \text{ kHz}$$

The ratio of the pulse width w to the PRT is known as the duty cycle. The duty cycle is normally expressed as a percentage.

$$\text{Duty cycle} = \frac{w(100)}{PRT}$$

A pulse width of 9 μs with a PRT of 120 μs produces a duty cycle of

$$\text{Duty cycle} = \frac{9(100)}{120}$$

$$= 7.5\%$$

Note in Fig. 10-36 the time between the end of the transmitted pulse and the beginning of the next pulse in sequence. It is during this interval that the echo will be received.

The duration of the transmitted pulse and the PRT are extremely critical in determining the performance of a radar and its application. Very short range radars use narrow pulses and short PRTs. If the object is only a short distance away, then the time for the echo will be relatively short. In short-range radars, the pulse width is made narrow to ensure that the pulse is terminated prior to the receipt of the echo of the target. If the pulse is too long, the return

signal may be masked or blanked by the transmitted pulse. On the other hand, long-range radars will typically use a longer PRT since it will take a greater time for the echo to return. This also permits a longer burst of energy to be transmitted, thereby ensuring a stronger return.

If the PRT is too short compared to the distance of the target, it is possible that the echo will not return during the time interval between two successive pulses. The return may occur after the second transmitted pulse. This is known as a double range or second return echo. Naturally, it will lead to ambiguous distance measurements.

Of course, the distance to the remote object is determined simply by measuring the time delay between the transmitted pulse and the received echo.

CW Radar

In *CW radar*, a constant-amplitude continuous microwave sine wave is transmitted. The echo, therefore, is also a constant-amplitude microwave sine wave of the same frequency but of lower amplitude. The question is then, "How can such a signal be useful in determining the range or other characteristics of a target?" The answer lies in the object itself which, in most cases, is moving with respect to the radar unit. When detecting a moving airplane, ship, missile, or automobile, the reflected signal undergoes a frequency change. It is this frequency change between the transmitted signal and the returned signal that is used to determine the speed of the target.

The frequency shift that occurs when there is relative motion between the transmitting station and a remote object is known as the *Doppler effect*. You have experienced the Doppler effect yourself with sound waves. For example, the horn on a car emits a tone at a fixed frequency. If the car is stationary, you hear this frequency. However, if the car is moving toward you while the horn is on, you will experience a tone of increased frequency. As the car moves closer to you, the sound waves are compressed, thereby giving the effect of a higher frequency. If the car is moving away from you with its horn on, the sound waves are stretched out, giving the effect of a lower frequency. This same effect works on both radio and light waves.

By measuring the amount of the frequency difference between the transmitted signal and the reflected signal, it is possible to determine

CW mode

Doppler effect

Speed measurement

the relative *speed* between the radar unit and the observed object.

$$V = 1.1 f \lambda$$

where f = frequency difference between transmitted and reflected signals, Hz
λ = wavelength of transmitted signal, m
V = relative speed between the two objects, m/s

Assume a frequency shift of 1000 Hz at a frequency of 10 GHz. A frequency of 10 GHz represents a wavelength of

$$\lambda = \frac{300}{f \text{ (in MHz)}}$$

$$= \frac{300}{10,000}$$

$$= 0.03 \text{ m}$$

Therefore the speed is

$$V = 1.1(1000)(0.03)$$

$$= 33 \text{ m/s}$$

Moving target indication (MTI)

In CW radar, it is the Doppler effect that provides frequency modulation of the carrier. In order for there to be a frequency change, the observed object must be moving toward or away from the radar unit. If the observed object moves parallel to the radar unit, there is no relative motion between the two objects, and no frequency modulation will occur. The great-

est value of CW radar is its ability to measure the speed of distant objects. The familiar police radar units use CW Doppler radar for measuring the speed of cars and trucks.

A popular variation of CW radar is FM CW radar. In these systems, the carrier is continuous but is frequency-modulated by a sawtooth or triangular wave. Since the frequency of the transmitted signal varies over time, the frequency of the echo signal can be compared to the frequency of the transmitted signal. The frequency difference will determine the time and, therefore, the distance between the transmitting station and the target. The frequency difference between the transmitted and echo signals is directly proportional to the range.

Some radar systems combine both pulse and Doppler techniques to improve performance and measurement capabilities. One such system evaluates successive echos to determine phase shifts that indicate when a target is moving. Such radars are said to incorporate *moving target indication (MTI)*. Through a variety of special signal-processing techniques, multiple moving targets can be distinguished not only from one another but from fixed targets as well.

Analysis of Radar System

Figure 10-37 shows a comprehensive block diagram of a typical radar unit. There are four

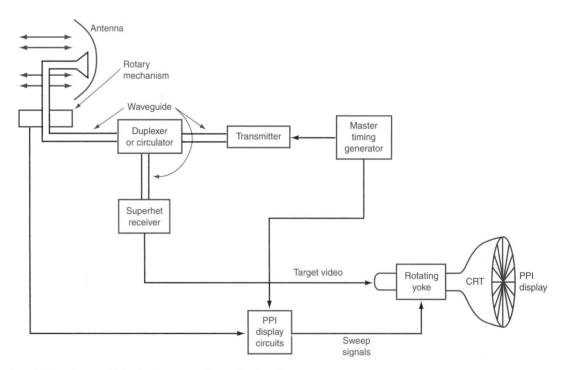

Fig. 10-37 General block diagram of a pulsed radar.

basic subsystems: the antenna, the transmitter, the receiver, and the display unit.

The transmitter in a pulsed radar system invariably uses a magnetron. Recall that a magnetron is a special high-power vacuum tube oscillator that operates in the microwave region. The cavity size of the magnetron sets the operating frequency. A master timing generator develops the basic pulses used for triggering the magnetron. The timing generator sets the pulse duration, PRT, and the duty cycle. The pulses from the timing network trigger the magnetron into oscillation, and the magnetron emits short bursts of microwave energy. Magnetrons are capable of extremely high powers, especially when operated on a pulse basis. Continuous average power may be low, but when pulsed, magnetrons can produce many megawatts of power for the short duration required by the application.

Although magnetrons are the most commonly used radar transmitter tube, both klystrons and TWTs are also used in some types of radars. These are more commonly found in CW Doppler radars. Low-power radars such as those used by the police for speed detection use Gunn diodes.

Referring to the block diagram, you can see that the transmitter output is then passed through a duplexer or circulator where it is then applied to the antenna. The duplexer is a special device that allows the transmitter and receiver to share a single antenna. The duplexer contains devices that prevent the high-power transmitted signal from getting into the receiver and damaging it.

A duplexer is a waveguide assembly containing special devices that prevent interference between the transmitter and receiver. The most commonly used device is a spark-gap tube. Different types of spark-gap tubes are used and are referred to as transmit-receive (TR) and anti-transmit-receive (ATR) tubes. The purpose of the TR tube is to prevent transmitter power from reaching the receiver. When RF energy from the transmitter is detected, the spark gap breaks down, creating a short circuit for the RF energy. The purpose of the ATR tube is effectively to disconnect the transmitter from the circuit during the receive interval. The TR and ATR tubes when combined with the appropriate quarter- and half-wavelength waveguides provide effective isolation between

transmitter and receiver. In lower-power radar systems, a standard circulator or isolator can be used for the same purpose.

The antenna system is typically a horn antenna with a parabolic reflector used to produce a very narrow beam width. A special waveguide assembly with a rotating joint allows the waveguide and horn antenna to be rotated continuously over a 360° angle.

The same antenna is also used for reception. During the pulse off time, the received signal passes through the antenna, the associated waveguide, and either the duplexer or circulator to the receiver. The receiver is a standard high-gain superheterodyne type. The signal is usually fed directly to a mixer in most systems, although some radars use an RF amplifier. The mixer is typically a single diode or diode bridge assembly. The LO feeds a signal to the mixer at the appropriate frequency so that an IF is developed. Both klystrons and Gunn diodes are used in LO applications. The IF amplifiers provide very high gain prior to demodulation. Some radar receivers may use double conversion.

The demodulator in a pulse radar system is typically just a diode detector since only the pulses must be detected. In Doppler and other more complex radars, some form of frequency- and phase-sensitive demodulator is used. Phase-locked loops are common in this application. The output of the demodulator is fed to a video amplifier which creates signals that will ultimately be displayed.

The display in most radar systems is a *cathode-ray tube (CRT)*. Various display formats can be used. The simplest form of display, known as an A scan, simply displays the transmitted and received pulses. The horizontal sweep on the oscilloscope is calibrated in yards or miles. See Fig. 10-38(*a*).

The most common type of CRT display is known as type P or plan position indicator (PPI). See Fig. 10-38(*b*). The advantage of the PPI display is that it shows both the range and azimuth of the target. The center of the display is assumed to be the location of the radar unit. Concentric circles indicate the range. The azimuth or direction is indicated by the position of the reflected target on the screen with respect to the vertical radius line. The targets show up as lighted blips on the screen.

The PPI display is developed by a sophisticated scanning system tied in to antenna

Cathode-ray tube (CRT)

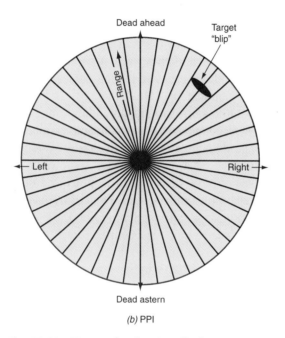

Phased array radars

(a) A scan

Dead ahead

Target "blip"

Range

Left

Right

Dead astern

(b) PPI

Fig. 10-38 Types of radar CRT displays.

rotation. As the antenna is rotated by a motor, an encoder mechanism sends signals to the PPI control circuits that designate the azimuth or direction of the antenna. At the same time, the horizontal and vertical deflection coils (called the yoke) around the neck of the CRT rotate in synchronism with the antenna. The electron beam in the CRT is swept from the center of the screen out to the edge. The sweep begins at the instant of pulse transmission and moves outward from the center. The beginning of each transmitted pulse begins another sweep of the electron beam. As the deflection yoke rotates, the beam moves in such a way that it appears that a radius from the center to the edge is continuously ro-

tating. If a target reflection occurs, a lighted blip will appear on the screen. Calibrations on the screen in the form of graticule markings or superimposed electron beam patterns permit the distance and azimuth reading to be read directly.

Phased Array Radar

An important newer type of radar provides greater flexibility in scanning narrow sectors and tracking multiple targets. Known as *phased array radars,* these high-tech units operate on the same principles but use a unique antenna. Instead of a single horn and parabolic reflector, multiple dipoles are used. Slots in a waveguide are also used. A half-wave dipole at microwave frequencies is very short. Therefore, many of them can be mounted together in a matrix or array. The result is a special collinear or end-fire array on a reflecting surface with very high gain. By using a system of separate feed lines and a variable phase shifter for each dipole, the beam width and directivity can be controlled electronically. This permits rapid scanning and on-the-fly adjustment of directivity. Phased array radars eliminate the mechanical systems needed for conventional radars.

Radar Applications

As for applications of radar, there are many. Their greatest use is in defense weapons systems and in safety and navigation applications. Search radars are used to locate enemy missiles, planes, and ships. Tracking radars are used on missiles and planes to acquire a target and zero in on them. Radars are widely used on planes and ships for navigating "blind" in fog or bad weather. Radars help ground controllers locate and identify nearby planes. Special radars assist planes in landing in bad weather when visibility is near zero.

Radars are also used as altimeters to measure height. High-frequency radars can actually be used to plot or map the terrain in an area. Special terrain-following radars allow high-speed jets to fly very close to the ground to avoid detection by enemy radar.

■ TEST

Answer the following questions.

80. When the speed of radio signals is known, radar can be used to determine the _____ of a target.

81. An echo is the _____ from a target.
82. The directivity of the radar antenna determines the _____ of a target.
83. The speed of a radar signal is _____ μs/nmi.
84. The elapsed time between radiating a radar signal and receiving its echo is 43.2 μs. The target distance in nautical miles is _____.
85. If a target is 1000 yd away, the time period between the transmission of the signal and receipt of the echo is _____ μs.
86. One nautical mile is equal to _____ statute miles.
87. For optimum reflection from a target, the wavelength of the radar signal should be _____ compared to the size of the target.
88. Radars operate in the _____ frequency range.
89. True or false. Both the transmitter and receiver in a radar share the same antenna.
90. The two main types of radar are _____.

91. A frequency shift of sound, radio, or light waves that occurs as the result of the relative motion between objects is called the _____ effect.
92. A CW radar operates at a frequency of 14 GHz. A frequency shift of 25 kHz is produced by a moving target. Its speed is _____ mi/h.
93. True or false. Distance can be measured with constant-frequency CW radar.
94. A waveguide assembly that lets the transmitter and receiver share an antenna is called a(n) _____.
95. The most commonly used component in a radar transmitter is a(n) _____ tube.
96. Both TR and ATR tubes incorporate a(n) _____.
97. The output display in a radar is usually a(n) _____.
98. A display that sweeps outward from the center of the screen while rotating is known as a(n) _____.
99. A radar using multiple antennas to shape the beam width and automatically adjust directivity is known as a(n) _____ radar.

Summary

1. Microwaves are radio signals in the frequency range from 1 to 300 GHz.

2. The RF spectrum below UHF is mostly already fully occupied leaving little or no room for the growth of new radio services.

3. At microwave frequencies, tremendous bandwidth is available for new radio services as well as for wide-bandwidth signals such as TV, multiplexed signals, or computer data.

4. The microwave frequencies are used primarily for telephone communications, radar, and satellite communications.

5. Other microwave applications include cable TV, space communications, radio astronomy, and heating.

6. The primary benefit of microwaves is wide-bandwidth availability.

7. The main disadvantages of microwaves are that they are limited to line-of-sight transmission distances, conventional components are not usable, and circuits are more difficult to analyze and design.

8. The allocation of the RF spectrum is handled by the Federal Communications Commission (FCC) in the United States.

9. Balanced transmission line is not used for microwaves because of radiation losses. Coaxial cable is not used because of its high attenuation.

10. The preferred transmission line for microwaves is waveguides.

11. Because of the short physical length of transmission lines at microwave frequencies, quarter- and half-wave lines are commonly used for tuned circuits and filters.

12. Two printed circuit board implementations of transmission lines, called stripline and microstrip, are widely used to create resonant circuits and filters.

13. A waveguide is a hollow metal pipe with a circular or rectangular cross section used for carrying microwave signals from one place to another.

14. A waveguide acts like a high-pass filter, passing all frequencies above its cutoff frequency and rejecting those below it.

15. The cutoff frequency f_{co} of a waveguide depends upon its physical size. For a rectangular waveguide, it is $300/2a$, where a is the wide dimension of the waveguide in meters.

16. The microwave signal carried by a waveguide is made up of electric (E) and magnetic (H) fields that bounce off the walls of the waveguide as they propagate along its length.

17. The modes of a waveguide describe the various patterns of electric and magnetic fields that are possible.

18. A transverse electric (TE) mode is one where the electric field is transverse or perpendicular to the direction of propagation.

19. A transverse magnetic (TM) mode is one where the magnetic field is perpendicular to the direction of propagation.

20. Waveguides are available in standard lengths and sizes, and special pieces are used for right-angle bends and 90° twists.

21. Half-wavelength sections of waveguides with shorted or closed ends are known as resonant cavities since they "ring" or oscillate at the frequency determined by their dimensions.

22. Cavity resonators are metallic chambers of various shapes and sizes that are used as parallel-tuned circuits and filters. They have a Q of up to 30,000.

23. Point-contact and Schottky or hot-carrier diodes are widely used as mixers in microwave equipment as they have low capacitance and inductance.

24. Varactor and snap-off diodes are widely used as microwave frequency multipliers. Multiplication factors of 2 and 3 are common with power levels of up to 50 W and efficiencies up to 80 percent.

25. A Gunn diode is a microwave semiconductor device used to generate microwave energy. When combined with a microstrip, stripline, or resonant cavity, simple low-power oscillators with frequencies up to 50 GHz are easily implemented.

26. Both IMPATT and TRAPATT diodes are GaAs devices operated with high reverse bias to produce avalanche breakdown. Both are used in microwave oscillators.

27. Both bipolar transistors and FETs are available to amplify microwave signals up to about 40 GHz. The MESFET has the best noise figure. Power amplification up to 150 W per transistor using class C can be achieved at low microwave frequencies. Monolithic microwave integrated circuits (MMICs) are available for amplification and mixing.

28. A klystron is a vacuum tube used for microwave amplification and oscillation.

29. Klystrons use a cavity resonator to velocity-modulate an electron beam which imparts energy to another cavity, producing power amplification. Klystrons are available which produce from a few to many thousands of watts.

30. A single-cavity reflex klystron is used as a microwave oscillator.

31. Klystrons are being gradually replaced by Gunn diodes and traveling-wave tubes.

32. A magnetron is a diode vacuum tube used as a microwave oscillator in radar and microwave ovens to produce powers up to the megawatt range.

33. In a magnetron, a strong magnetic field creates circular paths of electron flow to excite cavities into oscillation.

34. A traveling-wave tube (TWT) is a microwave power amplifier with very wide bandwidth.

35. A microwave signal applied to a helix around the TWT produces velocity and density modulation of the electron beam over a long distance which induces a higher-power signal in the helix.

36. The most commonly used microwave antenna is the horn, which is essentially a rectangular waveguide with a flared end.

37. A pyramidal horn flares in both waveguide dimensions. A sectoral horn flares in only one dimension.

38. Horn antennas are directional and produce a beam width in the 10° to 60° range with a gain in the 10- to 20-dB range, depending upon dimensions.

39. A parabolic or dish-shaped reflector is used with most microwave antennas to focus the RF energy into a narrow beam and increase gain.

40. The parabolic reflector usually has a diameter that is not less than 10 wavelengths at the operating frequency.

41. The gain and directivity of a parabolic reflector antenna is directly proportional to its diameter.

42. Parabolic reflector antennas are fed by placing a horn antenna at the focal point or by placing the horn at the center of the reflector and placing a small reflector at the focal point. The latter is known as Cassegrain feed.

43. A helical antenna is made up of six to eight turns of heavy wire or tubing to form a coil or helix. It is fed with coax and is backed up with a reflector.

44. Helical antennas are used at UHF and microwave frequencies and have a gain in the 12- to 20-dB range and a beam width in the 12° to 45° range.

45. Helical antennas produce circular polarization where the electric and magnetic fields rotate. The polarization may be right-hand or left-hand depending upon the direction in which the helix is wound.

46. Helical antennas can receive either vertically or horizontally polarized signals but can only receive a circularly polarized signal of the same direction.

47. Other popular microwave antennas include the bicone, slot, lens, and patch antennas.

48. Radar is the acronym for *ra*dio *d*etection *a*nd *r*anging.

49. Radar uses a reflected radio signal from a target to determine its distance, azimuth, elevation, and speed.

50. Radio waves travel at a speed of 186,000 mi/s or 162,000 nmi/s. Knowing the speed of radio waves permits the distance (range) of a remote target to be determined.

51. Radio signals travel at a speed of 5.375 μs/nmi or 6.18 μs/nmi.

52. The distance D in nautical miles to a target can be computed by knowing the total delay time T in microseconds for a signal to reach the target and return ($D = T/12.36$).

53. The strength of the reflection is a function of the wavelength of the radar signal and its relationship to the size of the target. Optimum reflection is obtained when the size of the target is one-quarter wavelength or larger of the signal frequency. Most radars operate in the microwave part of the spectrum.

54. The bearing or azimuth of a target with respect to the radar set is determined by a highly directional antenna. A narrow-beam-width antenna is

rotated continuously over 360° and the detection of a reflection from a target at a given bearing gives the target's direction.

55. There are two basic types of radar: pulse and continuous wave (CW).

56. In pulse radar, the transmitter emits microwave energy in the form of repetitive sine wave bursts or pulses.

57. The pulse repetition time (PRT) is the time interval between the beginning of successive pulses.

58. The pulse repetition frequency (PRF) or the pulse repetition rate in pulses per second is the reciprocal of PRT (PRF = 1/PRT).

59. The ratio of the pulse burst duration to the PRT is known as the duty cycle and is usually expressed as a percentage.

60. The PRT and duty cycle set the range of a radar: a short PRT and pulse width for short range, and a long PRT and pulse width for long range.

61. The reflection from the target occurs in the interval between successively radiated pulses.

62. In CW radar, a constant-amplitude–constant-frequency signal is continuously radiated. If the target is stationary, the reflected signal contains no distance information.

63. Continuous wave radar relies on the Doppler effect to produce frequency modulation of the carrier.

64. The Doppler effect is the change in frequency that occurs as the result of relative motion between the transmitter and a target. The Doppler effect occurs with sound, radio, and light signals.

65. The relative speed between transmitter and target is directly proportional to the amount of Doppler frequency shift.

66. The CW Doppler radar is used for speed measurement. Police radars are an example.

67. By frequency modulating a CW radar with a sawtooth or triangular wave, the frequency difference between the transmitted and received signals can be used to compute the distance or range to a target.

68. Radar sets consist of a transmitter, a receiver, an antenna, a master timing section, and a display.

69. Most radar transmitters use a magnetron oscillator, although some high-power radars use klystrons or TWTs. Low-power radars use a Gunn diode oscillator.

70. Radar receivers are superheterodynes with diode mixers.

71. The radar antenna is usually a horn with a parabolic reflector that rotates over a 360° angle.

72. A duplexer is a waveguide assembly that allows both transmitter and receiver to share the same antenna.

73. Spark-gap tubes called transmit-receive (TR) and anti-transmit-receive (ATR) tubes prevent high-power energy from the transmitter from getting into and damaging or desensitizing the receiver.

74. The radar display is normally a cathode-ray tube (CRT) that is calibrated to read out the range, bearing, and other data.

75. The most common CRT readout is the plan position indicator (PPI) where the radar is at the center and a radius rotates to reveal reflected targets as blips.

76. Phased array radars use a matrix of dipoles or slot antennas with variable phase shifters to permit automatic, high-speed, electronic beam switching; beam width changes, and sweeping or scanning.

Chapter Review Questions

Choose the letter which best answers each question.

10-1. The main benefit of using microwaves is
  a. Lower-cost equipment.
  b. Simpler equipment.
  c. Greater transmission distances.
  d. More spectrum space for signals.

10-2. Radio communications are regulated in the United States by the
  a. Federal Trade Commission.
  b. Congress.
  c. Federal Communications Commission.
  d. Military.

10-3. Which of the following is *not* a disadvantage of microwave?
  a. Higher-cost equipment
  b. Line-of-sight transmission
  c. Conventional components are not usable.
  d. Circuits are more difficult to analyze.

10-4. Which of the following is a microwave frequency?
a. 1.7 MHz
b. 750 MHz
c. 0.98 GHz
d. 22 GHz

10-5. Which of the following is *not* a common microwave application?
a. Radar
b. Mobile radio
c. Telephone
d. Spacecraft communications

10-6. Coaxial cable is not widely used for long microwave transmission lines because of its
a. High loss.
b. High cost.
c. Large size.
d. Excessive radiation.

10-7. Stripline and microstrip transmission lines are usually made with
a. Coax.
b. Parallel wires.
c. Twisted pair.
d. PCBs.

10-8. The most common cross section of a waveguide is a
a. Square.
b. Circle.
c. Triangle.
d. Rectangle.

10-9. A rectangular waveguide has a width of 1 in. and a height of 0.6 in. Its cutoff frequency is
a. 2.54 GHz.
b. 3.0 GHz.
c. 5.9 GHz.
d. 11.8 GHz.

10-10. A waveguide has a cutoff frequency of 17 GHz. Which of the signals will *not* be passed by the waveguide?
a. 15 GHz
b. 18 GHz
c. 22 GHz
d. 25 GHz

10-11. Signal propagation in a waveguide is by
a. Electrons.
b. Electric and magnetic fields.
c. Holes.
d. Air pressure.

10-12. When the electric field in a waveguide is perpendicular to the direction of wave propagation, the mode is said to be
a. Vertical polarization.
b. Horizontal polarization.
c. Transverse electric.
d. Transverse magnetic.

10-13. The dominant mode in most waveguides is
a. $TE_{0,1}$.
b. $TE_{1,2}$.
c. $TM_{0,1}$.
d. $TM_{1,1}$.

10-14. A magnetic field is introduced into a waveguide by a
a. Probe.
b. Dipole.
c. Stripline.
d. Capacitor.

10-15. A half-wavelength, closed section of a waveguide that acts as a parallel resonant circuit is known as a(n)
a. Half-wave section.
b. Cavity resonator.
c. *LCR* circuit.
d. Directional coupler.

10-16. Decreasing the volume of a cavity causes its resonant frequency to
a. Increase.
b. Decrease.
c. Remain the same.
d. Drop to zero.

10-17. A popular microwave mixer diode is the
a. Gunn.
b. Varactor.
c. Hot carrier.
d. IMPATT.

10-18. Varactor and step-recovery diodes are widely used in what type of circuit?
a. Amplifier
b. Oscillator
c. Frequency multiplier
d. Mixer

10-19. Which diode is a popular microwave oscillator?
a. IMPATT
b. Gunn
c. Varactor
d. Schottky

10-20. Which type of diode does *not* ordinarily operate with reverse bias?
a. Varactor
b. IMPATT
c. Snap-off
d. Tunnel

10-21. Low-power Gunn diodes are replacing
a. Reflex klystrons.
b. TWTs.
c. Magnetrons.
d. Varactor diodes.

10-22. Which of the following is *not* a microwave tube?
a. Traveling-wave tube
b. Cathode-ray tube
c. Klystron
d. Magnetron

10-23. In a klystron amplifier, velocity modulation of the electron beam is produced by the
a. Collector.
b. Catcher cavity.
c. Cathode.
d. Buncher cavity.

10-24. A reflex klystron is used as a(n)
a. Amplifier.
b. Oscillator.
c. Mixer.
d. Frequency multiplier.

10-25. For proper operation, a magnetron must be accompanied by a
a. Cavity resonator.
b. Strong electric field.
c. Permanent magnet.
d. High dc voltage.

10-26. The operating frequency of klystrons and magnetrons is set by the
a. Cavity resonators.
b. DC supply voltage.
c. Input signal frequency.
d. Number of cavities.

10-27. A magnetron is used only as a(n)
a. Amplifier.
b. Oscillator.
c. Mixer.
d. Frequency multiplier.

10-28. A common application for magnetrons is in
a. Radar.
b. Satellites.
c. Two-way radio.
d. TV sets.

10-29. In a TWT, the electron beam is density-modulated by a
a. Permanent magnet.
b. Modulation transformer.
c. Helix.
d. Cavity resonator.

10-30. The main advantage of a TWT over a klystron for microwave amplification is
a. Lower cost.
b. Smaller size.
c. Higher power.
d. Wider bandwidth.

10-31. High-power TWTs are replacing what in microwave amplifiers?

a. MESFETs
b. Magnetrons
c. Klystrons
d. IMPATT diodes

10-32. The most widely used microwave antenna is a
a. Half-wave dipole.
b. Quarter-wave probe.
c. Single loop.
d. Horn.

10-33. What happens when a horn antenna is made longer?
a. Gain increases
b. Beam width decreases
c. Both *a* and *b*
d. Neither *a* nor *b*

10-34. A pyramidal horn used at 5 GHz has an aperture that is 7 by 9 cm. The gain is about
a. 10.5 dB.
b. 11.1 dB.
c. 22.6 dB.
d. 35.8 dB.

10-35. Given the frequency and dimensions in Questions 10-34 above, the beam width is about
a. 27°.
b. 53°.
c. 60°.
d. 80°.

10-36. The diameter of a parabolic reflector should be at least how many wavelengths at the operating frequency?
a. 1
b. 2
c. 5
d. 10

10-37. The point where the antenna is mounted with respect to the parabolic reflector is called the
a. Focal point.
b. Center.
c. Locus.
d. Tangent.

10-38. Using a small reflector to beam waves to the larger parabolic reflector is known as
a. Focal feed.
b. Horn feed.
c. Cassegrain feed.
d. Coax feed.

10-39. Increasing the diameter of a parabolic reflector causes which of the following:
a. Decreased beam width
b. Increased gain

c. Increased beam width
d. *a* and *b*
e. *b* and *c*
f. None of the above.

10-40. A helical antenna is made up of a coil and a
a. Director.
b. Reflector.
c. Dipole.
d. Horn.

10-41. The output of a helical antenna is
a. Vertically polarized.
b. Horizontally polarized.
c. Circularly polarized.
d. Both *a* and *b*.

10-42. A common omnidirectional microwave antenna is the
a. Horn.
b. Parabolic reflector.
c. Helical.
d. Bicone.

10-43. The time from the transmission of a radar pulse to its reception is 0.12 ms. The distance to the target is how many nautical miles?
a. 4.85 nmi
b. 9.7 nmi
c. 11.2 nmi
d. 18.4 nmi

10-44. The ability of a radar to determine the bearing to a target depends upon the
a. Antenna directivity.
b. Speed of light.
c. Speed of the target.
d. Frequency of the signal.

10-45. The pulse duration of a radar signal is 600 ns. The PRF is 185 pulses per second. The duty cycle is
a. 1.1 percent.
b. 5.5 percent.
c. 31 percent.
d. 47 percent.

10-46. The Doppler effect is used to produce modulation of which type of radar signal?
a. Pulse
b. CW

10-47. The Doppler effect allows which characteristic of a target to be measured?
a. Distance
b. Azimuth
c. Altitude
d. Speed

10-48. The Doppler effect is a change in what signal characteristic produced by relative motion between the radar set and a target?
a. Amplitude
b. Phase
c. Frequency
d. Duty cycle

10-49. The most widely used radar transmitter component is a
a. Klystron.
b. Magnetron.
c. TWT.
d. Power transistor.

10-50. Low-power radar transmitters and receiver LOs use which component?
a. GaAs FET
b. Magnetron
c. Gunn diode
d. Klystron

10-51. What component in a duplexer protects the receiver from the high-power transmitter output?
a. Waveguide
b. Bandpass filter
c. Notch filter
d. Spark gap

10-52. Most radar antennas use a
a. Dipole.
b. Broadside array.
c. Horn and parabolic reflector.
d. Collinear array.

10-53. The most common radar display is the
a. A scan.
b. Color CRT.
c. Liquid-crystal display.
d. Plan position indicator.

10-54. A radar antenna using multiple dipoles or slot antennas in a matrix with variable phase shifters is called a(n)
a. A scan.
b. Phased array.
c. Broadside.
d. Circular polarized array.

10-55. Police radars use which technique?
a. Pulse
b. CW

10-56. Which of the following is a typical radar operating frequency?
a. 60 MHz
b. 450 MHz
c. 900 MHz
d. 10 GHz

Critical Thinking Questions

10-1. At microwave frequencies, the values of inductors and capacitors required for resonance, bypassing, and coupling become very small. Components like this are not available commercially. Explain how you could obtain or realize such components physically. (Hint: See Chap. 9.)

10-2. What is the main reason that vacuum tubes are still used in some microwave applications?

10-3. Assume that all of the radio and other communications signals in the spectrum from 100 kHz to 30 MHz are frequency-multiplexed on a single coaxial cable. Could this composite signal be used to frequency-modulate the carrier of a 30-GHz transmitter? Explain what the resulting output signal might look like in the frequency domain.

10-4. In electronic warfare, how does a plane's radar help and hurt in battle?

10-5. Explain how radar in an airplane could be used to measure its altitude.

10-6. A microwave oven generates hundreds of watts of power at 2.45 GHz. What type of microwave tube does it probably use?

10-7. If radar units were cheap enough (less than $100), what are some new applications to which they might be put?

10-8. A 12-GHz transmitter needs a beam width of less than 1°. What minimum diameter is needed for the parabolic dish? What is the gain of this antenna?

Answers to Tests

1. 1
2. bandwidth or spectrum
3. more difficult to analyze; different measurement techniques; resistors, capacitors, and inductors act like *LCR* circuits; conventional semiconductors do not work owing to internal capacitances and long transit time; special, expensive vacuum tubes are used for power amplification; line-of-sight transmission differences; excessive signal reflection and absorption
4. telephone, radar
5. television signal relay, space communications, radio telescopes, microwave heating
6. Federal Communications Commission
7. improved receiver selectivity, SSB, multiplexing, reduced FM deviation
8. internal capacitance, long transit time
9. *LCR*
10. 36, 3.6
11. 0.687 in., series resonant
12. attenuation
13. 6
14. waveguide
15. microstrip, stripline
16. probe, loop
17. high-pass
18. 4.921
19. (a) true, (b) false
20. H, E
21. $TE_{0,1}$
22. transverse magnetic
23. cavity resonator
24. parallel resonant
25. tuned circuits, filters
26. resonant frequency
27. high capacitance, high inductance, long transit time
28. P-type silicon, tungsten wire
29. Schottky barrier, or hot-carrier
30. mixer
31. varactor
32. varactor, snap-off or step-recovery diode
33. Gunn
34. IMPATT, TRAPATT, tunnel
35. negative resistance
36. reverse
37. forward
38. gain-bandwidth product, more figure
39. GaAs MESFET
40. 15
41. small-signal amplifiers, mixers
42. true
43. false
44. amplifier
45. cavity resonators
46. velocity
47. buncher, catcher
48. reflex

49. Gunn diodes, TWTs
50. permanent magnet
51. radar, microwave ovens
52. amplifier
53. helix
54. bandwidth
55. horn
56. increase, decrease
57. 11.8
58. 10, 60
59. 37.5, 13.3
60. wide
61. parabola
62. focal point
63. horn
64. 55.39, 0.242
65. increase, decrease
66. Cassegrain
67. simplicity, low cost
68. 12, 20
69. 12, 45
70. circularly
71. right-hand circular polarization
72. bicone
73. true
74. false

75. $\lambda/2$ wide
76. Above 40 GHz (millimeter waves). Attenuation is lower at these frequencies
77. printed circuit board, microstrip construction
78. circular, away from surface patch
79. higher gain, directivity
80. distance or range
81. reflection
82. bearing or azimuth
83. 6.18
84. 3.5
85. 6.1
86. 1.15
87. short
88. microwave
89. true
90. pulse, continuous wave
91. Doppler
92. 589
93. false
94. duplexer
95. magnetron
96. spark gap
97. CRT
98. plan position indicator (PPI)
99. phased array

Chapter 11

Introduction to Satellite Communications

Chapter Objectives

This chapter will help you to:

1. *Name* the most common communication satellite orbit and *tell* how it is used and how it is accomplished.
2. *State* the type of coordinates used in locating a communications satellite in orbit.
3. *Draw* a block diagram of the basic electronic unit in a satellite, *give* its name, and *tell* how it operates.
4. *Name* the six main subsystems of a satellite and *tell* what the function of each is.
5. *Draw* a basic block diagram of a satellite earth station, *name* the five basic subsystems, and *explain* the function of each.
6. *Name* four common applications for satellites.

A satellite is a physical object that orbits or revolves around some celestial body. Satellites occur in nature; our own solar system is a perfect example. The earth and other planets are satellites which revolve around the sun, and the moon is a satellite of the earth. A balance between the inertia of the revolving satellite and the gravitational pull of the orbited body keeps the satellite in orbit. Artificial satellites can be launched into orbit for a variety of purposes. One of these major applications is communications. In this chapter, we will discuss why satellites are such an important development in electronics communications and will briefly discuss how communications satellites work.

11-1 Satellite Orbits

In this section we will talk about the orbital dynamics of a satellite, that is, the forces that keep the satellite in orbit and the physical and mathematical laws that it follows. With this information you can better appreciate the technological achievement involved in putting a satellite into orbit and using it for communications and other purposes.

Centripetal acceleration
Centripetal force

Orbit Fundamentals

If a satellite were simply launched vertically from the earth and then released, it would fall back to earth because of gravity. In order for the satellite to go into orbit around the earth, it must have some forward motion. For that reason, the satellite is launched with both vertical and forward motion. The forward motion produces inertia which tends to keep the satellite moving in a straight line. However, gravity tends to pull the satellite toward the earth. The combined effect is called *centripetal acceleration*. The *centripetal force* is caused by the inertia of the satellite balanced by the earth's gravitational pull. If the velocity of the satellite was high enough, it would be carried away from the earth into space. It takes a speed of approximately 25,000 mi/h, the escape velocity of the earth, to cause a spacecraft to break the gravitational pull of the earth. At lower speeds gravity constantly pulls the satellite toward the earth. The acceleration of the satellite caused by the gravity pull exactly balances the effect of the satellite's own velocity.

The closer the satellite is to earth, the stronger the effect of the earth's gravitational pull. So in low orbits, the satellite must travel faster to avoid falling back to earth. The farther the satellite is from the earth, the lower its orbital speed. The lowest practical earth orbit is approximately 100 mi. At this height, the satellite's speed must be about 17,500 mi/h in order to stay in orbit. With this speed, the satellite orbits the earth in approximately $1^1/_2$ h. Communications satellites are usually much farther from earth. A typical distance is 22,300 mi. At this distance, a satellite need travel only about 6800 mi/h in order to stay in orbit. With this speed, the satellite revolves about the earth in approximately 24 h, the earth's own rotational time.

As it turns out, other factors beside just the velocity and gravitational pull determine a satellite's orbit. It is also affected by the satellite's own weight and the gravitational pulls of the moon and sun. Although these factors have less effect than the velocity and gravitational pull, over the long term these factors are important in determining the location and performance of a satellite.

Orbit Shape   A satellite revolves around the earth in either a circular or elliptical path as shown in Fig. 11-1. Circles and ellipses are, of

course, special geometric figures that can be accurately described mathematically. Figure 11-2 summarizes the mathematical characteristics of both circles and ellipses. Because the orbit is either circular or elliptical, it is possible to calculate the position of a satellite at any given time.

A satellite revolves in an orbit that forms a plane which passes through the center of gravity of the earth or the *geocenter*. This is illustrated in Fig. 11-3. In addition, the direction of the satellite's revolution may be either in the same direction as the earth's rotation or against the direction of the earth's rotation. In the former case, the orbit is said to be *posigrade* and in the latter case, *retrograde*. Most orbits are posigrade. In a circular orbit the speed of revolution is constant. However, in an elliptical orbit the speed changes depending upon the

Geocenter

Posigrade orbit
Retrograde orbit

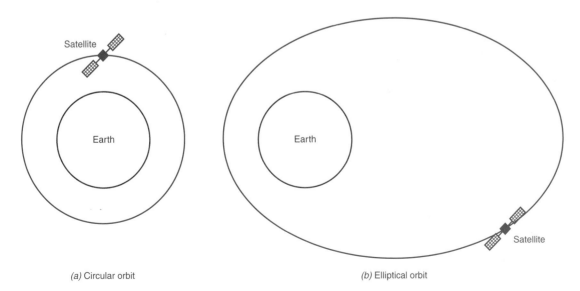

(a) Circular orbit

(b) Elliptical orbit

Fig. 11-1   Satellite orbits.

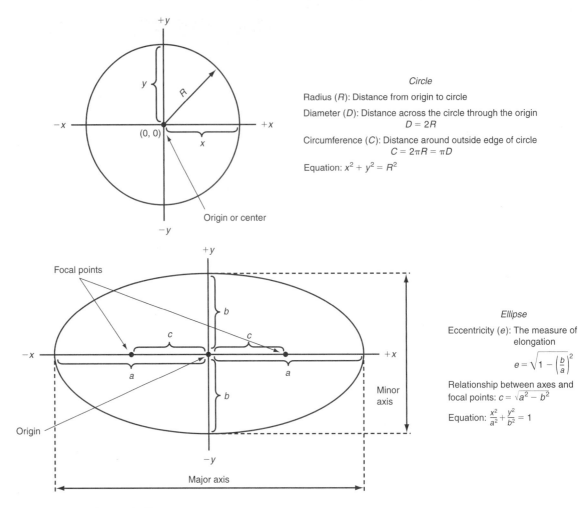

Fig. 11-2   Circles and ellipses.

Circle

Radius (*R*): Distance from origin to circle

Diameter (*D*): Distance across the circle through the origin
$$D = 2R$$

Circumference (*C*): Distance around outside edge of circle
$$C = 2\pi R = \pi D$$

Equation: $x^2 + y^2 = R^2$

Ellipse

Eccentricity (*e*): The measure of elongation

$$e = \sqrt{1 - \left(\frac{b}{a}\right)^2}$$

Relationship between axes and focal points: $c = \sqrt{a^2 - b^2}$

Equation: $\frac{x^2}{a^2} + \frac{y^2}{b^2} = 1$

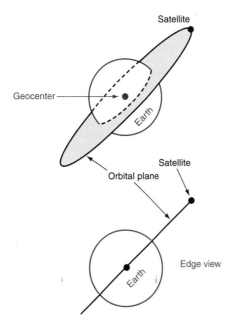

Fig. 11-3   The orbital plane passes through the geocenter.

height of the satellite above the earth. Naturally, the speed of the satellite is higher when it is close to the earth than when it is far away.

Now let's talk about some of the specific characteristics of a satellite orbit. Some of these characteristics include height, speed or period, angle of inclination, and angle of elevation.

In a circular orbit, the height is simply the distance of the satellite from the earth. However, in geometrical calculations, the height is really the distance between the center of the earth and the satellite. In other words, the distance includes the radius of the earth, which is generally considered to be approximately 3960 mi or 6370 km. A satellite that is 5000 mi above the earth in a circular orbit is 3960 + 5000 = 8960 mi from the center of the earth. See Fig. 11-4.

When the satellite is in an elliptical orbit, the center of the earth is one of the focal points of the ellipse. Refer to Fig. 11-5. In this case, the distance of the satellite from the earth varies depending upon its position. Typically the two

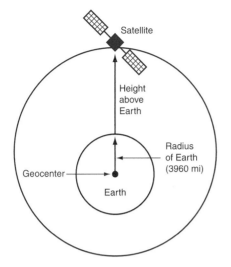

Fig. 11-4 Satellite height.

points of greatest interest are the highest and lowest points above the earth. The highest point is generally referred to as the *apogee,* and the lowest point is called the *perigee.* The apogee and perigee distances typically are measured from the geocenter of the earth and, therefore, include the earth's radius. To determine the height above the earth's surface, the radius must be subtracted from these figures.

Satellite Speed and Period The speed of a satellite is measured in either miles per hour, kilometers per hour, or knots (nautical miles per hour). As indicated earlier, the speed varies depending upon the distance of the satellite from the earth. For a circular orbit the speed is constant, but for

Communications satellites typically orbit the earth at a distance of about 22,300 mi at a speed of about 6800 mi/h.

an elliptical orbit the speed varies depending upon the height. Low earth satellites with a height of about 100 mi have a speed in the neighborhood of 17,500 mi/h. Very high satellites such as communications satellites that are approximately 22,300 mi out revolve much slower, a typical speed being in the neighborhood of 6800 mi/h.

The time that it takes for a satellite to complete one orbit is called the *sidereal period.* Some fixed or apparently motionless external object, such as the sun or a star, is used for reference in determining a sidereal period. The reason for this is that while the satellite is revolving around the earth, the earth itself is rotating.

Another method of expressing the time for one orbit is the *revolution* or *synodic period.* One revolution is the period of time that elapses between the successive passes of the satellite over a given meridian of earth longitude. Naturally, the synodic and sidereal periods differ from one another because of the earth's rotation. The time difference is determined by the height of the orbit, the angle of the plane of the orbit, and whether the satellite is in a posigrade orbit or retrograde orbit. The period is generally expressed in hours. Typical revolutional periods range from approximately

Sidereal period

Apogee
Perigee
Synodic period

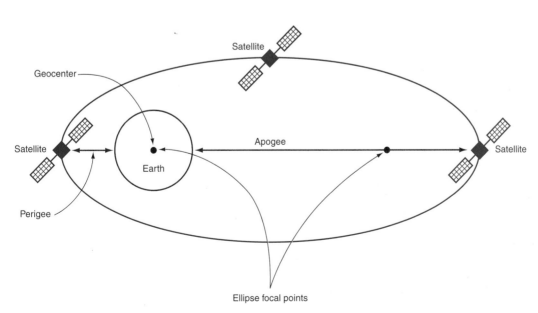

Fig. 11-5 Elliptical orbit showing apogee and perigee.

1¹/₂ h for a height of 100 mi to 24 h for a height of 22,300 mi.

Satellite Angles The *angle of inclination* of a satellite orbit is the angle formed between the line that passes through the center of the earth and the north pole and a line that passes through the center of the earth but which is also perpendicular to the orbital plane. This is illustrated in Fig. 11-6. Satellite orbits can have inclination angles of 0° through 180°.

Another definition of inclination is the angle between the equatorial plane and the satellite orbital plane as the satellite enters the northern hemisphere. This definition may be a little bit easier to understand, but it means the same thing as the preceding definition.

When the angle of inclination is 0° or 180°, the satellite will be directly above the equator. When the angle of inclination is 90°, the satellite will pass over both the north and south poles once during each orbit. Orbits with a 0° inclination are generally called *equatorial,* and orbits with inclinations of 90° are called *polar.*

The *angle of elevation* of a satellite is that angle that appears between the line from the earth station's antenna to the satellite and the line between the earth station's antenna and the earth's horizon. This is shown in Fig. 11-7. If the angle of elevation is too small, the signals between the earth station and the satellite have to pass through much more of the earth's atmosphere. Because of the very low powers used and the high absorption of the earth's atmosphere, it is desirable to minimize the amount of time that the signals spend in the atmosphere. Noise in the atmosphere also contributes to poor performance. The lower the angle of radiation, the more time this signal spends in the atmosphere. Five degrees is usu-

Angle of inclination

**Equatorial orbit
Polar orbit**

Angle of elevation

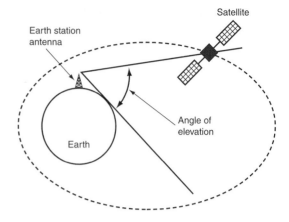

Fig. 11-7 Angle of elevation.

ally regarded as the minimum practical angle of elevation for good satellite performance. The higher the angle of elevation, the better.

Satellite Repeater To use a satellite for communications relay or repeater purposes, the ground station antenna must be able to follow or track the satellite as it passes overhead. Depending upon the height and speed of the satellite, the earth station will be able to use it for communications purposes only for that short period of time when it is visible. The earth station antenna will track the satellite from horizon to horizon, but at some point, the satellite will disappear around the other side of the earth. At this time, it can no longer support communications.

One solution to this problem is to launch a satellite with a very long elliptical orbit that allows the earth station to "see" the apogee. In this way the satellite stays in view of the earth station for most of its orbit, and the satellite is useful for communications for a longer period of time. It is only that short duration when the satellite disappears on the other side of the earth (perigee) that it can no longer be used.

Because of these orbital characteristics, intermittent communications is highly undesirable in many communications applications. Some means of providing continuous communication must be found. This can be done by using more than one satellite. Typically, three satellites, if properly spaced in the correct orbits, can provide continuous communications at all times. However, multiple tracking stations and complex signal switching or "handoff" systems between stations are required. This is expensive and inconvenient.

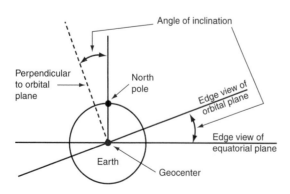

Fig. 11-6 Angle of inclination.

Geosynchronous Satellites

The solution to this problem is easy. Simply launch a synchronous or geostationary satellite. This is a satellite that orbits the earth around the equator at a distance of 22,300 mi or 35,860 km. A satellite at that distance revolves around the earth in exactly 24 h. In other words, the satellite revolves in exact synchronism with the earth's rotation. For this reason, it appears to be fixed or stationary, from which come the terms *synchronous, geosynchronous,* or *geostationary* orbit. Since the satellite remains apparently fixed, no special earth station tracking antennas are required. The antenna can simply be pointed at the satellite and remain in a fixed position. With this arrangement, continuous communications are possible. Most communications satellites in use today are of the geosynchronous variety. Approximately 40 percent of the earth's surface can be "seen" or accessed from such a satellite. Users inside that area can use the satellite for communications.

During an *eclipse* the earth or moon gets between the sun and the satellite. The shadow thus cast on the satellite causes sunlight to be blocked from the solar cell panels. Since these panels supply the main power to the satellite, an eclipse shuts off all power to the satellite.

Eclipses of geostationary satellites occur on the autumnal and vernal equinoxes, the forty-fourth day of fall and spring, respectively, and last from several minutes to over an hour. Occasionally, the moon's shadow will block sunlight to the satellite. During eclipse conditions, backup batteries take over so that operation can continue to take place.

Station Keeping

Once a satellite is in orbit, the forces acting on it tend to keep it in place. If the satellite's height and speed during launch are accurately controlled, the satellite will enter the proper orbit and remain there. However, even with a very good launch, the satellite will drift somewhat in its orbit. This drift is particularly undesirable in a geosynchronous satellite whose position is supposed to remain fixed for reliable continuous communications. *Orbital drift* is caused by a variety of forces. The gravitational pull of the sun and the moon affect the

satellite's position. Further, the earth's gravitational field is not perfectly consistent at all points on the earth. This is because the earth itself is not a perfect sphere. Instead, the earth is a little bit fatter around the equator and is flattened somewhat at the poles. So, even though the earth appears to be a nearly perfect sphere, it is more accurately described as oblate.

Because of this drift, the orbit of the satellite must be periodically adjusted. Most satellites contain small rockets or *thruster jets* for that purpose. These rockets, placed at various positions on the satellite, can be used to speed up or slow down the satellite for the purpose of compensating for orbital drift. Depending on how accurate the orbit must be, these rockets may be fired as often as every several weeks or as little as once per year.

The rockets can be conventional rocket motors of the type used to launch the satellite and put it into orbit. Smaller corrections are made with thrusters that simply cause motion by releasing a gas under pressure. The most common jet thruster uses a gas called hydrazine which, when released with a catalyst, causes an explosive jet force that moves the satellite. Most satellites have several thruster jets to make various satellite position adjustments possible. The process of firing the rockets under ground control to maintain or adjust the orbit is referred to as *station keeping.*

Attitude Control

In addition to maintaining the position of the satellite in orbit, some means must be provided to position the satellite for optimum performance. This is called *attitude control.* The attitude of the satellite must be controlled so that the antennas can be pointed toward the correct locations on earth. Attitude control is also necessary in some satellites to keep the satellite's solar panels pointed toward the sun so that maximum power is produced at all times. It may also be necessary in some applications where selected sensors or a TV camera must be pointed in the right direction. Attitude control is maintained by a combination of satellite stabilization techniques and jet thrusters for correctional purposes.

The attitude of the satellite is first determined right after the satellite is put into stable

Thruster

Geostationary

Eclipse

Station keeping

Attitude control

Orbital drift

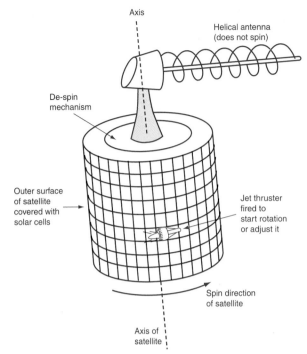

Axis

Helical antenna
(does not spin)

De-spin
mechanism

Outer surface
of satellite
covered with
solar cells

Jet thruster
fired to
start rotation
or adjust it

Spin direction
of satellite

Axis of
satellite

Fig. 11-8 A spin-stabilized satellite.

**Three-axis
stabilization**

Pitch

Roll

Yaw

**Spin
stabilization**

orbit. At some point, the various jet thrusters on the satellite are actuated to move the satellite in such a way that the correct attitude is assumed. For example, a thruster may be fired to rotate the satellite so that its antenna points toward earth. Once the initial attitude of the satellite is set, it must be maintained in this position. This is done by either of two stabilization methods. The most common stabilization method is spin stabilization.

Many satellites are cylindrical in shape as shown in Fig. 11-8. Such satellites may be made to spin on their axes. Once the satellite is in the proper orbit, a jet thruster is fired to begin spinning the satellite. A typical *spin-stabilized* satellite rotates at approximately 100 r/min. Once the satellite is spinning, it remains very stable. The spinning causes a gyroscopic or flywheel effect that keeps the satellite pointed in one direction. A typical position may be with the axis of the satellite pointed toward the earth. An antenna mounted at the axis will remain fixed in that position.

Sometimes the antenna system must be despun. This means that the portion of the satellite around the axis on the ends of the cylinder are independent of the outer spinning cylinder of the satellite. If the antennas themselves were attached to the spinning outer body, they would revolve and not provide a fixed position with

respect to the earth. Despinning allows the antennas to remain oriented to a fixed position on earth as the satellite spins. The gyroscopic effect holds the satellite in position.

Spin stabilization is also used on noncylindrical-shaped satellites. This is done by including a large flywheel at some point on the satellite body. Once the satellite is in position and its antennas, solar panels, and sensors are oriented, the flywheel is put into motion. Again, the gyroscopic effect of the flywheel keeps the satellite oriented with the proper attitude.

Another form of attitude control or stabilization is called *three axis stabilization.* The three axes are referred to as *pitch, roll,* and *yaw.* Any aircraft or spacecraft exhibits the properties of pitch, roll, and yaw. These axes are illustrated in Fig. 11-9. In a three-axis stabilized system, three heavy flywheels or reaction wheels, one for each axis, are spun by motors to provide a gyroscopic effect to stabilize the satellite. Any pitch, roll, or yaw is corrected for by firing thruster jets in the proper direction and by controlling flywheel motor speed. The three-axis stabilization system is far more accurate in attitude control and positioning than a spin-stabilized satellite. In those applications in which pinpoint accuracy of antenna pointing must be maintained, a three-axis stabilization system is used.

A three-axis stabilization system uses sensors to observe external reference points such as the sun, the earth, or some remote star. These optical or infrared sensors operate an electronic control system that determines when the attitude of the satellite needs adjustment. If adjustment is required, one or more thrusters are fired to make corrections.

In a three-axis stabilized satellite, attitude corrections are made automatically by the attitude-control subsystem of the satellite. However, most satellites can also be controlled from the ground station. By the use of command signals from the earth station, thrusters may be fired to make minor adjustments when required.

Satellite Position

In order to use a satellite, you must be able to locate its position in space. That position is usually predetermined by the design of the satellite and is achieved during the initial

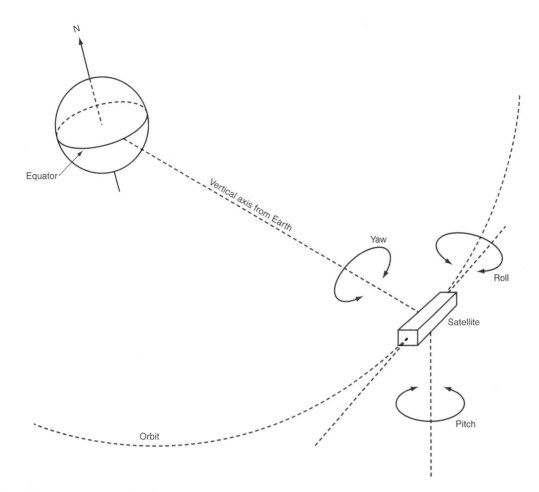

Fig. 11-9 Pitch, yaw, and roll.

launch and subsequent position adjustments. Once the position is known, the earth station antenna can be pointed at the satellite for optimum transmission and reception. For geosynchronous satellites, the earth station antenna will be adjusted once and will remain in that position except for occasional minor adjustments. For other satellites, their positions above the earth will vary depending upon their orbital characteristics. To use these satellites, special tracking systems must be used. A *tracking system* is essentially an antenna whose position can be changed in order to follow the satellite across the sky. To maintain optimum transmission and reception, the antenna must be continually pointed at the satellite as it revolves. We discuss methods of locating and tracking satellites in the following paragraphs.

The location of a satellite is generally specified in terms of latitude and longitude, as any point on earth can be. The satellite location is specified by a point on the surface of the earth directly below the satellite. This point is known as the *subsatellite point (SSP)*. The SSP is located using conventional latitude and longitude designations.

Latitude and longitude form a system for locating any given point on the surface of the earth. This system is widely used for navigational purposes. If you have ever studied a globe of the earth, you have seen the lines of latitude and longitude drawn on it. The lines of *longitude,* or *meridians,* are those lines drawn on the surface of the earth from the north pole to the south pole. Lines of *latitude* are those lines drawn on the surface of the earth from east to west parallel to the equator. The centerline of latitude is the *equator,* which separates the earth into north and south hemispheres.

Latitude is defined as the angle between the line drawn from a given point on the surface of the earth to the geocenter and the line between the geocenter and the equator. This is illustrated in Fig. 11-10. Zero-degree latitude would be at the equator, whereas 90° latitude would

Subsatellite point (SSP)

Latitude and longitude

Tracking system

Equator

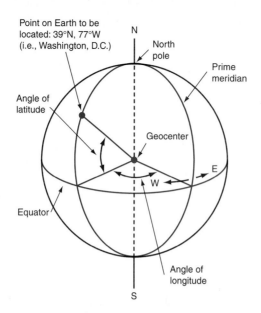

Fig. 11-10 Tracking and navigation by latitude and longitude.

where the meridian containing the green point of interest and the equator intersect. This is illustrated in Fig. 11-10. The designation east or west is usually added to the longitude angle to indicate whether the angle is being measured to the east or west of the prime meridian. As an example, the location of Washington, D.C., is given by the latitude and longitude of 39°N and 77°W.

To show how latitude and longitude are used to locate a satellite, refer to Fig. 11-11. This illustrates some of the many geosynchronous communications satellites serving the United States and other parts of North America. Since geosynchronous satellites orbit over the equator, their SSP is on the equator. For that reason, all geosynchronous satellites have a latitude of 0°.

Only geosynchronous satellites have a fixed SSP on the surface of the earth. The SSP of other satellites will move with respect to a given reference point on the earth. Their SSP traces a line on the earth known as the *subsatellite path* or *ground track*. The ground track for most satellites crosses the equator twice per orbit. The point where the SSP crosses the equator headed in the northerly direction is called the *ascending node*. The point where the SSP crosses the equator headed in the southerly direction is called the *descending node*. The ascending and descending nodes are designated by longitudinal angles. With these two points known, the satellite path can be traced across the surface of the earth between them. The ascending node is sometimes designated by the abbreviation EQX and is used as a reference point for locating and tracking a satellite.

**Subsatellite path
Ground track**

**Ascending node
(EQX)**

Prime meridian

Descending node

be at either the north or south pole. Usually an *N* or an *S* is added to the latitude angle in order to designate whether the point is in the northern or southern hemisphere.

A line drawn on the surface of the earth between the north and south pole is generally referred to as a meridian. One special meridian called the prime meridian is used as a reference point for measuring longitude. The *prime meridian* is the line on the surface of the earth drawn between the north and south poles that passes through Greenwich, England. The longitude of a given point is the angle between the line connecting the geocenter to the point where the prime meridian and equator intersect and the line connecting the geocenter and the point

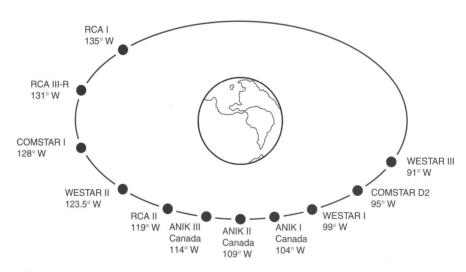

Fig. 11-11 The location of some of the many communications satellites in geosynchronous orbit.

The location of the satellite at any given time is specified by the SSP in terms of latitude and longitude. Except for geostationary satellites, this point changes continuously as the satellite orbits. The exact position of a satellite is usually designated by an orbit calendar. This is a listing that usually consists of the orbit number and the time of the occurrence of the ascending node (EQX). Usually the number of orbits that a satellite makes is tracked from the very instant it is put into orbit. By using various formulas involving the height, speed, and elliptical characteristics of the orbit, the time of occurrence of the ascending node can be computed for each orbit. With the orbital calendar and various maps and plotting devices, the ground track can be traced for each orbit. This allows the satellite user to determine whether or not the satellite is within a usable range.

Just knowing the location of the satellite is insufficient information for most earth stations that must communicate with the satellite. What the earth station really needs to know is the azimuth and elevation settings of its antenna to intercept the satellite. Most earth station satellite antennas are highly directional and must be accurately positioned to "hit" the satellite. The azimuth and elevation designations in degrees tell where to point the antenna.

Take a look at Fig. 11-12. *Azimuth* refers to the direction where north is equal to 0°. The azimuth angle is measured clockwise with respect to north. The angle of *elevation* is the angle between the horizontal plane and the pointing direction of the antenna.

Once the azimuth and elevation are known, the earth station antenna can be pointed in that direction. For a geosynchronous satellite, the antenna will simply remain in that position. For any other satellite, the antenna must then be moved as the satellite passes overhead.

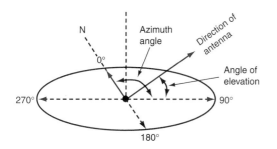

Fig. 11-12 Azimuth and elevation: azimuth = 90°, elevation = 40°.

For geosynchronous satellites, the angles of azimuth and elevation are relatively easy to determine. Because geosynchronous satellites are fixed in position over the equator, special formulas and techniques have been developed to permit easy determination of azimuth and elevation for any geosynchronous satellite for any point on earth. A chart used for this purpose is shown in Fig. 11-13. An example will illustrate its use.

Assume that a ground station is located in Washington, D.C. with coordinates of 39°N and 77°W and that the location of the satellite is 100°W longitude. Now take the difference between the longitude of the satellite and the longitude of Washington, D.C., in this case, 100 − 77 = 23°. This longitude and the latitude of Washington, D.C., 39°N, define a point on the chart in Fig. 11-13 (see Point A).

The lines drawn on the chart in Fig. 11-13 show various levels of both azimuth and elevation. The azimuth or radial lines originate at the lower left-hand 0° point and emerge upward and outward to the right. The other lines designate elevation. If the point defined falls within these sets of lines, then the satellite is within range of the ground station.

Next, note the elevation and radial values at the point defined on the chart. The angle of elevation is approximately 37°, and the radial value is approximately 146°. For a northern-hemisphere station, the azimuth reading is determined by subtracting the radial value from 360° or, in this case, 360° − 146° = 214°. If the earth station is in the southern hemisphere, the azimuth value is equal to 180° plus the radial value or 180° minus the radial value depending upon whether the satellite is east or west of the ground station. In this case, the earth station antenna will be pointed with an azimuth reading of 214° or approximately south-southwest. The angle of elevation is very low, being only about 37° above the horizon. Once the antenna has been set, minor adjustments can be made in both azimuth and elevation to optimize signal reception.

Tracking a nonsynchronous satellite is a more difficult job. Normally a computer is set up to calculate the orbit calendar, and various graphical devices are used to trace the ground track. From this information, azimuth and elevation angles for the tracking antenna can be determined.

Azimuth

Elevation

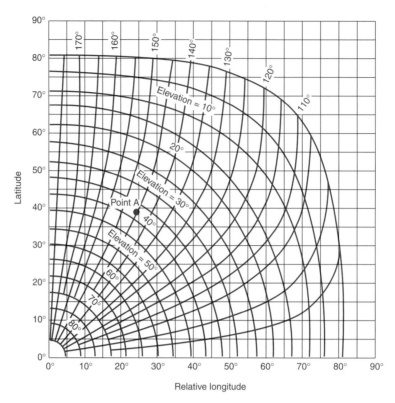

Fig. 11-13 Chart for obtaining azimuth and elevation directions from a ground station to a geostationary satellite.

Satellite Launching

Satellites are placed into their orbits by mounting them on top of rockets which literally shoot them into space. Most communications satellites today are large and heavy, weighing many thousands of pounds. A huge multistage rocket is required to put it into orbit. We will describe the launch procedure in detail so that you can understand the complex steps required to put a satellite into orbit about the earth. For this example, we will assume a communications satellite is being launched into geosynchronous orbit.

The satellite is mounted on top of the rocket and is covered by a fairing or shroud. The fairing is usually the top, pointed part of the rocket. Its purpose is to protect the satellite during the initial part of the flight and aerodynamically streamline the rocket.

Occasionally, the rocket will contain more than one satellite. The main satellite to be launched is called the *initial payload,* whereas other, smaller satellites are generally referred to as *secondary payloads.* The weight of the main payload and of any secondary payloads is accurately known. In this way, the rocket can be properly fueled and fine-tuned to achieve the desired thrust and speed. Since it is easier to adjust the weight of the payload rather than the thrust, ballast in the form of secondary payloads is added or subtracted as required.

The main booster rocket is ignited and lifts off. The rocket overcomes the earth's gravitational pull and rises slowly. Then a gyroscopic guidance system points the rocket in a specific direction. Typically, the rocket is not launched directly vertically upward. Instead, it is launched at a slight angle that will help the satellite achieve the desired orbit. During this time the rocket and its payload pass through the lower atmosphere of the earth.

If during the launch process the speed of the launch vehicle exceeds 25,000 mi/h, the satellite will not go into orbit. At this speed, the spacecraft will break away from the gravitational pull of the earth and go out into deep space. A spacecraft such as this, of course, is no longer called a satellite. Typically it is called a *space probe.* Examples of space probes are the special satellitelike craft that were constructed to survey the moon, Venus, Mars, and Saturn.

At approximately 50 to 100 mi above the earth, the large booster rocket flames out and

Payload

Space probe

is jettisoned. The second-stage rocket is ignited to carry the satellite farther upward into the higher parts of the atmosphere. At some point this rocket also burns out and is jettisoned. A third-stage rocket may also be required. This third and final stage, although not always used, is sometimes needed to achieve the final speed required to put the satellite into orbit. Also during this period the fairing is jettisoned. This exposes the satellite which no longer requires protection as the upper atmosphere is not harmful.

At some point the final rocket burns out and the satellite begins to coast, ultimately going into orbit. It enters the orbit near the perigee. Refer to Fig. 11-14. This is not the final orbit of the satellite. Instead, the satellite is first put into what is called a *transfer orbit,* a highly elliptical orbit that permits adjustments to the satellite to be made prior to its being placed into final position.

The satellite speed is adjusted to put it into a transfer orbit whose apogee is the final desired height for the geosynchronous satellite, usually 22,300 mi. The satellite is allowed to stay in this transfer orbit for several passes in order to permit various alignment maneuvers. For example, a geosynchronous satellite must be in position over the equator. The initial transfer orbit will not be over the equator because the launch site is typically not located on the equator. As a result, during the transfer orbit, an on-board rocket is fired at the appropriate time to put the orbit over the equator. Typically when a satellite is launched from Cape

Canaveral, it goes into an elliptical orbit with an inclination of 28.5° with respect to the equatorial plane.

After several revolutions in the transfer orbit, the satellite is placed into a circular orbit around the equator by firing the apogee kick motor. This is a small rocket motor built into the satellite and specifically designed to be fired for positioning purposes. The apogee kick motor is fired at the apogee of the elliptical transfer orbit. The satellite is oriented by ground station control with its jet thrusters so that the apogee kick motor fires in the proper direction to move the satellite in the direction that will achieve the desired equatorial orbit. The firing of the apogee kick motor removes the angle of inclination and circularizes the orbit. The satellite is allowed to drive in this orbit for several revolutions.

Transfer orbit

Finally, attitude stabilization is initiated. If spin stabilization is used, thrusters are fired to start the spinning. If three-axis stabilization is used, thrusters are fired until the sensors pick up the external reference points at which time the satellite's internal attitude-control system goes into operation to position the satellite properly. During this time, antennas are properly aligned and solar cell panels are adjusted to produce optimum power. The satellite is allowed to operate for several days and is tracked from the ground. At some point additional firing of the jet thrusters may be necessary to fine-tune the satellite in its final designated position.

NASA's highly successful space shuttle is now routinely used to put satellites into orbit. A separate rocket is not used. Instead, one or more satellites is loaded into the shuttle which is put into orbit. Astronauts deliver the satellites to their approximate location and then spin them out of the cargo bay into orbit. Ground stations make the final attitude and position adjustments.

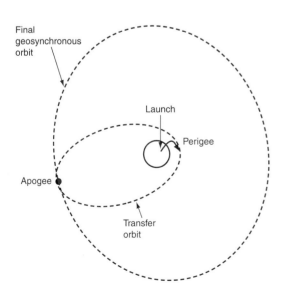

Fig. 11-14 Launching a satellite.

■ TEST _____

Answer the following questions.

1. A satellite is kept in orbit by a balance between two forces: _____.
2. A satellite that revolves in the same direction as the earth rotates is said to be in a _____ (posigrade, retrograde) orbit.
3. The geometric shape of a noncircular orbit is a(n) _____.

4. The center of gravity of the earth is called the _____.

5. The time for one orbit is called the _____.

6. The angle of inclination of a satellite is _____ degrees if it orbits over the equator and is _____ degrees if it orbits over the north and south poles.

7. To prevent excessive signal attenuation and noise in the atmosphere, satellite angles of elevation of less than _____ degrees should be avoided.

8. A satellite that rotates around the equator 22,300 mi from the earth is said to be in _____ orbit.

9. Small jet thrusters on a satellite are fired to correct the satellite's _____.

10. Name two ways satellites are stabilized in space.

11. The three axes of a satellite are _____.

12. The point on the earth directly below a satellite is called the _____.

13. Satellites are located by earth coordinates expressed in terms of _____.

14. The two angles used to point a ground station antenna are _____.

15. A satellite is put into final geosynchronous orbit from its transfer orbit by firing the _____.

Up link

Down link

11-2 Satellite Communications Systems

Transponder

Communications satellites are not originators of information to be transmitted. Although this is the case in some other types of satellites, it is not true for those used for communications purposes. Communications satellites are instead relay stations for other sources. If a transmitting station cannot communicate directly with one or more receiving stations because of line-of-sight restrictions, then a satellite can be used. The transmitting station sends the information to the satellite which, in turn, retransmits it to the receiving stations. The satellite in this application is what is generally known as a *repeater*.

Repeater

Transponders

Figure 11-15 shows the basic operation of a communications satellite. An earth station transmits information to the satellite. The satel-

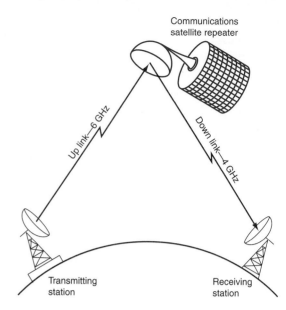

Fig. 11-15 Using a satellite as a microwave relay link.

lite contains a receiver which picks up the transmitted signal, amplifies it, and translates it to another frequency. This new frequency is then retransmitted to the receiving stations back on earth. The original signal being transmitted from the earth station to the satellite is called the *up link,* and the retransmitted signal from the satellite to the receiving stations is called the *down link.* Usually the down-link frequency is lower than the up-link frequency. A typical up-link frequency is 6 GHz, and a common down-link frequency is 4 GHz.

The transmitter-receiver combination in the satellite is known as a *transponder.* The basic function of a transponder is amplification and frequency translation. A simplified block diagram of a transponder is shown in Fig. 11-16. The reason for frequency translation is that the transponder cannot transmit and receive on the same frequency. The transmitter's strong signal would overload the receiver and block out

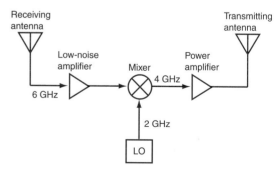

Fig. 11-16 A satellite transponder.

the very small up-link signal, thereby prohibiting any communications. By using widely spaced transmit and receive frequencies, no interference is encountered.

Transponders are also wide-bandwidth units so that they may receive and retransmit more than one signal. Any earth station signal within the receiver's bandwidth will be amplified, translated, and retransmitted on a different frequency.

Even though the typical transponder has a wide bandwidth, it is only used with a single signal to minimize interference and improve communications reliability. However, a satellite would not be an economical repeater if it were capable of handling only a single channel. As a result, most satellites contain multiple transponders each operating at a different frequency. A typical communications satellite has 12, 24, or more transponders. Each transponder represents an individual communications channel. Various multiplexing schemes are used so that each channel may carry multiple information transmissions.

Satellite Frequency Allocations

Most communications satellites operate in the microwave frequency spectrum. However, there are some exceptions to this. For example, many military satellites operate in the 200- to 400-MHz range which is generally considered to be in the UHF range. Also, the amateur radio OSCAR satellites operate in the VHF to UHF range.

The microwave spectrum is divided up into frequency bands which have been allocated to satellites as well as other communications services such as radar. These frequency bands are generally designated by a letter of the alphabet. The table in Fig. 11-17 shows the various frequency bands used in satellite communications.

The most widely used satellite communications band is the C band. The up-link frequencies are in the 5.925- to 6.425-GHz range. In any general discussion of the C band, the up-link frequency is generally said to be 6 GHz. The down-link frequencies are in the 3.7- to 4.2-GHz range, but again, in any general discussion of the C band, the down-link frequency is nominally said to be 4 GHz. Occasionally, you will see the C band referred to by the des-

ignation 6 GHz/4 GHz where the up-link frequency is given first.

Although most satellite communications activity takes place in the C band, there is a steady move toward the higher frequencies. In 1989 the Ku band received the most attention. The up-link frequencies are in the 14- to 14.5-GHz range, and the down-link frequencies are in the 11.7- to 12.2-GHz range. You will see the Ku band designated as 14/12 GHz.

Most new communications satellites will operate in this band. The reason for this shift upward in frequency is that the C band is becoming overcrowded. There are many communications satellites in orbit now, most of them operating in the C band. However, there is some difficulty with interference because of the heavy usage. The only way this interference will be minimized is to shift all future satellite communications to higher frequencies. Naturally, the electronic equipment needed to achieve these higher frequencies is more complex and expensive. Yet, the crowding and interference problems cannot be resolved in any other way. Further, for a given antenna size, the gain is higher in the Ku band than in the C band. This can improve communications reliability while decreasing antenna size and cost.

Two other bands of interest are the X and L bands. The military uses the X band for its satellites and radar, and the L band is used in

Frequency	Band
225–390 MHz	P
350–530 MHz	J
1530–2700 MHz	L
2500–2700 MHz	S
3400–6425 MHz	C
7250–8400 MHz	X
10.95–14.5 GHz	Ku
17.7–21.2 GHz	Kc
27.5–31 GHz	K
36–46 GHz	Q
46–56 GHz	V
56–100 GHz	W

Fig. 11-17 Frequency bands used in satellite communications.

marine and aeronautical communications and radar.

Satellite Bandwidth

Recall the frequencies designated for the C band up link and down link. These are 5925 to 6425 MHz and 3700 to 4200 MHz, respectively. You can see that the bandwidth between the upper and lower limits is 500 MHz. This is an incredibly wide band capable of carrying an enormous number of signals. In fact, 500 MHz covers all the radio spectrum we know so well from VLF through VHF and then some. Most communications satellites are designed to take advantage of this full bandwidth. This allows them to carry the maximum possible number of communications channels. Of course, this extremely wide bandwidth is one of the major reasons why microwave frequencies are so useful in communications. Not only can many communications channels be supported, but very high speed digital data requiring wide bandwidth can also be dealt with.

The transponder receiver "looks at" the entire 500-MHz bandwidth and picks up any transmission there. However, the input is "channelized" because the earth stations operate on selected frequencies or channels. The 500-MHz bandwidth is typically divided into 12 separate transmit channels, each 36 MHz wide. There are 4-MHz guard bands between channels that are used to minimize adjacent-channel interference. These communications channels are illustrated in Fig. 11-18. Note the center frequency for each channel. Remember that the up-link frequencies are translated by frequency conversion to the down-link channel. In both cases, the total bandwidth (500 MHz) and channel bandwidth (36 MHz) are the same. On board the satellite, a separate transponder is allocated to each of the 12 channels.

Although 36 MHz seems narrow compared to 500 MHz, each transponder bandwidth is still capable of carrying an enormous amount of information. For example, one typical transponder can handle up to 1000 one-way analog telephone conversations as well as one full-color TV channel. Each transponder channel can also carry high-speed digital data. Using certain types of modulation, a standard 36-MHz bandwidth transponder can handle digital data at rates up to 60 Mbits/s.

Increasing Channel Capacity

Although the transponders are quite capable, they nevertheless rapidly become overloaded

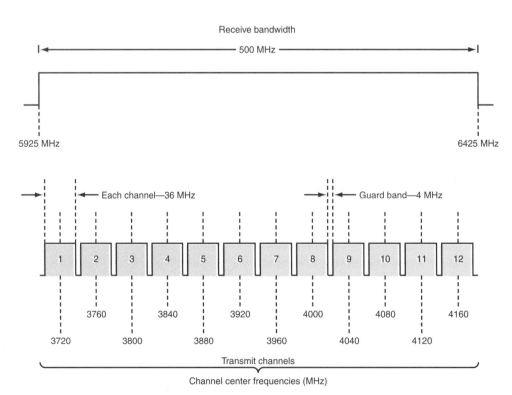

Fig. 11-18 Receive and transmit bandwidths in a C-band communications satellite.

with traffic. Further, there is at times more traffic than there are transponders to handle it. For that reason, numerous techniques have been developed to effectively increase the bandwidth and signal-carrying capacity of the satellite. Two of these techniques are known as frequency reuse and spatial isolation.

Frequency Reuse *Frequency reuse* effectively doubles the bandwidth and information-carrying capacity of a satellite. In this system, a communications satellite is provided with two identical sets of 12 transponders. The first channel in one transponder operates on the same channel as the first transponder in the other set, and so on. With this arrangement, the two sets of transponders transmit in the same frequency spectrum and, therefore, would appear to interfere with one another. However, this is not the case. The two systems, although operating on the same frequencies, are isolated from one another by special antenna techniques.

Recall that any radio signal is a combination of both electric and magnetic fields that exist in quadrature with one another. That is, the magnetic field is 90° offset from the electric field. Further, the electric field determines the so-called polarization of the signal. If the electric field is oriented vertically with respect to the earth, the signal is said to be vertically polarized. If the electric field is horizontal to the earth, the signal is said to be horizontally polarized. Of course, the positioning of the antenna determines whether the signal is vertically or horizontally polarized.

In theory, a signal that is transmitted with vertical polarization will not be received on an antenna that is horizontally polarized. The same is true of a horizontally transmitted signal, which will not be received by a vertically polarized antenna. In practice, perfect vertical or horizontal polarization can typically be received by antennas of either polarization. However, at microwave frequencies, the polarization can be more accurately controlled.

Circular polarization is also used in some types of antennas. There is left-hand and right-hand circular polarization. An antenna using one cannot work with the other.

By using transmitting and receiving antennas that are vertically or horizontally polarized or that use left- or right-hand polarization, two completely separate sets of transponders operating on the same frequency can be used simultaneously. One set of 12 transponders will have a vertically polarized or left-hand circular polarized antenna. The other set will use horizontal or right-hand circular polarization. By careful positioning and orientation of the antennas, one set of signals will not interfere with the other. For example, if two earth stations transmit on the same frequency but with different polarizations, one will be rejected by one transponder but picked up by another. Two transponders transmitting signals on the same frequency but with different polarizations will not interfere with one another as ground station antenna polarization will selectively sort them out. In this way, two 500-MHz bandwidths with signal-carrying capacity can be included in a satellite. The special antenna techniques that permit this will be discussed in more detail in another chapter.

Frequency reuse

Spatial Isolation *Spatial isolation* is another technique whereby additional information-carrying channels can be obtained. With this technique, extra sets of transponders are used. Special antenna techniques are used to isolate the inputs and outputs. In spatial isolation, very narrow beamwidth antennas are used to focus the down-link signals to specific areas of the earth. Such antennas are referred to as spot-beam antennas. By using such antennas on the spacecraft, the signals can be confined to a particular area. In this way, different earth stations can use the same frequencies. They do not interfere with one another because of the highly directional antennas. In this way, the total bandwidth or information-carrying capacity of the satellite can be doubled. For example, a satellite could contain up to four sets of 12 transponders each, all using the same frequencies. However, with frequency reuse techniques as well as spot beams, a total of 48 transponders can be used.

Spatial isolation

Polarization

About ⎯ Electronics

Magellan was the name of a famous explorer. It is also the name of a NASA spacecraft and of a GPS (global positioning system) receiver, now being used to help map the exact position of Darwin Island, one of the Galapagos Islands. The exact coordinates are: 01°40.855′ N, 91°59.963′ W.

Supply the missing information in each statement.

16. The most common use of a satellite is _____.
17. A communications satellite is basically a radio _____.
18. A(n) _____ transmits an up-link signal to the satellite.
19. The signal path from a satellite to a ground station is called the _____.
20. The basic communications electronics unit on a satellite is known as a(n) _____.
21. Most communications satellites operate in the _____ frequency range.
22. The most popular satellite frequency range is 4 to 6 GHz and is called the _____ band.
23. Military satellites often operate in the _____ band.
24. The Ku band extends from _____ to _____ GHz.
25. The bandwidth of a typical satellite is _____ MHz.
26. A typical C band transponder can carry _____ channels, each with a bandwidth of _____ MHz.
27. The term used to refer to the technique of using antenna polarization to separate signals on the same frequency to double the number of channels is _____.
28. Using very narrow beamwidth antennas to isolate signals on the same frequency is known as _____.

11-3 Satellite Subsystems

All satellite communications systems consist of two basic parts: the satellite or spacecraft and one or more earth stations. The satellite performs the function of a radio repeater or relay station. Two or more earth stations may communicate with one another through the satellite rather than directly point to point on the earth.

The heart of a communications satellite is the communications subsystem. This is a set of transponders that receive the up-link signals and retransmit them to earth. A transponder is a repeater that implements a wideband communications channel which can carry many simultaneous communications transmissions.

The transponders are supported by a variety of additional "housekeeping" subsystems. These include the power subsystem, the telemetry tracking and command subsystems, the antennas, and the propulsion and attitude stabilization subsystems. These are essential to the self-sustaining nature of the satellite.

Satellite Architecture and Organization

Figure 11-19 shows a general block diagram of a satellite. All the major subsystems are illustrated. The solar panels supply the electrical power for the spacecraft. They drive regulators that distribute dc power to all other subsystems. And they charge the batteries that operate the satellite during eclipse periods. Both dc-to-dc converters and dc-to-ac inverters are used to supply special voltages to some subsystems.

The communications subsystem consists of multiple transponders. These receive the up-link signals, amplify them, translate them in frequency, and amplify them again for retransmission as down-link signals. The transponders share a common antenna subsystem for both reception and transmission.

The telemetry, tracking, and command subsystem monitors on-board conditions, such as temperature and battery voltage, and transmits this data back to a ground station for analysis. The ground station may then issue orders to the satellite by transmitting a signal to the command subsystem, which then is used to control many spacecraft functions, such as firing the jet thrusters.

The jet thrusters and the apogee kick motor (AKM) are part of the propulsion subsystem. They are controlled by commands from the ground.

The attitude-control subsystem provides stabilization in orbit and senses changes in orientation. It fires the jet thrusters to perform attitude adjustment and stationkeeping maneuvers that keep the satellite in its assigned orbital position.

The main payload on a communications satellite, of course, is the communications subsystem that performs the function of a repeater or relay station. An earth station takes the signals to be transmitted, known as baseband signals, and frequency modulates a microwave

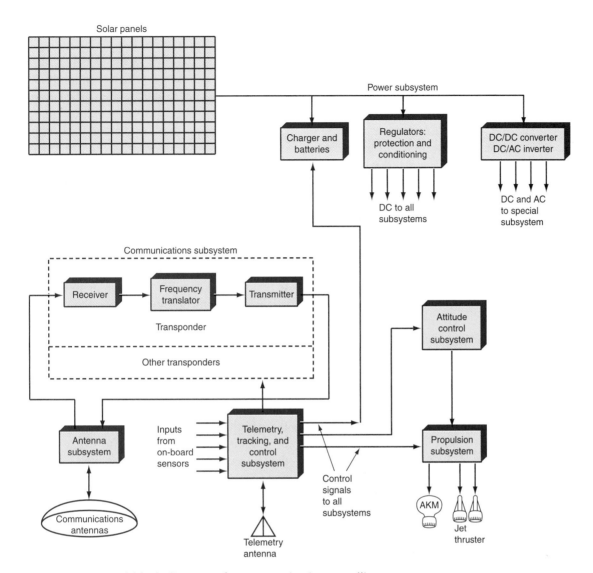

Fig. 11-19 General block diagram of a communications satellite.

carrier. The most common baseband signals are voice, video, and computer data. These up-link signals are then amplified, translated in frequency, and retransmitted on the down link to one or more earth stations. The component that performs this function is known as a transponder. Most modern communications satellites contain at least 12 transponders. More advanced satellites contain many more. These transponders operate in the microwave frequency range.

Communications Subsystem

The basic purpose of a transponder is simply to rejuvenate the up-link signal and retransmit it over the down link. In this role the transponder simply performs the function of an amplifier. By the time the up-link signal reaches the satellite, it is extremely weak. Therefore, it must be amplified before it can be retransmitted to the receiving earth station.

However, transponders are more than just amplifiers. An amplifier is a circuit that does nothing more than increase the voltage, current, or power level of a signal without changing its frequency or content. Such a transponder then would literally consist of a receiver and a transmitter that would operate on the same frequency. Because of the close proximity of the transmitter and the receiver in the satellite, the high transmitter output power for the down link would be picked up by the satellite receiver. Naturally, the up-link signal would be totally obliterated. Further, the transmitter output fed back into the receiver input would cause oscillation.

To avoid this problem, the receiver and transmitter in the satellite transponder are designed to operate at separate frequencies. In this way, they will not interfere with one another. The frequency spacing is made as wide as practical to prevent oscillation and to minimize the effect of the transmitter desensitizing the receiver. In many repeaters, even though the receive and transmit frequencies are different, the high output power of the transmitter can still affect the sensitive receiver input circuits and, in effect, desensitize them, making them less sensitive in receiving the weak up-link signals. The wider the frequency spacing between transmitter and receiver, the less of a problem this desensitizing is.

In typical satellites, the input and output frequencies are separated by huge amounts. At C band frequencies, the up-link signal is around 6 GHz, and the down-link signal is around 4 GHz. This 2-GHz spacing is sufficient to eliminate most problems. However, to ensure maximum sensitivity and minimum interference between up-link and down-link signals, the transponder contains numerous filters, which not only provide channelization, but also help to eliminate interference from external signals, regardless of their source.

There are three basic transponder configurations used in communications satellites. They are all essentially minor variations of one another, but each has its own advantages and disadvantages. These are the single-conversion, double-conversion, and regenerative transponders. Let's take a look at each.

Figure 11-20 shows a basic block diagram of a *single-conversion transponder*. The term "single conversion" refers to the fact that only a single-frequency translation process from the received signal to the transmitted signal takes place within the satellite.

The up-link signal is picked up by the receiving antenna and is first routed to a low-noise amplifier (LNA), usually a GASFET circuit. The signal is very weak at this point, even though it has been multiplied somewhat by the gain of the

receive antenna. Amplifiers with an extremely low noise figure or noise temperature must be used to increase the level of the signal.

Once the up-link signal has been amplified, it is translated in frequency. This is done in a mixer circuit. Any original modulation is retained during the frequency translation process. The mixer output is then amplified again and fed to a bandpass filter. The purpose of the bandpass filter is to remove all but the desired down-link signal of 4 GHz. Another function of the bandpass filter is to channelize the output. This means that only input signals on a specific frequency will be accepted by the bandpass filter. The LNA will increase their level, the mixer will translate them into the down-link frequency, and the amplifier following the mixer will further increase their level. However, if the signal falls out of the range of the filter, it will be rejected. The bandpass filters simply set up portions of the spectrum designated as channels.

Finally, the down-link signal is amplified by a high-power amplifier (HPA), usually a traveling-wave tube (TWT). Its output is further filtered at the channel frequencies to eliminate harmonics and intermodulation products. Intermodulation products are spurious, unwanted signals that are generated as the result of nonlinearities in the TWT. The input signals and any harmonics can be mixed together, producing many sum and difference frequencies unrelated to the desired signal. Filtering is used to eliminate them.

The resulting output is fed to the down-link antenna. In some cases, both the receive and transmit antennas are one and the same. A diplexer is used to keep the signals separate. The *diplexer* is a waveguide assembly that allows one antenna to be shared by a transmitter and a receiver.

Figure 11-21 shows a simplified block diagram of a *double-conversion transponder*. All the basic circuits of this system are similar in

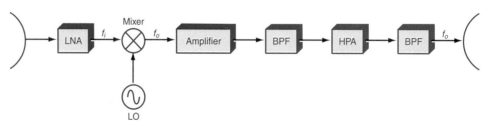

Fig. 11-20 A single-conversion transponder.

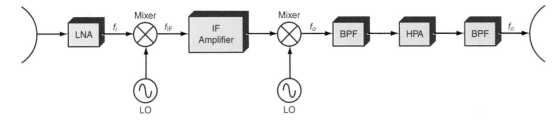

Fig. 11-21 A double-conversion transponder.

function to those of the single-conversion repeater. The only difference is that two frequency conversions are carried out.

Again, the up-link signal is picked up by the antenna, amplified by the LNA, and fed to the first mixer. The mixer translates the incoming signal into an IF. Typical IFs are 70 and 150 MHz. The IF signal out of the first mixer is fed to an IF amplifier where very high gain is achieved.

The output of the IF amplifier is fed to another mixer which translates the signal to the output frequency. Then the bandpass filter eliminates the unwanted mixed output and provides channelization. The HPA increases the signal level, and the output bandpass filter again provides for the elimination of harmonics and intermodulation components. Finally, the transmit antenna sends the signal on its way over the down link.

The main advantage of the double-conversion transponder is that greater flexibility in filtering and amplification can be achieved. Amplification and selectivity at the lower IF level is far easier to attain.

In some satellites using multiple transponders, there are switching arrangements for taking the input of one channel and cross-connecting it to the output of another channel. When this is done, it is frequently done at the IF level since switching is simpler at these lower frequencies.

A third configuration of a satellite transponder is the *regenerative repeater.* Like the other configurations, it performs the basic amplification and frequency translation functions. But in this case, the received signal is actually demodulated. Therefore, remodulation is necessary to create the down-link signal.

The basic block diagram of a regenerative transponder is shown in Fig. 11-22. The up-link signal is amplified and frequency-translated as before. However, the output is then demodulated. The output of the demodulator, of course, is the baseband signal.

A *baseband signal* refers to the basic information content transmitted on the up-link carrier. The baseband signal may be a voice telephone conversation, a TV signal, or digital data. The output of the demodulator, therefore, in the satellite is this baseband signal. Some amplification may be provided.

Next, the baseband signal is fed to a modulator along with a carrier at the down-link frequency. The modulation is usually FM. The output of the modulator is a new carrier containing the same information. The signal is then amplified, filtered, and transmitted over the down link.

This regenerative configuration has several basic advantages over the other configurations. First, it permits both the receiver and transmitter sections of the transponder to be optimized. In the other configurations, there is essentially no difference between the transmitter and the receiver. The transponder is a combination of both. You can't really tell where the receiver ends and the transmitter begins. In a regenerative repeater, the division line is clear since the receiver ends at the demodulator output. The

Baseband signal

Regenerative repeater

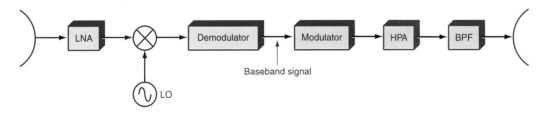

Fig. 11-22 A regenerative transponder.

transmitter begins at the modulator input. Clearly this separation permits each section to be fully optimized in performance without worrying about the interaction that normally occurs in more conventional configurations.

The amplification in a regenerative repeater is also easier to obtain at the very low baseband frequencies. The circuits are simpler and less expensive.

Further, the S/N ratio of the transponder overall can be improved using this technique. It has been shown that an S/N ratio increase of 2 to 3 dB is achievable over transponders with more conventional arrangements.

Finally, in multitransponder satellites with input and output switching, the regenerative arrangement is simpler and more flexible to implement. It is a lot easier to switch baseband signals than it is to switch higher-frequency signals. This flexibility allows a wider variety of switching options to be implemented.

Multichannel Systems

Virtually all modern communications satellites contain multiple transponders. This permits many more signals to be received and transmitted. A typical commercial communications satellite contains 12 transponders, or 24 if frequency reuse is incorporated. Military satellites often contain fewer transponders, whereas the newer, larger commercial satellites have provisions for up to 50 channels. Each transponder operates on a separate frequency, but its band-

width is wide enough to carry multiple channels of voice, video, and digital information.

There are two basic multichannel architectures in use in communications satellites. One is a broadband system, and the other is a fully channelized system.

As indicated earlier, a typical communications satellite spectrum is 500-MHz wide. This is usually divided into 12 separate channels, each with a bandwidth of 36 MHz. The center frequency spacing between adjacent channels is 40 MHz, thereby providing a 4-MHz spacing between channels to minimize adjacent-channel interference. Refer back to Fig. 11-18 for details. A wide-bandwidth repeater is designed to receive any signal transmitted within the 500-MHz total bandwidth. Such a transponder is shown in Fig. 11-23.

The receive antenna is connected to an LNA as in every transponder. Very wideband tuned circuits are used so that the entire 500-MHz bandwidth is received and amplified. A mixer translates all incoming signals into their equivalent lower down-link frequencies. In a C band communications satellite, the incoming signals are located between 5.925 and 6.425 MHz. An LO operating at the frequency of 2.225 GHz is used to translate the input signals down into the 3.7- to 4.2-GHz range. A wideband amplifier following the mixer amplifies this entire spectrum.

The channelization process occurs in the remainder of the transponder. For example, in a 12-channel satellite, 12 bandpass filters, each

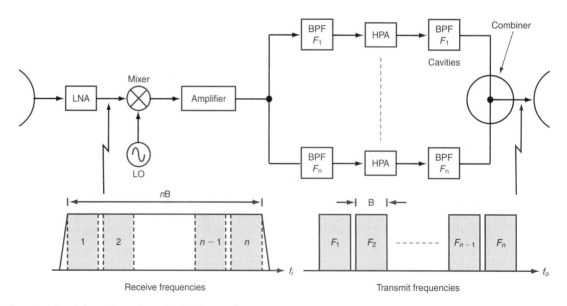

Fig. 11-23 A broadband multiple-channel repeater.

centered on one of the 12 channels, is used to separate all the various received signals. Figure 11-18 shows the 12 basic channels with their center frequencies, each having a bandwidth of 36 MHz. The bandpass filters separate out the unwanted mixer output signals and retain only the difference signals. Then, individual HPAs are used to increase signal level. These are usually TWTs. The output of each TWT amplifier is again filtered to minimize harmonic and intermodulation distortion problems. These filters are usually part of a larger assembly known as a multiplexer or combiner. In any case, it is a waveguide–cavity-resonator assembly that filters and combines all the signals for application to a single antenna.

It is logical to assume that if the receive function can be accomplished by wideband amplifier and mixer circuits, why isn't it possible to provide the transmit function in the same way? The answer is that it is generally not possible to generate the very high output power over such a wide bandwidth. In other words, there are no components and circuits currently available that do this well. The HPAs in most transponders are TWTs, which inherently have limited bandwidth. They operate well over a small range but cannot deal with the entire 500-MHz bandwidth allocated to a satellite. Therefore, in order to achieve the necessary high power levels, the channelization process is used.

Another transponder architecture used in communications satellites uses individual narrowband input channels instead of the single wideband input described previously. This arrangement is illustrated in Fig. 11-24. The receive antenna feeds a demultiplexer, a waveguide assembly, that separates the single wideband input into equal feeds for separate channels. Separate LNAs and bandpass filters centered on the designated channels are used at the inputs. The filters sort out the input signals, and the LNAs provide the desired amount of gain. Although each bandpass filter contributes to the noise level, this is offset by the narrowband operation. Keep in mind that the wider the bandwidth of a channel, the higher its noise content. By limiting the bandwidth, noise decreases. Therefore, with the individual channel arrangement, some improvement in noise temperature is achieved.

Each channel also has its own mixer, driven by a common LO that has a frequency of 2.225 GHz in the C band. The mixer outputs drive bandpass filters which, in turn, drive the HPA or TWT output amplifiers. Again, bandpass filters are used at the output of the TWTs. The output bandpass filters are part of a multiplexer-combiner unit which creates one feed for the transmitting antenna.

Power Subsystem

A key component of the satellite is its power subsystem. Figure 11-25 shows a block diagram of the *power subsystem.* Everything on board operates electrically or has electrical components in it. Most satellites, therefore, depend entirely upon their power supplies for success. The early satellites used on-board batteries to power everything, but batteries were

Power subsystem

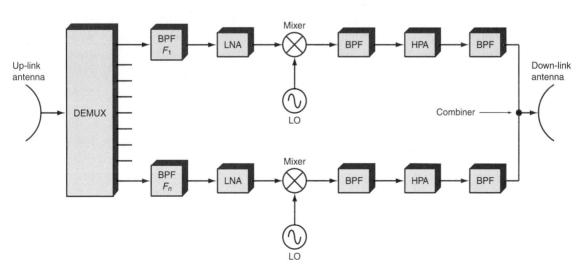

Fig. 11-24 Multichannel receiver transponder.

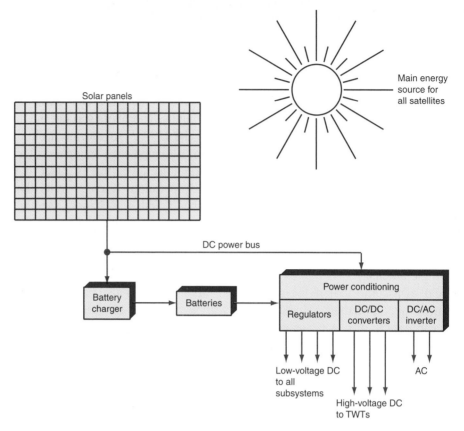

Solar panels

Main energy
source for
all satellites

DC power bus

Battery
charger

Batteries

Power conditioning

Regulators | DC/DC
converters | DC/AC
inverter

Low-voltage DC
to all
subsystems

High-voltage DC
to TWTs

AC

Fig. 11-25 Block diagram of the power subsystem.

quickly exhausted and could not be replaced. The life of a satellite is dependent strictly upon the quality of the batteries used and, of course, on the amount of power drain.

Today virtually every satellite uses solar panels for its basic power source. These solar panels are large arrays of photocells connected in various series and parallel circuits to create a powerful source of direct current. Early solar panels could generate hundreds of watts of power. Today huge solar panels are capable of generating many kilowatts. A key requirement is that the solar panels always be pointed toward the sun. The two basic satellite configurations shown in Fig. 11-26 use different methods. In cylindrically shaped satellites, the solar cells surround the entire unit and, therefore, some portion of them is always exposed to sunlight. In body-stabilized or three-axis satellites, individual solar panels are manipulated with various controls to ensure that they are correctly oriented with respect to the sun.

The solar panels generate a direct current which is used to operate the various components of the satellite. However, the dc power is

typically used to charge nickel-cadmium batteries that act as a buffer. At times when the satellite goes into an eclipse or when the solar panels are not properly positioned, the batteries take over temporarily and keep the satellite operating. The batteries aren't large enough to power the satellite for a long period of time; they are simply used as a backup system for eclipses, initial satellite orientation and stabilization, or emergency conditions.

The basic dc voltage from the solar panels is then conditioned in various ways. For example, it is typically passed through voltage regulator circuits before being used to power individual electronic circuits. Most electronic equipment works best with fixed, stable voltages; therefore, regulators are incorporated in most satellite systems.

Occasionally, voltages higher than those produced by the solar panels must also be generated. For example, the TWT amplifiers in most communications transponders require thousands of volts for proper operation. Special dc-to-dc converters are used to translate the lower dc voltage of the solar panels to the higher dc voltage required by the TWTs.

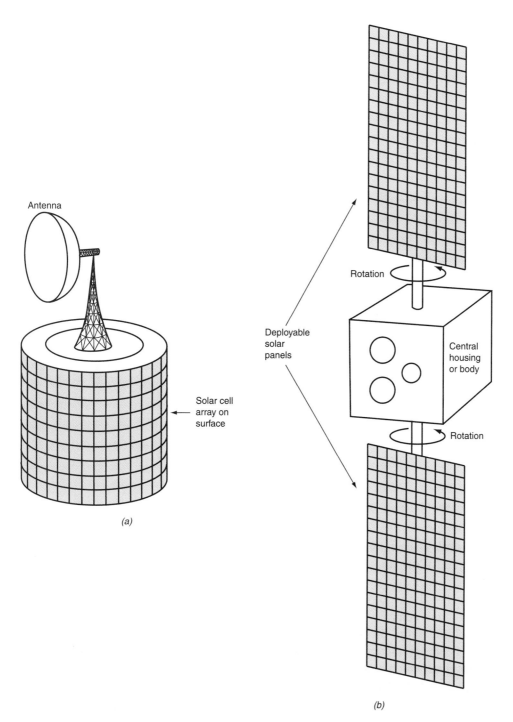

Antenna

Rotation

Deployable
solar
panels

Central
housing
or body

Solar cell
array on
surface

Rotation

(a)

(b)

Fig. 11-26 Satellite solar array configurations. (a) Spin-stabilized satellite, and
(b) body-stabilized satellite.

Telemetry, Tracking, and Control Subsystem

All satellites have a *telemetry, tracking, and control (TTC)* subsystem that allows a ground station to monitor and control conditions in the satellite. The telemetry system is used to report the status of the on-board subsystems to the ground station. A typical system is shown in Fig. 11-27. The telemetry system typically consists of various electronic sensors for measuring temperature, radiation level, power supply voltages, and other key operating characteristics. Both analog and digital sensors may be used. The sensors are selected by a multiplexer and are converted to a digital signal, which then modulates an internal transmitter. This transmitter sends the telemetry information

Telemetry, tracking and control (TTC)

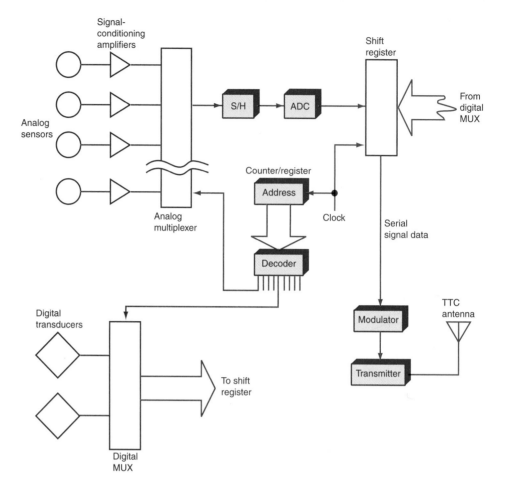

Fig. 11-27 General block diagram of a satellite telemetry unit.

back to the earth station where it is recorded and monitored. With this information, the ground station is then able to determine the operational status of the satellite at all times.

A command and control system permits the ground station to control the satellite. Refer to Fig. 11-28. Typically the satellite contains a command receiver which receives control signals from an earth station transmitter. The control signals are typically made up of various digital codes that tell the satellite what to do. Various commands may initiate a telemetry sequence, activate thrusters for attitude correction, reorient an antenna, or perform other operations as required by the special equipment specific to the mission. Usually, the control signals are processed by an on-board computer.

Most satellites now contain a small digital computer, usually microprocessor-based, that acts as a central control unit for the entire satellite. The computer contains a built-in ROM with a master control program. This master control program operates the computer and causes all other subsystems to be operated as required. The command and control receiver typically passes on new command codes that it receives from the ground station to the computer which then carries out the desired action.

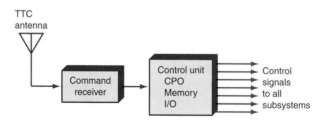

Fig. 11-28 The command receiver and controller.

The computer may also be used to make necessary computations and decisions. Information collected from the telemetry system may be first processed by the computer before it is sent to the ground station. The memory of the computer may also be used to store data temporarily before the data is processed or transmitted back to earth. The computer may also serve as an event timer or clock. The computer is a versatile control element that can typically be reprogrammed via the command and control system to carry out any additional functions that may be required, particularly those that were not properly anticipated by those designing the mission.

Attitude Control Subsystem

All satellites contain an attitude-stabilization subsystem with built-in jet thrusters for making corrections to the attitude of the satellite. By way of the command and control subsystem and the on-board computer, the *attitude-stabilization subsystem* receives signals that fire the appropriate thrusters at the proper time to achieve an adjustment in attitude. This may be to reorient antennas, maximize exposure of the solar panels to the sun, or make corrections indicated by photo- or infrared sensors that provide orientation input for a three-axis stabilized satellite. The attitude-stabilization system may also consist of gyroscopes, flywheels for spin stabilization, and antenna de-spin mechanisms.

In addition to the jet thrusters for attitude stabilization, most satellites also contain a *propulsion system.* A propulsion system is usually the apogee kick motor used to put the satellite into final orbit, or it may be one or more liquid or solid propellant rockets that could be used to change the orbit of a satellite or remove the satellite from orbit. The propulsion system is also operated by the on-board computer in response to the command control subsystem.

Antennas

All satellites have one or more antennas used for receiving signals from the ground station and transmitting information back to earth. Some antennas have a dual function. Numerous antennas are generally required because of the various satellite requirements. For example, one antenna may be used for communications relay purposes, and another antenna may be dedicated to the telemetry, tracking, and control functions. Most of these are highly directional gain antennas that must be accurately pointed. However, most satellites contain an omnidirectional antenna that is used on the command receiver so that the satellite may pick up ground control signals during launch, orbit transfer, and other periods prior to full attitude stabilization.

Applications Payload

All satellites have an applications subsystem made up of special components that make the satellite useful for the purpose intended. For a communications satellite, this subsystem is the transponders.

For a scientific satellite designed to gather information about space, a special instrumentation package will be included. This would be made up of various sensors, their signal-conditioning electronics, and additional telemetry equipment to transmit the information back to earth. The system may also include a tape recorder for temporarily storing information and transmitting it back to earth at a time to be determined later by the ground station. Nonsynchronous satellites often use recorders since data will be gathered during the times the ground station has no access to the satellite. The recorder captures data for later transmission.

An observation satellite such as those used for intelligence gathering or weather monitoring may use television cameras or infrared sensors to pick up various conditions on earth and in the atmosphere. This information is then transmitted back to earth by a special transmitter designed for this purpose. And, of course, there are many other variations of this subsystem, depending upon what it will be used for.

Attitude stabilization subsystem

Propulsion system

■ TEST

Answer the following questions.

29. The most common baseband signals handled by a satellite are _____.
30. The input circuit to a transponder is the _____.

31. List the three main functions of a transponder.
32. A(n) _____ circuit in the transponder performs the frequency conversion.
33. Power amplification in a transponder is usually provided by a(n) _____.
34. A(n) _____ transponder provides amplification at an IF of 70 or 150 MHz.
35. A transponder that demodulates the baseband signals and then remodulates a carrier is known as a(n) _____ transponder.
36. The circuit that provides channelization in a transponder is the _____.
37. The two basic architectures for multichannel transponders are _____.
38. The main power supplies in a satellite are the _____.
39. During an eclipse, the satellite is powered by _____.
40. The TTC subsystem is used to _____ operations on a satellite.
41. Attitude correction is made by firing _____.

11-4 Earth Stations

The earth station or ground station is the terrestrial base of the system. The earth station communicates with the satellite to carry out the designated mission. The earth station may be located at the end user's facilities or may be located remotely with ground-based intercommunication links between the earth station and the end user. In the early days of satellite systems, earth stations were typically situated in remote country locations. Because of their enormous antennas and other critical requirements, it was not practical to locate them in downtown or suburban areas. Today earth stations are much less complex and the antennas are smaller. Many earth stations are now located on top of tall buildings or in other urban areas directly where the end user resides. This offers the advantage of eliminating complex intercommunications systems between the earth station and the end user.

Ground Station Organization

Like the satellite, the earth station is made up of a number of different subsystems. The subsystems, in fact, generally correspond to those on board the satellite. However, most of the subsystems are larger and much more complex. Further, several additional subsystems exist at earth stations, that are not applicable to the satellite itself.

An earth station consists of five major subsystems. These are the antenna subsystem, the receive subsystem, the transmit subsystem, the ground communications equipment (GCE) subsystem, and the power subsystem. A general block diagram of these is illustrated in Fig. 11-29. Not shown here are the telemetry, control, and instrumentation subsystems.

The antenna subsystem, of course, consists of the parabolic reflector, the horn antennas, waveguide transmission line, and the related mounts. The receive subsystem consists of the LNA and filters as well as any power dividers and related circuitry. Many earth stations have multiple circuits for redundancy and reliability. These are switched on automatically if the main unit fails.

The transmit subsystem consists of the HPA and any intermediate driver stages plus related filters. The GCE consists of both receiving and transmitting circuits. The receiving portion consists of down converters, filters, demodulators, and demultiplexing equipment. The transmit portion of the GCE consists of multiplexers, modulators, up converters, and the related filters. Connections to the telephone system, terrestrial microwave relay links, computer interfaces, and so on are made through the GCE.

The power subsystem furnishes all the power to the other equipment. The primary sources of power are the standard ac power lines. The subsystem operates power supplies which distribute a variety of dc voltages to the other equipment. The power subsystem also consists of emergency power sources such as diesel generators, batteries, and inverters to ensure continuous operation during power failures.

Refer back to Fig. 11-29. The antenna subsystem usually includes a diplexer, a waveguide assembly that permits both the transmitter and the receiver to use the same antenna. The diplexer feeds a bandpass filter in the receiver section that ensures that only the received frequencies pass through to the sensitive receiving circuits. This bandpass filter blocks the high-power transmit signal which can occur simultaneously with reception. This prevents overload and damage to the receiver.

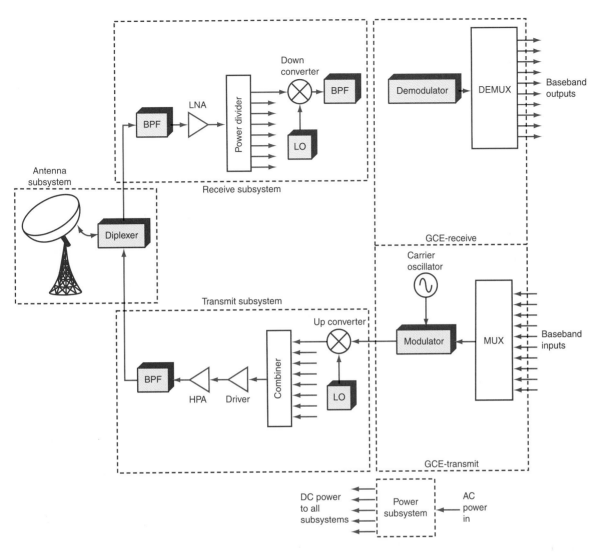

Fig. 11-29 General block diagram of an earth station.

The output of the bandpass filter feeds an LNA. This LNA drives a power divider. This is a waveguidelike assembly that splits the received signal into smaller but equal power signals. The power divider feeds several down converters. These are standard mixers fed by LOs that translate the received signals down to an IF, usually of 70 MHz. A bandpass filter ensures the selection of the proper sidebands out of the down converter.

The IF signal containing the data is then sent to the receive GCE where it is demodulated and fed to a demux where the original signals are finally obtained. The demux outputs are usually the baseband or original communications signals. In actual systems, several levels of demodulation and demultiplexing may have to take place to obtain the original signals.

In the GCE transmit subsystem, the baseband signals such as telephone conversations are applied to a multiplexer which permits multiple signals to be carried on a single channel. The multiplex signals are applied to a modulator along with the carrier oscillator. The modulator output is usually at the 70-MHz IF. Several levels of multiplexing and modulation may be used depending upon the application. This IF is fed to the transmit subsystem where it is translated to the final output frequency by way of an up-converter circuit. The up converter is a mixer with its LO tuned to generate the correct carrier channel. All the up-converted signals are then applied to a power combiner. This waveguidelike device equally combines all the RF signals into the final signal to be transmitted. This feeds a driver circuit which has sufficient power to operate the final amplifier or HPA. This is either a TWT or klystron. The output signal is applied to a bandpass filter and then to the diplexer and antenna.

All earth stations have a relatively large parabolic dish antenna that is used for sending and receiving signals to and from the satellite. Early satellites had very low-power transmitters, and, therefore, the signals received on earth were extremely small. Huge high-gain antennas were required to pick up minute signals from the satellite. The earth station "dishes" were 80 to 100 ft or more in diameter. Antennas of this size are still used in some satellite systems today, and even larger antennas have been used for deep-space probes.

Modern satellites now transmit with much more power. Advances have also been made in receiver components and circuitry. For that reason, smaller earth station antennas are now practical. In some applications, antennas as small as 2 or 3 ft in diameter can be used.

Typically, the same antenna is used for both transmitting and receiving. A diplexer, a special coupling device for microwave signals, permits a single antenna to be used for multiple transmitters and/or receivers. In some applications, a separate antenna is used for telemetry and control functions.

The antenna in an earth station must also be steerable. That is, it must be possible to adjust the antenna's azimuth and elevation so that the antenna can be properly aligned with the satellite. Earth stations supporting geosynchronous satellites can generally be fixed in position. However, azimuth and elevation adjustments are necessary to initially pinpoint the satellite and to permit minor adjustments over the satellite's life.

Satellites in nonstationary orbits must be tracked. Earth station antennas for such systems incorporate large, sophisticated tracking systems in which both the azimuth and elevation of the antenna is adjusted by large motors that are part of a servo system. The servo system is driven by the tracking system. The tracking system may be nothing more than a timer which operates in synchronism with the satellite orbit. The servo motors are driven continuously through gear trains so that the antenna is always pointed at the satellite. A computer may be used for the controlling purposes if it knows the satellite orbit characteristics. The tracking system can also be based on signals received by the satellite. When the satellite comes into view, its signal is picked up and its amplitude is used to drive the tracking system. The system is optimized so that the antenna adjustments are made to continuously seek the maximum amplitude signal from the satellite. A beacon transmitter on the satellite is used for transmitting the tracking signal.

Receive Substation

The down link is the receive subsystem of the earth station. It usually consists of very low noise preamplifiers that take the small signal received from the satellite and amplify it to a level suitable for further processing. The signal is then demodulated and sent on to other parts of the communications system.

The receive subsystem consists of the LNA, down converters, and related components. The purpose of the receive subsystem is to amplify the down-link satellite signal and translate it to a suitable IF. From that point, the IF signal is demodulated and demultiplexed as necessary to generate the original baseband signals.

Refer back to the general block diagram of the receive subsystem shown in Fig. 11-29. The output from the diplexer or circulator in the antenna subsystem feeds the RF signal to a bandpass filter (BPF). The purpose of this filter is to pass only the receive signal and to block any transmitted energy. Although the circulator or diplexer generally provides sufficient isolation between the received and transmitted signals, the filter provides additional protection of the receiver circuits.

The filter output is applied to an LNA. This provides initial amplification of the down-link signal. Typically, several stages of amplification are used to raise the tiny down-link signal to an amplitude suitable for use with the remaining circuits.

The output of the LNA is fed to a down-converter circuit consisting of a mixer and an LO. The down converter translates the RF signal to an IF where it will be dealt with further by the ground communications equipment.

The next stage in a satellite earth station receiver is a down converter which translates the microwave signal down to one or more IFs. The IF signal containing the original modulation is then sent to the GCE where it is demodulated and demultiplexed.

There are basically two types of down converters used in earth station receivers: single

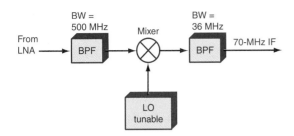

Fig. 11-30 Single-conversion down converter.

conversion and double conversion. A single-conversion circuit is shown in Fig. 11-30. The broadband down-link satellite signal is amplified by the LNA and is fed to a mixer circuit which is combined with an LO frequency. The resulting output is an IF of 70 MHz, the standard frequency used in most earth stations. The mixer itself may be a Schottky or a hot-carrier diode or in newer systems a GaAs MESFET. The LO is set to a frequency that will pick out one of the numerous channels in the wideband satellite signal. After the mixer is a bandpass filter which filters out any unwanted components from the mixer, including images. For example, to select 1 of the 12 channels in a standard 500-MHz bandwidth satellite signal, the LO would be tuned to a frequency of 70 MHz above or below the input channel carrier desired. The bandpass filter has a bandwidth of 36 MHz to pass the modulated and multiplexed signals on this channel.

In practice, single-conversion down converters are rarely used. The reason for this is image rejection problems and the difficulty of tuning or channel selection. To solve these problems, dual-conversion down converters are normally used.

Figure 11-31(a) shows a diagram of a typical dual-conversion down converter. The input bandpass filter, of course, passes the entire 500-MHz satellite signal. This is fed to a mixer along with an LO. The output of the mixer is an IF signal, usually 770 MHz. This is passed through a bandpass filter at that frequency with a bandwidth of 36 MHz.

The signal is then applied to another mixer. When combined with the LO frequency, the mixer output is the standard 70-MHz IF value. An IF of 140 MHz is used in some systems. After the mixer is a 36-MHz-wide bandpass filter that passes the desired channel.

In dual-conversion down converters, two different tuning or channel selection arrangements are used. One is referred to as RF tuning, and the other is referred to as IF tuning. In RF tuning, shown in Fig. 11-31(a), the first LO is made adjustable. Typically, a frequency synthesizer is used in this application. The frequency synthesizer generates a highly stable signal at selected frequencies whose increment can be changed. The synthesizer is set to a frequency that will select a desired channel. In RF tuning, the second LO has a fixed frequency to achieve the final conversion.

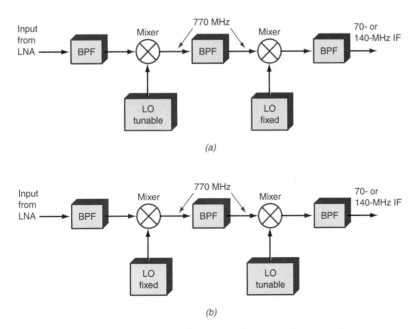

(a)

(b)

Fig. 11-31 Dual-conversion down converters. (a) RF tuning, and (b) IF tuning.

In IF tuning, the first LO is fixed in frequency, and the second is made tunable. Again, a frequency synthesizer is normally used for the tunable LO. See Fig. 11-31(b).

Figure 11-32 shows a full set of down converters as used in a standard earth station receiver. The amplified signal from the LNA is fed to a power divider circuit or branching network. This is either a waveguidelike device or in some cases a special stripline assembly that divides a single signal into multiple signals. The signals are identical to the received signal except that their amplitude is lower. The power dividers simply provide a means of allocating some of the amplified received signal to each of the independent down-converter circuits. It also provides necessary impedance matching.

The receiver GCE consists of one or more racks of equipment used for demodulating and demultiplexing the received signals. The down converters provide initial channelization by transponder, and the demodulators and demul-

tiplexing equipment processes the 70-MHz IF signal into the original baseband signals. Other intermediate signals may be developed as required by the application.

The outputs from the down converters are usually made available on a patch panel of coax connectors. See Fig. 11-32. These are interconnected via coax to the demodulators. The demodulators are typically packaged in a thin, narrow vertical module that plugs into a chassis in a rack. Many of these demodulators are provided. They are all identical in that they demodulate the IF signal. In FDM systems, each demodulator is an FM detector. The most commonly used type is the PLL discriminator. Equalization and de-emphasis are also taken care of in the demodulator.

In systems using TDM, the demodulators are typically those used to detect four-phase or quadrature PSK at 60 or 120 Mbits/s. (*Note:* These circuits are covered in Chap. 12.) The IF is usually at 140 MHz. Again, a patch panel

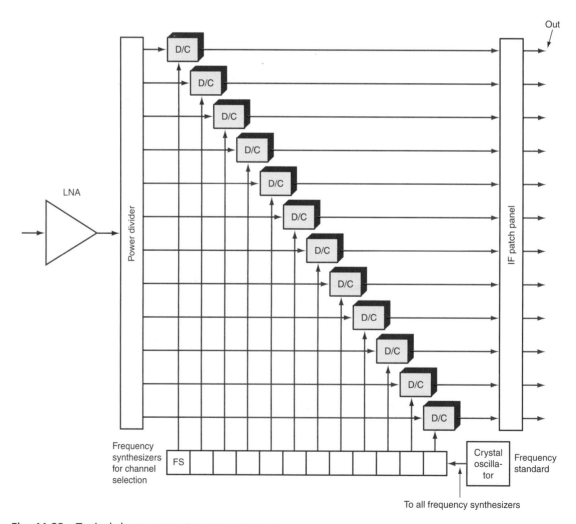

Fig. 11-32 Typical down-converter arrangement.

between the down converters and the demodulators permits flexible interconnection to provide any desired configuration. When video data is transmitted, the output of the frequency demodulator is the baseband video signal. It may then be transmitted by cable or used on premises as required.

If the received signals are telephone calls, then the demodulator outputs are sent to demultiplexing circuits. See Fig. 11-33. Again, in many systems, a patch panel between the demodulator outputs and the demultiplexer inputs is provided to make various connections as may be required. In FDM systems, standard frequency division multiplexing is used. This consists of additional SSB demodulators and filters. Depending upon the number of signals multiplexed, several levels of channel filters and SSB demodulators may be required to generate the original baseband voice signals. Once this is achieved, the signal may be transmitted over the standard telephone system network as necessary.

In TDM systems, time division demultiplexing equipment is used to reassemble the original transmitted data. The original baseband digital signals may be developed in some cases; in other cases these signals are used with modems as required for interconnecting the earth station with the computer that will process the data.

Transmit Substation

The up link is the transmitting subsystem of the earth station. It consists of all the electronic equipment that takes the signal to be transmitted, amplifies it, and sends it to the antenna. In a communications system, the signals to be sent to the satellite might be TV programs, multiple telephone calls, or digital data from a computer. These signals are used to modulate the carrier which is then amplified by a large TWT or klystron amplifier. Such amplifiers generally generate many hundreds of watts of output power. This is sent to the antenna by way of microwave waveguides, combiners, and diplexers.

The transmit subsystem consists of two basic parts: the up converters and the power amplifiers. The up converters translate the baseband signals modulated onto carriers up to the final up-link microwave frequencies. The

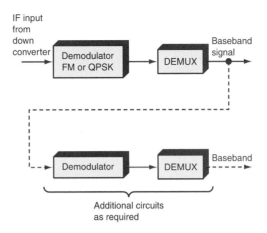

Fig. 11-33 The GCE for receiving.

power amplifiers generate the high-power signals that are applied to the antenna. The modulated carriers are created in the transmit GCE.

The transmit subsystem begins with the baseband signals. These are first fed to a multiplexer if multiple signals are to be carried by a single transponder. Telephone calls are a good example. Frequency or time division multiplexers are used to assemble the composite signal. The multiplexer output is then fed to a modulator. In analog systems, a wideband frequency modulator is normally used. It operates at a carrier frequency of 70 MHz with a maximum deviation of ± 18 MHz. Video signals are fed directly to the modulator; they are not multiplexed.

In digital systems, analog signals are first digitized with PCM converters. The resulting serial digital output is then used to modulate a QPSK modulator. The carrier frequency for these modulators is usually 140 MHz. Figure 11-34 shows the general transmit GCE configuration.

Once the modulated IF signals have been generated, up conversion and amplification will take place prior to transmission. Individual up converters are connected to each modulator output. Each up converter is driven by a frequency synthesizer which allows selection of the final transmitting frequency. The frequency synthesizer selects the transponder it will use in the satellite. The synthesizers are ordinarily adjustable in 1-kHz increments so that any up converter can be set to any channel frequency or transponder.

As in down converters, most modern systems use dual conversion. Both RF and IF

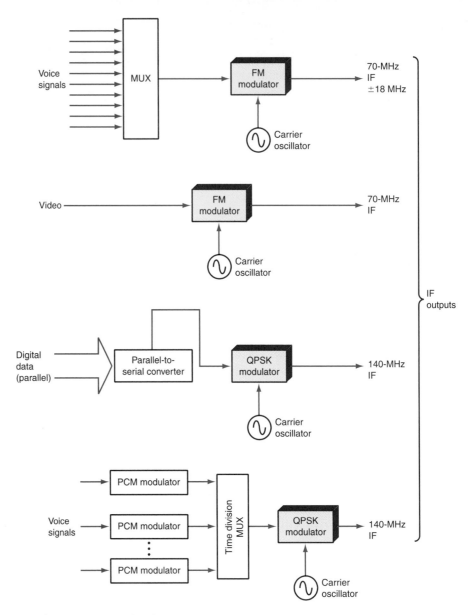

Fig. 11-34 Typical GCE transmit circuits.

tuning are used. Refer to Fig. 11-35. With IF tuning, a tunable carrier from a frequency synthesizer is applied to the mixer to convert the modulated signal to an IF level, which is usually 700 MHz. Another mixer fed by a fixed-frequency LO then performs the final up conversion to the transmitted frequency.

In RF tuning, a mixer fed by a fixed-frequency LO performs an initial up conversion to 700 MHz. Then, a sophisticated RF synthesizer applied to a second mixer provides up conversion to the final microwave frequency.

In some systems, all the IF signals at 70 or 140 MHz are combined prior to the up conversion. In this system, different carrier frequencies are used on each of the modulators to provide the desired channelization. The carrier frequencies when translated by the up converter will translate to the individual transponder center frequencies. A special IF combiner circuit mixes all the signals linearly and applies them to a single up converter. This up converter creates the final microwave signal.

In most systems, however, individual up converters are used on each modulated channel. At the output of the up converters, all the signals are combined in a microwave combiner which produces a single-output signal which will be fed to the final amplifiers. This arrangement is illustrated in Fig. 11-36.

The final combined signal to be transmitted to the satellite appears at the output of the RF

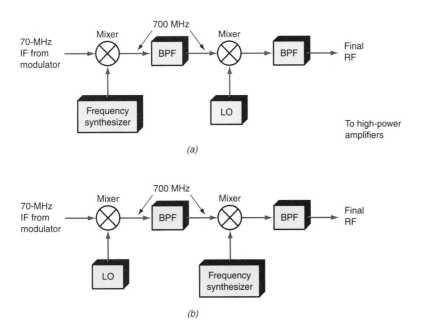

Fig. 11-35 Typical up-converter circuits. (a) IF tuning, and (b) RF tuning.

combiner, but it must first be amplified considerably before being sent to the antenna. This is done by the power amplifier. The power amplifier usually includes an initial stage called the intermediate power amplifier (IPA). This provides sufficient drive to the final, the HPA. See Fig. 11-36. Note also in Fig. 11-36 that redundant amplifiers are used. The main IPA and HPA are used until a failure occurs. Switches (SW) automatically disconnect the defective main amplifier and connect the spare to ensure

continuous operation. The amplified signal is then sent to the antenna via a waveguide, the diplexer, and a filter.

There are three types of power amplifiers used in earth stations: transistors, TWTs, and klystrons. Let's take a brief look at each. Transistor power amplifiers are used in small- and medium-size earth stations with low power. Powers up to 10 W are common. Normally, power GaAs FETs are used in this application. In some instances, a chain of varactor or

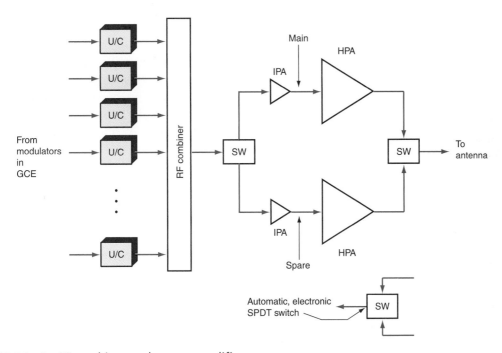

Fig. 11-36 An RF combiner and power amplifiers.

IMPATT diodes are used to create a string of frequency multipliers which translate an RF power signal generated by a bipolar transistor at a lower frequency to the desired final operating frequency. Again, powers of up to 10 W are achievable with these systems. The newer GaAs MESFET power devices when operated in parallel can achieve powers up to approximately 40 W. Further improvements are expected.

Most medium- and high-power earth stations use either TWTs or klystrons for the power amplifiers. There are two typical power ranges, one in the 200- to 400-W range and the other in the 2- to 3-kW range. The amount of power used depends upon the location of the station and its antenna size. Satellite transponder characteristics also influence the power required by the earth station. As improvements have been made in LNAs and satellites have been able to carry higher-power transponders, earth station transmitter power requirements have greatly decreased. This is also true for antenna sizes.

The majority of medium- and high-power earth stations use TWTs for power amplification. Traveling-wave tubes are available in sizes from a few watts to 20,000 W. Their greatest virtue is very wide bandwidth. A single TWT can transmit an entire 500-MHz bandwidth satellite signal.

On the down side, TWTs are extremely expensive. Further, they are very complex in operation. They require numerous expensive, regulated high-voltage power supplies. Finally, when they reach their maximum power saturation levels, TWTs produce considerable intermodulation products, spurious signals which affect the reliability of communications. When operating near maximum power, it is sometimes necessary to back off on input signal power to reduce the intermodulation signals.

The alternative to TWTs is klystrons. Klystrons are also available in a wide range of power sizes, from a fraction of a watt all the way to tens of thousands of watts. Klystrons are simpler in their construction and operation than TWTs and are, therefore, usually lower in cost. Less complex and less expensive power supplies are also required. Further, klystrons provide better intermodulation performance than TWTs.

However, the main disadvantage is that klystrons are basically narrowband devices. The maximum bandwidth for most high-power klystrons is approximately 150 MHz. This makes them suitable for transmitting only three adjacent 36-MHz-bandwidth channels. More klystrons must be used, therefore, to transmit all the desired signals. In some systems, narrowband klystrons are used one per channel. In such cases where multiple klystrons are used, the outputs are all combined in a single RF combiner to create a single signal that is fed to the antenna.

Regardless of whether klystrons or TWTs are used, if power levels in excess of 2 kW are required, liquid-cooling systems must typically be used. These are larger, more expensive devices and require the necessary hardware and plumbing to provide the liquid cooling.

The final HPA, whether it uses klystrons or TWTs, is usually driven by an IPA. This amplifier takes the low-level up-converted signals and amplifies them to a level capable of driving the final HPA devices. This IPA is typically a smaller TWT or klystron. The total power gain of the IPA and HPA combined is usually in the 60- to 70-dB range.

The output of the HPA located after any combiner is usually passed through a final filter to remove any undesirable harmonics and to provide final bandwidth shaping. The signal is then fed through a diplexer and the waveguide to the feed horns.

Power Substation

Most earth stations receive their power from normal ac sources. Standard power supplies convert the ac to the dc voltages required to operate all subsystems. However, most earth stations have backup power systems. Satellite systems, particularly those used for reliable communications of telephone conversations, TV programs, computer data, etc., must not "go down." The backup power system, which may consist of a diesel engine driving an ac generator, takes over if there is an ac power failure. When ac power fails, an automatic system starts the diesel engine. The generator creates the equivalent ac power which is automatically switched to the system. Smaller systems may use *uninterruptible power supplies (UPS)*, which derive their main power from batteries. Large battery arrays drive dc-to-ac inverters which produce the ac voltages for the system. Uninterruptible power supplies are not suitable

Uninterruptable power supplies (UPS)

for long power failures and interruptions because the batteries quickly become exhausted. However, for short interruptions of power, less than an hour, they are adequate.

Telemetry and Control Subsystem

The telemetry equipment consists of a receiver and the recorders and indicators that display the telemetry signals. The signal may be received by the main antenna or by a separate telemetry antenna. A separate receiver on a frequency different from the communications channels is used for telemetry purposes. The telemetry signals from the various sensors and transducers in the satellite are multiplexed onto a single carrier and are sent to the earth station. The earth station receiver demodulates and demultiplexes the telemetry signals into the individual outputs. These are then recorded and are also sent to various indicators, such as strip chart recorders, meters, and digital displays. Signals may be in digital form or converted to digital. They can also be sent to a computer where they can be further processed and stored. Mathematical operations on the signals may be necessary. For example, the number received from the satellite representing temperature may have to be processed by some mathematical formula to obtain the actual output temperature in degrees Celsius or Fahrenheit. The computer can also generate and format printed reports as required.

The control subsystem permits the ground station to control the satellite. This system usually contains a computer for entering the commands which modulate a carrier that is amplified and fed to the main antenna. The command signals can make adjustments in the satellite attitude, turn transponders off and on, actuate redundant circuits if the circuits they are backing up fail, and so on.

In some satellite systems where communications is not the main function, some instrumentation may be a part of the ground station. Instrumentation is a general term for all the electronic equipment that is used to deal with the information transmitted back to the earth station. It may consist of demodulators and demultiplexers, amplifiers, filters, A/D converters, or signal processors. The instrumentation subsystem is in effect an extension of the telemetry system. Besides relaying information about the satellite itself, the telemetry system may also be used to send back information related to various scientific experiments being conducted on the satellite. In satellites used for surveillance, the instrumentation may be such that it can deal with television signals sent back from an on-board TV camera. The possibilities are extensive, depending upon the actual satellite mission.

At the ground station, additional equipment may be required to process this information. The ground station equipment almost always includes tape recorders for recording the information, and the various pieces of equipment used to display it. Again, a computer may be part of the instrumentation for analyzing, processing, and reformatting the data.

Not all earth stations have the telemetry, control, and instrumentation subsystems; only those designated as control stations do. Most earth stations just send or receive.

■ TEST

Answer the following questions.

42. The signals to be communicated by the earth station to the satellite are known as _____ signals.
43. The receive or up-link subsystem uses a _____ to translate the satellite signal to an IF before demodulation and demultiplexing.
44. A common IF in an earth station receiver is _____ MHz.
45. The functions performed by the receive GCE section are _____.
46. The functions performed by the transmit GCE section are _____.
47. The two types of modulation used in the transmit GCE are _____.
48. A(n) _____ in the transmit subsection translates the modulated signal to the final transmission frequency.
49. _____ are often used to replace LOs for channel selection in earth station transmitters and receivers.
50. List the three main types of power amplifiers used in earth stations.
51. Most earth stations operate from the ac power line but have built-in _____ for emergency operation.
52. True or false. All earth stations contain telemetry, control, and instrumentation subsystems.

11-5 Applications Overview

Every satellite is designed to perform some specific task. Its predetermined application specifies the kind of equipment it must have on board and its orbit. Although our emphasis on satellites in this chapter is communications, satellites are useful for many other purposes. We have already mentioned some of them.

The main applications for satellites today is communications. Satellites used for this purpose act as relay stations in the sky. They permit reliable long-distance communications worldwide, and they solve many of the growing communications needs of industry and government. Communications applications will continue to dominate this industry.

Surveillance

A major application of satellites is surveillance or observation. From their vantage point high in the sky, satellites can look at the earth and transmit what they see back to ground stations for a wide variety of purposes. For example, military satellites are used to perform reconnaissance. On-board film cameras take photographs which can later be ejected from the satellite and brought back to earth for recovery. Television cameras can take pictures and send them back to earth as electrical signals. Infrared sensors detect heat sources, and small radars can profile earth features.

Intelligence satellites collect information about enemies and potential enemies. They permit monitoring for the purpose of proving compliance with nuclear test ban and missile treaties and agreed-upon conditions with other countries.

There are many different kinds of observation satellites. One special type is the meteorological or weather satellite. These satellites photograph cloud cover, and the pictures which are sent back to earth are used for determining and predicting the weather. Geodetic satellites photograph the earth for the purpose of creating more-accurate and more-detailed maps.

Satellites can also monitor the status of the earth's resources such as our land and oceans. They can observe crops, forests, lakes, and rivers. Satellites can spot diseased crop areas, mineral resources, sources of pollution, and other characteristics that are difficult, if not impossible, to detect in other ways.

Navigation

One of the newest and most useful satellite systems is the Global Positioning System (GPS). Its primary application is navigation. The system is a network of 24 low-earth-orbit satellites spaced equally around the world in overlapping patterns. Each satellite transmits a unique signal back to earth on low microwave frequencies such as 1.57542 and 1.2276 GHz. Receivers on the earth pick up transmissions from four satellites simultaneously. The receiver uses the signals in a microprocessor to compute the exact position of the receiver on earth. The receiver output is a display giving the latitude, longitude, and altitude of the receiver.

GPS was designed primarily as a navigation system for the military. It can also be used by individuals and industry for general navigational purposes. The military system has more accuracy than the civilian system. The military system error is less than 40 m, whereas the civilian system accuracy is less than 100 m from the actual location. Such accuracies are more than suitable in most air and sea navigation situations. However, greater accuracy may be required in land applications. Regardless of the accuracy, the real value of the system is that you can know your position on earth at any time by using a low-cost handheld microwave GPS receiver.

Each satellite is maintained in a *very* accurate orbit. Each satellite has a very accurate atomic clock. Time information is encoded on the transmitted signal. The receiver will receive a unique time reference from each satellite. The times are unique because of the different propagation times. For example, path 1 in Fig. 11-37 is the longest, and path 3 is the shortest. Signal 1 will be delayed the most. Since the velocity of propagation is known, the relative path lengths can be computed in the receiver microprocessor. Three (or more) simultaneous equations are then solved by the microprocessor to determine where the receiver is located.

TV Distribution

For years, TV signals have been transmitted through satellites for redistribution. Because of the very high frequency signals involved in TV transmission, the standard telephone system

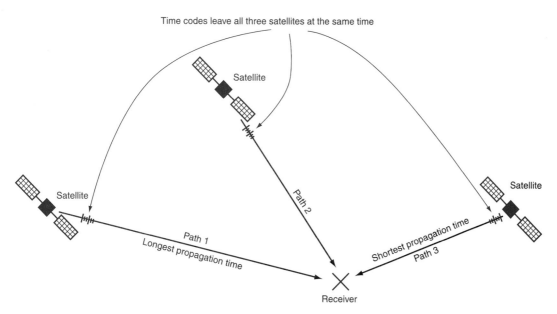

Time codes leave all three satellites at the same time

Satellite

Satellite

Satellite

Path 2

Path 1
Longest propagation time

Shortest propagation time
Path 3

Receiver

Fig. 11-37 Satellite identification information relates each satellite location with respect to points on the earth's surface.

cannot be used. Special coax as well as microwave relay links have been used to transmit TV signals from one place to another. However, with today's communications satellites, TV signals can be transmitted easily from one place to another. All the major TV networks and cable TV companies rely on communications satellites for TV signal distribution.

Consumers have discovered that with an appropriate home satellite antenna and receiver, they may pick up these TV signals. Satellites carrying TV broadcasts have been identified, and an entire industry has grown up around receiving these network distribution TV signals at home. Those who have them enjoy the broadest range of TV possible. These signals contain virtually everything that exists on the standard TV channels, much of which never reaches the public. The programs being transmitted are often uncut and unedited. However, many of these signals, especially those of the premium cable channels such as HBO and Showtime are scrambled. Viewers must buy a descrambler box and subscribe to receive these signals.

More recently, a new satellite TV service has been offered to consumers. The satellites transmit digitally encoded TV signals on the Ku band (12.5 to 18 GHz). Consumers pur-

chase a satellite receiver and 18-in. parabolic dish antenna that connects to a standard TV set. Several companies offer subscription services to several hundred channels of TV.

Satellite Telephones

Although the telephone system uses satellites for some long-distance calls, new satellite telephone systems are being developed. One example is Iridium, a worldwide cellular-type telephone system using a low-earth-orbit satellite (LEOS). The system uses 66 satellites 420 mi above the earth to communicate with handheld cell phones anywhere on earth. It uses the L band (1.6 GHz) and digital techniques for voice or data communications.

■ TEST

Answer the following questions.

53. The main use of satellites is for _____.
54. Name four types of surveillance sensors.
55. What are the name and the purpose of the 24-satellite system developed by the military?
56. Name three uses for surveillance satellites.
57. Consumers use satellites for _____.

Summary

1. A satellite is a physical object that orbits a celestial body.
2. A communications satellite containing electronic equipment acts as a repeater or relay station between two earth stations.
3. The basic component of a communications satellite is a receiver-transmitter combination called a transponder.
4. A satellite stays in orbit because the gravitational pull of the earth is balanced by the centripetal force of the revolving satellite.
5. Satellite orbits about the earth are either circular or elliptical.
6. Satellites orbit the earth from heights of 100 to 22,300 mi and travel at speeds of 6800 to 17,500 mi/h.
7. A satellite that orbits directly over the equator 22,300 mi from earth is said to be in a geostationary orbit. It revolves in synchronism with the earth's rotation, so it appears to be stationary when seen from points on the earth.
8. A satellite is stabilized in orbit by spinning it on its axis or building in spinning flywheels for each major axis (roll, pitch, yaw).
9. Attitude adjustments on a satellite are made by firing small jet thrusters to change the satellite's position or speed.
10. The location of a satellite is determined with latitude and longitude measurements that designate a point on the earth [subsatellite point (SSP)] directly below the satellite.
11. Azimuth and elevation angles determine where to point an earth station so that it intercepts the satellite.
12. Satellites are launched into orbit by rockets that give them vertical as well as forward motion.
13. A geosynchronous satellite is initially put into an elliptical orbit where its apogee is 22,300 mi high. The apogee kick motor is then fired to put the satellite into its final circular geostationary orbit.
14. Many satellites are put into orbit by launching them from NASA's space shuttle.
15. Most satellites operate in the microwave region.
16. Microwave satellites operate on assigned frequency bands designated by a letter. Common communications satellite bands are the C (3.4 to 6.425 GHz) and Ku (10.95 to 14.5 GHz) bands.
17. Satellite bands are typically 500 MHz wide and are divided into 12 segments, each 36 MHz wide. A transponder is used to cover each segment.
18. Frequency reuse is a technique that allows two sets of transponders to operate on the same frequency, thus doubling channel capacity. The two sets of channels use antennas of different polarizations to prevent interference with one another.
19. Spatial isolation is another technique for frequency sharing. It uses highly directional spot-beam antennas to prevent interference between stations on the same frequency.
20. The main subsystems in a satellite are the communications; power; telemetry; tracking, and control (TTC); propulsion; attitude stabilization; and antenna subsystems.
21. A transponder consists of a low-noise amplifier (LNA) that receives and amplifies the up-link signal, a mixer that converts the signal to another (lower) frequency, and a high-power amplifier that retransmits the signal on its new down-link frequency.
22. Double-conversion transponders use two mixers, one to translate the up-link signal to an IF where it is amplified and filtered, and another to translate the signal to its final down-link frequency.
23. Regenerative transponders demodulate the up-link signal to recover the baseband signals and then use them to remodulate a down-link transmitter. This improves the S/N ratio.
24. In a broadband transponder, a single mixer converts all channels within the 500-MHz bandwidth simultaneously to their down-link frequencies. These are selected by channel bandpass filters and then amplified by individual power amplifiers.

25. In a channelized transponder, each channel has its own LNA, bandpass filters, mixer, and high-power amplifier.

26. The power subsystem consists of solar panels, batteries, dc-to-dc converters, and regulators. The solar panels convert sunlight into dc power to operate all satellite electronics and to charge the batteries that take over when sunlight is blocked.

27. The TTC subsystem contains a receiver that picks up commands from a ground station and translates them into control signals that initiate some action on board. The telemetry system monitors physical conditions within the satellites and converts them into electrical signals that are transmitted back to earth.

28. The propulsion system consists of the apogee kick motor that puts the satellite into final orbit and the jet thrusters that are used for positioning and attitude control.

29. The stabilization subsystem for attitude control consists of spin components or three-axis flywheel gyros.

30. The antenna system consists of one or more highly directional horn or parabolic antennas and an omnidirectional TTC antenna.

31. Earth stations consist of transmit, receive, power, antenna, TTC, and ground control equipment (GCE) subsystems.

32. The transmit subsystem takes the baseband voice, video, or computer data signals; multiplexes them; and uses the composite signal to modulate a carrier. An up converter translates the signal to its final up-link frequency before it is amplified and transmitted.

33. The most common forms of modulation used are FM and QPSK.

34. Transistor power amplifiers are used in low-power earth stations; klystrons are used in high-power narrowband stations; and TWTs are used in high-power broadband stations.

35. Earth stations feature large parabolic dish antennas with high gain and directivity for receiving the weak satellite signal.

36. The receive subsystem in an earth station amplifies the signal with an LNA and then separates the channels with bandpass filters. Down converters translate the signals to a lower IF where they are demodulated and demultiplexed.

37. The GCE in an earth station interfaces the baseband signals to the transmit and receive subsystems. The receive GCE performs demodulation and demultiplexing. The transmit GCE performs modulation and multiplexing.

38. The most common application for satellites is communications.

39. Another major use of satellites is surveillance and reconnaissance.

40. Film cameras, TV cameras, infrared sensors, and radars are all used to observe a variety of conditions on earth from surveillance satellites.

41. Satellites play a major role in military and defense systems not only for communications but also for surveillance.

42. The 24-satellite Global Positioning System (GPS) makes accurate navigation possible anywhere on earth with a low-cost microwave receiver.

43. Consumers use satellite TV receivers to intercept TV signals transmitted by networks and cable TV companies.

44. A satellite TV system using digital transmission is available to consumers.

45. Satellites are also used in new portable telephone systems.

Chapter Review Questions

Choose the letter which best answers each question.

11-1. As the height of a satellite orbit gets lower, the speed of the satellite
  a. Increases.
  b. Decreases.
  c. Remains the same.
  d. None of the above.

11-2. The main function of a communications satellite is as a(n)
  a. Repeater.
  b. Reflector.
  c. Beacon.
  d. Observation platform.

11-3. The key electronic component in a communications satellite is the
  a. Telemetry.
  b. On-board computer.
  c. Command and control system.
  d. Transponder.

11-4. A circular orbit around the equator with a 24-h period is called a(n)
  a. Elliptical orbit.

b. Geostationary orbit.

c. Polar orbit.

d. Transfer orbit.

11-5. A satellite stays in orbit because the following two factors are balanced.

a. Satellite weight and speed

b. Gravitational pull and inertia

c. Centripetal force and speed

d. Satellite weight and the pull of the moon and sun

11-6. The height of a satellite in a synchronous equatorial orbit is

a. 100 mi.

b. 6800 mi.

c. 22,300 mi.

d. 35,860 mi.

11-7. Most satellites operate in which frequency band?

a. 30 to 300 MHz

b. 300 MHz to 3 GHz

c. 3 GHz to 30 GHz

d. Above 300 GHz

11-8. The main power sources for a satellite are

a. Batteries.

b. Solar cells.

c. Fuel cells.

d. Thermoelectric generators.

11-9. The maximum height of an elliptical orbit is called the

a. Perigee.

b. Apex.

c. Zenith.

d. Apogee.

11-10. Batteries are used to power all satellite subsystems

a. At all times.

b. Only during emergencies.

c. During eclipse periods.

d. To give the solar arrays a rest.

11-11. The satellite subsystem that monitors and controls the satellite is the

a. Propulsion subsystem.

b. Power subsystem.

c. Communications subsystem.

d. Telemetry, tracking, and command subsystem.

11-12. The basic technique used to stabilize a satellite is

a. Gravity-forward motion balance.

b. Spin.

c. Thruster control.

d. Solar panel orientation.

11-13. The jet thrusters are usually fired to

a. Maintain attitude.

b. Put the satellite into the transfer orbit.

c. Inject the satellite into the geosynchronous orbit.

d. Bring the satellite back to earth.

11-14. Most commercial satellite activity occurs in which band(s)?

a. L

b. C and Ku

c. X

d. S and P

11-15. How can multiple earth stations share a satellite on the same frequencies?

a. Frequency reuse

b. Multiplexing

c. Mixing

d. They can't

11-16. The typical bandwidth of a satellite band is

a. 36 MHz.

b. 40 MHz.

c. 70 MHz.

d. 500 MHz.

11-17. Which of the following is *not* usually a part of a transponder?

a. LNA

b. Mixer

c. Modulator

d. HPA

11-18. The satellite communications channels in a transponder are defined by the

a. LNA.

b. Bandpass filter.

c. Mixer.

d. Input signals.

11-19. The HPAs in most satellites are

a. TWTs.

b. Klystrons.

c. Vacuum tubes.

d. Magnetrons.

11-20. The physical location of a satellite is determined by its

a. Distance from the earth.

b. Latitude and longitude.

c. Reference to the stars.

d. Position relative to the sun.

11-21. The receive GCE system in an earth station performs what function(s)?

a. Modulation and multiplexing

b. Up conversion

c. Demodulation and demultiplexing

d. Down conversion

11-22. Which of the following types of HPA is *not* used in earth stations?

a. TWT

b. Transistor

c. Klystron

d. Magnetron

11-23. A common up-converter and down-converter IF is

a. 36 MHz.

b. 40 MHz.

c. 70 MHz.

d. 500 MHz.

11-24. The type of modulation used on voice and video signals is

a. AM.

b. FM.

c. SSB.

d. QPSK.

11-25. The modulation normally used with digital data is

a. AM.

b. FM.

c. SSB.

d. QPSK.

11-26. Which of the following is *not* a typical output from a GPS receiver?

a. Latitude

b. Speed

c. Altitude

d. Longitude

Critical Thinking Questions

11-1. What do you think are the primary factors that determine a satellite's useful life in space?

11-2. If you want an earth station to receive another satellite, what adjustments do you have to make and on what components?

11-3. What factors do you think determine an earth station's ability to select and distinguish one satellite from another closely adjacent satellite in orbit?

11-4. List as many civilian applications as you can think of for the Global Positioning System.

11-5. In what general direction would you point a dish antenna to receive a signal from a geosynchronous satellite if you are in the United States?

11-6. Why do satellites normally use UHF and higher frequencies?

11-7. Low earth orbit (LEO) satellites are routinely used for surveillance. Name several different types of sensors and input devices that might be used. How is this information sent back to earth?

Answers to Tests

1. centripetal force, gravitational pull
2. posigrade
3. ellipse
4. geocenter
5. period
6. 0, 90
7. 5
8. geosynchronous or geostationary
9. attitude
10. spin, three-axis stabilization
11. yaw, pitch, roll
12. subsatellite point
13. latitude, longitude
14. azimuth, elevation
15. apogee kick motor
16. communications
17. relay station
18. earth station
19. down link
20. transponder
21. microwave
22. C
23. X
24. 10.95, 14.5
25. 500
26. 12, 36
27. frequency reuse
28. spatial isolation
29. voice, video, computer or digital data
30. low-noise amplifier
31. signal amplification, frequency translation, power amplification
32. mixer
33. TWT
34. double-conversion
35. regenerative
36. bandpass filter
37. broadband, multichannel
38. solar panels
39. batteries

40. monitor, control
41. jet thrusters
42. baseband
43. down converter
44. 70
45. demodulation, demultiplexing
46. modulation, multiplexing
47. FM, QPSK
48. up converter
49. frequency synthesizers
50. transistor, TWT, klystron
51. backup power or uninterruptible power supply
52. false
53. communications
54. film cameras, TV cameras, infrared sensors, radar
55. Global Positioning System (GPS), navigation
56. weather forecasting, spying, earth resources, monitoring
57. TV reception

Chapter 12

Data Communications

Chapter Objectives

This chapter will help you to:

1. *Explain* the concepts of communications by serial binary data and *define* the terms ASCII, synchronous, asynchronous, and baud rate.
2. *State* the relationship between communications channel bandwidth and data rate in bits per second and *compute* one given the other.
3. *Define* the term modem and *explain* its basic function and application.
4. *Name* three common modem types in use today and *explain* their operation.
5. *Explain* the processes of modulation and demodulation with frequency-shift keying (FSK) and phase-shift keying (PSK).

6. *Explain* the need for and types of communications protocols and error-detection and error-correction schemes.
7. *Name* the three basic configurations of local networks and *discuss* the pros and cons of each.
8. *Explain* the operation and benefits of frequency hopping and direct-sequence spread spectrum systems.
9. *Define* the Internet and *explain* briefly how it works.

D ata communications is the transmission of binary or digital information from one point to another. Data communications systems permit the transfer of information between computers and also permit the remote operation of a computer from a terminal. Further, since any type of signal can be digitized, the transmission of voice, video, or other traditionally analog information in binary form is also known as data communications. Because of the low cost and effectiveness of digital techniques, data communications is becoming more widely used. This chapter introduces the basic concepts and equipment used in data communications and also provides a brief introduction to the Internet.

12-1 Digital Communications Concepts

Digital signals are binary pulses that have two distinct states, each represented by a voltage level. The pulses switch rapidly between these two levels. One level is referred to as a binary 0 or low, and the other as a binary 1 or high. Figure 12-1 shows a typical binary or digital signal. The binary 0 level might be 0 V or ground, and the binary 1 level might be +5 V. Any other two voltages can also be used.

Binary signals are easy to generate and process with electronic circuits. Binary pulses are usually produced by and manipulated with high-speed transistor switches. These switches are connected in a variety of configurations to form digital logic circuits. These digital logic circuits are available in IC form and can generate and process digital data at high speeds.

Usually, the binary signals represent data or information. Specifically, the binary signals are codes made up of groups or patterns of 0s and 1s. Each pattern represents a numerical

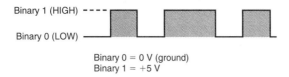

Binary 1 (HIGH) ----

Binary 0 (LOW) ———

Binary 0 = 0 V (ground)
Binary 1 = +5 V

Fig. 12-1   A binary signal.

value, a letter of the alphabet, or some special symbol, meaning, or message. The digital logic circuits process this data in various ways.

Data Codes

The earliest forms of data communications were binary in nature. The very first type of electronic communications in use was the *telegraph,* which was invented in 1832 and first demonstrated successfully in 1844 by Samuel F. B. Morse. A telegraph hand key acting as a switch connects to a remote receiver or sounder. The sounder is nothing more than a magnetic coil which attracts an armature that makes a clicking sound. Whenever the telegraph key is depressed, current flows through the coil producing a magnetic field that attracts the armature and makes a click. When the key is opened, the armature makes another click. The operation is strictly binary since there are only two states: off and on. For two-way communications, both stations have a key and a sounder so that either station may transmit or receive.

Morse created a special code using this off/on capability. Known as the *Morse code,* it is a series of dots and dashes that represent letters of the alphabet, numbers, and punctuation

Telegraph

Continuous wave (CW)

Baudot code

Morse code

marks. These are summarized in Fig. 12-2. When the armature is closed for a short duration, a dot is produced. When the armature is closed for a longer time, a dash is produced. With special training, humans can easily send and receive messages at high speeds. Speeds of 15 to 20 words per minute are easily mastered with the upper limit being 70 to 80 words per minute.

In the earliest radio communications the Morse code was used to send messages. A key turned the carrier of a transmitter off and on to produce the dots and dashes. These were detected at the receiver and converted by an operator back into letters and numbers making up the message. This type of radio communications is known as *continuous wave (CW).*

Although telegraph and CW communications are still used, for the most part they have been replaced by voice communications. However, the communications industry has rediscovered two-state or binary communications, and today more and more communications are taking place by that method. Even voice, video, and other analog signals are being converted into binary codes that can be transmitted by data communications techniques by using A/D and D/A conversion techniques.

One of the first binary codes was the *Baudot code,* used in the early teletype machines (the Baudot code is shown in Fig. 12-3). A teletype machine is a typewriterlike device that is used to send and receive coded signals over a communications link. Whenever a key on the typewriter keyboard is pressed, a unique code is generated and transmitted to the receiving machine which recognizes and then prints the corresponding letter, number, or symbol. Teletype machines were designed to eliminate the need for telegraph operators to learn the Morse code. Anyone who could type could send messages with a teletype machine.

The binary code used in teletype machines is a 5-bit code called the Baudot (pronounced "baw dough"). With 5 bits, $2^5 = 2 \times 2 \times 2 \times 2 \times 2 = 32$ different symbols can be represented. This, of course, is not enough to represent 26 letters of the alphabet, 10 numbers, and various punctuation marks. The Baudot code gets around this by using two "shift codes." Refer to Fig. 12-3 and note the figure-shift and letter-shift codes. If the message is preceded by the letter-shift code (11011), all of the

A	•—	T	—
B	—•••	U	••—
C	—•—•	V	•••—
D	—••	W	•——
E	•	X	—••—
F	••—•	Y	—•——
G	——•	Z	——••
H	••••	,	•—•—•—
I	••	.	•—•—•—
J	•———	1	•————
K	—•—	2	••———
L	•—••	3	•••——
M	——	4	••••—
N	—•	5	•••••
O	———	6	—••••
P	•——•	7	——•••
Q	——•—	8	———••
R	•—•	9	————•
S	•••	0	—————

Fig. 12-2   The Morse telegraph code. A dot (·) is a short click; a dash (—) is a long click.

Character Shift		Binary Code				
Letter	Figure	Bit: 4	3	2	1	0
A	—	1	1	0	0	0
B	?	1	0	0	1	1
C	:	0	1	1	1	0
D	$	1	0	0	1	0
E	3	1	0	0	0	0
F	!	1	0	1	1	0
G	&	0	1	0	1	1
H	#	0	0	1	0	1
I	8	0	1	1	0	0
J	'	1	1	0	1	0
K	(	1	1	1	1	0
L	)	0	1	0	0	1
M	.	0	0	1	1	1
N	,	0	0	1	1	0
O	9	0	0	0	1	1
P	0	0	1	1	0	1
Q	1	1	1	1	0	1
R	4	0	1	0	1	0
S	bel	1	0	1	0	0
T	5	0	0	0	0	1
U	7	1	1	1	0	0
V	;	0	1	1	1	1
W	2	1	1	0	0	1
X	/	1	0	1	1	1
Y	6	1	0	1	0	1
Z	"	1	0	0	0	1
Figure shift		1	1	1	1	1
Letter shift		1	1	0	1	1
Space		0	0	1	0	0
Line feed (LF)		0	1	0	0	0
Blank (null)		0	0	0	0	0

Fig. 12-3 The Baudot code.

following codes are interpreted as letters of the alphabet. Sending the figure-shift code (11111) causes all the following characters to be interpreted as numbers or punctuation marks. The Baudot code is still used in telegraph communications, but in most data communications, it has been replaced by codes that can represent more characters and symbols.

The most widely used data communications code is the *American Standard Code for Information Interchange,* abbreviated *ASCII* (pronounced "ass key"). This is a 7-bit binary code used for representing 128 numbers, letters, punctuation marks, and other symbols. Its big advantage over the Baudot code is that a sufficient number of code combinations is available to represent both upper- and lower-case letters of the alphabet. See Fig. 12-4.

Another code similar to ASCII is the *Extended Binary Coded Decimal Interchange Code,* abbreviated *EBCDIC* (pronounced "ebb see dick") developed by IBM. It is an 8-bit code allowing a maximum of 256 characters to be represented. Although it is an improvement over ASCII, its primary use is in IBM and IBM-compatible computing systems and equipment.

In data communications, messages are composed using one of these codes. The numbers and letters to be represented are coded usually by way of a keyboard, and the binary word representing each character is stored in a computer memory. The message may also be stored on magnetic tape or disk. The purpose of data communications is to transmit that data or information from one place to another.

Serial versus Parallel Transmission

As you know, there are two basic ways to transfer binary information from one place to another: parallel and serial. In *parallel data* transfers, all the bits of a code word are transferred simultaneously. This concept is illustrated in Fig. 12-5. The binary word to be transmitted is usually loaded into a register containing one flip-flop for each bit. The flip-flop outputs are each connected to a wire to carry that bit to the receiving circuit. The receiving circuit is also usually a storage register. As you can see, in parallel data transmission, there is one wire for each bit of information to be transmitted. This means a multiwire cable must be used. Multiple parallel lines that carry binary data are usually referred to as a *data bus.*

Parallel data transmission is extremely fast since all the bits of the data word are transferred simultaneously. The transmit speed is limited to the speed of the logic circuits used in the data transfer. Such data transfers can occur in only a few nanoseconds in many applications.

Parallel data transmission is not practical for long-distance communications. To transfer an 8-bit data word from one place to another, eight separate communications channels are needed, one for each bit. Although multiwire cables can be used over limited distances, for practical long-distance data communications they are impractical because of cost and signal attenuation. And, of course, parallel data transmission by radio would be even more complex and expensive.

Data transfers in communications systems are made serially. Each bit of the word is transmitted one after another. Figure 12-6 shows the concept of *serial* transmission. The figure shows the ASCII code for the letter J (1001010) being transmitted one bit at a time. The least significant

Parallel data

Data bus

American Standard Code for Information (ASCII)

Extended Binary Coded Decimal Interchange Code (EBCDIC)

Serial data

Bit:	6	5	4	3	2	1	0
NUL	0	0	0	0	0	0	0
SOH	0	0	0	0	0	0	1
STX	0	0	0	0	0	1	0
ETX	0	0	0	0	0	1	1
EOT	0	0	0	0	1	0	0
ENQ	0	0	0	0	1	0	1
ACK	0	0	0	0	1	1	0
BEL	0	0	0	0	1	1	1
BS	0	0	0	1	0	0	0
HT	0	0	0	1	0	0	1
NL	0	0	0	1	0	1	0
VT	0	0	0	1	0	1	1
FF	0	0	0	1	1	0	0
CR	0	0	0	1	1	0	1
SO	0	0	0	1	1	1	0
SI	0	0	0	1	1	1	1
DLE	0	0	1	0	0	0	0
DC1	0	0	1	0	0	0	1
DC2	0	0	1	0	0	1	0
DC3	0	0	1	0	0	1	1
DC4	0	0	1	0	1	0	0
NAK	0	0	1	0	1	0	1
SYN	0	0	1	0	1	1	0
ETB	0	0	1	0	1	1	1
CAN	0	0	1	1	0	0	0
EM	0	0	1	1	0	0	1
SUB	0	0	1	1	0	1	0
ESC	0	0	1	1	0	1	1
FS	0	0	1	1	1	0	0
GS	0	0	1	1	1	0	1
RS	0	0	1	1	1	1	0
US	0	0	1	1	1	1	1
SP	0	1	0	0	0	0	0
!	0	1	0	0	0	0	1
"	0	1	0	0	0	1	0
#	0	1	0	0	0	1	1
$	0	1	0	0	1	0	0
%	0	1	0	0	1	0	1
&	0	1	0	0	1	1	0
'	0	1	0	0	1	1	1
(	0	1	0	1	0	0	0
)	0	1	0	1	0	0	1
*	0	1	0	1	0	1	0

Bit:	6	5	4	3	2	1	0
+	0	1	0	1	0	1	1
,	0	1	0	1	1	0	0
-	0	1	0	1	1	0	1
.	0	1	0	1	1	1	0
/	0	1	0	1	1	1	1
0	0	1	1	0	0	0	0
1	0	1	1	0	0	0	1
2	0	1	1	0	0	1	0
3	0	1	1	0	0	1	1
4	0	1	1	0	1	0	0
5	0	1	1	0	1	0	1
6	0	1	1	0	1	1	0
7	0	1	1	0	1	1	1
8	0	1	1	1	0	0	0
9	0	1	1	1	0	0	1
:	0	1	1	1	0	1	0
;	0	1	1	1	0	1	1
<	0	1	1	1	1	0	0
=	0	1	1	1	1	0	1
>	0	1	1	1	1	1	0
?	0	1	1	1	1	1	1
@	1	0	0	0	0	0	0
A	1	0	0	0	0	0	1
B	1	0	0	0	0	1	0
C	1	0	0	0	0	1	1
D	1	0	0	0	1	0	0
E	1	0	0	0	1	0	1
F	1	0	0	0	1	1	0
G	1	0	0	0	1	1	1
H	1	0	0	1	0	0	0
I	1	0	0	1	0	0	1
J	1	0	0	1	0	1	0
K	1	0	0	1	0	1	1
L	1	0	0	1	1	0	0
M	1	0	0	1	1	0	1
N	1	0	0	1	1	1	0
O	1	0	0	1	1	1	1
P	1	0	1	0	0	0	0
Q	1	0	1	0	0	0	1
R	1	0	1	0	0	1	0
S	1	0	1	0	0	1	1
T	1	0	1	0	1	0	0
U	1	0	1	0	1	0	1

Bit:	6	5	4	3	2	1	0
V	1	0	1	0	1	1	0
W	1	0	1	0	1	1	1
X	1	0	1	1	0	0	0
Y	1	0	1	1	0	0	1
Z	1	0	1	1	0	1	0
[	1	0	1	1	0	1	1
\	1	0	1	1	1	0	0
]	1	0	1	1	1	0	1
^	1	0	1	1	1	1	0
—	1	0	1	1	1	1	1
`	1	1	0	0	0	0	0
a	1	1	0	0	0	0	1
b	1	1	0	0	0	1	0
c	1	1	0	0	0	1	1
d	1	1	0	0	1	0	0
e	1	1	0	0	1	0	1
f	1	1	0	0	1	1	0
g	1	1	0	0	1	1	1
h	1	1	0	1	0	0	0
i	1	1	0	1	0	0	1
j	1	1	0	1	0	1	0
k	1	1	0	1	0	1	1
l	1	1	0	1	1	0	0
m	1	1	0	1	1	0	1
n	1	1	0	1	1	1	0
o	1	1	0	1	1	1	1
p	1	1	1	0	0	0	0
q	1	1	1	0	0	0	1
r	1	1	1	0	0	1	0
s	1	1	1	0	0	1	1
t	1	1	1	0	1	0	0
u	1	1	1	0	1	0	1
v	1	1	1	0	1	1	0
w	1	1	1	0	1	1	1
x	1	1	1	1	0	0	0
y	1	1	1	1	0	0	1
z	1	1	1	1	0	1	0
{	1	1	1	1	0	1	1
;	1	1	1	1	1	0	0
}	1	1	1	1	1	0	1
~	1	1	1	1	1	1	0
DEL	1	1	1	1	1	1	1

NUL = null
SOH = start of heading
STX = start of text
ETX = end of text
EOT = end of transmission
ENQ = enquiry
ACK = acknowledge
BEL = bell
BS = back space
HT = horizontal tab
NL = new line

VT = vertical tab
FF = form feed
CR = carriage return
SO = shift-out
SI = shift-in
DLE = data link escape
DC1 = device control 1
DC2 = device control 2
DC3 = device control 3
DC4 = device control 4
NAK = negative acknowledge

SYN = synchronous
ETB = end of transmission block
CAN = cancel
SUB = substitute
ESC = escape
FS = field separator
GS = group separator
RS = record separator
US = unit separator
SP = space
DEL = delete

Fig. 12-4   The popular ASCII code.

bit (LSB) is transmitted first, the most significant bit (MSB) last. Each bit is transmitted for a fixed interval of time (t). The voltage levels representing each bit appear on a single data line one after another until the entire word has been transmitted. For example, the bit interval may be 1 ms, meaning that the voltage level for each bit in the word will appear for 1 ms. A 7-bit ASCII word would, therefore, take 7 ms to transmit.

Because of the sequential nature of serial data transmission, naturally it takes longer to send data this way than it does to transmit it by parallel means. However, with a high-speed logic circuit, even serial data transfers can take place at very high speeds.

In most computers and other digital equipment, the binary data is transferred between circuits in parallel format. When it is transmitted over long distances, it is sent in serial format. This means there must be techniques for converting between parallel and serial, and serial and parallel. Such data conversions are usually taken care of by shift registers. A typical circuit is shown in Fig. 12-7. A *shift register* is a sequential logic circuit made up of a number of flip-flops connected in a cascade. The flip-flops are capable of storing a multibit binary word. Usually this word is loaded in parallel into the transmitting register. When a clock pulse (CP) is applied to the flip-flops, the bits of the word

Shift registers

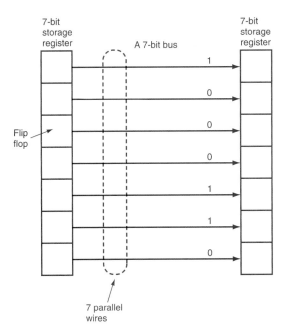

Fig. 12-5 A parallel data transfer of ASCII character F.

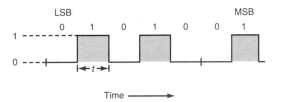

Fig. 12-6 Serial data transmission for the ASCII character J.

are shifted from one flip-flop to another in sequence. The last (right-hand) flip-flop in the transmitting register will ultimately store each of the bits in sequence as it is shifted out.

The serial data word is then transmitted over the communications link and is received by another shift register. The bits of the word are shifted into the flip-flops one at a time until the entire word is contained within the register. The flip-flop outputs may then be observed and the data stored in them transferred in parallel to other circuits.

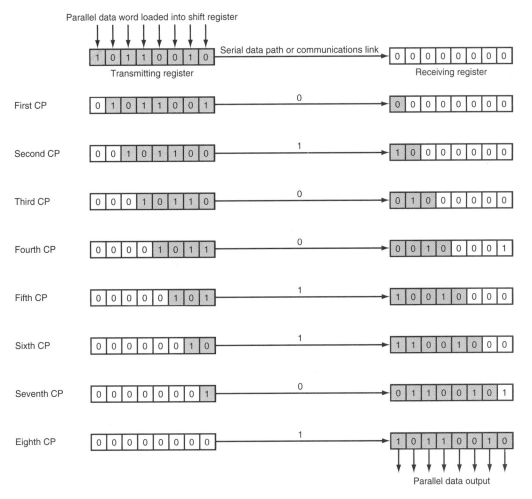

Fig. 12-7 Parallel-to-serial and serial-to-parallel data transfers with shift registers.

Serial Data Speed

The speed of data transfer is usually indicated by the number of bits per second (bits/s). Most data rates take place at relatively slow speeds, usually several thousand bits per second. The serial data rate in bits per second (bits/s) is the reciprocal of the time duration of one bit (T_b).

$$\text{Data rate (bits/s)} = \frac{1}{T_b}$$

If the bit duration is 2.5 ms, the data rate is:

$$\text{Data rate} = \frac{1}{2.5 \times 10^{-3}} = 400 \text{ bits/s}$$

A bit time of 34.7 μs gives a data rate of:

$$\text{Data rate} = \frac{1}{34.7 \times 10^{-6}} = 28,800 \text{ bits/s}$$

This can also be expressed as 28.8 kbits/s, where k = 1000. A bit time of 200 ms produces of data rate of

$$\text{Data rate} = \frac{1}{200 \times 10^{-9}} = 5,000,000 \text{ bits/s}$$

or 5 Mbps, where M = 1,000,000 bits/s.

If the data rate is known, the bit time can be determined.

$$T_b = \frac{1}{\text{bits/s}}$$

A 128-kbits/s rate has a bit time of:

$$T_b = \frac{1}{128 \times 10^3} = 7.81 \text{ } \mu s$$

However, in some data communications systems, bit rates as high as several hundred million bits per second are used.

Another term used to express the data speed in digital communications systems is baud rate. *Baud rate* is the number of signaling elements or symbols that occur in a given unit of time, such as 1 s. A signaling element or symbol in many cases is just a binary logic level, either 0 or 1. In that case, the baud rate is simply equal to the data rate in bits per second. However, in other cases the signaling element or symbol may be one of several discrete signal amplitudes or phase shifts, each of which represents two or more data bits. Several unique modulation schemes have been developed which use each symbol or baud to represent multiple bits. The number of symbol changes per unit of time is no higher than the straight binary bit rate, but more bits per unit time are transmitted. As a result, high bit rates can be transmitted over telephone lines or other severely bandwidth-limited communications channels which can-

not ordinarily handle them. Several of the modulation methods are discussed in Sec. 12-2.

An illustration will help make this clearer. Assume a system that represents 2 bits of data as different voltage levels. There are four possible combinations of 2 bits, and we assign a discrete level to each.

00	0 V	10	2 V
01	1 V	11	3 V

Here we have a system with four symbols, in this case four different voltage levels. Instead of transmitting a binary signal with only two levels, we transmit one of the four symbols or voltage levels. Assume we wanted to transmit the decimal number 201. This can be represented in binary by the 8-bit number 11001001. We could transmit the number serially as a sequence of equal-time-interval pulses that are either on (1) or off (0). Refer to Fig. 12-8(a). If each bit interval is 1 ms, then the bit rate is 1000 bits/s. The baud rate is also 1000 bits/s or 1000 baud.

Using our new four-level system, we could divide the word into 2-bit groups and transmit the appropriate voltage level representing each. Our number 11001001 would be divided into groups of 11/00/10/01. Thus the transmitted signal would be voltage levels of 3, 0, 2, and 1 V, respectively, each occurring for a fixed interval of say 1 ms. This is illustrated in Fig. 12-8(b). The baud rate is still 1000 baud because there is only one symbol per time interval (1 ms). However, the bit rate is 2000 bits/s because although only four symbols were

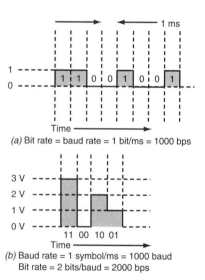

Fig. 12-8 The bit rate can be higher than the baud rate when each symbol represents two or more bits.

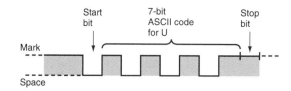

Fig. 12-9 Asynchronous transmission with start and stop bits.

transmitted, each symbol represents 2 bits. So, we have doubled the bit rate while keeping the baud rate constant. We have also shortened the total transmission time. It takes 8 ms to transmit the 8-bit binary word, but it only takes 4 ms to transmit the four-level signal.

Asynchronous Communications

There are two basic ways of transmitting serial binary data: asynchronous and synchronous. In *asynchronous* communications, each data word is accompanied by stop and start bits that identify the beginning and ending of the word. Each binary code word transmitted represents one character.

The use of *start and stop bits* to transmit an ASCII character is illustrated in Fig. 12-9. When no information is being transmitted, the communications line is usually high, or binary 1. In data communications terminology, this is referred to as a mark. To signal the beginning of a word, a start bit is transmitted. A start bit is a binary 0 or space as shown in Fig. 12-9. The start bit has the same duration as all other bits in the data word. The transition from mark to space indicates the beginning of the word and allows the receiving circuits to prepare themselves for the reception of the remainder of the bits.

After the start bit, the individual bits of the word are transmitted. In this case, the 7-bit ASCII code for the letter U is transmitted. Once the last code bit has been transmitted, a stop bit is included. The stop bit is the same duration as all other bits and again is a binary 1 or mark. In some systems, two stop bits are transmitted, one after the other, to signal the end of the word.

Most low-speed data communications, up to 115 K baud, take place using the asynchronous technique. It is extremely reliable, and the start and stop bits ensure that the sending and receiving circuits remain in step with one another.

The primary disadvantage of asynchronous communications is that the extra start and stop bits effectively slow down the data transmission. This is not a problem in low-data-volume, low-speed applications, but when huge volumes of information are to be transmitted, the start and stop bits represent a significant overhead; that is, they are a major percentage of the bits transmitted. For the transmission of a 7-bit ASCII character with start and stop bits, 9 bits are needed ($2 \div 9 = 22.2$ percent). By removing the start and stop bits and stringing the ASCII characters end to end, many more data words can be transmitted per second. In other words, the speed of transmission can be significantly increased. When large volumes of data are to be transmitted between computers, high speed is important to ensure minimum transmission time on expensive long-distance circuits.

Synchronous Communications

The technique of transmitting each data word one after another without start and stop bits is referred to as *synchronous* data communications. Data is usually transmitted in multiword blocks. To maintain synchronization between transmitter and receiver, a group of synchronization bits is placed at the beginning of the block and at the end of the block. Figure 12-10 shows one arrangement. Each block of data may represent hundreds or even thousands of characters. At the beginning of each block is a unique series of bits that identifies the beginning of the block. In Fig. 12-10, two 8-bit synchronization (SYN) codes signal the start of a transmission. The receiving equipment looks for these characters and then begins to receive the continuous data sent in sequential 8-bit words or bytes. At the end of the

Asynchronous

Stop and start bits

Synchronous

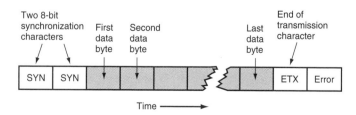

Fig. 12-10 Synchronous data transmission.

block is another special code (ETX) signaling the end of transmission. The receiving equipment will look for this code, detect it, and thereby recognize the end of the transmission. One or more error codes usually follow.

Although the special synchronization codes at the beginning and end do add extra bits to the transmission, overall they represent a very small increase in the total number of bits being transmitted. In any case, the number of bits used for synchronization purposes is small compared to the number of start and stop bits that would be transmitted if asynchronous transmission were used. Synchronous transmissions are much faster than asynchronous transmissions.

Relationship between Data Speed and Channel Bandwidth

An important concept in data communications, and analog communications for that matter, is that the amount of information that can be transmitted is proportional to the bandwidth of the communications channel and the time of transmission. The greater the bandwidth, the more information that can be transmitted in a given time. To transmit voice so that it is intelligible and recognizable, a bandwidth of only about 3 kHz is required. However, to transmit music with full fidelity, a bandwidth of 15 to 20 kHz is required because of the high frequencies and harmonics produced by the various musical instruments. Music inherently contains more information than voice and thus requires greater bandwidth.

Television is another example. Video requires a bandwidth of 4 MHz because of all the light and color detail in the scene or picture transmitted. A picture signal contains more information than a voice or music signal. Therefore, greater bandwidth is required to transmit it. A typical TV signal contains both voice and picture and is usually allocated 6 MHz of spectrum space.

It is also possible to transmit a picture in less bandwidth. This is done by transmitting it over a longer period of time. A black and white photo can be transmitted by TV in about 33 ms which to the human eye is virtually instantaneous. That same photo can also be transmitted over a voice-grade telephone line using a facsimile (fax) machine. The difference is that because the bandwidth is so much narrower, the transmission requires more time. It would take the fax machine 10 to 15 s to transmit the photo over the telephone line.

Just remember that the greater the bandwidth of a channel, the greater the amount of information you can transmit in a given time. You can still transmit the same amount of information over a narrower channel, but it will simply take longer. This general concept is known as *Hartley's law.* It is one of the many concepts that together form what is called information theory. *Information theory* is the study of the most efficient ways to transmit information.

Hartley's law also applies to binary data transmission. The greater the number of bits transmitted in a given time, the greater the amount of information that is conveyed. But the higher the bit rate, the wider the bandwidth needed to pass the signal without distortion. Narrowing the bandwidth of a channel will cause the harmonics in the binary pulses to be filtered out.

Binary pulses are rectangular waves made up of a fundamental sine wave plus many harmonics. The channel bandwidth must be wide enough to pass the harmonics to preserve the waveshape. On the other hand, most communications channels or media act like low-pass filters. Voice-grade telephone lines, for example, act as a low-pass filter with an upper cutoff frequency of about 3400 Hz. Naturally, harmonics higher in frequency than the cutoff will be filtered out. This produces signal distortion. Eliminating the harmonics rounds the signal off.

This filtering essentially converts the binary signal into its fundamental sine wave. The data is distorted but not lost. The information is still transmitted reliably but in the minimum possible bandwidth. The sine wave signal shape can easily be restored to a rectangular wave at the receiver with a Schmitt trigger, comparator, or other waveshaping circuit.

The upper cutoff frequency of the channel is approximately equal to the channel bandwidth. It is this bandwidth that determines the information capacity of the channel. The *channel capacity C* expressed in bits per second is twice the channel bandwidth B in hertz.

$$C = 2B$$

The bandwidth B is usually the same as the upper cutoff (3-dB-down) frequency of the channel. This is the maximum theoretical limit, and it assumes that no noise is present.

For example, assume a channel bandwidth of 10 kHz. The maximum theoretical bit capacity then is

$$C = 2B$$
$$= 2(10,000)$$
$$= 20,000 \text{ bits/s}$$

The channel capacity expression can be modified by taking into consideration multiple-level encoding schemes that permit more bits per symbol or baud to be transmitted. The new expression becomes

$$C = 2B \log_2 N$$

Here N is the number of different encoding levels per time interval. What this expression says is that for a given bandwidth, the channel capacity in bits per second can be higher if we use more than two levels or other symbols per time interval.

In our previous example, we used two levels per millisecond (0 or 1) in the binary signal of Fig. 12-8(a). The bit rate (and baud rate) is 1000 bits/s. The bandwidth needed to transmit this signal can be computed from the original expression.

$$C = 2B$$

Therefore,

$$B = \frac{C}{2}$$

With a 1000 bits/s channel rate, we will need an absolute minimum bandwidth of

$$B = \frac{1000 \text{ bits/s}}{2}$$
$$= 500 \text{ Hz}$$

The same result is obtained with the new expression. If $C = 2B \log_2 N$, then

$$B = \frac{C}{2 \log_2 N}$$

The logarithm of a number to the base 2 can be computed with the expression

$$\log_2 N = \frac{\log_{10} N}{\log_{10} 2}$$
$$= \frac{\log_{10} N}{0.301}$$
$$= 3.32 \log_{10} N$$

where N is the number whose logarithm is to be calculated. The base 10 or common logarithm can be computed on any scientific calculator. With two coding levels (binary 0 and 1), the bandwidth is

$$B = \frac{C}{2 \log_2 N}$$
$$= \frac{1000 \text{ bits/s}}{2(1)}$$
$$= 500 \text{ Hz}$$

Note that $\log_2 2$ for a binary signal is simply 1. That's how the basic expression is derived, or

$$C = 2B \log_2 N$$

For binary, $\log_2 N = \log_2 2 = 1$. Therefore,

$$C = 2B(1)$$
$$= 2B$$

Now, in Fig. 12-8(b), let's see what our multilevel coding scheme does.

$$B = \frac{C}{2 \log_2 N}$$

The channel capacity is still 1000 bits/s or 1000 baud as shown in Fig. 12-8(b) because each symbol (level) interval is 1 ms long. However, each symbol represents 2 bits. Therefore, the channel capacity is 2000 bits/s. The number of symbol levels $N = 4$. The bandwidth then is

$$B = \frac{2000 \text{ bits/s}}{2 \log_2 4}$$

and $\log_2 4 = 2$.

$$B = \frac{2000}{2(2)}$$
$$= \frac{2000}{4}$$
$$= 500 \text{ Hz}$$

Look what happened here. With a multilevel coding scheme, we can transmit at twice the speed in the same bandwidth.

Another important aspect of information theory is the impact of noise on the signal. As it turns out, the greater the amount of noise, the lower the channel capacity. The presence of noise reduces the amount of information that can be transmitted in a given bandwidth. Increasing the bandwidth increases the information rate. On the other hand, as you learned in a previous chapter, a wider bandwidth also allows more noise to pass. As you can see then, the choice of bandwidth is a compromise between information transmission rate and the amount of noise.

The relationship between channel capacity, bandwidth, and noise is summarized in what is known as the *Shannon-Hartley theorem*.

Shannon-Hartley theorem

$$C = B \log_2 (1 + \text{S/N})$$

where C = channel capacity, bits/s
B = bandwidth, Hz
S/N = signal-to-noise power ratio

To see how the Shannon-Hartley theorem works, assume we wish to calculate the maximum channel capacity of a voice-grade telephone line with its bandwidth of 3100 Hz. Assume an S/N ratio of 30 dB.

First, we must convert 30 dB into a power ratio. If

$$\text{dB} = 10 \log P$$

where P is the power ratio, then

$$P = \text{antilog} \frac{\text{dB}}{10}$$

Antilogs are easily computed on a scientific calculator. A 30-dB S/N ratio translates to a power ratio of

$$P = \text{antilog} \frac{30}{10}$$

$$= \text{antilog } 3$$

$$= 1000$$

We can now calculate the channel capacity.

$$C = B \log_2 (1 + \text{S/N})$$

$$= 3100 \log_2 (1 + 1000)$$

$$= 3100 \log_2 1001$$

The base 2 logarithm of 1001 is

$$\log_2 1001 = 3.32 \log_{10} 1001$$

$$= 3.32(3)$$

$$= 9.97$$

$$\approx 10$$

Therefore, the channel capacity is

$$C = 3100(10)$$

$$= 31{,}000 \text{ bits/s}$$

That is a surprisingly high bit rate for such a narrow bandwidth. In fact, it appears to conflict with what we learned earlier. That is, the maximum channel capacity is twice the channel bandwidth. If the bandwidth of the voice-grade line is 3100 Hz, then the channel capacity is only

$$C = 2B$$

$$= 2(3100)$$

$$= 6200 \text{ bits/s}$$

This is for a binary (two-level) system only, and it assumes no noise. How then can the Shannon-Hartley theorem predict that the channel capacity is 31,000 bits/s when noise is present?

The Shannon-Hartley expression says that we can theoretically achieve a 31,000-bits/s channel capacity on our 3100-Hz bandwidth line. What it doesn't say is that multilevel encoding is needed to do so. Let's go back to the basic channel capacity expression of

$$C = 2B \log_2 N$$

We know $C = 31{,}000$ bits/s and B is 3100 Hz. What we don't know is the number of coding or symbol levels N. Rearranging the formula, we get

$$\log_2 N = \frac{C}{2B}$$

$$= \frac{31{,}000}{2(3100)}$$

$$= \frac{31{,}000}{6200}$$

$$= 5$$

Therefore,

$$N = \text{antilog}_2 5$$

The antilog of a number is simply the value we get when we raise the base (2) to the number, in this case 5.

$$N = 2^5 = 32$$

What these calculations say is that we can achieve the channel capacity of 31,000 bits/s if we use a multilevel encoding scheme that uses 32 different levels or symbols per interval rather than pure two-level binary. The baud rate of the channel C is still 6200 or twice the bandwidth.

$$C = 2B$$

$$= 2(3100)$$

$$= 6200 \text{ baud}$$

But because we use a 32-level encoding scheme, the bit rate is 31,000 bits/s. As it turns out, we rarely attempt to use the maximum channel capacity. In practice, it is very difficult to achieve. Typical systems limit the channel capacity to one-third to one-half the maximum to ensure more reliable transmission in the presence of noise.

■ TEST

Answer the following questions.

1. Data communications is the transmission of _____ information.

2. True or false. Audio and video signals may be transmitted using data communications techniques.

3. The earliest form of data communications was the _____.

4. The earliest form of radio data communications is known as _____.

5. The Morse code is made up of combinations of _____ and _____ to represent characters.

6. Give the Morse code for the following characters (see Fig. 12-2): K, 8, ,.

7. The 5-bit code used in teletype systems is called the _____ code.

8. A system of data communications using typewriterlike devices to send and receive coded messages is known as _____.

9. If in the Baudot code the character 11011 is sent and then the code 00111, the character _____ is received.

10. The most widely used binary data code is _____ which uses 7 bits to represent characters.

11. An 8-bit data code used in IBM systems is known as _____.

12. Write the ASCII codes for the following characters: B, 3, ?.

13. What ASCII character is transmitted to ring a bell? BEL = _____.

14. True or false. Serial data transfers are faster than parallel transfers.

15. The digital circuit often used to perform parallel-to-serial and serial-to-parallel data conversions is the _____.

16. In asynchronous transmissions, _____ and _____ bits are used to signal the beginning and end of a character.

17. The speed of a serial data transmission is usually expressed in _____.

18. The number of symbols occurring per second in a data transmission is called the _____.

19. The data rate of a binary signal with a bit time of 115 μs is _____.

20. A binary signal with a data rate of 20 Mbits/s has a bit time of _____.

21. True or false. Asynchronous transmission is faster than synchronous transmission.

22. In synchronous data transmission, data is sent in _____ of characters with special codes to signal the beginning and end.

23. In serial data transmission, a binary 0 is called a(n) _____ and a binary 1 is called a(n) _____.

24. The study of efficient information transfer is known as _____.

25. The amount of information transmitted is proportional to the system _____ and the _____ allowed for transmission.

26. The channel capacity of a 5-kHz bandwidth binary system is _____ bits/s assuming no noise.

27. If an 8-level encoding scheme is used in a 10-kHz bandwidth system, the channel capacity is _____ bits/s.

28. True or false. Multilevel or multisymbol binary encoding schemes permit more data to be transmitted in less time, assuming a constant symbol interval.

29. By using a multisymbol encoding scheme, a higher bit rate can be achieved with _____ (more, less) bandwidth.

30. The channel capacity of a 6-MHz channel with an S/N ratio of 25 dB is _____ Mbits/s.

31. The minimum allowable bandwidth for a binary signal with a bit rate of 200 kbits/s is _____ kHz.

12-2 Modems

What is the fastest and easiest way to communicate electronically with anyone else in the United States or in the world? Radio communications is common and simple, but few people or organizations have radio transmitters and receivers. If they did, they would have to operate on your frequency, use your type of modulation, and be available at the appropriate time. Other conditions such as radio wave propagation also enter into the picture. So radio communications is simply not the most convenient.

Without belaboring the point further, the answer to the question is simple. The fastest and easiest way to communicate with anyone else in the world is by the telephone. The telephone network is a vast electronic communications system that is in place, and virtually everyone uses it. You can talk to anyone else in the world by simply dialing their number. Your voice signal then travels over hundreds or even thousands of miles of wire and even over microwave relay and satellite communications links. In newer long-distance telephone systems, your voice may be carried on a light beam in fiber-optic cable.

Although voice is the primary signal carried by the telephone system, this network is now widely used to carry digital information as well. When it is necessary for computers to communicate or for a personal computer or terminal to have access to a large mainframe computer, the telephone network serves as the communication medium.

There are two primary problems in transmitting digital data over the telephone network. The first problem is that binary signals are usually switched dc pulses. That is, the binary 1s and 0s are represented by pulses of a single polarity, usually positive. The telephone line is designed to carry only ac analog signals. Voice and other analog signals are processed by numerous components, circuits, and equipment along the way. Usually only ac signals of a specific frequency range are allowed. Voice frequencies in the 300- to 3000-Hz range are the most common. If a binary signal were applied directly to the telephone network, it simply would not pass. The transformers, capacitive coupling, and other ac circuitry virtually ensure that no dc signals get through. Further, binary data is usually transmitted at high speeds. This high-speed data would essentially be filtered out by the system with its limited bandwidth. The question then is: "Just how does digital data get transmitted over the telephone network?" The answer is by using modems.

A *modem* is a device containing both a *modulator* and a *demodulator*. A modem is used to convert binary signals into analog signals capable of being transmitted over the telephone lines and to demodulate such analog signals and reconstruct the equivalent binary output. Figure 12-11 shows two ways that modems are commonly used in digital data transmission. In Fig. 12-11(*a*), two mainframe computers exchange data by speaking through modems. While one modem is transmitting, the other is receiving. Full duplex operation is also possible. In Fig. 12-11(*b*), a remote video terminal or personal computer uses a modem to communicate with a large mainframe computer. Again, modems are used at each end of the communications system. The modem is the interface between these units.

A modem causes the binary dc pulses to modulate an analog sine wave carrier that is compatible with the telephone line. The modem makes the binary signals compatible with the analog telephone network. And even though the bandwidth of the telephone network is essentially restricted to 3 kHz, a variety of tricky modulation techniques permit serial data transmission rates as high as about 15,000 bits/s to be transmitted. In this section we take a look at the various modulation and demodulation techniques used in modern data communications modems. These include frequency-shift keying, phase-shift keying, and quadrature amplitude modulation.

FSK Modems

The oldest and simplest form of modulation used in modems is *frequency-shift keying*

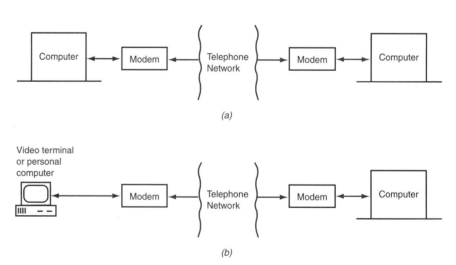

(a)

(b)

Fig. 12-11 How modems are used to permit digital data transmission on the telephone network.

(FSK). In FSK, two sine wave frequencies are used to represent binary 0s and 1s. For example, a binary 0, usually called a *space* in data communications jargon, has a frequency of 1070 Hz. A binary 1, referred to as a *mark,* is 1270 Hz. These two frequencies are alternately transmitted to create the serial binary data. The resulting signal looks something like that shown in Fig. 12-12. Note that both frequencies are well within the 300- to 3000-kHz bandwidth normally associated with the telephone system. This is illustrated in Fig. 12-13.

To permit simultaneous transmit and receive operations with a modem, known as *full-duplex operation,* another set of frequencies are defined. These are also indicated in Fig. 12-13. A binary 0 or space is 2025 Hz, and a binary 1 or mark is 2225 Hz. These tones are also within the telephone bandwidth but are spaced far enough from the other frequencies so that selective filters can be used to distinguish between the two. The 1070- and 1270-Hz tones are used for transmitting (originate) and the 2025- and 2225-Hz tones are used for receiving (answer).

A general block diagram of an FSK modem is illustrated in Fig. 12-14. Each modem contains an FSK modulator and an FSK demodulator so that both send and receive operations can be achieved. Bandpass filters at the inputs to each modem separate the two tones. For example, in the upper modem, a bandpass filter allows frequencies between 1950 and 2300 Hz to pass. This means that the 2025- and 2225-Hz tones will be passed but the 1070- and 1270-Hz tones generated by the internal modulator

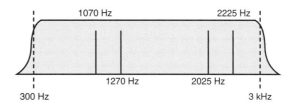

Fig. 12-13 The FSK signals within the telephone audio bandpass.

will be rejected. The lower modem has a bandpass filter that accepts the lower-frequency tones while rejecting the upper-frequency tones generated internally.

A wide variety of modulator and demodulator circuits are used to produce and recover FSK. Virtually all the circuits described in the chapter on FM can or have been used. A typical FSK modulator is simply an oscillator whose frequency can be switched between two frequencies, usually by switching in different capacitor values. Both *RC* and *LC* oscillators are used. At the lower audio frequencies, *RC* oscillators are preferred because of their simplicity. A variety of demodulators are also used. These include the PLL, pulse-averaging discriminator, and others.

Most of the newer modems use digital techniques because they are simpler and more adaptable to IC implementation. One type of digital FSK modulator is shown in Fig. 12-15. A clock oscillator generates a clock at a frequency of 271,780 Hz. It is applied to a binary frequency divider. This divider is usually some form of binary counter with various feedback logic gates to set the divide ratio. This frequency divider is set up so that it will divide by two different integer values. One divide ratio will produce the mark frequency, and the other will produce the space frequency.

To transmit a space at 1070 Hz, the divide ratio logic in Fig. 12-15 produces frequency division by 127. That is, when the serial binary input is zero, the frequency divider output will be $1/127$ of its input. This means that the output frequency will be 2140 Hz. This is fed to a single flip-flop that divides the frequency by 2, producing the desired 1070-Hz output. The flip-flop produces a 50 percent duty cycle square wave; that is, its on and off times are equal in length. The reason for this is that if the duty cycle of the frequency divider is other than 50 percent, the square wave when converted to a

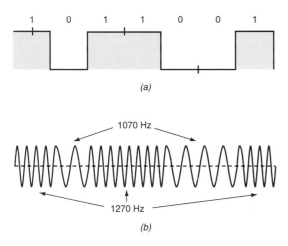

Fig. 12-12 Frequency-shift keying. (a) Binary signal, and (b) FSK signal.

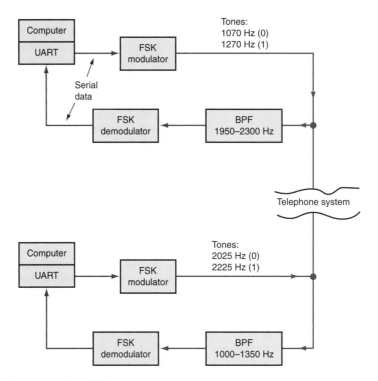

Fig. 12-14 Block diagram of an FSK modem.

sine wave will produce excessive distortion. The flip-flop output is passed through a low-pass filter which removes the higher odd harmonics producing a 1070-Hz sine wave tone.

When a binary 1 is applied to the divide ratio logic, the frequency divider will divide by 107. This produces an output frequency of 2540 Hz which when divided by 2 in the flip-flop produces the desired 1270-Hz output. The low-pass filter removes the higher-frequency harmonics, producing a sine wave output.

A digital method of FSK demodulation is shown in Fig. 12-16. The sine wave FSK signal is applied to a limiter which removes any amplitude variations and which shapes the signal into a square wave. The square wave is then applied to a gate circuit. The gate is used to turn a

1-MHz clock signal off and on. A binary counter counts or accumulates the 1-MHz clock pulses.

When a low-frequency space signal is applied, the period of the signal will be long, thereby allowing the gate to open and the binary counter to accumulate the clock pulses. The detection logic which is a set of binary gates determines whether the number in the binary counter is above or below some predetermined value. For the low-frequency tone input, the number in the counter will be higher than the counter value when the high-frequency or mark signal is applied to the input. The detection logic produces a binary 0 or binary 1 output depending upon whether the number in the counter is above or below a specific value between the two limits.

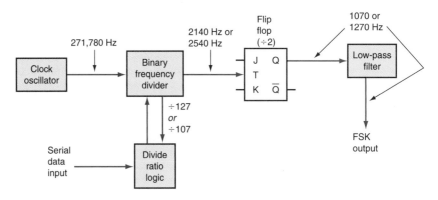

Fig. 12-15 A digital FSK modulator.

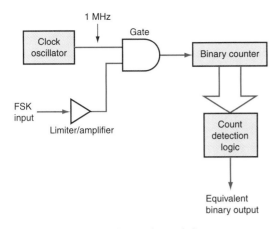

Fig. 12-16 A digital FSK demodulator.

Frequency-shift keying is used primarily in low-speed modems capable of transmitting data up to a speed of 300 baud or bits/s. This is a relatively slow data rate and is rarely used today. Higher speeds are used to shorten the transmission time for long messages or sets of data. Most communications data rates are 9600 baud or higher. Rates of 28.8, 33.6, and 56 kbits/s are common. To achieve these higher speeds, other forms of modulation are used. These include phase-shift keying and quadrature amplitude modulation.

The UART

Note that both modems in Fig. 12-14 contain a device called a UART which stands for *universal asynchronous receiver transmitter*. This is a digital IC used to perform parallel-to-serial and serial-to-parallel data transfers. Most data transfers and operations inside the computer are parallel, but as you know, data communications uses mainly serial binary data. The UART performs the necessary conversions.

Figure 12-17 shows a general block diagram of a UART. All this circuitry is typically contained within a single large-scale IC. Parallel data from the computer data bus is fed into and

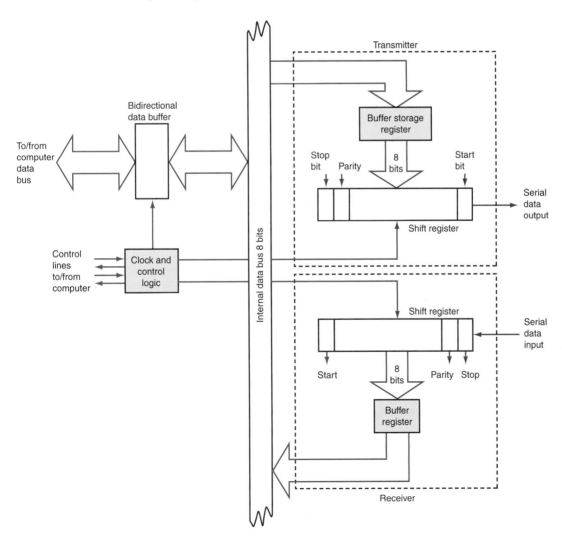

Fig. 12-17 General block diagram of a UART.

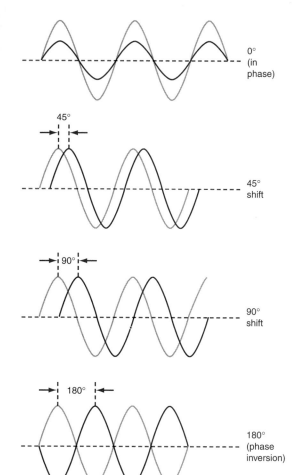

Fig. 12-18 Examples of phase shift.

out of a bidirectional data buffer which is usually a storage register with appropriate level-shifting circuits. The data, usually parallel 8-bit words, is put on an internal data bus. Before being transmitted, the data is stored first in a buffer storage register and then sent to a shift register. A clock signal shifts the data out serially 1 bit at a time. Note here that the internal circuitry adds start and stop bits and a parity bit. The start and stop bits signal the beginning and end of the word, and the parity bit is used to detect error. The resulting serial data word is transmitted 1 bit at a time.

The receive section of the UART is at the bottom of the diagram in Fig. 12-17. Serial data is shifted into a shift register. There the start, stop, and parity bits are stripped off. The remaining data is transferred to a buffer register, to the internal data bus, and through the bidirectional data buffer to the computer in parallel form. The clock and control logic circuits in the UART control all internal shifting and data transfer operations under the direction of control signals from the computer.

PSK Modems

Phase-shift keying (PSK)

In *phase-shift keying (PSK),* the binary signal to be transmitted changes the phase shift of a sine wave depending upon whether a binary 0 or binary 1 is to be transmitted. Recall that phase shift is a time difference between two sine waves of the same frequency. Figure 12-18 illustrates several examples of phase shift. Note that a phase shift of 180° represents the maximum difference and is also known as a phase reversal.

Figure 12-19 illustrates the simplest form of PSK known as *binary PSK (BPSK).* During the time that a binary 0 occurs, the carrier signal is transmitted with one phase, but when a binary 1 occurs, the carrier is transmitted with a 180° phase shift.

Binary PSK (BPSK)

Figure 12-20 shows one kind of circuit used for generating BPSK. It is a standard lattice ring modulator or balanced modulator used for generating DSB signals. The carrier sine wave, usu-

ally 1600 or 1700 Hz, is applied to the input transformer T_1, and the binary signal is applied to the transformer center taps. The binary signal provides a switching signal for the diodes. When a binary 0 appears at the input, A is positive and B is negative, so diodes D_1 and D_4 conduct. They act as closed switches, thereby connecting the secondary of T_1 to the primary of T_2. The windings are phased so that the BPSK output is in phase with the carrier input.

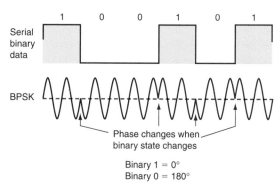

Fig. 12-19 Binary phase-shift keying (BPSK).

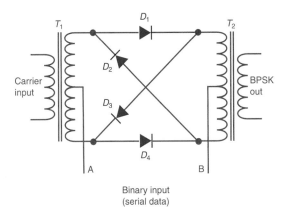

Fig. 12-20   A BPSK modulator.

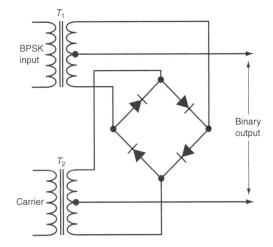

Fig. 12-21   A BPSK demodulator.

When a binary 1 appears at the input, A is negative and B is positive so diodes D_1 and D_4 are cut off while diodes D_2 and D_3 conduct. This causes the secondary of T_1 to be connected to the primary of T_2 but with the interconnections reversed. This introduces a 180° phase-shift carrier at the output.

Demodulation of BPSK signal is also done with a balanced modulator. A version of the diode ring or lattice modulator can be used as shown in Fig. 12-21. This is actually the same circuit as in Fig. 12-20, but the output is taken from the center taps. The BPSK and carrier signals are applied to the transformers. Integrated-circuit balanced modulators can also be used at the lower frequencies.

The key to demodulating BPSK is that a carrier with the correct frequency and phase relationship must be applied to the balanced modulator along with the BPSK signal. Typically the carrier is derived from the BPSK signal itself. This is done with a carrier recovery circuit shown in Fig. 12-22. A bandpass filter ensures that only the desired BPSK signal is passed. The signal is then squared or multiplied by itself by a balanced modulator or ana-

log multiplier which applies the same signal to both inputs. Squaring removes all the 180° phase shifts. The result is an output that is twice the input signal frequency or 2f. A bandpass filter set at twice the carrier frequency passes only this signal. The resulting signal is applied to the phase detector of a PLL. Note that a ×2 frequency multiplier is used between the VCO and phase detector. This ensures that the VCO frequency is at the carrier frequency. Because the PLL is used, the VCO will track any carrier frequency shifts. The result is a signal with the correct frequency and phase relationship for proper demodulation. The carrier is applied to the balanced modulator-demodulator along with the BPSK signal. The output is the recovered binary data stream.

In order to correctly recover the binary data transmitted by BPSK, the received signal must be fed to a balanced modulator along with a carrier whose frequency and phase is identical to that used on the original modulator. This ensures that the phase of the received signal is compared to the proper phase of the carrier.

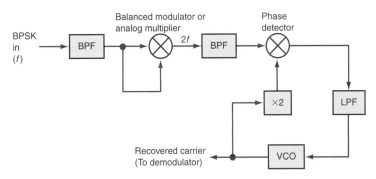

Fig. 12-22   A BPSK carrier recovery circuit.

Carrier recovery

The *carrier recovery circuit* just described produces a carrier of the correct phase at the receiver. However, the required circuitry is complex and expensive. To eliminate this problem and simplify the demodulation process, a version of BPSK called *differential PSK (DPSK)* can be used. In DPSK, there is no absolute carrier phase reference. Instead, the transmitted signal itself becomes the phase reference. In demodulating DPSK, the phase of the received bit is compared to the phase of the previously received bit.

Differential PSK (DPSK)

In order for DPSK to work, the original binary bit stream must undergo what is known as differential phase coding. This amounts to processing the serial bit stream through an inverted exclusive OR (XNOR) circuit as shown in Fig. 12-23(*a*). Note that the output of the XNOR circuit is applied to a 1-bit delay circuit before being applied back to the input. The 1-bit delay may simply be a clocked flip-flop or a delay line. The resulting bit pattern will permit the signal to be recovered by comparing the present bit phase with the previously received bit phase.

In Fig. 12-23(*b*), the input binary word to be transmitted is shown along with the output of the XNOR circuit. An XNOR circuit is simply a 1-bit comparator that produces a binary 1 output when both inputs are alike and a binary 0 output when the two bits are different. The output of the circuit is delayed for a 1-bit interval by storing it in a flip-flop. Therefore, the XNOR inputs are the current bit plus the previous bit. The XNOR signal is then applied to the balanced modulator along with the carrier to produce a BPSK signal.

Demodulation is extremely simple with the circuit shown in Fig. 12-24. The DPSK signal is applied to one input of the balanced modulator and a 1-bit delay circuit, either a flip-flop or a delay line. The output of the delay circuit is used as the carrier. The resulting output is filtered by a low-pass filter to recover the binary data. Typically the low-pass filter output is shaped with a Schmitt trigger or comparator to produce clean, high-speed binary levels.

The main problem with BPSK and DPSK is that the speed of data transmission is limited in a given bandwidth. One way to increase the binary data rate while not increasing the bandwidth required for the signal transmission is to encode more than 1 bit per phase change. There is a symbol change for each bit change so the baud (symbol) rate is the same as the bit rate. In BPSK and DPSK, each binary bit produces a specific phase change. It is also possible to use combinations of two or more bits to specify a particular phase shift. Therefore, a symbol change (phase shift) can represent multiple bits. We can encode more bits per baud. The bit rate of data transfer can be higher than the baud rate, while the signal occupies no more bandwidth.

QPSK

One commonly used system for doing this is known as quadrature, quaternary, or quadraphase PSK (*QPSK* or 4-PSK). In QPSK, each pair of successive digital bits in the transmitted word are assigned a particular phase as

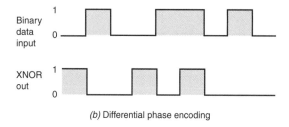

(a) DPSK modulator

(b) Differential phase encoding

Fig. 12-23 The DPSK process.

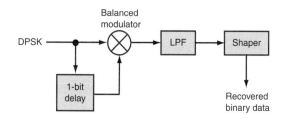

Fig. 12-24 A DPSK demodulator.

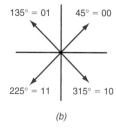

Dibit		Phase shift
0	0	45°
0	1	135°
1	1	225°
1	0	315°

135° = 01 | 45° = 00

225° = 11 | 315° = 10

(a)   *(b)*

Fig. 12-25 Quadrature PSK modulation. (*a*) Phase angle of carrier for different pairs of bits. (*b*) Phasor representation of carrier sine wave.

indicated in Fig. 12-25(*a*). Each pair of bits called a *dibit* is represented by a specific phase. A 90° phase shift exists between each pair of bits. Other phase angles can also be used as long as they have a 90° separation. For example, it is common to use phase shifts of 45°, 135°, 225°, and 315° as shown in Fig. 12-25(*b*).

A modulator circuit for producing QPSK is shown in Fig. 12-26. It consists of a 2-bit shift register implemented with flip-flops. This shift register is more commonly known as a bit splitter. The serial binary data train is shifted through this register. The bits from the two flip-flops are applied to balanced modulators. The carrier oscillator is applied to balanced modulator 1 and through a 90° phase shifter to balanced modulator 2. The outputs of the balanced modulators are linearly mixed to produce the QPSK signal.

The output from each balanced modulator is a BPSK signal as described before. With a binary 0 input, one phase of the carrier will be produced, and with a binary 1 input, the carrier phase will be shifted to 180°. The output of balanced modulator 2 will also have two phase states 180° out of phase with one another. Just keep in mind that the 90° carrier phase shift at the input causes the outputs from balanced modulator 2 to be shifted 90° from those of balanced modulator 1. The result is four different carrier phases which are combined two at a time in the linear mixer. The result is four unique output phase states.

Figure 12-27 shows the outputs of one possible set of phase shifts. Note that the carrier outputs from the two balanced modulators are shifted 90°. When the two carriers are algebraically summed in the mixer, the result is an output sine wave that has a phase shift of 225° which falls between the two balanced modulator signals.

Dibit

A demodulator for QPSK is illustrated in Fig. 12-28. The carrier recovery circuit is similar to the one described previously. The carrier is applied to balanced modulator 1 and is shifted 90° before being applied to balanced modulator 2. The outputs of the two balanced modulators are filtered and shaped into the dibit. The two bits are combined in a shift register and shifted out to produce the originally transmitted binary signal.

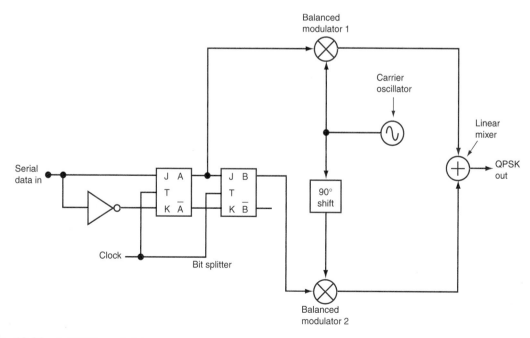

Fig. 12-26 A QPSK modulator.

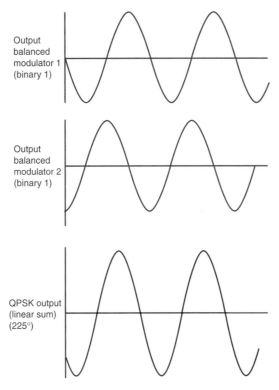

Quadrature amplitude modulation (QAM)

Output balanced modulator 1 (binary 1)

Output balanced modulator 2 (binary 1)

QPSK output (linear sum) (225°)

Fig. 12-27 How the modulator produces the correct phase by adding two signals.

More bits per phase change can also be encoded to produce even higher data rates. For example, three serial bits can be used to produce a total of eight different phase changes. This produces 8-PSK. Four serial input bits can produce 16 different phase changes. This results in 16-PSK which produces an even higher data rate. Again, the basic idea here is to encode more bits per carrier phase change or more bits per baud in order to produce a higher data rate within a restricted bandwidth.

QAM Modems

One of the most popular modulation techniques used in modems for encoding more bits per baud is *quadrature amplitude modulation (QAM)*. It uses both AM and PM of a carrier. In addition to producing different phase shifts, the amplitude of the carrier is also varied.

A popular version of QAM is known as 8-QAM. This system uses a total of four different phase shifts as in QPSK and two carrier amplitudes. With four possible phase shifts and two different carrier amplitudes, a total of eight different states can be transmitted. With eight states, 3 bits can be encoded for each baud or symbol transmitted. Each 3-bit binary word transmitted uses a different PM/AM combination.

One way to illustrate an 8-QAM signal is to use what is known as a constellation diagram, which shows all possible phase and amplitude combinations. An 8-QAM constellation diagram is shown in Fig. 12-29. The points in the diagram indicate the eight possible PM/AM combinations. Note that there are two amplitude levels for each phase position. Point A shows a low carrier amplitude with a phase shift of 135°. It represents 100. Point B shows a higher amplitude and a phase shift of 315°. This sine wave represents 011.

A block diagram of an 8-QAM modulator is shown in Fig. 12-30. The binary data to be transmitted is shifted serially into the 3-bit shift register. These bits are applied in pairs to two 2-to-4 level converters. A 2-to-4 level converter is a circuit that translates a pair of binary inputs into one of four possible dc output volt-

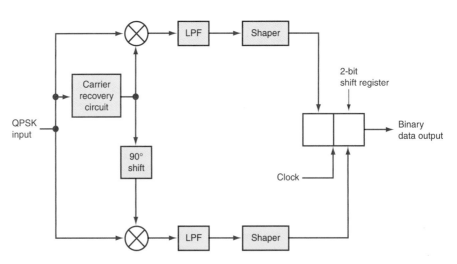

Fig. 12-28 A QPSK demodulator.

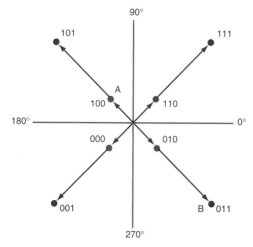

Note: Each vector has a specific amplitude and phase shift and represents one 3-bit word.

Fig. 12-29 A constellation diagram of QAM signal.

age levels. A 2-to-4 level converter is basically a simple D/A converter. The idea is to produce four voltage levels corresponding to the different combinations of two input bits. The result is four equally spaced voltage levels. These are applied to the two balanced modulators fed by the carrier oscillator and a 90° phase shifter as in a QPSK modulator. Each balanced modulator produces four different output PM/AM combinations. When these are combined in the linear mixer, eight different PM/AM combinations are produced. The result is 8-QAM as described previously. The most critical part of the circuit is the 2-to-4 level converters which must have very precise output amplitudes so that when they are combined in the linear summer,

the correct output and phase combinations are produced.

A 16-QAM signal can also be generated by encoding 4 input bits at a time. The result is 8 phase shifts and two amplitude levels, for a total of 16 different PM/AM combinations. Systems with 32- and 64-QAM signals are used.

The highest-speed modems now use QAM plus *trellis code modulation (TCM)*. TCM is a special digital encoding technique that permits many more constellation points to be mapped and recovered reliably. For example, a 28.8-kbits/s modem with TCM defines 768 different phase-amplitude constellation points.

Quadrature amplitude modulation is widely used in modems that transmit computer data over the telephone lines. Recall that the telephone system bandwidth is restricted to approximately 3 kHz. For that reason, it is extremely difficult to transmit binary data at speeds higher than approximately 1200 baud or bits/s. However, by using QPSK, QAM, or similar techniques, very high data rates can be achieved. For example, using QAM, data rates of up to 56 kbits/s are regularly achieved over the standard telephone network.

The BPSK, QPSK, QAM, and other techniques are also widely used to transmit digital data in microwave and satellite radio communications. With very high frequency carriers, bit rates of millions of bits per second can easily be achieved with minimum transmission error in a noisy environment.

Trellis code modulation (TCM)

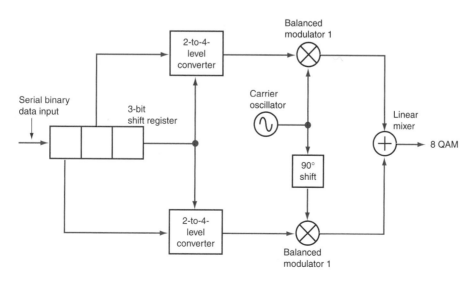

Fig. 12-30 An 8-QAM modulator.

ADSL Modems

Using current modem technology, QAM with trellis code modulation, the data rate is restricted to a maximum of 56 kbits/s over standard telephone lines. Although this speed (or less) is suitable for basic text applications, it is too slow for many applications using complex graphics or video. Downloading very large (megabytes) files of data or software can take many minutes or even hours at 56 kbits/s or less. Much higher rates are not only desirable but also necessary as the size of the data files and/or their complexity increases.

Asynchronous digital subscriber line (ADSL)

A recent development aimed at providing higher speed is a system called *asynchronous digital subscriber line (ADSL)*. This system permits downstream data rates up to 8 Mbits/s and upstream rates up to 640 kbits/s using the existing telephone lines. (*Asynchronous* means unequal upstream and downstream rates.)

Digital signal processor (DSP)

The connection between a telephone subscriber and the nearest telephone central office is twisted-pair cable using size 24 or 26 copper wire. Its length is usually anywhere between 9000 and 18,000 ft (2.7 to 5.5 km). This cable acts like a low-pass filter. Its attenuation to very high frequencies is enormous. Digital signals are seriously delayed and distorted by such a line. For this reason, only the lower 0- to 4-kHz bandwidth is used for voice. Traditional modems operate in the center of this voice band at 1700 or 1800 Hz using QAM.

ADSL employs some special techniques so that more of the line bandwidth can be used to increase data rates. Even though a 1-MHz signal may have an attenuation of up to 90 dB on an 18,000-ft line, special amplifiers and frequency compensation techniques make the line usable.

Discrete multitone (DMT)

The modulation scheme used with ADSL modems is called *discrete multitone (DMT)*. It divides the upper-frequency spectrum of the telephone line into 256 channels, each 4 kHz wide. See Fig. 12-31. Each channel, called a *bin,* is designed to transmit at speeds up to 15 kbits/s per baud or 60 kbits/s.

Each channel contains a carrier that is simultaneously phase-amplitude-modulated (QAM) by some of the bits to be transmitted. The serial data stream is divided up so that each carrier transmits some of the bits. All bits are transmitted simultaneously. Also, all the carriers are frequency-multiplexed into the line bandwidth above the normal voice telephone channel, as Fig. 12-31 shows.

The upstream signal uses the 4-kHz bins from 25.875 to 138.8 kHz, and the downstream signal uses bins in the 138-kHz to 1.1-MHz range.

The number of bits per baud and the data rate per bin are varied according to the noise on the line. The less noise there is in each bin, the higher the data rate. Very noisy bins will carry few or no bits, whereas quiet bins can accommodate the maximum 15 kbits/s per baud or 60 kbits/s.

This system is very complex and is implemented digitally with a high-speed microcomputer chip called a *digital signal processor (DSP)*. The DSP chip handles all modulation and demodulation functions by simulating them digitally.

Several different levels of ADSL are available. The data rate for each depends upon the length of the subscriber twisted pair. The shorter the cable, the higher the data rate. The highest standard rate is 6.144 Mbits/s downstream and 576 kbits/s upstream at a line distance not to exceed 9000 ft. The minimum rate is 1.536 Mbits/s downstream and 384 kbits/s upstream at line distances up to 18,000 ft (called ADSL Lite or G. Lite).

Although ADSL is not now widely implemented, it is expected that telephone companies will quickly adopt and implement ADSL systems. This will permit PC users to continue to use their existing telephone lines for data speeds of 1.536 Mbits/s or higher with an ADSL modem installed.

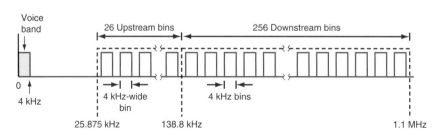

Fig. 12-31 Spectrum of telephone line used by ADSL.

Cable Modems

Many cable TV systems are set up to handle high-speed digital data transmission. The digital data is used to modulate a high-frequency carrier that is frequency-multiplexed onto the cable that also carries the TV signals.

Cable TV systems use a coaxial cable or combination fiber-optic cable plus coaxial cable system. The bandwidth of the cable is approximately 750 MHz. This spectrum is divided up into 6-MHz-wide TV channels containing the TV, voice, and video signals. The standard VHF and UHF frequencies normally assigned to wireless TV are used on the cable, along with some special cable frequencies. The TV signals are therefore frequency-division-multiplexed onto the cable.

Figure 12-32 shows the spectrum of the cable. Television channels extend from 50 MHz (Channel 2) up to 550 MHz. In this 500 MHz of bandwidth, up to 83 channels of 6 MHz can be accommodated.

The spectrum above the TV channels from 550 to 750 MHz is available for digital data transmission. Standard 6-MHz channels are used, giving approximately 33 channels. These channels are used for downstream data transmissions (from the remote computer down to the user).

The spectrum from 5 to 50 MHz is also unused, as you can see from Fig. 12-32. This region is divided into seven 6-MHz channels that are used for upstream data transmission (from the user up to the remote computer).

Cable modems use 64-QAM for downstream data. Using 64-QAM in a 6-MHz channel provides a data rate up to 30 Mbits/s. This method of modulation uses 64 different phase-amplitude combinations (symbols) to represent multiple bits.

Standard QPSK is used in the upstream channels to achieve a data rate of about 10 Mbits/s.

Cable modems provide significantly higher data rates than can be achieved over the standard telephone system. The primary limitation is the existence or availability of a cable TV system that offers such data transmission services.

■ TEST

Supply the missing information in each statement.

32. A modem contains both _____ and _____ circuits.
33. Modems convert _____ signals to _____ signals and vice versa.
34. The transmission medium most widely used with modems is the _____.
35. Modems are the interface between _____.
36. _____ modulation is used in low-speed modems.
37. In FSK, binary 0 and 1 levels are represented by different _____.
38. In an FSK modem, a high frequency represents a(n) _____ and a low frequency represents a(n) _____.
39. The typical frequencies used in a full-duplex FSK modem are as follows: originate 0 _____ Hz and 1 _____ Hz; answer 0 _____ Hz and 1 _____ Hz.
40. The IC used to perform serial-to-parallel and parallel-to-serial conversions and other operations in a modem is known as a(n) _____.
41. The modulation method that represents bits as different phase shifts of a carrier is known as _____.
42. In BPSK, phase shifts of _____ degrees and _____ degrees are used to represent binary 0 and binary 1, respectively.

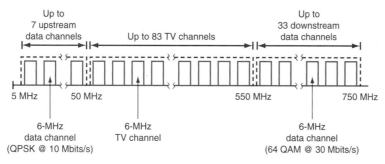

Fig. 12-32 Cable TV spectrum showing upstream and downstream data channels on the cable.

43. The circuit used to produce BPSK is a(n) _____.

44. The circuit used to demodulate BPSK is a(n) _____.

45. A(n) _____ circuit is used to generate the carrier to be used in demodulating a BPSK signal.

46. A carrier recovery circuit is not needed with _____ PSK modulation.

47. The key circuit used in a DPSK modulator is the _____.

48. The number of different phase shifts used in QPSK is _____.

49. In QPSK, how many bits are represented by each phase shift?

50. The number of bits represented by each phase shift in 16-PSK is _____.

51. The circuit used to create a dibit is known as a(n) _____.

52. A circuit that converts a 2-bit binary code into one of four dc voltage levels is known as a(n) _____.

53. Quadrature amplitude modulation is a combination of _____ modulation and _____ modulation.

54. Conventional telephone modems using QAM and TCM can achieve data rates up to _____.

55. The drawing showing all phase-amplitude combinations possible with QAM is called a(n) _____.

56. The upper speed of an ADSL modem is determined by _____.

57. The form of modulation used in ADSL is _____.

58. The upper speed limit of an ADSL modem is _____ downstream and _____ upstream.

59. The type of multiplexing used in ADSL and cable modems is _____.

60. A cable modem uses _____ modulation downstream and _____ modulation upstream.

61. The maximum downstream data rate on a cable modem is typically _____.

Determine whether each statement is true or false.

62. Conventional telephone modems must transmit signals in the 300-Hz to 3-kHz range.

63. Carrier recovery is not required in a QPSK demodulator.

64. When QPSK is used, the bit rate is faster than the baud (symbol) rate.

65. Each phase and/or amplitude change can represent only one bit.

66. With QAM, a 56 kbits/s signal can be transmitted within a 3000-Hz bandwidth.

12-3 Protocols and Error Detection and Correction

To communicate successfully using serial digital data, some guidelines must be established between the sender and the receiver. That is, specific rules and procedures must be agreed upon at the transmitting and receiving ends of the communications system. These rules and procedures are called *protocols*. A variety of different protocols are used in data communications.

When high-speed binary data is transmitted over a communications link, whether it is a cable or radio, errors will occur. These errors are changes in the bit pattern caused by interference, noise, or equipment malfunctions. Such errors will cause incorrect data to be received. To ensure reliable communications, schemes have been developed to detect such bit errors so that corrections can be made. In this section we will review some of the basic error-detection and error-correction schemes used in data communications.

Asynchronous Protocols

The simplest form of protocol is the use of stop and start bits in the transmission of individual characters using asynchronous methods. Refer back to Fig. 12-9. When no information is being transmitted, the line between the transmitter and receiver stays at the binary 1 or mark state. To begin transmission of a single 7- or 8-bit ASCII character, a start bit is generated. The start bit is a binary 0 or space transmitted for a 1-bit time. This signals the receiver that a character is about to be sent so that the receiver can then prepare itself for the forthcoming serial bits. The character bits are then transmitted. The character is followed by a stop bit that is a binary 1 or mark transmitted for 1- or 2-bit intervals. This signals the receiver that the character has been transmitted.

Even though this protocol is extremely simple, it illustrates the concept used in the transmission of all digital data. The minimum protocol, then, is a start bit and a stop bit framing a single character. But what happens when you want to transmit multiple characters in synchronous form?

In data communications, a message is more than one character. It is many characters composed of letters of the alphabet, numbers, punctuation marks, and other symbols in the desired sequence. A group of characters forms a *block.* When a message is sent, it is usually broken up into blocks, a block being a fixed number of characters. In synchronous data communications applications, the block is the basic transmission unit.

To identify a block, one or more special characters are transmitted prior to the block and after the block. These additional characters added to the beginning and end of the block perform a number of functions. These special characters are usually represented by 7- or 8-bit codes. Like the start and stop bits for a character, they signal the beginning and end of the transmission, but they are also used to identify a specific block of data and also provide means for error checking and detection.

Some of the characters at the beginning and end of each block are used for "handshaking" purposes. These are characters that give the transmitter and receiver status information. Figure 12-33 illustrates the basic *handshaking* process. For example, the transmitter may send a character indicating that it is ready to send data to the receiver. The receiver identifies that character and indicates its status. If the receiver is busy, it may indicate this by sending a character representing "busy" back to the transmitter. The transmitter will continue to send its ready signal until the receiver signals back that it is not busy or is ready to receive.

At that point, data transmission takes place. Once the transmission is complete, some additional handshaking takes place. The receiver acknowledges that it has received the information. The transmitter then sends a character indicating that the transmission is complete, and the receiver may acknowledge this.

To give you a more specific example, let's discuss the *Xmodem protocol,* which is widely used in data transmission between personal computers. Such data transmissions use asynchronous methods and the ASCII code.

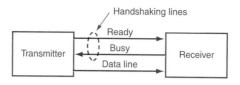

Fig. 12-33 The handshaking process in data communications.

The data transmission procedure begins with the receiving computer transmitting a negative acknowledge (NAK) character to the transmitter. The NAK is an ASCII character code of 7 bits that is transmitted serially back to the transmitter every 10 s until the transmitter recognizes it. Once the transmitter recognizes the NAK character, it will begin to send a 128-byte block of data. This block begins with a start of header (SOH) character, which is another ASCII character meaning that the transmission is beginning. A header is usually two or more characters preceding the actual data block which give auxiliary information. In the Xmodem protocol, the header consists of 2 bytes designating the block number. In most messages, several blocks of data are transmitted and each is numbered sequentially. The first byte is the block number in binary code. The second byte is the complement of the block number; that is, all bits have been inverted. Then, the 128-byte block is transmitted. At the end of the block, the transmitting computer sends a check sum byte. The check sum is a method of error detection that will be described later. It is the binary sum of all the binary information sent in the block.

The receiving computer looks at the block of data and also computes the check sum. If the check sum of the received block is the same as that transmitted, then it is assumed that the block was received correctly. If the block was received correctly, it sends an acknowledge (ACK) character back to the transmitter. Again, ACK is simply another ASCII code. Assuming ACK is received by the transmitter, then the next block of data is sent. However, if the block was not received correctly due to interference or equipment problems, the check sums will not match. As a result, the receiving computer will send the NAK codes back to the transmitter. If the transmitter receives NAK, it will automatically respond by sending the block again. This process is repeated until the entire block and the entire message is sent without errors.

Block

Handshaking

Xmodem protocol

When the entire message has been sent, the transmitting computer will send an end of transmission (EOT) character. The receiving computer replies with an ACK character, thus terminating the communication. Keep in mind that each character is sent along with its start and stop bits since this is an asynchronous protocol.

Synchronous Protocols

Error checking

In synchronous data communications, the protocols become even more complex. Like the protocols used in the asynchronous Xmodem system, various control characters are used for signaling purposes at the beginning and ending of the block of data to be transmitted.

Bisync protocol

An example of a typical synchronous protocol is IBM's *Bisync protocol,* which is widely used in computer communications. The protocol usually begins with the transmission of two or more sync (SYN) characters. See Fig. 12-34. These characters signal the beginning of the transmission and are also used to initialize the clock timing circuits in the receiving modem. This will ensure proper synchronization of the data transmitted 1 bit at a time.

Bit error rate (BER)

After the SYN characters, a start of header (SOH) character is transmitted. The header is a group of characters that usually identifies the type of message to be sent, a block number, a priority code, or some specific routing destination. The end of the header is signaled by a start of text (STX) character. At this point, the desired message is transmitted, 1 byte at a time. Remember that in synchronous transmission no start and stop bits are sent. The 7- or 8-bit words are simply strung together one after another. The receiver must sort them out into individual binary words which will be dealt with on a parallel basis later in the computer. A block usually consists of no more than 256 characters.

At the end of the block an end of block (ETB) character is transmitted. If this is the last block in a complete message, then an end

of text (ETX) character is transmitted. Finally, an end of transmission (EOT) character is transmitted signaling the end of the transmission. At the very end of the transmission, an error-detection code is appended. This is usually 1 or 2 bytes known as the binary check code (BCC), which will be discussed later.

Error Detection and Correction

As you have seen, the protocols contain some form of *error-checking method.* A parity bit is used in asynchronous character transmission. A BCC is used in both the Xmodem asynchronous and the Bisync synchronous protocols. Let's take a look at the error-detection and error-correction process in more detail.

The main objective in error detection and correction is to ensure a fully correct transmission of the message. When high data rates are used in a noisy environment, bit errors are inevitable. If the S/N ratio is good, the number of errors will be extremely small. The number of bit errors that occur for a given number of bits transmitted is referred to as the *bit error rate (BER).* The BER is similar to a probability in that it is the ratio of the number of bit errors to the total number of bits transmitted. If there is one error for 100,000 bits transmitted, the BER is $1/100,000 = 10^{-5}$. The bit rate may be higher or lower depending upon the equipment, the environment, and other considerations.

Many different methods have been used to ensure 100 percent accuracy in transmission. The simplest, of course, is simply to transmit each character or each message multiple times until it is properly received. This is known as redundancy. For example, one protocol may specify transmitting each character twice in succession. Entire blocks or messages may be treated in the same way.

Another approach to reliable transmission is to use special codes which permit automatic checking for accuracy. One such code is the ARQ exact-count code. This is a special 7-bit

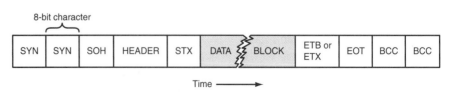

Fig. 12-34 Bisync synchronous protocol.

Bit:	1	2	3	4	5	6	7	Letter	Figure
	0	0	0	1	1	1	0	Letter shift	
	0	1	0	0	1	1	0	Figure shift	
	0	0	1	1	0	1	0	A	—
	0	0	1	1	0	0	1	B	?
	1	0	0	1	1	0	0	C	:
	0	0	1	1	1	0	0	D	(WRU)
	0	1	1	1	0	0	0	E	3
	0	0	1	0	0	1	1	F	%
	1	1	0	0	0	0	1	G	@
	1	0	1	0	0	1	0	H	£
	1	1	1	0	0	0	0	I	8
	0	1	0	0	0	1	1	J	(bell)
	0	0	0	1	0	1	1	K	(
	1	1	0	0	0	1	0	L	)
	1	0	1	0	0	0	1	M	.
	1	0	1	0	1	0	0	N	,
	1	0	0	0	1	1	0	O	9
	1	0	0	1	0	1	0	P	0
	0	0	0	1	1	0	1	Q	1
	1	1	0	0	1	0	0	R	4
	0	1	0	1	0	1	0	S	'
	1	0	0	0	1	0	1	T	5
	0	1	1	0	0	1	0	U	7
	1	0	0	1	0	0	1	V	=
	0	1	0	0	1	0	1	W	2
	0	0	1	0	1	1	0	X	/
	0	0	1	0	1	0	1	Y	6
	0	1	1	0	0	0	1	Z	+
	0	0	0	0	1	1	1		(blank)
	1	1	0	1	0	0	0		(space)
	1	0	1	1	0	0	0		(line feed)
	1	0	0	0	0	1	1		(carriage return)

Fig. 12-35 The ARQ exact-count code.

binary code used for representing letters of the alphabet, numbers, and other symbols. Refer to Fig. 12-35. Each 7-bit word contains exactly three binary 1s. One way to determine whether a character is received properly is to count the number of binary 1s in each character received. If the count is 3, then the character was most likely sent correctly. If noise occurs and causes one of the bits to change, the number of 1s would be a different value and an error would be indicated. This might signal the system to repeat the character or the entire block.

Parity

One of the most widely used systems of error detection is known as parity. *Parity* is a system in which each character transmitted contains one additional bit. That bit is known as a parity bit. The bit may be a binary 0 or binary 1, depending on the number of 1s and 0s in the character itself.

Two systems of parity are normally used: odd and even. Odd parity means that the total number of binary 1 bits in the character, including the parity bit, is odd. Even parity means that the number of binary 1 bits in the character, including the parity bit, is even. Examples of odd and even parity are indicated below. The seven left-hand bits are the ASCII character, and the right-hand bit is the parity bit.

Odd parity:	10110011
	00101001
Even parity:	10110010
	00101000

The parity bit of each character to be transmitted is produced by a parity generator circuit. The parity generator is made up of several levels of exclusive OR (XOR) circuits as shown in Fig. 12-36. Normally the parity generator circuit monitors the shift register in the computer or modem UART. Just prior to transmitting the data in the register by shifting it out, the parity generator circuit generates the correct parity value and inserts it as the last character in the data stream. In an asynchronous system, the start bit comes first, followed by the character bits, the parity bit, and finally one or more stop bits.

At the receiving modem or computer, the serial data word is transferred into a shift

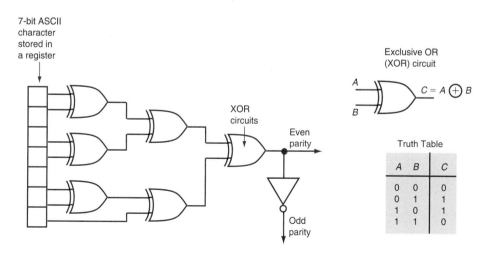

Fig. 12-36 A parity generator circuit.

register in the UART. A parity generator in the receiving UART produces the parity on the received character. If it matches the transmitted and received parity bit, then it is assumed that the character was transmitted correctly. If a parity error is detected, the system signals this to the computer. Then, depending upon the protocol or other procedures, the character may be retransmitted, the block may be transmitted, or the error may simply be ignored.

Although parity checking is simple and effective, it is useful only for detecting single bit errors. If two or more bit errors occur, the parity circuit may not detect it. If an even number of bit changes occurs, the parity circuit will not detect the error. In situations requiring more reliable transmission, more thorough schemes can be used.

Sometimes you will hear the individual character parity method of error detection referred to as the vertical redundancy check (VRC). In displaying characters transmitted in a data communications system, the bits are sometimes written vertically as illustrated in Fig. 12-37. The bit at the bottom is the parity or VRC bit as described earlier.

The term "vertical" also distinguishes the parity check from another method of error detection known as the horizontal or longitudinal redundancy check (LRC). An LRC is the process of exclusive-ORing all the characters in a particular block of transmitted data. For example, the top bit of the first word is exclusive-ORed with the top bit of the second word. The result of this operation is exclusive-ORed with the top bit of the third word, and so on until all the bits in a particular horizontal row have been exclusive-ORed. The final bit value then becomes 1 bit in a character known as the *block check character (BCC)* or the

block check sequence (BCS). Each row of bits is manipulated in the same way to produce the BCC. All the characters transmitted in the text as well as STX, ETX, ETB, EOT, and so on are included as part of the BCC. Typically, SYN and SOH characters are not included as part of the BCC.

The BCC is computed by circuits in the computer or modem as the data is transmitted, and it is then appended to the message. At the receiving end, the computer computes its own version of the BCC on the received data and compares it to the received BCC. Again, the two should be the same.

As it turns out, by knowing the parity on each character as well as the BCC, the exact location of a faulty bit can be determined. The individual character parity bits and the BCC bits provide a form of coordinate system that allows a particular bit error in one character to be identified. Once it is identified, the bit is simply complemented to correct it. The VRC identifies which character contains the bit error, while the LRC identifies which bit contains the error.

Another more reliable error-detection scheme is the *cyclic redundancy check (CRC)*. The CRC is a mathematical technique which is applied to the transmitted data. It is effective in catching 99.9 percent or more of transmission errors. The mathematical process is essentially a division. The entire string of bits in a block of data are considered to be one giant binary number which is divided by some preselected constant. The quotient resulting from the division is discarded, and any remainder is retained. This remainder is known as the CRC character.

Although the CRC mathematical process can actually be programmed using the computer's instruction set, in most applications it is computed by a special CRC hardware circuit. A typical circuit is shown in Fig. 12-38. It consists of several shift registers into which XOR gates have been inserted at specific points. The data to be checked is fed into the registers serially. The data is simply shifted in a bit at a time, and at the conclusion of the data, the contents will be the remainder of the division or the desired CRC value. Since 16 flip-flops total are used in the shift register, the CRC can be transmitted as two sequential 8-bit bytes. The CRC is computed as the data is transmitted, and the resulting CRC is appended to the block.

Cyclic redundancy check (CRC)

Block check character (BCC)

Character	D	A	T	A		C	O	M	LRC or BCC
(LSB)	0	1	0	1	0	1	1	1	1
	0	0	0	0	0	1	1	0	1
ASCII	1	0	1	0	0	0	1	1	0
	0	0	0	0	0	0	1	1	0
Code	0	0	1	0	0	0	0	0	1
	0	0	0	0	1	0	0	0	1
(MSB)	1	1	1	1	0	1	1	1	1
Parity R→ or VRC (odd)	1	1	0	1	0	0	0	1	0

Fig. 12-37 Vertical and horizontal redundancy checks.

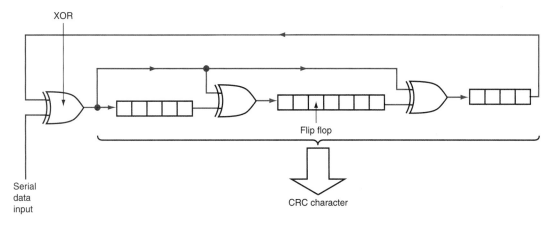

Fig. 12-38 A CRC error-detection circuit made with a 16-bit shift register and XOR gates.

At the receiving end, the CRC is computed by the receiving computer and compared to the received CRC characters. Assuming the two are alike, the message is correctly received. Any difference indicates an error, which then initiates retransmission or some other form of corrective action.

A number of error-correction schemes have also been devised to complement the error-detection methods just described. Typically these involve various mathematical processes on the characters to be transmitted which result in the addition of several bits of information to each character transmitted. The character and these extra bits can then be processed to determine whether an error occurs. If a bit error does occur, these error-correction codes will automatically identify it, making it possible for it to be corrected. The techniques for error correction involve complex computer and digital circuitry and are beyond the scope of this chapter.

■ TEST_____

Answer the following questions.

67. Rules and procedures that describe how data will be transmitted and received are referred to as _____.

68. List five methods of error detection and correction.

69. The process of exchanging signals between transmitter and receiver to indicate status or availability is called _____.

70. A popular protocol used in personal computer data communications is known as _____.

71. A string of characters making up a message or part of a message is referred to as a(n) _____ of data.

72. Stop and start bits are used with _____ (asynchronous, synchronous) data.

73. Synchronous protocols usually begin with the _____ character.

74. The characters that indicate message number, destination, or other facts are collectively referred to as the _____.

75. The _____ character designates the beginning of a message, while the _____ character indicates its end.

76. One simple but time-consuming way to ensure an error-free transmission is to send the data multiple times. This is called _____.

77. One type of special self-checking error codes is the _____ code.

78. Errors in data transmission are usually caused by _____.

79. The ratio of the number of bit errors to the total number of bits transmitted is known as the _____.

80. A bit added to the transmitted character to help indicate an error is called the _____ bit.

81. A number added to the end of a data block to assist in detecting errors is known as the _____.

82. Write the correct parity bits for each number.
    a. Odd: 0011000_____
    b. Even: 1111101_____
    c. Odd: 0101101_____
    d. Even: 1000110_____

83. Another name for parity is _____.

84. The basic building block of a parity generator circuit is the _____ gate.

85. Performing an exclusive OR on corresponding bits in successive data words produces a(n) _____ word.

86. An error-detection system that uses a BCC at the end of a block is known as a(n) _____.

87. Dividing the data block by a constant produces a quotient that is discarded and a remainder called the _____ character.

88. The CRC circuit is basically a(n) _____.

89. True or false. If a bit error can be identified, it can be corrected.

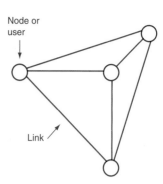

Fig. 12-39 A simple communications network.

12-4 Introduction to Networks

Up to now we have discussed communication as primarily a two-way channel where both parties can transmit and receive. Although this one-to-one mode of communications is relatively common, in many communications systems a user may communicate with many more than one additional station. In order for a station to communicate with other stations, some form of connection or link must be established. If wired communications is used, each station must be connected to all the other stations with which communication is desired. This interconnection of one station to many others is referred to as networking. A *network* is any interconnection of two or more stations that wish to communicate. Each station in a communications network is called a *node*. These nodes are interconnected with the desired wiring configuration. One possible arrangement of a network is shown in Fig. 12-39. Each node is connected to all other nodes. Many other organized interconnections are possible as you will see.

Today, electronic communications networks are very common. Many of our most familiar communications systems use network techniques. In this section, we will provide a brief introduction to communications networks.

Wide Area Networks

Perhaps the most familiar of all communications networks is the telephone system. The telephone system may indeed be the largest and most sophisticated network of all. Each person has a transmitter and receiver known as a telephone, and through many different networks that individual can communicate by telephone with virtually any other person on this planet. Large networks such as those that cover a complete country or state are generally referred to as *wide area networks (WANs)*. Most WANs are telephone systems. The long-distance systems of AT&T, MCI, GTE, and others are examples of WANs. These networks use cables of various types for their interconnection, and, in many cases, radio links using microwave relay stations or satellites also form part of the network. Most telephone networks, of course, carry voice communications, but as you have seen, when a modem is used, digital data can be converted into analog signals that are fully compatible with the voice-grade telephone system.

Metropolitan Area Networks

Smaller networks are also common. A typical example is a local cable television system. A cable TV company picks up programming from sources available on a satellite and rebroadcasts them via a system of coaxes to subscribers. In this case, the communications is only one-way, but the total interconnection is a network where each home is a node in the network. Such medium-sized networks are sometimes referred to as *metropolitan area networks (MANs)*.

Local Area Networks

Even smaller networks are used. Known as *local area networks (LANs)*, these networks

---

Wide area network (WAN)

Network

Node

Metropolitan area network (MAN)

Local area network (LAN)

interconnect multiple stations over a very small area. Most LANs have 10 to 100 users but never more than 1000. For example, LANs are widely used in companies to connect several offices within the same building or to connect different floors of the building. Somewhat larger LANs may interconnect several buildings within a complex, such as networks used on college campuses and in military installations. In a LAN, the nodes are usually computers or terminals. Local area networks made up of PCs are the most common type. Data transmitted over a LAN is typically digital rather than voice or video.

One of the most common types of LANs is a large mainframe or minicomputer connected to many remote terminals. Multiple users time-share the system. The remote nodes are dumb terminals or PCs set up to emulate a terminal. Users have access to all the system's capabilities and may communicate with one another by an electronic mail (e-mail) system.

Initially LANs were developed so that multiple PC users could share expensive or limited-use peripherals such as printers, hard disks, and modems. One computer in the network is set up as a controller through which other users have access to the peripherals. Such units are called servers, disk servers, printer servers, or modem servers. A variation of the disk server is a file server that combines a disk server with sophisticated software that lets users share data in a large database. The LANs also permit software and data sharing.

In addition to sharing peripherals, LANs permit users to talk with one another directly or through an electronic mail system. In the electronic mail system, one computer has a large storage unit in which users are assigned "mailboxes" where messages from other users are stored.

Network Topologies

Although many LANs stand alone, others are interconnected to other systems. Frequently a LAN is connected to a larger mainframe or minicomputer through one of the PCs in the network. This PC with its software and often special controller hardware is called a *gateway*.

A LAN is also sometimes connected to other LANs. A PC in each LAN acts as the link and controller and is called a *bridge*.

Bridge

There are many different ways in which stations or nodes can be interconnected to form a network. However, there are several popular configurations which are used again and again. These different network configurations or arrangements are known as *network topologies*. The more popular topologies are the star, the ring, the bus, and the tree. Let's take a look at each in more detail.

Network topologies

Star The *star network* is shown in Fig. 12-40. It consists of a central controller node and each individual station. In order for one station to communicate with another, the communication must take place through the central controller station. It is the central controller that allows any one user to be linked up with any other user.

Star network

An example of a star network is a mainframe or minicomputer that serves as the central controller for multiple video terminals with keyboards and CRT displays or PCs. Each station can communicate with the computer but must go through the computer to communicate with any other station.

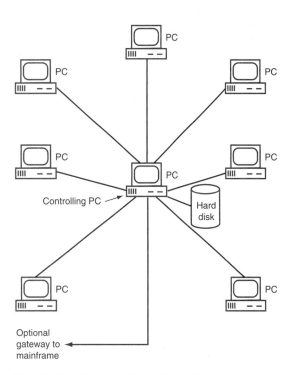

Fig. 12-40 A star LAN configuration.

Gateway

Another example of a star network is a *PBX system*. The acronym PBX stands for *private branch exchange*, which is the terminology used to describe a small telephone system within an office or building. The PBX allows any telephone in the building to be linked to any other telephone. The PBX is a large master switching circuit that automatically connects one station to another when the correct number is dialed. Although most PBXs are for voice applications, they are being more widely used for digital data.

Ring A *ring network* configuration is shown in Fig. 12-41. There is no central controller station since each station performs some control functions. All the stations are connected end to end to form a continuous loop, so each station is effectively connected to two other stations. Each small box in the diagram is the interface circuitry.

Stations in the ring can both send and receive. All nodes have the ability to recognize their assigned code or address and the capability of retransmitting a received signal. The message used to communicate between stations begins with the address of the station to whom the message is to be sent. The desired message then follows. The station originating the message sends it to the next station in line. In a ring, the direction of communication is usually unidirectional. If the message is not for the next station in line, the address will not be recognized, so that station will simply repeat the message and send it on to the next node. Each station in the ring performs that operation until the desired station is reached, at which time it recognizes the address and receives the message.

A good example of a ring LAN is IBM's popular Token Ring. It can accommodate hundreds of users. The data rate on early systems was 4 Mbits/s. Most systems have a data rate of 16 Mbits/s. Higher rates of 20 and 100 Mbits/s are also available. Shielded twisted pair cable is used to interconnect the PCs.

Bus Another widely used network configuration is the *bus* as shown in Fig. 12-42. Here all the stations effectively share a common cable called the bus. Communications is bidirectional on the bus; that is, any one station can talk to any other station. In a bus configuration, the sending station effectively broadcasts

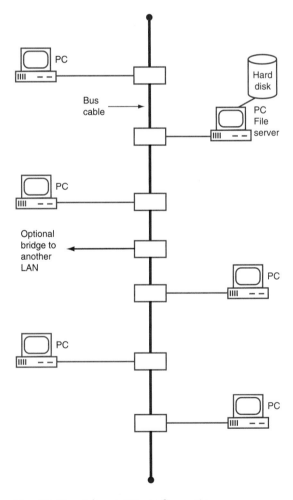

Fig. 12-42 A bus LAN configuration.

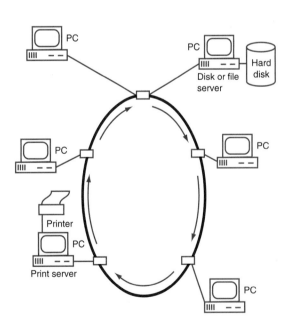

Fig. 12-41 A ring LAN configuration.

its message on the bus and all stations can receive the message.

The main advantage of the bus system is that it is much faster than any of the other topologies. In both the star and the ring configurations, other devices become involved in the transmission. In the bus system, the two stations desiring to communicate do so directly without interference from or assistance from any other station. Bus systems are fast and simple as a result.

The *tree topology* is essentially a variation of the bus system. The bus is extended into several branches so that more stations can be interconnected. In some large tree systems, a controller node is used.

The bus is the most widely used LAN configuration. An example of a bus system is *Ethernet,* which is used on more than 85 percent of the LANs in operation. The early Ethernet systems ran at 10 Mbits/s on coaxial cable. Today, most systems use twisted-pair cables. Ethernet is also available in versions that run at 100 and 1000 Mbits/s (1 Gbits/s). All use twisted-pair cable but fiber-optic cable is an option.

The three basic LAN topologies are compared and contrasted in the table of Fig. 12-43.

LAN Cabling

Three basic transmission media are used in interconnecting LANs: twisted-pair, coax, and fiber-optic cable. Twisted pair is a cable made of two no. 22-, 26-, or 28-gauge insulated solid-copper wires twisted loosely together. It is very inexpensive and easy to work with. It is excellent for short distances, no more than about 2 mi and usually less. It can accommodate data rates up to about 100 Mbits/s depending upon length. Twisted pair is popular in many simple LANs. The biggest disadvantage is that because it is unshielded, it can pick up noise, which can lead to bit errors. However, it is the most widely used LAN medium.

Coax is also widely used in LANs. Because it is shielded, noise is less of a problem. Coax can also be used over longer distances, up to several miles. Its wider bandwidth permits data rates up to 100 Mbits/s.

Fiber-optic cable is gaining in popularity as its cost comes down and its complexity decreases. Its main advantages are superior noise immunity and very high data rates, up to 100 Gbits/s. Distances up to about 5 mi can be accommodated.

In both wire and fiber-optic cable systems, repeaters may be used to extend the length of a network. A repeater receives, amplifies, and rejuvenates a weak and distorted signal and retransmits.

Broadband versus Baseband LANs

The LANs can also be categorized according to whether baseband or broadband operation is chosen. Baseband refers to the data or information signal whether it is binary or analog voice or video. In baseband systems, the digital data is applied directly to the medium. The entire bandwidth of the medium is used by the signal. Such systems are simple, inexpensive, and easy to work with, but only one signal can be carried on the medium at one time. Further, cable attenuation and distortion greatly limit the transmission distance, usually up to 1 mi with twisted pair or 5 mi with coax. Despite these disadvantages, most LANs are of the baseband type.

Broadband means that modulation techniques are used. Analog methods are used to transmit digital data. In this way, data signals can be translated up in frequency to specific channels. Frequency-division multiplexing techniques are used to create multiple channels for binary signals. Broadband systems use coax because it has a bandwidth of up to 300 to 450 MHz. As a result, an enormous number of high-speed channels can be created. In addition, transmission over longer distances (up to 10

Tree topology

Ethernet

Type	Advantages	Disadvantages
Star	1. Excellent control	1. Slow speed response during heavy use
	2. Facilities sharing	2. Limited number of nodes possible
	3. Failure of one node does not disable system	
Ring	1. Lowest cost of all topologies	1. Slow speed since nodes are accessed sequentially
	2. Ease of expansion	2. Failure of one node disables the system
Bus	1. Fastest speed due to direct access	
	2. Failure of one node will not disable the system	
	3. Ease of expansion	

Fig. 12-43 Comparison of LAN topologies.

mi) on coax can be achieved. Of course, broadband systems are more complex and expensive because modems are required at each node.

Ethernet

The best example of a modern LAN is Ethernet. More than 85 percent of all LANs use the Ethernet standard. Ethernet was developed by Xerox Corporation at Palo Alto Research Center in the 1970s. In 1980 Xerox joined with Digital Equipment Corporation and Intel to sponsor a joint standard for Ethernet. The collaboration resulted in a definition that became the basis for the IEEE 802.3 standard.

Topology Ethernet uses the bus topology. Network nodes simply tap into the bus cable (see Fig. 12-44). The bus may be a large coaxial cable (like RG-8/U), a small coaxial cable (like RG-58/U), or a twisted pair. Fiber-optic cable and wireless versions are also available. Information to be transmitted from one user to the other can move in either direction on the bus, but only one node can transmit at any given time.

Speed The standard transmission speed for Ethernet LANs is 10 Mbits/s. The time for each bit interval is the reciprocal of the speed, or $1/f$, where f is the transmission speed or frequency. With a 10-Mbits/s speed, the bit time is $1/10 \times 10^6 = 0.1 \times 10^{-6} = 0.1 \ \mu s$ (100 ns).

There are also 100-Mbits/s and 1-Gbits/s versions of Ethernet. The 10- and 100-Mbits/s versions are the most popular. The 1-Gbits/s system is relatively new and operates over

shorter distances than do the two slower versions.

Transmission Medium The standard transmission medium for Ethernet is coaxial cable. However, today twisted-pair versions of Ethernet are more common.

The two main types of coaxial cable used in Ethernet networks are RG-8/U and RG-58/U. RG-8/U cable has a characteristic impedance of 53 Ω, is approximately 0.4 in. in diameter, and is referred to as *thick cable*. It is usually bright yellow in color. N-type coaxial connectors are used to make the interconnections.

Ethernet systems using thick coaxial cable are generally referred to as *10Base5 systems,* where *10* means a 10-Mbits/s speed, *Base* means baseband operation, and *5* designates a 500-m maximum distance between nodes, transceivers, or repeaters. Ethernet LANs using thick cable are also referred to as *Thicknet.*

In 10Base5 Ethernet LANs, the cable is generally one long continuous bus that is routed from node to node. Depending on the distance between nodes, repeaters may have to be inserted to boost and rejuvenate the signal along the way. The bus is relatively easy to reconfigure; adding new nodes is easy. The bus cable can be cut and new connectors installed, allowing the bus to pass through while at the same time connecting to the new NICs. Special connectors referred to as *vampire* taps can pierce the coaxial cable without cutting it.

Ethernet systems implemented with thinner coaxial cable are known as *10Base2,* or *Thinnet* systems. Here the *2* indicates the maximum 200-m (actually, 185 m) run between nodes or repeaters. The most widely used thin cable is RG-58/U. It is slightly less than 0.25 in. in diameter and much more flexible and easier to work with than RG-8/U cable, but it has greater attenuation. BNC coax connectors are used.

More recent versions of Ethernet use twisted pair. The twisted-pair version of Ethernet is referred to as a *10Base-T network,* where *T* stands for twisted pair. The twisted pair used in a 10Base-T system is standard 22-, 24-, or 26-gauge solid copper wire with 8-pin RJ-45 modular connectors. The most commonly used cable is made up of four unshielded twisted pairs (UTPs). The wire size, insulation, and loose twist reduce cable capacitance, permitting higher speeds

Thicknet

Thinnet

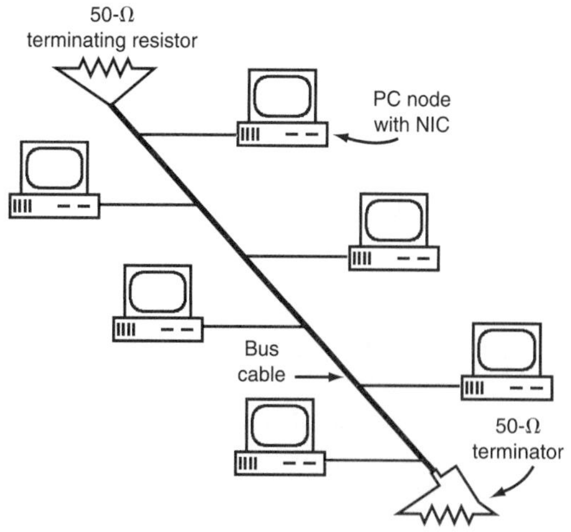

Fig. 12-44 The Ethernet bus.

and less attenuation. There are a variety of twisted-pair cable types. They are designated by a category number. Category 1 and 2 cables are used for common telephone wire carrying voice signals, not digital data. Category 3 cable was designed for digital baseband signals up to a rate of about 16 Mbits/s. Category 4 cable is rated to 20 Mbits/s. Both Category 3 and 4 are widely used with 10Base-T Ethernet and Token-Ring networks. Category 5 cable is the best and is capable of carrying digital data up to 100 Mbits/s or so depending upon its length. Twisted-pair lines act like transmission lines and have a characteristic impedance in the 100- to 150-Ω range. They must always be terminated in their characteristic impedance to ensure maximum power transfer and to minimize reflections.

The PC nodes connect to a hub, as shown in Fig. 12-45, which provides a convenient way to connect all nodes. Physically, a 10Base-T LAN looks like a star, but the bus is implemented inside the hub itself. It is usually easier and cheaper to install 10Base-T LANs than it is to install coaxial Ethernet systems, but the transmission distances are usually more limited.

Access Method *Access method* refers to the protocol used for transmitting and receiving information on a bus. Ethernet uses an access method known as *carrier sense multiple access with collision detection (CSMA/CD),* and the IEEE 802.3 standard is primarily devoted to a description of CSMA/CD.

Whenever one of the nodes on an Ethernet system has data to transmit to another node, the software sends the data to the NIC in the PC. The NIC builds a *packet,* a unit of data formed with the information to be transmitted. The completed packet is stored in RAM on the NIC while the sending node monitors the bus. Since all the nodes or PCs in a network monitor activity on the bus, whenever a PC is transmitting information on the bus, all the PCs detect (sense) what is known as the *carrier.* The carrier in this case is simply the Ethernet data being transmitted at a 10-Mbits/s (or 100-Mbits/s) rate. If a carrier is present, none of the nodes will attempt to transmit.

When the bus is free, the sending station initiates transmission. Then the transmitting node "broadcasts" the data on the bus so that any of the nodes can receive it. However, the packet contains a specific binary address defining the destination or receiving node. When the packet is received, it is decoded by the NIC and the recovered data is stored in the computer's RAM.

Although a node will not transmit if a carrier is sensed, at times two or more nodes may attempt to transmit at the same time or almost the same time. When this happens, a *collision* occurs. If the stations that are attempting to transmit sense the presence of more than one carrier on the bus, both will terminate transmission. The CSMA/CD algorithm calls for the sending stations to wait a brief interval of time and then attempt to transmit again. The waiting interval is determined randomly, making it statistically unlikely that both will attempt retransmission at the same time. Once a transmitting node gains control of the bus, no other station will

Access method

CSMA/CD

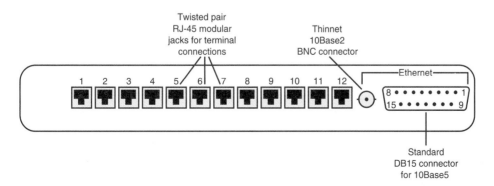

Fig. 12-45 Hub for Ethernet 10Base-T.

attempt to transmit until the transmission is complete. This method is called a *contention system* because the nodes compete, or contend, for use of the bus.

In Ethernet LANs with few nodes and little activity, gaining access to the bus is not a problem. However, the greater the number of nodes and the heavier the traffic, the greater the number of message transmission and potential collisions. The contention process takes time; when many nodes attempt to use a bus simultaneously, delays in transmission are inevitable. Although the initial delay might be only tens or hundreds of microseconds, delay times increase when there are many active users. Delay would become an insurmountable problem in busy networks if it were not for the packet system, which allows users to transmit in short bursts. The contention process is worked out completely by the logic in the NICs.

Packet Protocols Figure 12-46 shows the packet (frame) protocol defined by IEEE standard 802.3. The packet is made up of two basic parts: (1) the frame, which contains the data plus addressing and error detection codes, and (2) an additional 8 bytes (64 bits) at the beginning, which contain the preamble and the *start frame delimiter (SFD)*. The preamble consists of 7 bytes of data that help to establish clock synchronization with the NIC, and the SFD announces the beginning of the packet itself.

The *destination address* is a 6-byte, 48-bit code that designates the receiving node. This is followed by a 6-byte source address that identifies the sending node. Next is a 2-byte field that specifies how many bytes will be sent in the data field. Finally, the data itself is transmitted. Any integer number of bytes in the range from 46 to 1500 bytes can be sent in one packet. Longer messages are divided up

into as many separate packets as required to send the data.

Finally, the packet and frame end in a 4-byte frame-check sequence generated by putting the entire transmitted data block through a *cyclical redundancy check (CRC)*. The resulting 32-bit word is a unique number designating the exact combination of bits used in the data block. At the receiving end, the NIC again generates the CRC from the data block. If the received CRC is the same as the transmitted CRC, no transmission error has occurred. If a transmission error occurs, the software at the receiving end will be notified.

It is important to point out that data transmission in the Ethernet system is synchronous: The bytes of data are transmitted end to end without start and stop bits. This speeds up data transmission, but puts the burden of sorting the data on the receiving equipment. The clocking signals to be used by the digital circuits at the receiving end are derived from the transmitted data itself to ensure proper synchronization and counting of bits, bytes, fields, and frames.

Encoding *Encoding* refers to how the digital data is actually transmitted over the medium. Ethernet is a baseband technique in which the digital data is applied directly to the cable. In other words, no modulation is used. A special encoding method known as *Manchester encoding* is used. This method is shown in Fig. 12-47. The standard serial data format with common binary voltage levels is shown in Fig. 12-47(*a*). This method is usually referred to as the *non-return-to-zero (NRZ) method*. The Manchester encoded signal is shown in Fig. 12-47(*b*). Note that it is an ac signal, not a dc signal as in Fig. 12-47(*a*) since both positive and negative voltage levels are used. A binary 1 is represented by a positive pulse followed by a negative pulse. A binary 0

IEEE 802.3

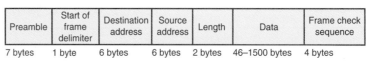

Preamble	Start of frame delimiter	Destination address	Source address	Length	Data	Frame check sequence
7 bytes	1 byte	6 bytes	6 bytes	2 bytes	46–1500 bytes	4 bytes

Fig. 12-46 Ethernet frame format.

is represented by a negative pulse followed by a positive pulse.

There are two benefits to this method. First, the positive and negative voltage levels average themselves out over time, resulting in a zero average voltage on the cable. Second, the transition that occurs at the center of each bit permits the clock signal to be derived from the data signal itself. This permits synchronous operation without transmitting the clock signal separately.

Most Ethernet systems use a dc version of the Manchester encoding as shown in Fig. 12-47(c). The voltages switch between zero and a negative value such as −3 V. This causes an average dc voltage of about −1.5 V to build up on the cable when data is being transmitted. This average voltage is used to detect a signal (carrier) on the cable during the access resolution stage of transmission.

100-Mbits/s Ethernet A more recent development in Ethernet technology is a modified version that permits a data rate of 100 Mbits/s, 10 times the normal Ethernet rate. Several versions have been developed. These are 100Base-T or 100Base-TX, also called Fast Ethernet, 100VG-AnyLAN, 100Base-T4, and 100Base-FX. All use twisted pair except the FX version, which uses fiber-optic cable. The TX and FX methods use the CSMA/CD access method and have the same packet size. The 100VG-AnyLAN and T4 versions use unique access methods. The 100VG-AnyLAN version has the IEEE standard number 802.12, and the other versions are subsets of the IEEE 802.3 standards. Many NICs support both 10- and 100-Mbits/s rates.

By far the most popular version of 100-Mbits/s Ethernet is 100Base-TX, or Fast Ethernet. It uses two unshielded twisted pairs instead of the single pair used in standard 10Base-T. One pair is used for transmitting, and the other is used for receiving permitting full-duplex operation, which is not possible with standard Ethernet. To achieve such high speeds on UTP, several important technical changes were implemented in Fast Ethernet. First, the cable length is restricted to 100 m of Category 5 UTP. This ensures minimal induc-

tance, capacitance, and resistance, which distort and attenuate the digital data.

Second, a new type of encoding method is used. Called *MLT-3*, this coding method is illustrated in Fig. 12-48. The standard NRZ binary signal is shown at (*a*), and the MLT-3 signal is illustrated at (*b*). Note that three voltage levels are used: +1, 0, and −1 V. If the binary data is a binary 1, the MLT-3 signal will change from the current level to the next level. If the binary input is 0, no transition takes place. If the signal is currently at +1, and a 1111 bit sequence occurs, the MLT-3 signal will change from +1 to 0 to −1 and then to 0 and then to +1 again. What this coding method does is to greatly reduce the frequency of the

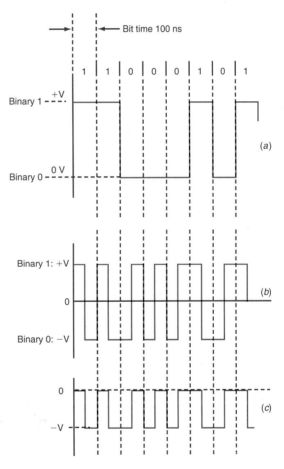

Fig. 12-47 Manchester encoding eliminates dc shift on baseband lines and provides a format from which a clock signal can be extracted. (*a*) Standard dc binary coding (unipolar NRZ). (*b*) Manchester ac encoding. (*c*) DC Manchester.

transmitted signal (to one-quarter of its original signal), making higher bit rates possible over UTP.

As for access method, 100Base-T uses CSMA/CD, described earlier. Most 100-Mbits/s Ethernet systems use the 100Base-T format described here.

100Base-T4 is a less-used form of 100Base-T. It uses Category 3 or 5 UTP up to 100 m from PC to hub. This version of Ethernet uses four twisted pairs. Three pairs are used to transmit the data in a parallel format, and the fourth pair is used for access determination. A version of CSMA/CD is used. With this arrangement, three pairs essentially carry 33 Mbits/s, each in parallel. Full duplex is not supported.

The 100VG-AnyLAN version of Ethernet was developed by Hewlett Packard. It is considerably different from all other versions of Ethernet. It uses Category 3 UTP up to 100 m long. Most versions use all four pairs in the cable, but some use just two pairs.

The topology of 100VG-AnyLAN is a star. All cables from PC go into a central hub that is controlled by a server computer. The access method is called *demand priority*. A proprietary encoding scheme is also used.

100Base-FX is a fiber-optic cable version of Ethernet. It uses two multimode fiber strands to achieve the 100-Mbits/s rate. The access method is CSMA/CD. This version of Ethernet is used primarily to interconnect individual LANs to one another over long distances. Full duplex operation can be achieved at distances up to 2 km.

Fiber-optic interrepeater links (FOIRLs) are fiber-optic communication channels with repeaters at each end that are designed to interconnect two Ethernet networks. If two networks that must work together are far apart, or if some of the nodes in a network exceed distance limits imposed by the network con-

figuration and hardware, FOIRLs may be the answer.

FOIRL repeaters are attached to an Ethernet bus, coaxial cable, or twisted pair, just as a regular node would be attached, and a fiber-optic cable links the FOIRL to a remote repeater in the other network on another mode. FOIRLs can be 500 to 100 m apart depending upon the configuration of the network. A FOIRL cable is actually two cables, one for transmitting and the other for receiving. Normally, the repeaters at each end of the cable attach to an Ethernet hub where all the other nodes terminate.

Gigabit Ethernet The latest version of Ethernet is Gigabit Ethernet, capable of achieving 1000 Mbits/s or 1 Gbit/s over Category 5 UTP. The 1-Gbit/s speed is more readily achieved with fiber-optic cable, but it is more expensive. This newest version of Ethernet uses existing lower-cost UTP to achieve a speed formerly reserved for fiber-optic cable. The most popular version of Gigabit Ethernet is defined by the IEEE standard 802.3z and is generally referred to as 1000Base-T.

The 1-Gbit/s rate is achieved by transmitting 1 byte of data at a time as if in a parallel data transfer system. This is done by using four UTP cables plus a coding scheme that transmits two bits per baud. The basic arrangement is shown in Fig. 12-49. One byte of data is divided into 2-bit groups, as it is in 8PSK. Each 2-bit group is sent to a D/A converter and an encoder that generates a five-level line code. The five levels are +2, +1, 0, −1, and −2 V. Each 2-bit group needs four coding levels. For example, 00 might be +2, 01 = +1, 10 = −1, and 11 = −2 V. The 0-V level is not used in coding but is used in clock synchronization and in an error-detection scheme. The resulting five-level code is fed to each twisted pair at a 125-Mbaud rate. Since 2 bits are transmitted for each baud or symbol level, the data rate on each twisted pair is 250 Mbits/s. Together, the four pairs produce a composite data rate of $4 \times 250 = 1000$ Mbits/s or 1 Gbit/s. Maximum distance is 100 m. Better and more reliable performance is achieved by restricting the distance from PC to hub to 25 m or less. The CSMA/CD access method is used. Several

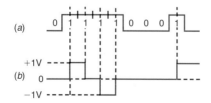

Fig. 12-48 MLT-3 encoding used with 100Base-T Ethernet. (*a*) NRZ. (*b*) MLT-3.

versions of Gigabit Ethernet are also available for fiber-optic cable.

Wireless Ethernet Several versions of Ethernet for transmission by radio have also been developed. This permits wireless LANs to be created. The cost and complexity of buying and installing cables are eliminated, and nodes can be relocated at any time without regard to where the cable is. Each PC is equipped with an NIC that incorporates a wireless transceiver. The most common application for wireless Ethernet is to interconnect two or more separate Ethernet LANs in different buildings that may be up to several miles apart. Such a wireless bridge allows the two LANs to communicate.

Most wireless Ethernet systems use direct-sequence spread spectrum (DSSS) in the 902- to 928-MHz or 2.4-GHz bands. Data speeds are generally limited to 2 to 6 Mbits/s at these frequencies. Faster versions using the 5.8-, 21-, and 38-GHz frequency ranges permit the conventional speeds of 10 and 100 Mbits/s to be obtained.

■ TEST

Answer the following questions.

90. The telephone system is an example of a _____.
91. True or false. A cable TV system is considered a network.
92. Most networks are smaller and cover shorter distances and are referred to as _____.
93. The acronym PBX means _____.
94. True or false. PBXs cannot be used with digital data.
95. The three main types of LAN topologies are _____.
96. A PBX is a type of _____ topology LAN.
97. A LAN that uses a central controller for multiple stations is the _____ topology.
98. One type of LAN connects multiple terminals and PCs to a large central _____ computer.
99. A LAN in which the message is passed from one station to the next until the destination is reached is called a(n) _____.
100. The stations or users in a network are referred to as _____.
101. A network in which all stations attach to a common cable is called a(n) _____.
102. The fastest network configuration is the _____.
103. In a(n) _____ LAN configuration, if one station fails, the whole system fails.
104. A connection from a LAN to a mainframe computer is known as a(n) _____.

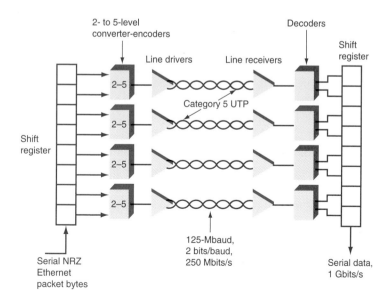

Fig. 12-49 Gigabit Ethernet.

105. A connection from one LAN to another is called a(n) _____.

106. The three transmission media used in LANs are _____.

107. An extended variation of the bus topology is called the _____.

108. The most widely used bus LAN is called _____. It permits three data rates of _____, _____, and _____.

Spread spectrum (SS)

109. _____ is the most widely used medium in LANs.

110. The medium that permits the highest data rates is _____.

111. The longest transmission distances can be achieved with _____ medium.

112. Give the average maximum data rates in Mbits/s for each medium: twisted-pair, coax, fiber-optic cable.

113. The digital, voice, or video voltages are referred to as _____ signals.

114. When the basic data signal is sent directly over the medium, the type of operation is said to be _____.

115. _____ operation uses modulation techniques and FDM techniques to increase speed and the number of operational channels.

116. Broadband operation requires the use of a(n) _____ at each node.

117. Most modern versions of Ethernet use _____ as the transmission medium.

118. The equipment used to connect all Ethernet nodes to the bus is called a(n) _____.

119. Write the transmission rates for each type of Ethernet: (a) 10Base-T _____. (b) 100Base-TX _____. (c) Gigabit Ethernet _____.

120. Data is carried over Ethernet in groups of bits called _____.

121. Twisted-pair cable capable of carrying data at rates of 100 Mbits/s is referred to as _____.

Code division multiple access (CDMA)

122. True or false. Ethernet can be carried over fiber-optic cable.

123. The access method used by Ethernet to resolve which node will transmit is _____.

124. State two ways used by Gigabit Ethernet to achieve such a high data rate on UTP. (a) _____ (b) _____.

12-5 Spread Spectrum

Spread spectrum (SS) is a modulation and multiplexing technique that distributes the signal and its sidebands over a very wide bandwidth. Traditionally, we evaluate the efficiency of modulation and multiplexing methods by how little bandwidth they consume. The continued increase in all types of radio communications, the resulting crowding, and the finite bounds of usable spectrum space have made everyone sensitive to how much bandwidth a given signal occupies. And designers of communications systems and equipment typically do all in their power to minimize the amount of bandwidth a signal takes. If that is the case, then how can a scheme that spreads the signal over a very wide piece of the spectrum be of value? The answer to this question and a discussion of the benefits and concepts of spread spectrum are the subjects of this section.

Applications

Spread spectrum is not new. It was developed primarily by the military after World War II. Its use has been restricted to military communications applications because it is a secure communications technique that is essentially immune to jamming. In the mid-1980s, the FCC authorized use of SS in civilian applications. Currently, unlicensed operation is permitted in the 902- to 928-MHz, 2.4- to 2.483-GHz, and 5.725- to 5.85-GHz ranges with 1 W of power. Spread spectrum on these frequencies is now being widely incorporated into a variety of commercial communications systems. One of the most popular new applications is wireless data communications. A variety of new wireless LANs and portable computer modems use SS techniques. A special version of SS known as *code division multiple access (CDMA)* is now widely used in the newer digital cell phones. Some cordless telephones also use SS. The use of SS is expected to grow because of the numerous benefits it offers.

Types of Spread Spectrum

There are two basic types of SS: *frequency hopping (FH)* and *direct sequence (DS)*. In frequency hopping, the frequency of the carrier of the transmitter is changed randomly at a rate higher than that of the serial binary data modulating the carrier. In direct sequence, the serial binary data is mixed with a higher-frequency pseudorandom binary code at a faster rate and the result is used to phase-modulate a carrier. Let's examine each technique in more detail.

FHSS

Figure 12-50 shows a block diagram of a frequency-hopping SS transmitter. The serial binary data to be transmitted is applied to a conventional FSK modulator. The two-tone FSK modulator output is applied to a mixer. Also driving the mixer is a frequency synthesizer. The output signal from the bandpass filter (BPF) after the mixer is the difference between one of the two FSK sine waves and the frequency of the frequency synthesizer. Note that the synthesizer is driven by a pseudorandom code generator. This is either a special digital circuit or the output of a specially programmed microprocessor.

The pseudorandom code is a pattern of binary 0s and 1s that changes randomly. For example, an 8-bit code word can make the synthesizer hop to 256 different frequencies. The randomness of the 1s and 0s makes the serial output of this circuit seem like digital noise. Sometimes the output of this generator is called *pseudorandom noise (PSN)*. The binary sequence is not totally random, because it does repeat after many bit changes. However, the randomness is sufficient to minimize the pos-

sibility of someone accidentally duplicating the code.

Pseudorandom noise sequences are usually generated by a shift register circuit similar to that shown in Fig. 12-51. Here eight flip-flops in the shift register are clocked by an external clock oscillator. The input to the shift register is derived by X-ORing two or more of the flip-flop outputs. It is this connection that produces the pseudorandom sequence. The output is taken from the last flip-flop in the register. By changing the number of flip-flops in the register and/or changing the outputs which are X-ORed and fed back, the code sequence can be changed. The serial PSN sequence can be converted to parallel codes for controlling the divide-by-N counter in the feedback loop of a phase-locked loop synthesizer. Or, the parallel codes can be converted to corresponding frequencies by means of direct digital synthesis. Alternatively, a microprocessor can be programmed to generate pseudorandom sequences.

In a frequency-hopping SS system, the rate of synthesizer frequency change is higher than the data rate. This means that although the data bit and the FSK tone it produces remain constant for one data interval, the frequency synthesizer will switch frequencies many times during this period. This is shown in Fig. 12-52. Here, the frequency synthesizer changes frequencies four times for each bit time of the serial binary data. The time that the synthesizer remains on a single frequency is called *dwell time*. The frequency synthesizer puts out a pseudorandom sine wave frequency to the mixer. The mixer creates a new carrier frequency for each dwell interval. The resulting signal, the frequency of which jumps around,

Frequency hopping (FH)

Direct sequence (DS)

Dwell time

Pseudorandom noise (PSN)

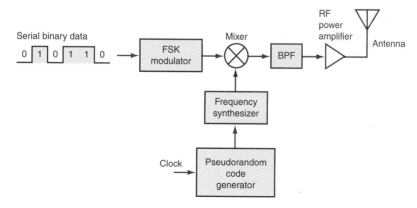

Fig. 12-50 A frequency-hopping SS transmitter.

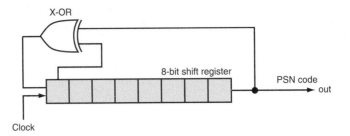

Fig. 12-51 A typical PSN code generator.

effectively scatters pieces of the signal all over the band. This wideband signal is amplified and fed to the antenna.

This is a signal that rapidly jumps around all over the place. Specifically, the carrier randomly switches between dozens or even hundreds of frequencies over a given bandwidth. The actual dwell time on any frequency varies with the application and data rate, but it may be as short as 10 milliseconds (ms). Currently, the FCC regulations specify that there must be a minimum of 50 hopping frequencies and that the dwell time must be no longer than 400 ms. Figure 12-53 shows what one frequency-hopping sequence might be. The horizontal axis is time divided into increments of the dwell time. The vertical axis is the transmitter output frequency divided into step increments of the phase-locked loop (PLL) frequency synthesizer.

As you can see, the signal is divided up and spread out over a very wide bandwidth. A signal that may occupy only a few kilohertz of spectrum is spread out over a range that is 10 to 10,000 times that wide. However, the signal does not remain on any one frequency for a long period of time. Because it jumps around randomly, the SS signal will not interfere with a traditional signal on any of the hopping frequencies. The SS signal actually appears to be more like noise to a conventional narrow bandwidth receiver. A conventional receiver picking up such a signal will not respond even to a signal with a duration of tens of milliseconds. In addition, the conventional receiver cannot receive the SS signal because it does not have a wide enough bandwidth and it cannot follow, or track, its random frequency changes. Thus the SS signal is as secure as if it were scrambled.

Also, consider the situation when there are two or more SS transmitters operating over the same bandwidth but with different pseudorandom codes. Since the transmitters will be hopping to different frequencies at different times, they will not typically occupy a given frequency simultaneously with any of the other signals. For this reason, the signals do not interfere with one another. Because of this unique operation, SS is also a kind of multiplexing, for it permits two or more signals to use a given bandwidth concurrently without interference. Therefore, this method makes it possible to pack more signals in a given band than would be possible with other methods.

Another distinct advantage of SS is its ability to mostly eliminate selective fading or multipath distortion and cancellation. Consider an FM telemetry transmitter operating at a narrow band of frequencies. Reflected and direct signals will cause nulls or dropouts as two waves arrive at the receiver out of phase. Spread spectrum is redundant (because of dwell time), and although a given receiver location might cancel one frequency, the others will get through.

As more and more signals use a band, the background noise produced by many signals switching increases, but not enough to prevent highly reliable communications.

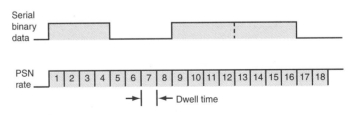

Fig. 12-52 Serial data and the PSN code rate.

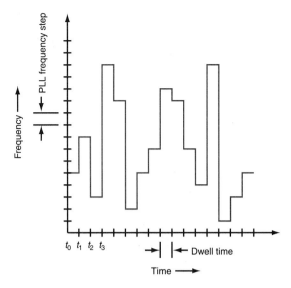

Fig. 12-53 A random frequency hop sequence.

A frequency-hopping receiver is shown in Fig. 12-54. The very-wide-band signal picked up by the antenna is applied to a broadband RF amplifier and then to a conventional mixer. The mixer, like that in any superheterodyne receiver, is driven by a local oscillator which in this case is a frequency synthesizer like the one used at the transmitter. To receive the signal, the receiver local oscillator must have the same pseudorandom code sequence so that it is tuned to the correct frequency to receive the signal on the correct frequency when it occurs. The signal is thus reconstructed as an IF signal which contains the original FSK data. The signal is then applied to an FSK demodulator which reproduces the original binary data train.

One of the most important parts of the receiver is the circuit that is used to acquire and synchronize the transmitted signal with the internally generated pseudorandom code. Even though the codes must be the same, it is still a problem to get the two codes into step with one another. This problem is solved by a preamble signal and code at the beginning of the transmission. Once synchronism has been established, the code sequences occur in step. Any receiver not having the correct code cannot receive the signal. You can see how secure an SS is because of this characteristic.

One way different stations are distinguished from one another is by their code sequence. Many stations can share a common band by using different codes. Instead of assigning each station a single frequency on which to operate, each is given a different pseudorandom code within the same band. This permits a transmitter to transmit selectively to a single receiver without other receivers in the band being able to pick up the signal.

DSSS Another type of SS is direct sequence (DS). A block diagram of a DS transmitter is shown in Fig. 12-55. The serial binary data is applied to an X-OR gate along with a serial pseudorandom code that occurs faster than the binary data. Figure 12-56 shows typical waveforms. One bit time for the pseudorandom code is called a *chip,* and the rate of the code is called the *chipping rate.* The chipping rate is faster than the data rate.

The signal developed at the output of the X-OR is then applied to a PSK modulator. Binary phase-shift keying (BPSK) is commonly used. The carrier phase is switched between 0° and 180° by the 1s and 0s of the X-OR output. Quadrature phase-shift keying (QPSK), quadrature amplitude modulation (QAM), and other forms of PSK may also be used. The PSK

Chip

Chipping rate

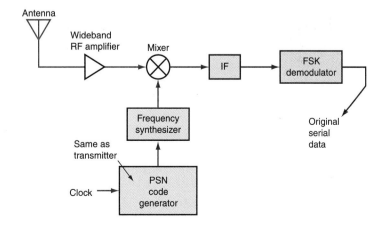

Fig. 12-54 A frequency-hopping receiver.

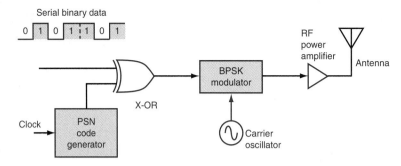

Fig. 12-55 A direct-sequence SS transmitter.

modulator is generally some form of balanced mixer as described earlier in this chapter. The signal phase-modulating the carrier, being much higher in frequency than the data signal, causes the modulator to produce multiple widely spaced sidebands whose strength is such that the complete signal takes up a great deal of the spectrum. Thus the signal is spread. And because of its randomness, the resulting signal appears to be nothing more than wideband noise to a conventional narrowband receiver.

One type of direct-sequence receiver is shown in Fig. 12-57. The broadband SS signal is amplified and then translated down to a lower IF frequency by mixing it in mixer 1 with a local oscillator signal. For example, the SS signal at an original carrier of 902 MHz might be translated down to an IF of 70 MHz. This IF signal is then compared to an IF signal that is produced in mixer 3 with the PSN sequence that is similar to that transmitted. This comparison takes place in mixer 2. The comparison process, called *correlation*, tells how alike

the two signals being compared are. If the two signals are identical, the correlation is 100 percent. If the two signals are not at all alike, the correlation is 0 percent. The correlation process in the mixer produces a signal that is averaged in the low-pass filter at the output of mixer 2. The output signal will be a high average value if the transmitted and received PSN codes are alike.

The signal out of mixer 2 is fed to a synchronization circuit. The sync circuit has the toughest job of any in the receiver. It must recreate the carrier in the exact frequency and phase so that demodulation can occur. The synchronizing circuit varies the clock frequency so that the PSN code output frequency varies in order to find the same chip rate as the incoming signal. The clock drives a PSN code generator containing the exact code used at the transmitter. The PSN code in the receiver is the same as that of the received signal, but the two are out of sync with one another. Adjusting the clock by speeding it up or slowing it down will

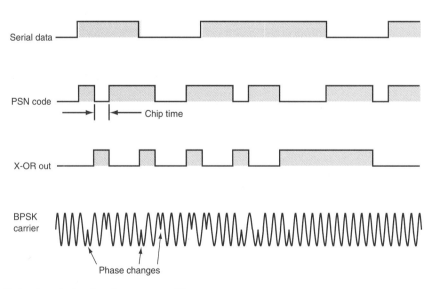

Fig. 12-56 Data signals in direct-sequence SS.

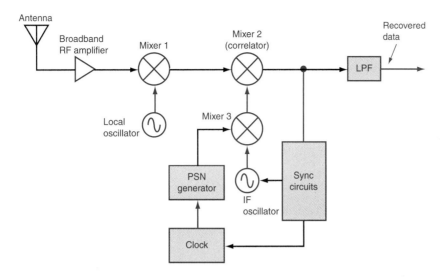

Fig. 12-57 A direct-sequence receiver.

eventually cause the two to come into synchronization.

The PSN code produced in the receiver is used to phase-modulate a carrier at the IF in mixer 3. Like all the other mixers, this one is usually of the double-balanced diode ring type. The output of mixer 3 is a BPSK signal similar to that being received. It is compared to the received signal in mixer 2, which acts as a correlator. The output of mixer 2 is then filtered to recover the original serial binary data. The received signal is said to be *despread.*

Direct sequence is also called *code division multiple access (CDMA)* or *spread spectrum multiple access.* The term *multiple access* refers to any technique that is used for multiplexing many signals on a single communications channel. CDMA is used in satellite systems so that many signals can use a transponder, thus making it more efficient and generating more income for the owner. CDMA is also one of the two new digital methods used in cellular telephone services because it permits more users to occupy a given band than other methods.

Although frequency hopping and direct sequence are the two most widely used types of SS, there are other methods. One example is called *chirp.* With this method, the signal modulates a carrier that is linearly swept over the band being used. This technique is used primarily in radar systems and in distance-ranging equipment.

Another type of SS is called *time-hopping.* In this system, the carrier is keyed on and off by the pseudorandom code sequence. The data signal modulates the carrier in the usual way.

Hybrid systems can also be created by combining two types of SS. Depending upon the application, such a combination will produce some additional benefits. The most common hybrid is a combination of frequency hopping and direct sequence.

Benefits of SS

What are the benefits of SS? First, the communications are secure; unauthorized listening is prevented. Unless the receiver has a very wide bandwidth and the exact pseudo-random code and type of modulation, the signal will not be received.

Second, SS signals are highly resistant to jamming. Jamming signals typically are restricted to a single frequency; jamming one frequency will not interfere with an SS signal.

In the same way, unintentional interference from a signal occupying the same band is greatly minimized and in most cases virtually eliminated.

A big advantage is that many users can share a single band with little or no interference. Because the signals are spread out and random, they will not interfere with one another. With very high usage, a slight increase in the background noise level may result, but in general it is not enough to create a problem. With SS, more signals can use a band than with any other type of modulation and multiplexing.

Another benefit is a resistance to fading. Frequency-selective fading occurs during signal

Despread

Code division multiple access (CDMA)

Chirp

Time-hopping

propagation because signals of different frequency arrive at the receiver at slightly different times. Wide variations of signal strength due to reflections and other phenomena during propagation are virtually eliminated by SS.

Spread spectrum also provides a way for the duration of transmission of a signal to be precisely timed. The pseudorandom code makes it possible to accurately determine the start and end of a transmission. Thus SS is a superior method for radar and other ranging applications that rely upon the time of transmission to determine distance.

Spread spectrum is a rapidly growing technique in transmissions. It is finding more and more applications in data communications as its benefits are discovered and as new components and equipment become available to implement it.

■ TEST_____

Answer the following questions.

125. Spread spectrum is both a _____ and a _____ technique.
126. The two main types of spread spectrum are _____ and _____.
127. In a frequency-hopping SS system, a _____ circuit generates the transmitter frequency.
128. In a frequency-hopping SS system, a _____ circuit selects the frequency produced by the synthesizer.
129. True or false. The hop rate is slower than the bit rate of the digital data.
130. To a narrowband receiver, an SS signal appears as a type of _____.
131. Two or more stations using SS and sharing a common band are identified and distinguished from one another by their unique _____.
132. The length of time a frequency-hopping SS transmitter stays on one frequency is called the _____.

133. In a direct-sequence SS transmitter, the data signal is mixed with a PSN signal in a(n) _____ circuit.
134. True or false. In a direct-sequence SS system, the chip rate is faster than the data rate.
135. The type of modulation used with direct-sequence SS is usually _____.

136. The most difficult part of SS communications is initiating and maintaining _____ prior to and during transmission.
137. The process of comparing one direct-sequence SS signal to another in an effort to obtain a match is called _____.
138. The two main benefits of SS are _____ and _____.
139. The main circuit used in making a PSN code generator is a _____.
140. The frequencies allocated to spread spectrum by the FCC for civilian applications are _____, _____, and _____ MHz.
141. Two growing applications for SS are _____ and _____.

12-6 The Internet

The Internet, sometimes called the Net, is a worldwide interconnection of computers by means of a complex network of many networks. Anyone can connect to the Internet for the purpose of communicating and sharing information with almost any other computer on the Internet. The Internet was established in the late 1960s under the sponsorship of the Department of Defense and later through the National Science Foundation. It provided a way for universities and companies doing military and government research to communicate and to share computer files and software. In the early 1990s, the Internet was privatized and opened to anyone. Today it is a system with tens of millions of users across the world.

Internet Applications

The primary applications of the Internet are e-mail, file transfer, the World Wide Web, e-commerce, and chat.

E-mail E-mail is the exchange of notes, letters, memos, and other personal communications by way of e-mail software and service companies. You write a message to a person and send it over the Internet via your e-mail provider, which, in turn, transfers it to the e-mail provider of the receiving person. That person retrieves the message at his or her convenience. Next to the telephone and the fax machine, e-mail is one of the most common communications methods in use today.

File Transfer *File transfer* refers to the ability of the Internet to transfer files of data or software from one computer to another. The file may be text, digitized photographic data, a computer program, and so on. A file transfer program (FTP) allows you to access files in remote computers and download them to your computer, where they may be used. Files can also be "attached" to e-mail messages. File transfer is an excellent research tool because it allows you to access massive amounts of data in the form of books, articles, newspapers, brochures, and data sheets, and hundreds of other sources.

World Wide Web Whenever most people refer to the Internet, they are really talking about the *World Wide Web* (WWW), or the Web for short. The Web is a specialized part of the Internet where companies, organizations, the government, or even individuals can post information for others to access and use. To do this, you establish a Web site, which is a computer that stores the information you wish to dispense. The information can be on your own computer or in the computer of a Web service provider. The information is in the form of pages, which may contain text, graphics, animations, sound, and even video.

You access these Web sites through the Internet by way of a special piece of software known as a *browser.* The browser allows you to call up a desired Web site by name or to search the Internet Web sites containing information you are interested in. The browser is the software that lets you navigate and explore the Web and to access and display the information. The two most widely used browsers are Netscape Navigator and Microsoft Internet Explorer.

A key feature of the Web is *hypertext.* Hypertext is a method that allows different pages or sites to be linked. When Web pages are created, usually with a language called *hypertext markup language (HTML),* programmers can insert links to other pages on the Web site, other parts of the same page, or even to different Web sites. For example, a page may contain a highlighted word (usually in blue letters), which means that it is a link to some related topic on a different page or site. Clicking with the mouse on that word automatically takes you to the related information. This ability to link to related or relevant information is one of the most powerful and useful features of the Web.

E-Commerce *E-commerce* or electronic commerce refers to doing business over the Net, usually buying and selling goods and services by way of the Web. Companies that have products or services to sell set up Web sites describing their wares and offering them for sale. Individuals may buy these items by giving a credit card number. The product is then shipped by mail or overnight carrier. On-line shopping is expected to grow significantly in the future.

Chat *Chat* is a system of initiating on-line conversations with individuals through specialized software. You can join chat groups in so-called chat rooms that discuss specific subjects. You type and transmit a message to the group; anyone connected to the chat room gets your message. The messages of all participants are displayed. All of this takes place in real time.

On-Line Services

Several large organizations offer an enormous amount of information in their own computers that you can access for a monthly fee. Some well-known examples of *on-line service companies* are America On-Line (AOL), CompuServe, and Prodigy. These companies are not the Internet, but they are connected to it. Almost any conceivable type of information is available including recent news, sports information, and data on many specialized subjects such as education, cars, travel, and restaurants. On-line companies also offer services such as e-mail and the ability to make purchases or airline reservations over the Internet, among many services. These companies also provide Internet access.

How the Internet Works

The Internet is so large, complex, and diverse that it is difficult to describe and explain. However, the purpose of this section is to give

File transfer

E-commerce

Chat
World Wide Web

Browser

On-line services

Hypertext

Hypertext markup language (HTML)

Uniform resource locator (URL)

Packets

Hypertext transfer protocol (http)

Domain

Asynchronous digital subscriber line (ADSL)

Integrated services digital network (ISDN)

Host

you a broad overview of how information is sent and received over the Internet. Virtually every conceivable type of data communications equipment and technique is used in the Internet. All the concepts you learned earlier in this chapter, plus many others, are used. Just keep in mind that the information is transmitted as serial binary pulses, usually grouped as bytes of data (8-bit chunks) within larger groups called *packets.* All the different types of communications media are used including twisted-pair cable, coax, fiber-optic cable, satellites, and other wireless connections.

Internet Addresses Each individual or computer on the Internet must have some kind of identifier or address. An addressing system for the Internet uses a simplified name-addressing scheme that defines a particular hierarchy. The upper level of the hierarchy is called a *domain.* A domain is a specific type of organization using the Internet. Such domains are assigned a part of the Internet address. The most common domains and their address segments are:

Domain	Address segment
Commercial companies	.com
Educational institutions	.edu
Nonprofit organizations	.org
Military	.mil
Government	.gov
Internet service providers	.net
Country	.us, .uk, .fr (United States, United Kingdom, France)

Another part of the address is the host name. The *host* refers to the particular computer connected to the Internet. The host name is often the name of the company, organization, or department sponsoring the computer. For example, IBM's host name is "ibm."

The first part of the address is the user's name or some abbreviation, concatenation, or nickname. You might use your name for an e-mail address or some made-up name that you would recognize. The complete address might look like this: ⟨billbob@xyz.net⟩.

The user name is separated from the host by the @ symbol. Note the dot between the host name and the domain name. This address gets converted into a series of numbers used by computers on the Net to identify and locate one another.

WWW Addresses To locate sites on the Web, you use a special address called a *uniform resource locator (URL).* A typical URL is: ⟨http://www.abs.com/newinfo⟩.

The first part of the URL specifies the communications protocol to be used, in this case *hypertext transfer protocol (http).* The www, of course, designates the World Wide Web. The abs.com part is the domain or the computer on which the Web site exists. The item after the slash(/) indicates a directory within the Web site software. Most Web sites have multiple directories, which are also usually further subdivided into pages.

Initial Connections A PC is connected to the Internet in a variety of ways. The most common method is through a modem that connects to the telephone system. There are traditional modems such as those described earlier and the newer *asynchronous digital subscriber line (ADSL)* and cable TV modems.

An alternative is an ISDN line. *Integrated services digital network (ISDN)* is a special digital communications service offered by most of the larger telephone companies. Communications are by means of baseband digital signals rather than by broadband techniques using modulation and demodulation.

A common way of connecting to the Internet is to use a LAN that your PC may be connected to. Most company and organization PCs are almost always connected to a LAN. The LAN has a server that handles the Internet connection, which may be by way of modem, ISDN interface, or even a T1 line to the telephone system.

The Role of the Telephone System The familiar telephone system is the first link to the

Internet. Because it is so large and so convenient, it is a logical way to make connections to remote computers. In many applications the telephone system is the only connection between computers. Many corporate computers are connected in this way, and you can connect directly to an on-line service provider this way. In any case, the telephone system is the first step in getting on the Net. Its primary function is to connect you to a facility known as an *Internet service provider (ISP)*.

Internet Service Provider An ISP is a company set up especially to tap into the network known as the Internet. The ISP has one or more servers to which are connected dozens, hundreds, or even thousands of modems, ISDN lines, or cable connections from subscribers. It is usually the ISP that provides you the software you will use in communications over the Internet. The ISP also usually provides your e-mail service. Some on-line service companies like AOL are also ISPs.

There are literally thousands of ISPs across the United States. Most are set up in the medium-size to larger cities. You access your ISP by dialing its local telephone number. If you live in a small town or a rural area, you can connect to an ISP in a nearby city, but you may have to make a long-distance call to do this. The ISP is connected to the Internet backbone by way of a fast digital interface such as T1 lines or by a faster method.

The Internet Backbone The *Internet backbone* is a collection of companies that install, service, and maintain large nationwide and even worldwide networks of high-speed fiber-optic cable. The companies own the equipment and operate it to provide universal access to the Internet. In many ways, these companies *are* the Internet, for they provide the basic communications medium, equipment, and software that permit any computer to access any other. Some of the larger backbone providers are MCI-WorldCom, GTE, Sprint, and PSINet.

Although each of the backbone providers has its own nationwide network, the providers are usually connected to one another to provide many different paths from one computer to another. There are about 40 to 50 of these interconnection points in the United States.

Known as *public peering points (PPPs),* these facilities provide the links between backbones.

Transmission Methods Data is transmitted over the Internet by a data communications method known as *packet switching.* The data to be transmitted is divided up into short chunks (usually less than 1500 bytes) called *packets.* When you are downloading a long file from some remote source, it will be packetized and sent in segments, one after another, and not necessarily in the order in which it is packetized. The packets may even take different routes through the backbone. If one line is busy, the system switches connections and sends the packet via another route. At the receiving end, the packets are reassembled in the correct order.

There are three key parts to the packet-switching system: switches, routers, and software called *TCP/IP.*

The packet-switching system is a network of exchanges using high-speed switches to connect multiple inputs to multiple outputs. Most exchanges are interlinked with other exchanges within the network. There are many possible paths through the various exchanges. Data is converted into packets and sent over this network.

There are three popular packet-switching systems: X.25, frame relay, and asynchronous transfer mode (ATM). The older X.25 system used extensive error detection and correction and as a result was very slow. This system has been abandoned in favor of frame relay, which is significantly faster. It does not use error correction, and it features a variable-length packet. ATM uses a fixed-length packet of 53 bytes. It can transmit at rates exceeding 1 Gbits/s over fiber networks. It can transmit data, voice, or video (and all at the same time using TDM) in digital form. Most Internet backbones use frame relay or ATM over the fiber backbone.

A router is a specialized computer that is used to sort out all the various signals and data on the Internet. The routers recognize and identify e-mail addresses and the host and domain addresses of the various sites on the Web. Routers examine all packets to see where they are going and then switch them to the right line for transmission.

The software that is used to packetize the data and format it for transmission and

Public peering points (PPPs)

Packet switching

Internet service provider (ISP)

Internet backbone

reception is transmission control protocol (TCP) and the Internet protocol (IP), better known as TCP/IP.

Tracing a Communications Session on the Internet

Now, let's consider a couple of examples of how you can access the Internet. The two most common applications, e-mail and Web access, are described in the next sections.

E-Mail Assume that you wish to send an e-mail message to a friend. You call up your Internet software and connect to your ISP. Then you write your message, using the e-mail software, and provide the address of your friend. You then send the message. You can even attach another file. The message will go from your PC through the modem to your local telephone office, where it is sent to your local ISP. The ISP server answers the call and stores your message. It then connects to the backbone and sends out the message. The various computers along the way sort out the host and domain names through routers. Next, the message arrives at the ISP used by your friend. The message then goes into your friend's hard drive of the server at the ISP. Your friend, at his or her leisure, accesses the e-mail account that says that mail has been received. If your friend replies, the process begins again.

Connecting to the Web To connect to the Web, you call up your browser and connect to your ISP. If you know the Web site, you type it into your browser. It will find the site and connect you. Web addresses look like this: ⟨http://www.sitename.host.domain⟩.

This format is referred to as the universal resource locator (URL). The letters http stand for "hypertext transfer protocol," which is the format used by most sites. The letters www, of course, stand for the Web, where the site is located. This is followed by the host computer and domain names as in other Internet addresses.

If you do not know the name of the site or if you just want to "surf" the Web and look for

sites, you can call up one of the special sites designed for searching the net. These sites use special search software called a *search engine* that queries other computers on the Internet for what you want. If you give the search site some key words to look for, it will provide you with a list of sites containing information you might be able to use. You can then link to them from the search site via your browser. Some of the most well known search sites are Yahoo, Excite, Lycos, Altavista, and Hot Bot. Most organizations "register" with the search sites so that their Web pages will be listed during a search.

Once you put the Web address into your browser, the ISP sends it out on the backbone, where computers and routers find the site and connect you. The Web site is located on a host server, which may be at the ISP or connected to an ISP. Again, all communications go through the backbone, and all the various computers and routers sort out the locations and paths.

■ TEST

Answer the following questions.

142. Name five interfaces that your PC can use to connect to the Internet.
143. What is the name of the facility that connects you to the Internet?
144. What communications medium do you usually use to connect to your ISP?
145. Briefly describe the Internet backbone.
146. What is the method by which data is transmitted on the backbone?
147. What is the name of the equipment that identifies, sorts, and connects computers on the Internet?
148. What is the name of the software and protocol used for communications on the Internet?
149. Define the World Wide Web.
150. What is the basic form and makeup of a Web site?
151. Name the software used to access the Web.
152. Name the three parts of an Internet address. Name the parts of a Web URL.
153. What is a search engine? What is it used for?

Summary

1. Data communications is the transmission and reception of binary data between computers and other digital equipment.
2. The earliest form of electronic communications, the telegraph, was a type of data communications.
3. Turning a carrier off and on in a code of dots and dashes is a kind of data communications known as continuous wave (CW).
4. Teletype is a form of telegraph that uses the 5-bit Baudot code to transmit between typewriterlike units.
5. The most widely used binary data communications code is the 7-bit American Standard Code for Information Interchange (ASCII).
6. Another popular code is the 8-bit Extended Binary Coded Decimal Interchange Code (EBCDIC) used mainly in IBM systems.
7. The two main methods of data transmission are serial and parallel. In serial transmission, each bit is transmitted sequentially. In parallel transmission, all bits are transmitted simultaneously.
8. Serial transfers are slower than parallel transfers but require only a single line or channel. Parallel transfers require multiple channels of lines called a bus.
9. The speed of data transmission is designated in terms of bits per second (bits/s) or baud.
10. Baud rate is the number of symbol changes per second. A symbol is an amplitude, frequency, or phase change.
11. The channel- or bit-rate capacity of a channel is directly proportional to the channel bandwidth and the time of transmission.
12. The channel capacity or binary signal transmission speed C in bits per second is equal to twice the channel bandwidth B when no noise is present ($C = 2B$).
13. When multiple levels or symbols are used to encode the data, the channel capacity C is greater for a given bandwidth B, or $C = 2B \log_2 N$ where N is the number of symbols used.
14. The channel capacity C in bits per second is proportional to the channel bandwidth B and the power S/N ratio, or

$$C = B \log_2 (1 + S/N)$$

15. The bit rate is higher than the baud (symbol) rate if multiple-level (symbol) encoding is used.
16. The two methods of data transmission are asynchronous and synchronous. In asynchronous transmission, data is sent as a continuous block of multiple characters framed with synchronization characters.
17. Synchronous transmission is faster than asynchronous transmission.
18. In data communications, a binary 1 is referred to as a mark and a binary 0 as a space.
19. Signals, whether voice, video, or binary, transmitted directly over a cable are known as baseband signals.
20. Voice and video signals are analog but may be converted to digital for data communications transmission.
21. Signals that involve a modulated carrier are called broadband signals.
22. Communications of binary data signals over the telephone network which is designed for analog signals is made possible by using a modem.
23. A modem is a modulator-demodulator unit that converts digital signals to analog and vice versa.
24. The most commonly used modulation techniques in modems are frequency-shift keying (FSK), phase-shift keying (PSK), and quadrature amplitude modulation (QAM).
25. Frequency-shift keying uses two frequencies for binary 0 and 1 (1070 and 1270 Hz or 2025 and 2225 kHz). It operates at speeds of 300 baud or less.
26. Modems capable of transmitting at standard higher rates of 1200, 2400, 4800, and 9600 bits/s use PSK and/or QAM.
27. Binary PSK (BPSK) uses a carrier of 1600 or 1700 Hz where a phase of 0° represents a binary of 0 and a 180° phase shift represents a binary 1, or vice versa.

28. Binary PSK is generated by a balanced modulator.

29. Binary PSK is demodulated by a balanced modulator.

30. To properly demodulate BPSK, the carrier at the demodulator must have exactly the same phase as the transmitting carrier.

31. A special carrier recovery circuit in the receiver produces the correct phase carrier from the BPSK signal.

32. Differential PSK eliminates the need for a special reference phase carrier by using a coding technique where the phase of each bit is referenced to the previous bit.

33. Quadrature PSK uses four equally spaced phase shifts of the carrier to represent two bits (dibit). For example, $00 = 45°$, $01 = 135°$, $11 = 225°$, $10 = 315°$.

34. In 8-PSK, 3 bits are coded per phase change. In 16-PSK, 4 bits are coded per phase change. Thus the bit rate is 3 or 4 times the symbol rate change or baud rate.

35. Quadrature amplitude modulation uses a combination of QPSK and two-level AM to code 3 bits per baud. Each of the eight possible 3-bit combinations is represented by a unique phase and amplitude signal.

36. Asynchronous digital subscriber line (ADSL) modems provide a way to use existing telephone lines to send high-speed data, as much as 8 Mbits/s downstream and 640 kbits/s upstream.

37. ADSL modems use a complex frequency-multiplexed set of QAM carriers to transmit the bits on a standard twisted-pair telephone line.

38. Cable modems use 6-MHz bands on a cable TV system cable to carry digital data. Rates up to 30 Mbits/s can be achieved with 64 QAM.

39. A protocol is a rule or procedure that defines how data is sent and received.

40. Protocols include "handshaking" signals between the transmitter and receiver that indicate the status of each.

41. The Xmodem protocol is widely used in personal computers.

42. In synchronous communications, a variety of special characters are sent before and after the block of data to ensure that the data is correctly received.

43. Bit errors that occur during transmission are caused primarily by noise.

44. The ratio of the number of bit errors that occur for a given number of bits transmitted is known as the bit error rate (BER).

45. Error-detection and -correction schemes have been devised to reduce bit errors and increase data accuracy.

46. One of the most widely used error-detection schemes adds a parity bit to each character transmitted, making the total number of binary 1s transmitted odd or even. If a bit error occurs, the parity bit derived at the receiver will differ from the one transmitted.

47. Parity generator circuits are made up of multiple levels of exclusive OR (XOR) gates.

48. Another name for parity is vertical redundancy check (VRC).

49. The longitudinal redundancy check (LRC) is another way to test for errors. Corresponding bits in adjacent data words are exclusive-ORed to generate a block check character (BCC) that is appended to the transmitted block.

50. The VRC and LRC provide a coordinate system that will identify the exact location of a bit error so that it may be corrected.

51. A widely used error-detection scheme is the cyclical redundancy check (CRC) where the data block is divided by a constant to produce a quotient and remainder. The remainder is the CRC character which is attached to the transmitted data block and compared to the CRC computed at the receiver.

52. A network is any interconnection of two or more stations that can communicate with one another.

53. An example of a wide area network (WAN) is the telephone system. An example of a metropolitan area network (MAN) is a cable TV system.

54. A local area network (LAN) is an interconnection of stations in a small area over short distances such as in an office building, on a military base, or on a college campus.

55. Local area networks were conceived to allow PC users to share expensive peripherals such as hard disks and printers but are now used for general communication and the sharing of software and data.

56. The most common physical configurations or topologies of LANs are the star, ring, and bus.

57. The bus is the fastest. The ring is the least expensive but is disabled if one station fails. The

star configuration is commonly used to connect many terminals and PCs to a larger mainframe or minicomputer.

58. The most widely used LAN type is Ethernet, which uses a coax or twisted-pair bus for speeds of 10 Mbits/s, 100 Mbits/s, or 1 Gbit/s.

59. The three most commonly used transmission media in networks are twisted-pair, coax, and fiber-optic cable.

60. Coaxial cable is also a widely used medium because of its high-speed capability and its shielding against noise.

61. Fiber-optic cable is growing in usage as its price declines. It has the highest speed capability of any medium.

62. Local area networks use both baseband and broadband techniques. Baseband refers to transmitting the information signal directly on the medium.

63. Broadband refers to using modulation techniques to transmit the data on carriers that can be assigned specific channels over a wide frequency range on a common medium.

64. Broadband systems require the use of modems at each node.

65. Spread spectrum (SS) is a modulation and multiplexing technique used primarily in data communications that deliberately spreads the signal out over a wide bandwidth rather than trying to restrict it to a narrow band.

66. The two most widely used types of spread spectrum are frequency hopping (FH) and direct sequence (DS).

67. Spread spectrum offers the benefits of privacy or security of communications, immunity to jamming, and lower sensitivity to frequency-selective fading.

68. In frequency-hopping SS, the serial binary data usually modulates a carrier by FSK. The FSK signal is mixed with a sine wave from a frequency synthesizer to form the final RF signal. The frequency synthesizer is switched at a rate of speed higher than the rate of the data signal, dwelling only briefly on each of many channel frequencies. Thus, the signal is broken up into small pieces and spread over a wide frequency range.

69. The frequency-hopping scheme is controlled by a pseudorandom binary code that switches at random from binary 0 to 1 and vice versa. The random nature of the signal causes the fre-

quency to jump all over the band, distributing pieces of the signal hither and yon. The pseudorandom code acts like digital noise and thus is called a pseudorandom noise (PSN) code.

70. In direct-sequence SS, the serial data is mixed with a higher-frequency PSN code in an X-OR circuit. The resulting higher-frequency binary signal produces more higher-frequency sidebands, thereby spreading the signal out over a wider bandwidth. The X-OR output usually phase-modulates the final carrier.

71. Because an SS signal is spread out randomly over a wide bandwidth, many signals can share a band without interference. A narrowband receiver will not respond to an SS signal except to interpret the random signals as a form of low-level noise.

72. The main problem in receiving SS signals is in acquiring the signal and synchronizing the transmitted PSN code to the same internally generated PSN code in the receiver. In direct-sequence SS, an electronic correlator circuit is responsible for achieving synchronism.

73. Different stations sharing a given band are defined and identified by their unique PSN code.

74. Spread spectrum was originally developed for and used in military equipment. In 1985, the FCC authorized SS for commercial or civilian use in the 902- to 928-, 2400- to 2483-, and 5725- to 5850-MHz bands.

75. The new applications for spread spectrum include wireless LANs, wireless computer modems, and digital cellular telephones.

76. The Internet is a worldwide network of computers linked to share information and software.

77. The most common applications of the Internet are e-mail, file transfer, e-commerce, chat, and the World Wide Web (WWW).

78. The WWW is that part of the Internet where people and organizations post Web sites containing information to be accessed by anyone with a PC.

79. The Web is accessed by way of software called a browser.

80. A PC's link to the Internet is by way of an interface such as conventional ADSL or cable modem, ISDN port, or LAN.

81. An organization known as an Internet service provider (ISP) uses a computer called a server to connect users to the Internet.

82. The link between most users and the ISP is through the telephone system.
83. The heart of the Internet is several backbones, which are fiber-optic cables with packet switching that provide multiple paths and links between ISPs.
84. Hardware units called routers identify and route signals over the backbone.
85. Software known as TCP/IP packetizes the data and facilitates its transmission over the backbone.
86. Internet addresses are of the form ⟨username@host.domain⟩. Web addresses or URLs are of the form ⟨http:www.username.host.domain⟩.

Chapter Review Questions

Choose the letter which best answers each question.

12-1. Data communications refer to the transmission of
   a. Voice.
   b. Video.
   c. Computer data.
   d. All the above.

12-2. Data communications uses
   a. Analog methods.
   b. Digital methods.
   c. Either of the above.
   d. Neither of the above.

12-3. Which of the following is *not* primarily a type of data communications?
   a. Telephone
   b. Teletype
   c. Telegraph
   d. CW

12-4. The main reason that serial transmission is preferred to parallel transmission is that
   a. Serial is faster.
   b. Serial requires only a single channel.
   c. Serial requires multiple channels.
   d. Parallel is too expensive.

12-5. Mark and space refer respectively to
   a. Dot and dash.
   b. Message and interval.
   c. Binary 1 and binary 0.
   d. On and off.

12-6. The number of amplitude, frequency, or phase changes that take place per second is known as the
   a. Data rate in bits per second.
   b. Frequency of operation.
   c. Speed limit.
   d. Baud rate.

12-7. Data transmission of one character at a time with start and stop bits is known as what type of transmission?
   a. Asynchronous
   b. Serial
   c. Synchronous
   d. Parallel

12-8. The most widely used data communications code is
   a. Morse.
   b. ASCII.
   c. Baudot.
   d. EBCDIC.

12-9. The ASCII code has
   a. 4 bits.
   b. 5 bits.
   c. 7 bits.
   d. 8 bits.

12-10. Digital signals may be transmitted over the telephone network if
   a. Their speed is low enough.
   b. They are converted to analog first.
   c. They are alternating current instead of direct current.
   d. They are digital only.

12-11. Start and stop bits, respectively, are
   a. Mark, space.
   b. Space, mark.
   c. Space, space.
   d. Mark, mark.

12-12. Which of the following is correct?
   a. The bit rate may be greater than the baud rate.
   b. The baud rate may be greater than the bit rate.
   c. The bit and baud rates are always the same.
   d. The bit and baud rates are not related.

12-13. A modem converts
   a. Analog signals to digital.
   b. Digital signals to analog.
   c. Both *a* and *b*.
   d. None of the above.

12-14. A carrier recovery circuit is *not* needed with
   a. BPSK.
   b. QPSK.
   c. DPSK.
   d. QAM.
12-15. The basic modulator and demodulator circuits in PSK are
   a. PLLs.
   b. Balanced modulators.
   c. Shift registers.
   d. Linear summers.
12-16. The carrier used with a BPSK demodulator is
   a. Generated by an oscillator.
   b. The BPSK signal itself.
   c. Twice the frequency of the transmitted carrier.
   d. Recovered from the BPSK signal.
12-17. A 56 kbits/s-rate signal can pass over the voice-grade telephone line if which kind of modulation is used?
   a. BPSK
   b. QPSK
   c. DPSK
   d. QAM/TCM
12-18. Quadrature amplitude modulation is
   a. Amplitude modulation only.
   b. QPSK only.
   c. AM plus PSK.
   d. AM plus FSK.
12-19. The transmission medium for an ADSL modem is
   a. coaxial cable.
   b. LAN wiring.
   c. fiber-optic cable.
   d. twisted-pair telephone line.
12-20. A 6-MHz bandwidth cable modem channel has an upper data rate of
   a. 1.536 Mbits/s.
   b. 6.144 Mbits/s.
   c. 8 Mbits/s.
   d. 30 Mbits/s.
12-21. A rule or procedure that defines how data is to be transmitted is called a(n)
   a. Handshake.
   b. Error-detection scheme.
   c. Data specification.
   d. Protocol.
12-22. A popular PC protocol is
   a. Parity.
   b. Xmodem.
   c. CRC.
   d. LRC.

12-23. A synchronous transmission usually begins with which character?
   a. SYN
   b. STX
   c. SOH
   d. ETB
12-24. The characters making up the message in a synchronous transmission are collectively referred to as a data
   a. Set.
   b. Sequence.
   c. Block.
   d. Collection.
12-25. Bit errors in data transmission are usually caused by
   a. Equipment failures.
   b. Typing mistakes.
   c. Noise.
   d. Poor S/N ratio at the receiver.
12-26. Which of the following is *not* a commonly used method of error detection?
   a. Parity
   b. BCC
   c. CRC
   d. Redundancy
12-27. Which of the following words has the correct parity bit? Assume odd parity. The last bit is the parity bit.
   a. 1111111 1
   b. 1100110 1
   c. 0011010 1
   d. 0000000 0
12-28. Another name for parity is
   a. Vertical redundancy check.
   b. Block check character.
   c. Longitudinal redundancy check.
   d. Cyclical redundancy check.
12-29. Ten bit errors occur in two million transmitted. The bit error rate is
   a. 2×10^{-5}.
   b. 5×10^{-5}.
   c. 5×10^{-6}.
   d. 2×10^{-6}.
12-30. The building block of a parity or BCC generator is a(n)
   a. Shift register.
   b. XOR.
   c. 2-to-4 level converter.
   d. UART.
12-31. A longitudinal redundancy check produces a(n)
   a. Block check character.
   b. Parity bit.

c. CRC.

d. Error correction.

12-32. Dividing the data block by a constant produces a remainder that is used for error detection. It is called the

a. Vertical redundancy check.

b. Horizontal redundancy check.

c. Block check character.

d. Cyclical redundancy check.

12-33. A CRC generator uses which components?

a. Balanced modulator

b. Shift register

c. Binary adder

d. Multiplexer

12-34. Which of the following is *not* a LAN?

a. PBX system

b. Hospital system

c. Office building system

d. Cable TV system

12-35. The fastest LAN topology is the

a. Ring.

b. Bus.

c. Star.

d. Square.

12-36. Which is *not* a common LAN medium?

a. Twin lead

b. Twisted pair

c. Fiber-optic cable

d. Coax

12-37. A mainframe computer connected to multiple terminals and PCs usually uses which configuration?

a. Bus

b. Ring

c. Star

d. Tree

12-38. A small telephone switching system that can be used as a LAN is called a

a. Ring.

b. WAN.

c. UART.

d. PBX.

12-39. Which medium is the least susceptible to noise?

a. Twin lead

b. Twisted pair

c. Fiber-optic cable

d. Coax

12-40. Which medium is the most widely used in LANs?

a. Twin lead

b. Twisted pair

c. Fiber-optic cable

d. Coax

12-41. Transmitting the data signal directly over the medium is referred to as

a. Baseband.

b. Broadband.

c. Ring.

d. Bus.

12-42. The technique of using modulation and FDM to transmit multiple data channels of a common medium is known as

a. Baseband.

b. Broadband.

c. Ring.

d. Bus.

12-43. What is the minimum bandwidth required to transmit a 56 kbits/s binary signal with no noise?

a. 14 kHz

b. 28 kHz

c. 56 kHz

d. 112 kHz

12-44. Sixteen different levels (symbols) are used to encode binary data. The channel bandwidth is 36 MHz. The maximum channel capacity is

a. 18 Mbits/s.

b. 72 Mbits/s.

c. 288 Mbits/s.

d. 2.176 Gbits/s.

12-45. What is the bandwidth required to transmit at a rate of 10 Mbits/s in the presence of a 28-dB S/N ratio?

a. 1.075 MHz

b. 5 MHz

c. 10 MHz

d. 10.75 MHz

12-46. Which circuit is common to both frequency-hopping and direct-sequence SS transmitters?

a. Correlator

b. PSN code generator

c. Frequency synthesizer

d. Sweep generator

12-47. Spread spectrum stations sharing a band are identified by and distinguished from one another by

a. PSN code.

b. Frequency of operation.

c. Clock rate.

d. Modulation type.

12-48. The type of modulation most often used with direct-sequence SS is

a. QAM.

b. SSB.

c. FSK.

d. PSK.

12-49. The main circuit in a PSN generator is a(n)
   a. X-OR.
   b. Multiplexer.
   c. Shift register.
   d. Mixer.

12-50. To a conventional narrowband receiver, an SS signal appears to be like
   a. Noise.
   b. Fading.
   c. A jamming signal.
   d. An intermittent connection.

12-51. Which of the following is *not* a benefit of SS?
   a. Jam-proof.
   b. Security.
   c. Immunity to fading.
   d. Noise proof.

12-52. Spread spectrum is a form of multiplexing.
   a. True.
   b. False.

12-53. The most critical and difficult part of receiving a direct-sequence SS signal is
   a. Frequency synthesis.
   b. Synchronism.
   c. PSN code generation.
   d. Carrier recovery.

12-54. The transmission medium over the Internet backbone is
   a. Fiber-optic cable.
   b. Twisted pair.
   c. Coaxial cable.
   d. Wireless links.

12-55. What is the name of the software that packetizes data and organizes its transmission over the Internet?
   a. Windows 98.
   b. TCP/IP.
   c. Browser.
   d. Search engine.

12-56. The actual link to the Internet is provided by
   a. The telephone company.
   b. A backbone company.
   c. The Internet service provider.
   d. The World Wide Web.

12-57. A key feature of the World Wide Web is
   a. Graphics, animation, sound, video, text.
   b. Hypertext links.
   c. Search capability.
   d. All of the above.

12-58. The most widely used Internet application is
   a. Chat.
   b. E-mail.
   c. FTP.
   d. The World Wide Web.

12-59. A URL is a(n)
   a. Internet e-mail address.
   b. Router code.
   c. Web address.
   d. File name.

12-60. What device recognizes addresses and sends data to the correct destination?
   a. Router
   b. Modem
   c. Server
   d. Packet switch

12-61. The basic structure of an Ethernet LAN is a
   a. Bus.
   b. Star.
   c. Tree.
   d. Ring.

12-62. Which of the following is *not* a common Ethernet transmission speed?
   a. 10 MHz
   b. 20 MHz
   c. 100 MHz
   d. 1 GHz

12-63. What is the name of the cable used in 100Base-T versions of Ethernet?
   a. RG-8/U
   b. RG-58/U
   c. Category 5 twisted pair
   d. Fiber-optic cable

12-64. What is the name of the connector used with UTP cable in Ethernet systems?
   a. R-11
   b. N-type
   c. BNC
   d. RJ-45

12-65. What is the name of the unit that consolidates all the wiring in a 10Base-T system by allowing all nodes to terminate at this common unit?
   a. Router
   b. Bridge
   c. Hub
   d. Bus strip

12-66. What is the maximum number of data bytes in an Ethernet packet?
   a. 46
   b. 256
   c. 1024
   d. 1500

12-67. What happens if two nodes try to transmit simultaneously on a common Ethernet cable?
   a. Both stop transmitting and randomly try again a short time later.
   b. Both stations continue to transmit without interference.

c. One station switches to another cable.

d. The stations take turns transmitting.

12-68. What type of error-detection scheme is used in an Ethernet packet?

    a. Parity

    b. Cyclical redundancy check

    c. Reed-Solomon

    d. Hamming codes

12-69. What is the name of the line code used with 10Base-T?

    a. 2QB1

    b. AMI

    c. Manchester

    d. MLT-3

12-70. What is probably the most widely used version of 100-Mbits/s Ethernet?

    a. 100Base-TX

    b. 100VG-AnyLAN

    c. 100Base-T4

    d. 100Base-FX

12-71. How does the MLT-3 line coding facilitate 100 Mbits/s operation?

    a. It codes 2 bits per baud by using three levels.

    b. It helps overcome noise by using the data to frequency-modulate a carrier.

c. It transmits the clock along with the signal, thus ensuring easier recovery.

d. It uses bipolar voltage levels to effectively reduce the frequency of the signals on the line.

12-72. Ethernet can use fiber-optic cable.

    a. True

    b. False

12-73. What is the IEEE designation for the Ethernet standard?

    a. RS-232

    b. 802.3

    c. FDDI

    d. TCP.IP

12-74. Which of the following techniques is *not* used in achieving 1 Gbits/s over UTP?

    a. Four twisted-pair cables

    b. Two bits per baud line coding

    c. Fiber-optic cable

    d. 125-Mbits/s clock

12-75. What is the name associated with 1-Gbits/s Ethernet?

    a. Fast Ethernet

    b. High-speed Ethernet

    c. HyperSpeed Ethernet

    d. Gigabit Ethernet

Critical Thinking Questions

12-1. How is voice intelligence transmitted via SS?

12-2. Could SS be used for satellite communications? Explain how and describe any benefits.

12-3. What are the three basic types of symbol changes that can be used to represent more bits per baud?

12-4. Explain how a wireless LAN works.

12-5. Which type of modem is fastest?

12-6. Describe how a wireless modem on a battery-operated laptop computer might work.

12-7. Assume that a 500-Hz square wave is applied to one end of a 10,000-ft-long telephone line. What is the equivalent bit rate in bits per second for this signal? Describe what the signal might look like at the end of the line.

12-8. Besides the telephone line and the cable TV connection, name two other possible ways in which your PC could be connected to the Internet.

12-9. Can voice be communicated over the Internet? Explain.

12-10. Describe a system or systems using data communications techniques that would allow you to monitor the temperature of a device at a point 400 ft from the PC on which the temperature would be recorded and displayed. Draw a block diagram of the major components.

12-11. Think up three new applications for the Internet that you have not heard of before.

12-12. Assume that you are building a Web site for yourself. What information would you put there that could be accessed by anyone?

1. binary or digital
2. true
3. telegraph
4. continuous wave (CW)
5. dots, dashes
6. K = $-\cdot-$; 8 $---\cdot\cdot$; , = $--\cdot\cdot--$
7. Baudot
8. teletype
9. M
10. ASCII
11. EBCDIC
12. B = 1000010; 3 = 0110011; ? = 0111111
13. 0000111
14. false
15. shift register
16. start, stop
17. bits per second (bits/s)
18. baud rate
19. 8.696 kbits/s
20. 50 ms
21. false
22. blocks
23. space, mark
24. information theory
25. bandwidth, time
26. 10,000
27. 60,000
28. true
29. less
30. 50
31. 100
32. modulator, demodulator
33. analog, digital (or binary)
34. telephone network
35. computers
36. frequency-shift keying
37. frequencies
38. mark (1), space (0)
39. 1070, 1270, 2025, 2225
40. universal asynchronous receiver transmitter (UART)
41. phase-shift keying
42. 0, 180
43. balanced modulator
44. balanced modulator
45. carrier recovery
46. differential
47. XNOR
48. four
49. 2 bits
50. four
51. bit splitter (or shift register)
52. 2- to 4-level converter
53. amplitude, PSK
54. 56 Kbits/s
55. constellation diagram
56. telephone line length
57. QAM
58. 8 Mbits/s, 640 kbits/s
59. frequency division multiplexing
60. 64 QAM, QPSK
61. 30 Mbits/s
62. true
63. false
64. true
65. false
66. true
67. protocols
68. redundancy, special error-checking codes, parity, block check code, cyclical redundancy check
69. handshaking
70. Xmodem
71. block
72. asynchronous
73. SYN or sync
74. header
75. STX, ETX, or ETB
76. redundancy
77. ARQ exact-count
78. noise
79. bit error rate (BER)
80. parity
81. block check character
82. a. 1
    b. 0
    c. 1
    d. 1
83. vertical redundancy check
84. exclusive OR
85. block check character
86. longitudinal redundancy check
87. cyclical redundancy check
88. shift register
89. true
90. wide area network
91. true
92. local area networks
93. private branch exchange
94. false
95. star, ring, bus

96. star
97. star
98. mainframe (or mini)
99. ring
100. nodes
101. bus
102. bus
103. ring
104. gateway
105. bridge
106. twisted pair, coax, fiber-optic cable
107. tree
108. Ethernet, 10 Mbits/s, 100 Mbits/s, 1000 Mbits/s
109. twisted pair
110. fiber-optic cable
111. fiber-optic cable
112. 100, 100, 1000
113. baseband
114. baseband
115. broadband
116. modem
117. Unshielded twisted pair (UTP)
118. hub
119. (a) 10 Mbits/s, (b) 100 Mbits/s, (c) 1 Gbit/s
120. packets
121. category 5
122. true
123. CSMA/CD
124. (a) 2- to 5-level encoding, (b) 4-UTP cables in parallel
125. modulation, multiplexing
126. frequency hopping, direct sequence
127. frequency synthesizer
128. pseudorandom noise code generator
129. false
130. noise
131. PSN code
132. dwell time
133. X-OR
134. true
135. PSK
136. synchronism
137. correlation
138. security, jam-proof
139. shift register
140. 902- to 928-, 2400- to 2483-, 5725- to 5850-MHz
141. wireless LANs, wireless computer modems
142. Standard analog modem, ADSL modem, cable modem, ISDN, LAN
143. Internet service provider
144. The telephone system
145. A high-speed fiber-optic network using packet switching.
146. Packet switching
147. Router
148. TCP/IP
149. A collection of Web sites where information is posted to access using text, graphics, sound, and video.
150. A Web site is made up of pages of information that can be viewed one screen at a time.
151. Browser.
152. User name, host, and domain. URL: http://www plus user name, host and domain names.
153. A search engine is a piece of software used by a search Web site to help you find Web sites of interest on a particular subject.

Chapter 13

Fiber-Optic Communications

Chapter Objectives

This chapter will help you to:

1. *Draw* a basic block diagram of a fiber-optic communications system and *tell* what each part of it does.
2. *State* eight benefits of fiber-optic cables over electrical cables for communications.
3. *Name* six typical communications applications for fiber-optic cable.
4. *Explain* how light is propagated through a fiber-optic cable.

5. *Name* the three basic types of fiber-optic cable and *state* the two materials from which they are made.
6. *Calculate* the transmission loss in decibels (dB) of fiber-optic cable over a distance.
7. *Name* the two types of optical transmitter components and the main operating frequency range.
8. *Explain* the operation of an optical detector and receiver.

The communications medium in most electronic communications systems is either a wire conductor cable or free space. But now, a new medium is growing in popularity, the fiber-optic cable. A fiber-optic cable is essentially a light pipe that is used to carry a light beam from one place to another. Light is an electromagnetic signal like a radio wave. It can be modulated by information and sent over the fiber-optic cable. Because the frequency of light is extremely high, it can accommodate very wide bandwidths of information and/or extremely high data rates can be achieved with excellent reliability. This chapter introduces you to the basic concepts and circuits used in light-wave communications.

13-1 Light-Wave Communications Systems

One of the main limitations of communications systems is their restricted information-carrying capabilities. In more specific terms what this means is that the communications medium can carry only so many messages. And, as you have seen, this information-handling ability is directly proportional to the bandwidth of the communications channel. In telephone systems, the bandwidth is limited by the characteristics of the cable used to carry the signals. As the demand for telephones has increased, better cables and wiring systems have been developed. Further, multiplexing techniques have been developed to transmit

multiple telephone conversations over a single cable. These techniques have the same effect as if the number of cables or channels of communications were greatly multiplied.

In radio communications systems, the information modulates a high-frequency carrier. The modulation produces sidebands, and therefore, the signal occupies a narrow portion of the RF spectrum, which we call a channel. However, the RF spectrum is finite. There is only so much space for radio signals. To increase the information capacity of a channel, the bandwidth of the channel must be increased. This reduces available spectrum space. Multiplexing techniques are used to send more signals in a given channel bandwidth, and methods have been developed to transmit more information in less bandwidth.

The information-carrying capacity of a radio signal can be increased tremendously if higher carrier frequencies are used. As the demand for increased communications capacity has gone up over the years, higher and higher RFs are being used. Today, microwaves are the preferred radio channels for this reason, but it is more complex and expensive to use these higher frequencies because of the special equipment required. Further, transmission distances are limited to the line of sight.

Laser

Today the use of microwave frequencies for communications has been greatly expanded. Although the microwave frequency spectrum is not totally occupied at this point, it is well used. A great deal of communication takes place by microwave signals transmitted through satellite relay stations. Satellites can receive and retransmit thousands of telephone calls to any place on earth or can carry very high speed digital data. Slowly but surely, however, we are reaching the limits of our radio spectrum.

Communicating by Light

One way to expand communications capability further is to use light as the transmission medium. Instead of using an electrical signal traveling over a cable or electromagnetic waves traveling through space, the information is put on a light beam and transmitted through space or through a special cable. In the late nineteenth century, Alexander Graham Bell, the inventor of the telephone, demonstrated that information could be transmitted by light. A special transmitter made up of a microphone attached to a mirror was used to transmit voice over a distance of several hundred feet using a light beam from the sun. The mirror was used to reflect the light from the sun to the receiv-

ing site. The microphone, mechanically attached to the mirror, caused the mirror to vibrate as words were spoken into it. The mirror vibrated at the voice frequency, thereby changing the amplitude of the reflected sunlight. At the receiving site, a special photocell picked up the reflected light rays. This photocell was a small batterylike device that produced a dc output voltage proportional to the amount of light received. As the reflected light varied, the voltage output from the photocell varied in the same way. The voltage could be connected to an earphone so that the voice would be accurately reproduced.

Communications by light beam in free space is impractical over very long distances. The main disadvantage is the great attenuation of the light due to atmospheric effects. Fog, haze, smog, rain, snow, and other conditions absorb, reflect, and refract the light, greatly attenuating it and thereby limiting the transmission distance. Artificial light beams used to carry information are also virtually obliterated during daylight hours by the sun.

Light-beam communications was made more practical with the invention of the *laser*. The laser is a special high-intensity, single-frequency light source. It produces a very narrow beam of brilliant light of a specific wavelength (color). Because of its great intensity, the laser beam can penetrate atmospheric obstacles better than other types of light, thereby making light-beam communication more reliable over longer distances. The primary problem with such free-space light-beam communication is that the transmitter and receiver must be perfectly aligned with one another.

Instead of using free space, some type of light-carrying cable can also be used. For centuries it has been known that light is easily transmitted through various types of transparent media such as glass and water, but it wasn't until the early 1900s that scientists were able to develop practical light-carrying media. By the mid-1950s, glass fibers were developed that permitted long light-carrying cables to be constructed. Over the years, these glass fibers have been perfected. Further, low-cost plastic fiber cable was also developed. Developments in these cables permitted them to be made longer with less attenuation of the light.

Today, these fiber-optic cables have been highly refined. Cables many miles long can be constructed and interconnected for the purpose of transmitting information on a light beam over very long distances. Thanks to these highly refined fiber-optic cables, a new transmission medium is now available. Its great advantage is that light beams have an incredible information-carrying capacity. Whereas hundreds of telephone conversations may be transmitted simultaneously at microwave frequencies, many thousands of signals can be carried on a light beam through a fiber-optic cable. Using multiplexing techniques similar to those used in telephone and radio systems, fiber-optic communications systems have an almost limitless capacity for information transfer.

A Fiber-Optic System

The components of a typical fiber-optic communications system are illustrated in Fig. 13-1. The information signal to be transmitted may be voice, video, or computer data. The first step is to convert the information into a form compatible with the communications medium. This is usually done by converting continuous analog signals such as voice and video (TV) signals into a series of digital pulses. An A/D converter is used for this purpose. Computer data is already in digital form.

These digital pulses are then used to flash a powerful light source off and on very rapidly. In simple low-cost systems that transmit over short

distances, the light source is usually a light-emitting diode (LED). This is a semiconductor device that puts out a low-intensity red light beam. Other colors are also used. Infra-red beams like those used in TV remote controls are also used in transmission. Another commonly used light source is the solid-state laser. This is also a semiconductor device that generates an extremely intense single-frequency light beam.

The light-beam pulses are then fed into a fiber-optic cable where they are transmitted over long distances. At the receiving end, a light-sensitive device known as a photocell or light detector is used to detect the light pulses. This photocell or photodetector converts the light pulses into an electrical signal. The electrical pulses are amplified and reshaped back into digital form. They are fed to a decoder, such as a D/A converter, where the original voice or video is recovered.

In very long transmission systems, repeater units must be used along the way. Since the light is greatly attenuated when it travels over long distances, at some point it may be too weak to be received reliably. To overcome this problem, special relay stations are used to pick up the light beam, convert it back into electrical pulses that are amplified, and then retransmit the pulses on another light beam. Several stages of repeaters may be needed over very long distances. But despite the attenuation problem, the loss is less than the loss that occurs with the electric cables.

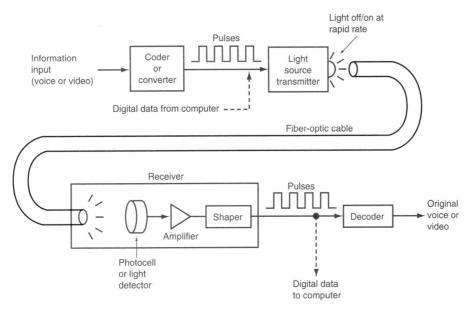

Fig. 13-1 Basic elements of a fiber-optic communications system.

Fiber-Optic Applications

Fiber-optic communications systems are being used more and more each day. Their primary use is in long-distance telephone systems. The fiber-optic cables are no more expensive or complex to install than standard electric cables, because their information-carrying capacity is many times greater. Most new long-distance telephone systems being installed use fiber-optic cables.

Fiber-optic communications systems are also being used in other applications. For example, they are being used to interconnect computers in networks within a large building, and they carry control signals in airplanes and ships. Many TV systems also use fiber-optic cable because of its wide bandwidth. In all cases, the fiber-optic cables replace conventional coax or twisted-pair cables. Figure 13-2 lists some of the applications in which fiber-optic cables are being used.

Benefits of Fiber-Optic Systems

We have already mentioned the main benefit of fiber-optic cables, that is, their enormous information-carrying capability. This is dependent upon the bandwidth of the cable. Bandwidth refers to the range of frequencies which a cable will carry. Electric cables such as coax do have a wide bandwidth, but the bandwidth of fiber-optic cable is much greater. Therefore, more signals can be carried.

But that's not all. Fiber-optic cable has many other benefits. These are summarized in Fig. 13-3. Is it any wonder that fiber-optic cables are becoming more widely used?

There are some disadvantages to fiber-optic cable. For example, its small size and brittle-

1. Wider bandwidth: Fiber-optic cables have higher information-carrying capability.
2. Lower loss: With fiber-optic cables, there is less signal attenuation over long distances.
3. Lightweight: Glass or plastic cables are much lighter than copper cables and offer benefits in those areas where low weight is critical (*i.e.*, aircraft).
4. Small size: Practical fiber-optic cables are much smaller in diameter than electrical cables. Therefore, more can be contained in a smaller space.
5. Strength: Fiber-optic cables are stronger than electrical cables and can support more weight.
6. Security: Fiber-optic cables cannot be "tapped" as easily as electrical cables, and they do not radiate signals that can be picked up for eavesdropping purposes. There is less need for complex and expensive encryption techniques.
7. Interference immunity: Fiber-optic cables do not radiate signals as some electrical cables do and cause interference to other cables. They are also immune to pickup of interference from other sources.
8. Greater safety: Fiber-optic cables do not carry electricity. Therefore, there is no shock hazard. They are also insulators so are not susceptible to lightning strikes as electrical cables are.

Fig. 13-3 Benefits of fiber-optic cables over conventional electric cables.

ness make it difficult to work with. Special, expensive tools and techniques are required. It is also more costly, at least for now. In time, with the continuously rising price of copper, fiber-optic cable will no doubt cost less than electric cable. Part of the high cost of fiber-optic cable is due to the difficulty of manufacturing it. As methods improve, costs will decline.

■ TEST

Answer the following questions.

1. Fiber-optic cables carry _____ rather than electrical signals.
2. The three main types of information carried by fiber-optic cables are _____, _____, and _____.
3. The major use of fiber-optic cables is in _____.
4. Fiber-optic cables are made of _____ or _____.
5. The main benefit of fiber-optic cable over electric cable is its _____.
6. True or false. Fiber-optic cable has more loss than electric cable over long distances.
7. True or false. Fiber-optic cable is smaller, lighter, and stronger than electric cable.
8. List the two main disadvantages of fiber-optic cable.
9. The two most commonly used light sources in fiber-optic systems are _____.

1. Local and long-distance telephone systems
2. TV studio-to-transmitter interconnection, eliminating microwave radio link
3. Closed-circuit TV systems used in buildings for security
4. Secure communications systems at military bases
5. Computer networks, wide area and local area
6. Shipboard communications
7. Aircraft communications
8. Aircraft controls
9. Interconnection of measuring and monitoring instruments in plants and laboratories
10. Data acquisition and control signal communications in industrial process control systems
11. Nuclear plant instrumentation
12. College campus communications
13. Utilities (electric, gas, etc.) station communications

Fig. 13-2 Applications of fiber-optic cables.

10. Voice and video signals are converted into _____ before being transmitted by a light beam.
11. The device that converts the light pulses into an electrical signal is a(n) _____.
12. Regenerative units called _____ are often used to compensate for signal attenuation over long distances.

13-2 How Fiber-Optic Cables Work

A fiber-optic cable is essentially a light pipe. It is not really a hollow tube carrying light, but it is a long thin strand of glass or plastic fiber. Most fiber cables have a circular cross section with a diameter of only a fraction of an inch. Some fiber-optic cables have a diameter only the size of a human hair. A light source is placed at the end of the fiber, and light passes through it and exits at the other end of the cable. How the light actually propagates through the fiber depends upon the laws of optics. In this section, we will review the basic fundamentals of light and the theories of optics that explain how light is contained within the light pipe.

The Spectrum of Light

Light is a kind of electromagnetic radiation. Another form of electromagnetic radiation that you are familiar with is a radio wave. Any electromagnetic wave is made up of both electric and magnetic fields that travel through space from one place to another. The basic characteristic of electromagnetic radiation is its frequency or wavelength. To put light waves into

A fiber-optic current sensor measures current without touching any wire in the circuit.

perspective with the lower radio and other frequencies, refer to the spectrum chart in Fig. 13-4. At the left end of the spectrum are very low frequency electrical signals such as 60-cycle ac power and audio or sound frequencies in the 20-Hz to 20-kHz range. These are low frequencies that have long wavelengths. Higher up the scale to the right is a wide range of radio frequencies. Radio frequencies range from approximately 10 kHz to 300 GHz. Microwaves extend from 1 GHz to 300 GHz.

Even further up the scale we find visible light. The frequency of the optical spectrum is in the range of 3×10^{11} to 3×10^{16} Hz. This includes both infrared and ultraviolet as well as the visible parts of the spectrum. The visible spectrum is from 4.3×10^{14} to 7.5×10^{14} Hz, but we rarely refer to the frequency of light. Instead, we state light in terms of wavelength. Recall that wavelength is a distance measured in meters between peaks of a wave. Light waves are very short and are usually expressed in nanometers (one billionth of a meter) or micrometers or microns (one millionth of a meter). Visible light is in the range of 400 to 700 nanometers (nm) or 0.4 to 0.7 micrometers (μm), depending upon the color of the light. Short-wavelength light is violet (400 nm), and red (700 nm) is long-wavelength light.

Right below visible light is a region known as infrared. Infrared rays cannot be seen. However,

Light

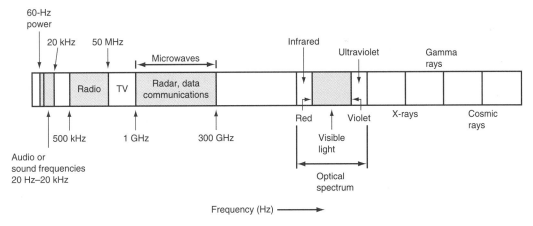

Fig. 13-4 Electromagnetic frequency spectrum.

they have the properties of light and, therefore, can be manipulated in the same way with mirrors, lenses, and other devices normally used to manipulate light. The source of most infrared is heat. Right above the visible spectrum is the ultraviolet range. Infrared and ultraviolet are included in what we call the optical spectrum. Beyond that are the x-rays, gamma rays, and cosmic rays, which we are not concerned with here.

Characteristics and Behavior of Light

Light waves travel in a straight line like microwaves do. Light rays emitted by a candle, light bulb, or other light source move out in a straight line in all directions. Like electricity, these light rays travel at the speed of light, which is generally considered to be 300,000,000 m/s or about 186,000 mi/s in free space.

Law of reflection

The speed of light depends upon the medium through which the light passes. The figures given above are correct for free space, meaning for light traveling in air or a vacuum. When light passes through another material such as glass, its speed is slower.

Light can be manipulated in many ways. For example, lenses are widely used to focus, enlarge, or decrease the size of light waves from some source. Lenses permit useful devices such as cameras, microscopes, binoculars, and telescopes to be constructed. Another simple way of manipulating light is to reflect it. When light rays hit a reflective surface, such as a mirror, the light waves are thrown back or deflected. By using mirrors, the direction of a light beam can be changed.

Refraction

Reflection

Reflection The *reflection* of light from a mirror follows a simple physical law. That is, the direction of the reflected light wave can be easily predicted if the angle of the light beam striking the mirror is known. Refer to Fig. 13-5. Assume an imaginary line that is perpendicular to the flat mirror surface. A perpendicular line, of course, makes a right angle with the surface as shown. This imaginary perpendicular line is referred to as the *normal*. The normal is usually drawn at the point where the mirror reflects the light beam.

If the light beam follows the normal, then the reflection will simply be back along the same path. The reflected light ray will exactly coincide with the original light ray.

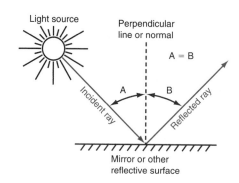

Fig. 13-5 The law of light reflection. The angle of reflection equals the angle of incidence.

If the light ray strikes the mirror at some angle A from the normal, then the reflected light ray will leave the mirror at the same angle B to the normal. This principle is known as the *law of reflection.* It is usually expressed in the following form: *The angle of incidence is equal to the angle of reflection.*

The light ray from the light source is normally called the incident ray. It makes an angle A with the normal at the reflecting surface. This is called the angle of incidence. The reflected ray is the light wave that leaves the mirror surface. Its direction is determined by the angle of reflection B which is exactly equal to the angle of incidence.

Refraction Another way that the direction of the light ray can be changed is by refraction. *Refraction* is the bending of a light ray that occurs when the light rays pass from one medium to another. In reflection, the light ray bounces away from the reflecting surface rather than being absorbed by or passing through the mirror. Refraction occurs only when light passes through transparent material such as air, water, or glass. This refraction or the bending of the light rays takes place at the point where two different substances meet. For example, where air and water meet, refraction will occur. The dividing line between the two different substances or media is known as the boundary or interface.

A simple example of refraction is to place a spoon or straw into a glass of water as shown in Fig. 13-6(*a*). If you observe the glass of water from the side, it will appear as if the spoon or straw is bent or offset at the surface of the water.

Another phenomenon caused by refraction is the effect that occurs whenever you observe an object under water. For example, you may be standing in a clear stream and observing a stone at the bottom. The stone is actually in a different position than it appears to be from your observation. See Fig. 13-6(*b*).

The refraction or bending of light occurs because light travels at different speeds in different materials. The speed of light in free space is typically much higher than the speed of light in water, glass, or other materials. The amount of refraction or bending of the light in a material is usually expressed in terms of the index of refraction *n*. This is the ratio of the speed of light in air to the speed of light in the substance.

$$\text{Index of refraction } (n)$$
$$= \frac{\text{speed of light in air}}{\text{speed of light in substance}}$$

Naturally, the index of refraction of air is 1, simply because 1 divided by itself is 1. The refractive index of water is approximately 1.3, whereas that of glass is 1.5.

To get a better understanding of this idea, let's consider a piece of glass with a *refractive index* of 1.5. What this means is that light will travel through 1.5 ft of air, but during that same time, the light will only travel 1 ft through the glass. The glass slows down the light wave considerably. The index of refraction is important because it tells exactly how much a light wave will be bent in various substances.

Refractive index

Whenever a light ray passes from one medium to the next, the light wave will be bent depending upon the index of refraction. In Fig. 13-7, you see a light ray passing through air. It makes an angle A with the normal. At the interface between the air and the glass, the direction of the light ray is changed. The speed

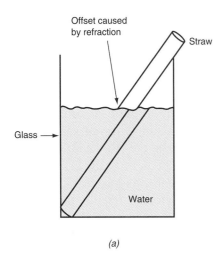

(a)

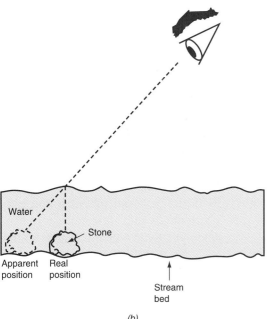

(b)

Fig. 13-6 Examples of the effect of refraction.

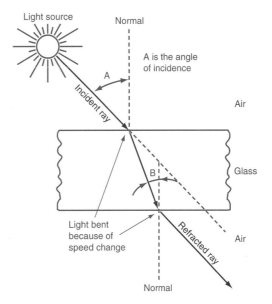

Fig. 13-7 How light rays are bent when passing from one medium to another.

of light is slower; therefore, the angle that the light beam makes to the normal B is different from the incident angle. If the index of refraction is known, the exact angle can be determined.

If the light ray then passes from the glass back into air, it will again change directions as Fig. 13-7 shows. The important point to note is that the angle of the refracted ray B is not equal to the angle of incidence A.

If you keep increasing the angle of incidence, at some point the angle of refraction will equal 90° to the normal as shown in Fig. 13-8(a). When this happens, the refracted light ray travels along the interface between the air and glass and the angle of incidence A is said to be the *critical angle*. The critical angle value depends upon the index of refraction of the glass.

Critical angle

Now, if you make the angle of incidence greater than the critical angle, the light ray will actually be reflected from the interface. See Fig. 13-8(b). When the light ray strikes the interface at an angle greater than the critical angle, the light ray does not pass through the interface into the glass. The effect is as if a mirror existed at the interface. When this occurs, the angle of reflection B is equal to the angle of incidence A as if a real mirror were used. This action is known as *total internal reflection*. Total internal reflection occurs only in

Total internal reflection

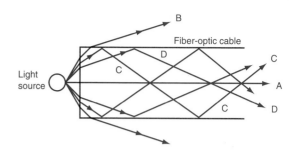

Fig. 13-9 Light rays in a fiber-optic cable.

materials in which the velocity of light is slower than in air. It is this basic principle that allows a fiber-optic cable to work.

How Light Travels in a Cable

Now let's see how this principle applies to the fiber-optic cable. Figure 13-9 shows a thin fiber-optic cable. A beam of light is focused on the end of the cable as shown. This beam of light can be positioned in a number of different ways so that the light enters the fiber at different angles. For example, light ray A enters the cable perpendicularly to the end surface. Therefore, the light beam simply travels straight down the fiber and exits at the other end.

The angle of light beam B is such that its angle of incidence is less than the critical angle, and, therefore, refraction takes place. The light wave passes through the fiber and exits the edge into the air at a different angle.

The angle of incidence of light beams C and D is greater than the critical angle. Therefore, total internal reflection takes place and the light beams are simply reflected off the surface of the fiber cable. The light beam bounces back and forth between the surfaces until it exits at the other end of the cable.

When the light beam reflects off the inner surface, the angle of incidence is equal to the angle of reflection. Because of this principle, light rays entering at different angles will take different paths through the cable. Since some paths will be longer than others, some light rays will exit sooner and some later.

In practice, the light source is placed so that the angle is such that the light beam passes directly down the center axis of the cable or so that the reflection angles are great. This prevents the light from being lost due to refraction at the interface. Because of total internal reflection, the light beam will continue to

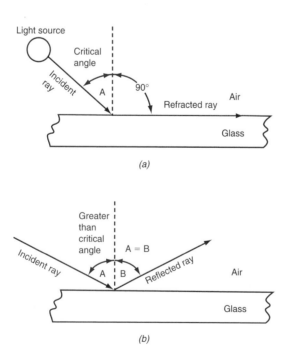

Fig. 13-8 Special cases of refraction. (a) Along the surface, and (b) reflection.

propagate through the fiber even though it is bent. Care must be taken in bending the cable so as not to break it and so that the bend angle is not great enough to cause the angle of incidence to change which, in turn, will cause the light to be lost due to refraction. With long slow bends, the light will stay within the cable.

■ TEST————————————————

Answer the following questions.

13. Light is a type of _____ radiation.
14. A fiber-optic cable can be viewed as a light _____.
15. Light travels in a
    a. Circle.
    b. Straight line.
    c. Curve.
    d. Random way.
16. The wavelength of light is usually expressed in _____ or _____.
17. The lowest-frequency visible light is
    a. Red
    b. Violet
18. A micrometer, or micron, is a length of _____ of a _____.
19. The wavelength range of visible light is _____ to _____ nm.
20. Infrared light has a wavelength that is
    a. Less than 400 nm.
    b. More than 700 nm.
    c. Less than 700 nm.
21. The optical spectrum is made up of what three parts?
22. The speed of light in air is _____ m/s or _____ mi/s.
23. True or false. The speed of light is slower in glass or water than it is in air.
24. The number that tells how fast light travels in a medium compared to air is the _____ of _____.
25. Light beams can be bounced or their direction can be changed by _____ with a(n) _____.
26. The bending of light rays due to speed changes when moving from one medium to another is called _____.
27. If a light ray strikes a mirror at an angle of 30° from the normal, it is reflected at an angle of _____° from the normal.
28. When the angle of refraction is 90° to the normal, the ray travels along the _____ between the two media.

Therefore, the incident ray strikes the surface at the _____ angle.
29. When the incident ray strikes the interface at an angle greater than the critical angle, _____ occurs.
30. The critical angle depends upon the value of the _____ of _____.

13-3 Fiber-Optic Cables

Just as standard electric cables come in a variety of sizes, shapes, and types, fiber-optic cables are available in different configurations. The simplest cable is just a single strand of fiber, whereas complex cables are made up of multiple fibers with different layers and other elements. In this section, we will introduce you to the various types of fiber-optic cables, show you how they are constructed, and explain the most popular types that you will encounter.

Fiber-Optic Construction

As indicated earlier, the portion of a fiber-optic cable that carries the light is made from either glass or plastic. Another name for glass is silica. Special techniques have been developed to create nearly perfect optical glass or plastic which is transparent to light. Such materials can carry light over a long distance. The glass or plastic is melted and pulled through a form to produce a fine threadlike fiber.

Glass has superior optical characteristics over plastic. However, glass is far more expensive and more fragile than plastic. Although plastic is less expensive and more flexible, its attenuation of light is greater. For a given intensity, light will travel a greater distance in glass than in plastic. For very long distance transmissions, glass is certainly preferred. For shorter distances, plastic is much more practical.

Step index

Core

Cladding

In an electric cable, the current-carrying element is a copper wire. However, copper wires are usually not used alone. Typically, most wires are covered with an insulation not only to protect the wire but to prevent electrical shorts between adjacent wires.

In the same way, a fiber-optic cable is rarely used alone. The fiber, which is called the *core,* is usually surrounded by a protective *cladding* as illustrated in Fig. 13-10. The cladding is also made of glass or plastic but has a lower index of refraction. This ensures that the proper interface is achieved so that the light waves remain within the core. In addition to protecting the fiber core from nicks and scratches, the cladding adds strength. Some fiber-optic cables have a glass core with a glass cladding. Others have a plastic core with a plastic cladding. Another common arrangement is a glass core with a plastic cladding. It is called plastic-clad silica (PCS) cable.

In observing a fiber-optic cable, you typically cannot tell the division between the core and the cladding. Since the two are usually made of the same types of material, to the naked eye it is not possible to see the difference. A plastic jacket similar to the outer insulation on an electric cable is usually put over the cladding.

Types of Fiber-Optic Cable

There are two basic ways of classifying fiber-optic cables. The first way is an indication of how the index of refraction varies across the cross section of the cable. The second way of classification is by mode. Mode refers to the various paths that the light rays can take in passing through the fiber. Usually these two methods of classification are combined to define the types of cable.

Step Index

There are two basic ways of defining the index of refraction variation across a cable. These are step index and graded index. *Step index* refers to the fact that there is a sharply defined step in the index of refraction where the fiber core and the cladding interface. It means that the core has one constant index of refraction N_1, while the cladding has another constant index of refraction N_2. Where the two come together, there is a distinct step as illustrated in Fig. 13-11. If you were to plot a curve showing how the index of refraction varies vertically as you move from left to right across the cross section of the cable, there would be a sharp increase in the index of refraction as the core is encountered and then a sharp decline in the index of refraction as the cladding is encountered.

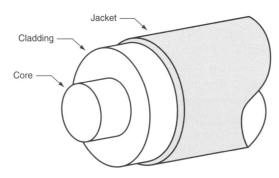

Fig. 13-10 Basic construction of a fiber-optic cable.

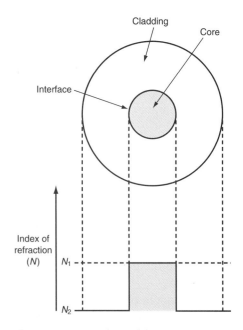

Fig. 13-11 A step-index cable cross section.

Graded Index

The other type of cable has a *graded index*. In this type of cable, the index of refraction of the core is not constant. Instead, the index of refraction varies smoothly and continuously over the diameter of the core as shown in Fig. 13-12. As you get closer to the center of the core, the index of refraction gradually increases, reaching a peak at the center and then declining as the other outer edge of the core is reached. The index of refraction of the cladding is constant.

Mode Mode refers to the number of paths for the light rays in the cable. There are two classifications: *single mode* and *multimode*. In single mode, light follows a single path through the core. In multimode, the light takes many paths through the core.

Each type of fiber-optic cable is classified by one of these methods of rating the index or mode. In practice, there are three commonly used types of fiber-optic cable. These are multimode step index, single-mode step index, and multimode graded index. Let's take a look at each of these types in more detail.

The *multimode step-index* fiber cable is probably the most common and widely used type. It is also the easiest to make and, therefore, the least expensive. It is widely used for short to medium distances at relatively low pulse frequencies.

The main advantage of a multimode step-index fiber is the large size. Typical core di-

ameters are in the 50- to 1000-μm range. Such large-diameter cores are excellent at gathering light and transmitting it efficiently. This means that an inexpensive light source such as an LED can be used to produce the light pulses. The light takes many hundreds or even thousands of paths through the core before exiting. You saw this earlier in Fig. 13-9. Because of the different lengths of these paths, some light rays take longer to reach the end of the cable than others. The problem with this is that it stretches the light pulses.

For example, in Fig. 13-13, a short light pulse is applied to the end of the cable by the source. Light rays from the source will travel in multiple paths. At the end of the cable, those rays which travel the shortest distance reach the end first. Other rays begin to reach the end of the cable later in time until the light ray with the longest path finally reaches the end, concluding the pulse. In Fig. 13-13 ray A reaches the end first, then B, then C. The result is a pulse at the other end of the cable that is lower in amplitude due to the attenuation of the light in the cable and increased in duration due to the different arrival times of the various light rays. This stretching of the pulse is referred to as *modal dispersion.*

Because the pulse has been stretched, input pulses cannot occur at a rate faster than the output pulse duration permits. Otherwise, the pulses will essentially merge, as shown in Fig. 13-14. At the output, one long pulse will occur and will be indistinguishable from the three separate pulses originally transmitted. This means that incorrect information will be received. The only cure for this problem is to

Graded index

Single mode

Multimode step index

Modal dispersion

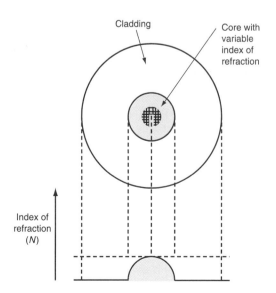

Fig. 13-12 Graded-index cable cross section.

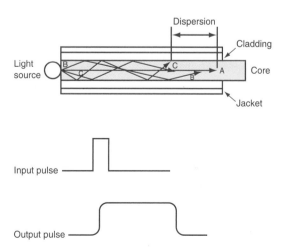

Fig. 13-13 A multimode step-index cable.

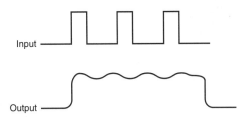

Fig. 13-14 The effect of modal dispersion on pulses occurring too rapidly in a multimode step-index cable.

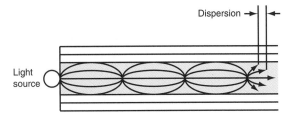

Fig. 13-16 A multimode graded-index cable.

reduce the pulse repetition rate or the frequency of the pulses. When this is done, proper operation occurs. But with pulses at a lower frequency, less information can be handled.

Single-mode step index

Single-Mode Step Index In a *single-mode,* or *mono-mode, step-index* fiber cable the core is so small that the total number of modes or paths through the core are minimized and modal dispersion is essentially eliminated. See Fig. 13-15. Typical core sizes are 2 to 15 μm. Virtually the only path through the core is down the center. With minimum refraction, no pulse stretching occurs. The output pulse has essentially the same duration as the input pulse.

The single-mode step-index fibers are by far the best since the pulse repetition rate can be high and the maximum amount of information can be carried. For very long distance transmission and maximum information content, single-mode step-index fiber cables should be used.

The main problem with this type of cable is that because of its extremely small size, it is difficult to make and is, therefore, very expensive. Handling, splicing, and making interconnections are also more difficult. Finally, for proper operation an expensive, superintense light source such as a laser must be used. For long distances, however, this is the type of cable preferred.

Multimode graded index

Multimode Graded Index *Multimode graded-index* fiber cables have several modes or paths

of transmission through the cable, but they are much more orderly and predictable. Figure 13-16 shows the typical paths of the light beams. Because of the continuously varying index of refraction across the core, the light rays are bent smoothly and converge repeatedly at points along the cable. The light rays near the edge of the core take a longer path but travel faster since the index of refraction is lower. All the modes or light paths tend to arrive at one point simultaneously. The result is that there is less modal dispersion. It is not eliminated entirely, but the output pulse is not nearly as stretched as in multimode step-index cable. The output pulse is only slightly elongated. As a result, this cable can be used at very high pulse rates and, therefore, a considerable amount of information can be carried on it.

This type of cable is also much wider in diameter with core sizes in the 50- to 100-μm range. Therefore, it is easier to splice and interconnect, and cheaper, less-intense light sources may be used.

Cable Variations

Most fiber-optic cables consist of more than the core, the cladding, and a jacket. They usually contain one or more additional elements. The simplest cable is the core with its cladding surrounded by a protective jacket. As in electric cables, this outer jacket or insulation is made of some type of plastic, typically polyethylene, polyurethane, or polyvinyl chloride (PVC). The main purpose of this outer jacket is to protect the core and cladding from damage. Usually fiber-optic cables are buried underground or are strung between supports. Therefore, the fibers themselves have to be protected from moisture, dirt, temperature variations, and other conditions. The outer jacket also helps minimize physical damage such as cuts, nicks, and crushing. The more complex cables may contain two or more fiber-optic

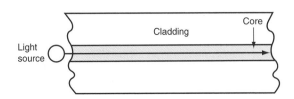

Fig. 13-15 Single-mode step-index cable.

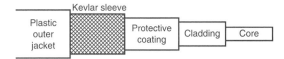

Fig. 13-17 Typical layers in a fiber-optic cable.

elements. Typical cables are available with 2, 6, 12, 18, and 24 fiber-optic cores.

There are many different types of cable configurations. Many have several layers of protective jackets. Some cables incorporate a flexible strength or tension element which helps minimize damage to the fiber-optic elements when the cable is being pulled or when it must support its own weight. Typically, this strength element is made of stranded steel or a special yarn known as Kevlar. Kevlar is strong and is preferred over steel because it is an insulator. In some cables, the Kevlar forms a protective sleeve or jacket over the cladding. Most claddings are covered with a clear protective coating for added strength and resistance to moisture and damage. Figure 13-17 shows one type of cable.

Fiber-optic cables are also available in a flat ribbon form as shown in Fig. 13-18. When working with multiple fibers, flat ribbon cable is generally easier to use because handling and identification of individual fibers is far easier. The flat cable is also more space-efficient for some applications.

Cable Attenuation

The main specification of a fiber-optic cable is its *attenuation*. Attenuation refers to the loss of light energy as the light pulse travels from one end of the cable to the other. The light pulse of a specific amplitude or brilliance is applied to one end of the cable, but the light pulse output at the other end of the cable will be much lower in amplitude. The intensity of the light at the output is lower because of various losses in the cable. The main reason for the loss in light intensity over the length of the cable is due to light absorption, scattering, and dispersion.

Absorption refers to how the light waves are actually "soaked up" in the core material due to the impurity of the glass or plastic. Scattering refers to the light lost because of light waves entering at the wrong angle and being lost in the cladding due to refraction. Dispersion, as you have already learned, refers to the pulse stretching caused by the many different paths through the cable. Although no light is lost as such in dispersion, the output is still lower in amplitude than the input but the length of the light pulse has increased in duration.

The amount of attenuation, of course, varies with the type of cable and its size. Glass has less attenuation than plastic. Wider cores have less attenuation than narrower cores. But more important, the attenuation is directly proportional to the length of the cable. It is obvious that the longer the distance the light has to travel, the greater the loss due to absorption, scattering, and dispersion. Doubling the length of a cable doubles the attenuation, and so on.

Attenuation

The attenuation of a fiber-optic cable is expressed in decibels per unit of length. The standard decibel formula used is

$$dB = 10 \log \frac{P_o}{P_i}$$

where P_o is the power out and P_i is the power in.

Figure 13-19 shows the percentage of output power for various decibel losses. For example, 3 dB represents half power. In other words, a 3-dB loss means that only 50 percent of the input appears at the output. The other 50 percent of the power is lost in the cable. The

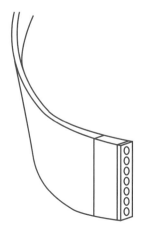

Fig. 13-18 Multicore flat ribbon cable.

Loss (dB)	Power output (%)
1	79
2	63
3	50
4	40
5	31
6	25
7	20
8	14
9	12
10	10
20	1
30	.1
40	.01
50	.001

Fig. 13-19 Decibel loss table.

higher the decibel figure, the greater the attenuation and loss. A 30-dB loss means that only one-thousandth of the input power appears at the end.

The standard specification for fiber-optic cable is the attenuation expressed in terms of decibels per kilometer. A kilometer is a smaller unit of distance than a mile. A mile is 5280 ft, whereas a kilometer is only 3281 ft. In other words, a kilometer is approximately 0.62 mi. That means that there are about 1.6 km per mile.

The attenuation ratings of fiber-optic cables vary over a considerable range. The finest single-mode step-index cables have an attenuation of only 1 dB/km. However, very large core plastic-fiber cables can have an attenuation of several thousand decibels per kilometer. A typical cable might have a standard loss of 10 to 20 dB/km. Typically, those fibers with an attenuation of less than 10 dB/km are called low-loss fibers, while those with an attenuation of between 10 and 100 dB/km are medium-loss fibers. High-loss fibers are those with over 100 dB/km ratings. Naturally, the smaller the decibel number, the less the attenuation and the better the cable.

You can easily determine the total amount of attenuation for a particular cable if you know the attenuation rating. For example, if a cable has an attenuation of 15 dB/km, then a 5-km length cable has a total attenuation of 5(15) = 75 dB. If two cables are spliced together and one has an attenuation of 17 dB and the other 24 dB, the total attenuation is simply the sum, or 17 + 24 = 41 dB.

When long fiber-optic cables are needed, two or more cables may be spliced together. The ends of the cable are perfectly aligned and then glued together with a special, clear, low-loss epoxy. Connectors are also used. A variety of connectors provide a convenient way to splice cables and attach them to transmitters, receivers, and repeaters.

Cable Splicing

Connectors are special mechanical assemblies that allow fiber-optic cables to be connected to one another. You might say they are a form of temporary splice that can be conveniently disconnected. Fiber-optic connectors are the optical equivalent of electrical plugs and sockets. They are mechanical assemblies that hold the ends of the cable and cause them to be accurately aligned with another cable. Most of them either snap together or have threads which allow the two pieces to be screwed together.

Connectors

Connectors force better alignment of the cables. The two ends of the cables must be aligned with precision so that excessive light is not lost. Otherwise, a splice or connection will introduce excessive attenuation. Figure 13-20 shows several ways that the cores can be misaligned. A connector corrects these problems.

A typical fiber-optic connector is shown in Fig. 13-21(*a*). One end of the connector called the ferrule holds the fiber securely in place. A matching fitting holds the other fiber securely in place. When the two are screwed together, the ends of the fibers touch, providing a low-loss coupling. Figure 13-21(*b*) shows how the connector aligns the fibers.

Dozens of different kinds of connectors are available for different applications. These connectors are used in those parts of the system

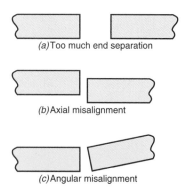

(a) Too much end separation

(b) Axial misalignment

(c) Angular misalignment

Fig. 13-20 Misalignment of fiber-optic cores causes excessive light loss.

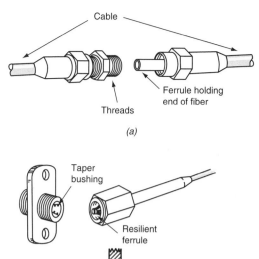

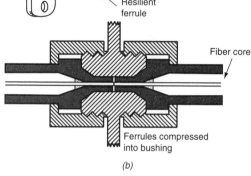

Fig. 13-21 Details of a fiber-cable connector.

where it may be desirable to occasionally disconnect the fiber-optic cable for making tests or repairs. Connectors are normally used at the end of the cable applied to the light source or at the end of the cable connected to the photodetector. Connectors are also used at the repeater units where the light is picked up, converted into an electrical pulse, amplified and reshaped, and then used to create a new pulse to continue the transmission over a long line.

The first step in installing a connector is to cleave the fiber so that it is perfectly square. Then the manufacturer's directions should be followed when attaching the connector to the fiber. Many manufacturers supply rolls of cable in different lengths with connectors already attached. In working with fiber-optic cable, connectors, and other accessories, you should always follow the manufacturer's directions.

■ TEST

Answer the following questions.

31. Which material has the best optical characteristics and lowest loss?
    a. Plastic
    b. Glass
    c. They are equal

32. The core is protected by the _____.
33. In PCS cable, the core is _____ and the cladding is _____.
34. The index of refraction is highest in the
    a. Core
    b. Cladding.
35. List the three main types of fiber-optic cable.
36. Stretching of the light pulse is called _____.
37. List the two types of cable in which light-pulse stretching occurs.
38. High-frequency pulses can be best transmitted over _____ cable.
39. Pulse stretching causes the information capacity of a cable to
    a. Increase.
    b. Decrease.
40. Graded index means that the _____ of _____ of the core varies over its cross section.
41. Single-mode step-index cable has a core diameter in the range of
    a. 100 to 1000 μm
    b. 50 to 100 μm
    c. 2 to 15 μm
42. A _____ is applied over the cladding to protect against moisture, damage, etc.
43. A common protective layer in a cable is made of _____ mesh.
44. Fiber-optic cables are available with the following number of cores: _____.
45. Light loss in a cable is called _____.
46. Light loss is caused by _____.
47. Light loss is measured in _____ per _____.
48. A cable with a loss of 5 dB will have _____ percent of the input appear at the output (see Fig. 13-19).
49. True or false. A kilometer is longer than a mile.
50. Which cable length will have the least attenuation?
    a. 40 ft
    b. 120 ft
    c. 1780 ft
    d. 1 km
51. Three cables with attenuations of 9, 22, and 45 dB are spliced together. The total attenuation is _____ dB.
52. True or false. Fiber-optic cables may be spliced.

RZ format

NRZ format

LED

53. To conveniently link and attach fiber-optic cables to one another and related equipment, _____ are used.

13-4 Optical Transmitters and Receivers

In an optical communications system, transmission begins with the transmitter, which consists of a modulator and the circuitry that generate the carrier. In this case, the carrier is a light beam that is modulated by digital pulses which turn it on and off. The basic transmitter is nothing more than a light source.

Optical Transmitters

Conventional light sources such as incandescent lamps cannot be used in fiber-optic systems. The reason for this is that they are simply too slow. An incandescent light source consists of a filament that heats up and emits light. Such a light source cannot be turned off and on fast enough because of the thermal delay in the filament. In order to transmit high-speed digital pulses, a very fast light source must be used. The two most commonly used light sources are LEDs and semiconductor lasers.

Infrared LEDs An *LED* is a PN junction semiconductor device that emits light when forward-biased. When a free electron encounters a hole in the semiconductor structure, the two combine, and in the process they give up energy in the form of light. Semiconductors such as gallium arsenide (GaAs) are superior to silicon in light emission. Most LEDs are GaAs devices optimized for producing red light. They are used for displays indicating whether a circuit is off or on, or for displaying decimal and binary data. However, because an LED is a fast semiconductor device, it can be turned off and on very quickly. Therefore, it is capable of transmitting the narrow light pulses required in a digital fiber-optic system.

Light-emitting diodes can be designed to emit virtually any color of light desired. Red LEDs are the most common, but yellow, green, and blue LEDs are also available. The LEDs used for fiber-optic transmission are usually in the red and low-infrared ranges. Typical wavelengths of LED light commonly used are 0.82, 0.94, 1.3, and 1.55 μm. These wavelengths are all in the near-infrared range just below red light. The light is not visible to the naked eye. These wavelengths have been chosen primarily because most fiber-optic cables have the lowest losses in these wavelength ranges. The most commonly used wavelength is 1.3 μm because many fiber-optic cables have minimum loss at that frequency.

Special LEDs are made just for fiber-optic applications. These units are made of GaAs indium phosphide (GaAsInP) and emit light at 1.3 μm. They come with a fiber-optic "pigtail" already attached for optimum coupling of light. The pigtail usually has a connector that attaches to the main cable.

There are two basic ways that digital data is formatted in fiber-optic systems. These are the *return-to-zero (RZ)* and *non-return-to-zero (NRZ)* formats. Both of these are illustrated in Fig. 13-22. The NRZ format at (*a*) is already generally familiar to you as we have discussed it previously in the chapter on data communications. Each bit occupies a separate time slot and is either a binary 1 or binary 0 during that time period. In the RZ format, the same time period is allotted for each bit but each bit is transmitted as a very narrow pulse (usually 50 percent of the bit time) or as an absence of a pulse as shown in Fig. 13-22(*b*). The digital data to be transmitted is converted into a serial pulse train and then into the desired RZ or

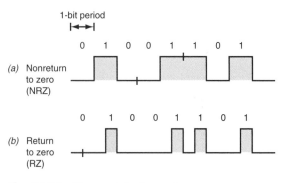
Fig. 13-22 Binary data formats.

NRZ format. These pulses are then applied to a light transmitter.

The light transmitter consists of the LED and its associated driving circuitry. A typical circuit is shown in Fig. 13-23. The binary pulses are applied to a logic gate which, in turn, operates a transistor switch Q_1 that turns the LED off and on. A positive pulse at the NAND gate input causes the NAND output to go to zero. This turns off Q_1, so the LED is forward-biased through R_2 and turns on. With zero input, the NAND output is 1, so Q_1 turns on and shunts current away from the LED. Very high current pulses are used to ensure a brilliant high-intensity light. High intensity is required if data is to be transmitted reliably over long distances. Most LEDs are capable of generating power levels up to approximately several hundred microwatts. With such low intensity, LED transmitters are good for only short distances. Further, the speed of the LED is limited. Turn-off and turn-on times are no faster than several nanoseconds, and, therefore, transmission rates are limited. Most LED-like transmitters are used for short-distance, low-speed digital fiber-optic systems.

Lasers The other commonly used light transmitter is a laser. A laser is a light source that emits coherent monochromatic light. Monochromatic means a single light frequency. Although an LED emits red light, that light covers a narrow spectrum of red frequencies. Monochromatic light has a pure single frequency. Coherent refers to the fact that all the light waves emitted are in phase with one another. Coherent light waves are focused into a narrow beam, which, as a result, is extremely intense. The effect is somewhat similar to that of using a highly directional antenna to focus radio waves into a narrow beam which also increases the intensity of the signal.

There are many different ways to make lasers. Some lasers are made from a solid rod of special material which is capable of emitting light when properly stimulated. Certain types of gases are also used. However, lasers used in fiber-optic systems are usually nothing more than specially made LEDs which are capable of operating as lasers.

The most widely used light source in fiber-optic systems is an injection laser diode (ILD). Like the LED, it is a PN junction diode usually made of GaAs. See Fig. 13-24. At some current level, it will emit a brilliant light. The physical structure of the ILD is such that the semiconductor structure is cut squarely at the ends to form internal reflecting surfaces. One of the surfaces is usually coated with a reflecting material such as gold. The other surface is only partially reflective. When the diode is properly biased, the light will be emitted and will bounce back and forth internally between the reflecting surfaces. The distance between the reflecting surfaces has been carefully measured so that it is some multiple of a half wave at the light frequency. The bouncing back and forth of the light waves causes their intensity to reinforce and build up. The structure is like a cavity resonator for light. The result is an incredibly high brilliance, single-frequency light beam that is emitted from the partially reflecting surface.

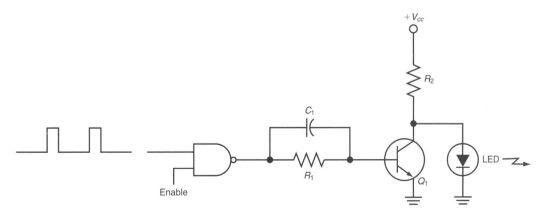

Fig. 13-23 Optical transmitter circuit using an LED.

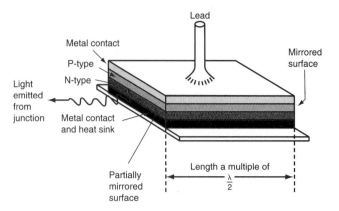

Fig. 13-24 An injection laser diode (ILD).

Injection laser diodes are capable of developing light power up to several watts. They are far more powerful than LEDs and, therefore, are capable of transmitting over much longer distances. Another advantage ILDs have over LEDs is their ability to turn off and on at a faster rate. High-speed laser diodes are capable of gigabit per second digital data rates.

Injection lasers dissipate a tremendous amount of heat and, therefore, must be connected to a heat sink for proper operation. Because their operation is heat-sensitive, most injection lasers are used in a circuit that provides some feedback for temperature control. This not only protects the laser, but also ensures proper light intensity and frequency.

A typical injection laser transmitter circuit is shown in Fig. 13-25. When the input is zero, the AND gate output is zero, so Q_1 is off and so is the laser. Capacitor C_2 charges through R_3 to the high voltage. When a binary 1 input occurs, Q_1 conducts, connecting C_2 to the ILD. Then C_2 discharges a very high current pulse

Photodiode

into the laser, turning it on briefly and creating an intense light pulse.

Optical Receivers

The receiver part of the optical communications system is relatively simple. It consists of a detector that will sense the light pulses and convert them into an electrical signal. This signal is then amplified and shaped into the original serial digital data. The most critical component, of course, is the light sensor.

The most widely used light sensor is a *photodiode*. This is a silicon PN junction diode that is sensitive to light. This diode is normally reverse-biased as shown in Fig. 13-26. The only current that flows through it is an extremely small reverse leakage current. Whenever light strikes the diode, this leakage current will increase significantly. It will flow through a resistor and develop a voltage drop across it. The result is an output voltage pulse.

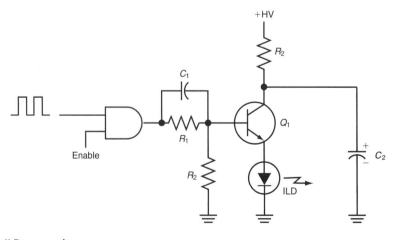

Fig. 13-25 An ILD transmitter.

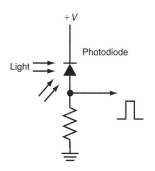

Fig. 13-26 How a photodiode converts light into voltage pulses.

The reverse current in a diode is extremely small even when the diode is exposed to light. The resulting voltage pulse is very small, so it must be amplified. This can be done by using a phototransistor. The base-collector junction is exposed to light. The base leakage current produced causes a larger emitter-to-collector current to flow. Thus the transistor amplifies the small leakage current into a larger, more useful output.

The sensitivity and response time of a photodiode can be increased by adding an undoped or intrinsic (I) layer between the P and N semiconductors to form a PIN diode. The I layer is exposed to the light. The diode is reverse-biased as previously discussed.

A widely used photosensor is the *avalanche photodiode (APD)*. It is the fastest and most sensitive photodiode available, but it is expensive and its circuitry is complex. Like the standard photodiode, the APD is reverse-biased. However, the operation is different. The APD uses the reverse breakdown mode of operation that is commonly found in zener and IMPATT microwave diodes. When a sufficient amount of reverse voltage is applied, an extremely high current will flow due to the avalanche effect.

Normally, several hundred volts of reverse bias, just below the avalanche threshold, is applied. When light strikes the junction, breakdown occurs and a large current flows. This high reverse current requires less amplification than the small current in a standard photodiode. The APDs are also significantly faster and are capable of handling the very high gigabit per second data rates possible in some systems.

Figure 13-27 shows the basic circuit used in most receivers. The current through the photodiode generated when light is sensed produces a current that is then amplified in an op amp. Following the amplifier is an op-amp comparator used as a shaping circuit that squares the pulses to ensure fast rise and fall times. The output is passed through a logic gate so that the correct binary voltage levels are produced.

Data Rate

The most important specification in a fiber-optic communications system is the data rate, that is, the speed of the optical pulses. For the very best systems using high-power ILDs and APD detectors, data rates of several billion bits per second are possible. This is known as a gigabit rate. Most systems have a data rate of 10 to 500 Mbits/s. Data rates up to 10 Gbits/s are available.

A system's performance is usually indicated by the product of the bit rate and the distance. This rating tells the fastest bit rate that can be produced over a 1-km cable. Assume a system with a 100 Mbits·km/s rating. This is a constant figure. If the distance increases, the bit rate decreases in proportion. In the above system, at 2 km the rate would drop to 50 Mbits/s. At 4 km, the rate is 25 Mbits/s, and so on.

Avalanche photodiode (APD)

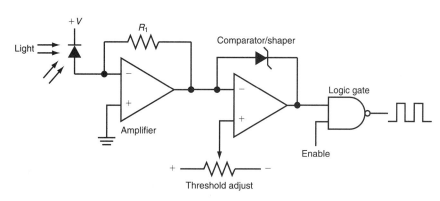

Fig. 13-27 A photodiode fiber-optic receiver.

Another important consideration is the maximum distance between repeaters. Obviously, the fewer the repeaters, the better. The average distance between repeaters is now in the 10- to 30-km range. The more components and systems improve, the greater this distance will become.

■ TEST

Answer the following questions.

54. The two most common light sources used in fiber-optic transmitters are _____.
55. The most popular light wavelength is _____ μm because fiber-optic cable attenuation is lowest at that wavelength.
56. True or false. The light from a 1.55-μm LED is visible.
57. Usually LEDs are made of _____.
58. Single-frequency light is called _____.
59. The condition of all emitted light waves being in phase is known as _____.
60. A single-frequency intense light source is known as a(n) _____.
61. The reflective surfaces on a laser diode structure form a(n) _____ that produces in-phase light waves.
62. For normal operation, LED and ILDs are _____ (reverse, forward)-biased.
63. Which is faster, an LED or ILD?
64. Which produces the brightest light, an LED or ILD?
65. During normal operation, all photodiodes are _____ (reverse, forward)-biased.
66. Light falling on the PN junction of a photodiode causes the diode's _____ current to increase.
67. The most sensitive and fastest light detector is the _____.
68. The two main circuits in a fiber-optic receiver are _____.
69. The product of the bit rate and distance of a system is 90 Mbits·km/s. The rating at 3 km is _____ Mbits/s.
70. In today's systems, the average maximum distance between repeaters is between _____ and _____ mi.

13-5 Fiber-Optic Data Communications Systems

Although fiber-optic cable can transmit analog signals such as TV video (by amplitude-

modulating the light beam), most fiber-optic systems transmit digital pulses. In other words, they are used for data communications. A sampling of some of the more common systems and applications is given here.

Local Area Networks

Although most LANs use twisted-pair cables (or coax), a growing trend is to run fiber-optic cable directly to the user's PC. The main benefit to this is faster network speeds.

Fiber-Distributed Data Interface (FDDI)
FDDI is a standard designed to connect up to 500 nodes (PCs or other computers) in a ring configuration over a distance of up to 60 mi (100 km). The data speed is 100 Mbits/s. A unique feature of the FDDI system is that it is implemented with a dual fiber cable. If one or both fibers are cut, the system is designed to automatically loop back and reconfigure itself to ensure uninterrupted operation. FDDI is most often used as the backbone for interconnecting several smaller LANs.

Ethernet The 10-, 100-, and 1000-Mbits/s versions of Ethernet can also use fiber-optic cable rather than twisted pair or coax. The operation of the network is the same, but the physical data transmission is by light rather than by high-speed electrical pulses. A fiber connection is more reliable, can operate over longer distances, and produces no electrical interference.

Fiber Channel *Fiber channel (FC)* is a flexible-fiber transmission standard that can be easily configured to almost any application requiring fast serial data transmission. FC can be connected directly between computers, in a loop, or in a switched network for multiple computers in a LAN. The speed is scalable to the application. FC offers data rates of 132.8, 265.6, 531.2, and 1062.5 Mbits/s.

FC is being used to replace some high-speed parallel data interfaces like the small computer systems interface (SCSI or "scuzzy").

Synchronous Optical Network (SONET)
SONET was developed to transmit digitized telephone calls in T1 format over fiber-optic

Fiber-distributed data interface (FDDI)

Fiber channel (FC)

Synchronous optical network (SONET)

cable at high speeds. Its primary use is to send time-multiplexed voice or data over switched networks. SONET is used between telephone central offices, between central offices and long-distance carrier facilities, and for long-distance transmission. It is also used for general digital transmission between computers.

SONET can be configured for different speed levels. The available speeds are 51.8, 155.5, 466.5, 622, and 933.1 Mbits/s as well as higher rates of 1.24, 1.86, 2.48, and 9.6 Gbits/s.

A system similar to SONET, known as *synchronous digital hierarchy (SDH),* is widely used in Europe.

Multiplexing on Fiber-Optic Cable Data is most easily multiplexed on fiber-optic cable by using TDM as in the T1 system. However, developments in optical components make it possible to use FDM on fiber-optic cable (called wavelength division multiplexing, or WDM), which permits multiple channels of data to operate over the cable's light-wave bandwidth.

WDM uses separate lasers to transmit serial digital data simultaneously on two or more different light wavelengths. Current systems use light in the 1550-nm range. A typical four-channel system uses laser wavelengths of 1534, 1543, 1550, and 1557.4 nm. Each laser is switched off and on with the desired data. The laser beams are then optically combined and transmitted over a single fiber cable.

Figure 13-28 illustrates the WDM system. A separate serial data source controls each laser. These may be single data sources or multiple TDM sources. At the receiving end of the cable, special optical filters are used to separate

the light beams into individual channels. Each light beam is detected with an optical sensor and then converted into the four individual data streams.

WDM significantly increases the data-handling capacity of fiber-optic cable. When WDM multiplexer/demultiplexer units are added to existing systems, more data channels and/or higher data speeds can be accommodated. Systems with 8, 16, or 32 channels are available.

Wavelength division multiplexing (WDM)

■ TEST

Answer the following questions.

71. True or false. Ethernet LANs may use fiber.
72. The topology of FDDI is _____, and its maximum speed is _____.
73. The maximum data rate of fiber channel is _____.
74. The primary application of SONET is _____.
75. The maximum speed of SONET is _____.
76. WDM is the same as _____ but at light frequencies.
77. Most WDM systems operate in the _____-nm range.

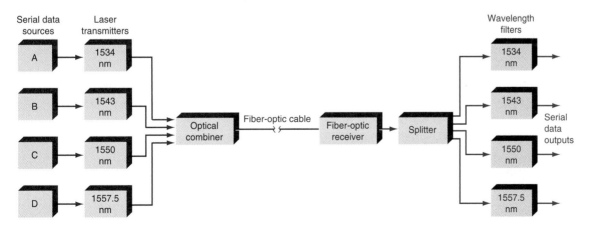

Fig. 13-28 A WDM system.

Summary

1. The information-carrying capacity of a cable or radio channel is directly proportional to its bandwidth.

2. The RF spectrum is heavily used and occupied. Only in the microwave region is there room for expansion.

3. Light is an electromagnetic signal like a radio wave but is much higher in frequency. It can be used as a carrier for information signals.

4. Because of the very high frequency of light compared to typical information signals, tremendous bandwidth is easily available.

5. Light waves carrying data signals can be transmitted in free space but are greatly attenuated by atmospheric effects and require pinpoint alignment.

6. Most light-wave communication is by way of a glass or plastic fiber cable that acts as a "light pipe" to carry light modulated by information signals.

7. The main components of a light-wave communications system are an A/D converter, a light source transmitter, a fiber-optic cable, a photodetector or light detector with amplifier and shaper, and a D/A converter.

8. Because of the great attenuation of light in a fiber-optic cable, repeater units are used to amplify and regenerate the signals over long distances.

9. The primary application of fiber-optic communications is in long-distance telephone systems.

10. The primary advantages of fiber-optic cables over conventional cables and radio are wider bandwidth, lower loss, light weight, small size, strength, security, interference immunity, and safety.

11. The main disadvantage of fiber-optic cable is that its small size and brittleness make it more difficult to work with.

12. Light waves, like radio waves, are a kind of electromagnetic radiation.

13. Light waves occur at very high frequencies in the range of 3×10^{11} to 3×16^{16} Hz.

14. Wavelength rater than frequency is used to express the place of light in the spectrum.

15. The wavelength of light is expressed in terms of nanometers (1 nm = 10^{-9} m) or micrometers (1 μm = 10^{-6} m). Micrometers are also called microns.

16. The visible light spectrum is from 700 nm (red) to 400 nm (violet).

17. The optical spectrum is made up of visible light, infrared at lower frequencies and ultraviolet at higher frequencies.

18. Infrared rays cannot be seen, but they act like light waves and can be manipulated in similar ways as with a lens or mirrors.

19. Light waves, like microwaves, travel in a straight line.

20. The angle at which light strikes a surface is called the angle of incidence. The angle at which light is reflected from a surface is called the angle of reflection. The angle of incidence is equal to the angle of reflection.

21. When a light ray passes from one medium to another, it is bent. This is called refraction.

22. The amount of refraction is called the index of refraction n and is the ratio of the speed of light in air to the speed of light in another medium, such as water, glass, or plastic ($n = 1$ in air, $n = 1.3$ in water, $n = 1.5$ in glass).

23. The angle of the incident light ray determines whether the ray will be reflected or refracted.

24. The critical angle is the angle of incidence that causes the refracted light to travel along the interface between two different media.

25. If the angle of incidence is made greater than the critical angle, reflection occurs instead of refraction.

26. Light entering a fiber-optic cable has an angle of incidence such that the light is reflected or bounded off the boundary between the fiber and the external media. This is called total internal reflection.

27. Fiber-optic cables are made from glass and plastic. Glass has the lowest loss but is brittle. Plastic is cheaper and more flexible, but has high attenuation.

28. A popular fiber-optic cable with a glass core and plastic cladding is called plastic-clad silica (PCS).

29. The cladding surrounding the core protects the core and provides an interface with a controlled index of refraction.

30. Step index means there is a sharp difference in the index of refraction between the core and cladding.

31. Graded index means that the index of refraction of the core varies over its cross section, highest in the center and lowest at the edges.

32. A single-mode cable is very small in diameter and essentially provides only a single path for light.

33. Multimode cores are large and provide multiple paths for the light.

34. Multiple light paths through a step-index core causes a light pulse to be stretched and attenuated. This is called modal dispersion, and it limits the upper pulse repetition rate and thus the information bandwidth.

35. Multiple light paths in a graded-index core are controlled so that they converge at multiple points along the cable. Modal dispersion does occur, but it is not as severe as that caused by a step-index core.

36. Modal dispersion does not occur in single-mode cores.

37. The three most widely used types of fiber-optic cables are multimode step-index, single-mode step-index, and multimode graded-index.

38. The primary specification of a fiber-optic cable is attenuation, which is usually expressed as the loss in decibels per kilometer.

39. Light loss in a fiber-optic cable is caused by absorption, scattering, and dispersion.

40. Cable attenuation is directly proportional to its length.

41. Cable losses range from 1 dB/km in glass single-mode step-index cable to 100 dB/km for plastic multimode step-index cable.

42. Fiber-optic cables can be spliced by gluing.

43. Special connectors are used to connect cables to one another and to the equipment.

44. Fiber-optic systems use light-emitting diodes (LEDs) and semiconductor lasers as the main light sources.

45. Light-emitting diodes are used in short-distance, low-speed systems. Injection laser diodes (ILDs) are used in long-distance, high-speed systems.

46. Most LEDs and ILDs emit light in the invisible near-infrared range (0.82 to 1.55 μm).

47. A popular operating frequency is 1.3 μm because fiber-optic cable has an attenuation null at this wavelength.

48. Laser diodes emit monochromatic or single-frequency light. The light waves are coherent, so they reinforce one another to create an intense and finely focused beam.

49. Intense laser light is produced by an ILD because reflecting surfaces in the structure form a cavity resonator for the light waves.

50. The most commonly used light sensor is a photodiode.

51. A photodiode is a PN junction that is reverse-biased and exposed to light. Light increases the leakage current across the junction. This current is converted into a voltage pulse.

52. PIN junction diodes are faster and more sensitive than conventional photodiodes.

53. The fastest and most sensitive light detector is the avalanche photodiode (APD).

54. The APD operates with a high reverse bias so that when light is applied, breakdown occurs and produces a fast, high-current pulse.

55. The receiver portion of a fiber-optic system is made up of a photodiode, amplifier, and shaper.

56. Fiber-optic systems are rated by the speed and the product of the bit rate and the distance.

57. A measure of the quality of a fiber-optic system is the maximum distance between repeaters.

58. Most fiber-optic systems transmit digital data.

59. Fiber-optic cable may be used in LANs. Some examples are FDDI and Ethernet.

60. SONET is a high-speed optical system used to transmit time-multiplexed voice signals at speeds up to 9.6 Gbits/s.

61. Wavelength division multiplexing (WDM) is the same as FDM on cable or wireless. Data can be transmitted on multiple light frequencies simultaneously over a single fiber.

Chapter Review Questions

Choose the letter which best answers each question.

13-1. Which of the following is *not* a common application of fiber-optic cable?
  a. Computer networks
  b. Long-distance telephone systems
  c. Closed circuit TV
  d. Consumer TV

13-2. Total internal reflection takes place if the light ray strikes the interface at an angle with what relationship to the critical angle?
  a. Less than
  b. Greater than
  c. Equal to
  d. Zero

13-3. The operation of a fiber-optic cable is based on the principle of
  a. Refraction.
  b. Reflection.
  c. Dispersion.
  d. Absorption.

13-4. Which of the following is *not* a common type of fiber-optic cable?
  a. Single-mode step-index
  b. Multimode graded-index
  c. Single-mode graded-index
  d. Multimode step-index

13-5. Cable attenuation is usually expressed in terms of
  a. Loss per foot.
  b. dB/km.
  c. Intensity per mile.
  d. Voltage drop per inch.

13-6. Which cable length has the highest attenuation?
  a. 1 km
  b. 2 km
  c. 95 ft
  d. 500 ft

13-7. The upper pulse rate and information-carrying capacity of a cable is limited by
  a. Pulse shortening.
  b. Attenuation.
  c. Light leakage.
  d. Modal dispersion.

13-8. The core of a fiber-optic cable is made of
  a. Air.
  b. Glass.
  c. Diamond.
  d. Quartz.

13-9. The core of a fiber-optic cable is surrounded by
  a. Wire braid shield.
  b. Kevlar.
  c. Cladding.
  d. Plastic insulation.

13-10. The speed of light in plastic compared to the speed of light in air is
  a. Less.
  b. More.
  c. The same.
  d. Zero.

13-11. Which of the following is *not* a major benefit of fiber-optic cable?
  a. Immunity from interference
  b. No electrical safety problems
  c. Excellent data security
  d. Lower cost

13-12. The main benefit of light-wave communications over microwaves or any other communications media is
  a. Lower cost.
  b. Better security.
  c. Wider bandwidth.
  d. Freedom from interference.

13-13. Which of the following is *not* part of the optical spectrum?
  a. Infrared
  b. Ultraviolet
  c. Visible color
  d. X-rays

13-14. The wavelength of visible light extends from
  a. 0.8 to 1.6 μm.
  b. 400 to 750 nm.
  c. 200 to 660 nm.
  d. 700 to 1200 nm.

13-15. The speed of light is
  a. 186,000 mi/h.
  b. 300,000 mi/h.
  c. 300,000 m/s.
  d. 300,000,000 m/s.

13-16. Refraction is the
  a. Bending of light waves.
  b. Reflection of light waves.
  c. Distortion of light waves.
  d. Diffusion of light waves.

13-17. The ratio of the speed of light in air to the speed of light in another substance is called the
  a. Speed factor.
  b. Index of reflection.
  c. Index of refraction.
  d. Speed gain.

13-18. A popular light wavelength in fiber-optic cable is
  a. 0.7 μm.
  b. 1.3 μm.

c. 1.5 μm.

d. 1.8 μm.

13-19. Which type of fiber-optic cable is the most widely used?

a. Single-mode step-index

b. Multimode step-index

c. Single-mode graded-index

d. Multimode graded-index

13-20. Which type of fiber-optic cable is best for very high-speed data?

a. Single-mode step-index

b. Multimode step-index

c. Single-mode graded-index

d. Multimode graded-index

13-21. Which type of fiber-optic cable has the least modal dispersion?

a. Single-mode step-index

b. Multimode step-index

c. Single-mode graded-index

d. Multimode graded-index

13-22. Which of the following is *not* a factor in cable light loss?

a. Reflection

b. Absorption

c. Scattering

d. Dispersion

13-23. A distance of 8 km is the same as

a. 2.5 mi.

b. 5 mi.

c. 8 mi.

d. 12.9 mi.

13-24. A fiber-optic cable has a loss of 15 dB/km. The attenuation in a cable 1000 ft long is

a. 4.57 dB.

b. 9.3 dB.

c. 24 dB.

d. 49.2 dB.

13-25. Fiber-optic cables with attenuations of 1.8, 3.4, 5.9, and 18 dB are linked together. The total loss is

a. 7.5 dB.

b. 19.8 dB.

c. 29.1 dB.

d. 650 dB.

13-26. Which light emitter is preferred for high-speed data in a fiber-optic system?

a. Incandescent

b. LED

c. Neon

d. Laser

13-27. Most fiber-optic light sources emit light in which spectrum?

a. Visible

b. Infrared

c. Ultraviolet

d. X-ray

13-28. Both LEDs and ILDs operate correctly with

a. Forward bias.

b. Reverse bias.

c. Neither *a* nor *b*.

d. Either *a* or *b*.

13-29. Single-frequency light is called

a. Pure.

b. Intense.

c. Coherent.

d. Monochromatic.

13-30. Laser light is very bright because it is

a. Pure.

b. White.

c. Coherent.

d. Monochromatic.

13-31. Which of the following is *not* a common light detector?

a. PIN photodiode

b. Photovoltaic diode

c. Photodiode

d. Avalanche photodiode

13-32. Which of the following is the fastest light sensor?

a. PIN photodiode

b. Photovoltaic diode

c. Phototransistor

d. Avalanche photodiode

13-33. Photodiodes operate properly with

a. Forward bias.

b. Reverse bias.

c. Neither *a* nor *b*.

d. Either *a* or *b*.

13-34. The product of the bit rate and distance of a fiber-optic system is 2 Gbits·km/s. What is the maximum rate at 5 km?

a. 100 Mbits/s

b. 200 Mbits/s

c. 400 Mbits/s

d. 1000 Gbits/s

13-35. Which fiber-optic system is better?

a. 3 repeaters

b. 8 repeaters

c. 11 repeaters

d. 20 repeaters

13-36. Which fiber-optic cable system is used to transmit time-multiplexed voice signals?

a. FDDI

b. SONET

c. Ethernet

d. Fiber channel

13-37. Wavelength division multiplexing means
    a. Time-division multiplexing with light.
    b. Frequency-modulating multiple light frequencies.
    c. Transmitting separate data sources on the same cable by using different light frequencies.
    d. Using multiple data sources to frequency-modulate sine wave carriers that are digitized and used to modulate the laser beam.

Critical Thinking Questions

13-1. Explain how you would transmit analog signals over a fiber-optic cable.
13-2. Name three potential new applications for fiber-optic communications that are not in Fig. 13-2 but take advantage of the benefits listed in Fig. 13-3.
13-3. Explain how a single fiber-optic cable can handle two-way communications, both half- and full-duplex.
13-4. Could an incandescent light be used for a fiber-optic transmitter? Explain its possible benefits and disadvantages.
13-5. In a 4-channel WDM system with a 2.48-Gbits/s rate on each, how many bits are transmitted per second? How many bytes are transmitted per second?
13-6. Suppose that a fiber-optic cable could be run to every home and business in the world. What current services could it replace? What new services could be added?
13-7. Consider wireless light-wave communications. State the major components of such a system. Give several possible applications. What might be the limiting factors of such a system?
13-8. Name one commonly used wireless light communications device.

Answers to Tests

1. light
2. voice, video, computer data
3. telephone systems
4. glass, plastic
5. wide bandwidth
6. false
7. true
8. brittleness, difficult to work with
9. LEDs, semiconductor lasers
10. binary or digital pulses
11. photocell or light detector
12. repeaters
13. electromagnetic
14. pipe
15. *b*
16. micrometers (microns), nanometers
17. *a*
18. one-millionth, meter
19. 400, 700
20. *b*
21. visible, infrared, ultraviolet
22. 300,000,000; 186,000
23. true
24. index, refraction
25. reflection, mirror
26. refraction
27. 30
28. critical
29. reflection
30. index, refraction
31. *b*
32. cladding
33. glass, plastic
34. *a*
35. multimode step-index, single-mode step-index, multimode graded-index
36. modal dispersion
37. multimode step-index, multimode graded-index
38. single-mode step-index
39. *b*
40. index, refraction
41. *c*
42. protective coating
43. Kevlar
44. 2, 6, 12, 18, 24
45. attenuation
46. dispersion, absorption, scattering
47. decibels, kilometer
48. 31
49. false
50. *a*
51. 76

52. true
53. connectors
54. LED, laser
55. 1.3
56. false
57. gallium arsenide
58. monochromatic
59. coherence
60. laser
61. resonant cavity
62. forward
63. ILD
64. ILD

65. reverse
66. leakage
67. avalanche photodiode
68. amplifier, comparator or shaper
69. 30
70. 6.17, 18.5 (10 and 30 km)
71. true
72. ring, 100 Mbits/s
73. 1062.5 Mbits/s
74. digital telephony
75. 9.6 Gbits/s
76. frequency division multiplexing
77. 1550

Chapter 14

Television

Chapter Objectives

This chapter will help you to:

1. *Describe* and give the specifications for a complete television signal including all its individual components.
2. *Explain* the process used by a television camera to convert a visual scene into a video signal.
3. *Draw* a simplified block diagram of the following units showing the main components and explaining their operation:
   a. TV transmitter
   b. TV receiver (conventional)
   c. Satellite TV receiver
4. *Explain* the operation of a typical infrared remote control.
5. *Draw* a block diagram of a cable TV system.
6. *Name* all the elements of and explain the operation of a cable TV system.
7. *Explain* the difference between a conventional satellite receiver and a DBS TV receiver.
8. *Define* digital television (DTV) and *state* the basic specifications of DTV receivers.

Video signal

This chapter introduces you to the subject of TV technology. It embodies almost all the principles and circuits covered elsewhere in this book. Studying TV is an excellent way to review communications fundamentals including modulation and multiplexing, transmitters and receivers, antennas and transmission lines, and even digital techniques.

In addition to standard audio transmission, TV systems use a TV camera to convert a visual scene into a voltage known as the *video signal*. The video signal represents the picture information and is used to modulate a transmitter. Both the picture and the sound signals are transmitted to the receiver, which demodulates the signals and presents the information to the user. The TV receiver is a special superheterodyne that recovers both the sound and the picture information. The picture is displayed on a picture tube.

This chapter emphasizes coverage of traditional analog TV but also introduces you to the new digital TV systems.

14-1 TV Signal

A considerable amount of intelligence is contained in a complete TV signal. As a result, the signal occupies a significant amount of spectrum space. The TV signal consists of two main parts: the sound and the picture. But it is far more complex than that. The sound today is usually stereo, and the picture carries color information as well as the synchronizing signals that keep the receiver in step with the transmitter.

Signal Bandwidth

The complete signal bandwidth of a TV signal is shown in Fig. 14-1. The entire TV signal occupies a channel in the spectrum with a bandwidth of 6 MHz. There are two carriers, one each for the picture and the sound.

Audio Signal The sound carrier is at the upper end of the spectrum. Frequency modulation is used to impress the sound signal on the carrier. The audio bandwidth of the

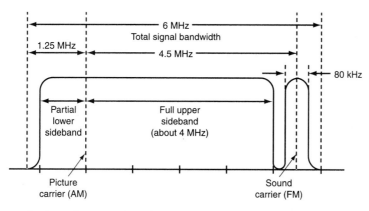

Fig. 14-1 Spectrum of a broadcast TV signal.

signal is 50 Hz to 15 kHz. The maximum permitted frequency deviation is 25 kHz, considerably less than the deviation permitted by conventional FM broadcasting. As a result, a TV sound signal occupies somewhat less bandwidth in the spectrum than a standard FM broadcast station. Stereo sound is also available in TV, and the multiplexing method used to transmit two channels of sound information is virtually identical to that used in stereo transmission for FM broadcasting.

Video Signal The picture information is transmitted on a separate carrier located 4.5 MHz lower in frequency than the sound carrier (Refer again to Fig. 14-1). The video signal derived from a camera is used to amplitude-modulate the picture carrier. Different methods of modulation are used for both sound and picture information so that there is less interference between the picture and sound signals. Further, amplitude modulation of the carrier takes up less bandwidth in the spectrum, and this is important when a high-frequency, content-modulating signal such as video is to be transmitted.

Note in Fig. 14-1 that vestigial sideband AM is used. The full upper sidebands of the picture information are transmitted, but a major portion of the lower sidebands is suppressed to conserve spectrum space. Only a vestige of the lower sideband is transmitted.

The color information in a picture is transmitted by way of FDM techniques. Two color signals derived from the camera are used to modulate a 3.85-MHz subcarrier which, in turn, modulates the picture carrier along with the main video information. The color subcarriers use double-sideband suppressed carrier AM.

The video signal can contain frequency components up to about 4.2 MHz. Therefore, if both sidebands were transmitted simultaneously, the picture signal would occupy 8.4 MHz. The vestigial sideband transmission reduces this excessive bandwidth.

TV Spectrum Allocation Because a TV signal occupies so much bandwidth, it must be transmitted in a very high frequency portion of the spectrum. TV signals are assigned to frequencies in the VHF and UHF range. U.S. TV stations use the frequency range between 54 and 806 MHz. This portion of the spectrum is divided into 68 6-MHz channels that are assigned frequencies (see Fig. 14-2). Channels 2 through 7 occupy the frequency range from 54 to 88 MHz. The standard FM radio broadcast band occupies the 88- to 108-MHz range. Aircraft, amateur radio, and marine and mobile radio communications services occupy the frequency spectrum from approximately 118 to 173 MHz. Additional TV channels occupy the space between 470 and 806 MHz. Figure 14-2 shows the frequency range of each TV channel.

To find the exact frequencies of the transmitter and sound carriers, use Fig. 14-2 and the spectrum outline in Fig. 14-1. To compute the picture carrier, add 1.25 MHz to the lower frequency of range given in Fig. 14-2. For example, for channel 6, the lower frequency is 82 MHz. The picture carrier is 82 + 1.25, or 83.25 MHz. The sound carrier is 4.5 MHz higher, or 83.25 + 4.5, that is, 87.75 MHz.

JOB TIP

Your employability will depend on keeping up with both analog and digital mixing technologies.

Channel	Frequency, MHz	Channel	Frequency, MHz
	Low-band VHF		UHF (cont.)
2	54–60	32	578–584
3	60–66	33	584–590
4	66–72	34	590–596
5	76–82	35	596–602
6	82–88	36	602–608
FM broadcast	88–108	37	608–614
Aircraft	118–135	38	614–620
Ham radio	144–148	39	620–626
Mobile or marine	150–173	40	626–632
		41	632–638
	High-band VHF	42	638–644
		43	644–650
7	174–180	44	650–656
8	180–186	45	656–662
9	186–192	46	662–668
10	192–198	47	668–674
11	198–204	48	674–680
12	204–210	49	680–686
13	210–216	50	686–692
	UHF	51	692–698
		52	698–704
14	470–476	53	704–710
15	476–482	54	710–716
16	482–488	55	716–722
17	488–494	56	722–728
18	494–500	57	728–734
19	500–506	58	734–740
20	506–512	59	740–746
21	512–518	60	746–752
22	518–524	61	752–758
23	524–530	62	758–764
24	530–536	63	764–770
25	536–542	64	770–776
26	542–548	65	776–782
27	548–554	66	782–788
28	554–560	67	788–794
29	560–566	68	794–800
30	566–572	69	800–806
31	572–578	Cellular telephone	806–902

Fig. 14-2 VHF and UHF TV channel frequency assignments.

Generating the Video Signal

The video signal is most often generated by a TV camera, a very sophisticated electronic device that incorporates lenses and light-sensitive transducers to convert the scene or object to be viewed into an electrical signal that can be used to modulate a carrier. All visible scenes and objects are simply light that has been reflected and absorbed and then transmitted to our eyes. It is the purpose of the camera to take the light intensity and color details in a scene and convert them into an electrical signal.

To do this, the scene to be transmitted is collected and focused by a lens upon a light-sensitive imaging device. Both vacuum tube and semiconductor devices are used for converting the light information in the scene into an electrical signal. Some examples are the *vidicon tube* and the *charged-coupled device (CCD)* so widely used in camcorders and all modern TV cameras.

The scene is divided into smaller segments that can be transmitted serially over a period of time. Again, it is the job of the camera to subdivide the scene in an orderly manner so that an acceptable signal is developed. This process is known as *scanning*.

Principles of Scanning Scanning is a technique that divides a rectangular scene into individual lines. The standard TV scene dimensions have an aspect ratio of 4:3; that is, the scene width is 4 units for every 3 units of height. To create a picture, the scene is subdivided into many fine horizontal lines called *scan lines*. Each line represents a very narrow portion of light variations in the scene. The greater the number of scan lines, the higher the resolution and the greater the detail that can be

Scanning

Scan lines

Vidicon tube

Charged-coupled device (CCD)

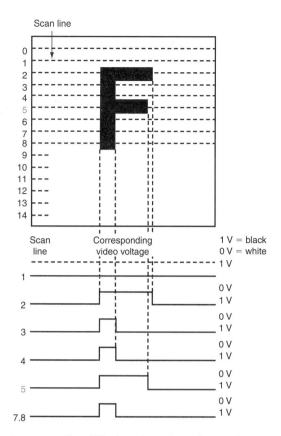

Fig. 14-3 Simplified explanation of scanning.

observed. U.S. TV standards call for the scene to be divided into a maximum of 525 horizontal lines.

Figure 14-3 is a simplified drawing of the scanning process. In this example, the scene is a large black letter F on a white background. The task of the TV camera is to convert this scene into an electrical signal. The camera accomplishes this by transmitting a voltage of 1 V for black and 0 V for white. The scene is divided into 15 scan lines numbered 0 through 14. The scene is focused on the light-sensitive area of a vidicon tube or CCD imaging device which scans the scene 1 line at a time, transmitting the light variations along that line as voltage levels. Figure 14-3 shows the light variations along several of the lines. Where the white background is being scanned, a 0-V signal occurs. When a black picture element is

encountered, a 1-V level is transmitted. The electrical signals derived from each scan line are referred to as the *video signal*. They are transmitted serially one after the other until the entire scene has been sent (see Fig. 14-4). This is exactly how a standard TV picture is developed and transmitted.

Since the scene contains colors, there are different levels of light along each scan line. This information is transmitted as different shades of gray between black and white. Shades of gray are represented by some voltage level between the 0- and 1-V extremes represented by white and black. The resulting signal is known as the *brightness*, or *luminance, signal* and is usually designated by the letter *Y*.

A more detailed illustration of the scanning process is given in Fig. 14-5. The scene is scanned twice. One complete scanning of the scene is called a *field* and contains $262\frac{1}{2}$ lines. The entire field is scanned in 1/60 of a second for a 60-Hz field rate. In color TV the field rate is 59.94 Hz. Then the scene is scanned a second time, again using $262\frac{1}{2}$ lines. This second field is scanned in such a way that its scan lines fall between those of the first field. This produces what is known as *interlaced scanning*, with a total of $2 \times 262\frac{1}{2} = 525$ lines. In practice, only about 480 lines show on the picture tube screen. Two interlaced fields produce a complete frame of video. With the field rate being 1/60 of a second, two fields produce a frame rate of 1/30 of a second, or 30 Hz. The frame rate in color TV is one-half the field rate, or 29.97 Hz. Interlaced scanning is used to reduce flicker, which is annoying to the eye.

Video signal

Brightness signal

Field

Interlaced scanning

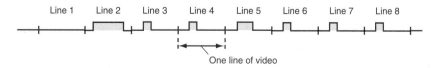

Fig. 14-4 The scan line voltages are transmitted serially. These correspond to the scanned letter F in Fig. 14-3.

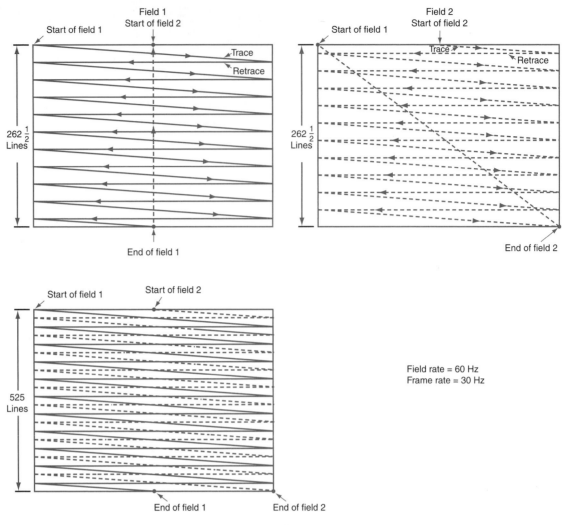

Field 1
Start of field 1
Start of field 2

Trace
Retrace

$262\frac{1}{2}$ Lines

End of field 1

Field 2
Start of field 1
Start of field 2

Trace
Retrace

$262\frac{1}{2}$ Lines

End of field 2

Field rate = 60 Hz
Frame rate = 30 Hz

Start of field 1
Start of field 2

525 Lines

End of field 1
End of field 2

Fig. 14-5 Interlaced scanning is used to minimize flicker.

This rate is also fast enough that the human eye cannot detect individual scan lines, which results in a stable picture.

The rate of occurrence of the horizontal scan lines is 15,750 Hz for monochrome, or black and white, TV and 15,734 Hz for color TV. This means that it takes about 1/15,734 s, or 63.6 μs, to trace out 1 horizontal scan line.

At the TV receiver, the picture tube is scanned in step with the transmitter to accurately reproduce the picture. To ensure that the receiver stays exactly in synchronization with the transmitter, special horizontal and vertical sync pulses are added to and transmitted with the video signal (see Fig. 14-6). After 1 line has been scanned, a horizontal blanking pulse comes along. At the receiver, the blanking pulse is used to cut off the electron beam in the picture tube during the time the beam must retrace from right to left to get ready for the next left to right scan line. The horizontal sync pulse is

used at the receiver to keep the sweep circuits that drive the picture tube in step with the transmitted signal. The width of the horizontal blanking pulse is about 10 μs. Since the total horizontal period is 63.6 μs, only about 53.5 μs is devoted to the video signal.

At the end of each field, the scanning must retrace from the bottom to the top of the scene so that the next field can be scanned. This is initiated by the vertical blanking and sync pulses. The entire vertical pulse blanks the picture tube during the vertical retrace. The pulses on top of the vertical blanking pulse are the horizontal sync pulses that must continue to keep the horizontal sweep in sync during the vertical retrace. The equalizing pulses help synchronize the half scan lines in each field. Approximately 30 to 40 scan lines are used up during the vertical blanking interval. Therefore, only 480 to 495 lines of actual video are shown on the screen.

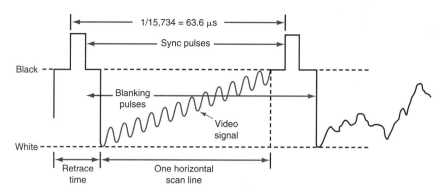

Fig. 14-6 Sync pulses are used to keep the receiver in step with the transmitter.

Relationship between Resolution and Bandwidth Scanning a scene or picture is a kind of sampling process. Consider the scene to be a continuous variation of light intensities and colors. To capture this scene and transmit it electronically, the light intensity and color variations must be converted into electrical signals. This conversion is accomplished through a process called *scanning,* whereby the picture is divided into many fine horizontal lines next to one another.

The *resolution* of the picture refers to the amount of detail that can be shown. Pictures with high resolution have excellent *definition,* or distinction of detail, and the pictures appear to be clearly focused. A picture lacking detail looks *softer,* or somewhat out of focus. The bandwidth of a video system determines the resolution. The greater the bandwidth, the greater the amount of definition and detail.

Resolution in a video system is measured in terms of the number of lines defined within the bounds of the picture. For example, the horizontal resolution (R_H) is given as the maximum number of alternating black-and-white vertical lines that can be distinguished. Assume closely spaced vertical black-and-white lines of the same width. When such lines are scanned, they will be converted into a square wave (50 percent duty cycle). One cycle, or period, t of this square wave is the time for one black and one white line. If the lines are very thin, the resulting period will be short and the frequency will be high. If the lines are wide, the period will be longer and the resulting frequency lower.

The *National Television Standards Committee (NTSC)* system restricts the bandwidth in the United States to 4.2 MHz. This translates into a period of 0.238 μs, or 238 ns. The width of a line is one-half this value, or 0.238/2 μs, or 0.119 μs. Remember that the horizontal sweep interval is about 63.6 μs. About 10 μs of this interval is taken up by the horizontal blanking interval, leaving 53.5 μs for the video. The displayed scan line takes 53.5 μs. With 0.119 μs per line, one horizontal scan line can resolve, or contain, up to 53.5/0.119, or 449.5, vertical lines. Therefore, the approximate horizontal resolution R_H is about 450 lines.

The vertical resolution R_V is the number of horizontal lines that can be distinguished. Only about 480 to 495 horizontal lines are shown on the screen. The vertical resolution is about 0.7 times the number of actual lines N_L:

$$R_V = 0.7\, N_L$$

If 485 lines are shown, the vertical resolution is 0.7×485, or 340, lines.

Color Signal Generation The video signal as described so far contains the video or luminance information, which is a black-and-white version of the scene. This is combined with the sync pulses. Now the color detail in the scene must somehow be represented by an electrical signal. This is done by dividing the light in each scan line into three separate signals, each representing one of the three basic colors, red, green, or blue. It is a principle of physics that any color can be made by mixing some combination of the three primary light colors (see Fig. 14-7).

In the same way, the light in any scene can be divided into its three basic color components by passing the light through red, green, and blue filters. This is done in a color TV camera, which is really three cameras in one (see Fig. 14-8). The lens focuses the scene on three separate light-sensitive devices such as a

Resolution

Definition

National Television Standards Committee (NTSC)

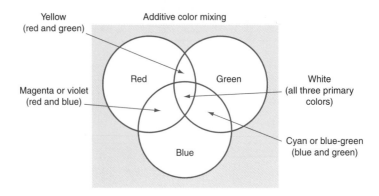

Fig. 14-7 Creating other colors with red, green, and blue light.

I and Q signals

vidicon tube or a CCD imaging device by way of a series of mirrors and beam splitters. The red light in the scene passes through the red filter, the green through the green filter, and the blue through the blue filter. The result is the generation of three simultaneous signals (R, G, and B) during the scanning process by the light-sensitive imaging devices.

The R, G, and B signals also contain the basic brightness or luminance information. If the color signals are mixed in the correct proportion, the result is the standard B&W video or luminance Y Signal. The Y signal is generated by scaling each color signal with a tapped voltage divider and adding the signals together as shown in Fig. 14-9(a). The Y signal is made up of 30 percent red, 59 percent green, and 11 percent blue. The resulting Y signal is what a black and white TV set will see.

Chrominance signal

The color signals must also be transmitted along with the luminance information in the same bandwidth allotted to the TV signal. This is done by a frequency-division multiplexing technique shown in Fig. 14-9(a). Instead of all three color signals being transmitted, they are

Quadrature modulation

combined into color signals referred to as the *I and Q signals*. These signals are made up of different proportions of the R, G, and B signals according to the following specifications:

$$I = 60\% \text{ red}, -28\% \text{ green}, -32\% \text{ blue}$$

$$Q = 21\% \text{ red}, -52\% \text{ green}, 31\% \text{ blue}$$

The minus signs in the above expressions mean that the color signal has been phase-inverted before the mixing process.

The I and Q signals are referred to as the *chrominance signals*. To transmit them, the they are *phase-encoded;* that is, they are used to modulate a subcarrier which is in turn mixed with the luminance signal to form a complete, or composite, video signal. These I and Q signals are fed to balanced modulators along with 3.58-MHz (actually 3.579545-MHz) subcarrier signals that are 90° out of phase [again refer to Fig. 14-9(a)]. This type of modulation is referred to as *quadrature modulation*, where *quadrature* means a 90° phase shift. The output of each balanced modulator is a double-

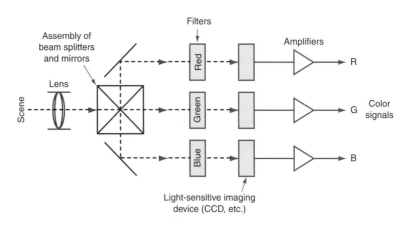

Fig. 14-8 How the camera generates the color signals.

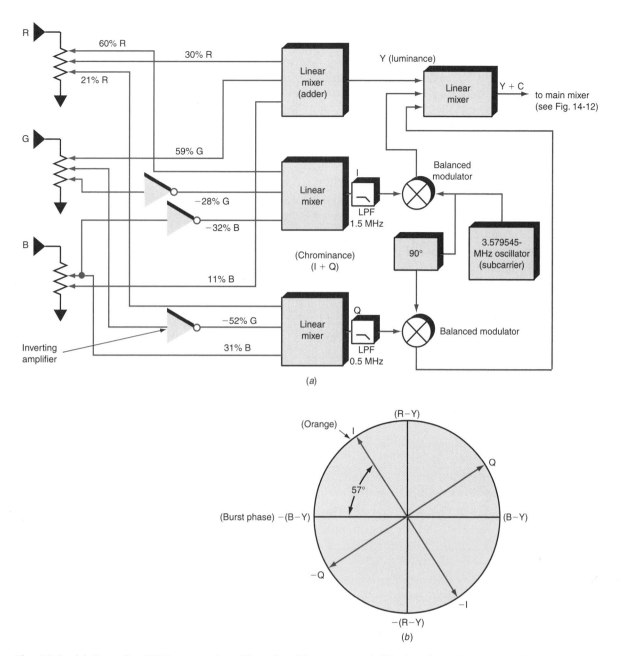

Fig. 14-9 (a) How the NTSC composite video signal is generated. (b) The chrominance signals are phase-encoded.

sideband suppressed carrier AM signal. The resulting two signals are added to the *Y* signal to create the composite video signal. The combined signal modulates the picture carrier. The resulting signal is called the *NTSC composite video signal*. This signal and its sidebands are within the 6-MHz TV signal bandwidth.

The *I* and *Q* color signals are also called the *R − Y* and the *B − Y signals* because the combination of the three color signals produces the effect of subtracting *Y* from the *R* or *B* signals. The phase of these signals with respect to the original 3.58-MHz subcarrier signal determines

the color to be seen. The color tint can be varied at the receiver so that the viewer sees the correct colors. In many TV sets an extra phase shift of 57° is inserted to ensure that maximum color detail is seen. The resulting *I* and *Q* signals are shown as phasors in Fig. 14-9(*b*). There is still 90° between the *I* and *Q* signals, but their position is moved 57°. The reason for this extra phase shift is that the eye is more sensitive to the color orange. If the *I* signal is adjusted to the orange phase position, better detail will be seen. The *I* signal is transmitted with more bandwidth than the *Q* signal, as can be seen by the response of the low-pass filters at the outputs of the *I* and *Q* mixers in Fig. 14-9(*a*).

The complete spectrum of the transmitted color signal is shown in Fig. 14-10. Note the color portion of the signal. Because of the frequency of the subcarrier, the sidebands produced during amplitude modulation occur in clusters that are interleaved between the other sidebands produced by the video modulation.

Remember that the 3.58-MHz subcarrier is suppressed by the balanced modulators and therefore is not transmitted. Only the filtered upper and lower sidebands of the color signals are transmitted. To demodulate these *double-sided (DSB) AM signals,* the carrier must be reinserted at the receiver. A 3.58-MHz oscillator in the receiver generates the subcarrier for the balanced modulator-demodulator circuits.

For the color signals to be accurately recovered, the subcarrier at the receiver must have a phase related to the subcarrier at the transmitter. To ensure the proper conditions at the receiver, a sample of the 3.58-MHz subcarrier signal developed at the transmitter is added to the composite video signal. This is done by gating 8 to 12 cycles of the 3.58-MHz subcarrier and adding it to the horizontal sync and blanking pulse as shown in Fig. 14-11. This is called the *color burst,* and it rides on what is called the *back porch* of the horizontal sync pulse. The receiver uses this signal to phase-synchronize the internally generated subcarrier before it is used in the demodulation process.

A block diagram of the TV transmitter is shown in Fig. 14-12. Note the sweep and sync circuits that create the scanning signals for the vidicons or CCDs as well as generate the sync

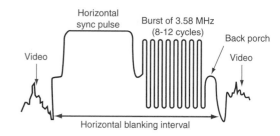

Fig. 14-11 The 3.58-MHz color subcarrier burst used to synchronize color demodulation at the receiver.

pulses that are transmitted along with the video and color signals. The sync signals, luminance *Y,* and the color signals are added to form the final video signal that is used to modulate the carrier. Low-level AM is used. The final AM signal is amplified by very high power linear amplifiers and sent to the antenna via a diplexer, which is a set of sharp bandpass filters that pass the transmitter signal to the antenna but prevent signals from getting back into the sound transmitter.

At the same time, the voice or sound signals frequency-modulate a carrier that is amplified by class C amplifiers and fed to the same antenna by way of the diplexer. The resulting VHF or UHF TV signal travels by line-of-sight propagation to the antenna and receiver.

■ TEST

Answer the following questions.

1. What is the bandwidth of a standard TV signal?
2. State the kind of modulation used on the video carrier and the sound carrier in a TV signal.

Double-sided (DSB) AM signal

Color burst

Back porch

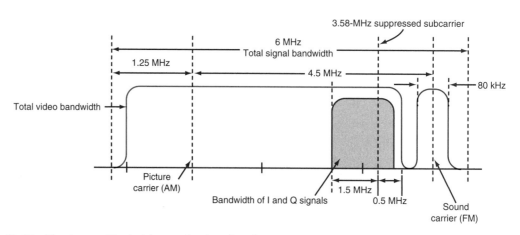

Fig. 14-10 The transmitted video and color signal spectrum.

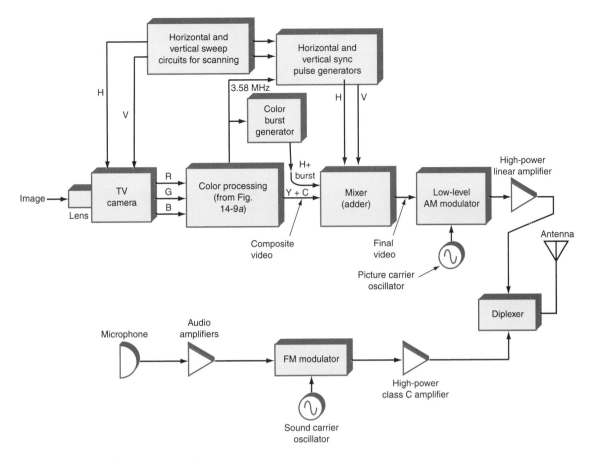

Fig. 14-12 Complete TV transmitter.

3. What is the frequency spacing between the sound and picture carriers?

4. What do you call the brightness signal produced by a monochrome video camera?

5. Name two widely used electronic imaging devices in TV cameras that convert light variations into a video signal.

6. What is the name of the process that breaks up a picture or scene into serially transmitted signals?

7. What are the three basic colors that can be used to produce any other color light?

8. What is the maximum number of scan lines used in an NTSC TV picture or frame of video?

9. How many scan lines make up one field of a TV picture?

10. What are the field and frame rates in NTSC color TV?

11. What is the rate of scanning one horizontal line in a color TV set?

12. What is the name of the circuit that lets the picture and sound transmitters use the same antenna?

13. The color camera signals are combined in a resistive matrix to produce the two composite color signals. What are they called?

14. How are the two color signals multiplexed and modulated onto the main video carrier?

15. What is the frequency of the subcarrier that the color signals modulate?

16. What characteristic of the composite color signal tells the receiver what the transmitted color is?

17. What type of modulation is used in the generation of the chrominance signals?

18. Compute the exact video and sound carriers for a channel 12 TV station.

19. What is the approximate upper frequency response of the video signal transmitted?

20. Using Carson's rule, calculate the approximate bandwidth of the sound spectrum of a TV signal.

21. How is the bandwidth of the video portion of a TV signal restricted to minimize spectrum bandwidth?

22. Describe the process by which the picture at the receiver is kept in step with the transmitted signal? What component of the TV signal performs this function?
23. Describe the process by which the color in a scene is converted into video signals. What signal is formed by adding the color signals in the following proportion: $0.11B + 0.59G + 0.3R$?

14-2 TV Receiver

The process involved in receiving a TV signal and recovering it to present the picture and sound outputs in a high-quality manner is complex. Over the course of the past 60 years since its invention, the TV set has evolved from a large vacuum tube unit into a smaller and more reliable solid-state unit made mostly with ICs.

A block diagram of a TV receiver is shown in Fig. 14-13. Although it is basically a superheterodyne receiver, it is one of the most sophisticated and complex electronic devices ever developed. Today, most of the circuitry is incorporated in large-scale ICs. Yet the typical TV receiver still uses many discrete component circuits.

The Tuner

The signal from the antenna or the cable is connected to the tuner, which consists of an RF amplifier, a mixer, and a local oscillator. The tuner is used to select which TV channel is to be viewed and to convert the picture and sound carriers plus their modulation to an *intermediate frequency (IF)*. As in most superheterodyne receivers, the local oscillator frequency is set higher than the incoming signal by the IF value.

Most TV set tuners are prepackaged in sealed and shielded enclosures. They are two tuners in one, one for the VHF signals and another for the UHF signals. The VHF tuner usually uses low-noise FETs for the RF amplifier and the mixer. UHF tuners use a diode mixer with no RF amplifier or a GaAs FET RF amplifier and mixer.

Tuning Synthesizer The local oscillators are *phase-locked loop (PLL) frequency synthesizers* set to frequencies that will convert the TV signals to the IF. Tuning of the local oscillator is typically done digitally. The PLL synthesizer

Surface acoustic wave (SAW) filter

Intermediate frequency (IF)

Phase-locked loop (PLL) frequency synthesizer

is tuned by setting the feedback frequency division ratio. In a TV set this is changed by a microprocessor which is part of the master control system. The interstage LC-resonant circuits in the tuner are controlled by varactor diodes. By varying the DC bias on the varactors, their capacitance is changed, thereby changing the resonant frequency of the tuned circuits. The bias control signals also come from the control microprocessor. Most TV sets are also tuned by IR remote control.

Video IF and Demodulation

The standard TV receiver IFs are 41.25 MHz for the sound and 45.75 MHz for the picture. For example, if a receiver is tuned to channel 4, the picture carrier is 67.25 MHz, and the sound carrier is 71.75 MHz (the difference is 4.5 MHz). The synthesizer local oscillator is set to 113 MHz. The tuner produces an output that is the difference between the incoming signal and local oscillator frequencies, or $113 - 67.25$ MHz, or 45.75 MHz, for the picture and $113 - 71.75$ MHz, or 41.25 MHz, for the sound. Because the local oscillator frequency is above the frequency of incoming signals, the relationship of the picture and sound carriers is reversed at the intermediate frequencies, the picture IF being 4.5 MHz above the sound IF.

The IF signals are then sent to the video IF amplifiers. Selectivity is usually obtained with a *surface acoustic wave (SAW) filter*. This fixed tuned filter is designed to provide the exact selectivity required to pass both of the IF signals with the correct response to match the vestigial sideband signal transmitted. Figure 14-14(*a*) is a block diagram of the filter. It is made on a piezoelectric ceramic substrate such as lithium niobate. A pattern of interdigital fingers on the surface convert the IF signals into acoustic waves that travel across the filter surface. By controlling the shapes, sizes, and spacings of the interdigital filters, the response can be tailored to any application. Interdigital fingers at the output convert the acoustic waves into electrical signals at the IF.

The response of the SAW IF filter is shown in Fig. 14-14(*b*). Note that the filter greatly attenuates the sound IF to prevent it from getting into the video circuits. The maximum response occurs in the 43- to 44-MHz range. The picture carrier IF is down 50 percent on the curve.

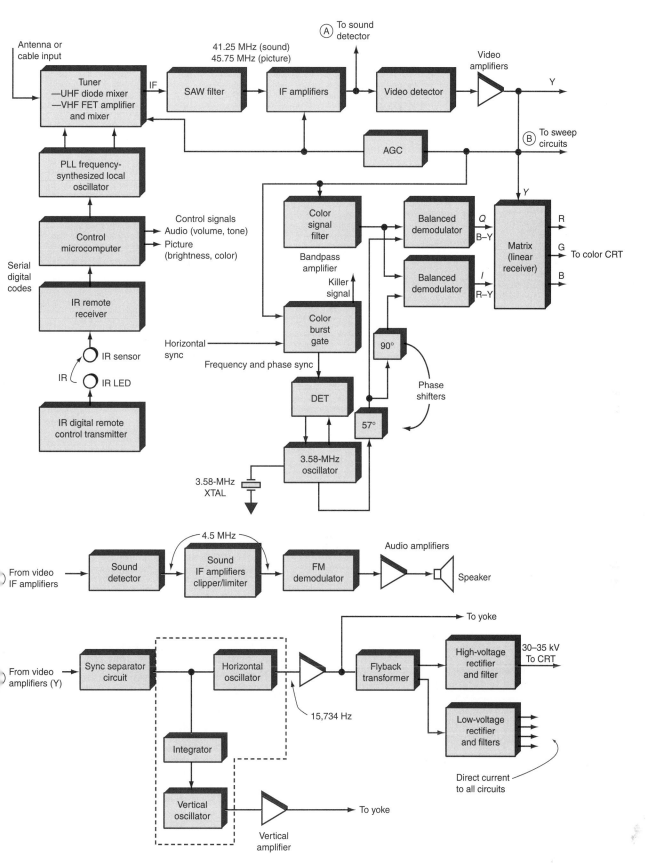

Fig. 14-13 Block diagram of TV receiver.

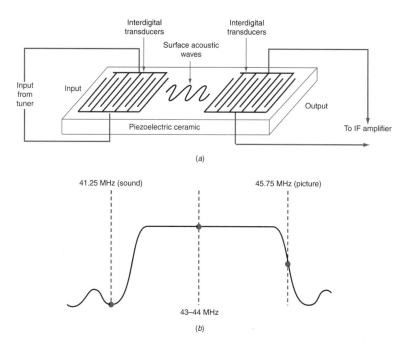

Interdigital transducers

Interdigital transducers

Surface acoustic waves

Input from tuner

Input

Piezoelectric ceramic

Output

To IF amplifier

(a)

41.25 MHz (sound)

45.75 MHz (picture)

43–44 MHz

(b)

Fig. 14-14 (a) Surface acoustic wave (SAW) filter. (b) Typical IF response curve.

Continue to refer to Fig. 14-13. The IF signals are next amplified by IC amplifiers. The video (luminance, or *Y*) signal is then recovered by an AM demodulator. In older sets, a simple diode detector was used for video detection. In most modern sets a synchronous balanced modulator type of synchronous demodulator is used. It is part of the IF amplifier IC.

The output of the video detector is the *Y* signal and the composite color signals, which are amplified by the video amplifiers. The *Y* signal is used to create an AGC voltage for controlling the gain of the IF amplifiers and the tuner amplifiers and mixers.

The composite color signal is taken from the video amplifier output by a filter and fed to color-balanced demodulator circuits. The color-burst signal is also picked up by a gating circuit and sent to a phase detector (*φ* DET) whose output is used to synchronize an oscillator that produces a 3.58-MHz subcarrier signal of the correct frequency and phase. The output of this oscillator is fed to two balanced demodulators that recover the *I* and *Q* signals. The carriers fed to the two balanced modulators are 90° out of phase. Note the 57° phase shifter used to correctly position the color phase for maximum recovery of color detail. The *Q* and *I* signals are combined in matrix with the *Y* signal, and out comes the three *R*, *G*, and *B* color signals. These are amplified

and sent to the picture tube, which reproduces the picture.

Sound IF and Demodulation

To recover the sound part of the TV signal, a separate sound IF and detector section are used. Continuing to refer to Fig. 14-13, note that the 41.25- and 45.75-MHz sound and picture IF signals are fed to a sound detector circuit. This is a nonlinear circuit that heterodynes the two IFs and generates the sum and difference frequencies. The result is a 4.5-MHz difference signal which contains both the AM picture and the FM sound modulation. This is the sound IF signal. It is passed to the sound IF amplifiers which also perform a clipping-limiting function which removes the AM, leaving only the FM sound. The audio is recovered with a quadrature detector or differential peak detector as described in Chap. 5. The audio is amplified by one or more audio stages and sent to the speaker. If stereo is used, the appropriate demultiplexing is done by an IC, and the left and right channel audio signals are amplified.

Synchronizing Circuits

A major part of the TV receiver is dedicated to the sweep and synchronizing functions that are unique to TV receivers. In other words, the receiver's job does not end with demodulation and recovery of the picture and sound. To display the picture on a picture tube, special sweep circuits are needed to generate the voltages and currents to operate the picture tube, and sync circuits are needed to keep the sweep in step with the transmitted signal.

The sweep and sync operations begin in the video amplifier. The demodulated video includes the vertical and horizontal blanking and sync pulses. The sync pulses are stripped off the video signal with a sync separator circuit and fed to the sweep circuits (refer to the lower part of Fig. 14-13). The horizontal sync pulses are used to synchronize a horizontal oscillator to 15,734 Hz. This oscillator drives a horizontal output stage that develops a sawtooth of current that drives magnetic deflection coils in the picture tube *yoke*. The yoke is placed around the neck of the tube to deflect the electron beam (see Fig. 14-16).

The horizontal output stage, which is a high-power transistor switch, is also part of a switching power supply. The horizontal output transistor drives a step-up–step-down transformer called the *flyback*. The 15.734-kHz pulses developed are stepped up, rectified, and filtered to develop the 30- to 35-kV-high direct current required to operate the picture tube. Step-down windings on the flyback produce lower-voltage pulses that are rectified and fil-

tered into low voltages which are used as power supplies for most of the circuits in the receiver.

The sync pulses are also fed to an IC that takes the horizontal sync pulses during the vertical blanking interval and integrates them into a 60-Hz sync pulse that is used to synchronize a vertical sweep oscillator. The output from this oscillator is a sawtooth sweep voltage at the field rate of 60 Hz (actually 59.94 Hz). This output is amplified and converted into a linear sweep current that drives the magnetic coils in the picture tube yoke. These coils produce vertical deflection of the electron beams in the picture tube.

In most modern TV sets, the horizontal and vertical oscillators are replaced by digital sync circuits (see Fig. 14-15). The horizontal sync pulses from the sync separator are normally used to phase-lock a 31.468-kHz *voltage-controlled oscillator (VCO)* that runs at two times the normal horizontal rate of 15.734 kHz. Dividing this by 2 in a flip-flop gives the horizontal pulses that are amplified and shaped in the horizontal output stage to drive the deflection coils on the picture tube. A digital frequency divider divides the 31.468-kHz signal by 525 to get a 59.94-Hz signal for vertical sync. This signal is shaped into a current sawtooth and amplified by the vertical output stage

Yoke

Voltage-controlled oscillator (VCO)

Flyback

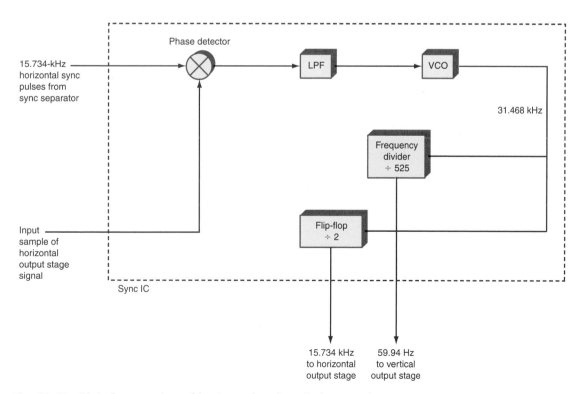

Fig. 14-15 Digital generation of horizontal and vertical sync pulses.

which drives the deflection coils on the picture tube.

Color Picture Tube

A picture tube is a vacuum tube called a *cathode-ray tube (CRT)*. Both monochrome (B&W) and color picture tubes are available. The CRT used in computer video monitors work like the TV picture tube described here.

Monochrome CRT The basic operation of a CRT is illustrated with a monochrome tube as shown in Fig. 14-16(a). The tube is housed in a bell-shaped glass enclosure. A filament heats a cathode that emits electrons. The negatively charged electrons are attracted and accelerated by positive-bias voltages on the elements in an electron gun assembly. The electron gun also focuses the electrons into a very narrow beam. A control grid that is made negative with

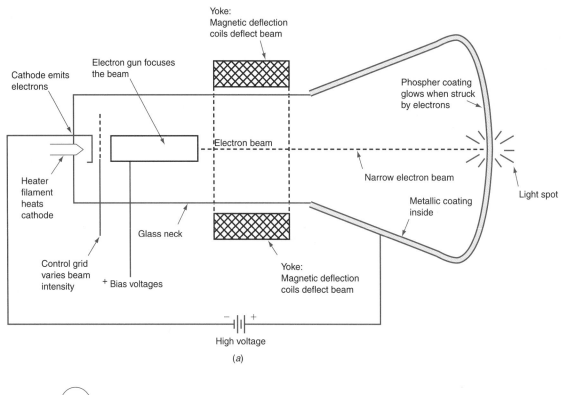

(a)

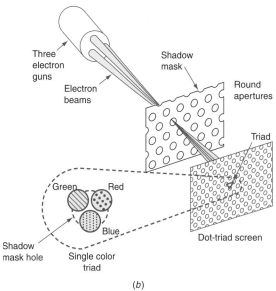

(b)

Fig. 14-16   (a) Basic construction and operation of a black and white (monochrome) cathode-ray tube (CRT). (b) Details of color picture tube.

respect to the cathode controls the intensity of the electron beam and the brightness of the spot it makes.

The beam is accelerated forward by a very high voltage applied to an internal metallic coating called *aquadag*. The *face,* or front, of the picture tube is coated internally with a phosphor that glows and produces white light when it is struck by the electron beam.

Around the neck of the picture tube is a structure of magnetic coils called the *deflection yoke.* The horizontal and vertical current linear sawtooth waves generated by the sweep and synchronizing circuits are applied to the yoke coils, which produce magnetic fields inside the tube that influence the position of the electron beam. When electrons flow, a magnetic field is produced around the conductor through which the current flows. In a CRT, the magnetic field that occurs around the electron beam is moved or deflected by the magnetic field produced by the deflection coils in the yoke. Thus, the electron beam is swept across the face of the picture tube in the interlaced manner described earlier.

As the beam is being swept across the face of the tube to trace out the scene, the intensity of the electron beam is varied by the luminance, or *Y,* signal, which is applied to the cathode or in some cases to the control grid. The *control grid* is an element in the electron gun that is negatively biased with respect to the cathode. By varying the grid voltage, the beam can be made stronger or weaker, thereby varying the intensity of the light spot produced by the beam when it strikes the phosphor. Any shade of gray, from white to black, can be reproduced in this way.

Color CRT   The operation of a color picture tube is similar to that just described. To produce color, the inside of the picture tube is coated with many tiny red, green, and blue phosphor dots arranged in groups of three called *triads.* Some tubes use a pattern of red, green, and blue stripes. These dots or stripes are energized by three separate cathodes and electron guns driven by the red, green, and blue color signals. Figure 14-16(*b*) shows how the three electron guns are focused so that they strike only the red, green, and blue dots as they are swept across the screen. A metallic plate with holes for each dot triad called a *shadow mask*

is between the guns and the phosphor dots to ensure that the correct beam strikes the correct color dot. By varying the intensity of the color beams, the dot triads can be made to produce any color. The dots are small enough so that the eye cannot see them individually at a distance. What the eye sees is a color picture swept out on the face of the tube.

Figure 14-17 shows how all the signals come together at the picture tube to produce the color picture. The *R, G,* and *B* signals are mixed with the *Y* signal to control the cathodes of the CRT. Thus the beams are properly modulated to reproduce the color picture. Note the various controls associated with the picture tube. The *R-G-B* screen, brightness, focus, and centering controls vary the dc voltages that set the levels as desired. The convergence controls and assembly are used to control the positioning of the three electron beams so that they are centered on the holes in the shadow mask and the electron beams strike the color dots dead center. The deflection yoke over the neck of the tube deflects all three electron beams simultaneously.

TV Remote Control

Almost every TV set sold these days, regardless of size or cost, has a wireless remote control. Other consumer electronic products have remote controls including VCRs, cable TV converters, CD players, stereo audio systems, and some ordinary radios. Generic remote controls are available to hook up to any device that you wish to control remotely.

All these devices work on the same basic principle. A small handheld battery-powered unit transmits a serial digital code via an infrared beam to a receiver that decodes it and carries out the specific action defined by the code. A TV remote control is one of the more sophisticated of these controls, for it requires many codes to perform volume control, channel selection, and other functions.

Figure 14-18 is a general block diagram of a remote control transmitter. In most modern units, all the circuitry, except perhaps for the IR LED driver transistors, is contained within a single IC. The purpose of the transmitter is to convert a keyboard entry into a serial binary code that is transmitted by IR to the receiver.

The keyboard is a matrix of momentary contact SPST pushbuttons. The arrangement

Aquadag

Deflection yoke

Control grid

Triad

Shadow mask

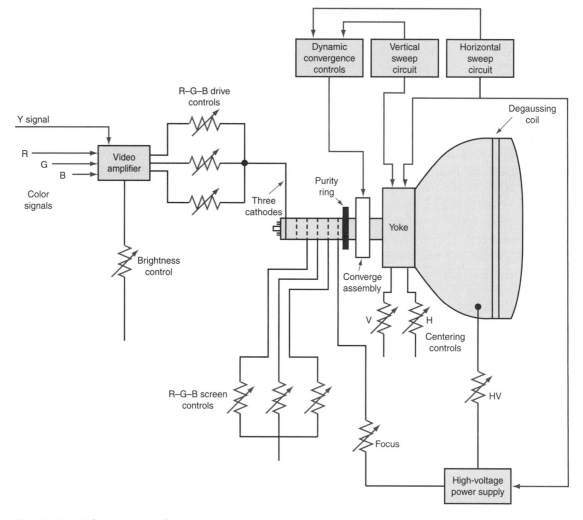

Fig. 14-17 Color picture tube circuits.

shown is organized as 8 rows and 4 columns. The row and column connections are made to a keyboard encoder circuit inside the IC. Pulses generated internally are applied to the column lines. When a key is depressed, the pulses from one of the column outputs are connected to one of the row inputs. The encoder circuit converts this input into a unique binary code representing a number for channel selection or some function such as volume control. Some encoders generate a few as 6 bits, and others generate up to 32 bit codes. Nine- and ten-bit codes are very common.

The serial output is generated by the shift register as data is shifted out. A standard NRZ serial code is generated. This is usually applied to a serial encoder to generate a standard biphase or Manchester code. Recall that the biphase code provides more reliable transmission and reception because there is a signal change for every 0-to-1 or 1-to-0 transition.

The actual bit rate is usually in the 30- to 70-kbits/s range.

The serial bit stream turns a higher-frequency pulse source off and on according to the code's binary 1s and 0s. The transmitter IC contains a clock oscillator that runs at a frequency in the 445- to 510-kHz range. A typical unit runs at 455 kHz using an external ceramic resonator to set the frequency. The serial data turns the 455-kHz pulses off and on. For example, a binary 1 generates a burst of 16 455-kHz pulses as shown in Fig. 14-19. When a binary 0 occurs in the data train, no pulses are transmitted. The figure shows a 6-bit code (011001) with a start pulse. The period T of the 455-kHz pulses is 2.2 μs. The pulse width is set for a duty cycle of about 25 percent, or $T/4$.

If 16 pulses make up a binary 1 interval, that duration is 16×2.2, or 35.2 μs. This translates into a code bit rate of $1/35.2 \times 10^{-6}$, or 28.410 kHz.

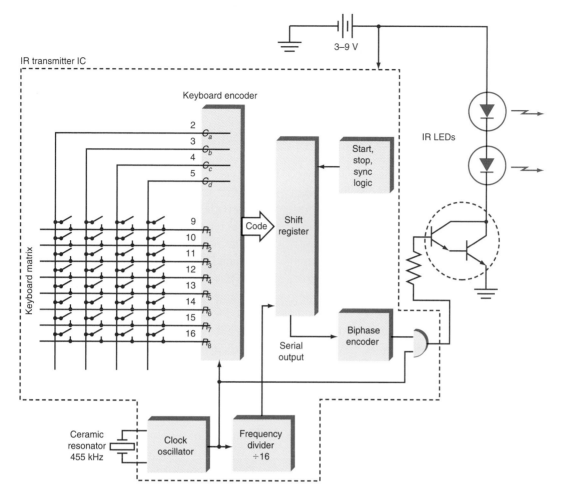

Fig. 14-18 IR TV remote control transmitter.

The 455-kHz pulses modulate the infrared light source by turning it off and on. The IR source is usually one or more IR LEDs. These are driven by a Darlington transistor pair external to the IC as shown in Fig. 14-18. Two or more LEDs are used to ensure a sufficient level of IR radiation to the receiver in the TV set. The LED current is usually very high, giving high IR output levels for reliable transmission to the receiver. Some remote units use three

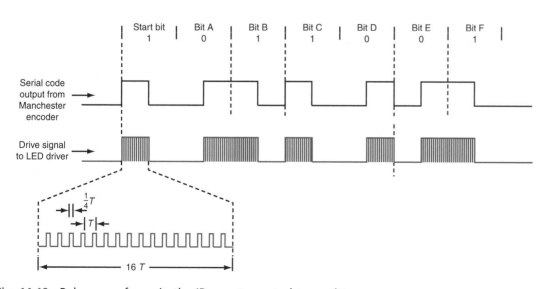

Fig. 14-19 Pulse wave forms in the IR remote control transmitter.

LEDs for a wide-angle transmission signal so that a high-amplitude signal will be received despite the direction in which the remote is pointed.

An IR receiver is shown in Fig. 14-20. The PIN IR photodiode is mounted on the front of the TV set, where it will pick up the IR signal from the transmitter. The received signal is very small despite the fact that the distance between the transmitter and receiver is only 6 to 15 ft on average. Two or more high-gain amplifiers boost the signal level. Most circuits have some form of AGC. The incoming pulses are then detected, shaped, and converted into the original serial data train. This serial data is then read by the control microcomputer that is usually part of the TV receiver.

The microcontroller is a dedicated microcomputer built into every TV set. A master control program is stored in a ROM. The microcomputer converts inputs from the remote control and front panel controls into output signals that control the various functions in a TV set such as channel selection and volume control.

The microcontroller inputs and decodes the incoming signal and then issues output control signals to all other circuits: the PLL frequency synthesizer that controls the TV tuner, the volume control circuits in the audio section, and in the more advanced receivers, chroma and video such as hue, saturation, brightness, and contrast. The microcontroller also generates, sometimes with the help of an external IC, the characters and simple graphics that can be displayed on the screen. Most microcontrollers also contain a built-in clock.

■ TEST

Answer the following questions.

24. What portion of a modern TV set uses digitally coded infrared signals to control channel selection, and volume level?
25. What is the name of the special filter that provides most of the selectivity for the TV receiver?
26. The picture and sound IFs are heterodyned together to form the sound IF. What is its frequency?
27. Name two common sound demodulators in TV receivers.
28. What is meant by a *quadrature 3.58-MHz subcarrier* as used in demodulation?
29. What circuit strips the horizontal sync pulses from the video detector output?
30. The horizontal sync pulses synchronize an internal sweep oscillator to what frequency in a color TV receiver?
31. What is the shape of the horizontal and vertical sweep signals which are currents applied to the horizontal and vertical deflection coils?
32. What is the name of the assembly around the neck of the picture tube to which the sweep signals are applied?
33. What element in the picture tube generates the electrons? What element in the picture tube focuses the electrons into a narrow beam?

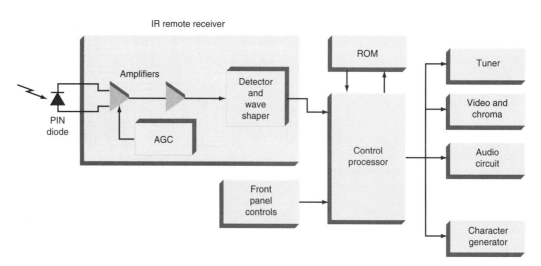

Fig. 14-20 The IR receiver and control microprocessor.

34. By what process is the electron beam deflected and swept across the face of the picture tube?

35. How many electron guns are used in a color CRT to excite the color dot triads on the face of the tube?

36. What is the name of the circuits that ensure the electron beams strike the correct color dots?

37. What stage in the TV receiver is used as the switching power supply to develop the HV that is required to operate the picture tube?

38. What is the name of the transformer used to step up and step down the horizontal sync pulses to produce horizontal sweep as well as most DC power supply voltages in a TV set?

39. How is channel selection in the tuner accomplished? What type of local oscillator is used, and how does it work?

40. A channel 33 UHF TV station has a picture carrier frequency of 585.25 MHz. What is the sound carrier frequency?

41. What are the TV receiver sound IF and the picture IF values?

42. What would be the local oscillator frequency to receive a channel 10 signal?

43. How is the spectrum space conserved in transmitting the video in a TV signal?

44. How is the composite chrominance signal demodulated?

45. How is the 3.85-MHz subcarrier oscillator in the receiver phase and frequency synchronized to the transmitted signal?

46. How is the vertical sweep oscillator synchronized to a sync pulse derived from the horizontal sync pulses occurring during the vertical blanking interval? What is the vertical sync frequency in a color TV set?

47. Explain how pressing a button on a TV remote control generates an IR signal.

48. What is the typical data rate of a TV remote control?

49. What frequency is used to modulate the IR beam?

50. What encoding method is commonly used in IR data transmission?

51. What component in the TV receiver actually decodes the serial data code sent by the remote control transmitter and initiates the intended action?

14-3 Cable TV

Cable TV, sometimes called *CATV*, is a system of delivering the TV signal to home receivers by way of a coaxial cable rather than over the air by radio wave propagation. A cable TV company collects all the available signals and programs and frequency-multiplexes them on a single coaxial cable that is fed to the homes of subscribers. A special cable decoder box is used to receive the cable signals, select the desired channel, and feed a signal to the TV set.

CATV Background

Many companies were established to offer TV signals by cable. They put up very tall high-gain TV antennas. The resulting signals were amplified and fed to the subscribers by cable. Similar systems were developed for apartments and condos. A single master antenna system was installed at a building, and the signals were amplified and distributed to each apartment or unit by cable.

Modern Cable TV Systems

Today, cable TV companies collect signals and programs from many sources, multiplex them, and distribute them to subscribers (see Fig. 14-21). The main building or facility is called the *headend*. The antennas receive local TV stations and other nearby stations plus the special cable channel signals distributed by satellite. The cable companies use parabolic dishes to pick up the so-called premium cable channels. A cable TV company uses many TV antennas and receivers to pick up the stations whose programming it will redistribute. These signals are then processed and then combined or frequency-multiplexed onto a single cable.

The main output cable is called the *trunk cable*. In older systems it is a large, low-loss coaxial cable, although newer cable TV systems use fiber-optic cable. The trunk cable is

Cable TV (CATV)

Headend

Trunk cable

About ⬤ Electronics

The Federal Communications Commission (FCC) regulates all broadcast transmissions. It assigns radio wavelength and television channels and licenses and regulates radio and television stations.

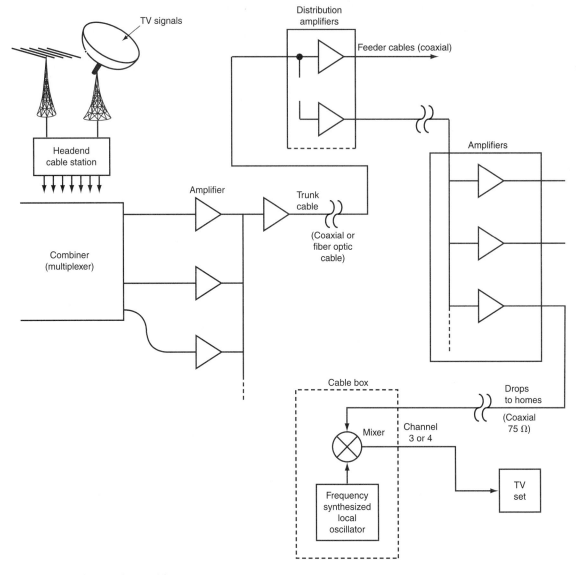

Fig. 14-21 The modem cable TV system.

usually buried and extended to surrounding areas. A junction box containing amplifiers takes the signal and redistributes it to smaller cables called *feeders,* which go to specific areas and neighborhoods. From there the signals are again rejuvenated with amplifiers and then sent to individual homes by coaxial cables called *drops.*

The coaxial cable (usually 75 Ω RG-59/U) comes into a home and is connected to a cable decoder box, which is essentially a special TV tuner that picks up the cable channels and provides a frequency synthesizer and mixer to select the desired channel. The mixer output is heterodyned to TV channel 3 or 4 and then fed to the TV set antenna terminals. The desired signal is frequency-translated by the cable box to channel 3 or 4 that the TV set can receive.

Cable TV is a popular and widely used service in the United States. More than 70 percent of U.S. homes have cable TV service. This service eliminates the need for antennas. And because of the direct connection of amplified signals, there is no such thing as poor, weak, noisy, or snowy signals. In addition, many TV programs are available only via cable, for example, the premium movie channels. The only downside to cable TV is that it is more expensive than connecting a TV to a standard antenna.

Signal Processing

The TV signals to be redistributed by the cable company usually undergo some kind of processing before they are put on the cable to the

Feeder

Drop

TV set. Amplification and impedance matching are the main processes involved in sending the signal to remote locations over what is sometimes many miles of coaxial cable. However, at the headend, other types of processes are involved.

Straight-Through Processors In early cable systems, the TV signals from local stations were picked up with antennas and the signal was amplified before being multiplexed onto the main cable. This is called *straight-through processing*. Amplifiers called *strip amplifiers* that are tuned to the received channels pass the desired TV signal to the combiner. Most of these amplifiers include some kind of gain control or attenuators that can reduce the signal level to prevent distortion of strong local signals. This process can still be used with local VHF TV stations, but today heterodyne processing is used instead.

Heterodyne Processors Heterodyne processing translates the incoming TV signal to a different frequency. This is necessary when satellite signals are involved. Microwave carriers cannot be put on the cable, so they are down-converted to some available 6-MHz TV channel. In addition, heterodyne processing gives the cable companies the flexibility of putting the signals on any channel they want to use.

The cable TV industry has created a special set of nonbroadcast TV channels, as shown in Fig. 14-22. Some of the frequency assignments correspond to standard TV channels, but others do not. Since all these frequencies are confined to a cable, there can be duplication of any frequency that might be used in radio or TV broadcasting. Note that the spacing between the channels is 6 MHz.

The cable company uses modules called *heterodyne processors* to translate the received signals to the desired channel (see Fig. 14-23). The processor is a small TV receiver. It has a tuner set to pick up the desired over-the-air channel. The output of the mixer is the normal TV IFs of 45.75 and 41.25 MHz. These pictures and sound IF signals are usually separated by filters, and they incorporate AGC and provide for individual gain control to make fine-tuning adjustments. These signals are then

Straight-through processing

Strip amplifier

Heterodyne processing

Channel		Frequency, Video Carrier, MHz	Channel		Frequency, Video Carrier, MHz
	Low-Band VHF			Superband (cont.)	
2		55.25	N		241.25
3		61.25	O		247.25
4		67.25	P		253.25
5		77.25	Q		259.25
6		83.25	R		265.25
	MidBand VHF		S		271.25
A-2		109.25	T		277.25
A-1		115.25	U		283.25
A		121.25	V		289.25
B		127.25	W		295.25
C		133.25			
D		139.25		Hyperband	
E		145.25	AA		301.25
F		151.25	BB		307.25
G		157.25	CC		313.25
H		163.25	DD		319.25
I		169.25	EE		325.25
	High-Band VHF		FF		331.25
7		175.25	GG		337.25
8		181.25	HH		343.25
9		187.25	II		349.25
10		193.25	JJ		355.25
11		199.25	KK		361.25
12		205.25	LL		367.25
13		211.25	MM		373.25
	Superband		NN		379.25
J		217.25	OO		385.25
K		223.25	PP		391.25
L		229.25	QQ		397.25
M		235.25	RR		403.25

Fig. 14-22 Special cable TV channels. Note that the video or picture carrier frequency is given.

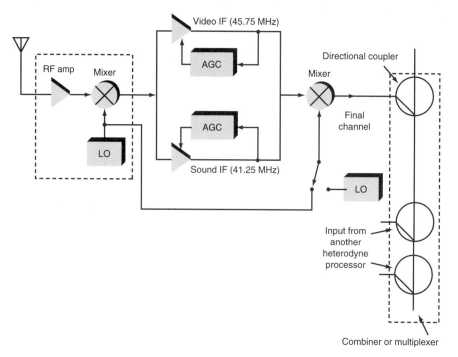

Video IF (45.75 MHz)

RF amp Mixer

AGC

LO

AGC

Sound IF (41.25 MHz)

Mixer

LO

Directional coupler

Final
channel

Input from
another
heterodyne
processor

Combiner or multiplexer

Fig. 14-23 A heterodyne processor.

Cable TV box

sent to a mixer where they are combined with a local oscillator signal to up-convert them to the final output frequency. A switch is usually provided to connect the input local oscillator to the output mixer. This puts the received signal back on the same frequency. In some cases that is done. However, setting the switch to the other position selects a different local oscillator frequency that will up-convert the signal to another desired channel frequency.

Some heterodyne processors completely demodulate the received signal into its individual audio and video components. This gives the cable company full control over signal quality by making it adjustable. In this way, the cable company could also employ scrambling methods if desired. The signals are then sent to a modulator unit that puts the signals on carrier frequencies. The resulting signal is then up-converted to the desired output channel frequency.

Combiner

All the signals on their final channel assignments are sent to a *combiner,* which is a large special-purpose linear mixer. Normally, directional couplers are used for the combining operation. Figure 14-23 shows how multiple directional couplers are connected to form the combiner or multiplexer. The result is that all the signals are frequency-multiplexed into a composite signal that is put on the trunk cable.

Cable TV Converter

The receiving end of the cable TV system at the customer's home is a box of electronics that selects the desired channel signal from those on the cable and translates it to channel 3 or 4, where it is connected to the host TV receiver through the antenna input terminals. The *cable TV box* is thus a tuner that can select the special cable TV channels and convert them to a frequency that any TV set can pick up.

Figure 14-24 shows a basic block diagram of a cable TV converter. The 75-Ω RG-59/U cable connects to a tuner made up of a mixer and a frequency synthesizer local oscillator capable of selecting any of the desired channels. The synthesizer is phase-locked and microprocessor controlled. Most control processors provide for remote control with a digital infrared remote control like that used on virtually every modern TV set.

The output of the mixer is sent to a modulator that puts the signal on either channel 3 or 4. The output of the modulator connects to the TV set antenna input. The TV set is then set to the selected channel and left there. All channel changing is done with the cable converter remote control.

Today, the cable converters have many advanced features, among them automatic identification and remote control by the cable

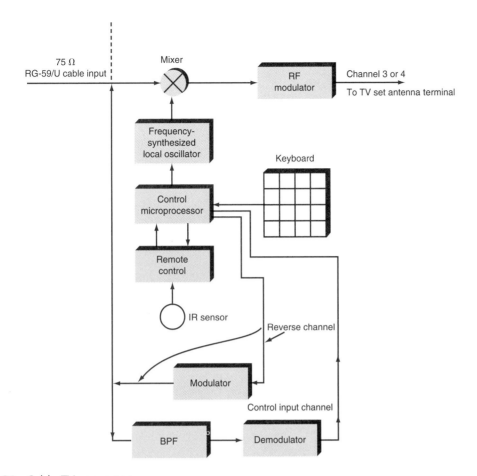

Fig. 14-24 Cable TV converter.

company. Each processor contains a unique ID code that the cable company uses to identify the customer. This digital code is transmitted back to the cable company over a special reverse channel. There are several 6-MHz channels below channel 2 that can be used to transmit special signals to or from the cable converter. The digital ID modulates one of these special reverse channels. These low channels can also be used by the cable company to turn on or disable a cable converter box remotely. A digital signal is modulated onto a special channel and sent to the cable converter. It is picked off by a special tuner or with a bandpass filter as shown in Fig. 14-21. The signal is demodulated, and the recovered signal is sent to the microprocessor for control purposes. It can be used to lock out access to any special channels for which the customer has not subscribed. The reverse channels can also be used for simple troubleshooting.

A cable modem is an extension of the cable TV converter. This device encodes and decodes digital data over one of the 6-MHz channels on the cable. The cable modem permits users to connect their personal computer to the Internet by way of the cable system. Because of the wide bandwidth of a cable TV channel, a very high data rate from 10 to 30 Mbits/s can be achieved. This is significantly higher than that obtainable on conventional telephone lines (see Chap. 12).

Some cable TV companies offer data services to users, and more are expected to offer this service in the future.

■ TEST

Answer the following questions.

52. What is the name given to the cable TV station that collects and distributes the cable signals?
53. What are the names of the main cables used to distribute the TV signals to subscribers?
54. What two types of cables are used for the main distribution of signals in a cable TV system?

55. Name the coaxial cable that feeds individual houses in a cable TV system. What is the type designation and impedance?

56. What is the name of the equipment at the cable station used to change the TV signal frequency to another frequency? Why is this done?

57. What local oscillator frequency would you use to translate the TV signal at its normal IF values to cable channel J?

58. What is the name of the circuit that is used to assemble or mix all the different cable channels together to form a single signal that is distributed by the cable?

59. Name the two main sections of a cable converter box used by the subscriber. What is a reverse channel on a cable TV converter box?

60. Describe the nature of the output signal developed by the cable converter box and where it is connected.

61. How is cable attenuation that occurs during distribution overcome?

62. Name one new nontelevision application of the cable TV system that is now becoming popular.

14-4 Satellite TV

One of the most common methods of TV signal distribution is via communications satellite. A communications satellite orbits around the equator about 22,300 mi out in space. It rotates in synchronism with the earth and therefore appears to be stationary. The satellite is used as a radio relay station (refer to Fig. 14-25). The TV signal to be distributed is used to frequency-modulate a microwave carrier, and then it is transmitted to the satellite. The path from earth to the satellite is called the *uplink*. The satellite translates the signal to another frequency and then retransmits it back to earth. This is called the *downlink*. A receive site on earth picks up the signal. The receive site may be a cable TV company or an individual consumer. Satellites are widely used by the TV networks, the premium channel companies, and the cable TV industry for distributing their signals nationally.

A newer form of consumer satellite TV is *direct broadcast satellite (DBS) TV.* The DBS systems are designed specifically for consumer reception directly from the satellite. The new DBS systems feature digitally encoded video and audio signals, which make transmission

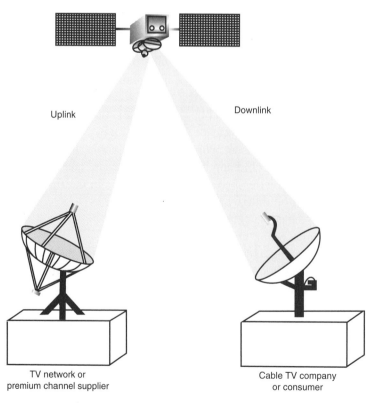

Uplink   Downlink

TV network or
premium channel supplier

Cable TV company
or consumer

Fig. 14-25 Satellite TV distribution.

and reception more reliable and provide outstanding picture and sound quality. By using higher-frequency microwaves, higher-powered satellite transponders, and very low noise GaAsFETs in the receiver, the customer's satellite dish can be made very small. These systems typically use an 18-in dish as opposed to the 5- to 12-ft-diameter dishes still used in many satellite TV systems.

Satellite Transmission

The TV signal to be uplinked to the satellite from a ground station is used to modulate a carrier in one of several available microwave satellite bands. The C band between approximately 3.7 to 4.2 GHz is the most commonly used. The video signal frequency-modulates the microwave carrier on one of 24 channel frequencies.

The audio accompanying the video frequency-modulates a subcarrier in the 5- to 8-MHz range. The 6.2- and 6.8-MHz subcarriers are the most common. Stereo sound is used to modulate the two subcarriers on 5.58 and 5.76 MHz. The video occupies the spectrum of approximately 0 to 5 MHz. The composite spectrum of the video and audio subcarrier signals used to frequency-modulate the uplink transmitter is illustrated in Fig. 14-26. This is the signal that must be recovered by the satellite receiver.

The signal is received by the satellite, and filters pass the signal through the selected transponder. In the transponder the signal is down-converted to a lower frequency, amplified, and retransmitted.

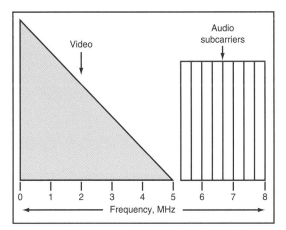

Fig. 14-26 Composite video-audio signal used to modulate the C-band uplink transmitter.

Satellite Receivers

A *satellite receiver* is a special subsystem designed to work with a consumer TV set. It consists of a parabolic dish antenna, a low-noise amplifier and down converter, an IF section with appropriate demodulators for both video and sound, and a method of interconnecting it to the conventional TV set. In addition, most satellite receivers contain circuitry for controlling the positioning of the satellite dish antenna. You will sometimes hear the satellite receiver referred to as *TVRO*, or *TV receive-only, system*. The following section describes the basic organization and operation of typical TVRO satellite receivers.

Antenna   The antenna is more critical in a satellite TV receiver than in any other kind of receiver. The signal from the satellite 22,300 mi away is extremely weak. In addition, there are hundreds of satellites in orbit above the earth, and their spacing is getting closer each year as the number of satellites continues to increase. A high-gain, highly directional parabolic dish antenna is used to select only the signal from the desired satellite and provide very high gain. Most satellite TV antennas range in size from approximately 6 to 15 ft. Over the years, lower-noise, higher-gain amplifiers have been developed using *gallium arsenide (GaAs) field effect transistors*. This has permitted dish antenna sizes to be reduced in size. In some of the newer systems, antennas as small as 3 to 4 ft in diameter are available. However, in most cases, the larger the dish, the higher the gain and the better the performance.

A part of determining antenna size is based on the location of the receiver in the United States. The signal strength of the satellite downlink signal varies considerably over the United States. In most cases, the signal strength is higher in the center of the country and considerably lower on the coasts. As a result, if the receiver is located on the east or west coast, higher gains and larger antennas are required

Satellite receiver

TV receive-only (TVRO) system

Gallium arsenide (GaAs) field effect transistor

If you are interested in a career in radio, visit the site for the National Association of Broadcasters: ⟨www.nab.org⟩

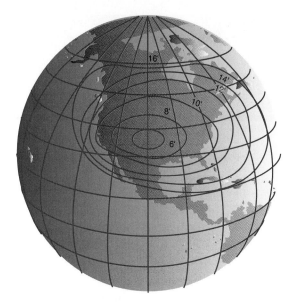

Fig. 14-27 Satellite "footprint" showing desired disk sizes for best reception.

for satisfactory reception. Figure 14-27 shows the "footprint" of the satellite antenna on earth. The contour lines indicate different signal levels, the strongest being in the center and the weakest being on the outside. Since the signal is the strongest in the center, smaller antennas can be used. Larger dishes must be used on the outer areas to receive an adequate signal level.

The antenna is a horn located at the focal point of a parabolic reflector (see Fig. 14-28). Signals picked up by the dish are focused on the horn, giving very high gain and exceptionally narrow directional characteristics. The antenna is built so that it can receive both horizontally and vertically polarized signals. The horn is usually coupled by a short piece of coaxial cable to the receiver input.

Receiver A satellite TV receiver is like any other communications receiver in that it is usually of the superheterodyne type. The basic configurations shown in Chap. 11 on satellite communications apply. See Fig. 11-31. A single-conversion receiver usually has a 70- or 140-MHz intermediate frequency (IF). A dual-conversion receiver uses a first IF of 770 MHz, although you may find other values in the 600- to 1500-MHz range. A second IF of 140 or 70 MHz follows. These values are chosen to minimize images.

If you are considering a career with a diverse electronics and communications corporation, you should take a look at SONY's Web site: ⟨www.sony.com⟩.

In some receivers the RF amplifier, first mixer, and local oscillator are located directly at the horn antenna on the dish. This is done to avoid the massive attenuation in a coaxial cable from the antenna to the receiver front end. Coaxial cable has a massive loss at 4 to 6 GHz (C band) and even more in the 11- to 18-GHz range (Ku band) used by direct broadcast satellites. With the first mixer at the antenna, the signal can be down-converted to a frequency that will produce less loss. Typically, a broadband converter is used to eliminate the need to tune the local oscillator. The entire bandwidth of the received signal is converted to a frequency that is usually in the 900- to 1400-MHz range, where coax attenuation is not so severe. This broadband signal enters the receiver and is further down-converted to the selected IF, either 70 or 140 MHz. From there it is demodulated and the resulting video and audio signals are used to remodulate a signal on VHF channel 3 or 4 to create a standard TV signal that is then connected to the antenna input of a conventional TV receiver.

Direct Broadcast Satellite System

The *direct broadcast satellite (DBS) system* is the newest form of satellite TV available to consumers. It was designed specifically to be an all-digital system in contrast to the analog systems currently in use. Data compression techniques are used to reduce the data rate required to produce high-quality picture and sound. The DBS system has almost totally replaced the older C-band TV receivers.

The DBS system features entirely new digital uplink ground stations and satellites. Since

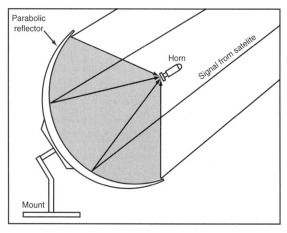

Fig. 14-28 Parabolic disk-horn antenna.

Direct broadcast satellite (DBS) system

the satellites are designed to transmit directly to the home, extra high power transponders are used to ensure a satisfactory signal level.

To receive the digital video from the satellite, a consumer must purchase a satellite TV receiver and antenna. These are similar to the satellite receivers just described; however, they work with digital signals and operate in the K_u rather than the C band. By using higher frequencies as well as higher-power satellite transponders, the necessary dish antenna can be extremely small. The new satellite DBS system antennas are only 18 in. in diameter. There are several special digital broadcast satellites in orbit, and some of the direct satellite TV sources include DirecTV, USSB, and PrimeStar. All provide full coverage of the major cable networks and the premium channels usually distributed to homes by cable TV, and all can be received directly. In addition to purchasing the receiver and antenna, the consumer must subscribe to one of the services supplying the desired channels.

Satellite Transmission The video to be transmitted must first be placed into digital form. To digitize an analog signal, it must be sampled a minimum of two times per cycle in order for sufficient digital data to be developed for reconstruction of the signal. Assuming that video frequencies of up to 4.2 Mbits/s are used, the minimum sampling rate is twice this, or 8.4 Mbits/s. For each sample, a binary number proportional to the light amplitude is developed. This is done by an A/D converter, usually with an 8-bit output. The resulting video signal, therefore, has a data rate of 8 bits times 8.4 Mbits/s, or 67.2 Mbits/s. This is an extremely high data rate. However, for a color TV signal to be transmitted in this way, there must be a separate signal for each of the red, green, and blue components making up the video. This translates to a total data rate of $3 \times$ 67.2, or 202 Mbits/s. Even with today's technology, this is an extremely high data rate that is hard to achieve reliably.

In order to lower the data rate and improve the reliability of transmission, the new DBS system uses compressed digital video. Once the video signals have been put into digital form, they are processed by *digital signal processing (DSP) circuits* to minimize the full amount of data to be transmitted. Digital compression greatly reduces the actual transmitting speed to somewhere in the 20- to 30-Mbits/s range. The compressed serial digital signal is then used to modulate the uplinked carrier using BPSK.

The DBS satellite uses the K_u band with a frequency range of 11 to 14 GHz. Uplink signals are usually in the 14- to 14.5-GHz range and the downlink signals usually cover the range of 10.95 to 12.75 GHz.

The primary advantage of using the K_u band rather than the C band is that the receiving antennas may be made much smaller for a given amount of gain. However, these higher frequencies are more affected by atmospheric conditions than the lower microwave frequencies. The biggest problem is the increased attenuation of the downlink signal caused by rain. Any type of weather involving rain or water vapor, such as fog, can seriously reduce the received signal. This is because the wavelength of K_u band signals is near that of water vapor. Therefore, the water vapor absorbs the signal. Although the power of the satellite transponder and the gain of the receiving antenna are typically sufficient to provide solid reception, there can be fadeout under heavy downpour conditions.

Finally, the digital signal is transmitted from the satellite to the receiver using circular polarization. The DBS satellites have *right-hand* and *left-hand circularly polarized (RHCP* and *LHCP) helical antennas.* By transmitting both polarities of signal, frequency reuse can be incorporated to double the channel capacity.

DBS Receiver A block diagram of a typical DBS digital receiver is shown in Fig. 14-29. The receiver subsystem begins with the antenna and its low-noise block converter. The horn antenna picks up the K_u band signal and translates the entire 500-MHz band used by the signal down to the 950- to 1450-MHz range, as explained earlier. Control signals from the receiver to the antenna select between RHCP and LHCP. The RF signal from the antenna is sent by coaxial cable to the receiver.

A typical DBS downlink signal occurs in the 12.2- to 12.7-GHz portion of the K_u band. Each transponder has a bandwidth of about 24 MHz. The digital signal is usually occurring at a rate of approximately 27 Mbits/s.

RHCP helical antenna

LHCP helical antenna

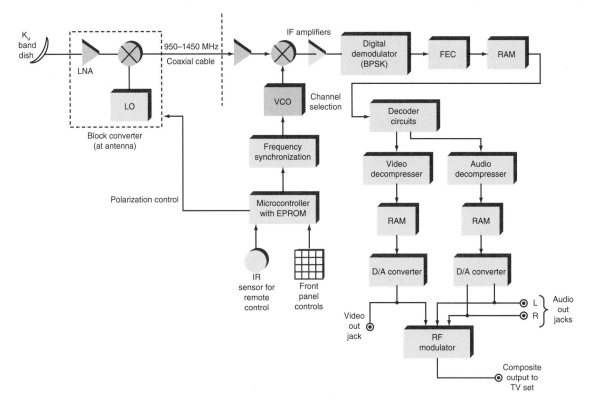

Fig. 14-29 Digital DBS TV receiver.

Figure 14-30 shows how the digital signal is transmitted. The digital audio and video signals are organized into data packets. Each packet consists of a total of 147 bytes. The first 2 bytes (16 bits) contain the *service channel identification (SCID) number.* This is a 12-bit number that identifies the video program being carried by the packet. The 4 additional bits are used to indicate whether the packet is encrypted and if so, which decoding key to use. One additional byte contains the packet type and a continuity counter.

The data block consists of 127 bytes, either 8-bit video signals or 16-bit audio signals. It may also contain digital data used for control purposes in the receiver. Finally, the last 17 bytes are the error-detection check codes. These 17 bytes are developed by an error-checking circuit at the transmitter. The ap-

pended bytes are checked at the receiver to detect any errors and correct them.

The received signal is passed through another mixer with a variable-frequency local oscillator to provide channel selection. The digital signal at the second IF is then demodulated to recover the originally transmitted digital signal, which is passed through a *forward-error correction (FEC) circuit.* This circuit is designed to detect bit errors in the transmission and to correct them on the fly. Any bits lost or obscured by noise during the transmission process are usually caught and corrected to ensure a near-perfect digital signal.

The resulting error-corrected signal is stored in *random access memory (RAM),* after which the signal is decoded to separate it into both the video and the audio portions. The resulting signals are then sent to the audio and video

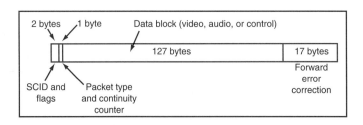

Fig. 14-30 Digital data packet format used in DBS TV.

decompression circuits. The DBS TV system uses digital compression-decompression standards referred to as an *MPEG2*. *MPEG* means "Motion Picture Experts Group," which is a standards organization that establishes technical standards for movies and video. MPEG2 is its latest and best compression method for video.

Although the new DBS digital systems will not replace cable TV, they provide the consumer with the capability of receiving a wide range of TV channels. The use of digital techniques provides an unusually high-quality signal.

Finally, the video and audio signals are converted to analog by D/A converters and sent to an RF modulator that develops a conventional TV signal.

■ TEST

Answer the following questions.

63. What is the name of the unit in a satellite that receives and then rebroadcasts the TV signal to be distributed?
64. What is the frequency range of standard satellites used for TV distribution, and what is the letter designation of that band?
65. In the United States, what direction must you point an antenna to "see" a satellite?
66. What is the approximate distance of the satellite from the earth?
67. What is the name given to a consumer satellite receiving station?
68. What kind of amplifiers are used in satellite TV receiver front-ends?
69. What allows an antenna to receive two different signals on the same frequency?
70. What kind of modulation is used to apply the video to the microwave uplink carrier?
71. What is a common intermediate frequency in a single-conversion satellite receiver?
72. What are two common first IFs used in double-conversion satellite receivers?
73. According to the satellite "footprint" in Fig. 14-27, what is the size of the antenna needed for good reception in south Florida?
74. What feature of the DBS system makes digital transmission and reception possible?

75. What natural occurrence can sometimes prohibit the reception of DBS signals?
76. Name two ways that a satellite receiver, either conventional or DBS, is connected to a conventional TV set.
77. Describe the antennas used in satellite transmission and reception.
78. Why are the amplifiers and mixers in a satellite receiver usually located at the antenna?
79. How is the audio signal modulated onto the microwave uplink carrier?
80. How is the tuning of a satellite receiver accomplished if the local oscillator is at the antenna?
81. Why are smaller antennas satisfactory in the central United States but not in Canada or Mexico?
82. Why are two demodulators needed in a satellite receiver? What form of modulation must be detected?
83. Describe how the antenna is oriented and positioned in a consumer satellite installation.
84. What is the main difference between conventional DBS satellite TV systems?
85. What is the operational frequency range and band designation of a DBS receiver?

14-5 Digital Television

Digital television (DTV), also known as *high-definition television (HDTV)*, will eventually replace the current NTSC system, which was invented in the 1940s and 1950s. The goal of DTV is to greatly improve the picture and sound quality.

After nearly a decade of evaluating alternative DTV systems, the FCC has finalized the standards and decreed that DTV will eventually become the U.S. TV standard by 2006. The first DTV stations began transmission in the ten largest U.S. cities on September 1, 1998. DTV sets can now be purchased by the consumer, but they are very expensive. As more DTV stations come on line and as more DTV programming becomes available, more consumers will buy DTV receivers and the cost will drop dramatically.

The DTV system is an extremely complex collection of

MPEG

DTV

For an overview of digital products and services, examine the site for ⟨www.digital.com⟩.

digital, communications, and computer techniques. A full discussion is beyond the scope of this book. However, this section is a brief introduction to the basic concepts and techniques used in DTV.

DTV Standards

DTV uses the scanning concept to paint a picture on the CRT. So you can continue to think of the DTV screen in terms of scan lines, as you would think of the standard NTSC analog screen. However, you should also view the DTV screen as made up of thousands of tiny dots of light called *pixels*. Each pixel can be any of 256 colors. These pixels can be used to create any image. The greater the number of pixels on the screen, the greater the resolution and the finer the detail that can be represented. Each horizontal scan line is divided up into hundreds of pixels. The format of a DTV screen is described in terms of the numbers of pixels per horizontal line by the number of vertical pixels (which is the same as the number of horizontal scan lines).

Pixel

One major difference between conventional NTSC analog TV and DTV is that DTV can use progressive line scanning rather than interlaced scanning. In progressive scanning each line is scanned one at a time from top to bottom. Since this format is compatible with computer video monitors, it is possible to display DTV on computer screens. Interlaced scanning can be used on one of the HDTV formats.

The FCC has defined a total of 14 different formats for DTV. Most are variations of the basic formats given in Table 14-1.

TABLE **14-1**

Standard	Aspect Ratio	Pixels/ Horiz. Line	Vert. Pixels*	Scan Rate, Hz
480p	4:3	640	480	24, 30, 60 Hz†
480i/p	4:3 or 16:9	704	480	24, 30, 60 Hz
720p	16:9	1280	720	24, 30, 60 Hz
1080i	16:9	1920	1080	24 or 30 Hz

*Number of scan lines.
†Standard PC VGA format.

The 480p (the *p* stands for "progressive") standard offers performance comparable to that of the NTSC system. It uses a 4:3 aspect ratio for the screen. The scanning is progressive. The vertical scan rate is selectable to fit the type of video being transmitted. This format is fully compatible with modern VGA computer monitors. The 704 × 480 format can use either progressive or interlaced scanning with either aspect ratio at the three vertical scan rates shown in Table 14-1.

The 720p format uses a larger aspect ratio of 16:9 (a 4:3 format is optional at this resolution also). This format is better for showing movies. Figure 14-31 shows the difference between the current and new DTV aspect ratios. The 1080i format uses the 16:9 aspect ratio but with more scan lines and more pixels per line. This format obviously gives the best

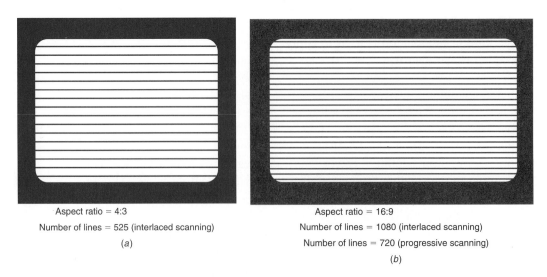

Aspect ratio = 4:3
Number of lines = 525 (interlaced scanning)

(a)

Aspect ratio = 16:9
Number of lines = 1080 (interlaced scanning)
Number of lines = 720 (progressive scanning)

(b)

Fig. 14-31 TV picture standards. (a) Current standard. (b) DTV standard.

resolution. The DTV set should be able to detect and receive any available format.

Digital TV Transmission Concepts

All the data communications concepts discussed in Chap. 12 are applicable to DTV. Obviously, in DTV both the video and the audio signals must be digitized by A/D converters and transmitted serially to the receiver. Because of the very high frequency of video signals, special techniques must be used to transmit the video signal over a standard 6-MHz bandwidth TV channel. And because both video and audio must be transmitted over the same channel, multiplexing techniques must be used. The FCC's requirement is that all of this information be transmitted reliably over the standard 6-MHz TV channels now defined for NTSC TV.

Assume that the video to be transmitted contains frequencies up to 4.2 MHz. For this signal to be digitized, it must be sampled at least two times per cycle or at a minimum sampling rate of 8.4 MHz. If each sample is translated into an 8-bit word (byte) and the bytes transmitted serially, the data stream would have a rate of 8×8.4 MHz or 67.2 MHz. This does not include the audio, which must also be digitized and multiplexed with the digital data. The resulting digital data rate is very high. To permit this quantity of data to be transmitted over the 6-MHz channel, special encoding and modulation techniques are used.

DTV Transmitter Figure 14-32 shows a block diagram of a DTV transmitter. The video from the camera consists of the R, G, and B signals that are converted into the luminance and chrominance signals. These are digitized by A/D converters. The luminance sampling rate is 14.3 MHz, and the chroma sampling rate is 7.15 MHz. The resulting signals are serialized and sent to a data compressor. The purpose of this device is to reduce the number of bits needed to represent the video data and therefore permit higher transmission rates in a limited-bandwidth channel. The data compression method used in DTV is referred to as MPEG2, which stands for the "Motion Picture Engineering Group-2" standard. This organization establishes standards for the compression of video and film pictures. The MPEG2 data compressor processes the data according to an algorithm that effectively reduces any redundancy in the video signal. For example, if the picture is one-half light blue sky, the pixel values will be the same for many lines. All of this data can be reduced to one pixel value

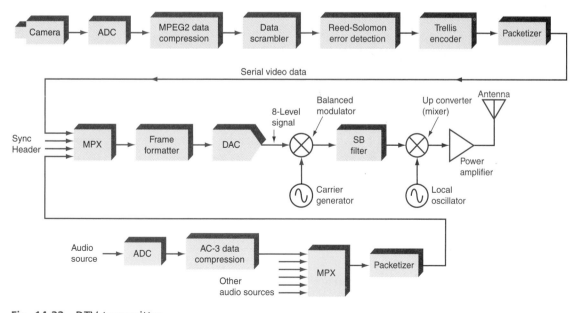

Fig. 14-32 DTV transmitter.

transmitted for a known number of times. The algorithm also uses fewer bits to encode the color than to encode the brightness because the human eye is much more sensitive to brightness than to color. The MPEG2 encoder captures and compares successive frames of video and compares them to detect the redundancy so that only differences between successive frames are transmitted.

The signal is next sent to a data randomizer. The randomizer scrambles or randomizes the signal. This is done to ensure that random data is transmitted even when no video is present or when the video is a constant value for many scan lines. This permits clock recovery at the receiver.

Next, the random serial signal is passed through a Reed-Solomon (RS) error detection and correction circuit. This circuit adds extra bits to the data stream so that transmission errors can be detected at the receiver and corrected. This ensures high reliability in signal transmission even under severe noise conditions. In DTV, the RS encoder adds 20-parity bytes per block of data that can provide correction for up to 10 byte errors per block.

The signal is next fed to a trellis encoder. This circuit further modifies the data to permit error correction at the receiver. Trellis encoding is widely used in modems. Trellis coding is not used in the cable TV version of DTV.

The audio portion of the DTV signal is also digital. It provides for compact disk (CD) quality audio. The audio system can accommodate up to six audio channels, permitting monophonic sound, stereo, and multichannel surround sound. The channel arrangement is flexible to permit different systems. For example, one channel could be used for a second language transmission or closed captioning.

Each audio channel is sampled at a 48 kbits/s rate, ensuring that audio signals up to about 24 kHz are accurately captured and transmitted. Each audio sample is converted into an 18-bit digital word. The audio information is time-multiplexed and transmitted as a serial bit stream at a frequency of 48 kbits/s $\times$ 6 channels $\times$ 18 bits = 5.184 Mbits/s. A data compression technique designated AC-3 is used to speed up audio transmission.

Next, the video and audio data streams are packetized; that is, they are converted into short blocks of data bytes that segment the video and

audio signals. These packets are then multiplexed along with some synchronizing signals to form the final signal to be transmitted. The result is a 188-bit packet containing both video and audio data plus four bytes of synchronizing bytes and a header. See Fig. 14-33. The header identifies the number of the packet and its sequence as well as the video format. Next, the packets are assembled into frames of data representing one frame of video. The complete frame consists of 626 packets transmitted sequentially. The final signal is sent to the modulator.

The modulation scheme used in DTV is 8-VSB or eight-level vestigial sideband amplitude modulation. The carrier is suppressed, and only the upper sideband is transmitted. The serial digital data is sent to a D/A converter where each sequential 3-bit group is converted into a discrete voltage level. This system encodes 3 bits per symbol, thereby greatly increasing the data rate within the channel. An example is shown in Fig. 14-34. Each 3-bit group is converted into a relative level of -7, -5, -3, -1, $+1$, $+3$, $+5$, or $+7$. This is the signal that amplitude-modulates the carrier. The resulting symbol rate is 10,800 symbols per second. This translates into a $3 \times 10,800 = 32.4$ Mbits/s data rate. Eliminating the extra RS and trellis bits gives an actual video/audio rate of about 19.3 Mbits/s.

A modified version of this format is used when the DTV signal is to be transmitted over a cable system. Trellis coding is eliminated and 16-VSB modulation is used to encode 4 bits per symbol. This gives double the data rate of terrestrial DTV transmission (38.6 Mbits/s).

The VSB signal can be created with a balanced modulator to eliminate the carrier and to generate the sidebands. One sideband is removed by a filter or by using the phasing system described earlier in the AM circuits chapter. The modulated signal is up-converted by a mixer to the final transmission frequency, which is one of the standard TV channels in the VHF or UHF range. A linear power amplifier is used to boost signal level prior to transmission by the antenna.

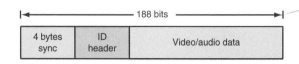

Fig. 14-33 Packet format for DTV.

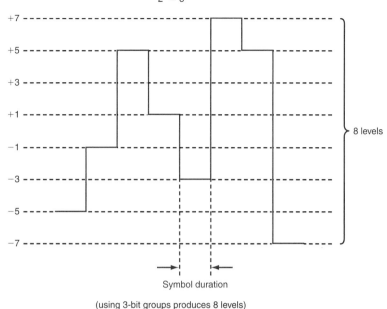

$2^3 = 8$

8 levels

Symbol duration

(using 3-bit groups produces 8 levels)

Fig. 14-34 Eight-level VSB signal.

DTV Receiver A DTV receiver picks up the composite signal and then demodulates and decodes the signal into the original video and audio information. A simplified receiver block diagram is shown in Fig. 14-35. The tuner and IF systems are similar to those in a standard TV receiver. From there the 8-VSB signal is demodulated using a synchronous detector into the original bit stream. A balanced modulator is used along with a carrier signal that is phase-locked to the pilot carrier to ensure accurate demodulation. A clock recovery circuit regenerates the clock signal that times all the remaining digital operations.

The signal then passes through an NTSC filter that is designed to filter out any one channel or adjacent channel interference from standard TV stations. The signal is also passed through an equalizer circuit that adjusts the signal to correct for amplitude and phase variations encountered during transmission.

The signals are demultiplexed into the video and audio bit streams. Next, the trellis decoder and RS decoder ensure that any received errors caused by noise are corrected. The signal is then descrambled and decompressed. The video signal is then converted back into the digital signals that will drive the D/A converters that, in turn, drive the red, green, and blue electron guns in the CRT. The audio signal is also demultiplexed and fed to AC-3 decoders.

The resulting digital signals are fed to D/A converters that create the analog audio for each of the six audio channels.

The State of DTV

As this is written, DTV is still relatively new, and few consumers own DTV sets. There is little original DTV programming, that is, programs that are shot with DTV cameras. Only some premium sports events and special programs are digitally transmitted. The currently available sets cost thousands of dollars. Most of these sets use large projection screens (up to 6 ft diagonal measurement) to implement the 16:9 video format. These expensive sets are used mainly by various types of establishments for sports event coverage.

As more programming becomes available, more sets will be sold, and as sales volume increases, prices will drop. It is expected that a converter box will be the first widely adopted DTV product. This is a DTV receiver just like the DBS satellite receiver that converts the signals into an NTSC format that can be displayed on a conventional TV set. Although the overall benefits of DTV will not be fully realized with this arrangement, it will make it possible to gradually phase out the old NTSC sets sooner. DTV VCRs are also expected to be available in the future.

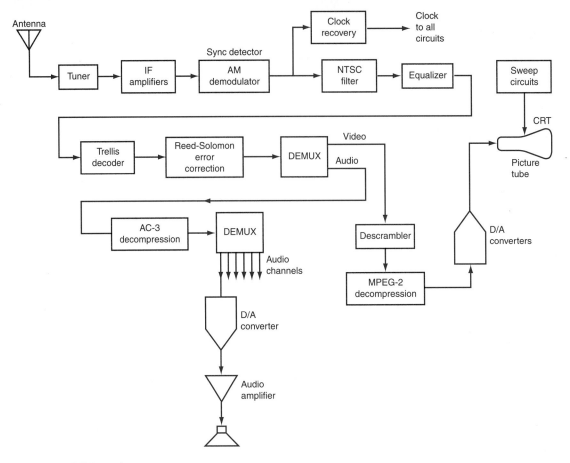

Fig. 14-35 DTV receiver.

■ TEST

Answer the following questions.

86. State the main reason for converting to a digital TV system.
87. What technique is used to reduce the data rate needed to transmit the digital TV signal in a limited bandwidth? What is the name of the video standard used? The audio standard used?
88. What is the bandwidth of a digital TV channel?
89. What is the name of the primary error-correction system used in DTV?
90. By what method is the number of bits per symbol increased to speed up data transmission?
91. Describe the type of modulation used in DTV.
92. What is the data rate through the TV channel?
93. How many audio channels are used? What is their sampling rate?
94. Calculate the total number of screen pixels in the 720p format.
95. What is the basic pack size of the DTV signal?
96. Define progressive scanning. Why is it used?

Summary

1. Television is the radio transmission of sound and pictures in the VHF and UHF ranges. The voice signal from a microphone frequency-modulates a sound transmitter. A camera converts a picture or scene into an electrical signal called the video or luminance Y signal, which amplitude-modulates a separate video transmitter. Vestigial sideband AM is used to conserve spectrum space. The picture and sound transmitter frequencies are spaced 4.5 MHz apart, with the sound frequency being the higher.

2. TV cameras use either a vacuum tube imaging device such as a vidicon or a solid-state imaging device such as the charged-coupled device (CCD) to convert a scene into a video signal.

3. A scene is scanned by the imaging device to break it up into segments that can be transmitted serially. The National Television Standards Committee (NTSC) standards call for scanning the scene in two $262\frac{1}{2}$ line fields which are interlaced to form a single 525-line picture called a frame. Interlaced scanning reduces flicker. The field rate is 59.94 Hz, and the frame or picture rate is 29.97 Hz. The horizontal line scan rate is 15,734 Hz or 63.6 μs per line.

4. The color in a scene is captured by three imaging devices which break a picture down into its three basic colors of red, green, and blue using color light filters. Three color signals are developed (R, G, B). These are combined in a resistive matrix to form the Y signal and are combined in other ways to form the I and Q signals. The I and Q signals amplitude-modulate 3.58-MHz subcarriers shifted 90° from one another in balanced modulators producing quadrature DSB suppressed signals that are added to form a carrier composite color signal. This color signal is then used to modulate the AM picture·transmitter along with the Y signal.

5. A TV receiver is a standard superheterodyne receiver with separate sections for processing and recovering the sound and picture. The tuner section consists of RF amplifiers, mixers, and a frequency-synthesized local oscillator for chan-

nel selection. Digital infrared remote control is used to change channels in the synthesizer via a control microprocessor.

6. The tuner converts the TV signals to intermediate frequencies of 41.25 MHz for the sound and 45.75 MHz for the picture. These signals are amplified in IC IF amplifiers. Selectivity is usually provided by a surface acoustic wave (SAW) filter. The sound and picture IF signals are placed in a sound detector to form a 4.5-MHz sound IF signal. This is demodulated by a quadrature detector or other FM demodulator to recover the sound. Frequency-multiplexing techniques similar to those used in FM radio are used for stereo TV sound. The picture IF is demodulated by a diode detector or other AM demodulator to recover the Y signal.

7. The color signals are demodulated by two balanced modulators fed with 3.58-MHz subcarriers in quadrature. The subcarrier is frequency- and phase-locked to the subcarrier in the transmitter by phase-locking to the color subcarrier burst transmitted on the horizontal blanking pulse.

8. To keep the receiver in step with the scanning process at the transmitter, sync pulses are transmitted along with the scanned lines of video. These sync pulses are stripped off the video detector and used to synchronize horizontal and vertical oscillators in the receiver. These oscillators generate deflection currents that sweep the electron beam in the picture tube to reproduce the picture.

9. The color picture tube contains three electron guns that generate narrow electron beams aimed at the phosphor coating on the inside of the face of the picture tube. The phosphor is arranged in millions of tiny red, green, and blue color dot triads or stripes. The electron beams excite the color dots or stripes in proportion to their intensity and generate light of any color depending upon the amplitude of the red, green, and blue signals. The electron beam is scanned or deflected horizontally and vertically in step

with the transmitted video signals. Deflection signals from the internal sweep circuits drive coils in a deflection yoke around the neck of the picture creating magnetic fields that sweep the three electron beams.

10. The horizontal output stage, which provides horizontal sweep, is also used to operate a flyback transformer that steps up the horizontal sync pulses to a very high voltage. These are rectified and filtered into a 30- to 35-kV voltage to operate the picture tube. The flyback also steps down the horizontal pulses and rectifies and filters them into low-voltage dc supplies that are used to operate most of the circuits in the TV set.

11. Cable television is the transmission of multiple TV signals which are frequency-multiplexed on a single cable to be used in lieu of over-the-air signal transmission. The headend of a cable TV station collects TV signals from local stations and from other sources by satellite and then modulates and multiplexes these signals on a cable that is sent to subscribers.

12. The main cable from the headend, called the trunk, is either large coaxial or fiber-optic cable, that takes the signals to other distribution points and to other cables called feeders, which transmit the signals to neighborhoods. The feeders are then tied to coaxial cable drops connected to each house. The cable signals are amplified at several points along the distribution path to maintain strong signals. A cable TV decoder box or tuner selects the desired channel and converts it into a channel 3 or 4 TV signal that connects to the TV receiver for presentation.

13. Cable TV systems are also being used for high-speed Internet access by using a special cable modem.

14. A cable modem uses one of the 6-MHz TV channels and modulation to send digital signals at a rate of 10 to 30 Mbits/s.

15. TV signals may also be accessed by satellite. Early satellite receivers used the microwave C band to pick up signals transmitted by networks and the premium program suppliers (HBO, CNN, etc.). These receivers are no longer used.

16. The direct broadcast satellite (DBS) system now beams TV signals to earth on the K_u microwave band. These signals are digital and can be received with a small (18 in.) dish antenna and a digital satellite receiver. The output of the digital receiver is converted to an analog TV signal that can be received by a standard TV set.

17. A new digital TV (DTV) system started broadcasts in 1998. Also known as high-definition television (HDTV), it produces much higher quality pictures and sound.

18. DTV uses the same 6-MHz TV channel but sends the video and sound signals in digital format. Data compression is used to reduce the data rate so that transmission can be through the standard 6-MHz channel. The video compression is designated MPEG2 and the audio compression is called AC-3.

19. The modulation is 8-level suppressed carrier signal (vestigial) sideband or 8 VSB. The data rate is 19.6 Mbits/s.

20. DTV is available in four basic screen formats: 640 pixels per line $\times$ 480 lines, 704 $\times$ 480, 1280 $\times$ 720, and 1920 $\times$ 1080. Aspect ratio is 4:3 for the first two formats, and 16:9 on the last three formats. These formats are compatible with computer video monitors when progressive scanning is used.

21. Digital audio is also used. Six channels are available with 48 kbits/s sampling rate, giving CD quality sound with stereo or surround sound.

22. DTV is transmitted in 188-bit packets that use error detection and correction methods to ensure data integrity.

23. DTV is not widely used now because little HDTV programming is available and receivers are very expensive. DTV is expected to be more widely adopted when more programming sources become available and the price declines.

Chapter Review Questions

Choose the letter that best answers each question.

14-1. The TV signal uses which types of modulation for picture and sound respectively?
  a. AM, FM
  b. DSB, FM
  c. FM, AM
  d. AM, DSB

14-2. If a TV sound transmitter has a carrier frequency of 197.75 MHz, the picture carrier is

a. 191.75 MHz.

b. 193.25 MHz.

c. 202.25 MHz.

d. 203.75 MHz.

14-3. The total bandwidth of an NTSC TV signal is

a. 3.58 MHz.

b. 4.5 MHz.

c. 6 MHz.

d. 10.7 MHz.

14-4. What is the total number of interlaced scan lines in one complete frame of an NTSC U.S. TV signal?

a. 262½

b. 525

c. 480

d. 625

14-5. What keeps the scanning process at the receiver in step with the scanning in the picture tube at the receiver?

a. Nothing

b. Color burst

c. Sync pulses

d. Deflection oscillators

14-6. What is the black-and-white or monochrome brightness signal in TV called?

a. RGB

b. Color subcarrier

c. Q and I

d. Luminance (Y)

14-7. What is the name of the solid-state imaging device used in TV cameras that converts the light in a scene into an electrical signal?

a. CCD

b. Photodiode matrix

c. Vidicon

d. MOSFET array

14-8. The I and Q composite color signals are multiplexed onto the picture carrier by modulating a 3.58-MHz subcarrier using

a. FM.

b. PM.

c. DSB AM.

d. Vestigial sideband AM.

14-9. The assembly around the neck of a picture tube that produces the magnetic fields that deflect and scan the electron beams is called the

a. Shadow mask.

b. Phosphor.

c. Electron gun.

d. Yoke.

14-10. The picture and sound carrier frequencies in a TV receiver IF are respectively

a. 41.25 and 45.75 MHz.

b. 45.75 and 41.25 MHz.

c. 41.75 and 45.25 MHz.

d. 45.25 and 41.75 MHz.

14-11. The sound IF in a TV receiver is

a. 4.5 MHz.

b. 10.7 MHz.

c. 41.25 MHz.

d. 45.75 MHz.

14-12. What type of circuit is used to modulate and demodulate the color signals?

a. Phase-locked loop

b. Differential peak detector

c. Quadrature detector

d. Balanced demodulator

14-13. What circuit in the TV receiver is used to develop the high voltage needed to operate the picture tube?

a. Low-voltage power supply

b. Horizontal output

c. Vertical sweep

d. Sync separator

14-14. What ensures proper color synchronization at the receiver?

a. Sync pulses

b. Quadrature modulation

c. 4.5-MHz carrier spacing

d. 3.58-MHz color burst

14-15. Which of the following is not a benefit of cable TV?

a. Lower-cost reception

b. Greater reliability

c. Less noise, stronger signals

d. Premium cable channels

14-16. What technique is used to permit hundreds of TV signals to share a common cable?

a. Frequency modulation

b. Mixing

c. Frequency-division multiplexing

d. Time-division multiplexing

14-17. A cable modem is used to transmit

a. TV signals.

b. Audio signals.

c. Two-way control signals.

d. Digital data.

14-18. The direct broadcast satellite (DBS) system uses the frequency range of

a. 4 to 6 GHz.

b. 7 to 9 GHz.

c. 11 to 14 GHz.

d. 18 to 22 GHz.

14-19. DBS TV is

a. Analog.

b. Digital.

14-20. The primary purpose of DTV is
  a. Higher resolution and higher-quality pictures.
  b. Higher-quality sound.
  c. Compatibility with personal computer video.
  d. All of the above.
14-21. Standard DTV uses the picture format of
  a. 640 × 480.
  b. 704 × 480.
  c. 1280 × 720.
  d. 1920 × 1080.
14-22. HDTV uses an aspect ratio of
  a. 1:1.
  b. 4:3.
  c. 7:5.
  d. 16:9.
14-23. High-definition picture formats use
  a. Progressive scan.
  b. Interleave scan.
14-24. What is the name of the video compression system used in DTV?
  a. AC-3
  b. JPEG
  c. MPEG2
  d. K6-2

14-25. How many audio channels are available in DTV?
  a. 1
  b. 2
  c. 4
  d. 6
14-26. What form of modulation is used in DTV?
  a. AM
  b. FM
  c. 8 VSB
  d. QPSK
14-27. What is the DTV data rate through the 6-MHz TV channel?
  a. 19.6 Mbits/s
  b. 44.1 Mbits/s
  c. 67.2 Mbits/s
  d. 128 Mbits/s
14-28. What is the name of the error-correction scheme used in DTV?
  a. Parity
  b. Hamming codes
  c. CRC
  d. Reed-Solomon

Critical Thinking Questions

14-1. Explain why color quadrature modulation is really a form of multiplexing.
14-2. You have probably heard of closed-circuit TV (CCTV), which is used for monitoring. Describe what you think such a system is made of. Is cable used? If so, what kind? Is modulation used? Is the system wireless?
14-3. How could digital data be multiplexed on a conventional NTSC TV signal?
14-4. Could telephone service be implemented on a cable TV system? Explain.
14-5. Name two or more ways that the high-definition picture is processed so that it can

be transmitted over a standard-bandwidth TV channel.
14-6. Could a TV signal be transmitted over the Internet? If so, describe how. What are the major problems and limitations of this kind of transmission?
14-7. Identify and discuss some of the problems you see in getting people to convert from their current analog TV sets to DTV sets.
14-8. Current VCRs record standard TV signals by having them frequency-modulate a carrier that is transferred to the tape. Will this system work for DTV?

Answers to Tests

1. 6 MHz
2. Video: AM, Audio: FM
3. 4.5 MHz
4. Luminance or Y
5. CCD and vidicon tube
6. Scanning
7. Red, green, blue (RGB)
8. 525
9. $262\frac{1}{2}$
10. Field rate: 60 Hz, Frame rate: 30 Hz (29.97 Hz)
11. $1/15734 = 63.6 \ \mu s$
12. Diplexer

13. *I* and *Q*
14. Quadrature amplitude modulation using suppressed carrier DSB
15. 3.58 MHz
16. Phase shift
17. DSB AM
18. $204 + 1.25 = 205.25$ MHz (video carrier), $205.25 + 4.5 = 209.75$ MHz (audio carrier)
19. 4.2 MHz
20. $BW = 2(25 + 15) = 80$ kHz
21. 3 MHz of the lower sideband is suppressed (vestigial sideband AM)
22. Synchronizing pulses are transmitted at the beginning and end of each horizontal scan line and during the time period between fields. These pulses are used to initiate scan lines in the receiver.
23. The *I* and *Q* color signals are converted back into the individual *R*, *G*, and *B* color signals, and these are applied to the red, green, and blue electron guns of the color CRT, which reproduces the color picture. The *Y* or luminance signal.
24. IR remote control
25. Surface acoustic wave (SAW) filter
26. 4.5 MHz
27. Quadrature and differential peak detector
28. Two 3.58-MHz sine waves shifted 90° from one another
29. Sync separator
30. 31.468 MHz
31. Linear current sweep
32. Yoke
33. Cathode. The plates in the electron gun focus the electrons into a beam.
34. Magnetic fields developed by the horizontal and vertical deflection coils in the yoke interact with the magnetic field produced by the electron beam to produce motion of the electron beams.
35. Three
36. Convergence
37. Horizontal (sweep) output stage
38. Flyback transformer
39. Varying the reverse bias voltage on a varactor diode changes the local oscillator frequency. The local oscillator is part of a phase-locked loop frequency synthesizer.
40. 589.75 MHz
41. 41.25 MHz (sound), 45.75 MHz (video)
42. 239 MHz
43. Part of the lower video sideband is filtered out.
44. Balanced modulators driven by quadrature (90° shift) local oscillators in a phase-locked loop

45. The 3.58-MHz local subcarrier oscillators are synchronized to the color-burst signal from the transmitter in a phase-locked loop.
46. The horizontal sync pulses are integrated in a long-time-constant low-pass filter to create the vertical sync pulse. 59.94 Hz.
47. Pressing a button on the remote control generates a binary code that modulates a light beam produced by an infrared light-emitting diode.
48. 30 to 70 kbits/s
49. 455 kHz
50. Manchester or biphase encoding is used.
51. A microcontroller in the tuner
52. Headend
53. The main cable is the "trunk" that sends signals to feeder cables, which, in turn, send signals to the "drop" cables to the individual homes.
54. Fiber-optic cables for the trunks and coaxial cables for the feeders and drops
55. The drop cable is typically RG-6/U or RG-59/U, 75-Ω cable.
56. A heterodyne processor is a mixer that converts the signal to another frequency in the cable's spectrum to position it as desired in the station's cable numbering system.
57. 263 MHz
58. Directional coupler, combiner, or multiplexer
59. The mixer and the synthesized, microprocessor-controlled local oscillator that interprets remote control signals for channel input. A reverse channel in a cable box allows the box circuitry to communicate back upstream to the headend.
60. The cable box is usually a standard TV signal on VHF channel 3 or 4 that is connected to the 75-Ω antenna input to the tuner on the TV set.
61. By using multiple amplifiers in the trunk, feeder, and drop cables
62. Internet access
63. Transponder
64. 4 to 6 GHz, C band
65. South (the equator is south of the United States)
66. 22,300 mi
67. TVRO (TV receive-only)
68. GaAs MESFET RF amplifiers
69. Horizontal and vertical polarization
70. FM
71. 70 or 140 MHz.
72. 770 MHz
73. 10 ft minimum; 11 or 12 ft is better.
74. Smaller but more powerful transponders with high output power to minimize receive antenna size
75. Rain

76. The video and audio are recovered and converted into a conventional analog TV signal on channel 3 or 4 that is connected to the standard TV receive antenna terminals.

77. A parabolic dish reflector with a horn antenna at the focal point is the standard satellite antenna. The size depends upon the frequency of operation.

78. To reduce noise and to minimize attenuation in the coaxial cable to the receiver.

79. The audio frequency-modulates a subcarrier in the 5- to 8-MHz range. These subcarriers are mixed with the video, and the composite signal modulates the uplink satellite signal.

80. A broadband mixer is used; therefore no tuning is necessary.

81. The satellite antennas have a stronger signal at the center for their footprint, which is centered on the United States. In outlying areas, the signal strength is less; thus an antenna with more gain for good reception is required.

82. Both audio and video demodulators are needed to recover the picture and sound. Both demodulators must detect FM.

83. The antenna is pointed south toward the equator and with an azimuth angle that corresponds to the satellite to be received. The antenna azimuth and elevation angles are set to the approximate values as determined by the location of the receiver. Then they are varied slowly until maximum signal strength is obtained.

84. Conventional satellite receivers (C band) are analog in nature and use FM for both picture and sound. Very large dish antennas (6 to 18 ft) are needed. DBS receivers are fully digital and use very small dish antennas (18 in.).

85. The frequency range of a DBS TV receiver is 10.95 to 12.75 GHz.

86. Improve picture and sound quality

87. Digital data compression. video: MPEG2, audio: AC-3

88. 6 MHz

89. Reed-Solomon

90. Multilevel digital signals are used to transmit multiple bits per symbol.

91. 8-VSB, eight-level vestigial sideband AM

92. 19.3 Mbits/s

93. Six audio channels with a sampling rate of 48 kbits/s

94. $1280 \times 720 = 921,600$ pixels

95. 188 bits

96. Progressive scanning uses successive scan lines rather than interlaced lines as in conventional NTSC TV. Progressive scanning is compatible with standard computer video monitors.

Chapter

15

The Telephone and Its Applications

Chapter Objectives

This chapter will help you to:

1. *Name* and *describe* the components in conventional and electronic telephones.

2. *Describe* the characteristics of the various signals used in telephone communications.

3. *State* the general operation of a cordless telephone.

4. *Explain* the general hierarchy of signal transmission within the telephone system.

5. *Explain* the operation of a facsimile machine.

6. *Explain* the operation of the cellular telephone system.

7. *Name* the four digital cell phone standards.

8. *Explain* the operation of a paging system.

9. *Draw* a block diagram of a basic paging receiver and *explain* its operation.

10. *Define* ISDN and explain its operation.

The telephone system is the largest and most complex electronic communications system in the world. It uses just about every type of electronic communications technique available including virtually all of those described in this book. The telephone communications system is so large and widely used that no text on electronic communications would be complete without a discussion of it. On the other hand, because the system is so large and complex, space is simply not available to cover it in great depth. However, a general discussion of the telephone system will better prepare readers for those communications applications that use the telephone system.

Although the primary purpose of the telephone system is to provide voice communications between individuals, it is also widely used for many other purposes. These include facsimile transmission and computer data transmission, as well as wireless transmission used in cordless cellular and paging telephone systems. Data transmission via modem was discussed in Chap. 12; therefore, it will not be repeated here. However, data transmission by the *integrated services digital network (ISDN) system,* which is part of the telephone system, will be discussed.

ISDN system

15-1 Telephones

The original telephone system was designed for full-duplex analog communications of voice signals. Today, the telephone system is still primarily analog in nature, but it employs a considerable number of digital techniques, not only in signal transmission but also in control operations.

The telephone system permits any telephone to connect with any other telephone in the world. This means that each telephone must have a unique identification code—the 10-digit telephone number assigned to each telephone. The telephone system provides a means of recognizing each individual number and switching systems that can connect any two telephones.

The Local Loop

Standard telephones are connected to the telephone system by way of a two-wire, twisted-pair cable that terminates at the local exchange or central office. As many as 10,000 telephone lines can be connected to a single central office (see Fig. 15-1). The connections from the central office go to the "telephone system" represented in Fig. 15-1 by the large cloud. This part of the system, which is mainly long distance, will be described in the next section. A call originating at telephone A will pass through the central office and then into the main system, where it is transmitted via one of many different routes to the central office connected to the desired location designated as B in Fig. 15-1. The connection between nearby local exchanges is direct rather than long distance.

The two-wire, twisted-pair connection between the telephone and the central office is referred to as the *local loop* or *subscriber loop*. The circuits in the telephone and at the central office form a complete electric circuit, or loop. This single circuit is analog in nature and carries both dc and ac signals. The dc power for operating the telephone is generated at the central office and supplied to each telephone over the local loop. The ac voice signals are transmitted along with the dc power. Despite the fact that only two wires are involved, full-duplex operation, that is, simultaneous send and receive, is possible. All dialing and signaling operations are also carried on this single twisted pair.

The basic telephone set and the local loop, standard and electronic telephone sets, and cordless telephones will be discussed in the next section.

The Telephone Set

A basic telephone or telephone set is an analog baseband transceiver. It has a handset which contains a microphone and a speaker, better known as a *transmitter* and a *receiver.* It also contains a ringer and a dialing mechanism. Overall, the telephone set fulfills the following basic functions:

The receive mode provides:

1. An incoming signal that rings a bell or produces an audio tone indicating that a call is being received.
2. A signal to the telephone system indicating that the signal has been answered.
3. Transducers to convert voice into electrical signals and electrical signals into voice.

The transmit mode:

1. Indicates to the telephone system that a call is to be made when the handset is lifted.
2. Indicates that the telephone system is ready to use by receiving a signal called the *dial tone.*
3. Provides a way of transmitting the telephone number to be called to the telephone system.
4. Receives an indication that the call is being made by receiving a ringing tone.
5. Provides a means of receiving a special tone indicating that the called line is busy.
6. Provides a means of signaling the telephone system that the call is complete.

All telephone sets provide these basic functions. Some of the more advanced electronic telephones have other features such as multiple line selection, hold, and speaker phone.

Local loop

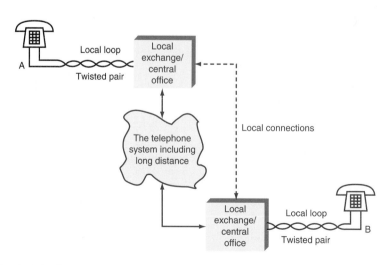

Fig. 15-1 The basic telephone system.

Figure 15-2 is a basic block diagram of a telephone set. The function of each block is described below. Detailed circuits for each of the blocks and their operation will be described later when the standard and electronic telephones are discussed in detail.

Ringer The *ringer* is either a bell or an electronic oscillator connected to a speaker. It is continuously connected to the twisted pair of the local loop back to the central office. When an incoming call is received, a signal from the central office will cause the bell or ringer to produce a tone.

Switch Hook A *switch hook* is a double-pole mechanical switch that is usually controlled by a mechanism actuated by the telephone handset. When the handset is "on the hook," the hook switch is open, thereby isolating all the telephone circuitry from the central office local loop. When a call is to be made or to be received, the handset is taken off the hook. This closes the switch and connects the telephone circuitry to the local loop. The direct current from the central office is then connected to the telephone, closing its circuits to operate.

Dialing Circuits The *dialing circuits* provide a way for entering the telephone number to be called. In older telephones, a pulse dialing system was used. A rotary dial connected to a switch produced a number of ON-OFF pulses corresponding to the digit dialed. These ON-OFF pulses formed a simple binary code for signaling the central office.

In most modern telephones, a tone dialing system is used. Known as the *dual-tone multi-frequency (DTMF) system,* this dialing method uses a number of pushbuttons that generate pairs of audio tones that indicate the digits selected.

Whether pulse dialing or tone dialing is used, circuits in the central office recognize both types of signals and make the proper connections to the dialed telephone.

Handset This unit contains a microphone for the transmitter and a speaker or receiver. When you speak into the transmitter, it generates an electrical signal representing your voice. When a received electrical voice signal occurs on the line, the receiver translates it into sound waves. The transmitter and receiver are independent units, and each has two wires connecting to the telephone circuit. Both of these connect to a special device known as the *hybrid.*

Hybrid The *hybrid* is a special transformer used to convert signals from the four wires from the transmitter and receiver into a signal suitable for a single two-line pair to the local loop. The hybrid permits *full-duplex,* that is, simultaneous send and receive, analog communications on the two-wire line. The hybrid also provides feedback from the transmitter to the receiver so that the speaker can hear his or her voice in the receiver. This feedback permits automatic voice-level adjustment.

The Standard Telephone and Local Loop

Figure 15-3 is a simplified schematic diagram of a conventional telephone and the local loop connections back to the central office. The circuitry at the central office will be discussed in more detail later. For now, note that the central

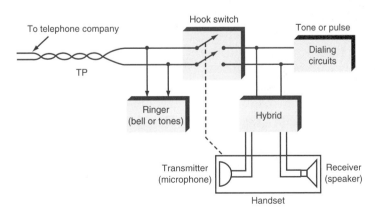

Fig. 15-2 Block diagram of a basic telephone.

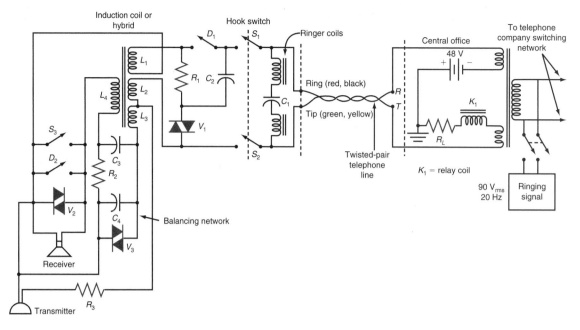

Fig. 15-3 Standard telephone circuit diagram showing connection to central office.

office applies a dc voltage over the twisted-pair line to the telephone. This dc voltage is approximately −48 V with respect to ground in the open-circuit condition. When a subscriber picks up the telephone, the switch hook closes, connecting the circuitry to the telephone line. The load represented by the telephone circuitry causes current to flow in the local loop and the voltage inside the telephone to drop to approximately 5 to 6 V.

The amount of current flowing in the local loop depends upon a number of factors. The dc voltage supplied by the central office may not be exactly −48 V. It can, in fact, vary several volts above or below the 48-V normal value.

As Fig. 15-3 shows, the central office also inserts some resistance R_L to limit the total current flow if a short circuit should occur on the line. This resistance can range from about 350 to 800 Ω. In Fig. 15-3, the total resistance is approximately 400 Ω.

The resistance of the telephone itself also varies over a relatively wide range. It can be as low as 100 Ω and as high as 400 Ω, depending upon the circuitry. The resistance varies because of the resistance of the transmitter element and because of the variable resistors called *varistors* used in the circuit to provide automatic adjustment of line level.

The local loop resistance depends considerably on the length of the twisted pair between the telephone and the central office. Although the resistance of copper wire in the twisted pair is relatively low, the length of the wire between the telephone and the central office can be many miles long. Thus the resistance of the local loop can be anywhere from 1000 to 1800 Ω, depending upon the distance.

Finally, the frequency response of the local loop is approximately 300 to 3400 Hz. This is sufficient to pass voice frequencies that produce full intelligibility. An unloaded twisted pair has an upper cutoff frequency of about 4000 Hz. But this cutoff varies considerably depending upon the overall length of the cable. When long runs of cable are used, special loading coils are inserted into the line to compensate for excessive roll-off at the higher frequencies.

The wires in Fig. 15-3 end at terminals on the telephone labeled *tip* and *ring*. These designations refer to the plug used to connect telephones to one another at the central office. At one time, large groups of telephone operators at the central office used plugs and jacks at a switchboard to connect one telephone to another manually. The jack is shown in Fig. 15-4. The tip and the ring are metallic contacts that are attached to the wires in the cable. They touch spring-loaded contacts in the jack to make the connection. Although such plugs and jacks are no longer used, the tip and ring designation is still used to refer to the two wires of the local loop.

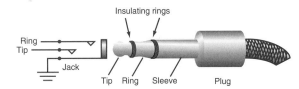

Fig. 15-4 Tip and ring designation on an old plug and jack.

The wires are also usually color-coded red and green. The tip wire is green and is usually connected to ground; the ring wire is red. Many telephone cables into a home or office also contain a second twisted pair if a separate telephone line is to be installed. These wires are usually color-coded black and yellow. Black and yellow correspond to ring and tip, respectively, where yellow is ground. Other color combinations are used in telephone wiring.

Now let's take a more detailed look at the operation of the various circuits in the telephone. The local loop circuitry will be discussed later.

Ringer In Fig. 15-3, the circuitry connected directly to the tip and ring local loop wires is the *ringer*. The ringer in most older telephones is an electromechanical bell. A pair of electromagnetic coils is used to operate a small hammer that alternately strikes two small metallic bells. Each bell is made of different materials so that each produces a slightly different tone. When an incoming call is received, a voltage from the central office operates the electromagnetic coils which, in turn, operate the hammer to ring the bells. The bells make the familiar tone produced by most standard telephones.

In Fig. 15-3, the ringer coils are connected in series with a capacitor C_1. This allows the ac ringing voltage to be applied to the coils but blocks the 48 V of direct current, thus minimizing the current drain on the 48 V of power supplied at the central office.

The ringing voltage supplied by the central office is a sine wave of approximately 90 V rms at a frequency of about 20 Hz. These are the nominal values, because the actual ringing voltage can vary from approximately 80 to 100 V rms with a frequency somewhere in the 15- to 30-Hz range. This ac signal is supplied by a generator at the central office. The ringing voltage is applied in series with the −48-Vdc signal from the central office power supply. The ringing signal is connected to the local loop line by way of a transformer T_1. The transformer couples the ringing signal into its secondary winding where it appears in series with the 48-Vdc supply voltage.

The standard ringing sequence is shown in Fig. 15-5. In U.S. telephones, the ringing voltage occurs for 1 s followed by a 3-s interval. Telephones in other parts of the world use different ringing sequences. For example, in the United Kingdom, the standard ring sequence is a higher-frequency tone occurring more frequently, and it consists of two ringing pulses 400 ms long, separated by 200 ms. This is followed by a 2-s interval of quiet before the tone sequence repeats.

Transmitter The *transmitter* is the microphone into which you speak during a telephone call. In a standard telephone, this microphone uses a carbon element that effectively

Ringer

Transmitter

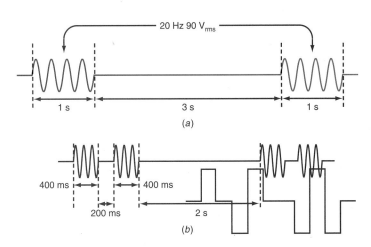

Fig. 15-5 Telephone ringing sequence. (*a*) United States and Europe. (*b*) United Kingdom.

translates acoustical vibrations into resistance changes. The resistance changes, in turn, produce current variations in the local loop representing the speaker's voice. A dc voltage must be applied to the transmitter so that current flows through it during operation. The 48 V from the central office is used in this case to operate the transmitter.

Figure 15-6 is a simplified diagram showing how a telephone transmitter works. The basic transmitter element is a small module containing fine carbon granules. One side of the module is a flexible diaphragm. Whenever you speak, you create acoustic energy in the form of the movement of air. The air molecules move in accordance with your voice frequency. The acoustic energy from the voice reaches the diaphragm and causes it to vibrate in accordance with the speech. An outward acoustic pressure wave causes the carbon granules to be compressed. Pushing the carbon granules closer together causes the overall resistance of the element to decrease. When the acoustic energy moves in the opposite direction, the carbon granules expand outward. Since they are less tightly compressed, their resistance increases. The transmitter element is in series with the telephone circuit, which includes the 48-V central office battery and the speaker in the remote handset. Speaking into the transmitter causes the current flow in the circuit to vary in accordance with the voice signal. The resulting ac voice signal produced on the telephone line is approximately 1 to 2 V rms.

Receiver The *receiver,* or earpiece, is basically a small permanent magnet speaker. A di-

aphragm is physically attached to a coil which rests inside a permanent magnet. Whenever a voice signal comes down a telephone line, it develops a current in the receiver coil. The coil produces a magnetic field that interacts with the permanent magnet field. The result is vibration of the diaphragm in the receiver, which converts the electrical signal into the acoustic energy that supplies the voice to the ear. As it comes in over the local loop lines, the voice signal has an amplitude of approximately 0.5 to 1 V rms.

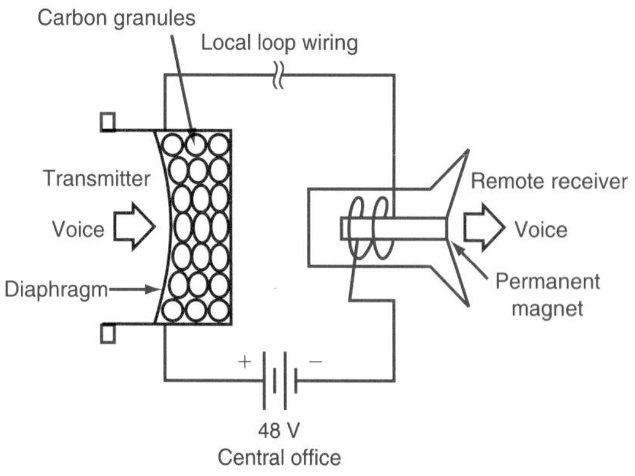

Hybrid The hybrid is a transformerlike device that is used to simultaneously transmit and receive on a single pair of wires. The hybrid, which is also sometimes referred to as an *induction coil,* is really several transformers combined into a single unit. The windings on the transformers are connected in such a way that signals produced by the transmitter are put on the two-wire local loop but do not occur in the receiver. In the same way, the transformer windings permit a signal to be sent to the receiver, but the resulting voltage is not applied to the transmitter.

In practice, the hybrid windings are set up so that a small amount of the voice signal produced by the transmitter does occur in the receiver. This provides feedback to the user so that he or she will speak with normal loudness. The feedback from the transmitter to the receiver is referred to as the *side tone.* If the side tone was not provided, there would be no signal in the receiver and the person speaking would have the sensation that the telephone was dead. By hearing his or her own voice in the receiver at a moderate level, the caller can speak at a normal level. Without the side tone, the speaker tends to speak more loudly, which is unnecessary.

Automatic Level Adjustment Because of the wide variation in the different loop lengths of the two telephones connected to one another, the circuit resistances will vary considerably, thereby causing a wide variation in the transmitted and received voice signal levels. All telephones contain some type of component or circuit that provides automatic voice-level adjustment so that the signal levels are approximately the same regardless of the loop lengths. In the standard telephone, this automatic loop

Induction coil

Receiver

Fig. 15-6 The transmitter and receiver in a telephone.

length adjustment is handled by components called *varistors*. These are labeled V_1, V_2, and V_3 in Fig. 15-3.

A varistor is a nonlinear resistance element whose resistance changes depending upon the voltage across it. When the voltage across the varistor increases, its resistance decreases. A decrease in voltage causes the resistance to increase.

The varistors are usually connected across the line. In Fig. 15-3, varistor V_1 is connected in series with a resistor R_1. This varistor automatically shunts some of the current away from the transmitter and the receiver. If the loop is long, the current will be relatively low and the voltage at the telephone will be low. This causes the resistance of the varistor to increase, thus shunting less current away from the transmitter and receiver. On short local loops, the current will be high and the voltage at the telephone will be high. This causes the varistor resistance to decrease; thus more current is shunted away from the transmitter and receiver. The result is a relatively constant level of transmitted or received speech.

Note that a second varistor V_3 is used in the balancing network. The balancing network (C_3, C_4, R_2) works in conjunction with the hybrid to provide the side tone discussed earlier. The varistor adjusts the level of the side tone automatically.

Tone Dialing Although some dial telephones are still in use and all central offices can accommodate them, most modern telephones use a dialing system known as *TouchTone*. It uses pairs of audio tones to create signals representing the numbers to be dialed. This dialing system is referred to as the *dual-tone multifrequency (DTMF) system*.

A typical DTMF keyboard on a telephone is shown in Fig. 15-7. Most telephones use a standard keypad with 12 buttons or switches for the numbers 0 through 9 and the special symbols * and #. The DTMF system also accommodates four additional keys for special applications.

In Fig. 15-7 numbers represent audio frequencies associated with each row and column of pushbuttons. For example, the upper horizontal row containing the keys for 1, 2, and 3 is labeled 697, meaning that when any one of these three keys is depressed, a sine wave of 697 Hz is produced. Each of the four horizontal rows produces a different frequency. The horizontal rows generate what is generally known as the *low group of frequencies.*

A higher group of frequencies is associated with the vertical columns of keys. For example, the keys for the numbers 2, 5, 8, and 0 produce a frequency of 1336 Hz when depressed.

If the number 2 is depressed, two sine waves are generated simultaneously, one at 697 Hz

Varistor

TouchTone

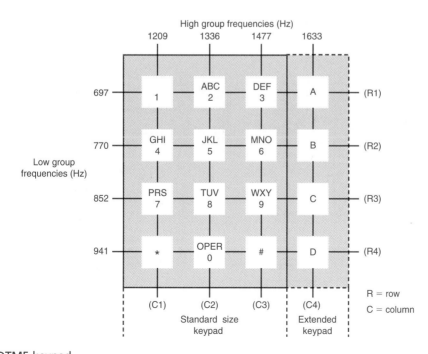

Fig. 15-7 DTMF keypad.

and the other at 1336 Hz. These two tones are linearly mixed. This combination produces a unique sound and is easily detected and recognized at the central office as the signal representing the dialed digit 2. The tolerance on the generated frequencies is usually within ±1.5 percent.

The earliest TouchTone telephones used a simple single transistor oscillator to generate two tones simultaneously. A typical circuit is shown in Fig. 15-8. The oscillator circuit consists of two LC-tuned circuits. The inductance is provided by tapped transformer windings. Winding A of transformer T_1 is the inductance for the low-frequency tones, and winding A on transformer T_2 provides the inductance for the high frequencies. The A windings are tapped to provide different inductance values for each of the various frequencies. A fixed capacitor C_1 resonates with the selected winding on T_1A, and capacitor C_2 resonates with the selected winding on T_2A. Whenever one of the push-buttons on the keypad is depressed, one of the switches in the low-frequency group and one

of the switches in the high-frequency group are closed simultaneously.

Referring to Fig. 15-8, note switch S_3. It is also actuated when any one of the keys on the keypad is depressed. A slight depression of the key establishes connection to the two resonant circuits as indicated above. During this time, the contacts on S_3 remain closed so that direct current flows through windings T_1A and T_2A. Further depression of the switch causes the contacts S_3 to open, breaking the current through windings T_1A and T_2A. Thus the two LC-resonant circuits are shocked into oscillation and two damped sine waves at the correct frequencies are generated.

The depression of S_3 causes the collector of transistor Q_1 to be connected to the dc voltage. Thus an electronic oscillator circuit is formed. Transformer windings T_1C and T_2C transmit some of the sine wave energy back to the base circuit of the transistor with the correct polarity to sustain continuous oscillation. Transformer windings T_1B and T_2B are connected in series with the emitter of the transistor. The resulting

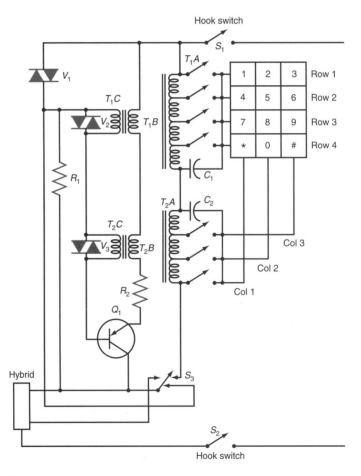

Fig. 15-8 DTMF oscillator.

two-tone signal is placed on the local loop where it will be received and detected by the central office. Some of the tone is also fed back through the hybrid to the receiver so that the calling party can hear it. The varistors V_2 and V_3 across windings T_1C and T_2C are used to automatically control the amplitude of the signal depending upon the amount of voltage on the local loop.

Electronic Telephones

When solid-state circuits came along in the late 1950s, an electronic telephone became possible and practical. Today, most new telephones are electronic, and they use integrated circuit technology.

The development of the microprocessor has also affected telephone design. Although simple electronic telephones do not contain a microprocessor, most multiple-line and full-featured telephones do. A built-in microprocessor permits automatic control of the telephone's functions and provides features such as telephone number storage and automatic dialing and redialing that are not possible in conventional telephones.

The variety of available electronic telephones is immense. The next section will give you an example of a sophisticated microprocessor-based electronic telephone.

Typical IC Electronic Telephone The major components of an electronic telephone circuit are shown in Fig. 15-9. Most of the functions are implemented with circuits contained within a single IC.

In Fig. 15-9, note that the TouchTone keypad drives a DTMF tone generator circuit. An external crystal or ceramic resonator provides an accurate frequency reference for generating the dual dialing tones.

The tone ringer is driven by the 20-Hz ringing signal from the phone line and drives a piezoelectric sound element.

The IC also contains a built-in line voltage regulator. It takes the dc voltage from the local loop and stabilizes it to provide a constant voltage to the internal electronic circuits. An external zener diode and transistor provide bias to the electret microphone.

The internal speech network contains a number of amplifiers and related circuits that fully duplicate the function of a hybrid in a standard telephone. This IC also contains a microcomputer interface. The box labeled MPU is a

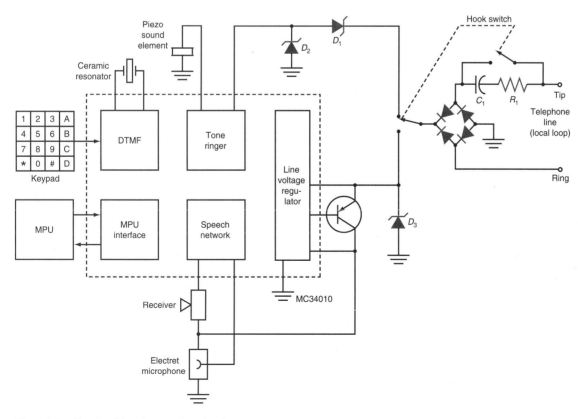

Fig. 15-9 Single-chip electronic telephone.

single-chip microprocessing unit. Although it is not necessary to use a microprocessor, if automatic dialing and other functions are implemented, this circuit is capable of accommodating them.

Finally, note the bridge rectifier and hook switch circuit. The twisted pair from the local loop is connected to the tip and ring connections. Both the 48-Vdc and 20-Hz ring voltages will be applied to this bridge rectifier. For direct current, the bridge rectifier provides polarity protection for the circuit, ensuring that the bridge output voltage is always a positive voltage. When the ac ringing voltage is applied, the bridge rectifier rectifies it into a pulsating dc voltage. The hook switch is shown with the telephone on the hook or in the "hung-up" position. Thus the dc voltage is not connected to the circuit at this time. However, the ac ringing voltage will be coupled through the resistor and capacitor to the bridge, where it will be rectified and applied to the two zener diodes D_1 and D_2 that drive the tone ringer circuit.

When the telephone is taken off the hook, the hook switch closes, providing a dc path around the resistor and capacitor R_1 and C_1. The path to the tone ringer is broken, and the output of the bridge rectifier is connected to zener diode D_3 and the line voltage regulator. Thus the circuits inside the IC are powered up and calls may be received or made.

Microprocessor Control   The more advanced electronic telephones also contain a built-in microcontroller. Like any microcontroller, it consists of the CPU, a ROM where a control program is stored, a small amount of random access read-write memory, and I/O circuits. The microcontroller, usually a single-chip IC, may be directly connected to the telephone IC, or some type of intermediate interface circuit may be used.

The function performed by the microcomputer is primarily that of storing telephone numbers and automatically redialing. Many advanced telephones have the capability of storing 12 commonly called numbers. The user puts the telephone into a program mode and uses the TouchTone keypad to enter the most frequently dialed numbers. These are stored in the microcontroller's RAM. To automatically dial one of the numbers, a pushbutton on the front of the telephone is depressed. This may be one of the TouchTone pushbuttons, or it

may be a separate set of pushbuttons provided for the purpose. When one of the pushbuttons is depressed, the microcontroller supplies a preprogrammed set of binary codes to the DTMF circuitry in the telephone IC. Thus the number is automatically dialed.

Line Interface   We have talked a great deal in general terms about how the telephone is connected to the local loop. Let's now examine the system more closely.

Most telephones are connected by way of a thin multiwire cable to a wall jack. A special connector on the cable, called an *RJ-11 modulator connector,* plugs into the matching wall jack. Two local loops are available if needed.

The wall jack is connected by wiring inside the walls to a central wiring point called the *subscriber interface.* Also known as the *wiring block* or *modular interface,* this is a small plastic housing containing all the wiring that connects the line from the telephone company to all the telephone wires in the house. Most modern houses and apartments are wired so that there is a wall jack in every room.

Figure 15-10 is a general diagram of the modular interface. The line from the telephone company usually passes through a protector which is for lightning protection. It then terminates at the interface box. An RJ-11 jack and plug are provided to connect to the rest of the wiring. This gives the telephone company a way to disconnect the incoming line from the rest of the house wiring to making testing and troubleshooting easier.

All the wiring is made by way of screw terminals. For a single-line house, the green and red tip and ring connections terminate at the terminals, and all wiring to the room wall jacks is connected in parallel at these terminals.

If a second line is installed, the black and yellow wires, which are the tip and ring connections, are also terminated at screw terminals. They are then connected to the inside house wiring.

Connections on the RJ-11 connector are shown in Fig. 15-11. The red and green wires terminate at the two center connections while the black and yellow wires terminate at the two outside connections. Most telephone wire and RJ-11 connectors have four wires and connections, but there are exceptions. With four wires a two-line phone can be accommodated.

RJ-11 modular connector

Subscriber interface

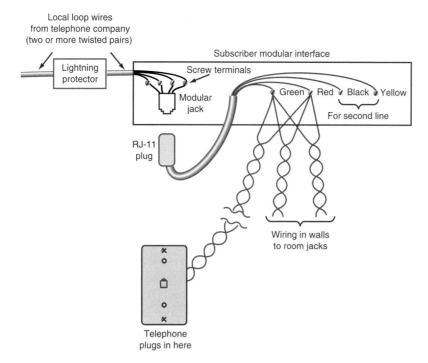

Local loop wires
from telephone company
(two or more twisted pairs)

Subscriber modular interface

Lightning
protector

Screw terminals

Modular
jack

Green Red Black Yellow

For second line

RJ-11
plug

Wiring in walls
to room jacks

Telephone
plugs in here

Fig. 15-10 Subscriber interface.

Cordless Telephones

Virtually all offices and most homes now have two or more telephones, most modern homes and apartments have a standard telephone jack in every room. This permits a single phone to be moved easily from one place to another, and it permits multiple (extension) phones. However, the ultimate convenience is a cordless telephone, which uses two-way radio transmission and provides total portability. Today, many homes have a cordless unit.

Cordless Telephone Concepts A cordless telephone is a full-duplex, two-way radio system made up of two units, the portable unit or handset and the base unit. The base unit is wired to the telephone line by way of a modular connector. It receives its power from the AC line. The base unit is a complete transceiver in that it contains a transmitter that sends the received audio signal to the portable unit and receives signals transmitted by the portable unit and retransmits them on the telephone line. It also contains a battery charger that rejuvenates the battery in the handheld unit.

The portable unit is also a full transceiver. It is battery powered and fully portable. This unit is designed to rest in the base unit where its battery can be recharged. Both units have an antenna.

The transceivers in both the portable and the base units use full-duplex operation. To achieve this, the transmitter and receiver must operate on different frequencies.

Figure 15-12 shows simplified block diagrams of the base and portable units of a typical cordless telephone. The base unit transmitter operates in the 43- to 46-MHz range and receives in the 49-MHz range. The portable unit has a receiver that operates in the 43- to 46-MHz range and a transmitter that operates in the 49-MHz range. These frequencies are far enough apart so that full simultaneous send and receive operations are possible at all times.

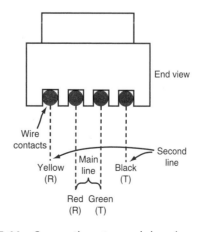

End view

Wire
contacts

Yellow
(R)

Main
line

Black
(T)

Second
line

Red Green
(R)   (T)

Fig. 15-11 Connections to modular plug.

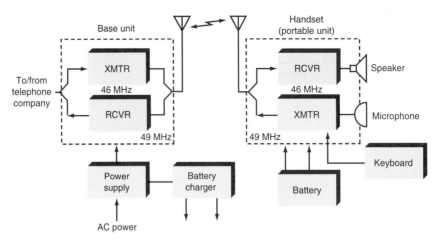

Fig. 15-12 Cordless telephone system.

Frequency Allocations The FCC has allocated 25 duplex channels for cordless telephone operation (see Fig. 15-13). Frequency modulation is used.

The transmitters in both the base station and the receiver are required to operate on the specific frequencies indicated in Fig. 15-13. Crystal control is used to set the transmit and

New cordless channels	
Base (TX)	**Handset (TX)**
43.72	48.76
43.74	48.84
43.82	48.86
43.84	48.92
43.92	49.02
43.96	49.08
44.12	49.10
44.16	49.16
44.18	49.20
44.20	49.24
44.32	49.28
44.36	49.36
44.40	49.40
44.46	49.46
44.48	49.50
Original cordless channels	
Base (TX)	**Handset (TX)**
46.61	49.67
46.63	49.845
46.67	49.86
46.71	49.77
46.73	49.875
46.77	49.83
46.83	49.89
46.87	49.93
46.93	49.99
46.97	49.97

Fig. 15-13 The FCC's 25-duplex channels for cordless analog telephones. All frequencies are in megahertz.

receive frequencies. Most cordless telephones operate on only one of the channels indicated. More sophisticated cordless telephones may have all 25 channels available so that the operating frequency can be changed if interference is occurring on one channel.

The FCC also regulates the transmitter power. There are rigid specifications regarding the maximum power output, which cannot exceed 500 mW. This requirement ensures that the transmitting distance is limited so that nearby cordless telephones will not interfere with one another. The maximum usable operating range is usually up to approximately 100 ft between the base station and portable unit. Operation over distances as great as 1000 ft has been achieved with some cordless telephones, although not typically. The maximum distance depends considerably upon the environment in which the telephone is used. The transmitting distance is extremely limited when the units are used in a concrete and steel building. Operation in a typical home is reliable at distances of about 100 ft. In most cases it is difficult to predict the actual operating range, and performance varies from one part of the house to another.

Despite the relatively limited distance over which a cordless telephone operates, there is still potential for interference with other nearby telephones. Under some conditions, you could receive signals from a cordless telephone at a neighbor's home, or your transmitter could interfere with the telephone usage of a neighbor. This problem is avoided in most modern telephones by the use of specific audio signaling tones that enable the transmitter and receiver circuitry in the telephone. If the correct tones are not transmitted and received between the portable and base units, the telephone will not operate. As a result, a cordless telephone will typically not interfere with a nearby unit, nor will the phone of a neighbor interfere with your own. Such tone signaling provides the privacy and security that you ordinarily expect from standard telephones.

Advanced Cordless Telephones Although cordless telephones work reliably, their range is limited, and signal quality is not quite up to standard telephone levels. Cordless telephones, particularly the portable units, are highly susceptible to all types of electrical noise and radio interference that may occur nearby and easily degrade the signal.

Because of this, new higher-quality cordless telephones have been developed. These phones use the *personal communications services (PCS) frequencies* of 902 to 928 MHz.

There are four basic types of advanced cordless telephones. The first type uses FM on channels in the 900-MHz range. This is still an analog phone, but the 900-MHz range allows greater power to be used and consequently longer-range transmission. Further, the 900-MHz signal is more "line-of-sight" in nature, which minimizes interference problems more common in the 46- to 48-MHz range.

The second type of cordless phone also operates in the 900-MHz range but is digital. The voice is converted to digital, and the resulting signal PSK modulates the carrier. An A/D conversion method known as *adaptive digital pulse code modulation (ADPCM)* is used. Digital phones have much greater clarity and are less sensitive to noise. They are more reliable over longer distances than their analog counterparts.

Another advanced cordless phone uses spread spectrum over the 902- to 928-MHz range. This phone has the best voice quality and is also secure because of the spread spectrum method. Many cordless phone users can operate simultaneously with no interference.

The fourth type uses spread spectrum in the 2.4-GHz range. It uses higher power for longer-distance transmission.

■ TEST

Answer the following questions.

1. Define specifically what is meant by the *local loop*.

2. What type of power supply is used to power a standard telephone, what are its specifications, and where is it located?
3. State the characteristics of the ringing signal supplied by the telephone company.
4. What is a hybrid?
5. True or false. Most telephone companies can still accommodate pulse dial telephones.
6. What type of transmitter (microphone) is used in a standard telephone, and how does it work?
7. Define what is meant by *tip and ring* and state what colors are used to represent them.
8. What is the name of the TouchTone dialing system?
9. What two-tone frequencies are generated when you press the star key on a TouchTone phone?
10. What is the name of the building or facility to which every telephone is connected?
11. What kind of microphone is used in an electronic telephone?
12. What is the purpose of the bridge rectifier circuit at the input to the connection of the telephone to the line to the telephone company?
13. Name one type of low-cost sounding device used to implement the bell or ringer in an electronic telephone.
14. True or false. In an electronic telephone, the hybrid is a special type of transformer.
15. Give two names or designations for the standard connector used on telephones.
16. State the two frequency ranges used by cordless telephones, and tell which is used for transmission and reception for the handheld and base units. How many channels are available?
17. Describe how one cordless telephone is prevented from interfering with another nearby cordless telephone.
18. State the basic specifications and benefits of the newer class of cordless telephone.

PCS frequencies

15-2 The Telephone System

Most of us take telephone service for granted, as we do other so-called utilities, for example, electric power. In the United States telephone

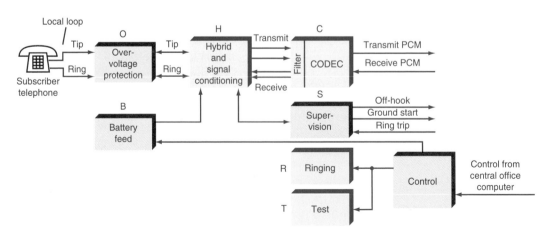
service is excellent. But this is certainly not the case in many other countries in the world.

When we refer to the "telephone system," we are talking about the organizations and facilities involved in connecting your telephone to the called telephone regardless of where it might be in the United States or anywhere else in the world. A number of different companies are involved in long-distance calls, although a single company is usually responsible for local calls in a given area. These companies make up the telephone system, and they design, build, maintain, and operate all the facilities and equipment used in providing universal telephone service. A vast array of equipment and technology is employed. Practically every conceivable type of electronic technology is used to implement worldwide telephone service.

The telephone, a small but relatively complex entity, is nothing compared to the massive system that backs it up. The telephone system can connect any two telephones in the world, and most people can only speculate on the method by which this connection takes place. It takes place on many levels and involves an incredible array of systems and technology. Obviously, it is difficult to describe such a massive system here. However, in this brief section, we will attempt to describe the technical complexities of interconnecting telephones, the central office and the subscriber line interface which connects each user to the telephone system, the hierarchy of interconnections within the telephone system, and the major elements and general operation of the telephone system. Long-distance systems are also discussed.

Subscriber Interface

Most telephones are connected to a local central office by way of the two-line, twisted-pair local loop cable. The central office contains all the equipment that operates the telephone and connects it to the "telephone system" that makes the connection to any other telephone.

Each telephone connected to the central office is provided with a group of basic circuits that power the telephone and provide all the basic functions such as ringing, dial tone, and dialing supervision. These circuits are collectively referred to as the *subscriber interface* or the *subscriber line interface circuits (SLIC)*. In older central office systems, the subscriber interface circuits used discrete components. Today, most functions of the subscriber line interface are implemented by one or perhaps two integrated circuits plus supporting equipment. The subscriber line interface is also referred to as the *line side interface*.

The SLIC provides seven basic functions generally referred to as *BORSCHT* (representing the first letters of the functions *battery, overvoltage protection, ringing, supervision, coding, hybrid,* and *test*). A general block diagram of the subscriber interface and BORSCHT functions is given in Fig. 15-14.

Battery The subscriber line interface at the central office must provide a dc voltage to the subscriber to operate the telephone. In the United States, this is typically -48 Vdc with respect to ground. The actual voltage can be anything between approximately -20 and -80 V when the phone is on the hook, that is, disconnected. The voltage at the telephone

Fig. 15-14 BORSCHT functions in the subscriber line interface at the central office.

drops to approximately 6 V when the phone is taken off the hook. The large differences between the on-hook and off-hook voltages has to do with the large voltage drop that occurs across the components in the telephone and the long local loop cable.

Overvoltage Protection   The circuits and components that protect the subscriber line interface circuits from electrical damage are referred to collectively as *overvoltage protection.* The phone lines are vulnerable to many types of electrical problems. Lightning is by far the worst threat, although other hazards exist, including accidental connection to an electrical power line or some type of misconnection that would occur during installation. Induced disturbances from other sources of noise can also cause problems. Overvoltage protection ensures reliable telephone operation even under such conditions.

Ringing   When a specific telephone is receiving a call, the telephone local office must provide a ringing signal. As indicated earlier, this is commonly a 90-V rms ac signal at approximately 20 Hz. The SLIC must connect the ringing signal to the local loop when a call is received. This is usually done by closing relay contacts that connect the ringing signal to the line. The SLIC must also detect when the phone is picked up (off hook) so that the ringing signal can be disconnected.

Supervision   *Supervision* refers to a group of functions within the subscriber line interface that monitor local loop conditions and provide various services. For example, the supervision circuits in the SLIC detect when a telephone is picked up to initiate a new call. A sensing circuit recognizes the off-hook condition and signals circuits within the SLIC to connect a dial tone. The caller then dials the desired number, which causes interconnection through the telephone system.

The supervision circuits continuously monitor the line during the telephone call. The circuits sense when the call is terminated and provide the connection of a busy signal if the called number is not available.

Coding   Coding is another name for A/D conversion and D/A conversion. Today, many telephone transmissions are made by way of serial digital data methods. The SLIC may contain codec that converts the analog voice signals into serial PCM format or convert received digital calls back into analog signals to be placed on the local loop. Transmission over trunk lines to other central offices or toll offices or for use in long-distance transmission is typically by digital PCM signals in modern systems.

Hybrid   Recall that in the telephone, a hybrid circuit (also known as a *two-wire* to *four-wire circuit*), usually a transformer, provides simultaneous two-way conversations on a single pair of wires. The hybrid combines the signal from the telephone transmitter with the received signal to the receiver on the single twisted pair. It keeps the signals separate within the telephone.

A hybrid is also used at the central office. It effectively translates the two-wire line to the subscriber back into four lines, two each for the transmitted and received signals. The hybrid provides separate transmit and receive signals. Although a single pair of lines is used in the local loop to the subscriber, all other connections to the telephone system treat the transmitted and received signals separately and have independent circuits for dealing with them along the way.

Test   In order to check the status and quality of subscriber lines, the phone company often puts special test tones on the local loop and receives resulting tones in return. These can give information about the overall performance of the local loop. The SLIC provides a way to connect the test signals to the local loop and to receive the resulting signals for measurement.

The basic BORSCHT functions are usually divided into two groups, high voltage and low voltage. The high-voltage parts of the system are the battery feed, the overvoltage protection, the ringing circuits, and the test circuits. The low-voltage group includes the supervision, coding, and hybrid functions. In older systems, all the functions were implemented with discrete component circuits. Today, these functions are generally divided between two ICs, one for the high-voltage functions and the other for the low-voltage functions.

The Telephone Hierarchy

Whenever you make a telephone call, your voice is connected through your local exchange to the telephone system. From there it passes through at least one other local exchange, which is connected to the telephone you are calling. Several other facilities may provide switching, multiplexing, and other services required to transmit your voice. The organization of this hierarchy in the United States is discussed in the next sections.

Central Office The central office or local exchange is the facility to which your telephone is directly connected by a twisted-pair cable. Also known as an *end office (EO),* the local exchange can serve up to 10,000 subscribers, each of whom are identified by a four-digit number from 0000 through 9999 (the last four digits of the telephone number).

The local exchange also has an exchange number. These are the three additional digits that make up a telephone number. Obviously, there can be as many as 1000 exchanges with numbers from 000 through 999. These exchanges become part of an area code region, which is defined by an additional three-digit number. Each area code is fully contained within one of the geographical areas assigned to one of the seven Bell regional operating companies.

Long-Distance Operation The United States is divided into seven telephone service regions. Service within each region is provided by one of the Regional Bell Operating Companies (RBOCs) that belong to one of the Regional Bell Holding Companies (RBHCs). Each of the RBHCs is typically divided into smaller operating companies serving one portion of the designated area. In all, there are 22 Bell operating companies that make up the seven large RBHC corporations.

In addition to the Bell operating companies, there are many other independent telephone companies. One of the largest is General Telephone and Electronics (GTE), which provides telephone service primarily in limited geographical areas around the country where the Bell operating companies do not serve. GTE and other smaller companies often provide telephone service in remote rural areas. All these Bell operating companies and any independent companies are referred to as *local exchange carriers* or *local exchange companies (LECs).*

The LECs provide telephone services to designated geographical areas referred to as *local access and transport areas (LATAs).* The United States is divided into approximately 200 LATAs. The LATAs are defined within individual states making up the seven operating regions. The LECs provide the telephone service for the LATAs within their regions but do not provide long-distance service for the LATAs.

Long-distance service is provided by long-distance carriers known as *interexchange carriers (IXCs).* The IXCs are the familiar long-distance carriers such as AT&T, MCI, and US Sprint. Long-distance carriers must be used for the interconnection for any inter-LATA connections. The LECs can provide telephone service within the LATAs that are part of their operating region, but links between LATAs within a region, even though they may be directly adjacent to one another, must be made through an IXC.

Each LATA contains a *serving,* or *point of presence (POP), office* that is used to provide the interconnections to the IXCs. The local exchanges communicate with one another via individual trunks. And all local exchanges connect to an LEC central office, which provides trunks to the POP. It is at the POP where the long-distance carriers can make their interface connections. The POPs must provide equal access for any long-distance carrier desiring to connect. Many POPs are connected to multiple IXCs, but in many areas, only one IXC serves a POP.

Figure 15-15 summarizes the hierarchy just discussed. Individual telephones within a LATA connect to the local exchange or central office by way of the two-wire local loop. The central offices within an LATA are connected to one another by trunks. These trunks may be standard baseband twisted-pair cables run underground or on telephone poles but they may also be coaxial cable, fiber-optic cable, or microwave radio links. In some areas, two or more central offices are located in the same building or physical facility. Trunk interconnections are usually made by cables.

The local exchanges are also connected to an LEC central office when a connection cannot be made between two local exchanges that are not directly trunked. The call passes from the local exchange to the LEC central office,

LATA

IXC

POP

LEC

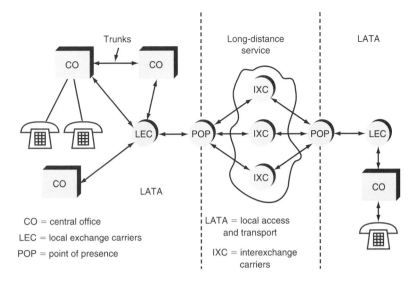

Fig. 15-15 Organization of the telephone system in the United States.

CO = central office
LEC = local exchange carriers
POP = point of presence

LATA = local access and transport
IXC = interexchange carriers

where the connection is made to the other local exchange.

The LEC central office is also connected to the POP. Depending upon the organization of the LEC within the LATA, the LEC central office may contain the POP.

Note in Fig. 15-15 that the POP provides the connections to the long-distance carriers, or IXCs. (The IXCs are the interexchange carriers.) The long-distance network connects to the remote POPs, which in turn are connected to other central offices and local exchanges.

Today, AT&T has implemented what it calls *dynamic nonhierarchical routing (DNHR)*. In most cases, the types of switching offices and trunk paths available make it possible to establish a long-distance connection with two or fewer switching centers.

Most other long-distance carriers have their own specific hierarchical arrangements. A variety of switching offices across the country are linked by trunks using fiber-optic cable or microwave relay links. Multiplexing techniques are used throughout to provide many simultaneous paths for telephone calls.

In all cases, the various central offices and routing centers provide switching services. The whole idea is to permit any one telephone to directly connect with any other specific telephone. The purpose of all the different levels in the telephone system hierarchy is to provide the interconnecting trunk lines as well as switching equipment that makes the desired interconnection.

DNHR

15-3 Facsimile

Facsimile, or *fax,* is an electronic system for transmitting graphical information by wire or radio. Facsimile is used to send printed material by scanning it and converting it into electronic signals that modulate a carrier to be transmitted over the telephone lines. Since modulation is involved, fax transmission can also take place by radio. With facsimile, documents such as letters, photographs, line drawings, or any printed information can be converted into an electrical signal and transmitted with conventional communications techniques. The components of a fax system are illustrated in Fig. 15-16.

Although facsimile is used to transmit pictures, it is not TV because it does not transmit sound messages or live scenes and motion. However, it does use scanning techniques that

Fax

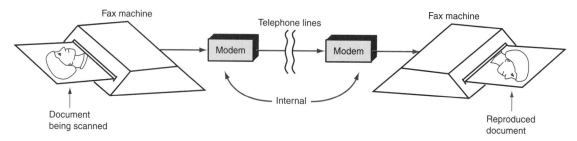

Fig. 15-16 Components of a facsimile system.

are in some way generally similar to those used in TV. A scanning process is used to break a printed document up into many horizontal scan lines which can be transmitted and reproduced serially.

History of Facsimile

Like most other communications technologies, facsimile is not new. It was invented in 1842 by a Scottish inventor named Alexander Bain. Later, the original process was improved in England in the 1850s by Fredrick Bakewell. A German scientist, Arthur Korn, incorporated photoelectric scanning into the fax process in 1907, which made fax truly practical. In the 1920s, fax development was expedited, and commercial products were developed. The first radio fax transmissions were made in the 1930s.

The earliest users of fax were the newspapers and wire service companies which transmitted news information and photos by fax. It was also used by telegram companies, the government, and military, and in many cases, commercial business organizations. Today fax is widely used to transmit weather photos from satellites to ground stations.

The fax machine has long been recognized as a major communications convenience. But its cost remained high until the 1970s, when major breakthroughs were made in the development of

semiconductor devices such as photosensors and microprocessors. The development of data compression techniques further improved transmission times. Today, the fax machine is just about as common in offices as the telephone and the personal computer.

How Facsimile Works

The early facsimile machines were electro-mechanical devices consisting of a rotary scanning drum and some form of printing mechanism. The typical facsimile machine was capable of both sending and receiving. Figure 15-17 shows the basic scanning mechanism. The document to be transmitted was wrapped around and tightly affixed to the drum. A motor rotated the drum through a gear train mechanism. The motor was designed to operate at an accurate speed so that the scanning rate was precise. A precision tuning fork oscillator was used to control the scanning drum motor for precise speed value.

Scanning of the document was done with a light and photocell arrangement. A lead screw driven by the gear train moved a scanning head consisting of a light source and a photocell. An incandescent light source, focused to a tiny

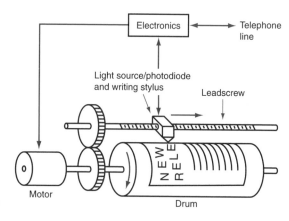

Fig. 15-17 Drum scanner used in early fax machines.

point with a lens system, was used to scan the document. The lens was also used to focus the reflected light from on the document onto the photocell. As the light scanned the letters and numbers in a typed or printed document or the gray scale in a photograph, the photocell produced a varying electronic signal whose output amplitude was proportional to the amount of reflected light. This baseband signal was then used to amplitude or frequency-modulate a carrier in the audio frequency range. This permitted the signal to be transmitted over the telephone lines.

Figure 15-18 shows how a printed letter might have been scanned. Assume the letter F is black on a white background. The output of a photodetector as it scans across line *a* is shown in Fig. 15-18(*a*). The output voltage is high for white and low for black. The output of the photodetector is also shown for scan lines *b* and *c*. The output of the photodetector is used to modulate a carrier, and the resulting signal is put on the telephone line.

The resolution of the transmission is determined by the number of scan lines per vertical

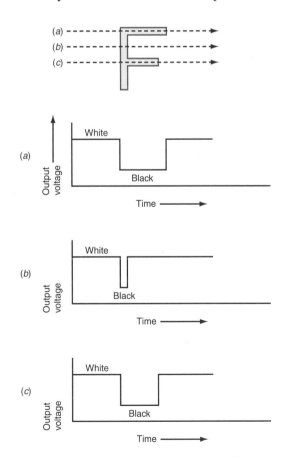

Fig. 15-18 Output of a photosensitive detector during different scans.

inch. The greater the number of lines scanned, the finer the detail transmitted and the higher the quality of reproduction. Older systems had a resolution of 96 lines per inch (LPI), and the new systems use 200 LPI.

On the receiving end, a demodulator recovered the original signal information, which was then applied to a stylus. The purpose of the stylus was to redraw the original information on a blank sheet of paper. A typical stylus converted the electrical signal into heat variations that burned the image into heat-sensitive paper. Other types of printing mechanisms were also used.

In the printing process, a blank sheet of paper was wrapped around a cylindrical drum and rotated at the same speed as the transmitting drum. The drum rotation was precisely controlled by a precision tuning fork oscillator. The drum rotations in both the transmitter and receiver were synchronized so that the information would be reproduced simultaneously with transmission. Because of the frequency spectrum limitations of the telephone lines and radio channels, scanning speed was slow to ensure the production signals that were within the available bandwidth. Transmission times of many minutes per page were typical.

Today's modern fax machine is a high-tech electro-optical machine. Scanning is done electronically, and the scanned signal is converted into a binary signal. Then digital transmission with standard modem techniques is used.

Figure 15-19 is a block diagram of a modern fax machine. The transmission process begins with an image scanner that converts the document into hundreds of horizontal scan lines. Many different techniques are used, but they all incorporate a photo- (light-) sensitive device to convert light variations along one scanned line into an electrical voltage. The resulting signal is then processed in various ways to compress the data for quicker transmission. The resulting signal is sent to a modem where it modulates a carrier set to the middle of the telephone voice spectrum bandwidth. The signal is then transmitted to the receiving fax machine over the public-switched telephone network.

The receiving fax machine's modem demodulates the signal that is then processed to recover the original data. The data is decompressed and then sent to a printer, which

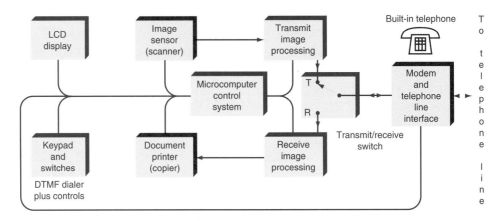

Fig. 15-19 Block diagram of modem fax machine.

Transceiver

reproduces the document. Since all fax machines can transmit as well as receive, they are referred to as *transceivers*. The transmission is half duplex because only one machine may transmit or receive at a time.

Most fax machines have a built-in telephone, and the printer can also be used as a copy machine. An embedded microcomputer handles all control and operation including paper handling.

Image Processing

CCD

Most fax machines use *charged coupled devices (CCDs)* for scanning. A CCD is a light-sensitive semiconductor device that converts varying light amplitudes into an electrical signal. The typical CCD device is made up of many tiny reverse-biased diodes that act like capacitors which are manufactured in a matrix on a silicon chip (see Fig. 15-20). The base forms one large plate of a capacitor which is electrically separated by a dielectric from many thousands of tiny capacitor plates as shown. When the CCD is exposed to light, the CCD capacitors charge to a value proportional to the light intensity. The capacitors are then scanned or sampled electronically to determine their charge. This creates an analog output signal that accurately depicts the image focused on the CCD.

A CCD is actually a device that breaks up any scene or picture into *individual picture elements,* or *pixels.* The greater the number of CCD capacitors, or pixels, the higher the resolution and the more faithfully a scene, photograph, or document may be reproduced. CCD devices are available with a matrix of

Pixel

many thousands of pixels, thereby permitting very high resolution picture transmission. CCDs are widely used in modern video cameras in place of the more delicate and more expensive vidicon tubes. In the video camera (camcorder), the lens focuses the entire scene on a CCD matrix. This same approach is used in some fax machines. In one type of fax machine, the document to be transmitted is placed face down as it might be in a copy machine. The document is then illuminated with brilliant light from a xenon or fluorescent bulb. A lens system focuses the reflected light on a CCD. The CCD is then scanned, and the resulting output is an analog signal whose amplitude is proportional to the amplitude of the reflected light.

In most desktop fax machines, the entire document is not focused on a single CCD. Instead, only a narrow portion of the document is lighted and examined as it is moved through the fax machine with rollers. A complex system of mirrors is used to focus the lighted area on the CCD (see Fig. 15-21).

The more modern fax machines use another type of scanning mechanism that does not use lenses. The scanning mechanism is an assembly made up of an LED array and a CCD array. These are arranged so that the entire width of a standard $8 \frac{1}{2} \times 11$ in. page is scanned simultaneously one line at a time. The LED array illuminates a narrow portion of the document. The reflected light is picked up by the CCD scanner. A typical scanner has 2048 light sensors forming one scan line. Figure 15-22 shows a side view of the scanning mechanism. The 2048 pixels of light are converted into voltages proportional to the light variations on

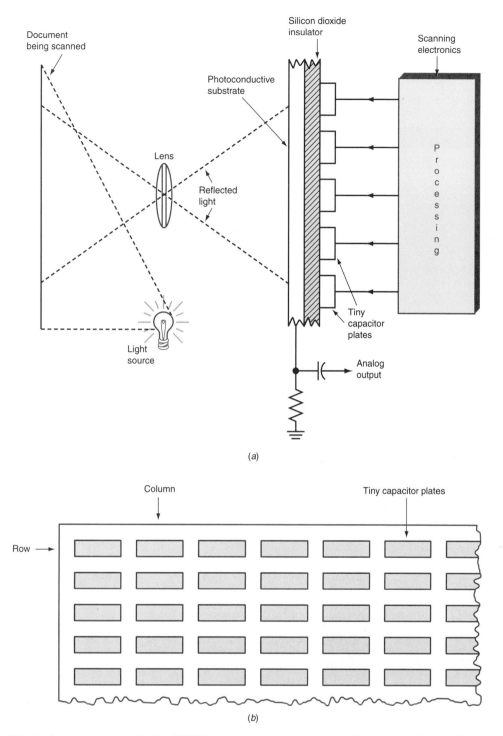

Fig. 15-20 A charge coupled device (CCD) is used to scan documents in modern fax machines. (a) Cross section. (b) Detail of capacitor matrix.

one scanned line. These voltages are converted from a parallel format to a serial voltage signal. The resulting analog signal is amplified and sent to an AGC circuit and an S/H amplifier. The signal is then sent to an A/D converter where the light signals are translated into binary data words for transmission.

Data Compression

An enormous amount of data is generated by scanning one page of a document. A typical $8\frac{1}{2} \times 11$ in. page represents about 40,000 bytes of data. This can be shortened by a factor of 10 or more with data compression techniques. Furthermore, because of the narrow bandwidth

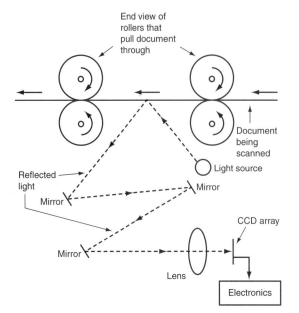

Fig. 15-21 Scanning mechanism in a fax machine.

Data compression is a digital data processing technique that looks for redundancy in the transmitted signal. White space or continuous segments of the page that are the same shade produce continuous strings of data words that are the same. These can be eliminated and transmitted as a special digital code that is significantly faster to transmit. Other forms of data compression use various mathematical algorithms to reduce the amount of data to be transmitted.

The data compression is carried out by a *digital signal processing (DSP) chip*. This DSP chip is a super high speed microprocessor with embedded ROM containing the compression program. The digital data from the A/D converter is passed through the DSP chip, from which comes a significantly shorter string of data that represents the scanned image. This is what is transmitted, and in far less time than the original data could be transmitted.

At the receiving end, the demodulated signal is decompressed. Again, this is done through a DSP chip especially programmed for this function. The original data signal is recovered and sent to the printer.

Modems

Every fax machine contains a built-in modem that is similar to a conventional data modem

of telephone lines, data rates are also limited. That is why it takes so long to transmit one page of data. Developments in high-speed modems have helped reduce the transmission time, but the most important developments are data compression techniques that reduce the overall amount of data, which significantly decreases the transmission time and telephone charges.

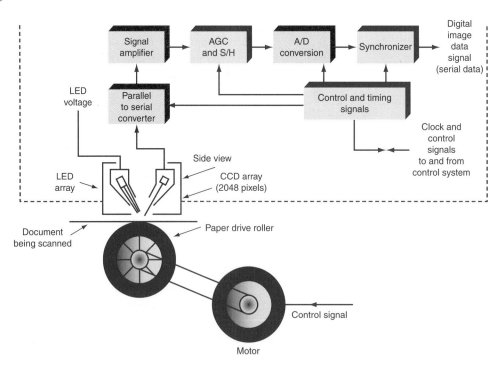

Fig. 15-22 LED/CCD scanner mechanism in a modern fax machine.

for computers. These modems are optimized for fax transmission and reception. And they follow international standards so that any fax machine can communicate with any other fax machine.

A number of different modulation schemes are used in fax systems. Analog fax systems use AM or FM. Digital fax uses PSK or QAM. To ensure compatibility between fax machines of different manufacturers, standards have been developed for speed, modulation methods, and resolution by the *International Telegraph and Telephone Consultative Committee,* better known by its French abbreviation *CCITT.* The CCITT is now known as the *ITU-T,* or *International Telecommunications Union.* The ITU-T fax standards are divided into four groups:

1. *Group 1 (G1 or GI):* Analog transmission using frequency modulation where white is 1300 Hz and black is 2100 Hz. Most North American equipment uses 1500 Hz for white and 2300 Hz for black. The scanning resolution is 96 lines per inch (LPI). Average transmission speed is 6 minutes per page ($8^1/_2 \times 11$ in., or A4 metric, size which is slightly longer than 11 in.).
2. *Group 2 (G2 or GII):* Analog transmission using FM or vestigial sideband AM. The vestigial sideband AM uses a 2100-Hz carrier. The lower sideband plus part of the upper sideband are transmitted. Resolution is 96 LPI. Transmission speed is 3 min or less for an $8^1/_2 \times 11$ in. or A4 page.
3. *Group 3 (G3 or GIII):* Digital transmission using PCM black and white only or up to 32 shades of gray. PSK or QAM to achieve transmission speeds of up to 9600 baud. Resolution 200 LPI. Transmission speed of less than 1 minute per page with 15 to 30 s being typical.
4. *Group 4 (G4 or GIV):* Digital transmission, 56 kbits/s, resolution up to 400 LPI, and speed of transmission less than 5 s.

The older G1 and G2 machines are no longer used. The most common configuration is Group 3. Most G3 machines can also read the G2 format.

G4 machines are not yet widely used. They are designed to use digital transmission only with no modem over very wide band dedicated digital-grade telephone lines. G4 machines will become popular when the new integrated services digital network (ISDN) telephone system goes into use in the future. Both G3 and G4 formats also employ digital data compression methods that shorten the binary data stream considerably, thereby speeding up page transmission. This is important because shorter transmission times cut long-distance telephone charges and reduce operating costs.

CCITT

Fax Machine Operation

Figure 15-23 is a block diagram of the transmitting circuits in a modern G3 fax transceiver. The analog output from the CCD array is serialized and fed to an A/D converter which translates the continuously varying light intensity into a stream of binary numbers. Sixteen gray scale values between white and black are typical. The binary data is sent to a DSP digital data compression circuit. The binary output in serial data format modulates a carrier which is transmitted over the telephone lines. The techniques are similar to those employed in modems. Speeds of 2400/4800 and 7200/9600 baud are common. Most systems use some form of PSK or QAM to achieve very high data rates on voice-grade lines.

In the receiving portion of the fax machine, the received signal is demodulated and then sent to DSP circuits where the data compression is removed and the binary signals are restored to their original form. The signal is then applied to a printing mechanism. In most older fax machines, thermal printers are used. Thermal printers use a special heat-sensitive paper. The *printhead,* or *writing stylus* as it is sometimes called, contains tiny heating elements that are turned off and on by the received signal. A typical printhead has 2048 thick-film resistors that are heated individually to reproduce the transmitted pixels. The heat darkens the paper at the appropriate point to re-create the original image. Some newer thermal printers use a heating element printhead to melt ink on a special ribbon onto plain 20-lb bond paper.

New fax machines use an ink-jet printer or a laser printer. In these machines, laser scanning of an electrosensitive drum, similar to the drum

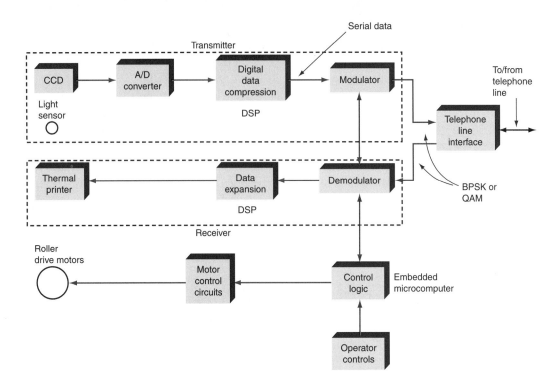

Fig. 15-23 Block diagram of a facsimile machine.

used in laser printers, produces output copies using the proven techniques of xerography.

The control logic in Fig. 15-23 is usually an embedded microcomputer. Besides all of the internal control functions it implements, it is used for "handshaking" between the two machines that will communicate. This ensures compatibility. Handshaking is usually carried out by exchanging different audio tones. The called machine responds with tones designating its capability. The calling machine compares this to its own standards and then either initiates the transmission or terminates it because of incompatibility. If the transmission proceeds, the calling machine sends synchronizing signals to ensure that both machines start at the same time. The called machine acknowledges the receipt of the sync signal, and transmission begins. All the protocols for establishing communications and sending and receiving the data are standardized by the ITU-T. Transmission is half duplex.

As improvements have been made in picture resolution quality, transmission speed, and cost, facsimile machines have become much more popular. The units can be easily attached with standard RJ-11 modular connectors to any telephone system. In most business applications, the fax machine is typically dedicated to

a single line. Most fax machines feature fully automatic operation with microprocessor-based control. A document can be sent to a fax machine automatically. The sending machine simply dials the receiving machine and initiates the transmission. The receiving machine answers the initial call and then reproduces the document before hanging up.

Most fax machines have a built-in telephone and are designed to share a single line with conventional voice transmission. The built-in telephone usually features TouchTone dialing and number memory plus automatic redial and other modern telephone features. Most fax machines also have automatic send and receive features for fully unattended operation. Smaller portable fax machines are available. Fax machines may also be used with standard cellular telephone systems in automobiles.

Another popular variation is the *fax modem,* an internal modem designed to be plugged into a personal computer. It allows standard modem operation for connection to online services and any desired remote computer usage. However, this device also contains the circuits for a fax modem but without the image scanning and printing.

A document produced with a word processor is usually stored in ASCII format as a file

Fax modem

on disk. This file can be transmitted serially to the fax modem, which compresses it and sends it over the telephone lines. The receiving computer demodulates the signal, decompresses it, and stores it in RAM and then on disk. The resulting document can be read on the video monitor or printed by the usual means.

■ TEST

Answer the following questions.

23. True or false. Fax can transmit photos and drawings as well as printed text.
24. What is the most common transmission medium for fax signals? What other medium is commonly used?
25. True or false. Facsimile was invented before radio.
26. Who sets the standards for fax transmission?
27. Vestigial sideband AM is used in what group type fax machines?
28. What is the name of the semiconductor photosensitive device used in most modern fax machines to convert a scanned line into an analog signal?
29. What is the group designation given to most modern fax machines?
30. To ensure compatibility between sending and receiving fax machines, the control logic carries out a procedure using audio tones to establish communications. What is this process called?
31. What circuit in the fax machine makes the fax signal compatible with the telephone line?
32. What is the upper speed limit of a G3 fax machine over the telephone lines?
33. What is the resolution of a G3 fax machine in lines per inch?
34. Explain the process and hardware used to convert images to be transmitted into electrical signals in a fax machine.
35. Describe two methods of scanning used in modern fax machines.
36. What is the most commonly used type of printer in a fax machine? How does it work?
37. Describe how fax signals are processed to speed up transmission.
38. Fax transmission is usually
    a. Full duplex.
    b. Half duplex.

39. Fax signals representing the image to be transmitted before they are prepared for the telephone lines are
    a. Analog.
    b. Digital.
40. True or false. Group 4 transmissions do not use the standard telephone lines. What are the speed and resolution of Group 4 fax transmissions?

15-4 Cellular Telephone Systems

A *cellular radio system* provides standard telephone operation by full-duplex two-way radio at remote locations. Cellular radios or telephones can be installed in cars or trucks and are also available in handheld models. Each cellular telephone permits the user to link up with the standard telephone system that permits calls to any part of the world.

Cellular radio system

The Telephone Company division of AT&T developed the cellular radio system during the late 1970s and fully implemented it during the early 1980s. Today, cellular radio telephone service is available nationwide in most medium-size to large cities and along most major highways. The original system was known as the *advanced mobile phone service,* or *AMPS.* Although this analog system is still widely used, it is rapidly being replaced by improved digital systems. Both types are covered here.

AMPS

Cellular Concepts

The basic concept behind the cellular radio system is that rather than serving a given geographical area with a single transmitter and receiver, the system divides the service area into many smaller areas known as *cells,* as shown in Fig. 15-24. The typical cell covers only several square miles and contains its own receiver and low-power transmitter. The cell site is designed to reliably serve only vehicles in its small cell area.

About ⟸ Electronics

Low-earth-orbit (LEO) satellite telecommunications systems have the potential for increased coverage, compared to cellular networks.

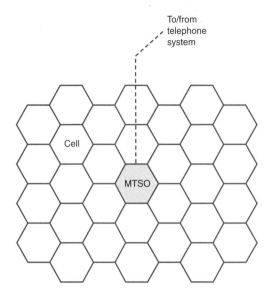

To/from telephone system

Cell

MTSO

Fig. 15-24 The area served by a cellular telephone system is divided into small areas called *cells. Note:* Cells are shown as ideal hexagons, but in reality they will have circular or other geometric shapes. These areas may overlap, and the cells may be of different sizes.

MTSO

Each cell is connected by telephone lines or a microwave radio relay link to a master control center known as the *mobile telephone switching office (MTSO).* The MTSO controls all the cells and provides the interface between each cell and the main telephone office. As the vehicle containing the telephone passes through a cell, it is served by the cell transceiver. The telephone call is routed through the MTSO and to the standard telephone system. As the vehicle moves, the system automatically switches from one cell to the next. The receiver in each cell station continuously monitors the signal strength of the mobile unit. When the signal strength drops below a desired level, it automatically seeks a cell where the signal from the mobile unit is stronger. The computer at the MTSO causes the transmission from the vehicle to be switched from the weaker cell to the stronger cell. This is called a *handoff.* All of this takes place in a very short period of time and is completely unnoticeable to the user. The result is that optimum transmission and reception are obtained.

The cellular system operates in the 800- to 900-MHz range, previously reserved for the higher UHF TV channels 68 through 83, which were rarely used. Originally there were 666 30-kHz-wide full-duplex channels available for

communications. Today, 832 channels are used. The cellular system also uses what is known as *frequency reuse,* which allows cells within the system to use the same frequency channel. Because the cells are physically small, and low-power transmitters are used, and the cell sites use directional antennas, the signal does not stray beyond the cell boundaries. This allows other cells within the system to share the same frequency channel without interference. Frequency reuse tremendously increases the number of available channels. Obviously, to prevent interference, adjacent cells are not permitted to use the same channel.

Another feature of the cellular system is that different cell sizes can be accommodated. In low-usage areas, the cells can be large. As the number of users increases, the large cells can be divided into smaller cells. This provides ease of expansion as the number of users grows. A typical system can serve up to about 50,000 subscribers, and 10,000 of them may use the system simultaneously.

The newer digital systems have even greater capacity. Some of these systems operate in the 1.7- 1.8-GHz bands.

A Cellular Telephone Unit

Figure 15-25 is a general block diagram of a cellular mobile radio unit. This applies to either

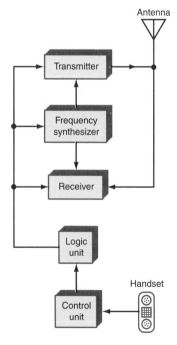

Fig. 15-25 General block diagram of a cellular radio.

the larger units designed to be mounted in a vehicle or the smaller handheld devices. The two types of units are identical in operation, but the larger units used in vehicles have a higher power transmitter and are inherently larger.

The unit consists of five major sections: transmitter, receiver, synthesizer, logic unit, and control unit. Mobile radios derive their operating power from the car battery. Portable units contain built-in rechargeable batteries. The transmitter and receiver share a single antenna. The sections are discussed below.

Transmitter The transmitter block diagram is shown in Fig. 15-26. It is a low-power FM unit operating in the frequency range of 825 to 845 MHz. There are 666 30-kHz transmit channels. Channel 1 is 825.03 MHz, channel 2 is 825.06 MHz, and so on up to channel 666 on 844.98 MHz. The carrier furnished by a frequency synthesizer is a phase modulated by the voice signal. The phase modulator produces a deviation of ±12 kHz. Preemphasis is used to help minimize noise. The modulator output is translated up to the final transmitter frequency by a mixer whose second input also comes from the frequency synthesizer. The mixer output is fed to class C power amplifier stages where the output signal is developed. The final amplifier stage is designed to supply 3 W to the antenna of a vehicle-mounted unit but only about 500 mW in a handheld unit.

A unique feature of the higher-power transmitter is that its output power is controllable by the cell site and MTSO. Special control signals picked up by the receiver are sent to an *automatic power control (APC) circuit* that sets the transmitter to one of 8 power output levels. The APC circuit can introduce power attenuation in steps of 4 dB from 0 dB (3 W) to 28 dB (4.75 mW). This is done by controlling the supply voltage to one of the intermediate-power amplifier stages.

The output power of the transmitter is monitored internally by built-in circuits. A microstrip directional coupler taps off an accurate sample of the transmitter output power and rectifies it into a proportional dc signal. This signal is used in the APC circuit and is transmitted back to the cell site permitting the MTSO to know the present power level.

This automatic power control feature permits optimum cell site reception with minimal power. It also helps to minimize interference from other stations in the same or adjacent cells.

The transmitter output is fed to a duplexer circuit or isolator that allows the transmitter and receiver to share the same antenna. Since cellular telephone units use full-duplex operation, the transmitter and receiver will operate simultaneously. The transmit and receive frequencies are spaced 45 MHz apart to minimize interference. However, an isolator is still needed to keep transmitter power out of the sensitive receiver. The duplexer consists of two very sharp bandpass filters, one for the transmitter and one for the receiver. The transmitter output passes through this filter to the antenna.

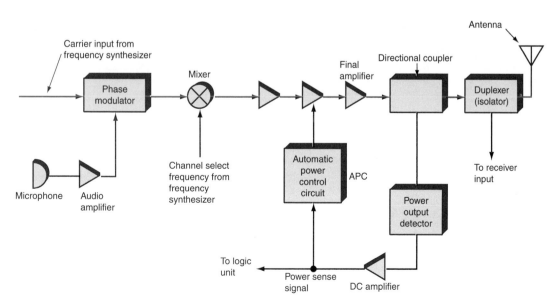

Fig. 15-26 Cellular transmitter.

Receiver The receiver is a dual-conversion superheterodyne (refer to Fig. 15-27). An RF amplifier boosts the level of the received cell site signal. The receiver frequency range is 870.03 to 889.98 MHz. There are 666 receive channels spaced 30 kHz apart. The first mixer translates the incoming signal down to a first IF of 82.2 MHz. Some receivers use a 45-MHz first IF. The local oscillator signal for the mixer is derived from the frequency synthesizer. The local oscillator frequency sets the receive channel. The signal passes through IF amplifiers and filters to the second mixer, which is driven by a crystal controlled local oscillator. The second IF is usually either 10.7 MHz or 455 kHz. The signal is then demodulated, de-emphasized, filtered, and amplified before being applied to the output speaker in the handset.

The output of the demodulator is also fed to other filter circuits that select out the control audio tones and digital control data stream sent by the cell site to set and control both the transmitter and the receiver. The demodulator output is also filtered into a DC level whose amplitude is proportional to the strength of the received signal. This is the *receive signal strength indicator (RSSI) signal* that is sent back to the cell site so that the MTSO can monitor the received signal from the cell and make decisions about switching to another cell.

Frequency Synthesizer The frequency synthesizer section develops all the signals used by

the transmitter and receiver (see Fig. 15-28). It uses standard PLL circuits and a mixer. A crystal controlled oscillator provides the reference for the PLLs. One PLL incorporates a VCO (number 2) whose output frequency is used as the local oscillator for the first mixer in the receiver. This signal is mixed with the output of a second PLL VCO to derive the transmitter output frequency.

As in other PLL circuits, the output VCO frequency is determined by the frequency division ratio of the divider in the feedback path between the VCO and the phase detector. In a cellular radio, this frequency division ratio is supplied by the MTSO via the cell site. When a mobile unit initiates or is to receive a call, the MTSO computer selects an unused channel. It then transmits a digitally coded signal to the receiver containing the frequency division ratios for the transmitter and receiver PLLs. This sets the transmit and receive channel frequencies.

Logic Unit The logic unit shown in Fig. 15-29 contains the master control circuitry for the cellular radio. It is made up of a microprocessor with both RAM and ROM plus additional circuitry used for interpreting signals from the MTSO/cell site and generating control signals for the transmitter and receiver.

All cellular radios contain a *programmable read-only memory (PROM) chip* called the *number assignment module (NAM)*. The NAM

NAM

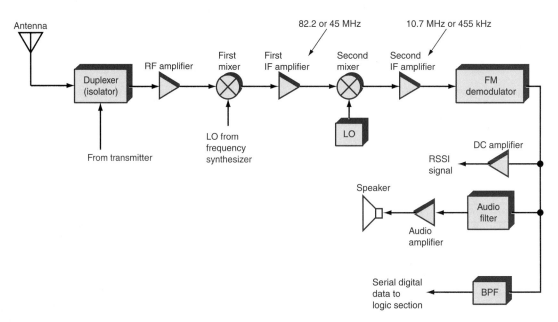

Fig. 15-27 Cellular receiver.

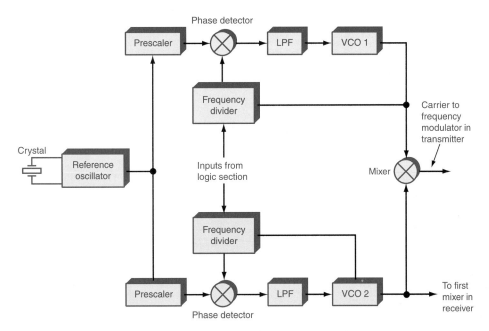

Fig. 15-28 Frequency synthesizer.

contains the *mobile identification number (MIN)*, which is the telephone number assigned to the unit. The NAM PROM is "burned" when the cellular radio is purchased and the MIN assigned. This chip allows the radio to identify itself when a call is initiated or when the radio is interrogated by the MTSO.

All cellular mobile radios are fully under the control of the MTSO through the cell site. The MTSO sends a serial digital data stream at 10K bps through the cell site to the radio to control the transmit and receive frequencies and transmitter power. The MTSO monitors the received cell signal strength at the cellular

MIN

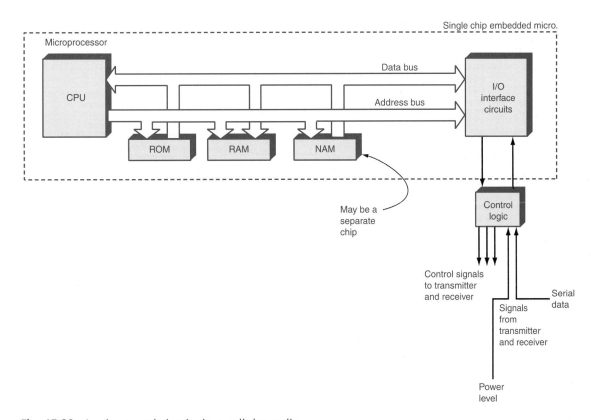

Fig. 15-29 Logic control circuits in a cellular radio.

radio by way of the RSSI signal, and it monitors transmitter power level. These are transmitted back to the cell site and MTSO. Audio tones are also used for signaling purposes.

Control Unit The control unit contains the handset with speaker and microphone. This may be a standard handset as used in a regular telephone on a mobile unit. However, these circuits are built into the handheld units. The main control unit contains a complete TouchTone dialing circuit (see Fig. 15-30). The control unit is operated by a separate microprocessor that drives the LCD display and other indicators. It also implements all manual control functions. The microprocessor memory permits storage of often-called numbers and an auto-dial feature.

Operational Procedure

Described below is the sequence of operations that occur when a person initiates a cellular telephone call:

1. The operator applies power to the unit which turns on the transmitter and receiver.
2. The receiver seeks an open control channel. Twenty-one control or paging channels are used to establish initial contact with a cell and the MTSO. When contact is made, the cell site reads the NAM data and the MTSO computer verifies that it is a valid number.
3. The operator enters the number to be called via a keyboard.
4. The operator sends the telephone number to be called by pressing a send or call button.
5. The cell and MTSO search for an open channel and send the frequency data to a cellular transceiver.
6. The RSSI signal is read to determine the optimum cell selection. The transmitter power is adjusted.
7. Handshake signals are exchanged, signifying that contact has been established.
8. The MTSO calls the designated number. Conversation takes place.
9. If the mobile unit passes from one cell to another, the MTSO senses the RSSI signal and "hands off" the mobile signal from one cell to another to maintain maximum signal strength.
10. The call is terminated.

When a mobile unit is to receive a call, the MTSO and cell site transmit a call signal containing the MIN over a control channel. The transceiver monitors the control channels, iden-

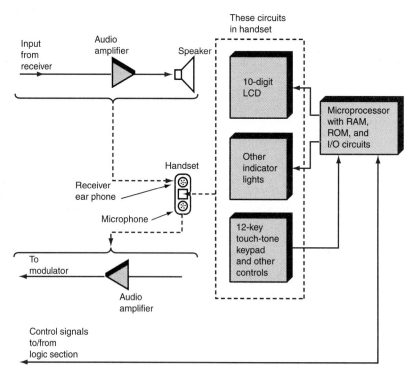

Fig. 15-30 Control unit with handset.

tifies its MIN, and turns on. From that point on, the sequence beginning at step 5 above is similar.

Keep in mind that although most cellular telephones carry voice communications, like other telephones they can also carry data. Special modems are available that permit a personal computer to be connected to a cellular telephone for online communications. Even a fax machine can be connected to a cellular telephone.

Digital Cellular Telephone Systems

The cellular telephone system has been far more successful than ever expected. When cellular service was first started during the mid-1980s, few expected it to be as commonly used as it is. Industry analysts predicted that fewer than 1 million cellular telephones would be in use by 1995. However, by 1998, that figure was exceeded by a factor of 20 or more.

Thanks to rapid technological developments, the cost and size of cellular telephones have dropped considerably, from as high as $3000 per unit in the early days to less than $100 for a simple unit today. However, the costs of cellular service have remained high and in fact have increased because of the high cost of system expansion and maintenance. Yet, heavy competition keeps prices reasonable. Today cellular telephone service is available in most areas of the United States and Europe. In addition, many systems are at full capacity, although there is still room for some growth in most areas. Overall, the systems are reaching their limits of growth. This has mainly to do with the lack of available spectrum space for expansion.

Some of the new cellular systems are expected to use the microwave bands. The FCC has set aside the 1.7- to 1.9-GHz band for the new cellular telephone and other wireless services. All new systems will be digital. The current AMPS system is fully analog except for the 10 kbits/s data channel used by the current system to establish communications and perform the control functions. The newer systems digitize the voice signal and transmit it as a digital bit system, which modulates a carrier using BPSK or some variation thereof.

The advantage of an all-digital system is a significant improvement in noise rejection and reliability of communications. Further, digital techniques facilitate the use of multiplexing methods, which will increase the number of telephone calls that can be handled within a given spectrum space.

Three digital cellular telephone systems are in use today: the Groupe Special Mobile *(GSM)* or *Global System for Mobile Communications* used in Europe and the IS-54 and IS-95 systems used in the United States. Each of these systems is described below.

GSM System   This system is widely used throughout Europe. It uses the 890- to 915-MHz band for uplink signals (signals from the cell user to the cell site) and the 935- to 960-MHz band for downlink signals (from cell site to user). In newer GSM systems, referred to as DCS-1800, the uplink range is 1.71 to 1.785 GHz and the downlink frequencies are 1.805 to 1.88 GHz. The spacing between the uplink and downlink signals is 45 MHz on the lower frequencies and 95 MHz on the upper frequencies. Each channel has a bandwidth of 25 kHz. Remember, in order to achieve full duplex operation, the uplink and downlink signals occur simultaneously in their own 25-kHz channel. The carrier spacing of the channels is 200 kHz. There are 124 channels in the lower frequency band, and 374 channels in the upper frequency band.

GSM uses time-division multiplexing to allow eight simultaneous telephone calls to use each channel. This system is known as *time-division multiple access (TDMA)*. The voice signals are digitized by an A/D converter and converted to serial format. The basic digitized bit rate is 13 kbits/s. The serial voice data is time-multiplexed into eight channels. Each voice channel contains one digital sample of the audio signal. Some channels may be blank if they are not being used.

The type of modulation used in GSM is called *Gaussian minimum shift keying (GMSK)*. Minimum shift keying (MSK) is a form of FSK in which the two frequencies selected to represent binary 0 and 1 are related in such a way that their zero crossing times are the same. When switching between frequencies takes place, it occurs at the zero crossing points. This reduces the number of harmonics. Therefore, the bandwidth used by an MSK signal is narrower than a standard FSK signal. In this

GSM system

TDMA

GMSK

context "Gaussian" reference means that a special type of low-pass filter with a Gaussian response filters the serial digital data to reduce its bandwidth prior to modulation. This combination produces a very narrow band digital signal.

IS-54

IS-54 System The *IS-54 system* is one of two different North American digital systems. It is a TDMA system, but its specifications differ from those of GSM. The IS-54 system was designed so that it could share the same channels allocated to the existing analog AMPS system. This was done to provide a path for eventual changeover from full analog to full digital operation. The IS-54 signals are designed to fit within the 30-kHz channels using exactly the same uplink and downlink frequencies as the AMPS system. The simultaneous duplex channels are also spaced 45 MHz.

The IS-54 TDMA system permits three time-multiplexed voice signals per 30-kHz channel. Future variations will allow six voice signals per channel. The digital bit rate of the signal is 48.6 kbits/s. The modulation scheme is referred

DQPSK

to as *DQPSK (differential quadrature phase shift keying)* with improved synchronization features and narrower bandwidth. It also includes the prefilter of the digital bit stream to further reduce overall signal bandwidth without materially affecting intelligibility.

IS-95 System This is North America's second digital system. It is a spread spectrum system

CDMA

called *code division multiple access (CDMA)*. Recall that spread spectrum is both a modulation scheme and a multiplexing method. The

IS-95 system

IS-95 system was also designed to use the same frequency ranges as AMPS and IS-54. However, it divides the space differently. It allows a total of 20 channels spaced at 1.25-MHz intervals. A maximum of 64 voice signals can use each channel. A wideband CDMA (W-CDMA) system is being developed to permit more channels spread over twice the bandwidth.

The IS-95 system uses a chip rate of 1.23 Mbits/s with direct sequence. The modulation is QPSK. The CDMA system allows many more voice signals with a given frequency range making it more efficient than TDMA. However, the IS-54 and IS-95 systems will coexist rather than one being replaced by the other. The FCC provides for dividing the frequency ranges so that all of the various forms

of cellular telephone systems can operate with minimum interference to one another.

■ TEST

Answer the following questions.

41. What do you call the small zones into which the area to be served by a cellular telephone system is divided?
42. What is the master control station for a group of cells called?
43. How many cellular channels were initially available, and how many are available now?
44. True or false. Cellular telephone radios operate full duplex.
45. What type of modulation is used in a cellular radio?
46. What is the maximum power output of a large mobile cellular transmitter? A hand-held unit?
47. What is the transmit frequency range?
48. What is the receive frequency range?
49. What is the channel spacing between cellular stations?
50. What is the frequency separation between send and receive frequencies?
51. Name common values for first IFs and second IFs in a cellular receiver.
52. What circuit in the cellular telephone allows a transmitter and receiver to share an antenna?
53. What is the source of the frequency divider ratios in the frequency synthesizer PLLs in the cellular telephone?
54. Name two signals that are transmitted back to the cell site and monitored by the MTSO.
55. Name three conditions in the transceiver controlled by the MTSO.
56. What is the name of the section of the cellular transceiver that interprets the serial digital data from the cell site and MTSO?
57. What is the NAM, and where is it kept?
58. What is AMPS?
59. Give two reasons why there are more channels available with cellular systems than in older mobile telephone systems.
60. How are the transmit and receive frequencies used by a cellular telephone determined?
61. What is the purpose of the APC circuit? What operates this circuit?

62. A receiver has a first IF of 82.2 MHz. The second IF is 456 kHz. What is the local oscillator frequency on the second mixer?
63. Explain the concept of a communications channel using TDMA.
64. Name the European digital cellular standard. Name the two American digital standards.
65. Which access mode (multiplex) is used by IS-95? IS-54? GSM?
66. Name the method of modulation used by IS-95, IS-54, and GSM.
67. Why is a digital cell phone system superior to an analog system?
68. Describe the CDMA system operation.

15-5 Paging Systems

Paging is a radio communications system designed to signal individuals wherever they may be. Paging systems operate in the simplex mode, for they broadcast signals or messages to individuals who carry small battery-operated receivers. Millions of people carry paging receivers. Typically they work in jobs that require maintaining constant communications with their employer and/or customers. The paging receiver operates continuously. To contact an individual with a pager, all you need to do is make a telephone call. A paging company will send a radio signal that will be received by the pager. The paging receiver usually has a built-in audible signaling device that informs the person that he or she is being paged. The signal may be as simple as an audio tone that indicates that the individual should call a telephone number to receive a message or otherwise make contact. Alternatively, the paging company can transmit a short printed message to the paging receiver. Some paging receivers have a small LCD screen on which a telephone number is displayed. This tells the paged individual which number to call. Some paging receivers have larger LCD screens that are capable of displaying several lines of alphanumeric text information.

Paging System Operation

Although the paging business is not owned by telephone companies, it is closely allied with the telephone business, because the telephone system provides the initial and final communications process. The most common paging process is described below (see Fig. 15-31).

To contact a person who has a pager, an individual dials the telephone number assigned to that person. The call is received at the office of the paging company. The paging company responds with one or more signaling tones that tell the caller to enter the telephone number which the paged person should call. Once the number is entered, the caller presses the pound sign key to signal the end of the telephone entry. The calling party then hangs up.

The paging system records the telephone number in a computer and translates this number into a serial binary-coded message. A unique protocol is used. The message is transmitted as a data bit stream to the paging receiver. The serial binary-coded message modulates the carrier of a radio transmitter. Paging systems usually operate in the VHF and

Paging

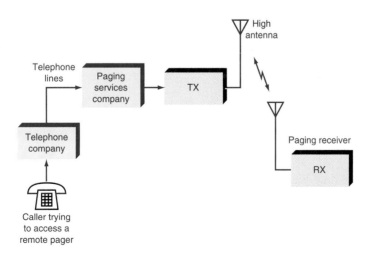

Fig. 15-31 The paging system.

UHF frequency ranges. A variety of bands have been assigned by the FCC specifically for paging purposes.

The paging company usually has a large antenna system mounted on a tower or the top of a tall building so that its communications range is considerable. Most paging systems can locate an individual within a 30-mi radius.

When the signal is transmitted, the paging receiver picks it up on its assigned frequency. The paged individual receives the message, as described above, and responds.

POCSAG

As indicated earlier, modern paging systems can transmit complete messages to an individual. These messages are usually entered on a personal computer and transmitted through a modem over the telephone system to the paging company. In this case the message is received, stored in a computer, and formulated into the correct protocol for transmission to the pager. The length of the message is typically restricted to several lines of text. A typical paging receiver may be able to display four lines of 20 characters each.

FLEX

All pagers emit an audio beep when the message is received. Some pagers vibrate rather than emit a beep.

Paging Formats

The earliest paging systems used a tone signaling system. Each paging receiver is assigned a special code called a *cap code,* which is a sequence of numbers or a combination of letters and numbers. The code is broadcast over the paging region. If the pager is within the region, it will pick up and recognize its unique code. The cap code is encoded using audio tones. Early systems used two-tone signaling. Later systems used a five-tone code.

The tones frequency-modulated the transmitter carrier in a fixed protocol or sequence. All paging receivers pick up every transmission, but they recognize only their own code. When a code is recognized, the beeper in the receiver goes off, informing the user that a call has been received.

Figure 15-32 shows a typical digital protocol. It begins with a preamble packet of bits that is the dotting sequence of clock pulses that help establish synchronization in the receiver. Motorola, the leading pager manufacturer, calls this sequence *dotting comma.* A unique pream-

Fig. 15-32 A digital protocol for pagers.

ble binary word is also transmitted multiple times. This transmission is followed by a sync packet that provides further dotting and a sequence of sync words.

Digital paging systems use one of two common protocols, POCSAG and FLEX. *POCSAG* stands for Post Office Code Standard Advisory Group. It uses a two level ASK-PSK modulation scheme to transmit data at a rate of up to 2.4 kbits/s over a 25-kHz channel in the 150-, 470-, or 900-MHz range.

The *FLEX* protocol was developed by Motorola, the world leader in pager design and manufacturing. It uses a two- or four-level ASK-FSK combination modulation scheme to send data at a rate from 1.6 to 6.4 kbits/s. The rate is self-adjusting to the channel conditions. It also uses CRC error detection and a forward error-correcting scheme to ensure data integrity even on noisy channels.

A Paging Receiver

A typical FLEX paging receiver is shown in Fig. 15-33. It uses three integrated circuits. The first is the RF portion of the receiver with a low-noise amplifier (LNA), a mixer and local oscillator, and an A/D converter (ADC). Most modern paging receivers are a special version of a superheterodyne called a *direct conversion* or *zero IF (ZIF) receiver.* In this type of receiver, the local oscillator is set to the same frequency as the incoming signal. The sum and difference frequencies produced by the mixer are twice the operating frequency and zero frequency. In other words, the IF is zero. The double frequency is filtered out. Since the original modulation produces sidebands that are above and below the operating frequency, these heterodyne with the local oscillator to produce the original modulating signal at the output of the mixer. In this type of receiver, the mixer is also the demodulator. A low-pass filter at the output of the mixer is all that is needed to recover the original signal. An A/D converter is used to translate this signal into the original serial bit stream.

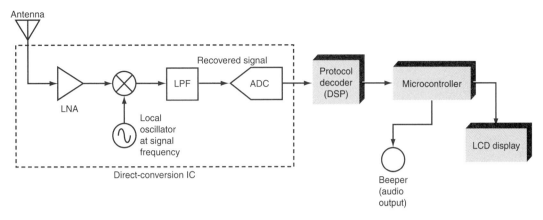

Fig. 15-33 A digital paging receiver.

The serial data is then sent to a protocol decoder chip. This is a specialized digital signal processor (DSP) chip that recognizes the protocol and recovers the message that is sent on to the third chip, a general-purpose microcontroller that operates the liquid crystal display (LCD).

■ TEST

Answer the following questions.

69. Name the major parts of a paging system, and briefly explain its operation.
70. Name the two main digital protocols used in paging systems.
71. What is dotting, and why is it needed?
72. What is the output of a paging receiver?
73. Explain the operation of a direct-conversion receiver.

15-6 Integrated Services Digital Network

Integrated services digital network (ISDN) is a digital communications interface designed to replace the local analog loop now used in the public switched network. It supports digital voice telephones as well as fax machines, computers, video, and other digital data sources. It allows computers to access online services directly without a modem. ISDN is a telephone company service that is available in many major cities. It is used just like a standard telephone service. It was created so that the installed base of twisted-pair cable used for analog loops could be used for digital transmission. Its primary application is high-speed Internet access.

The ISDN Interface

Although ISDN service is available from local telephone companies, all-new equipment or special interfaces are required to use it. Once in place, ISDN can be used for voice telephone communications, fax, digital video, and all types of data transmissions. Figure 15-34 shows the interface connections to an ISDN line. The NT_1 box connects to the telephone company lines via twisted-pair, called the *U interface*, which uses multilevel encoding. The NT_1 box connects to the NT_2 box which provides the interface to the user's equipment. The circuitry for NT_1 and NT_2 is usually in the same enclosure, and this piece of equipment is owned by the customer. The NT_1 interface provides encoding and decoding, multiplexing, and timing. The NT_2 interface provides for connection to LAN, PBX, and other devices.

The communications devices connect to the NT_2 circuits over what is called the *S bus*. The S bus is twisted pair using special three-level modified AMI encoding. ISDN–compatible devices are called *terminal equipment*. Examples are special ISDN telephones, digital fax machines, or personal computer ISDN interfaces. These are designated as TE_1 devices. Equipment that is not ISDN compatible is designated TE_2. To use such equipment on an ISDN line, a *terminal adapter (TA)* is required. Different types of TAs are used depending upon the type of device. Standard analog telephones and computers with non-ISDN I/O interfaces are instances of such equipment.

ISDN

Basic Rate Interface

The two basic types of ISDN connections are the *basic rate interface (BRI)* and the *primary*

BRI

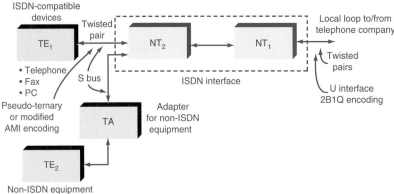

ISDN-compatible devices

Twisted pair

Local loop to/from telephone company

TE₁

• Telephone
• Fax
• PC

Pseudo-ternary or modified AMI encoding

S bus

NT₂

NT₁

Twisted pairs

U interface 2B1Q encoding

ISDN interface

TA

Adapter for non-ISDN equipment

TE₂

Non-ISDN equipment
• Telephone (analog)
• PC (RS-232, etc.)

Fig. 15-34 The ISDN interface.

PRI

rate interface (PRI). The basic rate interface is made up of a single twisted pair. Two bearer (B) channels for voice and one data control (D) channel are time-multiplexed on the line. Its designation is 2B + D. The B channels are capable of 64 kbits/s data transmission. The D channel can operate at 16 kbits/s. The two B channels are used for 64 kbits/s baseband transmissions of any kind of data. One channel is normally used for transmitting, and the other for receiving. The 16-kbits/s channel is used for signaling and control functions between the user and the telephone company such as dialing and busy signals.

Because multiple channels are provided in an ISDN cable, various configurations of data transmissions are possible: simplex, half duplex, or full duplex. ISDN interfaces permit reconfiguration "on the fly," depending upon the application. For instance, the data to be transmitted can be partitioned into 2-bit words which can be transmitted simultaneously at 64 kbits/s. This is the equivalent of transmitting at a rate of 128 kbits/s. The total information bandwidth is the sum of the individual line data rates, in this case, 64 + 64 + 16 = 144 kbits/s. Considering the framing and auxiliary bits also used, the total data rate is 192 kbits/s. Most ISDN interfaces will be of this type.

Primary Rate Interface

The primary rate interface is called 23B + D and is made up of 23 64 kbits/s B channels and a 64 kbits/s D channel. Again, a single twisted pair is used with all channels being time-multiplexed. The total data bandwidth is 24 × 64 kbits/s, or 1536 kbits/s, or 1.536 Mbits/s. Special designations are given to the various popular combinations such as H0 for 6 64-kbits/s channels used at 384 kbits/s, H10 for 23 64 kbits/s channels used for 1472-kbits/s transmission, and H11 for 24 channels at 1536 kbits/s.

Primary rate ISDN is primarily for larger companies that need to accommodate a PBX system or needs internetwork connections to remote points. The primary rate service is T1 compatible.

ISDN Signals

Since the standard twisted pair used in the local loop is bandwidth limited to about 4 kHz, special techniques are needed to increase the speed of data transmission. ISDN uses a special multilevel digital signal on the U interface to the central office.

ISDN uses baseband communications methods; that is, the digital signal is connected directly to the line without modulation. However, the standard serial two-level binary signal is not used. A special multivoltage signal is used. Known as 2B1Q, this signal uses four voltage levels (-3, -1, $+1$, and $+3$ V) to represent pairs of bits according to the scheme listed below:

$$+1 \text{ V} = 11$$
$$+3 \text{ V} = 10$$
$$-1 \text{ V} = 01$$
$$-3 \text{ V} = 00$$

The data to be transmitted is split up into 2-bit sections and encoded as the appropriate

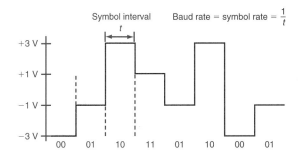

Fig. 15-35 2B1Q ISDN signal.

voltage level. This is the same technique used in modems to increase the data rate. Each level represents 1 symbol or baud. And since 2 bits per baud are being encoded, the data rate is increased over the standard binary rate possible. Figure 15-35 shows a 2B1Q or 2 bits per quaternary symbol.

Figure 15-36 shows the frame format for the basic rate ISDN signal. This is the format of the data at the S bus. There are two different but similar formats, one for network to subscriber transmission and the other for subscriber to network transmission. Each frame contains 48 bits that occur in 250 μs. This gives a bit time of 250/48, or 5.208 μs. This translates to a bit rate of $1/5.208 \times 10^{-6}$, or 192,000 bits/s, or 192 kbits/s.

In each frame there are 2 bytes (also called *octets*) of data for the B_1 bearer channel and 2 bytes of data for the B_2 bearer channel. The spacing between bytes is 125 μs, which translates to an 8-kHz sampling rate, which is standard for digital conversion of voice in the telephone system. There are 4 D channel bits per frame, but they are distributed rather than clustered. The other bits are used for framing, control, and signaling functions. DC balancing bits are added to ensure that the number of positive bits equals the number of negative bits so that the line remains DC balanced, that is, that there is no net positive or negative charge over one frame. Special multiplexing and demultiplexing circuits at each end of the system format the data into frames or extract it for use.

The encoding method used for this S-bus signal is a special pseudoternary, or three-level, binary signal as shown in Fig. 15-37. A binary 1 or mark is represented as a 0-V level, and a binary 0 or space is represented as a positive or negative voltage pulse (approximately 2 V). This is a modified form of the *alternate mark inversion (AMI) format* used in the T1 system. Consecutive binary 0s are transmitted as alternating polarity pulses to ensure that no dc offset occurs on the line. A 0 followed by another 0 of the same polarity is a code violation. Code

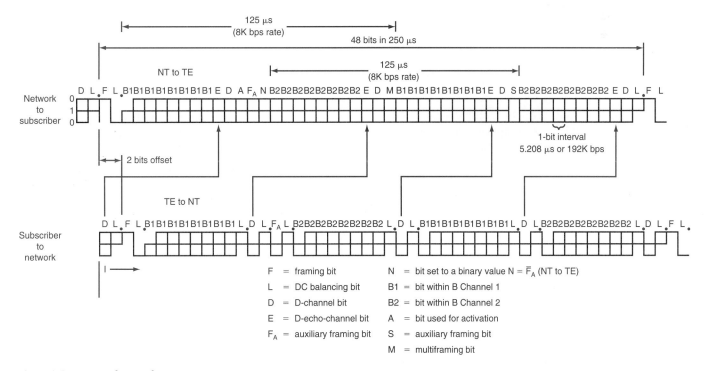

F = framing bit   N = bit set to a binary value N = $\bar{F}_A$ (NT to TE)

L = DC balancing bit   B1 = bit within B Channel 1

D = D-channel bit   B2 = bit within B Channel 2

E = D-echo-channel bit   A = bit used for activation

F_A = auxiliary framing bit   S = auxiliary framing bit

M = multiframing bit

Fig. 15-36 ISDN frame formats.

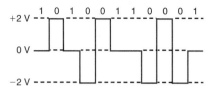

1 0 1 0 0 1 1 0 0 0 1

+2 V

0 V

−2 V

Fig. 15-37 S-bus line encoding. Pseudoternary or modified AMI format. Number of pulses equals number of negative pulses per frame to maintain 0 Vdc average on the line.

violations are used to identify the start and stopping points in a frame.

Wireless Local Loop

The local loop consists of a twisted-pair cable between a subscriber's telephone and the nearest central office. Usually, the cable contains several twisted pairs, but only one is used to make the connection. The others are for future expansion for additional voice lines, fax machines, and modems. The local loop cable is usually anywhere from several thousand feet long to as much as 18,000 ft long. Such a cable is expensive to purchase, install, and maintain. Yet there are hundreds of millions of them in place—and more on the way.

Local loop cables have proved to be very reliable and more than adequate for voice transmission. But today, the local loop is increasingly used for other applications such as fax transmissions and Internet access. Because of the poor frequency response of a local loop cable, the speed of data transmission is severely limited. Further, it does not lend itself to multiplexing multiple signals on a single line. These limitations greatly restrict the uses to which the telephone and its accessories can be put. Advances in modem design have helped increase data speed over the local loop. Technologies like ISDN and ADSL also extend the usefulness of the local loop. However, further improvements in cable bandwidth would open the possibilities for improved telecommunication services. Some possibilities are fiber-optic cable local loop connections and wireless.

Recent developments in cellular telephone technology have made it possible to build wireless local loops at a reasonable cost. Obviously, no cable is used. The link between the sub-

scriber and the central office is radio. An example is Lucent Technologies' *AirLoop*. It consists of a subscriber transceiver connected to the telephone. Provisions for other connections such as fax and PC are also provided. A small antenna at the subscriber's home or office transmits and receives signals to and from the central office base station. This base station connects to the main telephone network through the usual switching systems. The operation is "seamless" in that the user never knows that the local loop is radio.

AirLoop operates in the microwave frequency bands and uses a special wide-bandwidth version of code-division multiple access (W-CDMA), the digital spread-spectrum system widely used in some digital cell phone systems. It accommodates standard voice transmission but also provides high-speed digital data connections for fax and Internet access. Data rates up to 144 kbits/s for regular service and 512 kbits/s for packet transmission are available. And Lucent is working to significantly increase these rates for future versions.

The primary benefits of a wireless local loop over a conventional local loop are:
1. Full equivalency of a wired loop with all features and services
2. Lower cost (no cable installation and maintenance costs)
3. Faster installation
4. Flexibility (permits changes required by the customer to be made easily and quickly)
5. Wider bandwidth, permitting faster data rates

Although wireless local loop is not yet widely available, it is growing in popularity and usage. As more subscribers discover its benefits and more local telephone companies offer the service, wireless local loop is sure to grow.

■ TEST

Answer the following questions.

74. What is ISDN?
75. What type of wiring does ISDN use?
76. What type of signals can an ISDN system carry?
77. State the specifications for basic rate ISDN.
78. State the specifications for primary rate ISDN.

79. Define the designations TE_1, TE_2, TA, NT_1, and NT_2.

80. How can data rates exceeding the bit rate capacity of a single cable be achieved with ISDN?

81. How can non-ISDN devices be used with an ISDN system?

82. List five possible applications for ISDN.

83. What is the name of the signal encoding used in ISDN to achieve high data rates on a low-data-rate line? State the specifications of the coding method.

84. What is the basic frequency response of the telephone local loop? Can it carry digital as well as analog signals?

Summary

1. The telephone is the most widely used electronic communications device.
2. Standard subscriber telephones are connected to a central office or local exchange by a twisted-pair cable called the local loop.
3. Standard telephones are powered by 48 Vdc and a 20-Hz sine wave ringing signal supplied by the telephone company.
4. Modern telephones use electronic circuits.
5. The "dialing" system used on telephones is called dual-tone multifrequency (DTMF) because two different sine wave tones are generated for each key pressed.
6. Most modern telephones are connected to the telephone line with twisted-pair cable terminated in a modular connector designated RJ-11.
7. Cordless telephones are full-duplex two-way radios operating on a pair of frequencies.
8. The original cordless telephones used FM on 25 channels in the 44-, 46-, and 49-MHz range.
9. The newer cordless telephones use the 900-MHz frequency range. Some use FM, but the better phones use digital signals including spread spectrum.
10. Facsimile or fax is an electronic communication technique for transmitting printed documents including text, line drawings, photos, and other graphical information via telephone lines or radio.
11. Fax machines scan the document to be transmitted using photo-optical techniques to create a baseband signal that modulates a carrier prior to transmission.
12. The receiving fax machine demodulates the carrier to recover the baseband signal which is sent to a printer where the original document is faithfully reproduced.
13. The quality of reproduction is a function of the scanning resolution in lines per inch.
14. The greater the number of scan lines, the finer the definition.
15. The most widely used device to convert the scanned lines into an electrical signal is a light-sensitive semiconductor component known as a charge coupled device (CCD).
16. The CCD contains thousands of tiny capacitors that charge to a value proportional to the light intensity. The capacitors are sequentially sampled. Their charges are read out, creating an analog signal corresponding to the lines scanned.
17. The baseband signal modulates a carrier prior to transmission; AM, FM, and PSK are commonly used.
18. Facsimile standards for modulation, transmission speed, and other factors are set by the International Telegraph and Telephone Consultative Committee (CCITT). There are four basic standards designated group 1 through 4.
19. Group 1 fax machines are analog and use FM where black is 1500 Hz (or 1300 Hz) and white is 2300 Hz (or 2100 Hz). Resolution is 96 lines per inch, and transmission speed is 6 minutes per page. Group 1 machines are no longer used.
20. Group 2 machines are analog and use FM or vestigial sideband AM with a 2100-Hz carrier. Resolution is 96 lines per inch, and transmission speed is 3 minutes per page or less.
21. Group 3 machines use digital techniques and PSK or QAM to achieve speeds of up to 9600 baud over the telephone lines. Resolution is 200 lines per inch, and transmission times are less than 1 min, with less than 30 s being common. Most modern fax machines are of the group 3 type.
22. Group 4 fax machines are digital and are designed for wideband telephone lines. Digital transmission rates are 56 kbits/s. Resolution is 400 lines per inch. Few group 4 machines are in use yet.
23. Most fax printers are of the thermal type and use special heat-sensitive paper, although laser printers using xerography are used in some higher-priced machines.
24. Fax is widely used by newspapers, business, and the military. Radio fax is used for transmitting weather satellite photos to earth.
25. Cellular radios provide telephone service in vehicles and are used as portable telephone units.

26. The area served by cellular telephones is divided into small zones called cells.

27. Each cell is served by a repeater containing a low-power transmitter and a receiver. The cells are connected by wire to a computer-controlled master station called the mobile telephone switching office (MTSO). The MTSO links to the telephone system.

28. By operating at high frequencies in the spectrum and through the application of frequency reuse, many channels are available to users.

29. There are 666 full-duplex telephone channels in most service areas. The receive frequencies are in the 870- to 890-MHz range.

30. The transmit frequencies are in the 825- to 845-MHz range.

31. Channel spacing is 30 kHz.

32. Spacing between the simultaneously used transmit and receive frequencies is 45 MHz.

33. Cellular radios use FM with a maximum deviation of $\pm$ 12 kHz.

34. Mobile cellular transmitters have a maximum output power of 3 W which can be decreased in steps under the control of the MTSO to minimize adjacent cell interference.

35. Mobile cellular receivers are of the dual-conversion type with a 45- or 82.2-MHz first IF and a 10.7-MHz or 455-kHz second IF.

36. Both transmit and receive frequencies are determined by PLL frequency synthesizers that are set by the MTSO to a clear channel.

37. The MTSO monitors received cell signal strength [received signal strength indicator (RSSI)] and transmitter power output and makes decisions about when to "hand off" the mobile unit to another cell to maintain optimum signal strength.

38. The MTSO controls transmitter power as well as transmit and receive frequencies. Serial digital data containing this information is transmitted to the mobile unit whose logic section interprets it and effects the changes.

39. Each cellular radio contains a PROM called the number assignment module (NAM) which stores the unit's telephone number referred to as the mobile identification number (MIN).

40. Most cellular radios contain two microprocessors, one for controlling the logic section and another for operating the displays and dialing circuits in the handset and control section.

41. Most newer cellular telephones are digital. These telephones offer the benefits of less sensitivity to noise, more telephone calls per channel, and easier computer control.

42. There are three popular digital cell phone standards. The GSM (Global System for Mobile Communications) is used in Europe. It uses time-division multiple access to put eight telephone calls on each channel.

43. The GSM system uses the 890- to 915- and 935- to 960-MHz frequency range. There are 124 25-kHz channels spaced at 200-kHz intervals. The modulation is GMSK, a narrow-band form of FSK.

44. There are two digital cell phone standards used in the United States, IS-54 and IS-95. Both are designed to use the same frequency range now used by the AMPS system. The goal is to eventually replace the AMPS system completely in the future.

45. The IS-54 system uses TDMA with three calls per 30-kHz channel. The modulation is DQPSK.

46. The IS-95 system uses CDMA or spread spectrum. It uses 20 channels spaced at 1.25-MHz intervals. It uses direct sequence spread spectrum with a chip rate of 1.23 MHz. The modulation is QPSK.

47. A paging system allows a customer to call or page a user by sending a message by standard Touch-Tone telephone using a simplex transmitter that broadcasts the signal to an individual's paging receiver operating in the 150-, 470-, or 900-MHz range.

48. The paging receiver is a small battery-operated superheterodyne with a tone beeper, an LCD alphanumeric display, or a vibrator.

49. Older pagers used a system of audio tones for signaling. Most newer pagers use digital techniques.

50. Two common digital paging protocols are PSC-SAG and FLEX. Data rates are in the 1.6- to 6.4-kbits/s range.

51. Typical pager receivers use direct conversion techniques in which the local oscillator frequency is set to the incoming signal frequency, thereby producing a zero IF. Such a system directly recovers the signal.

52. A digital signal processor detects and decodes the protocol for display.

53. The Integrated System Digital Network (ISDN) was designed to transmit telephone calls, fax

messages, or digital data by digital means on a twisted-pair line. Most telephone companies in the larger cities offer ISDN services.

54. The most common application of ISDN is high-speed Internet access.

55. ISDN is a baseband system in which digital signals are transmitted on the twisted-pair line without modulation. A special four-level (symbol) code called 2B1Q permits 2 bits to be encoded per symbol and thus produces high-speed digital transmission of speeds up to 192 kbits/s.

56. The basic rate ISDN line has two data channels and one signaling and control channel. This is known as 2B + D. Each bearer or data channel can transmit at 64 kbits/s. The control channel transmits at 16 kbits/s. The two data lines may be combined to transmit at 128 kbits/s.

57. The primary rate ISDN interface has 23 data channels and one control channel.

58. ISDN uses time-division multiplexing to put multiple channels on a single twisted pair.

Chapter Review Questions

Choose the letter which best answers each question.

15-1. The local loop is
  a. An antenna used for telephone communications.
  b. The connection from the user's telephone to the central office.
  c. A cell site.
  d. A ring network used to connect users to the telephone office.

15-2. A telephone is
  a. Full duplex.
  b. Half duplex.
  c. Simplex.
  d. Multiplex.

15-3. The type of cable used between the telephone subscriber and the central office is
  a. Coaxial cable.
  b. Fiber-optic cable.
  c. Twin lead.
  d. Twisted pair.

15-4. Standard telephones receive their power from
  a. An internal ac power supply.
  b. A battery.
  c. 48 Vdc supplied by the telephone company.
  d. 90 Vac supplied by the telephone company.

15-5. The ringing voltage supplied by the telephone company is
  a. A 20-Hz sine wave.
  b. A 60-Hz sine wave.
  c. 48 Vdc.
  d. A sine wave of any audio frequency.

15-6. What is the name of the "dialing" system used in modern telephones?
  a. Rotary pulse
  b. Audio pulse
  c. Multitone alphanumeric
  d. Dual-tone multifrequency

15-7. What two sine wave frequencies are produced when the 8 key is pressed?
  a. 697 and 1477 Hz
  b. 852 and 1336 Hz
  c. 770 and 1209 Hz
  d. 941 and 1336 Hz

15-8. What is the designation given to the modular plug used on telephone cable?
  a. BNC
  b. F-connector
  c. RJ-11
  d. RJ-45

15-9. What type of modulation is used by a standard analog cordless telephone?
  a. AM
  b. FM
  c. PSK
  d. QAM

15-10. What is the frequency range of the newer digital cordless telephones?
  a. 43–49 MHz
  b. 150 MHz
  c. 470 MHz
  d. 900 MHz

15-11. What is the designation of a long-distance carrier?
  a. LEC
  b. POP
  c. IXC
  d. LATA

15-12. Printed documents to be transmitted by fax are converted into a baseband electrical signal by the process of
  a. Reflection.
  b. Scanning.
  c. Modulation.
  d. Light variation.

15-13. The most commonly used light sensor in a modern fax machine is a
a. Phototube.
b. Phototransistor.
c. Liquid-crystal display.
d. Charge coupled device.

15-14. In FM fax, the frequencies for black and white are
a. 1500 and 2300 Hz.
b. 2300 and 1500 Hz.
c. 1300 and 2400 Hz.
d. 1070 and 1270 Hz.

15-15. Which resolution produces the best quality fax?
a. 96 lines per inch
b. 150 lines per inch
c. 200 lines per inch
d. 400 lines per inch

15-16. Group 2 fax uses which modulation?
a. SSB
b. FSK
c. Vestigial sideband AM
d. PSK

15-17. The most widely used fax standard is
a. Group 1.
b. Group 2.
c. Group 3.
d. Group 4.

15-18. Group 3 fax uses which modulation?
a. QAM
b. FSK
c. Vestigial sideband AM
d. FM

15-19. Older fax printers are of which type?
a. Impact
b. Thermal
c. Electrosensitive
d. Laser xerographic

15-20. Facsimile standards are set by the
a. FCC.
b. DOD.
c. ITV-T.
d. IEEE.

15-21. What type of graphics are commonly transmitted by radio fax?
a. Newspaper text
b. Architectural drawings
c. Cable movies
d. Satellite weather photos

15-22. The transmission speed of group 4 fax is
a. 4800 baud.
b. 9600 baud.
c. 56 kbits/s.
d. 192 kbits/s.

15-23. The master control center for a cellular telephone system is the
a. Cell site.
b. Mobile telephone switching office.
c. Central office.
d. Branch office.

15-24. Each cell site contains a
a. Repeater.
b. Control computer.
c. Direct link to a branch exchange.
d. Touch-tone processor.

15-25. Multiple cells within an area may use the same channel frequencies.
a. True
b. False

15-26. Cellular telephones use which type of operation?
a. Simplex
b. Half duplex
c. Full duplex
d. Triplex

15-27. The maximum frequency deviation of an FM cellular transmitter is
a. 6 kHz.
b. 12 kHz.
c. 30 kHz.
d. 45 kHz.

15-28. The maximum output power of a cellular transmitter is
a. 4.75 mW.
b. 1.5 W.
c. 3 W.
d. 5 W.

15-29. Receive channel 22 is 870.66 MHz. Receive channel 23 is
a. 870.36 MHz.
b. 870.63 MHz.
c. 870.96 MHz.
d. 870.69 MHz.

15-30. A transmit channel has a frequency of 837.6 MHz. The receive channel frequency is
a. 729.6 MHz.
b. 837.6 MHz.
c. 867.6 MHz.
d. 882.6 MHz.

Note: The LO frequency is usually higher than the receive frequency.

15-31. A receive channel frequency is 872.4 MHz. To develop an 82.2 MHz IF, the frequency synthesizer must supply an LO signal of
a. 790.2 MHz.   c. 954.6 MHz.
b. 827 MHz.   d. 967.4 MHz.

15-32. The output power of a cellular radio is controlled by the
  a. User or caller.
  b. Cell site.
  c. Called party.
  d. MTSO.

15-33. When the signal from a mobile cellular unit drops below a certain level, what action occurs?
  a. The unit is "handed off" to a closer cell.
  b. The call is terminated.
  c. The MTSO increases power level.
  d. The cell site switches antennas.

15-34. In a cellular radio, the duplexer is a
  a. Ferrite isolator.
  b. Waveguide assembly.
  c. Pair of TR/ATR tubes.
  d. Pair of sharp bandpass filters.

15-35. The digital cell phone system used in Europe is called
  a. IS-54.
  b. IS-95.
  c. AMPS.
  d. GSM.

15-36. What type of modulation is used in the IS-54 digital cell phones?
  a. QAM
  b. DQPSK
  c. BPSK
  d. GMSK

15-37. Which type of multiple access does the IS-95 digital cell phone use?
  a. CDMA
  b. FDMA
  c. TDMA
  d. Spatial

15-38. How many telephone calls can be accommodated on each channel of the U.S. TDMA system?
  a. 2
  b. 3
  c. 8
  d. 10

15-39. Which digital cell phone system uses spread spectrum?
  a. AMPS
  b. GSM
  c. IS-95
  d. IS-54

15-40. Which of the following is *not* one of the advantages of a digital cell phone over an analog cell phone?
  a. Longer transmission distances.
  b. Less interference from noise.
  c. Easier computer control.
  d. More transmissions per channel.

15-41. What is the frequency range of the digital cell phones in the United States?
  a. 902–928 MHz
  b. 890–915 and 935–960 MHz
  c. 1.71–1.785 and 1.805–1.88 GHz
  d. 825–845 and 870–890 MHz

15-42. A direct conversion receiver is one with
  a. Dual conversion.
  b. Zero IF.
  c. TRF features.
  d. Feedback.

15-43. Most modern pagers are
  a. Analog.
  b. Digital.

15-44. Which of the following is *not* a type of output on a paging receiver?
  a. Vibration
  b. Audible beep
  c. LCD display
  d. Video

15-45. ISDN uses which type of cable?
  a. Coaxial
  b. Fiber optic
  c. Twisted pair
  d. Speaker

15-46. How many data channels does basic rate ISDN have?
  a. 1
  b. 2
  c. 3
  d. 24

15-47. What is the data rate of a single data channel of ISDN?
  a. 64 kbits/s
  b. 128 kbits/s
  c. 192 kbits/s
  d. 1.536 Mbits/s

15-48. The primary application of ISDN is
  a. Local area networks.
  b. Long-distance telephone service.
  c. Local loop.
  d. Internet access.

15-49. Which type of system is ISDN?
  a. Baseband
  b. Broadband

15-50. ISDN uses
  a. Frequency-division multiplexing.
  b. Time-division multiplexing.
  c. Code-division multiplexing.

Critical Thinking

15-1. What must be done to a primary rate ISDN signal to make it compatible with a T1 system?

15-2. Discuss how it would be possible to use the standard ac power lines for telephone voice transmission. What circuits might be needed, and what would be the limitations of this system?

15-3. Can TV signals be transmitted over the telephone lines? Explain how this might be done.

15-4. Can color pictures be transmitted by fax? Explain how.

15-5. Can digital data be transmitted by cell phone? Explain how this might be done.

15-6. Caller ID is a telephone service that can display the telephone number of a caller. Explain how this works.

15-7. Name three systems that give faster Internet service than a conventional telephone modem.

15-8. Could you make a cell phone that would accommodate all four of the major formats? Explain how this could be done.

Answers to Tests

1. The local loop is the connection between a telephone subscriber and the central office. It is a twisted-pair cable ranging from 8000 to 18,000 ft in length.

2. A 48-Vdc supply located at the central office

3. A 90- to 100-V 20-Hz sine wave

4. A hybrid is a circuit or component (usually a transformer in a telephone) that allows a single line to carry both incoming and outgoing voice signals simultaneously.

5. true

6. A carbon microphone is the standard in most conventional phones. When you speak into a carbon microphone, a vibrating diaphragm compresses and expands a carbon resistor element in accordance with the voice. The varying resistance varies the current in a circuit that becomes the voice signal.

7. The tip and ring refer to electrical connections on an old-style telephone plug used by operators. These connections are made to the twisted-pair local loop. The tip is usually a green wire and the ring is red. Alternatively, yellow and black wires are used.

8. Dual-tone multifrequency (DTMF)

9. 941 and 1209 Hz

10. Central office or local exchange

11. Electret

12. To ensure that the polarity of the direct current from the local exchange is of the correct polarity

13. Piezoelectric sound element

14. false

15. Modular connector or RJ-11

16. Most cordless phones use FM in the 43- to 49-MHz frequency band with 25 channels. The 900-MHz band is also used for cordless phones. Analog phones use FM and digital phones use some form of PSK and/or spread spectrum.

17. Cordless phones use low-frequency audio tones to trigger and enable the telephone circuitry if a call is received. If no tone is received from the base unit, the phone will not ring or work.

18. The newer cordless phones use digital audio. One type uses an analog-to-digital method known as ADPCM and PSK. Another type uses direct-sequence or frequency-hopping spread spectrum. Both operate in the 902- to 928-MHz band. Spread-spectrum phones are also available in the 2.4-GHz band. Digital phones have better clarity, transmit over longer distances, and are less susceptible to noise.

19. The basic BORSCHT functions are battery, overvoltage protection, ringing, supervision, coding, hybrid, and test.

20. The first three numbers are the area code, which designates specific areas of the country. The second three digits are the exchange code, which defines a specific central office. The last four digits define one of 10,000 subscriber numbers connected to the exchange.

21. Links between central offices can be twisted pair, coaxial cable, or fiber-optic cable. In rare cases, microwave links may be used.

22. LATA means local access and transport area. LATAs are set up by LECs which define areas within a state or region that are made up of local exchanges connected to one another. An LEC (local exchange carrier) is the telephone company that is set up to serve several states or a large geographical area. POP (point of presence) is a connection point between the LATAs and

long-distance transmission systems. IXCs (interexchange carriers) are the long-distance companies.

23. true
24. The telephone system, wireless (radio, satellite, etc.)
25. true
26. ITU
27. Group 2
28. CCD
29. G3
30. Handshaking
31. The modem
32. 9600 baud
33. 96 lines per inch
34. Reflected light from a page is converted into an electrical signal by a photocell or a CCD. The resulting signal usually has pulses corresponding to black and white areas on the page. This signal modulates an audio frequency carrier that is connected to the telephone line.
35. A photocell can be moved to scan one thin line on a page, or a group of LEDs are used to illuminate one scan line on the page. The reflected light is converted into black and white or gray areas by a CCD.
36. The most common and most economical printer is a thermal printer that uses a heat-sensitive paper that is burned by a special print head to form the characters.
37. The digital data resulting from the scanning is compressed into smaller groups of bits that can be transmitted faster.
38. *b.*
39. *a.*
40. true; G4 speed is 56 kbits/s
41. Cells
42. Mobile telephone switching office (MTSO)
43. 666, 832
44. true
45. FM
46. 3 W, 500 mW
47. 825–845 MHz
48. 870–890 MHz
49. 30 kHz
50. 45 MHz
51. 45 or 82.2 MHz for first IF, 10.7 MHz or 455 kHz for second IF
52. Duplexer
53. Microcontroller
54. 10-kbits/s digital control signal and the received signal strength indicator (RSSI)
55. Transmit and receive frequencies and power level
56. Control logic, usually a single-chip microcomputer.
57. NAM means number assignment module. It is a PROM.
58. AMPS means advanced mobile phone service and is the name given to the original analog cell phone system still in operation.
59. Higher frequencies are used where more spectrum space is available, and the cellular system permits frequency reuse that multiplies the number of channels available.
60. The MTSO assigns the cell phones to an available channel.
61. APC means automatic power control. It is a circuit in the transmitter section of a cell phone that is under the control of the MTSO via the 10-kbits/s digital control channel.
62. 82.656 MHz
63. A TDMA system is a digital system that uses a single-frequency channel to transmit multiple voice signals by digitizing them and transmitting them serially and sequentially time-multiplexed.
64. GSM is the European standard. The American standards are IS-95 and IS-54.
65. IS-95 uses spread spectrum, IS-54 uses TDMA, and GSM uses TDMA.
66. IS-95 uses direct-sequence spread spectrum and QPSK. IS-54 uses DQPSK. GSM uses GMSK.
67. Digital systems are less susceptible to noise and are more reliable.
68. CDMA (code-division multiple access) is another name for spread spectrum. The voice signal is digitized and then passed through an XOR with a digital chipping signal that creates a direct-sequence spread-spectrum signal. The fast digital signal is used to modulate a carrier that usually uses a form of PSK. This results in a very broad band signal.
69. The main parts of any paging system are (a) the originating telephone, (b) the telephone system, (c) a paging services company with a transmitter and antenna, and (d) the user with a paging receiver.
70. POCSAG and FLEX.
71. Dotting refers to the transmission of a short sequence of clock pulses that the receiver uses to synchronize itself with the transmitted signal.
72. The output of a paging signal may be one or more of the following: audio tone (beep), a vibration, and a message on an LCD screen.
73. A direct-conversion receiver has a local oscillator frequency that is equal to the carrier frequency of the transmitted signal. This results in a zero IF. In

this type of receiver, the output of the first mixer is the original data signal. Only a low-pass filter is needed to recover it.

74. Integrated services digital network, an all-digital baseband connection to the local exchange designed to transmit digitized voice or data.

75. Twisted pair

76. Any type of signal that can be put into digital form such as computer data, voice, video, or any analog control or measurement signals.

77. The basic rate refers to two 64-kbits/s data channels plus one 16-kbits/s control/signaling channel. The two 64-kbits/s channels may be combined into one 128-kbits/s channel.

78. The primary rate channel provides for 23- to 64-kbits/s data channels plus one 64 kbits/s signaling channel. The maximum data rate is 1.536 Mbits/s.

79. A TE1 device is ISDN-compatible and may be a telephone, fax machine, or PC interface. A TE2 device is non-ISDN-compatible but can have a digital output that could be transmitted over the ISDN line. A TA unit is hardware that connects a noncompliant TE2 device to the ISDN line. The NT1 and NT2 units together form the ISDN interface. The NT1 unit provides encoding, decoding, multiplexing, and demultiplexing. The NT2 device provides the physical interface to the line and the local loop.

80. Special line coding techniques and transmitting multiple bits per symbol boost the bit rate capability of the line.

81. A TE2 interface box can make a non-ISDN device compatible with the line.

82. Computer data, voice, video, fax, and control/measurements.

83. 2B1Q. Four voltage levels of -3, -1, $+1$, $+3$ are used to transmit 2 bits per symbol or voltage level.

84. The basic frequency response of a local loop connection is up to 4 kHz. It can carry voice without special conditions. It can carry low-bit-rate baseband digital signals or high-bit-rate digital signals if a modem is used.

Base rate interface (BRI), 509–510
Basic groups, 198
Batteries:
    for satellite systems, 326, 330, 338–339
    for telephone systems, 488–489
Baudot code, 348–349
Baud rates, 352, 356
BCC (binary check code), 372
BCC (block check character), 374
BCS (block check sequence), 374
Beacon transmitters, 332
Beam antennas, 242–243
Beam width of antennas, 241–242, 280–282
Bearing with radar, 290
Beat frequency oscillator (BFO) circuits, 173–174
Bell, Alexander Graham, 408
BER (bit error rate), 372
Bessel functions, 74
BFO (beat frequency oscillator) circuits, 173–174
Biasing of class C amplifiers, 119–122
Bicone antennas, 285
Bidirectional antennas, 241
Bifilar winding, 119
Binary check code (BCC), 372
Binary PSK (BPSK), 362–364, 389
Binary signals, 347–349, 383
Bins, 368
Bisync protocol, 372
Bit error rate (BER), 372
Bit splitter, 365
Block check character (BCC), 374
Block check sequence (BCS), 374
Blocks, error checking in transmission of, 371
Body-stabilized satellites, 326–327
Booster rockets, 314–315
BORSCHT, 488, 489
BPF (*see* Bandpass filters)
BPSK (binary PSK), 362–364, 389
Bridge circuit, 46
Brightness signal, 437
Broadband, 5
Broadband amplifiers, 134
Broadband LANs, 379–380
Broadband satellite systems, 324–325
Broadside arrays, 244
Browsers, Internet, 393, 396
Buffer amplifiers, class A, 115
Buncher cavity, 275
Buses, data, 349
Bus networks, 378–379, 380–385

C
Cables, fiber-optic, 415–421
Cable television, 7, 453–457
    cable modems, 369
    as network, 376–377
    satellites for, 341
Cable TV box, 456–457
Capacitance:
    and L networks, 129
    and microwaves, 271
Cap code, 508
Capture effect, 78
Capture range for PLLs, 104

Carrier recovery circuit, 364
Carriers, 5, 9–11, 22, 382
    and DSB, 33, 37, 54, 364
    power of, 31–33
    for SSB selection, 56
    (*See also* Single-sideband suppressed carrier signals)
Carrier sense multiple access with collision detection (CSMA/CD), 382, 384
Carrier suppression, 52–53
Carson's rule, 75–76
Cassegrain feeds, 284
Cathode-ray tubes (CRTs), 293–294, 436, 448–449
Cavity resonators, 268–269
C band, 317–318, 324–325, 460
CB receivers and transmitters, 4, 8, 12, 37
    frequency synthesizers for, 181
    IF for, 155, 171
    power of, 111
CCD (charge-coupled devices), 436–437, 439–440, 442, 494–495
CCITT, facsimile standards by, 497
Cellular radio systems, 499–506
Celsius scale, 158
Central office, telephone, 490
Centripetal acceleration, 304
Centripetal force, 304
Ceramic filters for SSB, 60, 167
Channels, 3, 260, 324–325
    bandwidth of, 14, 354–356
    capacity of, 318–319, 354–355
Characteristic impedance:
    of folded dipoles, 239
    of transmission lines, 227–228, 245–246
Charge-coupled devices, 436–437, 439–440, 442, 494–495
Chat, 393
Chip, 389
Chipping rate, 389
Chirp, 391
Chrominance signals, 440, 441
Circular orbits, 305–308, 315
Circular polarization, 285–286, 319
Citizens band radio, 4, 8, 12, 37
    (*See also* CB receivers and transmitters)
Cladding, 416
Class A amplifiers:
    for AM modulators, 48
    for AM transmitters, 112
    efficiency of, 80, 122
    for IF, 152
    for RF, 115–116, 122, 272–273
Class AB amplifiers:
    for AM modulators, 48
    for RF, 115
Class B amplifiers:
    for AM modulators, 48
    efficiency of, 80, 122
    for RF, 115, 117–119, 122
Class C amplifiers:
    for AM modulators, 48–50
    efficiency of, 80, 122
    as frequency multipliers, 113, 122
    harmonics from, 121
    in phase modulators, 93–95
    RF, 115, 119–122, 132
    for transmitters, 111

Clippers, diode, 136–137
Clipping, sideband, 122
Clock pulse, 351
Clock waveforms for PAM systems, 208–210
Coaxial cable, 225
    baluns from, 246
    for cable television, 369, 453–454
    characteristic impedance of, 227–228
    for LANs, 379, 380–381
    for microwaves, 263
    velocity factor of, 226–227
Codec ICs, 217
Code division multiple access (CDMA), 386, 391, 506, 512
Coherent light waves, 423–424
Collector modulator circuits, 48–49
Collinear arrays, 243–244
Collisions, 382
Color burst, 446
Color signals, 439–442, 449
Combiners for satellites, 325, 336–337
Communications:
    applications of, 6, 7–8
    importance of, 1–3
    systems for, 3–4
    types of electronic, 4–6
    (*See also* Data communications)
Communications Act of 1934, 15
Communication satellites, 340–341
Companding circuits, 214–216
Compression:
    with facsimiles, 495–496
    speech, 135–136, 137
Compression amplifier, 215
Computers:
    electronic telephones and, 484
    fax modems, 493–494, 496–497, 498–499
    fiber optics to interconnect, 409
    satellite control by, 328–329, 339
    (*See also* Data communications; Internet)
Conical horn antennas, 279
Connectors, fiber-optic, 420–421
Constant-bandwidth FM telemetry channels, 196–197
Constellation design, 366
Contention systems, 382
Continuous tone control squelch (CTCS) systems, 172–173
Continuous-wave transmission, 110–111
    BFO circuits in, 173–174
    for radar systems, 291–292
    telegraph as, 348
    transceivers using, 179
Control grid, television, 449
Control systems for satellites, 309–313, 314–315, 320–321, 327–329, 339
Converters:
    cable television, 456–457
    of frequency, 122–123, 151–152, 332–334
    parallel and serial data, 350–352
    in superheterodyne receivers, 150
    (*See also* Analog-to-digital converters; Digital-to-analog converters; Mixers)
Cordless telephones, 485–487
Core, 416